THE GEORGE FISHER BAKER
NON-RESIDENT LECTURESHIP
IN CHEMISTRY AT
CORNELL UNIVERSITY

GENOMICS

GENOMICS

The Science and Technology Behind the Human Genome Project

Charles R. Cantor
Cassandra L. Smith
Center for Advanced Biotechnology
Boston University
Boston, Massachusetts

A Wiley-Interscience Publication

JOHN WILEY & SONS, INC.

New York • Chichester • Weinheim • Brisbane • Singapore • Toronto

Library of Congress Cataloging-in-Publication Data
Cantor, Charles R.
    Genomics : the science and technology behind the human genome
  project / Charles R. Cantor and Cassandra Smith.
        p.   cm.
      Includes index.
      ISBN 0-471-59908-5 (alk. paper)
      1. DNA--Analysis.   2. Nucleotide sequence.   3. Gene mapping.
  I. Smith, Cassandra.   II. Title.
  QP624.C36   1999
  572.8′ 6--dc21   98-40448

Printed in the United States of America.

10 9 8 7 6 5 4 3

Dedicated to
Charles DeLisi, who started it.
Rick Bourke, who made it so much fun
to explain it.

CONTENTS

7   Cytogenetics and Pseudogenetics   **208**

8   Physical Mapping   **234**

9   Enhanced Methods for Physical Mapping   **285**

10   DNA Sequencing: Current Tactics   **325**

11   Strategies for Large-Scale DNA Sequencing   **361**

PREFACE

This book is an outgrowth of the George Fisher Baker Lecture series presented by one of us (C.R.C.) at Cornell University in the fall of 1992. This author is tremendously grateful to all those at Cornell who made this occasion truly memorable, personally, and most productive, intellectually. Included especially are Jean Fréchet, Barbara Baird, John Clardy, Jerrold Meinwald, Fred McLafferty, Benjamin Widom, David Usher, and Quentin Gibson, among many others. He also wants to express great thanks to Rick Bourke, who, several times provided extraordinary help, without which lectures would have been missed or required rescheduling.

The book also builds in part on the research experience of both authors over the past 15 years. Without the help of numerous talented co-workers and collaborators, we would never have achieved the accomplishments that form some of the work described herein. No book is produced without heroic efforts on the part of support staff who handle endless chores, even in this era of efficient word processing. We are especially grateful to Michael Hirota for hours of patient deciphering of handwritten scribble. A major part of the effort in completing any substantial publication is receiving and dealing with constructive criticism from those who have read early drafts. Joel Graber and Chandran Sabanayagam helped with the illustrations and many other details; Scott Downing, Foual Siddiqi, and many others used preliminary versions as a textbook in courses over the past few years. To them we owe our thanks for taking the time to reveal ambiguities and painful lapses in our logic. Correcting these has, we hope, made the book far more accurate and comprehensible.

Inevitably in a fast-moving field like genome research, parts of this book were out of date even at the time it was submitted for publication. Furthermore, like many books based on a lecture series, this one relies too heavily on the authors' own work to provide a truly fair representation of many parallel or competing efforts in the field. No attempt was made to be comprehensive or representative in citing sources and additional readings. Despite these limitations, by concentration on fundamentals, the authors hope that a vast majority of workers in the field will find this text a useful companion.

CHARLES R. CANTOR
CASSANDRA L. SMITH

Boston Massachusetts

INTRODUCTION

WHY DNA IS INTERESTING

The topic of this book is the analytical chemistry of DNA. DNA is, arguably, the most interesting of all molecules. Certainly it is the largest, well-defined, naturally occurring molecule. The peculiar nature of DNA—its relatively regular structure consisting of a sequence of just four different kinds of bases, the intense specificity of base-pairing interactions, its high charge density, and the existence of biological machinery to duplicate it and copy it into RNA—has led to a powerful arsenal of methods for DNA analysis. These include procedures for detecting particular DNA sequences, for purifying these sequences, and for immortalizing these sequences.

It probably seems strange for most readers to hear the term analytical chemistry applied to DNA. However, this is what many studies of DNA, including the massive current efforts to map and sequence the human genome, are all about. The basic chemical structure and biological properties of DNA are mostly well understood. However, the complexity of DNA in most samples of biological interest can be staggering. Many experimental efforts, ranging from clinical diagnoses to gene identification, ultimately come down to asking whether a particular DNA sequence is present, where it is located along a larger DNA molecule, and, sometimes, how many copies are present. Such questions are at the heart of traditional analytical chemistry. However, the methods used for DNA are mostly unique to it.

Some might say that the basic chemistry of DNA is rather uninteresting and mostly well worked out. It might seem that DNA bases and the ribose-phosphate backbone are not fertile ground for exotic or innovative chemical modification or manipulation. However, recent focus on these very basic aspects of DNA chemistry has led to the development of a number of novel compounds with interesting chemical characteristics as well as unique potentials for detection and manipulation of nucleic acids.

DNA itself has thus far shown only modest evidence of possessing any intrinsic catalytic activities, although the prospect that more will be discovered in the future is surely plausible. Rather, the fascination of DNA lies in the enormous range of activities made possible by its function as the cell's primary storage of genetic information. In the DNA sequence lies the mechanism for the specific interaction of one DNA (or RNA) molecule with another, or the interaction of DNAs with proteins.

WHAT THIS BOOK IS ABOUT

In this book we first review the basic chemical and biological properties of DNA. Our intention is to allow all readers to access subsequent chapters, whether they begin this journey as biologists with little knowledge of chemistry, chemists with equally little knowledge of biology, or others (engineers, computer scientists, physicians) with insufficient grounding or recent exposure to either biology or chemistry. We also review the basic biological and chemical properties of chromosomes, the nucleoprotein assemblies that package DNA into manipulatable units within cells. Methods for fractionation of the chromosomes of both simple and complex organisms were one of the key developments that opened up the prospect of large-scale genome analysis.

Next the principles behind the three major techniques for DNA analysis are described. In hybridization, a specific DNA sequence is used to identify its complement, usually among a complex mixture of components. We understand hybridization in almost quantitative detail. The thermodynamics and kinetics of the process allow accurate predictions in most cases of the properties of DNAs, although there are constant surprises. In the polymerase chain reaction (PCR) and related amplification schemes, short DNA segments of known sequence are used to prime the synthesis of adjacent DNA where the sequence need not be known in advance. This allows us, with bits of known sequence and synthetic DNA primers, to make essentially unlimited amounts of any relatively short DNA molecules of interest. In electrophoresis, DNA molecules are separated primarily, if not exclusively, on the basis of size. Electrophoresis is the primary tool used for DNA sequencing and for most DNA purifications and analytical methods. The high charge density, great length, and stiffness of DNA lends it particularly amenable to high-resolution length separations, and some variants of electrophoresis, like pulsed field gel electrophoresis (PFG), seem uniquely suited to DNA analysis. Frequently these three basic techniques are combined such as in the electrophoretic analysis of a PCR reaction or in the use of PCR to speed up the rate of a subsequent hybridization analysis.

We next describe DNA mapping. Maps are low-resolution structures of chromosomes or DNA molecules that fall short of the ultimate detail of the full DNA sequence. Many different types of maps can be made. Cytogenetic maps come from examination of chromosomes, usually as they appear in the light microscope after various treatments. Interspecies cell hybrids formed from intact chromosomes or chromosomes fragmented by radiation are also useful mapping tools. Closer to the molecular level of DNA are restriction maps, which are actually bits of known DNA sequence spaced at known intervals, or ordered libraries, which are sets of overlapping clones that span large DNA regions. Methods for producing all of these types of maps will be described and evaluated.

Totally different flavors of maps emerge from genetic or pseudogenetic analyses. Even with no knowledge about DNA structure or sequence, it is possible to place genes in order along a chromosome by their pattern of inheritance. In the human species direct breeding experiments are impossible for both moral and practical reasons. However, retrospective analysis of the pattern of inheritance in respectable numbers of families is frequently an acceptable alternative. For genetic disease searches, it is particularly useful to study geographically isolated, inbred human populations that have a relatively large preponderance of a disease. Some of these populations, like families studied in Finland, have the added advantage of having religious or government records that detail the genetic history of the population.

Genetically mapped observable traits like inherited diseases or eye color are related, ultimately to DNA sequence, but at the current state of the art they afford a really different view of the DNA than direct sequence analysis. An analogy would be a person whose sense of smell or taste might be so accomplished that these could determine the components of a perfume or a sauce without the need for any conventional chemical analysis. Following the impact of the genes gives us direct functional information without the need for any detailed chemical knowledge of the sequence that is ultimately responsible for it. Occasional properties are inherited through the pattern of modification of DNA bases rather than through the direct base sequence of DNA that we ordinarily consider when talking about genes. As a footnote, one must also keep in mind that it is only the DNA in germ (sperm and egg) cells that is heritable. This means that DNA in somatic (nongermline) cells may undergo nonheritable changes.

Standard techniques for the determination of the base sequence of DNA molecules are well developed and their strengths and weaknesses well understood. Through automated methodology, raw DNA sequence data can be generated by current procedures at a rate of about 10^5 base pairs per day per scientist. Finished DNA sequence, assembled and reduced in errors by redundancy, accumulates at a tenfold slower rate by standard strategies. Such figures are impressive, and laboratories interested in small DNA targets can frequent accumulate sequence data faster than they can provide samples worth sequencing, or faster than they can analyze the resulting data in a useful fashion. The situation is different when a large-scale sequencing effort is contemplated, encompassing thousands of continuous kilobases (kb) of DNA. Here new strategies are being formulated and tested to make such larger-scale projects efficient.

The task of determining the DNA sequence of even one human genome with 3×10^9 base pairs is still daunting to any existing technologies. Perhaps these techniques can be optimized to gain one or two orders of magnitude of additional speed, and progress in this direction is described. However, it seems likely that eventually new approaches to DNA sequencing will appear that may largely surpass today's methods, which rely on gel electrophoresis, an intrinsically slow process. Among these methods, mass spectrometry and sequencing by hybridization (SBH, in which data are read essentially by words instead of letters) are described in greatest detail, because they seem highly likely to mature into generally useful and accessible tools.

The ultimate purpose behind making DNA maps and determining DNA sequences is to understand the biological function of DNA. In current terms this means acquiring the ability to find the position of genes of interest with sufficient accuracy to actually possess these genes on DNA fragments suitable for further study and manipulation. Our ability to do this is improving rapidly. One of the basic rationales behind the human genome project is that it will systematically reveal the locations of all genes. This will be far more efficient and cost effective than a continued search for genes one by one. In addition to gene location and identity, DNA can be used to study the occurrence of mutations caused by various environmental agents or by natural, spontaneous, biological processes. This task is still a formidable one. We may need to search for a single altered base among a set of 10^8 normally inherited ones. The natural variation among individuals confounds this task, because it must be first filtered out before any newly created differences, or mutations, can be identified.

In a typical higher organism, more than 90% of the DNA sequence is not, apparently, translated into protein sequence. Such material is frequently termed *junk*. Sydney Brenner was apparently the first to notice that junk is a more appropriate term than garbage because junk is kept while garbage is discarded, and this noncoding component of the genome has remained with us. A great challenge for the future will be to dissect out any important functional elements disguised within the junk. Much of this so-called junk is repeated DNA sequence, either simple tandem repeats like $(AC)_n$ or interspersed, longer (and less perfect) repeats. Although from a biological standpoint much of the genome appears to be junk, for DNA analysis it is frequently extremely useful because it can be exploited to look at whole classes of DNAs simultaneously. For example, human-specific repeats allow the human DNA component of an interspecies cell hybrid to be detected selectively.

By taking advantage of the properties of particular DNA sequences, it is often possible to perform sequence-specific manipulations such as purification, amplification, or cloning of desired single species or fractions from very complex mixtures. A number of relatively new techniques have been developed that facilitate these processes. Some have implications far beyond the simple chemical analysis of DNA. This feature of DNA analysis, the ability to focus on one DNA species in 10^{10}, on a desired class of species which represents, say 10^{-5} of the whole, or on a large fraction of a genome, say 10%, is what really sets DNA analysis apart from most other systematic analytical protocols.

As large amounts of DNA sequence data accumulate, certain global patterns in these data are emerging. Today more than 1,000 megabases (Mb) of DNA sequence are available in publicly accessible databases. This number will spiral upward to levels we cannot even guess at today. Some of the patterns that have emerged in the existing data are greatly facilitating our ability to find the locations of previously unknown genes, just from their intrinsic sequence properties. Other patterns exist that may have little to do with genes. Some are controversial; others we simply have no insight on at all, yet.

Like it or not, a DNA revolution is upon us. Pilot projects in large-scale DNA sequencing of regions of model organisms have been successfully concluded, and the complete DNA sequence of many bacterial species and one yeast species have been published. Considerable debate reigns over whether to proceed with continued large-scale genomic DNA sequence or just concentrate on genes themselves (isolated through complementary DNA copies [cDNAs] of cellular mRNAs). Regardless of the outcome of these debates, and largely independent of future further advances in DNA sequence analysis, it now seems inevitable that the next half decade will see the successful sequencing and discovery of essentially all human genes. This has enormous implications for the practice of both biological research and medical care. We are in the unique position of knowing, years ahead of time, that ongoing science will have a great, and largely positive, impact on all humanity. However, it behooves us to look ahead, at this stage, contemplate, and be prepared to deal with the benefits to be gained from all this new knowledge and cope with the few currently perceptible downside risks.

GENOMICS

1 DNA Chemistry and Biology

BASIC PROPERTIES OF DNA

DNA is one of the fundamental molecules of all life as we know it. Yet many of the features of DNA as described in more elementary sources are incorrect or at least misleading. Here a brief but fairly rigorous overview will be presented. Special stress will be given to try to clarify any misconceptions the reader has based on prior introductions to this material.

COVALENT STRUCTURE

The basic chemical structure of DNA is well-established. It is shown in Figure 1.1. Because the phosphate-sugar backbone of DNA has a polarity, at each point along a polynucleotide chain the direction of that chain is always uniquely defined. It proceeds from the 5′-end via 3′- to 5′-phosphodiester bonds until the 3′-end is reached. The structure of DNA shown in Figure 1.1 is too elaborate to make this representation useful for larger segments of DNA. Instead, we abbreviate this structure by a series of shorthand forms, as shown in Figure 1.2. Because of the polarity of the DNA, it is important to realize that different sequences of bases, in our abbreviations, actually correspond to different chemical structures (not simply isomers). So ApT and TpA are different compounds with, occasionally, rather different properties. The simplest way to abbreviate DNA is to draw a single polynucleotide strand as a line. Except where explicitly stated, this is always drawn so that the left-hand end corresponds to the 5′-end of the molecule.

RNA differs from DNA by having an additional hydroxyl at the 2′-position of the sugar (Fig. 1.1). This has two major implications that distinguish the chemical and physical properties of RNA and DNA. The 2′-OH makes RNA unstable with respect to alkaline hydrolysis. Thus RNA is a molecule intrinsically designed for turnover at the slightly alkaline pH's normally found in cells, while DNA is chemically far more stable. The 2′-OH also restricts the range of energetically favorable conformations of the sugar ring and the phosphodiester backbone. This limits the range of conformations of the RNA chain, compared to DNA, and it ultimately restricts RNA to a much narrower choice of helical structures. Finally the 2′-OH can participate in interactions with phosphates or bases that stabilize folded chain structures. As a result an RNA can usually attain stable tertiary structures (ordered, three-dimensional, relatively compact structures) with far more ease than the same corresponding DNA sequence.

DOUBLE HELICAL STRUCTURE

The two common base pairs A–T and G–C are well-known, and little evidence for other base interactions within the DNA's double helix exists. The key feature of double-helical DNAs (duplexes), which dominates their properties, is that an axis of symmetry relates the two strands (Box 1.1).

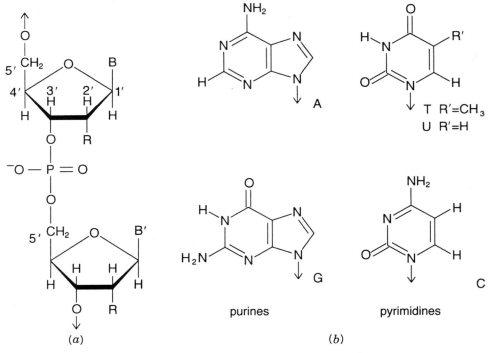

Figure 1.1   Structure of the phosphodiester backbone of DNA and RNA, and the four major bases found in DNA and RNA: the purines, adenine (A) and guanine (G), and the pyrimidines, cytosine (C) and thymine (T) or uracil (U). *(a)* In these abbreviated structural formulas, every vertex not occupied by a letter is a carbon atom. R is H in DNA, OH in RNA; B and B′ indicate one of the four bases. *(b)* The vertical arrows show the base atoms that are bonded to the C1′ carbon atoms of the sugars.

In ordinary double-helical DNA this is a pseudo C2 axis, since it applies only to the backbones and not to the bases themselves, as shown in Figure 1.3*a*. However, certain DNA sequences, called *self-complementary,* have a true C2 symmetry axis perpendicular to the helix axis. We frequently abbreviate DNA duplexes as pairs of parallel lines, as shown in Figure 1.3*b*. By convention, the top line almost always runs from 5′-end to 3′-end.

abbreviations

$$5'\ \ B_pB'\ \ 3' \qquad \neq\ \ B_p'B$$

$$5'\ \ B\,B'\ \ 3' \qquad \neq\ \ B'B$$

$$5'\ \ \rule{3em}{0.4pt}\ \ 3'$$

Figure 1.2   Three ways in which the structures of DNA and RNA are abbreviated. Note that BpB′ is not the same chemical compound as B′pB.

BOX 1.1
C2 SYMMETRY

C2 symmetry implies that a structure is composed of two identical parts. An axis of rotation can be found that interchanges these two parts by a 180-degree rotation. This axis is called a *C2 axis*. An example of a common object that has such symmetry is the stitching of a baseball, which is used to bring two figure-8 shaped structures together to make a spherical shell. An example of a well-known biological structure with C2 symmetry is the hemoglobin tetramer which is made up of two identical dimers each consisting of one alpha chain and one beta chain. Another example is the streptavidin tetramer which contains four copies of the same subunit and has three different C2 axes. One of these passes horizontally right through the center of the structure (shown in Fig. 3.26).

Helical symmetry means that a rotation and a translation along the axis of rotation occur simultaneously. If that rotation is 180 degrees, the structure generated is called a *twofold helix* or a *pleated sheet*. Such sheets, called *beta structures,* are commonly seen in protein structure but not in nucleic acids. The rotation that generates helical symmetry does not need to be an integral fraction of 360 degrees; DNA structures have 10 to 12 bases per turn; under usual physiological conditions DNA shows an average of about 10.5 bases per turn. In the DNA double helix, the two strands wrap around a central helical axis at each turn.

Pseudo C2 symmetry means that some aspects of a structure can be interchanged by a rotation of 180 degrees, while other aspects of the structure are altered by this rotation. This process might be imagined as a disk painted with the familiar yin and yang symbols of the Korean flag. Then, except for a color change, a C2 axis perpendicular to the yin and yang exchanges them. The pseudo C2 axes in DNA are perpendicular to the helix axis. They occur in the plane of each base pair (this axis interchanges the position of the two paired bases) and between each set of adjacent base pairs (this axis interchanges a base on one strand with the nearest neighbor of the base to which it is paired to the other strand). Thus for DNA with 10 bases per turn there are 20 C2 axes per turn.

The antiparallel nature of the DNA strands imposed by the pseudosymmetry of their structure means that the bottom strand in this simple representation runs in the opposite direction to our usual convention. Where the DNA sequence permits it, double helices can also be formed by the folding back of a single strand upon itself to make structures called *hairpins* (Fig. 1.3c) or more complex topologies such as structures called *pseudoknots* (Fig. 1.3d).

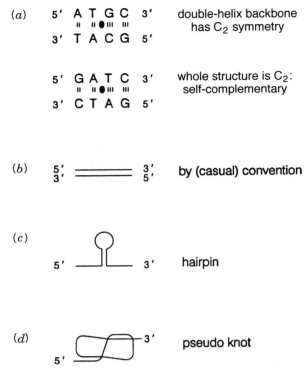

Figure 1.3 Symmetry and pseudosymmetry in DNA double helices. *(a)* The vertical lines indicate hydrogen bonds between base pairs. The base pairs and their attached backbone residues can be flipped by a 180° rotation around an axis through the plane of the bases and perpendicular to the helix axis (shown as a filled lens-shaped object). *(b)* Conventional way of representing a double-stranded DNA. *(c)* Example of a DNA hairpin. *(d)* Example of a DNA pseudoknot.

In a typical DNA duplex, the phosphodiester backbones are on the outside of the structure; the base pairs are internal (Fig. 1.4). The structure of the double helix appears to be regular because the A–T base pair fills a three-dimensional space in a manner similar to a G–C base pair. The spaces between the two backbones are called *grooves*. Usually one groove is much broader (the major groove) than the other (the minor groove). The structure appears as a pair of wires held fairly close together and wrapped loosely around a cylinder. Three major classes of DNA helical structures (secondary structures) have been found thus far. DNA B, the structure first analyzed by Watson and Crick, has 10 base pairs per turn. DNA A, which is very similar to the structure almost always present in RNA, has 11 base pairs per turn. Z DNA has 12 base pairs per turn; unlike DNA A and B, it is a left-handed helix. Only a very restricted set of DNA sequences appears able to adopt the Z helical structure. The biological significance of the range of structures accessible to particular DNA sequences is still not well understood. Elementary texts often focus on hydrogen bonding between the bases as a major force behind the stability of the double helix. The pattern of hydrogen bonds in fact is responsible for the specificity of base–base interactions but not their stability. Stability is largely determined by electrostatic and hydrophobic interactions between parallel overlapping base planes which generate an attractive force called *base stacking*.

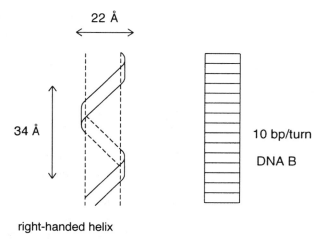

22 Å

34 Å

10 bp/turn

DNA B

right-handed helix

Figure 1.4   Schematic view of the three-dimensional structure of the DNA double helix. Ten base pairs per turn fill a central core; the two backbone chains are wrapped around this in a right-handed screw sense. Note that the A–T, T–A, G–C, and C–G base pairs all fit into *exactly* the same space between the backbones.

METHYLATED BASES

DNA from most higher organisms and from many lower organisms have additional analogues of the normal bases. In bacteria these are principally *N6*-methyl A and 5-methyl C (or 4-methyl C). Higher organisms contain 5-methyl C. The presence of methylated bases has a strong biological effect. Once in place, the methylated bases are apt to maintain their methylated status after DNA replication. This is because the hemi-methylated duplex produced by one round of DNA synthesis is a far better substrate for the enzymes that insert the methyl groups (methylases) than the unmethylated sequence. The result is a form of inheritance (epigenetic) that goes beyond the ordinary DNA sequence: DNA has a memory of where it has been methylated.

The role of these modified bases is understood in some detail. In bacteria these bases mostly arise by a postreplication endogenous methylation reaction. The purpose of methylation is to protect the cellular DNA against endogenous nucleases, directed against the same specific DNA sequences as the endogenous methylases. It constitutes a cellular defense or restriction system that allows a bacterial cell to distinguish between its own DNA, which is methylated, and the DNA from an invading bacterial virus (bacteriophage) or plasmid, which is unmethylated and can be selectively destroyed. Unless the host DNA is replicating too fast for methylation to keep up, it is continually protected by methylation shortly after it is synthesized. However, once lytic bacteriophages are successfully inside a bacterial cell, their DNA will also become methylated and protected from destruction. In bacteria particular methylated sequences function to control initiation of DNA replication, DNA repair, gene expression, and movement of transposable elements.

In bacteria, methylases and their cognate nucleases recognize specific sequences that range in size from 4 to 8 base pairs in length. Each site is independently methylated. In higher organisms the principal, if not the exclusive, site of methylation is the sequence CpG which is converted to ^mCpG. Although the eukaryotic methylases recognize only a dinucleotide sequence, methylation (or unmethylated) at CpGs appears to be regionally specific, suggesting that nearby ^mCpG sequences interact. This plays a role in allowing cells with the same exact DNA sequence to maintain stable, different patterns of gene expression (cell differentiation), and it also allows the contributions of the genomes of two parents to be distinguished in offspring, since they often have different patterns of DNA methylation.

The fifth base in the DNA of humans and other vertebrates is 5-methyl C. Its presence has profound consequences for the properties of these DNA. The influence of ^mC on DNA three-dimensional structure is not yet fully explored. We know, however, that it favors the formation of some unusual DNA structures, like the left-handed Z helix. Its biological importance remains to be elucidated. However, it is on the level of the DNA sequence, the primary structure, where the effect of ^mC is most profoundly felt. To understand the impact, it is useful to consider why DNA contains the base T (5-methyl U) instead of U, which predominates, overwhelmingly, in RNA. While the base T conveys a bit of extra stability in DNA duplexes because of interactions between the methyl group and nearby bases, the most decisive influence of the methyl group of T is felt in the repair of certain potential mutagenic DNA lesions. By far the most common mutagenic DNA damage event in nature appears to be deamination of C. As shown in Figure 1.5, this yields, in duplex DNA, a U mispaired with a G. Random repair of this lesion would result, 50% of the time, in replacement of the G by an A, a mutation, instead of replacement of the U by a C, restoring the original sequence.

Apparently the intrinsic rate of the C to U mutagenic process is far too great for optimum evolution. Some rate of mutation is always needed; otherwise, a species could not adapt or evolve. Too high a rate can lead to deleterious mutations that interfere with reproduction. Thus the mutation rate must be carefully tuned. Nature accomplishes this for deamination of C by a special repair system that recognizes the G–U mismatch and selectively excises the U and replaces it with a C. This system, centered about an enzyme called *Uracil DNA glycosylase,* biases the repair process in a way that effectively avoids most mutations. However, a problem arises when the base 5-^mC is present in the DNA. Then, as shown in Figure 1.5, deamination produces T which is a normally occurring base. Although the G–T mismatch is still repaired with a bias toward restoring the presumptive original ^mC, the process is not nearly as efficient (nor should it be, since some of the time the G–T mismatch will have come by misincorporation of a G for an A). The result is that ^mC represents a mutation hotspot within DNA sequences that contain it.

In the DNA of vertebrates, all known ^mC occur in the sequence ^mCpG. About 80% of this sequence occurs in the methylated form (on both strands). Strikingly the total occurrence of CpG (methylated or not) is only 20% of that expected from simple binomial statistics based on the frequency of occurrence of the four bases in DNA:

$$\frac{X_{CpG}}{X_C X_G} = 0.2$$

where X indicates the mole fraction. The remainder of the expected CpG has apparently been lost through mutation. The occurrence of the product of the mutation, TpG, is elevated, as expected. This is a remarkable example of how a small bias, over the evolutionary time scale, can lead to dramatic alteration in properties. Presumably the rate of mutation of ^mCpG's continues to slow as the target size decreases. There must also be considerable functional constraints on the remaining ^mCpG's that prevent their further loss.

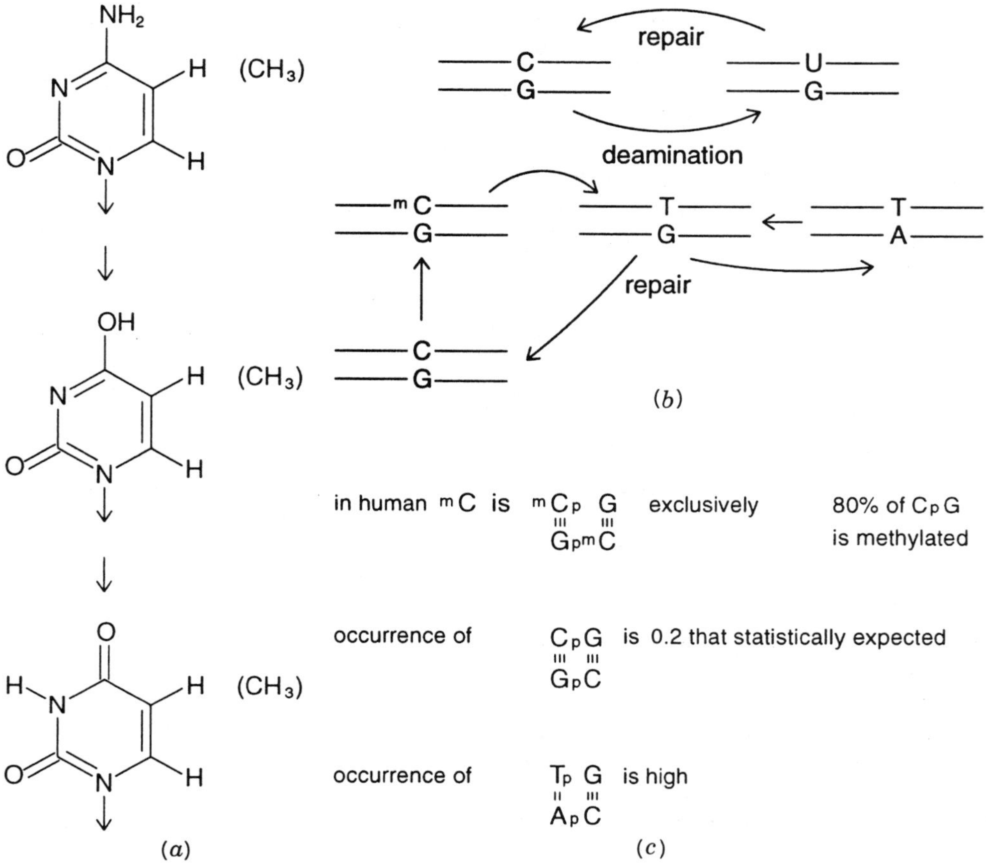

Figure 1.5 Mutagenesis and repair processes that alter DNA sequences. *(a)* Deamination of C and ᵐC produce U and T, respectively. *(b)* Repair of uracil-containing DNA can occur without errors, while repair of mismatched T–G pairs incurs some risk of mutagenesis. *(c)* Consequences of extensive ᵐCpG mutagenesis in mammalian DNA.

The vertebrate immune system appears to have learned about the striking statistical abnormality of vertebrates. Injections of DNA from other sources with a high G + C content, presumably with a normal ratio to CpG, act as an adjuvant; that is, these injections stimulate a generally heightened immune response.

PLASTICITY IN DNA STRUCTURE

Elementary discussions of DNA dwell on the beauty and regularity of the Watson-Crick double helix. However, the helical structure of DNA is really much more complex than this. The Watson-Crick structure has 10 base pairs per turn. DNA in solution under physiological conditions shows an average structure of about 10.5 base pairs per turn, roughly halfway between the canonical Watson-Crick form and the A-type helix with a larger diameter and tilted base pairs which are characteristic of RNA. In practice, these are just average forms. DNA is revealed to be fairly irregular by methods that do not have to average over long expanses of structure. DNA is a very plastic molecule with a backbone easily distorted and with optimal geometry very much influenced by its local sequences. For example, certain DNA sequences, like properly spaced sets of ApA's promote extensive curvature of the backbone. Thus, while base pairs predominate, the angle between the base pairs, the extent of their stacking (which holds DNA together) above and below neighbors, their planarity, and their disposition relative to helix axis can vary substantially. Almost all known DNA structures can be viewed in detail by accessing the Nucleic Acid Database, NDB <http://ndbserver.rutgers.edu/>.

We do not really know enough about the properties of proteins that recognize DNA. One extreme view is that these proteins look at the bases directly and, if necessary, distort the helix into a form that fits well with the structure of protein residues in contact with the DNA. The other extreme view has a key role played by the DNA structure with proteins able to recognize structural variants, without explicit consideration of the sequence that generated them. These views have very different implications for proteins that might recognize classes of DNA sequences rather than just distinct single sequences. We are not yet able to decide among these alternative views or to adopt some sort of compromise position. The structures of the few protein-nucleic acid complexes known can be viewed in the NDB.

DNA SYNTHESIS

Our ability to manufacture specific DNA sequences in almost any desired amounts is well developed. Nucleic acid chemists have long learned and practiced the powerful approach of combining chemical and enzymatic syntheses to accomplish their aims. Automated instruments exist that perform stepwise chemical synthesis of short DNA strands (oligonucleotides) principally by the phosphoramidite method. Synthesis proceeds from the 3'-end of the desired sequence using an immobilized nucleotide as the starting material (Fig. 1.6a). To this are added, successively, the desired nucleotides in a blocked, activated form. After each condenses with the end of the growing chain, it is deblocked to allow the next step to proceed. It is a routine procedure to synthesize several compounds 20 nucleotides long in a day. Recently instruments have been developed that allow almost a hundred compounds to be made simultaneously. Typical instruments produce about a thousand times the amount of material needed for most biological experiments. The cost is about $0.50 to $1.00 per nucleotide in relatively efficient settings. This arises primarily from the costs of the chemicals needed for the synthesis. Scaling down the process will reduce the cost accordingly, and efforts to do this are a subject of intense interest. For certain strategies of large-scale DNA analysis, large numbers of different oligonucleotides are required. The resulting cost will be a significant factor in evaluating the merits of the overall scheme. The currently used synthetic schemes make it very easy to incorporate unusual or modified nucleotides at desired places in the sequence, if appropriate derivatives are available. They also make it very easy to add, at the ends of the DNA strand, other functionalities like chemically reactive alkyl amino or thiol groups or useful biological ligands like biotin, digoxigenin, or fluorescein. Such derivatives have important uses in many analytical application, as we will demonstrate later.

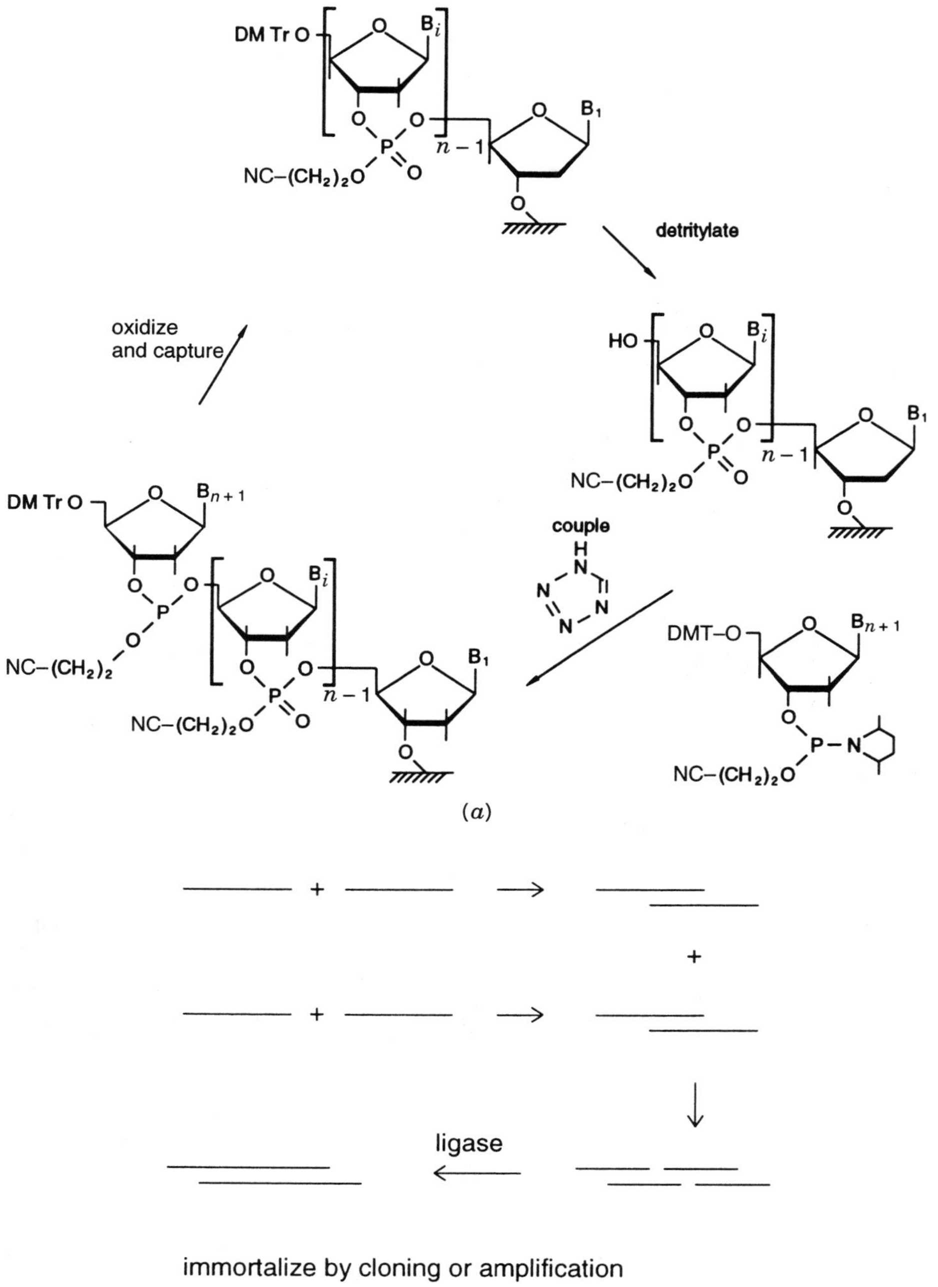

Figure 1.6 DNA synthesis by combined chemical and enzymatic procedures. (*a*) Phosphoramidite chemistry for automated solid state synthesis of DNA chains. (*b*) Assembly of separately synthesized chains by physical duplex formation and enzymatic joining using DNA ligase.

BOX 1.2
SIMPLE ENZYMATIC MANIPULATION OF DNAs

The structure of a DNA strand is an alternating polymer of phosphate and sugar-based units called nucleosides. Thus the ends of the chain can occur at phosphates (p) or at sugar hydroxyls (OH).

Polynucleotide kinase can specifically add a phosphate to the 5′-end of a DNA chain.

$$5' \text{ HO–ApTpCpG–OH } 3' \xrightarrow[\text{kinase}]{\text{ATP}} 5' \text{ pApTpCpG–OH } 3'$$

Phosphatases remove phosphates from one or both ends.

$$5' \text{ pApTpCpGp } 3' \xrightarrow[\substack{\text{alkaline} \\ \text{phosphatase}}]{} 5' \text{ HO–ApTpCpG–OH } 3'$$

DNA ligases will join together two DNA strands that lie adjacent along a complementary template. These enzymes require that one of the strands have a 5′-phosphate:

```
                    HO p
       5' GpCpCpT GpTpCpCpA  3'  →  5' GpCpCpTpGpTpCpCpA  3'
       3' CpGpGpApCpApGpGpA  5'      3' CpGpGpApCpApGpGpA  5'
```

DNA ligase can also fuse two double-stranded DNAs at their ends provided that 5′-phosphates are present:

```
  5' ——— 3'   5'p ——— 3'      5' ——— p ——— 3'
            +                →
  3' ——— p5'   3' ——— 5'      3' ——— p ——— 5'
```

This reaction is called blunt-end ligation. It is not particularly sensitive to the DNA sequences of the two reactants.

Restriction endonucleases cleave both strands of DNA at or near the site of a specific sequence. They usually cleave at all sites with this particular sequence. The products can have blunt-ends, 3′-overhangs, or 5′-overhangs, as shown by the examples below:

```
  5'—pApCpGpTp—3'      5'—pApC–OH          pGpTp—3'
                    →                 +
  3'—pTpGpCpAp—5'      3'—pTpGp        HO–CpAp—5'


  5'—pApCpGpTp—3'      5'—pApCpGpT–OH          p—3'
                    →                    +
  3'—pTpGpCpAp—5'      3'—p              HO–TpGpCpAp—5'


  5'—pApCpGpTp—3'      5'—OH              pApCpGpT—3'
                    →                +
  3'—pTpGpCpAp—5'      3'—pTpGpCpAp       HO—5'
```

Restriction enzymes always leave 5′-phosphates on the cut strands. The resulting fragments are potential substrates for DNA ligases. Most restriction enzymes cleave at sites with C2 symmetry like the examples shown above.

(continued)

BOX 1.2 *(Continued)*

Exonucleases remove nucleotides from the ends of DNA. Most cleave single nucleotides one at a time off one end of the DNA. For example, Exonuclease III carries out steps like

$$5'\ \text{ApCpGp} \quad\text{------}\quad \text{pTpApA} \quad 3'$$
$$3'\ \text{TpGpCp} \quad\text{------}\quad \text{pApTpT} \quad 5'$$

$$5'\ \text{Ap}\ \ \text{CpGp} \text{------} \text{pTpA--OH}\ 3'$$
$$3'\ \text{HO--GpCp} \text{------} \text{pApTpT} \quad 5'$$

$$5'\ \text{ApCpGp} \quad\text{------}\quad \text{pT--OH} \quad 3'$$
$$3'\ \text{HO--}\ \text{Cp} \quad\text{------}\quad \text{pApTpT} \quad 5'$$

and continues until there are no double-stranded regions left.

Terminal transferases add nucleotides at the 3'-end of a DNA strand. They require that this end not have a 3'-phosphate. No template is required; hence the sequence of the incorporated residues is random. However, if only a single type of pppdN is used, then a set of homopolymeric products is produced

$$\text{GpApTpCpA} + \text{pppdT} \xrightarrow[\text{tranferase}]{} \text{GpApTpCpA (pT)}n$$

where n is variable. Some DNA polymerases (Box 1.6) also have a terminal transferase activity (see Box 1.5).

Separately synthesized DNA strands can be combined to yield synthetic double helices. The specificity and strength of base pairing will ensure the formation of the correct duplexes under almost all circumstances if the two strands can form 8 or more base pairs. As shown in Figure 1.6*b*, synthetic DNA duplexes can be strung together, and nicks in the polynucleotide backbone can be sealed with the enzyme DNA ligase. This enzyme requires a 5'-phosphate group and a free 3'-OH in order to form a phosphodiester bond (Box 1.2). There are two commonly used forms of the enzyme: the species isolated from *Escherichia coli* uses NAD as an energy source, while the bacteriophage T7 enzyme requires ATP. Thermostable ligases isolated from thermophilic organisms are also available. Their utility will be demonstrated later.

Once a duplex DNA is available, it can be immortalized by cloning it into an appropriate vector (see Box 1.3). Duplexes (or single strands) can also be amplified in vitro by methods such as the polymerase chain reaction (Chapter 4). The result is that for any desired species, a chemical synthesis needs only to be done once. Thereafter biological or enzymatic methods usually suffice to keep up the supply of product. It remains to be seen in the future, as larger stocks of specific synthetic DNA sequences accumulate worldwide, whether it is worthwhile to set up a distribution system to supply the potentially millions of different compounds to interested users, or whether it will be simpler and cheaper to manufacture a needed compound directly on site.

By taking advantage of the specificity of base pairing, it is possible to synthesize and assemble more complex DNA structures than those that predominate in nature. Naturally occurring DNAs are either circles with no ends or linear duplexes with two protected ends.

BOX 1.3
CLONING DNAs IN BACTERIAL PLASMIDS
AND BACTERIOPHAGE

Cloning is the process of making many identical cells (or organisms) from a single precursor. If that precursor contains a target DNA molecule of interest, the cloning process will amplify that single molecule into a whole population. Most single-cell organisms are clonal. Their progeny are identical replicas. Cloning can also be done by manipulating single, immortal, or immortalizable cells of higher organisms including plants and animals. Here we concentrate on bacterial cloning, which can be carried out in a large number of different species but often involves the favorite laboratory organism *Escherichia coli*.

Cloning DNA requires three components: the target DNA of interest, a vector DNA that will accommodate the target as an insertion and that contains a replication origin for it to be propagated indefinitely, and a host cell that will replicate the vector and its insert. The simplest early cloning systems used two types of vectors, either plasmids, which are small DNAs that can replicate independently of the host cell DNA, or bacteriophages, which also have independently replicating DNA systems. Plasmids are usually double-stranded circular DNAs. Bacteriophages can be circular or linear, single stranded, or double stranded. Here we illustrate the steps in cloning double strands.

The vector is usually designed so that it has a unique cutting site for a particular restriction enzyme. Digestion linearizes the vector and produces two ends, usually with the same overhanging sequence. The target is cut with the same enzyme, or alternatively, with a different enzyme that nevertheless leaves the same overhangs. A complex target will yield many fragments, only a few of which may be of interest.

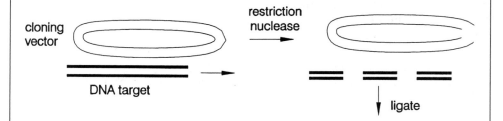

When the vector and target are mixed, they can anneal and form suitable substrates of DNA ligase. Depending on the relative concentrations used, and whether the 5'-phosphate ends produced by the restriction nuclease are left intact or are removed with a phosphatase, a variety of different products combining targets and vectors will be formed. A number of tricks exist to bias the ligation in favor of particular 1:1 target-vector adducts. For example, in the scheme above a single target and vector can come together in two different polarities. More complex schemes allow the polarity to be preselected.

(continued)

BOX 1.3 *(Continued)*

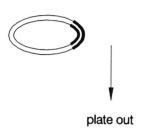

plate out

Next the vector-target adduct must be reintroduced into the bacterial cells. Depending on the vector, different procedures are used. If the vector is a plasmid, the bacterial cells are made permeable by chemical, enzymatic, electrical, or mechanical procedures. When mixed with DNA, these bacteria take up some of the vector before the gaps in their surfaces can be resealed. The process is called *transformation* or *transfection.* Very similar procedures can be used to introduce DNA into other kinds of cells. The efficiencies vary widely depending on the type of vector, type of cell, and the procedures used for permeabilization.

If the vector is a bacteriophage, it is usually preferable to repackage the vector-target adduct in vitro and then allow the assembled cells to infect the host cells naturally. This increases the efficiency of delivering DNA into the host cells, both because bacteriophage insertion is often quite effective and because there is no need to permeabilize the host cells, a process that frequently kills many of them.

Traditionally microbiological screening systems are used to detect host cells propagating cloned DNA. If a dilute suspension of bacteria is allowed to coat the surface of a culture dish, colonies will be observed after several days of growth, each of which represents cloned progeny arising from a single cell. The cloning process can be used to introduce markers that aid in the selection or screening of bacteria carrying plasmids of potential interest. For example, the plasmid may carry a gene that confers resistance to an antibiotic so that only host cells containing the plasmid will be able to grow on a culture medium in which that antibiotic is placed. The site at which the vector is cut to introduce the target can disrupt a gene that yields a colored metabolic product when the precursor of that product is present in the medium. Then vectors that have been designated as having no target insert will still yield colored colonies, while those containing a target insert will appear white.

When a bacteriophage vector is used, the same sets of selection and screening systems can be used. If the bacteriophage shows efficient lysis, that is, if infected cells disrupt

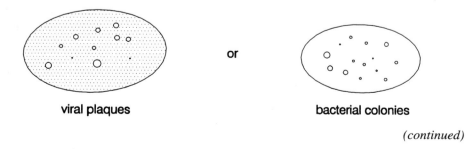

viral plaques   or   bacterial colonies

(continued)

BOX 1.3 *(Continued)*

and release bacteriophage, then a convenient method of cloning is to grow a continuous layer of susceptible bacteria and then cover the culture with a thin layer of dilute bacteriophage suspension. Individual bacteriophage will infect cells, leading to continual cycles of bacteriophage growth and cell lysis until visible holes in the bacterial lawn, called bacteriophage plaques, are observed. Individual colonies or plaques are selected for further study by picking, literally by sampling with a sterile toothpick or other sharp object, and then reinnoculated, respectively, into a permissive growth medium or strain.

A key variable in the choice of vector is the range of target sizes that can be accommodated. Plasmids generally allow a broad range of target sizes, while bacteriophages, because of constraints in the packaging process, generally accept only a narrow range of target sizes. The earliest cloning vectors were efficient for relatively small targets, under 10 kb. Progressively interest has turned to ever larger targets, specialized vector/host systems needed to propagate such targets are described later (Boxes 2.3 and 8.2).

A second key variable in the choice of cloning vector is the copy number. Some plasmids and bacteriophages can grow in tens of thousands of copies per *E. coli* cell. This greatly facilitates recovery and purification of cloned DNA inserts. However, if DNA fragments toxic to the host are the desired targets, a high copy number vector system will make it more difficult for them to be propagated.

The major advantages of cloning, as compared to the in vitro replication systems discussed in Chapter 4, are the ability to handle very large numbers of samples in parallel and the absence of a need for any prior knowledge about the desired target sequences. The major disadvantages of cloning are that it is a relatively time-consuming process difficult to automate, and some DNA sequences are not cloneable either because they are toxic to particular hosts, or they are unstable in those hosts and fragments rearrange or delete.

However, sequences can be designed that associate to give specific structures with three, four, or more ends. These are called DNA junctions (Fig. 1.7). Similar four-stranded structures are actually seen in cells as intermediate steps in genetic recombination. However, these so-called Holliday structures are unstable, since the participating strands are pairs of identical sequences. The location of the junction can move around by a process called *branch migration*. Synthetic junctions can be designed that are stable because their DNA sequences do not allow branch migration. At low salt concentrations these structures become relatively flat, as indicated in Figure 1.7; at higher salt concentrations they form an X-shaped structure with one pair of co-axially stacked helices crossing over a second pair of co-axially stacked helices.

Still other types of DNA structures can be formed by taking advantage of the fact that in the double helix, specific hydrogen bond donor and acceptor sites on the bases remain exposed in the major groove of the double helix. Under appropriate conditions this allows the association of a third DNA strand in a sequence-specific manner to form a DNA triplex. It produces a DNA structure with four ends. The great utility of triplexes, which we will exploit in considerable detail later (Chapter 14), is that triplexes can be formed

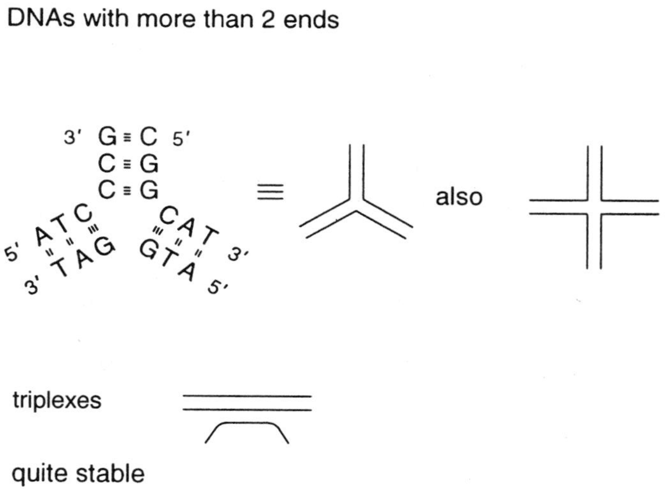

DNAs with more than 2 ends

Figure 1.7   DNA structures with more than two ends.

and broken under conditions that do not particularly alter the stability of duplexes but facilitate the sequence-specific manipulation of DNA.

DNA AS A FLEXIBLE SET OF CHEMICAL REAGENTS

Our ability to synthesize and replicate almost any DNA is encouraging novel applications of DNAs. Two examples will be sketched here: the use of DNAs as aptamers and the use of DNAs in nanoengineering. Aptamers are molecules selected to have high affinity binding to a preselected target. The plasticity of single-stranded nucleic acids polymers combined with the affinity of complementary bases to base pair provides an enormous potential for variation in three-dimensional (tertiary) structure. These properties have been taken advantage of to identify nucleic acids that bind tightly to specific ligands. In this approach random DNA or RNA libraries are made by synthesizing 60 to 100 base variable compositions. DNAs and RNAs are ideal aptamers because powerful methods exist to work with very complex mixtures of species and to purify from these mixtures just the molecules with high affinity for a target from which to characterize the common features of these classes of molecules. These cycles of design and optimization form the heart of any engineering process. (See Chapter 14 for a detailed discussion.)

The goal, in nanoengineering is to make machines, motors, transducers, and tools at the molecular level. It is really the ultimate chemical synthesis, since entire arrays of molecules (or atoms) must be custom designed for specific mechanical or electromagnetic properties. The potential advantages of using DNAs for such purposes is due to several factors: the great power to synthesize DNAs of any desired length and sequence, the ability to make complex two- and three-dimensional arrays using Holliday junctions, and the formation of structures of accurate length by taking advantage of the great stiffness of duplex DNA. (See Box 1.4.) Proteins can be anchored along the DNA at many points to create arrays with more complex properties. The disadvantages of using DNA and proteins for nanoengineering is that these structures are mechanically easier to deform than typical atomic or molecular solids; they usually require an aqueous environment, and they have relatively limited electrical or mechanical properties. There are two potential ways to circumvent the potential disadvantages of DNA in nanoengineering. One is to use DNA (and proteins) to direct the assembly of other types of molecules with the desired properties needed to make engines or transducers. A second is to use the DNA as a resist: to cast a solid surface or volume around it, remove the DNA, and then use the resulting cavity as a framework for the placement of other molecules. In both applications DNA is really conceived of as the ultimate molecular scaffold, with adjustable lengths and shapes. Time will tell if this fantasy can be ever realized in practice. A simple test of the use of DNA to control the spacing of two proteins is described in Box 1.4.

Our powerful ability to manipulate DNA is really just hinted at by some of the above discussion. We know most of the properties of short DNA molecules; our ability to make quantitative predictions about the properties of unknown sequences is not bad. What makes the problem of managing cellular DNAs difficult is their enormous size. The DNAs of small viruses have thousands of base pairs. Bacterial chromosomal DNA molecules are typically 1 to 10 million base pairs (0.3–3.0 mm) long. Human chromosomal DNAs range in size from about 0.5 to 2.5×10^8 base pairs. The largest DNA molecules in nature may approach a billion base pairs, which corresponds to molecular weights of almost a trillion Daltons. These are indeed large molecules; just describing their sequence in any detail, if we knew it, would be a formidable task. What is impressive is that all DNAs, from the largest to the smallest, can be analyzed by a small number of very similar physical and genetic techniques (Chapters 3–5). However, it is the recent development of some physical methods that have moved biological experimentation beyond the study of the relatively few organisms with well-developed genetic systems. The results of these experiments indicate that the biology of a large number of organisms seems to fall along very similar lines.

BOX 1.4
POTENTIAL USE OF DNA IN NANOENGINEERING

As a test case for nanoengineering with DNA, we have explored the use of DNA as a spacer of known length between two sites with biological functionality. The immunological problem which motivated this work was a desire to understand the complex set of molecular interactions that occurs when an antigen-presenting cell is recognized by a T lymphocyte. Antigen is presented as peptide bound to a cell surface molecule known as a major histocompatibility molecule, in complex with associated accessory membrane proteins. The detection is done by the clonetypic T cell receptor: a complex membrane protein that also can associate with other accessory proteins. Our current picture of this interaction is summarized in Figure 1.8. To refine this picture, we need to determine the structure of the complex, to decide what elements in this structure pre-exist on the cell surface in associated form in each of the participating cells before they become engaged, and to monitor the fate of the components during and after engagement. Because the complex of proteins involved in antigen presentation is so

(continued)

BOX 1.4 *(Continued)*

large, it is difficult to study cellular interaction by short-range methods such as energy transfer or crosslinking. Thus we must seek to make longer molecular rulers that can span distances of up to 100 Å or more.

The idea is to make pairs of antibodies or antibody combining sites separated by DNA spacers. Monoclonal antibodies exist for many chemical structures (epitopes) on the molecules present on the surfaces of T cells or antigen-presenting cells. The two approaches used to link such antibodies are shown schematically in Figure 1.9. The key point in both approaches is that separate, complementary DNA single strands are conjugated to particular antibodies or fragments. Then pairs of these conjugates are mixed, and the double-helix formation directs the specific production of heterospecific antibody conjugates. The length of the DNA double helix is 3.4 Å per base pair. Thus a 32-base DNA duplex will be about 100 Å long. It is expected to behave as an extremely rigid structure with respect to lateral bending, but it can undergo torsional twisting. Thus, while the distance between the tethered antibodies should be relatively fixed, the angle between them may be variable. This should help both sites reach their targets simultaneously on the same cell surface. The actual utility of such compounds remains to be established, but it is interesting that it is possible to make such constructs relatively easily.

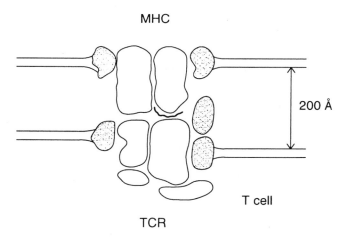

questions about assemblies:
structure
existence prior to engagement
co-internalization

Figure 1.8   Schematic illustration of the recognition of an antigen-presenting cell by a T lymphocyte, a key step in the immune response. The antigen-presenting cell has on its surface molecules of the major histocompatibility complex (MHC) which have bound an antigen (black chain). The T cell has specific receptors (TCR's) that recognize a particular MHC molecule and its bound peptide. Both cells have accessory proteins (dotted) that assist the recognition.

(continued)

BOX 1.4 *(Continued)*

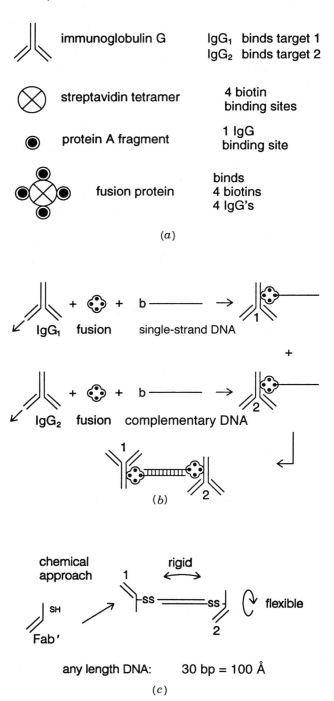

Figure 1.9   Use of DNA molecules as spacers to construct long molecular rulers. *(a)* Symbols used. *(b)* Noncovalent coupling. *(c)* Covalent coupling.

BASIC DNA BIOLOGY

DNA is the essential store of genetic information in the cell. That information must be duplicated to be passed to daughter cells by the process known as *replication*. It must be read out as an RNA copy by the process known as *transcription* so that this copy (after editing in many cases) can be used to direct the synthesis of proteins.

DNA replication is limited by the fact that all of the known enzymes that can faithfully copy the information encoded in DNA have very restricted properties. Such enzymes are called *DNA polymerases*. They all require a 3′-ended nucleic acid (or a protein substituting for this 3′-end) as a primer, and a template strand, which is copied. They cannot initiate synthesis de novo but can only elongate a strand defined by a primer (Figure 1.10). They can, for the most part, copy a DNA strand all of the way to its end. Accurate replication cannot depend on base pairing specificity alone because the thermodynamic difference between the stability of a single base pair and a mismatch is not very great (Chapter 3). Editing is used to correct any misincorporated bases. Some of this editing is carried out by the DNA polymerase itself. A newly incorporated mispaired base can be excised by the 3′-exonuclease activity of most DNA polymerases. Some enzymes also have a 5′-exonuclease activity that degrades any DNA strands in front of a wave of new synthesis. This process is called *nick translation*. Other enzymes that lack this activity will displace one strand of a duplex in order to synthesize a new strand. This process is called *strand displacement*. These enzyme activities are illustrated in Box 1.5.

Special procedures are used to synthesize the primers needed to start DNA replication. For example, RNA primers can be used and then degraded to result in a complete DNA strand. Both strands of DNA must be replicated. The unidirectional mode of DNA polymerases makes this process complicated, as shown in Figure 1.11. In a single-replication fork the synthesis of one strand (the leading strand), once initiated, can proceed in an uninterrupted fashion. The other strand (the lagging strand) must be made in a retrograde fashion from periodically spaced primers, and the resulting fragments are then stitched together. Most replication processes employ not a single fork but a bidirectional pair of replication forks, as shown in Figure 1.11.

Any DNA replication process poses an interesting topological problem, whether it is carried out by a cell or in the laboratory. Because the two DNA strands are wound about a common axis, the helix axis, they are actually also twisted around each other. They cannot be unwound without rotation of the structure, once for each helix turn. Thus DNA must spin as it is being replicated. In the bacterium *E. coli,* for example, the rate of replication is fast enough to make a complete copy of the chromosomal DNA in about 40 minutes. Since the chromosome contains almost 5 million base pairs of DNA, the required replication rate is 1.2×10^5 base pairs per minute. Since *E. coli* uses a pair of bidirectional replication forks, each must move at 6×10^4 bases per minute. Thus each side of the replication fork must unwind 6×10^3 helical turns per minute: it must rotate at 6×10^3 rpm.

DNA polymerase only goes one way

$$5' \xrightarrow{\hspace{3cm}} 3'$$
$$3' \text{———————} 5'$$

Figure 1.10 Primer extension by DNA polymerase in a template-directed manner. The newly synthesized strand is shown as a dashed arrow.

BOX 1.5
PROPERTIES OF DNA POLYMERASES

DNA polymerases play the central role in how cells replicate accurate copies of their DNA molecules. These enzymes also serve to illustrate many of the basic properties of other enzymes that make an RNA copy of DNA, a DNA copy of an RNA, or replicate RNAs. In the laboratory DNA polymerases are an extraordinarily useful tool for many of the most common ways in which DNAs are manipulated or analyzed experimentally. All DNA polymerases can extend a primer along a template in a sequence specific manner.

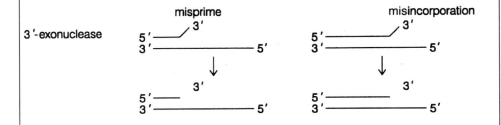

Most polymerases have one or more additional activities. A 3'-exonuclease activity will prevent the use of mispaired primers. This activity, essentially looks backward as the polymerase proceeds, and if an incorrect base is inserted, the polymerase pauses to remove it. In the absence of pppdN's, the 3'-exonuclease activity will progressively shorten the primer by removing nucleotides one at a time, even though they are correctly base-paired to the template.

Many DNA polymerases have a 5'-exonuclease activity. This degrades any DNA ahead of the site of chain extension. The result is the progressive migration of a nick in the strand complementary to the template. Thus this activity is often called nick translation.

(continued)

BOX 1.5 *(Continued)*

One function of nick translation is to degrade primers used at earlier stages of DNA synthesis. Frequently these primers are RNA molecules. Thus the nick translation, 5′-exonuclease activity ensures that no RNA segments remain in the finished DNA. This activity also is used as part of the process by which DNA damaged by radiation or chemicals is repaired.

The final DNA polymerase activity commonly encountered is strand displacement. This is observed in some mechanisms of DNA replication.

strand displacement

$$5'\underbrace{\hspace{2cm}}^{3'\,5'}\hspace{-0.5cm}3'\qquad\qquad 5'\underbrace{\hspace{3cm}}_{}3'\nearrow^{5'}$$
$$3'\underline{\hspace{4cm}}5'\qquad\qquad 3'\underline{\hspace{4cm}}5'$$

In most DNA replication, however, a separate enzyme, DNA helicase, is used to melt the double helix (separate the base-paired strands) prior to chain extention. Some DNA polymerases also have a terminal transferase activity. For example, *Taq* polymerase usually adds a single nontemplated A onto the 3′-ends of the strands that it has synthesized.

A large library of enzymes exists for manipulating nucleic acids. Although several enzymes may modify nucleic acids in the same or a similar manner, differences in catalytic activity or in the protein structure may lead to success or failure in a particular application. Hence the choice of a specific enzyme for a novel application may require an intimate knowledge of the differences between the enzymes catalyzing the same reaction. Given the sometimes unpredictable behavior of enzymes, empirical testing of several similar enzymes may be required.

More details are given here about one particularly well-studied enzyme. DNA polymerase I, isolated from *Escherichia coli* by Arthur Kornberg in 1963, established much of the nomenclature used with enzymes that act on nucleic acids. This enzyme was one of the first enzymes involved in macromolecular synthesis to be isolated, and it also displays multiple catalytic activities. DNA polymerase I replicates DNA, but it is mostly a DNA repair enzyme rather than the major DNA replication enzyme in vivo. This enzyme requires a single-stranded DNA template to provide instructions on which DNA sequence to make, a short oligonucleotide primer with a free 3′-OH terminus to specify where synthesis should begin, activated precursors (nucleoside triphosphates), and a divalent cation like $MgCl_2$. The primer oligonucleotide is extended at its 3′-end by the addition, in a 5′- to 3′-direction, of mononucleotides that are complementary to the opposite base on the template DNA. The activity of DNA polymerase I is distributive rather than processive, since the enzyme falls off the template after incorporating a few bases.

(continued)

BOX 1.5 *(Continued)*

Besides the polymerase activity, DNA polymerase I has two activities that degrade DNA. These activities are exonucleases because they degrade DNA from either the 5′- or the 3′-end. Both activities require double-stranded DNA. The 3′-exonuclease proofreading activity is the reverse reaction of the 5′- to 3′-polymerase activity. This activity enhances the specificity of extension reaction by removing a mononucleotide from the 3′-primer end when it is mismatched with the template base. This means that the specificity of the DNA polymerase I extension reaction is enhanced from $\sim 10^{-8}$ to $\sim 10^{-9}$. Both the extension and 3′-exonuclease activity reside in the same large, proteolytic degradation fragment of DNA polymerase, also called the Klenow fragment.

The 5′- to 3′-exonuclease activity is quite different. This activity resides in the smaller proteolytic degradation fragment of DNA polymerase I. It removes oligonucleotides containing 3–4 bases, and its activity does not depend on the occurrence of mismatches. A strand displacement reaction depends on a concerted effort of the extension and 5′-exonuclease activity of DNA polymerase I. Here the extension reaction begins at a single-stranded nick; the 5′-exonuclease activity degrades the single-stranded DNA annealed to the template ahead of the 3′-end being extended, thus providing a single-stranded template for the extension reaction. The DNA polymerase I extension reaction will also act on nicked DNA in the absence of the 5′-exonuclease activity. Here the DNA polymerase I just strand displaces the annealed single-stranded DNA as it extends from the nick.

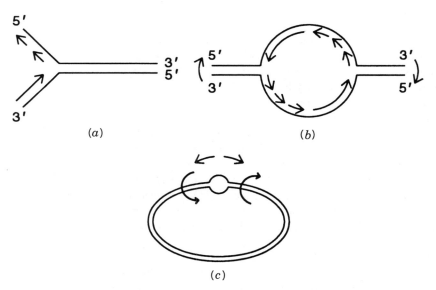

(a)   (b)

(c)

Figure 1.11   Structure of replication forks. *(a)* A single fork showing the continuously synthesized leading strand and discontinuously synthesized lagging strand. *(b)* A double fork, common in almost all genomic DNA replication. *(c)* Topological aspects of DNA replication. Thin arrows show translational motion of the forks; bold arrows show the required DNA rotation around the forks.

Unfortunately, the rotations generated by the two forks do not cancel each other out. They add. If this rotation were actually allowed to occur across massive lengths of DNA, the cell would probably be stirred to death. Instead, optional topoisomerases are used to restrict the rotation to regions close to the replication fork. Topoisomerases can cut and reseal double-stranded DNA very rapidly without allowing the cut ends to diffuse apart. Thus the rotations can let the torque generated by helix unwinding to be dissipated. (The actual mechanism of topoisomerases is more complex and indirect, but the outcome is the same as we have stated.)

The information stored in the sequence of DNA bases comprises inherited characteristics called genes. Early in the development of molecular biology, observed facts could be accounted for by the principle that one gene (i.e., one stretch of DNA sequence) codes for one protein, a linear polymer containing a sequence composed of up to 20 different amino acids, as shown in Figure 1.12a. A three-base sequence of DNA directs the incorporation of one amino acid; hence the genetic code is a triplet code. The gene would define both the start and stop points of actual transcription and the ultimate start and stop points of the translation into protein. Nearby would be additional DNA sequences that regulate the nature of the gene expression: when, where, and how much mRNA should be made. A typical gene in a prokaryote (a bacterium or any other cell without a nucleus) is one to two thousand base pairs. The resulting mRNA has some upstream and downstream untranslated regions; it encodes (one or more) proteins with several hundred amino acids.

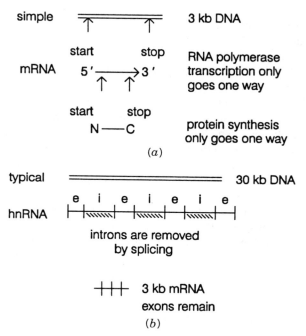

Figure 1.12 What genes are. *(a)* Transcription and translation of a typical prokaryotic gene. N and C indicate the amino and carboxyl ends of the peptide backbone of a protein. *(b)* Transcription and translation of a typical eukaryotic gene. Introns (i) are removed by splicing leaving only exons (e).

multiple reading frames

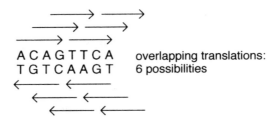

Figure 1.13 Six possible reading frames (arrows) for a stretch of DNA sequence.

The genes of eukaryotes (and even some genes in prokaryotes) are much larger and more complex. A typical mammalian gene might be 30,000 base pairs in length. Its regulatory regions can be far upstream, downstream, or buried in the middle of the gene. Some genes are known that are almost 100 times larger. A key difference between prokaryotes and eukaryotes is that most of the DNA sequence in eukaryotic genes is not translated (Fig. 1.12b). A very long RNA transcript (called *hnRNA,* for heterogeneous nuclear RNA) is made; then most of it is removed by a process called *RNA splicing.* In a typical case several or many sections of the RNA are removed, and the remaining bits are resealed. The DNA segments coding for the RNA that actually remain in the mature translated message are called *exons* (because they are expressed). The parts excised are called *introns.* The function of introns, beyond their role in supporting the splicing reactions, is not clear. The resulting eukaryotic mRNA, which codes for a single protein, is typically 3 kb in size, not much bigger than its far more simply made prokaryotic counterpart.

Now that we know the DNA structure and RNA transcription products of many genes, the notion of one gene one protein has to be broadened considerably. Some genes show a pattern of multiple starts. Different proteins can be made from the same gene if these starts affect coding regions. Quite common are genes with multiple alternate splicing patterns. In the simplest case this will result in the elimination of an exon or the substitution of one exon for another. However, much more complicated variations can be generated in this way. Finally DNA sequences can be read in multiple reading frames, as shown by Figure 1.13. If the sequence allows it, as many as six different (but not independent) proteins can be coded for by a single DNA sequence depending on which strand is read and in what frame. Note that since transcription is unidirectional, in the same direction as replication, one DNA strand is transcribed from left to right as the structures are drawn, and the other from right to left.

Recent research has found evidence for genes that lie completely within other genes. For example, the gene responsible for some forms of the disease neurofibromatosis is an extremely large one, as shown in Figure 1.14. It has many long introns, and they are tran-

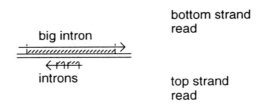

Figure 1.14 In eukaryotes some genes can lie within other genes. A small gene with two introns coded for by one of the DNA strands lies within a single intron of a much larger gene coded for by the other strand. Introns are shown as hatched.

scribed off of the opposite strand used for the transcription of the type one neurofibromatosis gene. The small gene is expressed, but its function is unknown.

GENOME SIZES

The purpose of the human genome project is to map and sequence the human genome and find all of the genes it contains. In parallel, the genomes of a number of model organisms will also be studied. The rationale for this is clear-cut. The human is a very poor genetic organism. Our lifespan is so long that very few generations can be monitored. It is unethical (and impractical) to control breeding of humans. As a result one must examine inheritance patterns retrospectively in families. Typical human families are quite small. We are a very heterogeneous outbred species, with just the opposite genetic characteristics of highly inbred, homogeneous laboratory strains of animals used for genetic studies. For all these reasons experimental genetics is largely restricted to model organisms. The gold standard test for the function of a previously unknown gene is to knock it out and see the resulting effect, in other words, determine the phenotype of a deletion. For organisms with two copies of their genome, like humans, this requires knocking out both gene copies. Such a double knockout is extremely difficult without resorting to controlled breeding. Thus model organisms are a necessary part of the genome project.

Considerable thought has gone into the choice of model organisms. In general, these represent a compromise between genome size and genetic utility. *E. coli* is the best-studied bacterium; its complete DNA sequence became available early in 1997. *Saccharomyces cerevisiae* is the best studied yeast, and for that matter the best studied single-cell eukaryotic organism. Its genetics is exceptionally well developed, and the complete 12,068 kb DNA sequence was reported in 1996, the result of a worldwide coordinated effort for DNA sequencing. *Caenorhabditis elegans,* a nematode worm has very well-developed genetics; its developmental biology is exquisitely refined. Every cell in the mature organism is identified as are the cell lineages that lead up to the mature adult. The last invertebrate canonized by the genome project is the fruit fly *Drosophila melanogaster.* This organism has played a key role in the development of the field of genetics, and it is also an extraordinarily convenient system for studies of development. The fruit fly has an unusually small genome for such a complex organism; thus the utility of genomic sequence data is especially apparent in this case.

For vertebrates, if a single model organism must be selected, the mouse is the obvious choice. The size of the genome of *Mus musculus* is similar to that of humans. However, its generation time is much shorter, and the genetics of the mouse is far easier to manipulate. A number of inbred strains exist with relatively homozygous but different genomes; yet these will crossbreed in some cases. From such interspecific crosses very powerful genetic mapping tools emerge, as we will describe in Chapter 6. Mice are small, hence relatively inexpensive to breed and maintain. Their genetics and developmental biology are relatively advanced. Because of efforts to contain the projected costs of the genome project, no other "official" model organisms exist. However, many other organisms are of intense interest for genome studies; some of these are already under active scrutiny. These include maize, rice, *Arabidopsis thaliana,* rats, pigs, cows, as well as a number of simpler organisms.

In thinking of which additional organisms to subject to genome analysis, careful attention must be given to what is called the G-value paradox. Within even relatively similar classes of organisms, the genome size can vary considerably. Furthermore, as the data in Table 1.1 reveal, there is not a monotomic relationship between genome size and our

TABLE 1.1   Genome Sizes (base pairs)

Bacteriophage lambda	5.0×10^4
Escherichia coli	4.6×10^6
Yeasts	12.0×10^6
Giardia lamblia	14.0×10^6
Drosophila melanogaster	1.0×10^8
Some hemichordates	1.4×10^8
Human	3.0×10^9
Some amphibians	8.0×10^{11}

Note: These are haploid genome sizes. Many cells will have more than one copy of the haploid genome.

view of how evolutionarily advanced a particular organism is. Thus, for example, some amphibians have genomes several hundred times larger than the human. Occasional organisms like some hemichordates or the puffer fish have relatively small genomes despite their relatively recent evolution. The same sort of situation exists in plants.

In planning the future of genome studies, as attention broadens to additional organisms, one must decide whether it will be more interesting to examine closely related organisms or to cast as broad a phylogenetic net as funding permits. Several organisms seem to be of particular interest at the present time. The fission yeast *Schizosaccharomyces pombe* has a genome the same size as the budding yeast *S. cerevisiae*. However, these two organisms are as far diverged from each other, evolutionarily, as each is from a human being. The genetics of *S. pombe* is almost as facile as that of *S. cerevisiae*. Any features strongly conserved in both organisms are likely to be present throughout life as we know it. Both yeasts are very densely packed with genes. The temptation to compare them with full genomic sequencing may be irresistible. Just how far genome studies will be extended to other organisms, to large numbers of different individuals, or even to repeated samplings of a given individual will depend on how efficient these studies eventually become. The potential future need for genome analysis is almost unlimited, as described in Box 1.6.

BOX 1.6
GENOME PROJECT ENHANCEMENTS

DNA Sequencing Rate:

bp Per Person Per Day	Accessible Targets
10^6	One human, five selected model organisms
	Organisms of commercial value
10^7	Selected diagnostic DNA sequencing
10^8	Human diversity (see Chapter 15)
	5×10^9 individuals $\times$ 6 to 12×10^6 differences = 3 to 6×10^{16}
	Full diagnostic DNA sequencing
10^9	Environment exposure assessment

NUMBERS OF GENES

It has been estimated that half of the genes in the human genome are central nervous system specific. For such genes, one must wonder how adequate a model the mouse will be for the human. Even if there are similar genes in both species, it is not easy to see how the counterparts of particular human phenotypes will be found in the mouse. Do mice get headaches, do they get depressed, do they have fantasies, do they dream in color? How can we tell? For such reasons it is desirable, as the technology advances to permit this, to bring into focus the genomes of experimental animals more amenable to neurophysiological and psychological studies. Primates like the chimp are similar enough to the human that it should be easy to study them by starting with human material as DNA probes. Yet the differences between humans and chimps are likely to be of particular interest in defining the truly unique features of our species. Other vertebrates, like the rat, cat, and dog, while more distant from the human, may also be very attractive genome targets because their physiologies are very convenient to study, and in some cases they display very well-developed personality traits. Other organisms, such as the parasitic protozoan, *Giardia lamblia* or the blowfish, fugu, are eukaryotes of particular interest because of their comparatively small genome sizes.

The true goal of the genome project is to discover all of the genes in an organism and make them available in a form convenient for future scientific study. It is not so easy, with present tools and information, to estimate the number of genes in any organism. The first complete bacterial genome to be sequenced is that of *H. influenzae* Rd. It has 1,830,137 base pairs and 1743 predicted protein coding regions plus six sets of three rRNA genes and numerous genes for other cellular RNAs like tRNA. *H. influenzae* is not as well studied as *E. coli,* and we do not yet know how many of these coding regions are actually expressed. For the bacterium *E. coli,* we believe that almost all genes are expressed and translated to at least a detectable extent. In two-dimensional electrophoretic fractionations of *E. coli* proteins, about 2500 species can be seen. An average *E. coli* gene is about 1 to 2 kb in size; thus the 4.6 Mb genome is fully packed with genes. Yeasts are similarly packed. Further details about gene density are given in Chapter 15.

In vertebrates the gene density is much more difficult to estimate. An average gene is probably about 30 kb. In any given cell type, 2d electrophoresis reveals several thousand protein products. However, these products are very different in different cell types. There is no way to do an exhaustive search. Various estimates of the total number of human genes range from 5×10^4 to 2×10^5. The true answer will probably not be known until long after we have the complete human DNA sequence, because of the problems of multiple splicing patterns and genes within genes discussed earlier. However, by having cloned and sequenced the entire human genome, any section of DNA suspected of harboring one or more genes will be easy to scrutinize further.

SOURCES AND ADDITIONAL READINGS

Alivisatos, A. P., Jonsson, K. P., Peng, X., Wilson, T. E., Loweth, C. J., Bruchez, M. P., and Schultz, P. G. 1996. Organization of "nanocrystal molecules" using DNA. *Nature* 382: 609–611.

Berman, H. M. 1997. Crystal studies of B-DNA: The answers and the questions. *Biopolymers* 44: 23–44.

Berman, H. M., Olson, W. K., Beveridge, D. L., Westbrook J., Gelbin, A., Demeny, T., Hsieh, S.-H., Srinivasan, A. R., and Schneider, B. 1992. The Nucleic Acid Database: A comprehensive relational database of three-dimensional structures of nucleic acids. *Biophysical Journal* 63: 751–759.

Cantor, C. R., and Schimmel, P. R. 1980. *Biophysical Chemistry.* San Francisco: W. H. Freeman, ch. 3 (Protein structure) and ch. 4 (Nucleic acid structure).

Garboczi, D. N., Ghosh, P., Utz, U., Fan, Q. R., Biddison, W. E, and Wiley, D. C. 1996. Structure of the complex between human T-cell receptor, viral peptide and HLA-A2. *Nature* 384: 134–141.

Hartmann, B., and Lavery, R. 1996. DNA structural forms. *Quarterly Review of Biophysics* 29: 309–368.

Klinman, D. A., Yi, A., Beaucage, S., Conover, J., and Krieg, A. M. 1996. CpG motifs expressed by bacterial DNA rapidly induce lymphocytes to secrete IL-6, IL-12, and IFN-g. *Proceeding of the National Academy of Sciences USA* 93: 2879–2883.

Lodish, H., Darnell, J., and Baltimore, D. 1995. *Molecular Cell Biology,* 3rd. ed. New York: Scientific American Books.

Mao, C., Sun, W., and Seeman, N. C. 1997. Assembly of Borromean rings from DNA. *Nature* 386: 137–138.

Mirkin, C. A., Letsinger, R. L., Mucic, R. C., and Storhoff, J. J. 1996. A DNA-based method for rationally assembling nanoparticles into macroscopic materials. *Nature* 382: 607–609.

Niemeyer, C. M., Sano, T., Smith, C. L., and Cantor, C. R. 1994. Oligonucleotide-directed self-assembly of proteins: Semisynthetic DNA-streptavidin hybrid molecules as connectors for the generation of macroscopic arrays and the construction of supramolecular bioconjugates. *Nucleic Acids Research* 22: 5530–5539.

Saenger, W. 1984. *Principles of Nucleic Acid Structure.* New York: Springer-Verlag.

Timsit, H. Y., and Moras, D. 1996. Cruciform structures and functions. *Quarterly Review of Biophysics* 29: 279–307.

2 A Genome Overview at the Level of Chromosomes

BASIC PROPERTIES OF CHROMOSOMES

Chromosomes were first seen by light microscopy, and their name reflects the deep color they take on with a number of commonly used histological stains. In a cell a chromosome consists of a single piece of DNA packaged with various accessory proteins. Chromosomes are the fundamental elements of inheritance, since it is they that are passed from cell to daughter cell, from parent to progeny. Indeed a cell that did not need to reproduce would not have to keep its DNA organized into specific large molecules. Some single-cell organisms, like the ciliate Tetrahymena, actually fragment a working copy of their DNA into gene-sized pieces for expression while maintaining an unbroken master copy for reproductive purposes.

BACTERIAL CHROMOSOMES

Bacteria generally have a single chromosome. This is usually a circular DNA duplex. As shown in Figure 2.1, the chromosome has at least three functional elements. The replication origin (ori) is the location of the start of DNA synthesis. The termination (ter) region provides a mechanism for stopping DNA synthesis of the two divergent replication forks. Also present are *par* sequences which ensure that chromosomes are partitioned relatively uniformly between daughter cells. A description of a bacterial chromosome is complicated by the fact that bacteria are continually replicating and transcribing their DNA.

In rapidly growing organisms, a round of replication is initiated before the previous round is completed. Hence the number of copies of genomic DNA depends on how rapidly the organism is growing and where in the genome one looks. Genes near the origin are often present at several times the copy number of genes near the terminus. In general, this seems to have little effect on the bacterium. Rather, bacteria appear to take advantage of this fact. Genes whose products are required early in the replication cycle, or in large amounts (like the ribosomal RNAs and proteins), are located in the early replicated regions of the chromosomes. Although the bacterial chromosome is relatively tolerant of deletions (involving nonessential genes) and small insertions, many large rearrangements involving inversion or insertions are lethal. This prohibition appears to be related to conflicts that rise between convergent DNA replication forks and the transcription machinery of highly expressed genes.

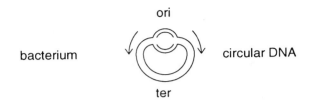

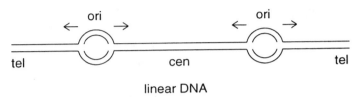

Figure 2.1   Basic functional elements in chromosomes: ori (replication origin), tel (telomere), cen (centromere), ter (termination region). Little is known about eukaryotic termination regions.

Bacteria will frequently harbor additional DNA molecules smaller than their major chromosome. Some of these may be subject to stringent copy number control and to orderly partitioning, like the major chromosome. These low copy number plasmids use the same DNA replication machinery as the chromosome, and they are present in a copy number that is equal to the chromosomal copy number. An example is the naturally occurring F+ DNA of *E. coli* which represents about 2% of the DNA present in the cell.

In some bacteria essential genes are located on two genetic elements. For instance, in *Rhodobacteria sphaeroides* the genes encoding the ribosomal RNAs (rRNAs) are located on a 0.9 Mb chromosome, whereas the remainder of the genome is located on a 3 Mb chromosome. In Pseudomonas species, many genes that code for catabolic (degradative) enzymes are located on large extrachromosomal plasmids. Plasmids containing antibiotic resistance genes have been isolated from many species. In part, it is the rapid transfer of plasmids, sometimes even between different genera, that accounts for the rapid development of large populations of antibiotic resistant bacteria.

The control of replication of other, usually smaller plasmids, is more relaxed. These plasmids can have a very high intracellular copy number, and they are usually the focus of recombinant DNA cloning experiments. These plasmids use a DNA replication mechanism that is distinct from that used by the chromosomes. In fact selective inhibition of chromosomal replication machinery focuses the cell replication machinery on producing more plasmid such that it is possible to increase the copy number of these plasmids to about 1000 copies per cell. Selection of growth conditions, which depend on genes carried by the plasmid, can also be used to increase or decrease its copy number. Some plasmids do not contain a par functioning region and are not partitioned in an orderly fashion to daughter cells. This means that their inheritance is subject to statistical fluctuations that may lead to significant instabilities especially with low copy plasmids. Since the genome is defined as all of the DNA in a cell, plasmids and other extrachromosomal DNA elements must be counted as part of it.

Bacterial chromosomes contain bound protein molecules that are essential for normal growth. Some of these promote an organization or packaging of DNA similar to that seen in higher organisms. However, the proteins that appear to be responsible for packaging are present in such small amounts that can only interact with about 20% of the genomic DNA. Furthermore the packaging does not seem to be as orderly or as stable as the packaging of DNA in eukaryotic chromosomes. Bacterial chromosomal DNA is organized into topological constrained domains (Box 2.1) that average 75 kb in size. The way in which this occurs, and its functional consequences, are not yet understood in a rigorous way.

BOX 2.1
TOPOLOGICAL PROPERTIES OF DNA

Because the two strands of DNA twist around a common axis, for unbent DNA, each turn of the helix is equivalent to twisting one strand 360° around the other. Thus, when the DNA is circular, if one strand is imagined to be planar, the other is wrapped around it once for each helix turn. The two circular strands are thus linked topologically. They cannot be pulled apart except by cutting one of the strands. A single nick anywhere in either strand removes this topological constraint and allows strand separation. The topological linkage of the strands in circular DNA leads to a number of fascinating phenomena, and the interested reader is encouraged to look elsewhere for detailed descriptions of how these are studied experimentally and how they are analyzed mathematically. (Cantor and Schimmel, 1980; Cozzarelli and Wang, 1990).

For the purpose of this book, the major thing the reader must bear in mind is that physical interactions between DNA strands can be blocked by topological constraints. If a linear strand of DNA contacts a surface at two points, the region between these points is topologically equivalent to a circle that runs through the molecule and then through the surface. Hybridization of a complementary strand to this immobilized strand will require twisting the former around the latter. This may be difficult if the latter is close to the surface, and it will be impossible if the former is circular itself.

Cells deal with the topological constraints of DNA double helices by having a series of enzymes that relax these constraints. Type I topoisomerases make a transient single-stranded nick in DNA, which allows one strand to rotate about the other at that point. Type II topoisomerases make a transient double-strand nick and pass an intact segment of the duplex through this nick. Though it is less obvious, this has the effect of allowing the strands to rotate 720° around each other. When DNA is replicated or transcribed, the double helix must be unwound ahead of and rewound behind the moving polymerases. Topoisomerases are recruited to enable these motions to occur.

In condensed chromatin, loops of 300 Å fiber are attached to a scaffold. Although the DNA of mammalian chromosomes is linear, these frequent attachment points make each constrained loop topologically into a circle. Thus topoisomerases are needed for the DNA in chromatin to function. Type II topoisomerases are a major component of the proteins that make up the chromosome scaffold.

CHROMOSOMES OF EUKARYOTIC ORGANISMS

All higher organisms usually have linear chromosomal DNA molecules, although circles can be produced under special circumstances. These molecules have a minimum of three functional features, as shown in Figure 2.1. Telomeres are specialized structures at the ends of the chromosome. These serve at least two functions. They provide a mechanism by which the ends of the linear chromosomes can be replicated. They stabilize the ends. Normal double-strand DNA ends are very unstable in eukaryotic cells. Such ends could be the result of DNA damage, such as that caused by X rays, and could be lethal events. Hence very efficient repair systems exist that rapidly ligate ends not containing telomeres together. If several DNAs are broken simultaneously in a single cell, the correct fragment pairs are unlikely to be reassembled, and one or more translocations will result. If the ends are not repaired fast enough by ligation, they may invade duplex DNA in order to be repaired by recombination. The result, even for a single, original DNA break, is a re-arranged genome.

Centromeres are DNA regions necessary for precise segregation of chromosomes to daughter cells during cell division. They are the binding site for proteins that make up the kinetochore, which in turn serves as the attachment site for microtubules, the cellular organelles that pull the chromosomes apart during cell division.

Another feature of eukaryotic chromosomes is replication origins. We know much less about the detailed structural properties of eukaryotic origins than prokaryotic ones. What is clear from inspecting the pattern of DNA synthesis along chromosomes is that most chromosomes have many active replication origins. These do not necessarily all initiate at the same time, but a typical replicating chromosome will have many active origins. The presence of multiple replication origins allows for complete replication of entire human genome in only 8 hours. It is not known how these replication processes terminate.

CENTROMERES

In the yeast, *S. cerevisiae,* the centromere has been defined by genetic and molecular experiments to reside in a small DNA region, about 100 bp in length. This region contains several A–T rich sequences that may be of key importance for function. The centromere of *S. cerevisiae* is very different in size and characteristics than the centromeres of more advanced organisms, or even the yeast *S. pombe.* This is perhaps not too surprising in view of the key role that the centromere plays in cell division. Unlike these other species, *S. cerevisiae* does not undergo symmetrical cell division. Instead, it buds and exports one copy of each of its chromosomes to the daughter cell. In species that produce two nominally identical daughter cells, centromeres appear to be composed mostly of DNA with tandemly repeating sequences. The most striking feature of these repeats is that the number of copies can vary widely. A small repeating sequence, $(GGAAT)_n$, has recently been found to be conserved across a wide range of species. This conservation suggests that the sequence may be a key functional element in centromeres. While not yet proved, physical studies on this DNA sequence reveal it to have rather unusual helical properties that at least make it an attractive candidate for a functional element. As shown in Figure 2.2, the G-rich strand of the repeat has a very stable helical structure of its own, although the detailed nature of this structure is not yet understood.

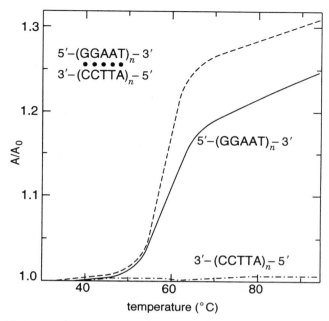

Figure 2.2   Evidence for the presence of some kind of unusual helical structure in a simple centromeric repeating sequence. The relative amount of absorbance of 260 nm UV light is measured as a function of temperature. Shown are results for the two separated strands of a duplex, and the duplex itself. What is unusual is that one of the separated strands shows a change in absorbance that is almost as sharp and as large as the intact duplex. This indicates that it, alone, can form some sort of helical structure. (From Moyzis et al., 1988.)

The other repeats in the centromeres are much longer tandemly repeated structures. In the human these were originally termed *satellite DNA* because they were originally seen as shoulders or side bands when genomic DNA fragments were fractionated by base composition by equilibrium ultracentrifugation on CsCl gradients (see Chapter 5). For example, the alpha satellite of the african green monkey is a 171 base pair tandem repeat. Considerable effort has gone into determining the lengths and sequences of some of these satellites, and their organization on the chromosome. The results are complex. The repeats are often not perfect; they can be composed of blocks of different lengths or different orientation. The human alpha satellite, which is very similar in composition (65% identity) to the african green monkey satellite, does not form a characteristic separate band during density ultracentrifugation. Thus the term satellite has evolved to now include tandemly repeated DNA sequences which may be of the same composition of the majority of genome. An example of a satellite sequence is shown in Figure 2.3.

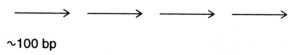

~100 bp

Figure 2.3   Example of a tandemly repeating DNA sequence in centromeric DNA.

The implications of the specific sequences of the repeats, and their organization, for cen-tromere function are still quite cloudy.

The size of centromeres, judged by the total length of their simple sequence blocks, varies enormously among species, and even among the different chromosomes contained in one cell, without much indication that this is biologically important. Centromeres in *S. pombe* are only 10^4 to 10^5 bases in size, while those in the human can be several Mb. Centromeres on different human chromosomes appear to be able to vary in size widely, and within the population there is great heterogeneity in the apparent size of some cen-tromeres. It is as though nature, having found a good thing, doesn't care how much of it there is as long as it is more than a certain minimum.

TELOMERES

Telomeres in almost all organisms with linear chromosomes are strikingly similar. They consist of two components as shown in Figure 2.4. At the very end of the chromosome is a long stretch of tandemly repeating sequence. In most organisms the telomere is domi-nated by the hexanucleotide repeat $(TTAGGG)_n$. The repeating pattern is not perfect; sev-eral other sequences can be interspersed. In a few species the basic repeat is different, but it always has the characteristic that one strand is T and G-rich. The G + T rich strand is longer than the complementary strand, and thus the ends of the chromosome have a pro-truding 3'-end strand of some considerable length. This folds back on itself to make a sta-ble helical structure. The best evidence suggests that this structure is four stranded, in analogy to the four-strand helix made by aggregates of G itself. Pairs of telomeres might have to associate in order to make this structure. Alternatively, it could be made by loop-ing the 3'-end of one chromosome back on itself three times. The details not withstand-ing, these structures are apparently effective in protecting the ends of the chromosomes from attack by most common nucleases.

Next to the simple sequence telomeric repeats, most chromosomes have a series of more complex repeats. These frequently occur in blocks of a few thousand base pairs. Within the blocks there may be some tandemly repeated sequence more complex than the hexanucleotide telomeric repeat. The blocks themselves are of a number of different types, and these are distributed in different ways on different chromosomes. Some unique sequences, including genes, may occur between the repeating sequences. It is not clear if any chromosome really has a unique telomere, or if this matters. Some researchers feel that the sub-telomeric repeats may play a role in positioning the ends of the chromosomes at desired places within the nucleus. Whether and how this information might be coded by the pattern of blocks on a particular chromosome remains to be determined. At least some subtelomeric sequences vary widely from species to species.

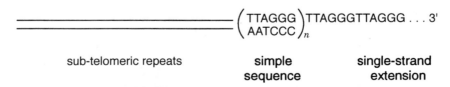

sub-telomeric repeats   simple   single-strand
                               sequence   extension

Figure 2.4   Structure of a typical telomere.

DYNAMIC BEHAVIOR OF TELOMERES

The actual length of the simple telomeric repeating DNA sequence is highly variable both within species and between species. This appears to be a consequence, at least in part, of the way in which telomeres are synthesized and broken down. Telomeres are not static structures. If the growth of cells is monitored through successive generations, telomeres are observed to gradually shrink and then sometimes lengthen considerably. Nondividing cells and some cancer cells appear to have abnormal telomere lengths. At least two mechanisms are known that can lead to telomere degradation. These are shown in Figure 2.5a. In one mechanism the single-strand extension is subject to some nuclease cleavage. This shortens that strand. The alternate mechanism is based on the fact that the 5′-ended strand must serve as a starting position for DNA replication. This presumably occurs by the generation of an RNA primer that is then extended inward by DNA replication. Because of the tandem nature of the repeated sequence, the primer can easily be displaced inward before synthesis continues. This will shorten the 5′-strand. Both of these mechanisms seem likely to occur in practice.

A totally different mechanism exists to synthesize telomeres and to lengthen existing telomeres (Fig. 2.5b). The enzyme telomerase is present in all cells with telomeres. It is a ribonucleoprotein. The RNA component is used as a template to direct the synthesis of the 3′-overhang of the telomere. Thus the telomere is lengthened by integral numbers of repeat units. It is not known how the complex and subtle variations seen in this simple sequence arise in practice. Perhaps there is a family of telomerases with different templates. More likely, some of the sequences are modified after synthesis, or the telomerase may just be sloppy.

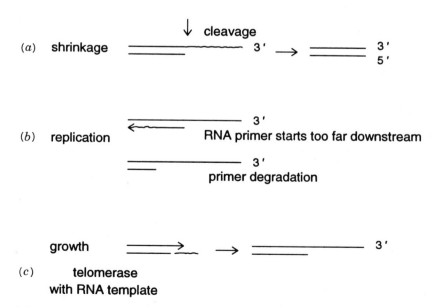

Figure 2.5 Dynamics of telomeric simple repeating sequences. Mechanisms for telomere shrinkage: *(a)* Nuclease cleavage; *(b)* downstream priming by telomerase; *(c)* mechanism of telomere growth.

The total length of telomeric DNA in the human ranges from 10 to 30 kb. Because of its heterogeneity, fragments of DNA cut from any particular telomere by restriction enzymes have a broad size range and appear in gel electrophoretic size fractionations as broad, fuzzy bands. This is sufficiently characteristic of telomere behavior to constitute reasonable evidence that the band in question is telomeric. In mice, the length of telomeres is much longer, typically 100 kb. We don't know what this means. In general, the message from both telomeres and centromeres is that eukaryotic cells apparently feel no pressure to minimize the size of their genome or the sizes of these important functional elements. Unlike viruses, which must fit into small packages for cell escape and reinfection, and bacterial cells which are under selection pressure in rich media to replicate as fast as they can, eukaryotic cells are not mean and lean.

CHROMATIN AND THE HIGHER-ORDER STRUCTURE OF CHROMOSOMES

A eukaryotic chromosome is only about half DNA by weight. The remainder is protein, mostly histones. There are five related, very basic histone proteins that bind tightly to DNA. The rest is a complex mixture called, loosely, *nonhistone chromosomal proteins.* This mixture consists of proteins needed to mediate successive higher-order packaging of the DNA, proteins needed for chromosome segregation, and proteins involved in gene expression and regulation. An enormous effort has gone into characterizing these proteins and the structures they form. Nevertheless, for the most part their role in higher-order structure or function is unknown.

At the lowest level of chromatin folding, 8 histones (two each of four types) assemble into a globular core structure that binds about 140 bp of DNA forming it into a coiled structure called the nucleosome (Fig. 2.6). Not only are nucleosomes very similar in all organisms from yeast to humans but in addition the four core histones are among the most evolutionarily conserved proteins known. Nucleosomes pack together to form a filament that is 100 Å in diameter and is known by this name. The details of the filament are different in different species because the lengths of the spacer DNA between the nucleosomes varies. In turn the 100 Å filament is coiled upon itself to make a thicker structure, called the 300 Å fiber. This appears to be solenoidal in shape. Stretches of solenoid containing on average 50 to 100 kb of DNA are attached to a protein core. In condensed metaphase chromosomes, this core appears as a central scaffold of the chromosome; it can be seen when the chromosome is largely stripped of other proteins and examined by electron microscopy (Fig. 2.7). A major component of this scaffold is the enzyme topoisomerase II (Box 2.1), which probably serves a role analogous to DNA gyrase in *E. coli* of acting as a swivel to circumvent any topological problems caused by the interwound nature of DNA strands.

At other stages in the cell cycle, the scaffold proteins may actually attach to the nuclear envelope. The chromosome is then suspended from this envelope into the interior of the nucleus. During mitosis, if this picture is correct, the chromosomes are then essentially turned inside out, separated into daughters, and reinverted. The topological domains created by DNA attachment to the scaffold at 50 to 100 kb intervals are similar in size to those seen in bacteria, where they appear to be formed by much simpler structures.

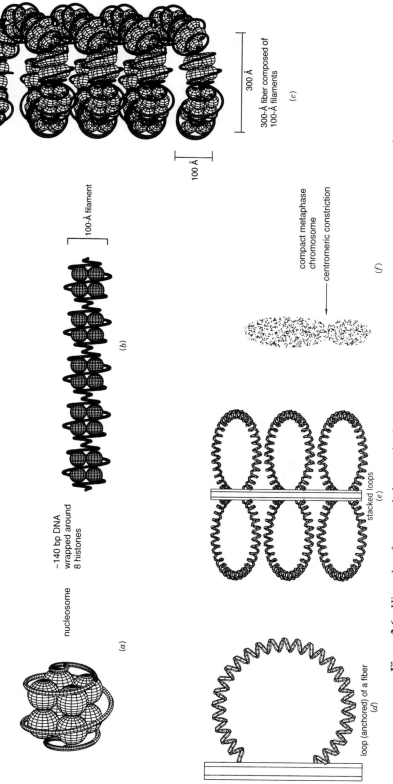

Figure 2.6 Hierarchy of structural elements in chromatin and chromosomes: (*a*) The nucleosome; (*b*) a 100 Å filament; (*c*) a 300 Å solenoid. (*d*) chromosome loop anchored to a protein scaffold; (*e*) successive loops stacked along the scaffold; (*f*) the appearance of a condensed metaphase chromosome. The path of the scaffold is unknown, but it is not straight.

hierarchy of structures

nucleosome

~140 bp DNA wrapped around 8 histones

(*a*)

100-Å filament

(*b*)

300-Å fiber composed of 100-Å filaments

300 Å

100 Å

(*c*)

loop (anchored) of a fiber

(*d*)

stacked loops

(*e*)

compact metaphase chromosome

centromeric constriction

(*f*)

The most compact form of chromosomes occurs in metaphase. This is reasonable because it is desirable to pass chromosomes efficiently to daughter cells. Surely this is facilitated by having more compact structures than the unfolded objects seen in Figure 2.7. Metaphase chromosomes appear to consist of stacks of packed 300 Å chromatin fiber loops. Their structures are quite well controlled. For example, they have a helical polarity that is opposite in the two sister chromatids (the chromosome pairs about to separate in cell division). When the location of specific genes on metaphase chromosomes is examined (as described later), they appear to have a very fixed position within the morphologically visible structure of the chromosome. A characteristic feature of all metaphase chromosomes is that the centromere region appears to be constricted. The reason for this is not known. Also for unknown reasons some eukaryotic organisms, such as the yeast *S. cerevisiae* do not have characteristic condensed metaphase chromosomes.

The higher-order structure of chromatin and chromosomes poses an extraordinary challenge for structural biologists because they are so complex and because these structures are so large. The effort needed to reveal the details of these structures may not be worthwhile. It remains to be shown whether the details of much of the structure actually matter for any particular biological function. One extreme possibility is that this is all cheap packaging; that most of it is swept away whenever the underlying DNA has to be uncovered for function in gene expression or in recombination. We do not yet know if this is the case, but our ability to manipulate DNAs and chromosomes has grown to the point where it should soon be possible to test such notions explicitly. It would be far more elegant and satisfying if we uncover sophisticated mechanisms that allow DNA packaging and unpackaging to be used to modulate DNA function.

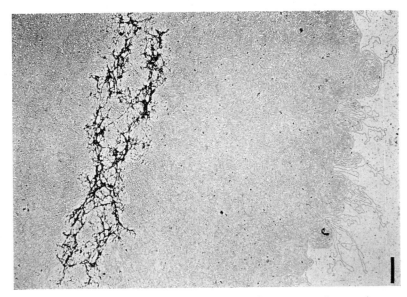

Figure 2.7 Electron micrograph of a single mammalian chromosome, denatured to remove most of the protein and allow the DNA to expand. The X-shaped structure is the protein scaffold that defines the shape of the condensed chromosome. The DNA appears as a barely resolvable mass of fiber covering almost the entire field.

CHROMOSOMES IN THE CELL CYCLE

Bacteria are organisms without nuclei. They are continuously synthesizing DNA and dividing if food is plentiful. In contrast, nucleated cells are often quiescent. Sometimes they are even frozen forever in a nondividing state; examples are cells in the brain or heart muscle. Most eukaryotic cells proceed through a similar cycle of division and DNA synthesis, illustrated in Figure 2.8. Cell division is called *mitosis*. It occurs at the stage labeled M in the figure. After cell division, there is a stage, G1, during which no DNA synthesis occurs. Initiation of DNA synthesis, triggered by some stimulus, transforms the cell to the S phase. Not all of the DNA is necessarily synthesized in synchrony. Once synthesis is completed, another resting phase ensues, G2. Finally, in response to a mitogenic stimulus, the cell enters metaphase, and mitosis occurs in the M stage. In different cell types the timing of the cycle, and the factors that induce its progression, can vary widely.

Only in the M stage are the chromosomes compact and readily separable or visualizable under the light microscope. In other cell cycle stages, most portions of chromosomes are highly extended. The extended regions are called *euchromatin*. Their extension appears to be a prerequisite for active gene expression. This is reasonable considering the enormous steric and topological barriers that would have to be overcome to express a DNA sequence embedded in the highly condensed chromatin structure hierarchy. There are regions, called *heterochromatin*, unusually rich in simple repeated sequences that do not decondense after metaphase but instead remain condensed throughout the cell cycle. Heterochromatin is characteristic of centromeres but can occur to different extents in other regions of the genome. It is particularly prevalent on the human Y chromosome which contains the male sex determining factor but relatively few other genes. Heterochromatic regions are frequently heterogeneous in size within a species. They are generally sites where little or no gene expression occurs.

In most cases the level of expression of a gene does not depend much on its position in the genome, so long as the cis-acting DNA regions needed for regulation are kept reasonably near the gene. In fact typical eukaryotic genes are bracketed by sequences, such as enhancers or nuclear scaffold sites that eliminate transcriptional cross talk between adjacent genes. However, there are some striking exceptions to this rule, called *positional variation*. Using genetic or molecular methods, genes can be moved from euchromatic regions to heterochromatic regions. This usually results in their being silenced. Silencing also sometimes occurs when genes are placed near to telomeres (Fig. 2.9). The mechanism of this latter silencing is not understood.

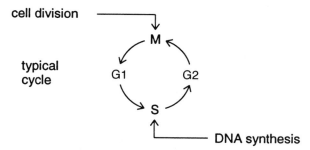

Figure 2.8   The cell division cycle in almost all eukaryotic cells. M means a cell undergoing mitosis; S means a cell in the act of DNA synthesis.

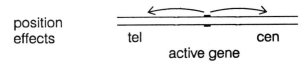

Figure 2.9 Differences in chromatin affect gene expression. Genes transposed to heterochromatic or telomeric regions are sometimes silenced.

GENOME ORGANIZATION

There are very visible patterns of gene organization. Many genes occur in families, such as globins, immunoglobulins, histones, or zinc finger proteins. These families presumably arose mostly by gene duplications of a common precursor. Subsequent evolutionary divergence led to differences among family members, but usually sufficient traces remain of their common origin through conserved sequences to allow family members to be identified. An alternative mechanism for generating similar sets of genes is convergent evolution. While examples of this are known, it does not appear to be a common mechanism.

The location of gene families within a genome offers a fascinating view of some of the processes that reshape genomes during evolution. Some families are widely dispersed throughout the genome such as zinc finger proteins, although these may have preferred locations. Other families are tightly clustered. An example is the globin genes shown in Figure 2.10. These lie in two clusters: one on human chromosome 11 and one on human chromosome 16. Each cluster has several active genes and several pseudogenes, which may have been active at one time but now are studded with mutations that make them unable to express functional protein. Some families like the immunoglobulins are much more complex than the globin family.

When metaphase chromosomes are stained in various different ways and examined in the light microscope, a distinct pattern of banding is seen. An example is shown in Figure 2.11a for human chromosomes. The same bands are usually seen with different stains, implying that this pattern is a reflection of some general intrinsic property of the chromosomes rather than just an idiosyncratic response to a particular dye or a particular staining protocol. Not all genomes show such distinct staining patterns as the human, but most higher organisms do.

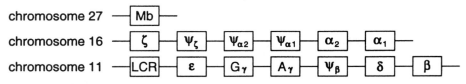

Figure 2.10 Genomic organization of the human globin gene family. Hemoglobins expressed in the adult are alpha and beta; hemoglobins expressed in the embryo are gamma and delta; hemoglobins expressed in the early embryo are zeta and eta. Myoglobin is expressed throughout development. Gene symbols preceded by psi are pseudogenes, no longer capable of expression.

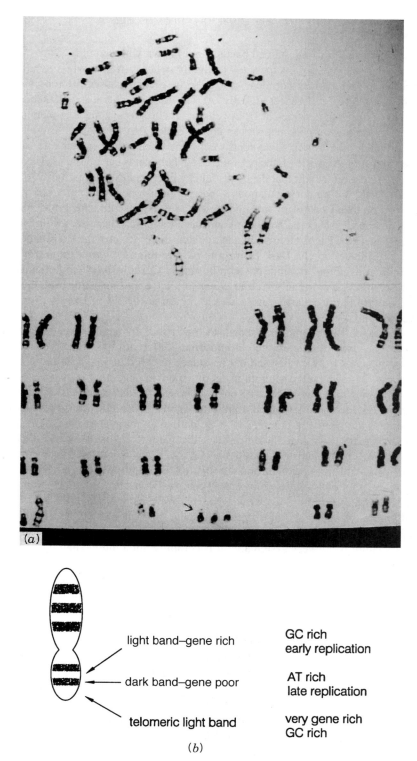

light band–gene rich   GC rich
early replication

dark band–gene poor   AT rich
late replication

telomeric light band   very gene rich
GC rich

(b)

Figure 2.11 Chromosome banding. *(a)* A typical preparation of banded chromosomes. Cells are arrested in metaphase and stained with Geimsa. Individual chromosomes are identified by the pattern of dark and light bands, and rearranged manually for visual convenience. The particular individual in this case is a male because there is one X and one Y chromosome, and he has Down's syndrome because there are three copies of chromosome 21. *(b)* Typical properties of bands.

The molecular origin of the stained bands is not known with certainty. Dark bands seen with a particular stain, Geimsa, appear to be slightly richer in A + T, while light bands are slightly richer in G + C. It is not clear how these base composition differences can yield such dramatic staining differences directly. At one time the light and dark bands were thought to have different densities of DNA packing. As progress is made in mapping extensive regions of the human genome, we can compare the DNA content in different regions. Thus far, although there is still some controversy, not much strong evidence for significant differences in the DNA packing density of light and dark bands can be found. The most tenable hypothesis that remains is that the bands reflect different DNA accessibility to reagents, perhaps as a result of different populations of bound nonhistone proteins.

While the physical origin of chromosome bands is obscure, the biological differences that have been observed between bands are dramatic. There are two general phenomena (Fig. 2.11b). Light Geimsa bands are rich in genes, and they replicate early in the S phase of the cell cycle. Dark Geimsa bands are relatively poor in genes, and they are late replicating. Finer distinctions can be made. Certain light bands, located adjacent to telomeres, are extremely rich in genes and have an unusually high G + C content. An example is the Huntington's disease region at the tip of the short arm of human chromosome 4.

The appearance of chromosome bands is not fixed. It depends very much on the method that was used to prepare the chromosome. Different procedures focused on analyzing chromosomes from earlier and earlier stages in cell division yield more elongated chromosomes that reveal increasing numbers of bands. In general, it is customary to work with chromosomes that show a total of only about 350 bands spanning the entire human genome because the more extended forms are more difficult to prepare reproducibly. Some examples are shown in Figure 2.12. One particular annoyance in studying chromosome banding patterns is that it complicates the naming of bands. Unfortunately, the nomenclature in common use is based on history. Early workers saw few bands and named them outward from the centromere as p1, p2, etc., for the short (p = petit) arm and q1, q2, etc., for the long arm (q comes after p in the alphabet). When a particular band could be resolved into multiplets, its components were named q21, q22, etc. If in later work, with more expanded chromosomes, additional sub-bands could be seen, these were renamed as q21.1, q21.2, etc. More expansion led to more names as in q21.11, q21.12. This nomenclature is not very systematic; it is certainly not a unique naming system, and it risks obfuscating the true physical origins of the bands. However, we appear to be stuck with it. Like the Japanese system for assigning street addresses in the order in which houses were constructed, it is wonderful for those with a proper historical perspective, but treacherous for the newcomer.

Humans have 22 pairs of autosomes and two sex chromosomes (XX or XY). Their DNAs range in size from chromosome 1, the largest with 250 Mb to chromosome 21, the smallest, with 50 Mb. One of each pair of autosomes and one sex chromosome is inherited from each parent. The overall haploid DNA content of a human cell is 3×10^9 bp. At 660 Da per base pair, this leads to a haploid genome molecular weight of about 2×10^{12}. The chromosomes are distinguishable by their size and unique pattern of stained bands. A schematic representation of each, in relatively compact form, is given in Figure 2.13. There are a few interesting generalizations from this genome overview. All human telomeres, except Yq, 19q, and 3p, are Geimsa light bands. The ratio of light to dark banding on different chromosomes can vary quite a bit from 19 which is mostly light, and appears to have a very large number of genes, to chromosomes 3 and 13 which are mostly dark, and are presumably relatively sparse in genes.

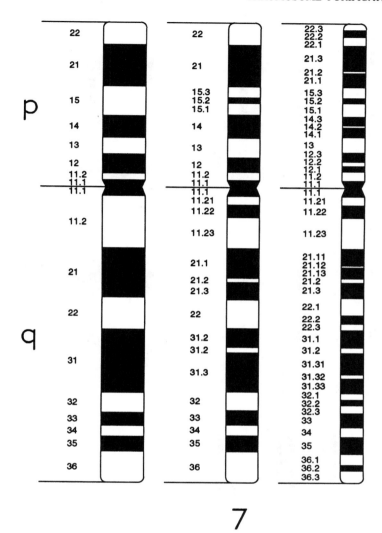

Figure 2.12 Example of how different chromosome preparations change the number of bands visible and the appearance of these bands. Three different levels of band resolution for human chromosome 7 are shown schematically. Also illustrated is the way these bands are numbered.

CHROMOSOME PURIFICATION

The past decade has seen tremendous strides in our ability to purify specific human chromosomes. Early attempts, using density gradient sedimentation, never achieved the sort of resolution necessary to become a routine analytical or preparative technique. The key advance was the creation of fluorescence activated flow sorters with sufficient intensity to allow accurate fluorescence determinations on single metaphase chromosomes. The fluorescence activated flow sorter originally was developed for intact cells, hence the name FACS (fluorescence activated cell sorter). However, it was soon found to be applicable for chromosomes, especially if more powerful lasers were used, and these were focused more tightly.

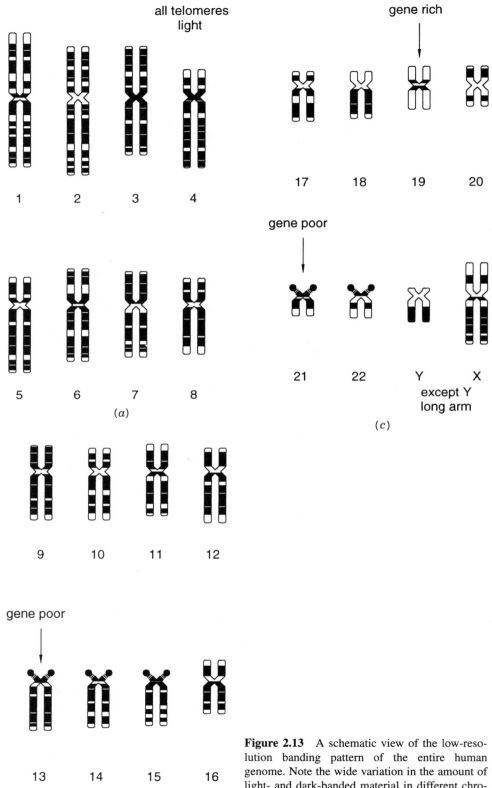

Figure 2.13 A schematic view of the low-resolution banding pattern of the entire human genome. Note the wide variation in the amount of light- and dark-banded material in different chromosomes.

FACS instruments can be used to determine a profile of chromosome sizes or other characteristics, by pulse height analysis of the emission from large numbers of chromosomes, or they can be used for actual fractionation, one chromosome at a time, as shown schematically in Figure 2.14.

In FACS, fluorescently stained metaphase chromosomes are passed in a collimated flowing liquid stream, one at a time past a powerful focused laser beam. After passing the laser, the stream is broken into uniform droplets by ultrasonic modulation. Each emission pattern is captured and integrated, and the resulting pulse height is stored as an event. If the resulting signal falls between certain preset limits, a potential is applied to the liquid stream just before the chromosome-containing droplet breaks off. This places a net charge on that droplet, and its path can then be altered selectively by an electric field. The result is the physical displacement of the droplet, and its chromosome, to a collection vessel. The circuitry must be fast enough to analyze the emission pattern of the chromosomes and relay this information before the droplet containing the desired target is released. In practice, more than one colored dye is used, and the resulting emission signal is detected at several different wavelengths and angles and analyzed by several-parameter logic. This produces an improved ability to resolve the different human chromosomes. The ideal pattern expected from single parameter analysis is shown in Figure 2.15. Each peak should show the same area, since (neglecting sex chromosomes) each is present in unit stoichiometry.

Real results are more complex as shown by the example in Figure 2.16. Some chromosomes are very difficult to resolve, and appear clustered together in an intense band. The most difficult to distinguish are human chromosomes 9 to 12. In general, larger chromosomes are more fragile and more easily broken than smaller chromosomes. Thus they appear in substoichiometric amounts, and debris from their breakage can contaminate fractions designed to contain only a particular small chromosome. The other limitation of chromosome purification by FACS is that it is a single molecule method. Typical sorting rates are a few thousand chromosomes per second. Even if the yield of a particular chromosome were perfect, this would imply the capture of only a few hundred per second. In practice, observed yields are often much worse than this. Several high-speed sorters have been constructed that increase the throughput by a factor of 3 to 5.

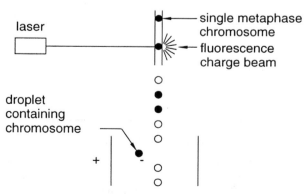

Figure 2.14 Schematic illustration of the purification of metaphase chromosomes (shown as black dots) by fluorescence activated flow-sorting (FACS).

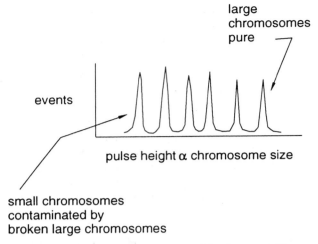

Figure 2.15 Ideal one-dimensional histogram expected for flow-sorted human chromosomes.

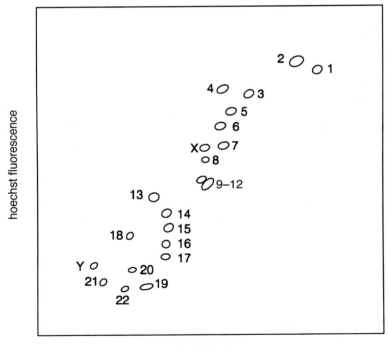

Figure 2.16 An example of actual flow analysis of human chromosomes. Two different fluorescence parameters have been used to try to resolve the chromosomes better. Despite this, four chromosomes, 9 through 12, are unresolved and appear as a much more intense peak than their neighbors. (Provided by the Lawrence Livermore National Laboratory, Human Genome Center.)

However, these are not yet generally available instruments. Thus FACS affords a way of obtaining fairly pure single chromosome material, but usually not in as large quantities as one would really like to have. Alternatives to FACS purification of chromosomes are still needed. One possibility is discussed in Box 2.2.

The problem of contamination of small chromosomes with large, and the problem of resolution of certain chromosomes, can be circumvented by the use of rodent-human hybrid cells. These will be described in more detail later. In ideal cases they consist of a single human chromosome in a mouse or hamster background. However, even if more than one human chromosome is present, they are usually an improved source of starting material. FACS is performed on hybrids just as on pure human cells. Windows (bounds on particular fluorescence signals) are used to select the desired human chromosome. Although this will be contaminated by broken chromosome fragments, these latter will be of rodent origin. The degree of contamination can be easily assessed by looking for rodent-specific DNA sequences.

FACS-sorted chromosomes can be used directly by spotting them onto filters, preparing DNA in situ, and using the resulting filters as hybridization targets for particular DNA sequences of interest (Chapter 3). In this way the pattern of hybridization of a particular DNA sequence allows its chromosome assignment. This procedure is particularly useful when the probe of interest shows undesirable cross-hybridization with other human or with rodent chromosomes. However, for most applications, it is necessary to amplify the flow-sorted human chromosome material. This is done either by variants of the polymerase chain reaction, as described in Chapter 4, or by cloning the DNA from the sorted chromosome into various vectors. Plasmids, bacteriophages like lambda (Box 1.2), P1, or cosmids (Box 2.3), and bacterial or yeast artificial chromosomes (BACs or YACs, Box 8.1) have all been used for this purpose. Collections of such clones are called single chromosome libraries. While early libraries were often heavily contaminated and showed relatively uneven representation of the DNA along the chromosome, more recently-made libraries appear to be much purer and more representative.

Single-chromosome libraries represent one of the most important resources for current genome studies. They are readily available in the United States from the American Type Culture Collection (ATCC). The first chromosome-specific libraries consisting of small clones were constructed in plasmid vectors. A second set of chromosome-specific libraries consists of larger 40 kb cosmid clones. One way in which such libraries are characterized is by their coverage, the probability that a given region is included on at least one clone. If the average insert size cloned in the library is N base pairs, the number of clones is n, and the size of the chromosome is C base pairs, the redundancy of the library is just Nn/C. Assuming that the library is a random selection of DNA fragments of the chromosome, one can compute from the coverage the probability that any sequence is represented in the library. Consider a region on the chromosome. The probability that it will be contained on the first clone examined is N/C. The probability that it will not be contained on this clone is $1 - N/C$. After n clones have been picked at random the probability that none of them will contain the region selected is $(1 - N/C)n$. Thus we can write that the fraction, f, of the chromosome covered by the library is

$$f = 1 - \left(1 - \frac{N}{C}\right)^n$$

BOX 2.2
PROSPECTS FOR ELECTROPHORETIC PURIFICATION
OF CHROMOSOMES

In principle, it should be possible to use agarose gel electrophoresis to purify chromosomes. DNA molecules up to about 50,000 bp in size are well resolved by ordinary agarose electrophoresis; while larger DNAs, up to about 10 Mb in size, can be fractionated effectively by pulsed field gel (PFG) electrophoresis. Secondary pulsed electrophoresis (SPFG), where short intense pulses are superimposed on the normally slowly varying pulses in PFG, expands the fractionation range of DNA even further (see Chapter 5). An ideal method of chromosome fractionation would be fairly general; it would allow one to capture the chromosome of interest and discard the remainder of the genome. One approach to such a scheme exploits the fact that genetic variants can be found in which the desired chromosome is a circular DNA molecule. Bacterial chromosomes are naturally circular. Eukaryotic chromosomes can become circles by recombination between the simple telomeric repeating sequences or subtelomeric repeats, as shown in Figure 2.17. Many cases of individuals with circular human chromosomes are picked up by cytogenetic analysis. In most cases the circle produces no direct deleterious phenotype because all that is lost is telomeric sequence.

It has been known for a long time that DNA circles larger than 20 kb have a very difficult time migrating in agarose gels under conventional electrophoretic conditions. The explanation is presumably entrapment of the DNA on agarose fibers, as shown in Figure 2.18. At the typical field strengths used for electrophoresis, a linear molecule, once entrapped, can slip free again by moving along its axis, but a circle is permanently trapped because of its topology. Changing field directions helps larger circles to move, which is consistent with this picture. Thus field strengths can be found where all of the linear chromosomes in a sample will migrate fairly rapidly through the gel, while circles stay at the origin. For example, in PFG the 4.6 Mb circular *E. coli* chromosomal DNA does not move, but once the chromosome is linearized by a single X-ray break, it moves readily. A mutant circular chromosome II of *S. pombe,* which is 4.8 Mb in size, does not move, while the normal linear chromosome moves readily at low electrical field strengths.

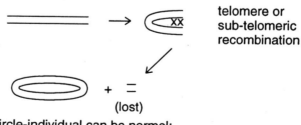

Figure 2.17   Generation of circular chromosomal DNA molecules by homologous recombination at telomeric or sub-telomeric repeated DNA sequences.

(continued)

BOX 2.2 *(Continued)*

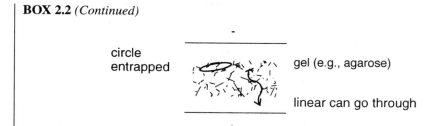

circle
entrapped

gel (e.g., agarose)

linear can go through

Figure 2.18 Entrapment of a circular DNA molecule on agarose fibers.

The linear DNA molecules that make up intact human chromosomes are so large that they do not appear able to enter agarose at all under any PFG conditions so far tried. It will be shown in Chapter 5 that these molecules do apparently enter the gel under SPFG conditions. No size fractionation is seen, but the molecules migrate well. We reasoned that under these conditions a circular human chromosomal DNA would be unable to enter agarose; if this were the case we would have a simple bulk procedure for human chromosome purification. Thus far we have experimented to no avail with a cell line containing a chromosome 21 circle. This material seems to co-migrate in the gel with ordinary linear chromosome 21 DNA. The most likely explanation is that under the conditions we used, the molecule has been fragmented—by physical forces during the electrophoresis itself, by nuclease contamination, or, much less likely, the molecule (unlike the morphological appearance of the chromosome) was never a circle to begin with. We will need to explore a wider range of conditions, and look at other circular human chromosomes.

We can arrange this to solve for n and thus determine the number of clones needed to achieve a fractional coverage of f.

$$n = \frac{log(1 - f)}{log(1 - N/\text{C})}$$

Typical useful libraries will have a redundancy of two- to tenfold.

In practice, however, most libraries are shown to be over-represented in some genome regions, under-represented in others, and totally missing certain DNA segments. Cloning biases can arise from many reasons. Some bacterial strains carry restriction nucleases that specifically degrade some of the methylated DNA sequences found in typical mammalian cells. Some mammalian sequences if expressed produce proteins toxic to the host cell. Others may produce toxic RNAs. Strong promoters, inadvertently contained in a high-copy number clone, can sequester the host cell's RNA polymerase, resulting in little or no growth. DNA sequences with various types of repeated sequences can recombine in the host cell, and in many cases this will lead to loss of the DNA stretch between the repeats. The inevitable result is that almost all libraries are fairly biased.

BOX 2.3
PREPARATION OF SINGLE CHROMOSOME LIBRARIES IN COSMIDS AND P1

Because flow-sorting produces only small amounts of purified single chromosomes, proce-dures for cloning this material must be particularly efficient. The ideal clones will also have relatively large insert capacities so that the complexity of the library, namely the number of clones needed to represent one chromosome equivalent of insert DNA, can be kept within reasonable bounds. The earliest single-chromosome libraries were made in bacteriophage or plasmid vectors (see Box 1.3), but these were rapidly supplanted by cosmid vectors. Libraries of each human chromosome in cosmids have been made and distributed by a col-laboration between Lawrence Livermore National Laboratory and Los Alamos National Laboratory. These libraries are available today at a nominal cost from the American Type Culture Collection. Gridded filter arrays of clones from most of these libraries have also been made by various genome centers. Interrogation of these filters by hybridization with a DNA probe or interrogation of DNA pools with PCR primers will identify clones that con-tain specific DNA sequences. The use of the same arrays by multiple investigators at differ-ent sites facilitates coordination of a broad spectrum of genome research.

Cosmids are chosen as vectors for single chromosome libraries because they have relatively large inserts. A cosmid clone consists of two ends of bacteriophage lambda DNA (totaling about 10 kb in length) with all of the middle of the natural vector re-moved. The ends contain all of the sequence information needed to package DNA into viruses. Hence a cloned insert can replace the central 40 kb of lambda. Recombinant molecules are packaged in vitro using extracts from cells engineered to contain the proteins needed for this reaction. Lambda DNA packaged into a bacteriophage head is a linear molecule, but the ends have 12 base complementary 5'-extensions. Once the virus infects an *E. coli* cell, the two ends circularize and are ligated together as the first step in the viral life cycle. The 5'-extensions are called *COS sites,* for cohesive ends, and the name of this site has been carried over to the vectors that contain them (COSmids). Cosmids propagate in *E. coli* cells as low-copy plasmids.

Bacteriophage P1 offers another convenient large insert cloning system. P1 pack-ages its DNA by a headful mechanism that accommodates about 90 kb. Hence, if the target DNA is much larger than 90 kb, it will be cut into adjacent fragments as it is packaged. A typical P1 cloning vector is shown below. The vector is equipped with bacteriophage SP6 and T7 promoters to allow strand-specific transcription of the in-sert. The resulting clones are called P1 artificial chromosomes (PACs).

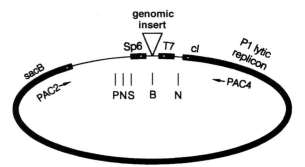

PAC cloning vector. In the circle are shown cutting sites for some restriction nucleases and also locations of known sequences, PAC2, PAC4 suitable for PCR amplifications of the insert (see Chapter 4).

CHROMOSOME NUMBER

There are two determinants of the total number of chromosomes in a cell. The first is the number of different chromosomes. The second is the number of copies of a particular chromosome. In general, in a species the number of different autosomes (excluding sex chromosomes) is preserved, and the number of copies of each chromosome is kept in a constant ratio, although the absolute number can vary in different cells. We use the term *ploidy* to refer to the number of copies of the genome. Yeast cells are most typically haploid with one genome copy, but after mating they can be diploid, with two copies. Human gametes are haploid; somatic cells vary between diploid and tetraploid depending on what stage in the cell cycle they are in.

When actively growing eukaryotic cells are stained for total DNA content and analyzed by FACS, a complex result is seen, as shown in Figure 2.19. The three peaks represent diploid G1 cells, tetraploid G2 and M cells, and intermediate ploidy for cells in S phase. Plant cells frequently have much higher ploidy. In specialized tissues much higher ploidy is occasionally seen in animal cells as, for example, in the very highly polyploid chromosomes of Drosophila salivary glands. When such events occur, it is a great boon for the cytogeneticist because it makes the chromosomes much easier to manipulate and to visualize in detail in the light microscope.

Aneuploidy is an imbalance in the relative numbers of different chromosomes. It is often deleterious and can lead to an altered appearance, or phenotype. Ordinarily gene dosage is carefully controlled by the constant ratios of different segments of DNA. An extra chromosome will disturb this balance because its gene products will be elevated. In the human the most commonly seen aneuploidy is Down's syndrome: trisomy chromosome 21. The result is substantial physical and mental abnormalities, although the individuals survive. Trisomy 13 is also seen, but this trisomy leads to even more serious deformations and the individuals do not survive long beyond birth. Other trisomies are not seen in live births because fetuses carrying these defects are so seriously damaged that they do not survive to term.

In the human (and presumably in most other diploid species) monosomies, loss of an entire chromosome, are almost always fatal. This is due to the presence of recessive lethal alleles. For example, imagine a chromosome that carries a deletion for a gene that codes for an essential enzyme. As long as the corresponding gene on the homologous chromosome is intact, there may be little phenotypic effect of the haploid state of that gene.

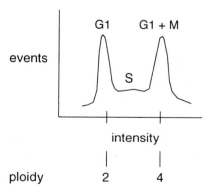

Figure 2.19 FACS analysis of the amount of DNA in a population of rapidly growing and dividing cells. Peaks corresponding to particular cell cycle stages are indicated.

However, monosomy will reveal the presence of any recessive lethal allele on the remaining chromosome. The result is not compatible with normal development.

Cell ploidy is maintained by mitosis, for somatic cells (see Box 2.4), and it is altered by meiosis (see Box 6.1) during gametogenesis. Errors during these processes or errors made during DNA damage and repair can lead to cells that themselves are aneuploid or that produce aneuploid offspring. Cells with defects in their division cycle frequently accumulate abnormal chromosome complements. This is particularly dramatic in many late-stage tumors which frequently have large numbers of different chromosome abnormalities.

BOX 2.4
MITOSIS

During cell division, events must be carefully controlled to ensure that each daughter cell receives one copy of each of the two homologous chromosomes, namely receives one copy of the paternal genome and one copy of the maternal genome. An obvious potential source of confusion is the similarity of the two parental genomes. The way in which this confusion is effectively avoided is shown by the schematic illustration of some of the steps in mitosis in Figure 2.20. Our example considers a cell with only two chromosome types. At the G1 phase the diploid cell contains one copy of each parental chromosome. These are shown condensed in the figure, for clarity, but remember that they are not condensed except at metaphase. After DNA synthesis the cell is tetraploid; there are now two copies of each parental genome. However, the two copies are paired; they remain fused at the centromere. We call these structures *sister chromatids*. Hence there is no chance for the two parental genomes to mingle. During mitosis the paired sister chromatids all migrate to the metaphase plate of the cell. Microtubules form between each centromere and the two centrioles that will segregate into the two daughter cells. As the microtubules shrink, each pair of sister chromatids is dragged apart so that one copy goes to each daughter cell.

Errors can occur; one type is called nondisjunction. The sister chromatids fail to separate so that one daughter gets both sister chromatids; the other gets none. Usually such an event will be fatal because of recessive lethal alleles present on the sole copy of that chromosome present in the daughter. One additional complication bears mention. While sister chromatids are paired, a process called sister chromatid exchange can occur. In this form of mitotic recombination, DNA strands from one sister invade the other; the eventual result is a set of DNA breaks and reunions that exchanges material between the two chromatids. Thus each final product is actually a mosaic of the two sisters. Since these should be identical anyway (except for any errors made in DNA synthesis) this process has no phenotypic consequences. We know of its existence most compellingly through elegant fluorescence staining experiments conceived by Samuel Latt at Harvard Medical School. He used base analogues to distinguish the pre-existing and newly synthesized sister chromatids, and a fluorescent stain that showed different color intensities with the two base analogues. Thus each chromatid exchange point could be seen as a switch in the staining color, as shown schematically in Figure 2.21.

(continued)

BOX 2.4 *(Continued)*

cell division
with only 2 chromosomes,
1 from each parent

(not really condensed
except at M)

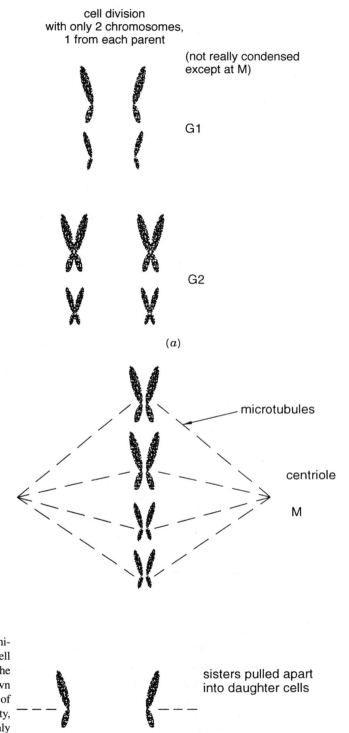

G1

G2

(a)

microtubules

centriole

M

sisters pulled apart
into daughter cells

Figure 2.20 Steps in mitosis, the process of cell division. For simplicity the chromosomes are shown condensed at all stages of the cell cycle. In actuality, they are condensed only during metaphase.

(b)

(continued)

BOX 2.4 (*Continued*)

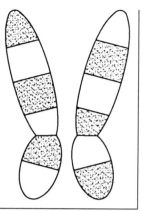

Figure 2.21   Visualization of sister chromatid exchange, by fluorescence quenching. Newly synthesized DNA was labeled with 5-bromoU which quenches the fluorescence of the acridine dye used to stain the chromosomes at metaphase.

Partial aneuploidy can arise in a number of different ways. The results are often severe. One common mechanism is illustrated in Figure 2.22. Reciprocal chromosome translocations are fairly common, and they are discovered by genetic screening because the offspring of such individuals frequently have genetic abnormalities. The example, shown in the figure, is a reciprocal translocation between chromosome 5 and chromosome 20. Such translocations can occur by meiotic or mitotic recombination. Unless the break points interrupt vital genes, the translocation results in a normal phenotype because all of the genome is present in normal stoichiometry. This illustrates once again that the arrangement of genes on the chromosomes is not usually critical.

Now consider the result of a mating between the individual with a reciprocal translocation and a normal individual. Fifty percent of the children will have a normal dosage of all of their chromosomes. Half of these will have a totally normal genotype because they will have received both of the normal homologs originally present in the parent with the reciprocal translocation. Half will have received both abnormal chromosomes from that parent; hence their genome will still be balanced. The remaining 50% of the offspring will show partial aneuploidy. Half of these will be partially trisomic for chromosome 20, partially monosomic for chromosome 5. The other half will be partially monosomic for chromosome 20, partially trisomic for chromosome 5.

UNUSUAL CHARACTERISTICS OF SEX CHROMOSOMES AND MITOCHONDRIA

In mammals a female carries two copies of the X chromosome; males have one X and one Y. However, this simple difference in karyotype (the set of chromosomes) has profound effects that go beyond just the establishment of sex. The first thing to consider is why we need sex at all. In species with just one sex, each organism can reproduce clonally. The offspring of that organism may be identical. If the organism inhabits a wide ecological range, different selection processes will produce a geographical pattern of genetic differences, but there is no rapid way to combine these in response to a shifting environment. Sex, on the other hand, demands continual outbreeding, so it leads to much more efficient mixing of the gene pool of a species.

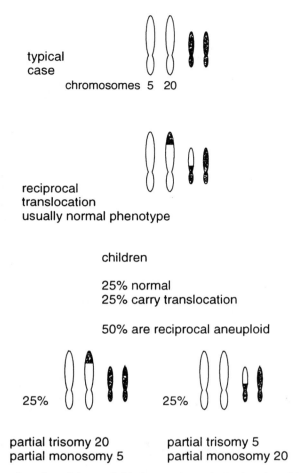

typical
case

chromosomes 5   20

reciprocal
translocation
usually normal phenotype

children

25% normal
25% carry translocation

50% are reciprocal aneuploid

25%   25%

partial trisomy 20   partial trisomy 5
partial monosomy 5   partial monosomy 20

Figure 2.22   Generation of partial aneuploidy by a reciprocal translocation, followed by segregation of the rearranged chromosomes to gametes and, after fertilization, generation of individuals
that are usually phenotypically abnormal.

We are so used to the notion of two sexes, that it often goes unquestioned why two and
only two? Current speculation is that this is the result of the smallest component of mammalian genomes, the mitochondrial DNA. Mitochondria have a circular chromosome,
much like the typical bacterial genome from which they presumably derived. This
genome codes for key cellular metabolic functions. In the human it is 16,569 kb in size,
and the complete DNA sequence is known. A map summarizing this sequence is shown in
Figure 2.23. Mitochondria have many copies of this DNA. What is striking is that all of
an individual's mitochondria are maternally inherited. The sperm does contain a few mitochondria, and these can enter the ovum upon fertilization, but they are somehow
destroyed.

Bacterial DNAs carry restriction nucleases that can destroy foreign (or incompatible) DNA
(see Box 1.2). Perhaps, from their bacterial origin, mitochondria also have such properties.

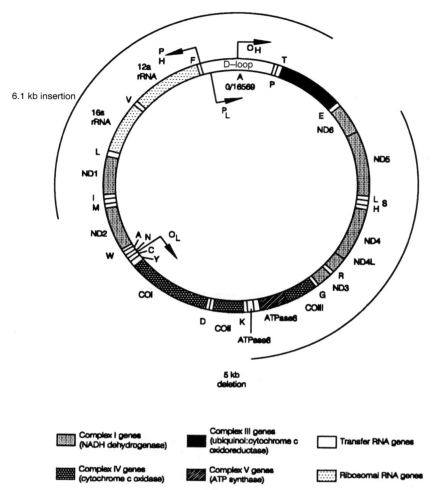

Figure 2.23   Map of human mitochondrial DNA. The tRNAs are indicated by their cognate amino acid letter code. The genes encoded by the G-rich heavy (H) strand are on the outside of the circle, while those for the C-rich light (L) strand are on the inside. The H- and L-strand origins (O_H and O_L) and promoters (P_H and P_L) are shown. The common 5-kb deletion associated with aging is shown outside the circle. (Adapted from Wallace, 1995.)

If they do, this would explain why one sex must contribute all of the mitochondria. It can be used as an argument that there should only be two sexes. In fact, however, cases are known where organisms have more than two sexes. The slime mold, *Physarum polycephalum,* has 13 sexes. However, these turn out to be hierarchical. When two sexes mate, the higher one on the hierarchy donates its mitochondria to the offspring. This ensures that only one parental set of mitochondria survive. So one important thing about sex is who you get your mitochondria from.

In the human and other mammals, the Y chromosome is largely devoid of genes. The long arm is a dark G band (Fig. 2.13), and the short arm is small. However, an exception is the gene-rich tip of the short arm, which is called the *pseudoautosomal region.*

BOX 2.5
MORE ON MITOCHONDRIAL DNA

The pure maternal inheritance of mitochondria makes it very easy to trace lineages in human populations, since all of the complexities of diploid genetics are avoided. The only analogous situation is the Y chromosome which must be paternally inherited. One region of the mitochondrial DNA, near the replication origin, codes for no known genes. This region shows a relatively fast rate of evolution. By monitoring the changes in the DNA of this region, Allen Wilson and his coworkers have attempted to trace the mitochondrion back through human prehistory to explore the origin of human ethnic groups and their geographic migrations. While considerable controversy still exists about some of the conclusions, most scientists feel that they can trace all existing human groups to a single female progenitor who lived in Africa some 20,000 years ago.

The mitochondrion has recently been implicated in studies by Norman Arnheim, Douglas Wallace, and their coworkers as a major potential site of accumulated damage that results in human aging. In certain inherited diseases a large deletion occurs in mitochondrial DNA. This deletion drops out more than 5 kb of the genome between two repeated sequence elements (Fig. 2.23). It presumably arises by recombination. Small amounts of similar deletions have been detected in aging human tissue, particularly in cells like muscle, heart, and brain that undergo little or no cell division. While the full significance of these results remains to be evaluated, on the surface these deletions are striking phenomena, which provide potential diagnostic tools for what may be a major mechanism of aging and a way to begin to think rationally about how to combat it. The mitochondrion is the site of a large amount of oxidative reactions; these are known to be able to damage DNA and stimulate repair and recombination. Hence it is not surprising that this organelle should be a major target for DNA aging.

This region is homologous to the tip of the short arm of the X chromosome. A more detailed discussion of this region will be presented in Chapter 6. There are also a few other places on X and Y where homologous genes exist. Beyond this, most of the X contains genes that have no equivalent on the Y. This causes a problem of gene dosage. The sex chromosomes are unbalanced because a female will have two copies of all these X-linked genes while the male will have only one. The gene dosage problem is solved by the process known as X-inactivation.

Mature somatic cells of female origin have a densely staining condensed object called a Barr body. This object is absent in corresponding cells of male origin. Eligibility of female athletes competing in the Olympics used to be dependent on the presence of a Barr body in their cells. Mary Lyon first demonstrated that the Barr body is a highly condensed X chromosome. Since we know that condensed chromatin is inactive in expression, this suggests that in the female one of the two X chromosomes is inactivated. This process, X-inactivation, occurs by methylation of C. It covers the entire X chromosome except for the pseudoautosomal region and other genes that are homologous on X and Y. The exact mechanism is still not understood in detail, but it seems to be a process that is nucleated at some X-specific sequences and then diffuses (except where barriers limit its spread). Cells with translocations between the X chromosome and autosomes are known; in these cases the inactivation can spread onto part of the adjacent autosome fragment.

If X-inactivation occurred at the single-cell stage of an embryo; one of the two parental Xs would be lost, and males and females would have similar sex-linked genetic properties. However, X-inactivation occurs later in embryogenesis. When it occurs, the two parental X chromosomes have an equal probability of inactivation. The resulting female embryo then becomes a mosaic with half the cells containing an active paternal X chromosome, half an active maternal X chromosome. When these two chromosomes carry distinguishable markers, this mosaicism is revealed in patterns of somatic differences in clones of cells that derive from specific embryonic progenitors. One spectacular example is the tortoise shell cat (Fig. 2.24). This X chromosome of this animal can carry two different color coat alleles. The male is always one color or the other because it has one allele or the other. The female can have both alleles and will inactivate each in a subset of embryonic ectodermal cells. As

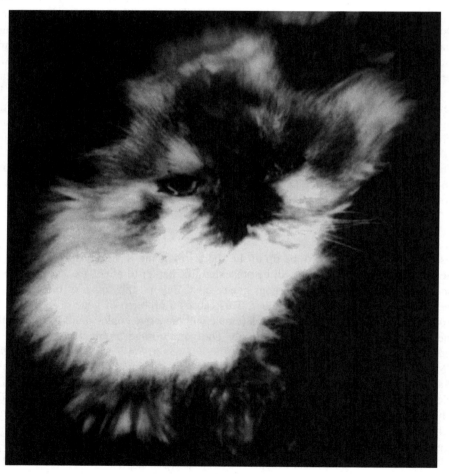

Figure 2.24 A tortoise shell Himalayan female cat. A gene responsible for overall development of skin pigment is temperature sensitive. As a result, pigmentation occurs only in regions where the animal is normally cold such as the tips of the ears, nose, and paws. These colored regions are mottled because the cat is a mosaic of different color alleles because of random X-inactivation early in development. (Photograph courtesy of Chandran Sabanayagam.)

these cells multiply and differentiate, a two-colored animal emerges; the pattern of color distribution reveals the clonal origin of the cells that generated it. The tortoise shell phenotype is only seen in the female. Thus there are truly fundamental differences between the sexes; females are mosaics in all their tissues for most of the genes on the X chromosome; males are monosomic.

Abnormalities in sex chromosome balance reveal the interplay between factors directed by genes on the X and Y chromosomes. An individual with only one X chromosome is called XO. This individual is female, but she is short and has other marked abnormalities. It is not clear how much of this is due to deleterious recessive alleles on the X and how much is due to gene dosage. The abnormality is called Turner's syndrome. An XXY individual is male, but he has some feminine characteristics such as breasts. This syndrome is called Kleinfelter's. This individual will have a Barr body because X-inactivation occurs in cells with more than one X chromosome. An individual with XYY is a male, mostly normal, but tall.

SYNTENY

In general, from some of the items discussed in the last few sections, we can conclude that the relative number of chromosomes (or even parts of chromosomes) matters a great deal, but the actual number of different chromosomes is not terribly important. A classic case is two closely related deer that are phenotypically extremely similar. The Reeves Muntjak has only 3 autosomes plus sex chromosomes. These autosomes are enormous. The Indian Muntjak has 22 autosomes plus sex chromosomes, a number far more typical of other mammals. The two deer must have very similar genomes, but these are distributed differently. The chromosome fusion events that resulted in the Reeves Muntjak must be fairly recent in evolution. Their major consequence is that the two species cannot interbreed because the progeny would have numerous chromosome imbalances.

The Muntjak example suggests that chromosome organization can vary, in a dramatic way, superficially, without much disruption of gene organization and gene content. This notion is fully borne out when the detailed arrangement of genes is observed in different species. For example, in most closely related species, like human and chimp, the detailed appearance of chromosome bands is nearly identical. Humans and chimps in fact show only two regions of significant chromosome morphological differences in the entire genome. These regions are clearly of some interest for future study, since they could reveal hints of genes that differ significantly in the two species.

When more distantly related organisms are compared, their chromosomes appear, superficially, to be very different. For example, the mouse has 19 autosomes; the human has 22. All of the mouse chromosomes are acrocentric: Their centromeric constrictions occur at one end of the chromosome. In essence each mouse chromosome has only one arm. In contrast, most of the human chromosomes have an internally located centromere; in many cases it is near the center and the two arms are comparable in size. Only five human chromosomes are acrocentric, and each of these is a special case: Chromosomes 13, 14, 15, 21, and 22 all have tiny short arms containing large numbers of tandemly repeated rDNA genes (see Fig. 2.13). However, when the detailed arrangement of genes along chromosomes is compared in the mouse and human, a much more conserved structural pattern is seen.

Synteny is the term used to describe similar arrangements of genes on maps, whether these are genetic maps or chromosome morphology. Most regions of the mouse and human genomes display a high degree of synteny. This is summarized in Figure 2.25,

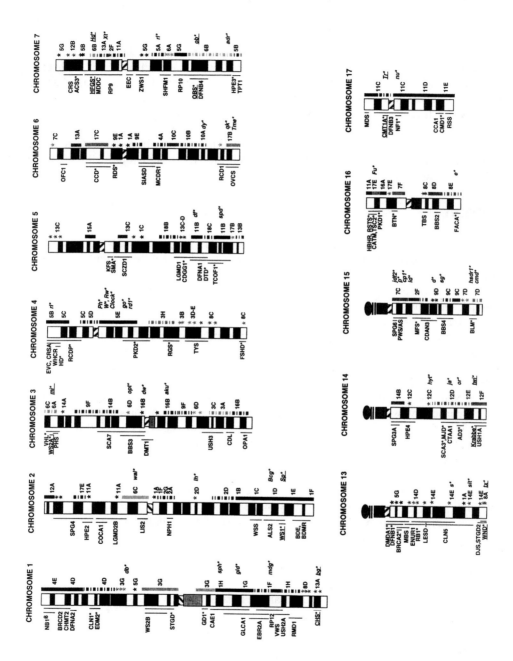

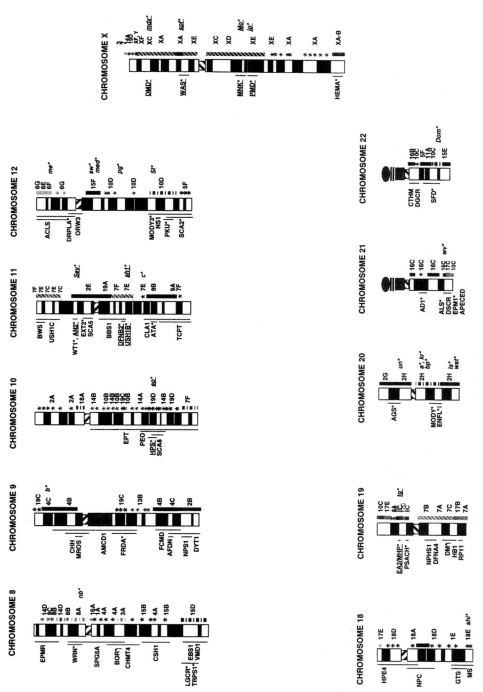

Figure 2.25   Correspondences between human and mouse chromosomes. Corresponding mouse regions are indicated on each human chromosome. (Kindly provided by Lisa Stubbs; Carver and Stubbs, 1997.)

which reveals that a typical mouse chromosome has genes that map to two to four human chromosomes, and vice versa. If we look at genes that are neighbors in the mouse, these are almost always neighbors in the human. Figure 2.26 illustrates this for mouse chromosome 7: The corresponding human 19q regions contain blocks of the same genes in the same order. This is extremely useful because it means that genetic or physical mapping data in one species can usually be applied to other species. It speeds, enormously, the search for corresponding genes in different species. The superficial differences between the mouse and human genome have presumably been caused largely by gross chromo-

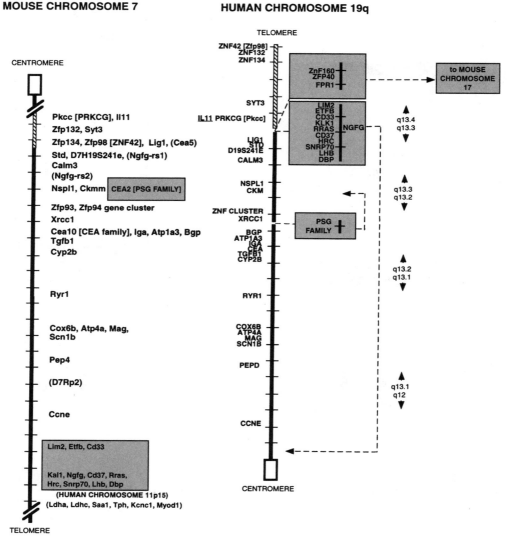

Figure 2.26 An example of a detailed synteny map. Shown are sections of mouse chromosome 7 corresponding to regions of human chromosome 19q which contain similarly ordered blocks of markers. (Kindly provided by Lisa Stubbs; Stubbs et al., 1996.)

some rearrangements in both species. It has been estimated that only a few hundred such rearrangements will ultimately account for all of the differences we see in the contemporary versions of these chromosomes.

SOURCES AND ADDITIONAL READINGS

Buckle, V. J., Edwards, J. H., Evans, E. P., Jonasson, J. A., Lyon, M. F., Peters, J., and Searle, A. G. 1984. Chromosome maps of man and mouse II. *Clinical Genetics* 1: 1–11.

Cantor, C. R., and Schimmel, P. R. 1980. *Biophysical Chemistry.* San Francisco: W. H. Freeman, chapter 24.

Carver, E. A., and Stubbs, L. 1997. Zooming in on the human-mouse comparative trap: Genome conservation re-examined on a high resolution scale. *Genome Research* 7: 1123–1137.

Cozzarelli, N. R., and Wang, J. C., eds. 1990. *DNA Topology and Its Biological Effects.* Cold Spring Harbor, NY: Cold Spring Harbor Laboratory Press.

Harrington, J. J., Van Bokkeln, G., Mays, R. W., Gustashaw, K., and Willard, H. F. 1997. Formation of de novo centromeres and construction of first-generation human artificial microchromosomes. *Nature Genetics* 15: 345–355.

Holmquist, G. P. 1992. Review article: Chromosome bands, their chromatin flavors, and their functional features. *American Journal of Human Genetics* 51: 17–37.

König, P., and Rhodes, D. 1997. Recognition of telomeric DNA. *Trends in Biochemical Sciences* 22: 43–47.

Levy, M. Z., Allsopp, R. C., Futcher, A. B., Greider, C. W., and Harley, C. B. 1992. Telomere end-replication problem and cell aging. *Journal of Molecular Biology* 225: 951–960.

Manuelidis, L. 1990. A view of interphase chromosomes. *Science* 250: 1533–1540.

Moyzis, R. K., Buckingham, J. M, Cram, L. S., Dani, M., Deaven, L. L., Jones, M. D., Meyne, J., Ratliff, R. L., and Wu, J. R. 1988. A highly conserved repetitive DNA sequence, (TTAGGG)n, present at the telomeres of human chromosomes. *Proceedings of the National Academy of Sciences USA* 85: 6622–6626.

Niklas, R. B. 1997. How cells get the right chromosomes. *Science* 275: 632–637.

Therman, E. 1986. *Human Chromosomes: Structure, Behavior, Effects,* 2nd ed. New York: Springer-Verlag.

Saccone, S., De Sario, A., Wiegant, J., Raap, A. K., Valle, G. D., and Bernardi, G. 1993. Correlations between isochores and chromosomal bands in the human genome. *Proceedings of the National Academy of Sciences USA* 90: 11929–11933.

Stubbs, L., Carver, E. A., Shannon, M. E., Kim, J., Geisler, J., Generoso, E. E., Stanford, B. G., Dunn, W. C., Mohrenweiser, H., Zimmermann, W., et al. 1996. Detailed comparative map of human chromosome 19q and related regions of the mouse genome. *Genomics* 35: 499–508.

Wallace, D. C. 1995. Mitochondrial DNA variation in human evolution, degenerative disease, and aging. *American Journal of Human Genetics* 57: 201–223.

3 Analysis of DNA Sequences by Hybridization

BASIC REQUIREMENTS FOR SELECTIVITY AND SENSITIVITY

The haploid human genome is 3×10^9 base pairs, and a typical human cell, as described in the last chapter, is somewhere between diploid and tetraploid in DNA content. Thus each cell has about 10^{10} base pairs of DNA. A single base pair is 660 Da. Hence the weight of DNA in a single cell can be calculated as $10^{10} \times 660 / (6 \times 10^{23}) = 10^{-11}$ g or 10 pg. Ideally we would like to be able to do analyses on single cells. This means that if only a small portion of the genome is the target for analysis, far less than 10 pg of material will need to be detected. By current methodology we are in fact able to determine the presence or absence of almost any 20-bp DNA sequence within a single cell, such as the sequence ATTGGCATAGGAGCC-CATGG. This analysis takes place at the level of single molecules. Two requirements must be met to perform such an exquisitely demanding analysis. There must be sufficient experimental sensitivity to detect the presence of the sequence. This sensitivity is provided by either chemical or biological amplification procedures or by a combination of these procedures. There must also be sufficient experimental selectivity to discriminate between the desired, true target sequence and all other similar sequences, which may differ from the target by as little as one base. That specificity lies with the intrinsic selectivity of DNA base pairing, itself.

The target of a 20-bp DNA sequence is not picked casually. Twenty bp is just about the smallest DNA length that has a high probability, a priori, of being found in a single copy in the human genome. This can be deduced as follows from simple binomial statistics (Box 3.1).

For simplicity, pretend that the human genome contains equal amounts of the four bases, A, T, C, and G, and that the occurrences of the bases are random. (These constraints will be relaxed elsewhere in the book when some of the unusual statistical properties of natural DNAs need to be considered explicitly. Then the expected frequency of occurrence of any particular stretch of DNA sequence, such as n bases beginning as ATCCG . . ., is 4^{-n}. The average number of occurrences of this particular sequence in the haploid human genome is $3 \times 10^9 \times 4^{-n}$. For a sequence of 16 bases, $n = 16$, the average occurrence is $3 \times 10^9 \times 4^{-16}$ which is about 1. Thus such a length will tend to be seen as often as not by chance; it is not long enough to be a unique identifier. There is a reasonable chance that the sequence 16 bases long will occur several times in different places in the genome. Choosing $n = 20$ gives an average occurrence of about 0.3%. Such sequences will almost always be unique genome landmarks. One corollary of this simple exercise is that it is a very futile exercise to look at random for the occurrence of particular 20-mers in the sequence of a higher organism unless there is good a priori reason for suspecting the presence of these sequences. This means that sequences of length 20 or more can be used as unique identifiers (see Box 3.2).

BOX 3.1
BINOMIAL STATISTICS

Binomial statistics describe the probable outcome of events like coin flipping, events that depend on a single random variable. While a normal coin has a 50% chance of heads or tails with each flip, we will consider here the more general case of a weighted coin with two possible outcomes with probabilities p (heads) and q (tails). Since there are no other possible outcomes $p + q = 1$. If N successive flips are executed, and the outcome is a particular string, such as *hhhhttthhh,* the chance of this particular outcome is $p^n q^{N-n}$, where n is the number of times heads was observed. Note that all strings with the same numbers of heads and tails will have the same a priori probability, since in binomial statistics each event does not affect the probability of subsequent events. Later in this book we will deal with cases where this extremely simple model does not hold. If we care only about the chance of an outcome with n heads and $N - n$ tails, without regard to sequence, the number of such events is $N!/(n!)(N - n)!$, and so the fraction of times this outcome will be seen is $(p^n q^{N-n})N!/(n!)(N - n)!$

A simple binomial model can also be used to estimate the frequency of occurrence of particular DNA base sequences. Here there are four possible outcomes (not quite as complex as dice throwing where six possible outcomes occur). For a particular string with n_A A's, n_C C's, n_G G's and n_T T's, and a base composition of X_A, X_C, X_G, and X_T the chance occurrence of that string is $X_A^{n_A} X_C^{n_C} X_G^{n_G} X_T^{n_T}$. The number of possible strings with a particular base composition is $N!/(n_A! n_C! n_G! n_T!)$, and by combining this with the previous term, the probability of a string with a particular base composition can easily be computed. Incidentally, the number of possible strings of length N is 4^N, while the number of different base compositions of this length is $(N + 3)!/(N!3!)$.

The same statistical models can be used to make estimates that two people will share the same DNA sequences. Such estimates are very useful in DNA-based identity testing. Here we consider just the simple case of two allele polymorphisms. In a particular place in the genome, suppose that a fraction of all individuals have one base, f_1, while the remainder have another, f_2. The chance that two individuals share the same allele is $f_1^2 + f_2^2 = g^2$. If a set of M two-allele polymorphisms $(i, j, k, \ldots)$ is considered simultaneously, the chance that two individuals are identical for all of them is $g_i^2 g_j^2 g_k^2 \ldots$ By choosing M sufficiently large, we can clearly make the overall chance too low to occur, unless the individuals in question are one and the same. However, two caveats apply to this reasoning. First, related individuals will show a much higher degree of similarity than predicted by this model. Monozygotic twins, in principle, should share an identical set of alleles at the germ-line level. Second, the proper allele frequencies to use will depend on the racial, ethnic, and other genetic characteristics of the individuals in question. Thus it may not always be easy to select appropriate values. These difficulties notwithstanding, DNA testing offers a very powerful approach to identification of individuals, paternity testing, and a variety of forensic applications.

BOX 3.2
DNA SEQUENCES AS UNIQUE SAMPLE IDENTIFIERS

The following table shows the number of different sequences of length n and compares these values to the sizes of various genomes. Since genome size is virtually the same as the number of possible short substrings, it is easy to determine the lengths of short sequences that will occur on average only once per genome. Sequences a few bases longer than these lengths will, for all practical purposes, occur either once or not at all, and hence they can serve as unique identifiers.

LENGTH	NUMBER OF SEQUENCES	GENOME, GENOME SIZE (BP)
8	6.55×10^4	Bacteriophage lambda, 5×10^4
9	2.60×10^5	
10	1.05×10^6	
11	4.20×10^6	*E. coli*, 4×10^6
12	1.68×10^7	*S. cerevisiae, 1.3 $\times 10^7$*
13	6.71×10^7	
14	2.68×10^8	All mammalian mRNAs, 2×10^8
15	1.07×10^9	
16	4.29×10^9	Human haploid genome, 3×10^9
17	1.72×10^{10}	
18	6.87×10^{10}	
19	2.75×10^{11}	
20	1.10×10^{12}	

DETECTION OF SPECIFIC DNA SEQUENCES

DNA molecules themselves are the perfect set of reagents to identify particular DNA sequences. This is because of the strong, sequence-specific base pairing between complementary DNA strands. Here one strand of DNA will be considered to be a target, and the other, a probe. (If both are not initially available in a single-stranded form, there are many ways to circumvent this complication.) The analysis for a particular DNA sequence consists in asking whether a probe can find its target in the sample of interest. If the probe does so, a double-stranded DNA complex will be formed. This process is called *hybridization,* and all we have to do is to discriminate between this complex and the initial single-stranded starting materials (Fig. 3.1*a*).

The earliest hybridization experiments were carried out in homogeneous solutions. Hybridization was allowed to proceed for a fixed time period, and then a physical separation was performed to capture double-stranded material and discard single strands. Hydroxyapatite chromatography was used to do this discrimination because conditions could be found in which double-stranded DNA bound to a column of hydroxylapatite, while single strands were eluted (Fig. 3.1*b*). The amount of double-stranded DNA could be quantitated by using a radioisotopic label on the probe or the target, or by measuring the bulk amount of DNA captured or eluted. This method is still used today in select cases, but it is very tedious because only a few samples can be conveniently analyzed simultaneously.

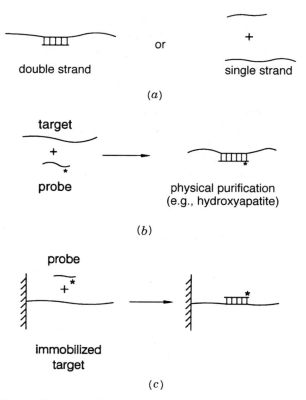

Figure 3.1 Detecting the formation of specific double-stranded DNA sequences. *(a)* The problem is to tell whether the sequence of interest is present in single-stranded (s.s.) or double-stranded (d.s.) form. *(b)* Physical purification by methods such as hydroxyapatite chromatography. *(c)* Physical purification by the use of one strand attached to an immobilized phase. A label is used to detect hybridization of the single-stranded probe.

Modern hybridization protocols immobilize one of the two DNAs on a solid support (Fig. 3.1c). The immobilized phase can be either the probe or the target. The complementary sample is labeled with a radioisotope, a fluorescent dye, or some other specific moiety that later allows a signal to be generated in situ. The amount of color, or radioactivity, on the immobilized phase is measured after hybridization, for a fixed period, and subsequent washing of the solid support to remove adsorbed, but nonhybridized, material. As will be shown later, an advantage of this method is that many samples can be processed in parallel. However, there are also some disadvantages that will become apparent as we proceed.

EQUILIBRIA BETWEEN DNA DOUBLE AND SINGLE STRANDS

The fraction of single- and double-stranded DNA in solution can be monitored by various spectroscopic properties that effectively average over different DNA sequences. Such measurements allow us to view the overall reaction of DNA single strands.

Ultraviolet absorbance, circular dichroism, or the fluorescence of dyes that bind selectively to duplex DNA can all be used for this purpose. If the amount of double-stranded (duplex) DNA in a sample is monitored as a function of temperature, the results typically obtained are shown in Figure 3.2. The DNA is transformed from double strands at low temperature, rather abruptly at some critical temperature, to single strands. The process, for long DNA, is usually so cooperative that it can be likened to the melting of a solid, and the transition is called *DNA melting*. The midpoint of the transition for a particular DNA sample is called the melting temperature, T_m. For DNAs that are very rich in the bases G + C, this can be 30 or 40°C higher than for extremely (A + T)-rich samples. It is such spectroscopic observations, on large numbers of small DNA duplexes that have allowed us to achieve a quantitative understanding of most aspects of DNA melting.

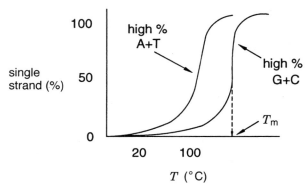

Figure 3.2 Typical melting behavior for DNA as a function of average base composition. Shown is the fraction of single-stranded molecules as a function of temperature. The midpoint of the transition is the melting temperature, T_m.

The goal of this section is to define conditions that allow sequence-specific analyses of DNA using DNA hybridization. Specificity means the ratio of perfectly base-paired duplexes to duplexes with imperfections or mismatches. Thus high specificity means that the conditions maximize the amount of double-stranded perfectly base-paired complex and minimize the amount of other species. Key variables are the concentrations of the DNA probes and targets that are used, the temperature, and the salt concentration (ionic strength).

The melting temperature of long DNA is concentration *in*dependent. This arises from the way in which T_m is defined. Large DNA melts in patches as shown in Figure 3.3a. At T_m, the temperature at which half the DNA is melted, (A + T)-rich zones are melted, while (G + C)-rich zones are still in duplexes. No net strand separation will have taken place because no duplexes will have been completely melted. Thus there can be no concentration dependence to T_m.

In contrast, the melting of short DNA duplexes, DNAs of 20 base pairs or less, is effectively all or none (Fig. 3.3b). In this case the concentration of intermediate species, partly single-stranded and partly duplex, is sufficiently small that it can be ignored. The reaction of two short complementary DNA strands, *A* and *B*, may be written as

$$A + B = AB$$

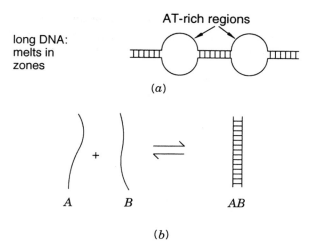

(a)

(b)

Figure 3.3   Melting behavior of DNA. (a) Structure of a typical high molecular weight DNA at its melting temperature. (b) Status of a short DNA sample at its melting temperature.

The equilibrium (association) constant for this reaction is defined as

$$K_a = \frac{[AB]}{[A][B]}$$

Most experiments will start with an equal concentration of the two strands (or a duplex melted to give the two strands). The initial total concentration of strands (whatever their form) is C_T. If, for simplicity, all of the strands are initially single stranded, their concentrations is

$$[A]_o = [B]_o = \frac{C_T}{2}$$

At T_m half the strands must be in duplex. Hence the concentrations of the different species at T_m will be

$$[AB] = [A] = [B] = \frac{C_T}{4}$$

The equilibrium constant at T_m is

$$K_a = \frac{[AB]}{[A][B]} = \frac{C_T/4}{(C_T/4)^2} = \frac{4}{C_T}$$

Do not be misled by this expression into thinking that the equilibrium constant is concentration dependent. It is the T_m that is concentration dependent. The equilibrium constant is temperature dependent. The above expression indicates the value seen for the equilibrium constant, $4/C_T$, at the temperature, T_m. This particular T_m occurs when the equilibrium is observed at the total strand concentration, C_T.

A special case must be considered in which hybridization occurs between two short single strands of the same identical sequence, C. Such strands are self-complementary. An example is GGGCCC which can base pair with itself. In this case the reaction can be written

$$2C = C_2$$

The equilibrium constant, K_a, becomes

$$K_a = \frac{[C_2]}{[C]^2}$$

At the melting temperature half of the strands must be duplex. Hence

$$[C_2] = [2C] = \frac{C_T}{4}$$

where C_T as before is the total concentration of strands. Thus we can evaluate the equilibrium expression at T_m as

$$K_a \frac{C_T/4}{(C_T/2)^2} = \frac{1}{C_T}$$

As before, what this really means is that T_m is concentration dependent. In both cases simple mass action considerations ensure that T_m will increase as the concentration is raised.

The final case we need to consider is when one strand is in vast excess over the other instead of both being at equal concentrations. This is frequently the case when a trace amount of probe is used to interrogate a concentrated sample or, alternatively, when a large amount of probe is used to interrogate a very minute sample. The formation of duplex can be written as before as $A + B = AB$, but now the initial starting conditions are

$$[B]_o \gg [A]_o$$

Effectively the total strand concentration, C_T, is thus simply the initial concentration of the excess strand: B_o. At T_m,

$$[AB] = [A]$$

Thus the equilibrium expression at T_m becomes

$$K_a = \frac{[AB]}{[A][B]} = \frac{1}{C_T}$$

The melting temperature is still concentration dependent.

The importance of our ability to drive duplex formation cannot be underestimated. Figure 3.4 illustrates the practical utility of this ability. It shows the concentration dependence of the melting temperature of two different duplexes. We can characterize each re-

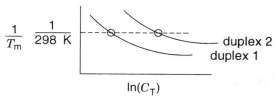

Figure 3.4   The dependence of the melting temperature, T_m, of two short duplexes on the total concentration of DNA strands, C_T.

action by its melting temperature and can attempt to extract thermodynamic parameters like the enthalpy change, ΔH, and the free energy change, ΔG, for each reaction. However, with the T_m's different for the two reactions, if we do this at T_m, the thermodynamic parameters derived will refer to reactions at two different temperatures. There will be no way, in general, to compare these parameters, since they are expected to be intrinsically temperature dependent. The concentration dependence of melting saves us from this dilemma. By varying the concentration, we can produce conditions where the two duplexes have the same T_m. Now thermodynamic parameters derived from each are comparable. We can, if we wish, choose any temperature for this comparison. In practice, 298 K has been chosen for this purpose.

THERMODYNAMICS OF THE MELTING OF SHORT DUPLEXES

The model we will use to analyze the melting of short DNA double helices is shown in Figure 3.5. The two strands come together, in a nucleation step, to form a single pair. Double strands can form by stacking of adjacent base pairs above or below the initial nucleus until a full duplex has zippered up. It does not matter, in our treatment, where the initial nucleus forms. We also need not consider any intermediate steps beyond the nucleus and the fully duplex state. In that state, for a duplex of n base pairs, there will be $n - 1$ stacking interactions (Fig. 3.6). Each interaction reflects the energetics of stacking two adjacent base pairs on top of each other. There are ten distinct such interactions, as ApG/CpT, ApA/TpT, and so on (where the slash indicates two complementary antiparallel strands). Because their energetics are very different, we must consider the DNA sequence explicitly in calculating the thermodynamics of DNA melting.

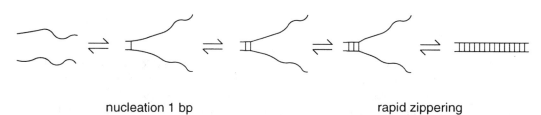

nucleation 1 bp   rapid zippering

Figure 3.5   A model for the mechanism of the formation of duplex DNA from separated complementary single strands.

oligomer
duplex

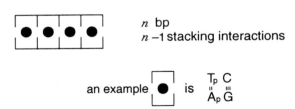

Figure 3.6   Stacking interactions in double-stranded DNA.

For each of the ten possible stacking interactions, we can define a standard ΔG_s^0 for the free energy of stacking, a ΔH_s^0 for the enthalpy of stacking, and a ΔS_s^0 for the entropy of stacking. These quantities will be related at a particular temperature by the expression

$$\Delta G_s^0 = \Delta H_s^0 - T\Delta S_s^0$$

For any particular duplex DNA sequence, we can compute the thermodynamic parameters for duplex formation by simply combining the parameters for the competent stacking interactions plus a nucleation term. Thus

$$\Delta G^0 = \Delta G_{nuc}^0 + \Sigma\Delta G_s^0 + \Delta g_{sym}$$

$$\Delta H^0 = \Delta H_{nuc}^0 + \Sigma\Delta H_s^0$$

and similarly for the entropy, where the sums are taken over all of the stacking interactions in the duplex, where $\Delta g_{sym} = 0.4$ kcal/mol if the two strands are identical; otherwise, $\Delta g_{sym} = 0$. The equilibrium constant for duplex formation is given by

$$K = ks_1 s_2 s_3 \ldots s_{n-1}$$

where k is the equilibrium constant of nucleation, related to the ΔG_{nuc}^0 by

$$\Delta G_{nuc}^0 = -RT\ln k$$

and each s_i is the microscopic equilibrium constant for a particular stacking reaction, related to the ΔG_i^0 for that reaction by

$$\Delta G_i^0 = -RT\ln s_i$$

The key factor involved in predicting the stability of DNA (and RNA) duplexes is that all of these thermodynamic parameters have been measured experimentally. One takes advantage of the enormous power available to synthesize particular DNA sequences, combines complementary pairs of such sequences, and measures their extent of duplex as a function of temperature and concentration. For example, we can study the properties of

A_8/T_8 and compare these with A_9/T_9. Since the only difference between these complexes is an extra ApA/TpT stacking interaction, the differences will yield the thermodynamic parameters for that interaction. Other sequences are more complex to handle, but this has all been accomplished.

It is helpful to choose a single standard temperature and set of environmental conditions for the tabulation of thermodynamic data: 298 K has been selected, and ΔG_s^0's at 298 K in 1 M NaCl are listed in Table 3.1. Enthalpy values for the stacking interactions can be obtained in two ways: either by direct calorimetric measurements or by examining the temperature dependence of the stacking interactions. ΔH_s^0's are also listed in Table 3.1. So are ΔS_s^0's, which can be calculated from the relationship $\Delta G_s^0 = \Delta H_s^0 - T\Delta S_s^0$. From these data, thermodynamic values at other temperatures can be estimated as shown in Box 3.3. The effects of salt are well understood and are described in detail elsewhere (Cantor and Schimmel, 1980).

The results shown in Table 3.1 make it clear that the effects of the DNA sequence on duplex stability are very large. The average ΔH_s^0 is -8 kcal/mol; the range is -5.6 to -11.9 kcal/mol. The average ΔG_s^0 is -1.6 kcal/mol with a range of -0.9 to -3.6 kcal/mol. Thus the DNA sequence must be considered explicitly in estimating duplex stabilities. The two additional parameters needed to do this concern the energetics of nucleation. These are relatively sequence independent, and we can use average values of $\Delta G_{nuc}^0 = +5$ kcal/mol (except if no G–C pairs are present, then $+6$ kcal/mol should be used) and $\Delta H_{nuc}^0 = 0$.

For estimating the stability of perfectly paired duplexes, these nucleation parameters and the stacking energies in Table 3.1 are used. Table 3.2 shows typical results when calculated and experimentally measured ΔG^0's are compared. The agreement in almost all cases is excellent, and the few discrepancies seen are probably within the range of typical experimental errors. The approach described above has been generalized to predict the thermodynamic properties of triple helices, and presumably it will also serve for four-stranded DNA structures.

TABLE 3.1   Nearest-neighbor Stacking Interactions in Double-stranded DNA

	Nearest-neighbor Thermodynamics		
Interaction	$-\Delta H°$ (kcal/mol)	$-\Delta S°$ (cal/Kmol)	$-\Delta G°$ (kcal/mol)
AA/TT	9.1	24.0	1.9
AT/TA	8.6	23.9	1.5
TA/AT	6.0	16.9	0.9
CA/GT	5.8	12.9	1.9
GT/CA	6.5	17.3	1.3
CT/GA	7.8	20.8	1.6
GA/CT	5.6	13.5	1.6
CG/GC	11.9	27.8	3.6
GC/CG	11.1	26.7	3.1
GG/CC	11.0	26.6	3.1

Source: Adapted from Breslauer et al. (1986).

BOX 3.3
STACKING PARAMETERS AT OTHER TEMPERATURES

In practice, it turns out to be a sufficiently accurate approximation to assume that the various ΔH_s^0's are independent of temperature. This is a considerable simplification. It allows direct integration of the van't Hoff relationship to compute thermodynamic parameters at any desired temperature from the parameters measured at 298 K. Starting from the usual relationship for the dependence of the equilibrium constant on temperature,

$$\frac{d(\ln K)}{d(1/T)} = \frac{\Delta H^0}{R}$$

we can integrate this directly:

$$\int d(\ln K) = \frac{\Delta H^0}{R} \int d\left(\frac{1}{T}\right)$$

With limits of integration from T_0, which is 298 K, to any other melting temperature, T_m, the result is

$$\ln K(T_m) - \ln K(T_0) = \frac{\Delta H^0}{R}\left(\frac{1}{T_m} - \frac{1}{T_0}\right)$$

However, the first term of the left-hand side of the equation, for a pair of complementary oligonucleotides, is just $4/C_T$. The second term on the left-hand side is equal to $\Delta G(T_0)/RT_0$. Inserting these values and rearranging gives a final, useful expression for computing the T_m of a duplex at any temperature from data measured at 298 K. All of the necessary data are summarized in Table 3.1.

$$\frac{T_m}{T_0} = \frac{\Delta H^0}{\Delta H^0 - \Delta G^0(T_0) + RT_0 \ln(C_T/4)}$$

THERMODYNAMICS OF IMPERFECTLY PAIRED DUPLEXES

In contrast to the small number of discrete interactions that must be considered in calculating the energetics of perfectly paired DNA duplexes, there is a plethora of ways that duplexes can pair imperfectly. We have available model compound data on most of these so that estimates of the energetics can be made. However, the large number of possibilities precludes a complete analysis of imperfections in the context of all possible sequences, at least for the present.

The simplest imperfection is a dangling end, as shown in Figure 3.7a. If both ends are dangling, their contributions can be treated separately. From available data it appears that a dangling end contributes -8 kcal/mol on average to the overall ΔH of duplex formation, and -1 kcal/mol to the overall ΔG of duplex formation. The large enthalpy arises

**TABLE 3.2   Predicted and Observed Stabilities
(free energy of formation) of Various
Oligonucleotide Duplexes**

Comparison of Calculated and Observed ΔG (kcal/mol)		
Oligomeric Duplex	$-\Delta G$pred	$-\Delta G$obs
1 GCGCGC CGCGCG	11.1	11.1
2 CGTCGACG GCAGCTGC	11.2	11.9
3 GAAGCTTC CTTCGAAG	7.9	8.7
4 GGAATTCC CCTTAAGG	9.3	9.4
5 GGTATACC CCATATGG	6.7	7.4
6 GCGAATTCGC CGCTTAAGCG	16.5	15.5
7 CAAAAAG GTTTTTC	6.1	6.1
8 CAAACAAAG GTTTGTTTC	9.3	10.1
9 CAAAAAAAG GTTTTTTTC	9.9	9.6
10 CAAATAAAG GTTTATTTC	8.5	8.5
11 CAAAGAAAC GTTTCTTTC	9.3	9.5
12 CGCGTACGCGTACGCG GCGCATGCGCATGCGC	32.9	34.1

Note: Calculations use the equations given in the text and the measured thermodynamic values given in Table 3.1.

Source: Adapted from Breslauer et al. (1986).

because the first base of the dangling end can still stack on the last base pair of the duplex. Note that there are two distinct types of dangling ends: a 3′-overhang and a 5′-overhang. At the current level of available information, we treat these as equivalent. Simple dangling ends will arise whenever a target and a probe are different in size.

The next imperfection, which leads to considerable destabilization, is an internal mismatch. As shown in Figure 3.7b, this leads to the loss of two internal ΔH_s^0's and two internal ΔG_s^0's. Apparently this is empirically compensated by some residual stacking either between the bases that are mispaired and the duplex borders or between the bases themselves. Whatever the detailed mechanism, the result is to gain back about -8 kcal/mol in ΔH. There is no effect on the ΔG. A larger internal mismatch is called an *internal loop*. Considerable data on the thermodynamics of such loops exist for RNA, and much less for DNA. In general, such structures will be far less stable than the perfect duplex because of the loss of additional free energies and enthalpies of stacking.

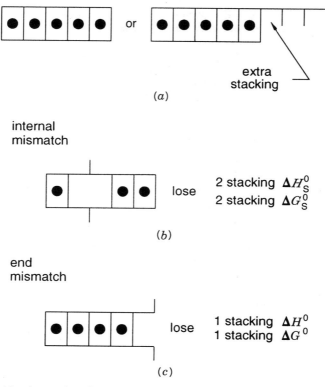

Figure 3.7   Stacking interactions in various types of imperfectly matched duplexes. *(a)* Dangling ends. *(b)* Internal mismatch. *(c)* Terminal mismatch.

A related imperfection is a bulge loop in which two strands of unequal size come together so that one is perfectly paired while the other has an internal loop of unpaired residues. There can also be internal loops in which the single-stranded regions are of different length on the two strands. Again the data describing the thermodynamics of such structures are available mostly just for RNAs (Cantor and Schimmel, 1981). There are many complications. See, for example, Schroeder et al. (1996).

A key imperfection that needs to be considered to understand the specificity of hybridization is a terminal mismatch. As shown in Figure 3.7c, this results in the loss of one ΔH_s^0 and one ΔG_s^0. Such an external mismatch should be less destabilizing than an internal mismatch because one less stacking interaction is disrupted. It is not clear, at the present time, how much any stacking of the mismatch to the end of the duplex partially compensates for the lost duplex stacking. It may be adequate to model an end-mismatch like a dangling end. Alternatively, one might consider it like an internal mismatch with only one lost set of duplex stacking interactions. Both of these cases require no correction to the predicted ΔG of duplex formation, once the lost stacking interactions have been accounted for. Therefore at present it is simplest to concentrate on ΔG estimates and ignore ΔH estimates. More studies are needed in this area to clear up these uncertainties. In Chapter 12 we will discuss attempts to use hybridization of short oligonucleotides to infer

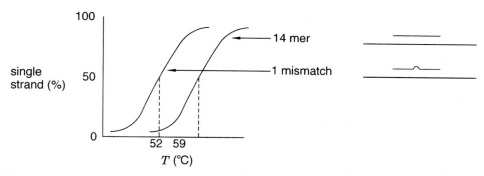

Figure 3.8 Equilibrium melting behavior of a perfectly matched oligonucleotide duplex and a duplex with a single internal mismatch.

sequence information about a larger DNA target. As interest in this approach becomes more developed, the necessary model compound data will doubtlessly be accumulated. Note that there are 12 distinct terminal mismatches, and their energetics could depend not only on the sequence of the neighboring duplex stack but also on whether one or both strands of the duplex are further elongated. A large number of model compounds will need to be investigated.

Despite the limitations and complications just described, our current ability to predict the effect of a mismatch on hybridization is quite good. A typical example is shown in Figure 3.8. Note that the melting transitions of oligonucleotide duplexes with 8 or more base pairs are quite sharp. This reflects the large ΔH_s^0's and the all-or-none nature of the reaction. The key practical concern is finding a temperature where there is good discrimination between the amount of perfect duplex formed and the amount of mismatched duplex. The results shown in Figure 3.8 indicate that a single internal mismatch in a 14-mer leads to a 7°C decrease in T_m. Because of the sharpness of the melting transitions, this can easily be translated into a discrimination of about a factor of ten in amount of duplex present.

KINETICS OF THE MELTING OF SHORT DUPLEXES

The major issue we want to address here is whether, by kinetic studies of duplex dissociation, one can achieve a comparable or greater discrimination between the stability of different duplexes than by equilibrium melting experiments. In practice, equilibrium melting experiments are technically difficult because one must discriminate between free and bound probe or target. If the analysis is done spectroscopically, it can be done in homogeneous solution, but this usually requires large amounts of sample. If it is done with a radioisotopic tag (or a fluorescent label that shows no marked change on duplex formation), it is usually done by physical isolation of the duplex. This assumes that the rate of duplex dissociation is slow compared with the rate of physical separation, a constraint that usually can be met.

In typical hybridization formats a complex is formed between probe and target, and then the immobilized complex is washed to remove excess unbound probe (Fig. 3.9).

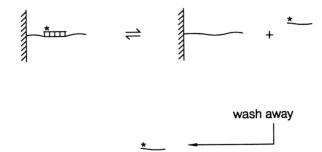

Figure 3.9 Schematic illustration of the sort of experiment used to measure the kinetics of duplex melting.

The complex is stored under conditions where little further dissociation occurs, and afterward the amount of probe bound to the target is measured. A variation on this theme is to allow the complex to dissociate for a fixed time period and then to measure the amount of probe remaining bound. It turns out that this procedure gives results very similar to equilibrium duplex formation. The reason is that the temperature dependence for the reaction between two long single strands to form a duplex is very slight. The activation energy for this reaction has been estimated to be only about 4 kcal/mol. Figure 3.10 shows a hypothetical reaction profile for interconversion between duplex and single strands. If we assume that the forward and reverse reaction pathways are the same, then the activation energy for the dissociation of the double strand will be just $\Delta H + 4$ kcal/mol. Since ΔH is much larger than 4 kcal/mol for most duplexes, the temperature dependence of the reaction kinetics will mirror the equilibrium melting of the duplex.

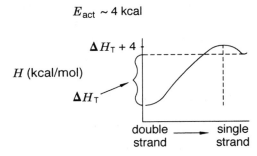

Figure 3.10 Thermodynamic reaction profile for oligonucleotide melting. Shown is the enthalpy, H, as a function of strand separation.

It is simple to estimate the effect of temperature on the dissociation rate. At T_m we expect an equation of the form

$$k_m = A \exp\left(-\frac{[\Delta H + 4]}{RT_m}\right)$$

to apply, where A is a constant, and the exponential term reflects the number of duplexes that have enough energy to exceed the activation energy. At any other temperature, T_d,

$$k_d = A \exp\left(-\frac{[\Delta H + 4]}{RT_d}\right)$$

The rate constants k_m and k_d are for duplex melting at temperatures T_m and T_d, respectively. We define the time of half release at T_d to be t_{wash}, and the time of half release at T_m to be $t_{1/2}$. Then simple algebraic manipulation of the above equations yields

$$\ln\left(\frac{t_{wash}}{t_{1/2}}\right) = \frac{(\Delta H + 4)(1/T_d - 1/T_m)}{R}$$

This expression allows us to calculate kinetic results for particular duplexes once measurements are made at any particular temperature.

KINETICS OF MELTING OF LONG DNA

The kinetic behavior of the melting of long DNA is very different from that of short duplexes in several respects. Melting near T_m cannot normally be considered an all-or-none reaction. Melting experiments at temperatures far above T_m reveal an interesting complication. The base pairs break rapidly, and if the sample is immediately returned to temperatures below T_m, they reform again. However, if a sufficient time at high temperature is allowed to elapse, now the renaturation rate can be many orders of magnitude slower. What is happening is shown schematically in Figure 3.11. Once the DNA base pairs are broken, the two single strands are still twisted around each other. The twist represented by each turn of the double helix must still be unwound. As the untwisting begins, the process greatly slows because the coiled single strands must rotate around each other. Experiments show that the untwisting time scales as the square of the length of the DNA. For molecules 10 kb or longer, these unwinding times can be very long. This is probably one of the reasons why people have trouble performing conventional polymerase chain reaction (PCR) amplifications on high molecular weight DNA samples. The usual PCR protocols allow only 30 to 60 seconds for DNA melting (as discussed in the next chapter.) This is far too short for strand untwisting of large DNAs.

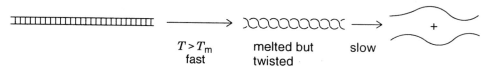

$T > T_m$ melted but   slow
fast   twisted

Figure 3.11   Kinetic steps in the melting of very high molecular weight DNA.

KINETICS OF DOUBLE-STRAND FORMATION

The kinetics of renaturation of short duplexes are simple and straightforward. The reaction is essentially all or none; nucleation of the duplex is the rate-limiting step, and it comes about by intermolecular collision of the separated strands. Duplex renaturation kinetic measurements can be carried out at any temperature sufficiently below the melting temperature so that the reverse process, melting, does not interfere. The situation is much more complicated when longer DNAs are considered. This is illustrated in Figure 3.12*a,* where the absorbance increase upon melting is used to measure the relative amounts of single- and double-stranded structures. If a melted sample, once the strands have physically unwound, is slowly cooled, full recovery of the original duplex can be achieved. However, depending on the sample, the rate of duplex formation can be astoundingly slow, as we will see. If the melted sample is cooled too rapidly, the renaturation is not complete. In fact, as shown in Figure 3.12*b,* true renaturation has not really occurred at all. Instead, the separated single strands start to fold on themselves to make hairpin duplexes, analogous to the secondary structures seen in RNA. If the temperature falls too rapidly, these structures become stable; they are kinetically trapped and, effectively, can never dissociate to allow formation of the more stable perfect double-stranded duplex.

To achieve effective duplex renaturation, one needs to hold the sample at a temperature where true duplex formation is favored at the expense of hairpin structures so that these rapidly melt, and the unfolded strands are free to continue to search for their correct partners. Alternatively, hairpin formation is kept sufficiently low so that there remains a sufficiently unhindered DNA sequence to allow nucleation of complexes with the correct partners.

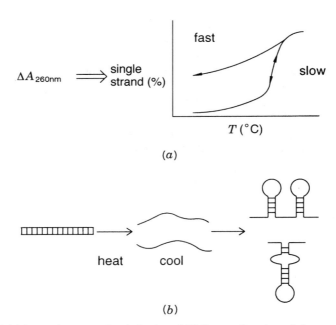

Figure 3.12 Melting and renaturation behavior of DNA as a function of the rate of cooling. *(a)* Fraction of initial base pairing as a function of temperature. *(b)* Molecular structures in DNA after rapid heating and cooling.

In practice, renaturations are carried out most successfully when the renaturation temperature is $(T_m - 25)°C$. In free solution the simple model, shown in Figure 3.5, is sufficient to account for renaturation kinetics under such conditions. It is identical to the picture used for short duplexes. The rate-limiting step is nucleation, when two strands collide in an orientation that allows the first base pair to be formed. After this, rapid duplex growth occurs in both directions by a rapid zippering of the two strands together. This model predicts that the kinetics of renaturation should be a pure second-order reaction. Indeed, they are. If we call the two strands A and B we can describe the reaction as

$$A + B \rightarrow AB$$

The kinetics will be governed by a second-order rate constant k_2:

$$\frac{d[AB]}{dt} = k_2[A][B]$$

$$\frac{-d[A]}{dt} = \frac{-d[B]}{dt} = k_2[A][B]$$

In a typical renaturation experiment, the two strands have the same initial concentration: $[A]_o = [B]_o = C_s$, where C_s is $\frac{1}{2}$ the initial total concentration of DNA strands,

$$C_s = \frac{[A]_o + [B]_o}{2}$$

At any subsequent time in the reaction, simple mass conservation indicates that the concentrations of single strands remaining will be $[A] = [B] = C$, where C is the total instantaneous concentration of each strand at time t. This allows us to write the rate equation for the renaturation reaction as in terms of the disappearance of single strands as

$$\frac{-dC}{dt} = k_2 C^2$$

We can integrate this directly from the initial concentrations to those at any time t,

$$-\int \frac{dC}{C^2} = \int_0^t k_2 dt$$

which becomes

$$\frac{1}{C} - \frac{1}{C_s} = k_2 t$$

and can be rearranged to

$$\frac{C_s}{C} = 1 + k_2 t C_s$$

It is convenient to rearrange this further to

$$\frac{C}{C_s} = f_s = \frac{1}{(k_2 t C_s + 1)}$$

where f_s is the fraction of original single strands remaining.

When $f_s = 0.5$, we call the time of the reaction the half time, $t_{1/2}$. Thus

$$k_2 t_{1/2} C_s = 1 \quad \text{or} \quad t_{1/2} = \frac{1}{k_2 C_s}$$

The key result, for a second-order reaction, is that we can adjust the halftime to any value we choose by selecting an appropriate concentration.

The concentration that enters the equations just derived is the concentration of unique, complementary DNA strands. For almost all complex DNA samples, this is not a concentration we can measure directly. Usually the only concentration information readily available about a DNA sample is the total concentration of nucleotides, C_0. This is obtainable by direct UV absorbance measurements or by less direct but more sensitive fluorescence measurements on bound dyes. To convert C_0 to C_s, we have to know something about the sequence of the DNA. Suppose that our DNA was the intact genome of an organism, with N base pairs, as shown schematically in Figure 3.13. The concentration of each of the two strands at a fixed nucleotide concentration, C_0, is just

$$C_s = \frac{C_0}{2N}.$$

Now suppose that the genome is divided into chromosomes, or the chromosomes into unique fragments, as shown in Figure 3.13. This does not change the concentration of any of the strands of these fragments. (An important exception occurs if the genome is randomly broken instead of uniquely broken; see Cantor and Schimmel, 1980.) Thus we can write the product of k_2, the concentration and half-time, as

$$k_2 t_{1/2} C_s = \frac{k_2 t_{1/2} C_0}{2N} = 1$$

and this can be rearranged to

$$C_0 t_{1/2} = \frac{2N}{k_2}$$

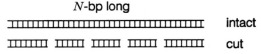

N-bp long

intact

cut

Figure 3.13   Schematic illustration of intact genomic DNA, or genomic DNA digested into a set of nonoverlapping fragments.

This key result means that the rate of genomic DNA reassembly depends linearly on the genome size. Since genome sizes vary across many orders of magnitude, the resulting renaturation rates will also. Here we have profound implications for the design and execution of experiments that use genomic DNA. Molecular biologists have tended to lump the two variables C_0 and $t_{1/2}$ together, and they talk about the parameter $C_0 t_{1/2}$ as "a half cot." We will use this term, but don't try to sleep on it.

Some predicted kinetic results for the renaturation of genomic DNA that illustrate the renaturation behavior of the different samples are shown in Figure 3.14. These results are calculated for typical renaturation conditions with a nucleotide concentration of 1.5×10^{-4} M. Note that under these conditions a 3-kb plasmid will renature so quickly that it is almost impossible to work with the separate strands under conditions that allow renaturation. In contrast, the 3×10^9 bp human genome is predicted to require 58 days to renature under the very same conditions. In practice, this means never; much higher concentrations must be used to achieve the renaturation of total human DNA.

The actual renaturation results seen for human DNA are very different from the simple prediction we just made. The reason is a significant fraction of human DNA is highly repeated sequences. This changes the renaturation rate of these sequences. The concentration of strands of a sequence repeated m times in the genome is $mC_0/2N$. Thus this sequence will renature much faster:

$$C_0 t_{1/2} = \frac{2N}{mk_2}$$

The same equation also handles the case of heterogeneity in which a particular sequence occurs in only a fraction of the genomic DNA under study. In this case $m < 1$, and the renaturation proceeds slower than expected. Typical renaturation results for human (or any mammalian) DNA are shown in Figure 3.15. There are three clearly resolved kinetic

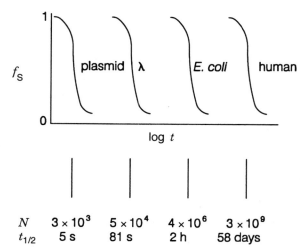

Figure 3.14 Renaturation kinetics expected for different DNA samples at a total initial nucleotide concentration of 1.5×10^{-4} M at typical salt and temperature conditions. The fraction of single strand remaining, f_s, is plotted as a function of the log of the time, t. Actual experiments conform to expectations except for the human DNA sample. See Figure 3.15.

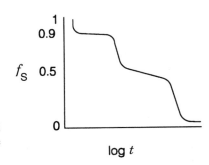

Figure 3.15 Renaturation kinetics actually seen for total human genomic DNA. The parameters plotted are the same as in Figure 3.14.

phases to the reaction. About 10% of the DNA renatures almost immediately. It turns out this is very simple sequence DNA, some of which is self-complementary, and can renature by folding back on itself to make a hairpin. Other single-stranded components of this class form novel DNA ordered structures like some triplet-repeating sequences (see Chapter 13). About 40% of the DNA renatures 10^5 to 10^6 times as fast as expected for single-copy human DNA. This consists mostly of highly repeated sequences like the human *Alu* repeat. The effective $C_0t_{1/2}$ for this DNA is about 1. The remaining half of the genome does show the expected renaturation times for single-copy DNA.

 The repeated sequences that renature fast, but intermolecularly, are mostly interspersed with single-copy DNA. Thus the length of the DNA fragments used in a renaturation experiment will have a profound effect on the resulting products that one might purify at various times during the reaction. Repeats that come together fast will usually be flanked by single-copy sequences that do not correspond. If one purifies double stranded material, it will be a mixture of duplex and single strands, as shown in Figure 3.16. Upon further renaturation the dangling single strands will find their partners, but only at the cost of generating a complex crosslinked network of DNA (Fig. 3.16*b*). Eventually this will act as such a barrier to diffusion that some of the strands will never find their mates.

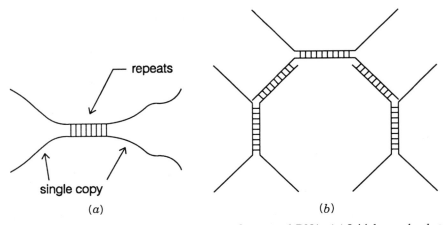

Figure 3.16 Effects of repeats on the structure of renatured DNA. *(a)* Initial complex between two DNAs that share partial sequence complementarity at a repeat. *(b)* Subsequent complexes formed at C_0t's sufficiently large to allow duplex formation between single copy sequences.

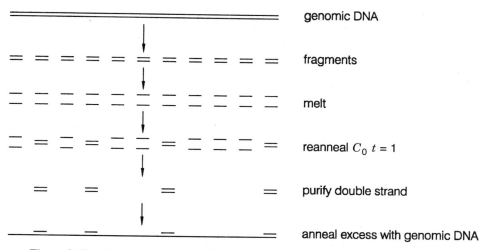

Figure 3.17   Blocking repeats by prehybridization with excess purified repeated DNA.

There are two solutions to this problem. One is to cut the genomic DNA down to sizes of a few hundred base pairs so that most single-copy sequences are essentially removed from the repeats. The other alternative, if larger DNA is needed for a particular application, is to purify repeated DNA as small fragments and pre-anneal it at high concentrations to single-stranded genomic DNA. This will block the repeat sites, as shown in Figure 3.17, and then the flanking single strands will be free to find their complements without network formation.

One could ask why not cut genomic DNA down to tiny fragments for all reannealing or hybridization experiments. Indeed many current schemes for analyzing DNA with hybridization do use short oligonucleotide probes. These have one disadvantage. Their intrinsic k_2 is small. Recall that k_2 really represents the nucleation reaction for duplex formation. This is the rate of formation of the first correct base pair that can zip into the duplex structure. The number of potential nucleation sites increases as L, the length of the strands. There is a second way in which length affects k_2; this is the excluded volume of the coiled DNA single strands which retards the interpenetration needed for duplex formation. This depends on $L^{-1/2}$. When these two effects are combined, the result is that k_2 depends on $L^{1/2}$. This discourages the use of small fragments for hybridization unless they are available at very high concentrations. Ultimately, if too small fragments are used, there will also be an effect on the stability of the resulting duplexes.

COMPLEXITY

In thinking about DNA hybridization, it is useful to introduce the notion of DNA complexity. This is the amount of different DNA sequence present in a sample. For a DNA sample with repeated sequence and heterogeneities, the complexity is

$$\frac{\Sigma_i N_i}{m_i}$$

where the sum is taken over every type of DNA species present; m_i is the number of times a species of size N_i is repeated. The complexity will determine the hybridization kinetics of the slowest renaturing species in the sample.

HYBRIDIZATION ON FILTERS

In contemporary DNA work, hybridization of probes in solution to targets immobilized on filters is much more common than duplex formation in homogeneous solution. Two basic formats are most frequently used; these are summarized in Figure 3.18. In a dot blot, a sample of DNA from a cell, a viral plaque, or a sorted chromosome is immobilized (nominally) on the surface of a filter. It is anchored there by heating or UV crosslinking. The exact mechanism of this immobilization is unknown, but it is important, especially for procedures in which the same filter is going to be subjected to many serial hybridizations. Generally, the DNA sample is denatured by alkali or heating to temperatures near 100°C just before immobilization so that what is stuck to the filter is largely single-stranded material. A number of different types of filters are in common use; the most frequent are nitrocellulose and nylon.

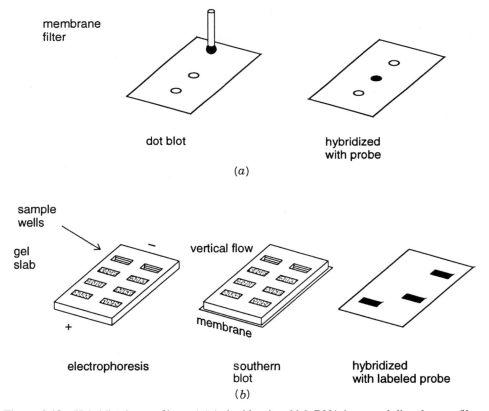

Figure 3.18 Hybridization on filters. *(a)* A dot blot, in which DNA is spotted directly on a filter. *(b)* A Southern blot in which a filter replica is made of DNA after a gel-electrophoretic separation.

In the Southern blot format, a mixture of DNA species is size-separated by electrophoresis on polyacrylamide or agarose gels (see Chapter 5). A replica of the gel-separated material is made by transfer of the DNA through the plane of the separation gel to a parallel filter, as shown in Figure 3.18. In this way the targets of the subsequent hybridization reaction can be characterized by their electrophoretic properties. Several methods exist for transfer of the gel sample to the filter. These include bulk liquid flow through the capillary network of a stacked set of paper sheets, vacuum transfer of the fluid phase of the gel, pressure transfer of the fluid phase of the gel, or electrophoretic transfer of the DNA by using electrodes perpendicular to the plane of the original gel separation. All of these methods work.

Several procedures exist where the transfer step can be avoided. One is prehybridization of labeled probes to samples before electrophoresis. Another is drying the separation gel and using it directly as a target for hybridization with radiolabeled probes. These procedures have never achieved the popularity of the Southern blot, mainly because they are less amenable to successful serial probings.

There are two basic advantages of filter hybridization compared with homogeneous phase hybridization. The target samples are highly confined. Therefore many different samples can be reacted with the same probe simultaneously. The second advantage is that the original target molecules are confined in separate microscopic domains on the filter surface. Thus they cannot react with each other. This largely avoids the formation of entangled arrays, and it simplifies some of the problems caused by interspersed repeated DNA sequences. Hybridization at surfaces does, however, produce some complications. The immobilized target DNA is presumably attached to the surface at several places. This introduces a number of obstacles to hybridization. The first of these is the potential for nonspecific attraction or repulsion of the probe DNA to the surface. More serious are the topological boundaries imposed by the points of surface-DNA attachment. These greatly reduce the effective length of the target DNA. As measured by the relative kinetics of solution and immobilized targets, the latter appear to have an effective length of only ten bases. The topological barriers will also affect the ability of DNA's with nonlinear topologies to hybridize (Fig. 3-19).

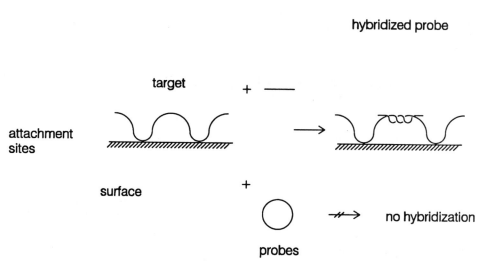

Figure 3.19 Topological obstacles to filter hybridization as a result of frequent DNA attachment points.

Formation of double strands requires twisting of one strand around the other (see Box 2.1). Thus a circular probe (e.g., circular single-stranded DNA from a bacteriophage M13 clone) will be incapable of forming extensive double helix with an immobilized DNA target. An ingenious use of this topological constraint to improve the detection of specific hybridization signals is the padlock probe approach described in Box 3.4.

BOX 3.4
PADLOCK PROBES

Ulf Landegren and his collaborators recently demonstrated that DNA topology can be used to improve the stringency of detection of correctly hybridized DNA probes. A schematic view of the procedure that they have developed is shown in Figure 3.20. Consider the hybridization of a linear single-stranded probe to a single-stranded circular target. The probe is specially designed so that it's 3'- and 5'-terminal sequences are complementary to a continuous block of DNA sequence in the target. Probes like this can always be made synthetically (or they can be made by circularization with an arbitrary flanking sequence by ligation and then cleavage within the probe sequence in a

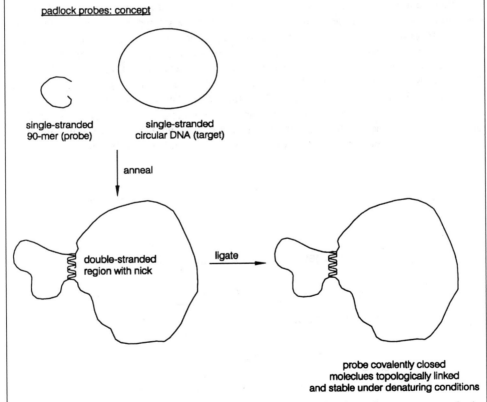

Figure 3.20 Example of the use of padlock probes. Here ligation is used to create a topologically linked probe-target complex in solution. The same approach can be used to link probes onto DNA targets immobilized on filters. (Drawing provided by Fouad Siddiqi.)

(continued)

BOX 3.4 *(Continued)*

way similar to the production of jumping probes described in Chapter 8). When the probe is hybridized to the target, a substrate for DNA ligase is created. After ligation the probe and target form a continuous stretch of double-stranded DNA. If this duplex is greater than 10 bp in length, the probe and target strands are topologically linked. Thus, as described in Box 2.1, although strong denaturants can melt the double-stranded region, the two resulting single strands cannot physically separate.

When padlock probes are used in a filter hybridization format, a single-stranded circular target is not required. Because the target is fixed on the filter in multiple places, as shown in Figure 3.19, each interior segment is topologically equivalent to circular DNA. Thus padlock probes that are ligated after hybridization to such segments will become topologically linked with them. This permits very stringent washing conditions to be used to remove any probe DNA that is hybridized but not ligated, or probe DNA that is nonspecifically adsorbed onto the surface of the filter.

One variant on blot hybridization is the reverse dot blot. Here the probe is immobilized on the surface while the labeled target DNA is introduced in homogeneous solution. This really does not change the advantages and disadvantages of the blot approach very much. Another variation on hybridization to blotted DNA samples is in-gel hybridization. Here DNA samples are melted and reannealed in hydrated gels, sometimes in between two distinct electrophoresis fractionations. Such hybridization schemes have been developed to specifically detect DNA fragments that have rearranged in the genome, by Michio Oishi, and to isolate classes of DNA repeats, by Igor Roninsen.

The kinetics of hybridization of DNA on filters or in dried gels (Fig. 3.21) can be described by a variation of the same basic approach used for solution hybridization kinetics. The basic mechanism of the reaction is still bimolecular. At long distances the diffusion of probe DNA in solution toward the target will be more efficient than the diffusion of the immobilized target toward the probe, but these bulk motions do not really matter that much to the rate. What matters is diffusion once the two DNAs are in each others' neighborhood, and here we can think of them as still showing equivalent behavior. Usually, however, the relative concentrations of the probe and sample are very different. Most often the initial probe concentration, $[P]_0$, is much greater that the initial target concentration $[T]_0$, and we say that the probe drives the hybridization. The amount of free probe remains essentially constant during the reaction. Starting from the same basic second-order rate equation, we can write

$$\frac{-d[T]}{dt} = k_2[P][T] = k_2[P]_0[T]$$

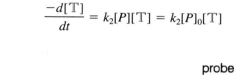

Figure 3.21   Kinetics of hybridization of a probe to a target DNA immobilized on a filter at initial concentrations of $[P]_0$ and $[T]_0$, respectively.

Because $[P]_0$ is constant, the resulting kinetics appear as pseudo–first order rather than second order. The rate equation can be integrated as

$$\int \frac{-d[T]}{[T]} = k_2[P]_0 \int_0^t dt$$

from the initial concentration of target, $[T]_0$, at time zero, to the concentration $[T]$, at any later time t, to give

$$-\ln\left[\frac{[T]}{[T]_0}\right] = k_2[P]_0 t \quad \text{or}$$

$$\frac{[T]}{[T]_0} = \exp\left(-k_2[P]_0 t\right)$$

The halftime of the reaction is defined as the time at which $[T]/[T]_0 = \frac{1}{2}$. This can be written as

$$t_{1/2} = \frac{\ln 2}{k_2[P]_0} = \frac{\ln 2}{k_2 m C_0/2N}$$

where N is the genome size of the probe, C_0 is the probe concentration in nucleotides, and m reflects any repeated sequences or heterogeneity. Note that if the probe is single stranded, the factor of two should be removed from the denominator. This allows us to write

$$C_0 t_{1/2} = 2 \ln 2 \left(\frac{N}{mk_2}\right)$$

The equation is almost the same as the definition of $C_0 t_{1/2}$ for ordinary homogeneous solution hybridization.

Occasionally the probe contains a trace radiolabel. The target is in great excess, and it drives the hybridization reaction. The above equations are exactly the same except that the role of target and probe are reversed, and $mC_0/2N$ refers to *target* DNA.

SENSITIVE DETECTION

Hybridization frequently involves the detection of very small amounts of target DNA. Thus it is essential that the probe be equipped with a very sensitive detection system. The most conventional approach is to use a radiolabeled probe. [32]P is the most common choice because its short half-life and high energy create a very intense radiodecay signal. The key to the use of ^{32}P is to incorporate the label at very high specific activity at many places in the probe. High specific activity means that many of the phosphorous atoms in the probe are actually ^{32}P and not [31]P. The most common way to do this is random priming of DNA polymerase by short oligonucleotides to incorporate [alpha ^{32}P]-labeled deoxynucleoside triphosphates ([^{32}P]dpppN's) at many places in the chain. There is a trade-off between priming at too many places, which gives probes too short to hybridize well, and priming too infrequently, which runs the risk of not sampling the probe uniformly. In principle, the highest specific activity probes would be made by using all four [^{32}P]dpppN's. In practice, however, only one or two [^{32}P]dpppN's are usually employed. Attempts to make even hotter probes are usually plagued by technical difficulties.

The short half-life of ^{32}P is responsible for its intense signal, but it also leads the major disadvantage of this detection system: The probes, once made, are rapidly decaying. A second disadvantage of very high energy radioisotopes like ^{32}P is that the decay produces a track in a photographic film or other spatially resolved detectors that can be significantly longer than the spatial resolution of the target sample. This blurs the pattern seen in techniques like autoradiographic detection of high-resolution DNA sequencing gels or in situ hybridization analysis (Chapter 7) of gene location on chromosomes. To circumvent this problem, lower-energy radioisotopes like ^{35}S-labeled thio-derivatives of dpppN's can be incorporated into DNA. However, to avoid the disadvantages of radioisotopes altogether, many alternative detection systems have been explored. A few will be discussed here. The principal concern with each is the ratio of the target signal to the background signal. The advantage of radioactive detection is that there is no intrinsic background from the sample (just a counting background due to cosmic rays and the like). This advantage disappears when chemical or spectroscopic detection methods are used.

After ^{32}P, fluorescence is currently the most popular method for DNA detection, with chemiluminescent methods (Box 3.5) rapidly developing as strong alternatives. The ultimate sensitivity of fluorescent detection allows one to work with single fluorescent molecules. To do this, the dye label is pumped continually with an intense exciting beam of light, until it is finally bleached in some photochemical reaction. Clearly the more times the dye can absorb before it is destroyed, and the higher a fraction of dye decays that emit a photon, the greater is the sensitivity. Both of these factors indicate that the ideal dye will have a fluorescence quantum yield (probability of de-excitation by light emission) as close to 1.0 as possible and a photodestruction quantum yield as low as possible.

BOX 3.5
CHEMILUMINESCENT DETECTION

In chemiluminescence, a chemical reaction is used to generate a product that is in an electronically excited state. The product subsequently relaxes to the ground state by emitting a photon. The photon is detected, and the accumulation of such events can be used to make an image of the target. The advantage of this approach is that since there is no exciting light, there is no background of stray light from the exciting beam, nor is there any background of induced fluorescence from the filter or other materials used in the hybridization and blotting. An example of some of the chemical steps used in current chemiluminescent detection of DNA is shown in Figure 3.22. These methods have sensitivity comparable to those with the highest specific activity radioisotopes used, and they are more rapid. However, they often do not allow convenient serial reprobings of the filter; this is a distinct disadvantage, and the possibility for multicolor detection, well developed with fluorescence, is not yet common with chemiluminescence.

(continued)

BOX 3.5 (*Continued*)

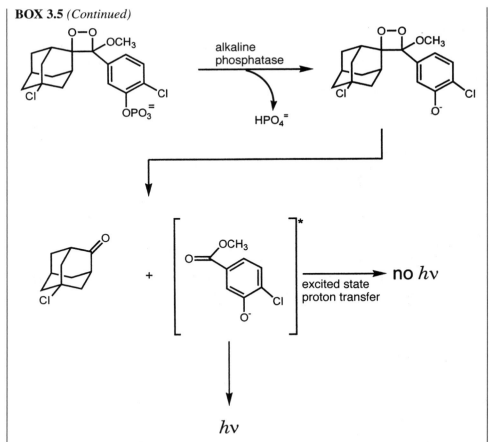

Figure 3.22   Chemistry used in the chemiluminescent detection of DNA. An alkaline phosphatase-conjugated probe is used for hybridization. Then it is presented with the substrate CDP-star™. Upon dephosphorylation of the substrate by alkaline phosphatase, a metastable phenolate anion intermediate is formed that decomposes and emits light at a maximum wavelength of 466 nm.

While such single molecule detection is achievable, it is intrinsically noisy, as is any single molecule method. For this reason it is helpful to supplement fluorescent detection schemes by intermediate chemical amplification steps. Thus a single target molecule produces or is coupled to a large number of intermediates, and each of these in turn becomes the target for a fluorescence detection system. Two general schemes for chemical amplification are shown in Figure 3.23. Catalytic amplification uses an enzyme coupled to the probe. After hybridization the enzyme is allowed to catalyze a reaction that releases a fluorescent or chemiluminescent product, or an intensely colored product. This must be done under circumstances where the products do not diffuse too far away from the site of their creation; otherwise, the spatial resolution of the blot will be degraded. In stoichiometric amplification, potential diffusion problems are avoided by building up a physical complex of molecules attached to the original probe. One way to accomplish this is by successive rounds of complementary polyvalent regents, sites, or ligands. This eventually generates a large network that can be coupled to many detector molecules.

Figure 3.23 Two types of chemical amplification systems. Catalytic amplification; stoichiometric amplification using two complementary multivalent reagents. D is a molecule that allows detection, such as a fluorescent dye.

There are a number of different ligands and polyvalent reagents that have been widely used for nucleic acid analysis. Several of the reagents are illustrated in Figure 3.24. Potentially the most powerful, in several respects, is the combination of biotin and streptavidin or avidin. Biotin can be attached to the 5'- or 3'-ends of nucleic acids, as discussed earlier, or it can be incorporated internally by various biotinylated dpppN analogues. Streptavidin and avidin bind biotin with a binding constant K_a approaching 10^{15} under physiological conditions. The three-dimensional structure of streptavidin is known from two X-ray crystallographic studies (Figure 3.25). Since avidin and streptavidin are tetravalent, there is a natural amplification route. After attaching single streptavidins to biotinylated probes, the three remaining biotin-binding sites on each protein molecule can be coupled to a polyvalent biotinylated molecule such as a protein with several biotins. Then more streptavidin can be added, in excess; the network is grown by repeats of such cycles (Fig. 3.26a). In practice, for most nucleic acid applications, streptavidin, the protein product from the bacterium *Streptomyces avidinii,* is preferred over egg white avidin because it has an isoelectric point near pH 7, which reduces nonspecific binding to DNA, and because it has no carbohydrate, which reduces nonspecific binding to a variety of components and materials in typical assay configurations.

Most other stoichiometric amplification systems use monoclonal antibodies with high affinity for haptens that can easily be coupled to DNA. Digoxigenin and fluorescein are the most frequent choices. After the antibody has been coupled to the hapten-labeled DNA, amplification can be achieved by sandwich techniques in which a haptenated second antibody, specific for the first antibody, is allowed to bind, and then additional cycles of monoclonal antibody and haptenated second antibody are used to build up a network. An example in the case of fluorescein is shown in Figure 3.26b. Other haptens that have proved useful for such amplification systems are coumarin, rhodamine, and dinitrophenyl. In all of these cases a critical concern is the stability of the network. While anti-

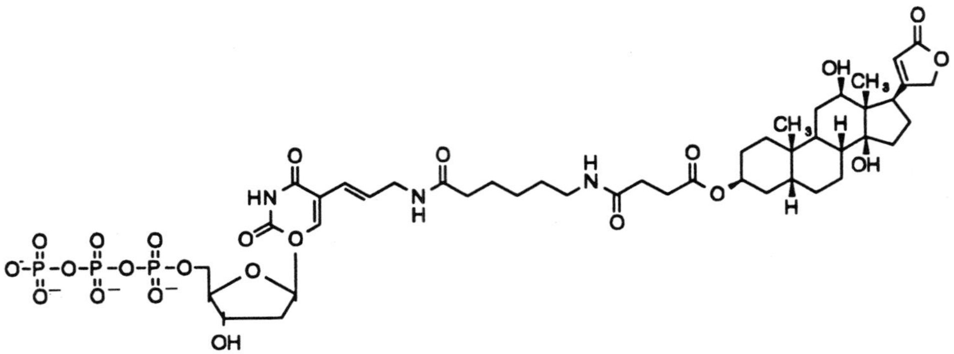

Figure 3.24 Three ligands used in stoichiometric amplification systems. *(a)* Digoxigenin. *(b)* Biotin. *(c)* Fluorescein. All are shown as derivatives of dpppU, but other derivatives are available.

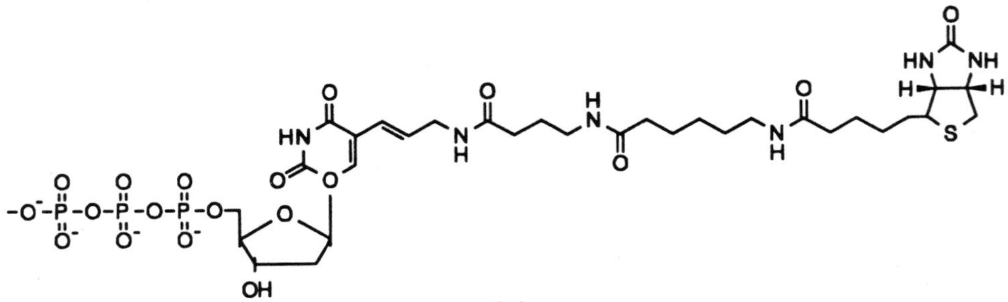

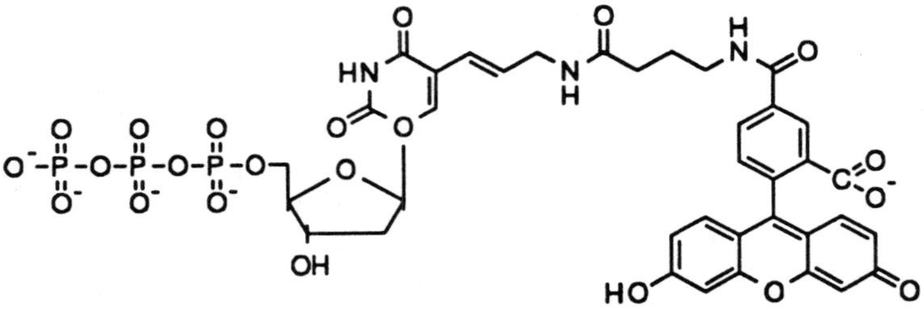

bodies and their complexes with haptens are reasonably stable, streptavidin-biotin complexes are much more stable and can survive extremes of temperature and pH in ways comparable to DNA. Parenthetically, a disadvantage of the streptavidin system is that the protein and its complexes are so stable that it is very difficult to reverse them to generate free DNA again, if this is needed. Even greater degrees of signal amplification can be achieved by using dendrimers as described in Box 3.6.

All of these amplification systems work well, but they do not have the same power of sample multiplication that can be achieved when the amplification is carried out directly at the DNA level by enzymatic reactions. Such methods are the subject of the next chapter. Ultimately sample amplification systems can be combined with color-generating amplification systems to produce exquisitely sensitive ways of detecting multiple DNA samples, sometimes in multiple colors.

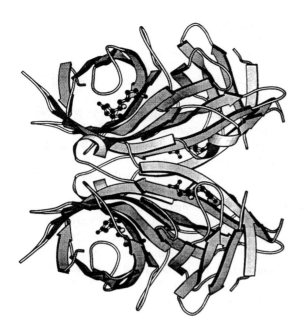

Figure 3.25 The three-dimensional structure of streptavidin. Four bound biotin molecules are shown in boldface. (Illustration created by Sandor Vajda using protein coordinates provided by Wayne Hendrickson.)

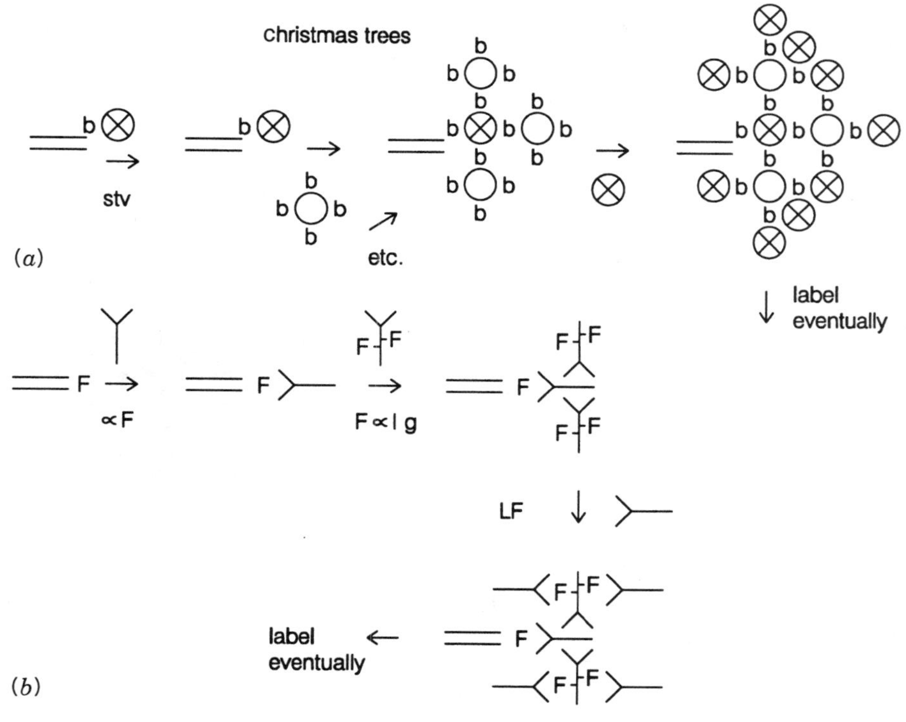

christmas trees

(a)

stv

etc.

↓ label eventually

(b)

∝F

F∝Ig

LF

label eventually

Figure 3.26 Detailed structural intermediates formed by two methods for stoichiometric amplification. *(a)* Streptavidin and some other protein containing multiple attached biotin (b) residues. *(b)* A monoclonal antibody directed against fluorescein (F) and a fluorescinated polyclonal antibody specific for the monoclonal antibody.

BOX 3.6
DENDRIMERIC DNA PROBES

Dendrimers are a chemical amplification system that allows large structures to be constructed by systematic elaboration of smaller ones. A traditional dendrimer is formed by successive covalent additions of branched reactive species to a starting framework. Each layer added grows the overall mass of the structure considerably. The process is a polyvalent analogue of the stoichiometric amplification schemes described in Figure 3.26.

Recently schemes have been designed and implemented to construct dendrimeric arrays of DNA molecules. Here branched structures are used to create polyvalency, and base-pairing specificity is used to direct the addition of each successive layer. The types of structures used and the complexity of the products that can be formed are illustrated schematically in Figure 3.27. These structures are designed so that each layer presents equal amounts of two types of single-stranded arms for further complexation. Ultimately one type of arm is used to identify a specific target by base pairing, while the other type of arm is used to bind molecules needed for detection. Dendrimers could be built on a target layer by layer, or they can be preformed with specificity selected for each particular target of interest. The latter approach appears to offer a major increase in sensitivity in a range of biological applications including Southern blots, and in situ hybridization.

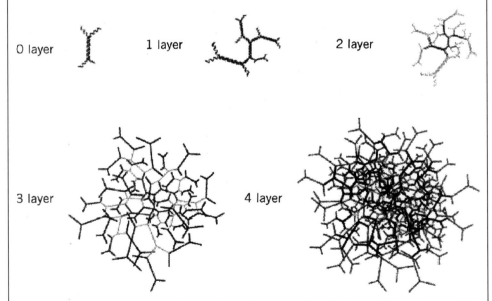

Figure 3.27   Dendrimer layer growth. Figure also appears in color insert. (Illustration provided by Thor Nilsson.)

SOURCES AND ADDITIONAL READINGS

Breslauer, K. J., Franz, R., Blöcker, H., and Marky, L. A. 1986. Predicting DNA duplex stability from the base sequence. *Proceedings of the National Academy of Sciences USA* 83: 3746–3750.

Cantor and Schimmel. 1980. *Biophysical Chemistry* III. San Francisco: W. H. Freeman, pp. 1226–1238.

Nilsson, M., Malmgren, H., Samiotaki, M., Kwiatkowski, M., Chowdhary, B. P., and Landegren, U. 1994. Padlock probes: Circularizing oligonucleotides for localized DNA detection. *Science* 265: 2085–2088.

Yokota, H., and Oishi, M. 1990. Differential cloning of genomic DNA: Cloning of DNA with an altered primary structure by in-gel competitive reassociation. *Proceedings of the National Academy of Sciences USA* 87: 6398–6402.

Roberts, R. W., and Crothers, D. M. 1996. Prediction of the Stability of DNA triplexes. *Proceedings of the National Academy of Sciences USA* 93: 4320–4325.

Roninson, I. B. 1983. Detection and mapping of homologous, repeated and amplified DNA sequences by DNA renaturation in agarose gels. *Nucleic Acids Research* 11: 5413–5431.

SantaLucia Jr., J., Allawi, H. T., and Seneviratne, P. A. 1996. Improved nearest-neighbor parameters for predicting DNA duplex stability. *Biochemistry* 35: 3555–3562.

Schroeder, S., Kim, J., and Turner, D. H. 1996. G–A and U–U mismatches can stabilize RNA internal loops of three nucleotides. *Biochemistry* 35: 16015–16109.

Sugimoto, N., Nakano, S., Yoneyama, M., and Honda, K. 1996. Improved thermodynamic parameters and helix initiation factor to predict stability of DNA duplexes. *Nucleic Acids Research* 24: 4501–4505.

Sugimoto, N., Nakano, S., Katoh, M., Matsumura, A., Nakamuta, H., Ohmichi, T., Yoneyama, M., and Sasaki, M. 1995. Thermodynamic parameters to predict stability of RNA/DNA hybrid duplexes. *Biochemistry* 34: 11211–6.

Wetmur, J. G. 1991. DNA probes: Applications of the principles of nucleic acid hybridization. *Critical Reviews in Biochemistry and Molecular Biology* 26: 227–259.

4 Polymerase Chain Reaction and Other Methods for In Vitro DNA Amplification

WHY AMPLIFY DNA?

The importance of DNA signal amplification for sensitive detection of DNA through hybridization was discussed in the previous chapter. Beyond mere sensitivity, there are two basic reasons why direct amplification of DNA is a vital part of DNA analysis. First, DNA amplification provides a route to an essentially limitless supply of material. When amplification is used this way, as a bulk preparative procedure, the major requirement is that the sample be uniformly amplified so that it is not altered, distorted, or mutated by the process of amplification. Only if these constraints can be met, can we think of amplification as a true immortalization of the DNA.

The second rationale behind DNA amplification is that selective amplification of a region of a genome, chromosome, or sample provides a relatively easy way to purify that segment from the bulk. Indeed, if the amplification is sufficient in magnitude, the DNA product becomes such an overwhelming component of the amplification mixture that the starting material is reduced to a trivial contaminant for most applications.

Amplification can be carried out in vivo by growing living cells (Box 1.2) or in vitro by using enzymes. There are several overwhelming advantages of in vitro amplification. Any possible toxic effects of a DNA target on the host cell are eliminated. There is no need to purify the amplified material away from the host genome or the vector used for cloning. Base analogues can be used that would frequently be unacceptable to a living cell system. Samples can be manipulated by automated methods that are far easier to implement in vitro than in vivo. The major limitation of existing in vitro amplification methods until recently is that they were restricted to relatively short stretches of DNA, typically less than 5 kb. New long polymerase chain reaction (PCR) procedures have extended the range of in vitro amplification up to about 20 kb. For longer targets than this, in vivo cloning methods must still be used.

BASIC PRINCIPLES OF THE POLYMERASE CHAIN REACTION (PCR)

What makes PCR a tool of immense power and flexibility is the requirement of DNA polymerases for pre-existing DNA primers. Thus DNA polymerases cannot start DNA chains de novo; a primer can be used to determine where, along a DNA template, the synthesis of the complement of that stand begins. It is this primer requirement that allows the selective amplification of any DNA region by using appropriate, specific DNA primers.

Once started, a DNA polymerase like *E. coli* DNA polymerase I (pol I) will proceed in the 3′- to 5′-direction until it has copied all the way to the 5′-end of the template. The simplest amplification scheme for in vivo DNA amplification is successive cycles of priming, chain extension, and product denaturation, shown schematically in Figure 4.1. If these steps are carried out efficiently, the result is a linear increase in the amount of product strand with increasing cycles of amplification. This scheme, called *linear amplification,* is very useful in preparing DNA samples for DNA sequencing. Here chain terminating analogues of dpppN's are added in trace amounts to produce a distribution of products with different chain lengths (see Chapter 10). When linear amplification is used in this context, it is called *cycle sequencing.*

A step up in complexity from linear amplification is the use of two antiparallel primers, as shown in Figure 4.2*a*. These primers define the region to be amplified, since after the first few cycles of DNA synthesis, the relative amount of longer DNA sequences that contain the region spanned by the primer becomes insignificant. The target DNA is denatured, and two antiparallel primers are added. These must be in sufficient excess over target that renaturation of the original duplex is improbable, and essentially all products are primers annealed to single-stranded templates. The first cycle of DNA synthesis copies both of the original template strands. Hence it doubles the number of targets present in the starting reaction mixture. Each successive round of DNA denaturation, primer binding, and chain extension will, in principle, produce a further doubling of the number of target molecules. Hence the amount of amplified product grows exponentially, as 2^n, where n is the number of amplification cycles. This is the basic design of a typical PCR procedure. Note that only the DNA sequence flanked by the two primers is amplified (Fig. 4.2).

Early PCR protocols employed ordinary DNA polymerases such as the Klenow fragment of *E. coli* DNA polymerase I (a truncated version of the natural protein with its 5′-exonuclease activity removed). The difficulty with this approach is that these enzymes

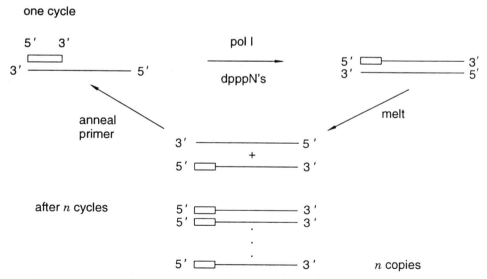

Figure 4.1   Linear amplification of DNA by cycles of repeated in vitro synthesis and melting.

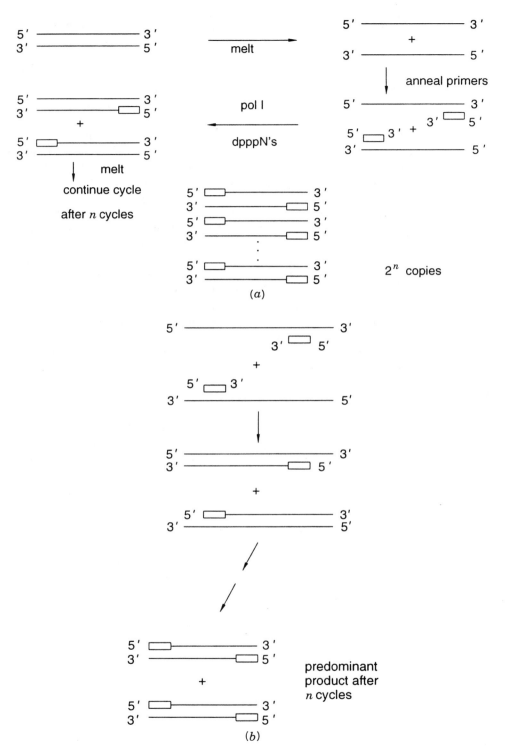

Figure 4.2 Exponential amplification of DNA in vitro by the use of two antiparallel primers. *(a)* Polymerase chain reaction (PCR) with successive cycles of DNA synthesis and melting. *(B)* Only the DNA flanked by the primers is amplified.

are easily denatured by the high temperatures needed to separate the two strands of duplex DNA. Unlike DNA, however, proteins like DNA polymerases, once denatured, are generally reluctant to renature. Thus, with each cycle of amplification, it was necessary to add fresh polymerase for efficient performance. This problem was relieved when DNA polymerases from thermophilic organisms became available. The most widely used of these enzymes in PCR is from *Thermis acquaticus,* the *Taq* polymerase. This enzyme has an optimal temperature for polymerization in the range of 70 to 75°C. It extends DNA chains at a rate of about 2 kb per minute. Most important, it is fairly resistant to the continual cycles of heating and cooling required for PCR. The half-life for thermal denaturation of *Taq* polymerase is 1.6 h at 95°C, but it is less than 6 minutes at 100°C. When very high denaturation temperatures are needed, as in the PCR amplification of very G + C rich DNA, it is sometimes useful to employ even more thermal-stable polymerases such as enzymes isolated from organisms that grow in the superheated water (temperature above 100°C) near geothermal vents. Such enzymes are called *vent polymerases.* Examples are the enzyme from *Thermococcus litoralis* with a half-life of 1.8 h at 100°C and the enzyme from *Pyrococcus furiosis* with a half-life of 8 h at 100°C. However, these enzymes are not as processive as *Taq* polymerase. This means that they tend to fall off their template more readily, which makes it more difficult to amplify long templates.

With *Taq* polymerase typical PCR cycle parameters are

DNA denaturation	92–96°C for 30 to 60 seconds
Primer annealing	55–60°C for 30 seconds
Chain extension	72°C for 60 seconds

The number of cycles used depends on the particular situation and sample concentrations, but typical values when PCR is efficient are 25 to 30 cycles. Thus the whole process takes about an hour. Typical sample and reagent concentrations used are

Target	$< 10^{-15}$ mol, or down to as little as 1 mol
Primer	2×10^{-11} mol
dpppN's	2×10^{-8} mol of each

These concentrations imply that the reaction will eventually saturate once the primer and dpppN's are depleted.

There are a few peculiarities of the *Taq* polymerase that must be taken into consideration in designing PCR procedures or experimental protocols based on PCR. *Taq* polymerase has no 3'-proofreading exonuclease activity. Thus it can, and does, misincorporate bases. We will say more about this later. *Taq* polymerase does have a 5'-exonuclease activity. Thus it will nick translate, which is sometimes undesirable. However, mutants exist that remove this activity, and they can be used if necessary. The most serious problem generally encountered is that *Taq* polymerase can add an extra nontemplate-coded A to the 3'-end of DNA chains, as shown in Figure 4.3*a*. It can also, perhaps using a related activity, make primer dimers, which may or may not contain additional uncoded residues as shown in Figure 4.3*b*. This is a serious problem because once such dimeric primers are created, they are efficient substrates for further amplification. These primer dimers are a major source of artifacts in PCR. However, they are usually short and can be removed by gel filtration or other sizing methods to prevent their interference with subsequent uses of the PCR reaction product.

Figure 4.3 Artifacts introduced by the use of *Taq* DNA polymerase. *(a)* Nontemplated terminal adenylation; *(b)* primer dimer formation.

(a)   *(b)*

The efficiency of typical PCR reactions can be impressive. Let us consider a typical case where one chooses to amplify a short, specific region of human DNA. This sort of procedure would be useful, for example, in examining a particular region of the genome for differences among individuals, without having to clone the region from each person to be tested. To define this region, one needs to know the DNA sequence that flanks it, but one does not need know anything about the DNA sequence between primers. For unique, efficient priming, convergent primers about 20 bp long are used on each side of the region. A typical amplification would start with $1\mu g$ of total human genomic DNA. Using Avogadro's number, 6×10^{23}, the genome size of 3×10^9 bp, and the 660 Da molecular weight of a single base pair, we can calculate that this sample contains

$$\frac{10^{-6} \times 6 \times 10^{23}}{3 \times 10^9 \times 6.6 \times 10^2} = 3 \times 10^5 \text{ copies}$$

of the genome or 3×10^5 molecules of any single-copy genomic DNA fragments containing our specific target sequence. If PCR were 100% efficient, and 25 cycles of amplification were carried out, we would expect to multiply the number of target molecules by $2^{25} = 6.4 \times 10^7$. Thus, after the amplification, the number of targets should be

$$6.4 \times 10^7 \times 3 \times 10^5 = 2 \times 10^{13} \text{ molecules}$$

If the product of the DNA amplification is 200 bp, it will weigh

$$200 \text{ bp} \times \frac{(6.6 \times 10^2 \text{ Da/bp}) \times (2 \times 10^{13})}{6 \times 10^{23} \text{ g/Da}} = 4\mu g$$

If such efficient amplification could be achieved, the results would be truly impressive. This means that one would be sampling just a tiny fraction of the genome, and in a scant hour of amplification, the yield of this fraction would be such as to amount to 80% of the total DNA in the sample. Thus one would be able to purify any DNA sample.

In practice, the actual PCR efficiencies typically achieved are not perfect, but they are remarkably good. We can define the efficiency, E, for n cycles of amplification by the ratio of product, P, to starting material, S, as

$$\frac{P}{S} = (1 + E)^n$$

Actual efficiencies turn out to be in the range of 0.6 to 0.9; this is impressive. Such high efficiencies immediately raise the notion of using PCR amplification as a quantitative tool to measure not just the presence of a particular DNA sequence in a complex sample but to determine its amount. In practice, this can be done, but it is not always reliable; it usually requires coamplification with a standard sample, or competition with known amounts of a related sample.

NOISE IN PCR: CONTAMINATION

As in any high-gain amplification system, any fluctuations in conditions are rapidly magnified, especially when they occur early in the reaction. There are many sources of noise in PCR. One extremely common source is variations in the temperature at different positions in typical thermal cycling blocks. A typical apparatus for PCR uses arrays of tube holders (or microtitre plate holders) that can be temperature controlled by heating elements and a cooling bath, by thermoelectric heating and cooling, by forced convection, or by switching among several pre-equilibrated water baths. A number of different fundamental designs for such thermal cyclers are now available, and the serious practitioner would be well advised to look carefully at the actual temperature characteristics of the particular apparatus used. For extremely finicky samples, such as those that need very G + C rich primers, it may be necessary to use the same sample well each time to provide reproducible PCR. It is for reasons like this that PCR, while it has revolutionized the handling of DNA in research laboratories, has not yet found broad acceptance in clinical diagnostic laboratories despite its potential power.

A major source of PCR noise appear to lie with characteristics of the samples themselves. The worst problem is undoubtedly sample contamination. If the same sample is used repetitively in PCR assays, the most likely source of contamination is the PCR reaction product from a previous assay. With PCR we are dealing with a system that amplifies a DNA sequence by 10^7 fold. Thus, even if there is one part per million in carry over contamination; it will completely dominate the next round of PCR. A second major source of contamination is DNA from organisms in dust or from DNA shed by the experimenter, in the form of dander, hair follicles, sweat, or saliva. This problem is obviously of greatest significance when the samples to be amplified contain human sequences.

The basic cure for most contamination is to carry out PCR under typical biological containment procedures and use good sterile technique including plugged pipetmen tips, to prevent contamination by aerosols, and laminar flow hoods, to minimize the ability of the investigator to inadvertently contaminate the sample. However, these approaches are not always sufficient to deal with the contamination caused by previous PCR experiments on similar samples. This situation is extremely frustrating because it is not uncommon for neophytes to have success with procedures that then progressively deteriorate as they gain more experience, since the overall level and dispersal of contaminant in the laboratory keeps rising. There are general solutions to this problem, but they have not yet seen widespread adoption. One helpful procedure is to UV irradiate all of the components of a PCR reaction before adding the target. This will kill double-stranded DNA contaminants in the polymerase and other reagents. A more general solution is to exploit the properties of the enzyme, Uracil DNA glycosylase (DUG), which we described in Chapter 1. This enzyme degrades DNA at each incorporated dU. Normally these incorporations are rare mutagenic events, and the lesion introduced into the DNA is rapidly repaired.

DUG is used in PCR by carrying out the initial PCR amplification using dpppU instead of dpppT. The amplification product can be characterized in the normal way; the properties of DNA with T fully substituted by dU are not that abnormal except for a lower melting temperature and inability to be recognized by many restriction nucleases. Next, when a subsequent PCR reaction needs to be done, the sample, including the target, is treated with DUG prior to thermal cycling. This will destroy any carryover from the previous PCR because there will be so many dU's removed that the resulting DNA will be incapable of replication. If the second PCR is also performed with dpppU, its subsequent carryover can also be prevented by a DUG treatment, and this procedure can be repeated *ad libertum*.

PCR NOISE: MISPRIMING

Typical PCR conditions with two convergent primers offer a number of possible unintended primed amplifications. These are illustrated in Figure 4.4. If the primers are not chosen wisely, one of the two primers may be able to act alone to amplify DNA as shown in Figure 4.4*b*. Alternatively, the two convergent primers may have more than one site in the target that allows amplification. There is no way to plan for these events, unless the entire sequence of the sample is known. However, the chances of such accidental, unintended, but perfect priming can be minimized by using long enough primers so that the probability of such a coincidental match in DNA sequence is very small.

A more serious and more common occurrence is mispriming by inexact pairing of the primer with the template. Note that if the 3′-end of the primer is mispaired with the template, this is unlikely to lead to amplification, and thus there will be little harm. However, if the 5′-end of the primer is mispaired, the impact is much more serious, as shown in Figure 4.4*c* and *d*. If primer annealing is carried out under insufficiently stringent conditions, once elongation is allowed to start, a 5′ mispaired primer may still be able to lead to DNA synthesis. In the next round of PCR the incorrect elongated product from this synthesis will serve as a template if it contains a sequence complementary to any of the primers in the solution. However, when this synthesis extends past the original mispaired primer, the sequence that is made is now the precise complement of that primer. From this round on, no stringency conditions will discriminate between the desired product and the misprinted artifact. Thus more than one product will amplify efficiently in subsequent steps. The key is to prevent the mispriming in the first place. If the misprimed sequence is nearly identical to the primed sequence, the most obvious way to solve this problem is to change primers.

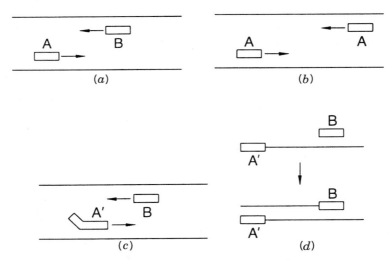

Figure 4.4 Effects of mispriming on PCR reaction products. *(a)* Desired product. *(b)* Product formed by an inverted repeat of one primer. *(c)* Product formed by nonstringent annealing. *(d)* After the second round of DNA synthesis, the product in *(c)* is now a perfect match to the primer for subsequent rounds of amplification.

Hopefully, another sequence near by on the desired template will not have its near mate somewhere else in the sample. However, there are more general cures for some mispriming as shown below.

In typical PCR reactions the denatured target and all of the other components are mixed at room temperature, and then the cycles of amplification are allowed to start. This has the risk that polymerase extensions of the primers may start before the reaction is heated to the optimal temperature for the elongation step. If this occurs, it enhances the risk of mispriming, since room temperature is far from a stringent enough annealing temperature for most primers. This problem can easily be avoided by what is called *hot start PCR*. Here the temperature is kept above the annealing temperature until all the components, including the denatured target, have been added. This avoids most of the mispriming during the first step; and it is always the first step in PCR that is most critical for the subsequent production of undesired species.

A very powerful approach to favoring the amplification of desired products and eliminating the amplification of undesired products is nested PCR. This can be done whenever a sufficient length of known sequence is available at each end of the desired target. The process is shown schematically in Figure 4.5. Two sets of primers are used; one is internal to the other. Amplification is allowed to proceed for half the desired rounds using the external primers. Then the primers are switched to the internal primers, and amplification is allowed to continue for the remaining rounds. The only products that will be present at high concentration at the end of the reaction are those that can be amplified by both sets of primers. Any sequence that can inadvertently be amplified by one set is most unlikely to be a target for the second set, since, in general, there is no relationship or overlap of the sequences used as the two sets of primers. With nested priming it is possible to carry out many more than 30 rounds of amplification with relatively little background noise. Hence this procedure is to be especially recommended when very small samples are used and large numbers of amplifications are needed to produce desired amounts of product.

Instead of full nesting, sometimes it is desirable or necessary to use a dual set of primers on one side of the target but only a single set on the other. This is called *hemi-nesting,* and it is still much safer and generally yields much cleaner products than no nesting at all. Particularly elegant versions of hemi-nesting have been demonstrated where the two nested primers can both be introduced at the start of the PCR at a temperature at which only the external primer functions well. Halfway through the amplification cycles, the annealing temperature is shifted so that now the internal primer becomes by far the favored one.

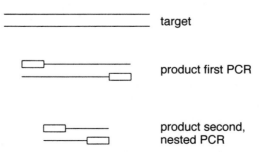

Figure 4.5   The use of nested primers to increase the specificity of PCR amplification.

MISINCORPORATION

We mentioned earlier that *Taq* polymerase has no 3' editing exonuclease. Because of this it has a relatively high rate of misincorporation when compared with many other DNA polymerases. Thus the products of PCR reactions accumulate errors. The misincorporation rate of *Taq* polymerase has been estimated as 1.7×10^{-4} to 5×10^{-6} per nucleotide per cycle. The error rate depends quite a bit on the reaction conditions, especially on the concentration of dpppN's used, and on the sequence of the target. The impact of these mispairing rates on the product are straightforward to calculate. At any site in the target, the fraction of correct bases after n cycles of amplification will be

$$X_{corr} = (1 - X_{mis})^n \sim 1 - nX_{mis}$$

where X_{mis} is the fraction of misincorporation rate at that site for a single cycle. For 30 cycles the fraction of misincorporated bases at any site will be $30X_{mis} = 5.1 \times 10^{-3}$ to 1.5×10^{-4}, using the numbers described above. Thus at any site the correct sequence is still overwhelmingly predominant.

However, if one asks, instead, for the amplification of a DNA of length, L, how many incorrect bases will each DNA product molecule have after n steps, the number is LnX_{mis}. With $L = 1000$, this means that products will have from 0.15 to 5 incorrect bases. How serious a problem is this? It depends on the use to which the DNA will be put. As a hybridization probe, these errors are likely to be invisible. If one sequences the DNA directly, the errors will still be invisible (except for the rare case where a misincorporation occurred in the first round or two of the amplification and then was perpetuated). This is because the errors are widely distributed at different sites on different molecules, and sequencing sees only the average occupant of each site. However, if the PCR products are cloned, the impact of misincorporation is much more serious. Now, since each clone is the immortalization of a single DNA molecule, it will contain whatever particular errors that molecule had. In general, it is a hazardous idea to clone PCR products and then sequence them. The sequences will almost always have errors. Similarly PCR starting from single DNA molecules is fine for most analyses, but one cannot recommend it for sequencing because, once again, the products are likely to show a significant level of misincorporation errors.

LONG PCR

A number of factors may limit the ability of conventional PCR to amplify long DNA targets. These include depurination of DNAs at the high temperatures used for denaturation, inhibition of the DNA polymerase by stable intramolecular secondary structure in nominally single-stranded templates, insufficient time for strand untwisting during conventionally used denaturation protocols, as described in Chapter 3, and short templates. The first problem can be reduced by using increased pH's to suppress purine protonation, a precursor to depurination. The second problem can be helped somewhat by adding denaturants like dimethyl sulfoxide (DMSO). The third problem can be alleviated by using longer denaturation times. The last problem can be solved by preparing DNA in agarose (Chapter 5). However, the most serious obstacle to the successful PCR amplification of long DNA targets rests in the properties of the most commonly used DNA polymerase, the *Taq* polymerase.

Taq polymerase has a significant rate of misincorporation as described above. However, it lacks a 3′ proofreading exonuclease activity. Once a base is misincorporated at the 3′-end, the chances that the extension will terminate at this point become markedly enhanced. This premature chain termination ultimately leads to totally ineffective PCR amplification above DNA sizes of 5 to 10 kb. To circumvent the problem of premature chain termination, Wayne Barnes and coworkers have added trace amounts of a second thermally stable DNA polymerase like *Pfu,* Vent, or Deep Vent that possesses a 3′-exonuclease activity. This repairs any terminal mismatches left by *Taq* polymerase, and then the latter can continue chain elongation. With such a two-enzyme procedure, successful PCR amplification of DNA targets in the 20 kb to 40 kb range are now becoming common.

INCORPORATING EXTRA FUNCTIONALITIES

PCR offers a simple and convenient way to enhance or embroider the properties of DNA molecules. As shown in Figure 4.6, primers can be pre-labeled with radioisotopes, biotin, or fluorescent dyes. Thus the ends of the amplified targets can be derivatized as an intrinsic part of the PCR reaction. This is extremely convenient for many applications. Since two primers are chosen, two different labels or tags can be used. One frequent and every effective strategy is to put a capture tag on one primer—like a biotin—and a detection tag on the other—like a fluorophore. After the amplification, the product is captured and analyzed. Only double-stranded material that is the result of amplification that incorporated both of the primers should be visible.

PCR can also be used to modify the DNA sequence at the ends of the target. For example, as shown in Figure 4.6*b*, the primer can overhang the ends of the desired target. As successive amplification cycles are carried out, the DNA duplexes that accumulate will contain the target sequence flanked by the additional segments of primer sequence. This has a number of useful applications. It allows any restriction sites needed to be built into the primer. Then, as shown in the figure, after the PCR reaction the product can be cleaved at these sites for subsequent ligation or cloning steps.

Another use for overhanging primers arises in circumstances where the original amount of known sequence is too short or too imperfect to allow efficient amplification. This problem arises, for example, when a primer is made to an imperfectly repeating sequence. The usual desire in such experiments is to amplify many different copies of the repeat (e.g., to visualize human DNAs among a background of rodent DNA in a hybrid cell as illustrated in Chapter 14), but few of the repeats match the primer well enough to really give good amplification. By having an overhanging primer, after the first few rounds of amplification, the complements to the primer sequence now contain the extra overhang (Fig. 4.6*c*). The resulting template-primer complexes are much more stable and amplify much more effectively.

SINGLE-SIDED PCR

A major limitation in conventional PCR is that known DNA sequence is needed on both sides of the desired target. It is frequently the case, as shown in Figure 4.7*a*, that a known sequence is available only at one place within the desired target, or at one end of it.

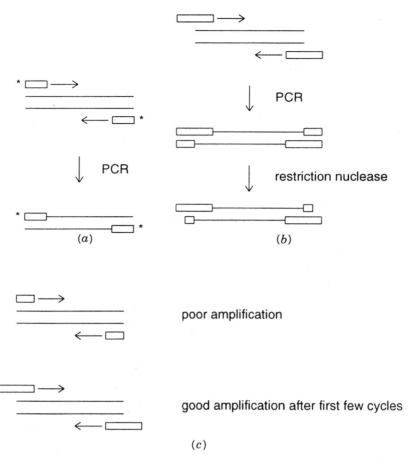

Figure 4.6 Introducing extra functionalities by appropriately designed primers. *(a)* Incorporation of a 5'-terminal label (asterisk). *(b)* Incorporation of flanking restriction sites, useful for subsequent cloning. *(c)* Compensating for less than optimal length initial sequence information.

This problem typically arises when one is trying to walk by from a known region of the genome into flanking unknown regions. Then one starts with a bit of known sequence at the extreme edge of the charted region, and now the goal is to make a stab into the unknown. There is still not a consensus on the best way to do this, but there have been a few successes, and many failures.

Consider the simple scheme shown in Figure 4.7*b*. The known sequence is somewhere within a restriction fragment of a length suitable for PCR. One can ligate arbitrary bits of known DNA sequence, called *splints* or *adapters,* onto the ends of the DNA fragment of interest. Now, in principle, the use of the one known primer and one of the two ligated splint primers will allow selective amplification of one side of the target. This amplification process will work as expected. However, the complication arises not from the target but from all the other molecules in the sample. They too are substrates for amplification using two copies of the splint primer (Fig. 4.7*c*).

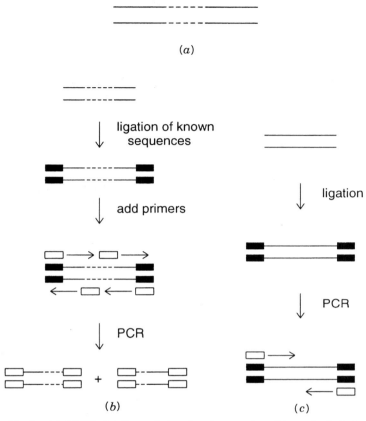

Figure 4.7   Single-sided PCR. *(a)* General situation that requires this method; dashed line represents the only known sequence in the target. *(b)* A potentially simple scheme. *(c)* Unwanted amplification products that defeat the scheme in *(b)*.

Thus the desired product will be overwhelmed with undesirable side products. One must either separate these away (e.g., by using a capture tag on the known sequence primer) or find a way of preventing the undesired products from amplifying in the first place (like suppression PCR, discussed later in this chapter). Sometimes it is useful to do both, as we will illustrate below.

In capture PCR, one can start with the very same set of primers shown for the unsuccessful example in Figure 4.7. However, the very first cycle of the amplification is performed using only the sequence-specific primer with a capture tag, like biotin. Then, before any additional rounds of amplification are executed, the product of the first cycle is physically purified using streptavidin-coated magnetic beads or some other strepatvidin-coated surface (Fig. 4.8). Now the splint primer is added, along with nonbiotinylated specific primer, and the PCR amplification is allowed to proceed normally. When the procedure is successful, very pure desired product is achieved, since all of the potential side product precursors are removed before they are able to be amplified.

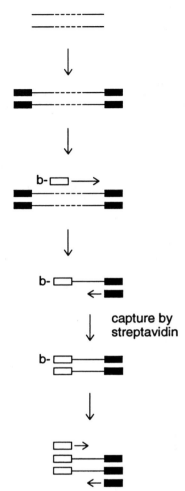

Figure 4.8 Capture PCR, where the known sequence is used to purify the target DNA from a complex sample before further rounds of amplification.

An alternative, general approach to single-sided PCR is to use a splint and primer combination designed so that the primer will work only after the splint has been replicated once, and the splint can only be replicated by synthesis initiated at the known bit of sequence. Two versions of this are shown in Figure 4.9. In one case a splint is used that is dephosphorylated, so it cannot ligate to itself, and the double-stranded region is short and A + T rich (Fig. 4.9a). After this is ligated to the target, the sample is heated to melt off the short splint strand. Next one cycle of DNA synthesis is allowed, using only the primer in the known target sequence. This copies the entire splint and produces a template for the second splint-specific primer. Both primers are now added, and ordinary PCR is allowed to proceed. However, only those molecules replicated during the first PCR cycle will be substrates for subsequent cycles. In the second version, called *bubble PCR*, the splint contains a noncomplementary segment of DNA which forms an interior loop (Fig. 4.9b). As before, the first round of PCR uses only the sample-specific primer. This copies the mispaired template strand faithfully so that when complementary primer is added to it, normal PCR can ensue. These procedures are reported to work well. It is worth noting that they can easily be enhanced by attaching a capture tag to the target-specific primer in the

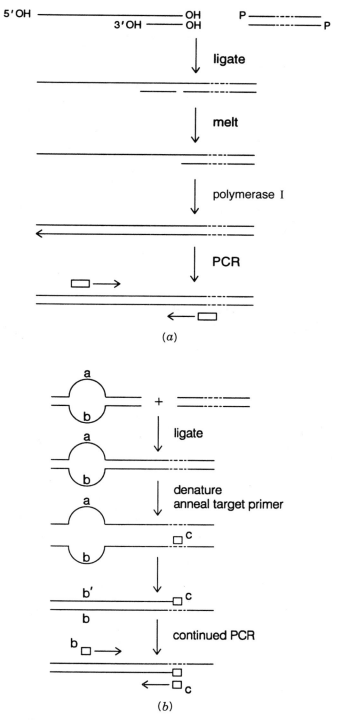

Figure 4.9   Two single-sided PCR schemes that use a linker that must be replicated before the complementary primer will function; dashed lines indicate known target sequence. *(a)* use of an appropriately designed dephosphorylated linker. *(b)* Bubble PCR.

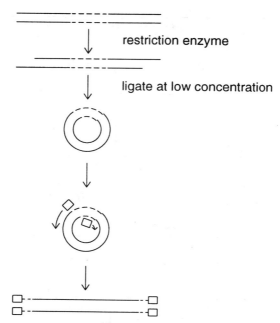

Figure 4.10   Inverse PCR, using primers designed to be extended outward on both sides of the known sequence (dashed lines).

first step. This will allow purification, as in capture PCR, before the reaction is allowed to continue.

A third variation on the single-sided PCR theme is inverse PCR. This is extremely elegant in principle. Like many such novel tricks, it was developed independently and simultaneously by several different research groups. Inverse PCR is shown schematically in Figure 4.10. Here the target is cut with a restriction enzyme to generate ends that are easily ligatable. Then it is diluted to very low concentrations and exposed to DNA ligase. Under these conditions the only products expected are DNA circles. To perform PCR, two primers are chosen within the known sequence, but they are oriented to face outward. Successful PCR with these primers should produce a linear product in which two, originally separate, segments of the unknown sequence are now fused at a restriction site and lie in between the two bits of known sequence. In practice, this procedure has not seen widespread success. One difficulty is that it is not easy to obtain good yields of small double-stranded circles by ligation. A second problem is the topological difficulties inherent in replication of circular DNA. The polymerase complex must wind through the center of the DNA circle once for each turn of helix. In principle, this latter problem could be removed by cleaving the known sequence with a restriction enzyme to linearize the target prior to amplification.

REDUCING COMPLEXITY WITH PCR

PCR allows any desired fraction of the genome to be selectively amplified if one has the primers that define that fraction. The complexity of a DNA sample was defined in

Chapter 3. It is the total amount of different DNA sequence. For single-copy DNA the complexity determines the rate of hybridization. Thus it can be very useful to selectively reduce the complexity of a sample, since this speeds up subsequent attempts to analyze that sample by hybridization. The problem is that in general, one rarely has enough information about the DNA sequences in a sample to choose a large but specific subset of it for PCR amplification.

A powerful approach has been developed to use PCR to selectively reduce sample complexity without any prior sequence knowledge at all. This approach promises to greatly facilitate genetic and physical mapping of new, uncharted genomes. It is based on the use of short, random (arbitrary) primers. Consider the use of a *single* oligonucleotide primer of length n. As shown in Figure 4.11, this primer can produce DNA amplification only if its complementary sequence exists as an inverted repeat, spaced within a distance range amenable to efficient PCR. The inverted repeat requires that we specify a DNA sequence of $2n$ bases. For a statistically random genome of N base pairs, the probability of this occurring at any particular place is $N4^{-2n}$, which is quite small for almost any n large enough to serve as an effective primer. However, any placement close enough for PCR will yield amplification products of the two primer sites. If L is the maximum practical PCR length, the probability that some observable PCR product will be seen is $LN4^{-2n}$. It is instructive to evaluate this expression for a mammalian genome with $N = 3 \times 10^9$ bp. For $L = 2000$ the results are

OLIGONUCLEOTIDE LENGTH	NUMBER OF PCR PRODUCTS	TOTAL AMOUNT OF AMPLIFIED DNA
8	1500	1.5×10^6 bp
9	100	1.0×10^5 bp
10	6	6.0×10^3 bp

These results make it clear that by using single arbitrary short primers, we can sample useful discrete subsets of a genome. Each different choice of primer will presumably give a largely nonoverlapping subset. The complexity of the reaction products can be controlled by the primer length to give simple or complex sets of DNA probes. This method has been used, quite successfully, to search for new informative polymorphic genetic markers in plants. It has been called *RAPD mapping,* which is short for randomly amplified polymorphic DNA. The idea is to amplify as large a number of bands as can be clearly analyzed by a single electrophoretic lane and then to compare the patterns seen in a diverse set of individuals.

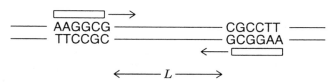

Figure 4.11 Sampling a DNA target (reducing complexity) with a single short oligonucleotide primer (RAPD method).

ADDITIONAL VARIANTS OF THE BASIC PCR REACTION

Here we will illustrate a number of variations on the basic PCR theme that increase the utility of the technique. The first of these is a convenient approach to the simultaneous analysis of a number of different genomic targets. This approach is called *multiplex PCR,* and it is carried out in a reverse dot blot format. The procedure is illustrated schematically in Figure 4.12. Each target to be analyzed is flanked by two specific primers. One is ordinary; the other is tagged with a unique 20-bp overhanging DNA sequence. PCR can be carried out separately (or in one pot if conditions permit), and then the resulting products pooled. The tagged PCR products are next hybridized to a filter consisting of a set of spots, each of which contains the immobilized complementary sequence of one of the tags. The unique 20-bp duplex formed by each primer sequence will ensure that the corresponding PCR products become localized on the filter at a predetermined site. Thus the overall results of the multiplex PCR analysis will be viewed as positive or negative signals at specific locations on the filters. This approach, where amplified products are directed to a known site on a membrane for analysis, dramatically simplifies the interpretation of the results and makes it far easier to automate the whole process.

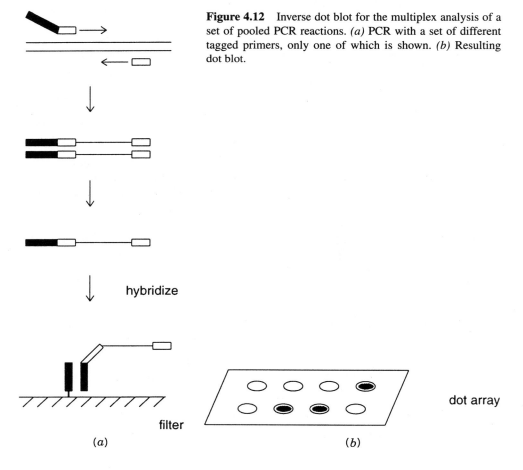

Figure 4.12   Inverse dot blot for the multiplex analysis of a set of pooled PCR reactions. *(a)* PCR with a set of different tagged primers, only one of which is shown. *(b)* Resulting dot blot.

For the approach shown in Figure 4.12, it would be far more useful to produce single-stranded product from the PCR reaction, since only one of the two strands of the amplified target can bind to the filter; the other strand will actually act as a competitor. Similarly, for DNA sequencing, it is highly desirable to produce a single-stranded DNA product. Presence of the complementary strand presents an unwanted complication in the sequencing reactions, and it can also act as a potential competitor. To produce single strands in PCR, a very simple approach called *asymmetric PCR* can be used. Here ordinary PCR is carried out for a few less than the usual number of cycles, say 20. Then one primer is depleted or eliminated, and the other is allowed to continue through an additional 10 cycles of linear PCR. The result is a product that is almost entirely single stranded. Clearly one can have whichever strand one wants by the appropriate choice of primer.

For diagnostic testing with PCR, one needs to distinguish different alleles at the DNA sequence level. A general approach for doing this is illustrated in Figure 4.13. It is called *allele-specific PCR*. In the case shown in Figure 4.13, we have a two-allele polymorphism. There is a single base difference possible, and we wish to know if a given individual has two copies of one allele, or the other, or one copy of each. The goal is to distinguish among these three alternatives in a single, definitive test. To do this, two primers are constructed that have 3'-ends specific for each of the alleles. A third general primer is used somewhere downstream where the two sequences are identical. The key element of allele-specific PCR is that because *Taq* polymerase does not have a 3'-exonuclease, it cannot use or degrade a primer with a mispaired 3'-terminus. Thus the allele-specific primers will only amplify the allele to which they correspond precisely. Analysis of the results is simplified by using primers that are tagged either by having different lengths or by having different colored fluorescent dyes. With length tagging, electrophoretic analysis will show different size bands for the two different alleles, and the heterozygote will be revealed as a doublet. With color tagging, the homozygotes will show, for example, red or green fluorescence, while the heterozygote will emit both red and green fluorescence, which our eye detects as yellow.

Color can also be used to help monitor the quantitative progress of a PCR reaction. Colored primers would not be useful unless their color were altered during the chain extension process. A similar problem holds for the potential use of colored dpppN's. The best approach to date is a slightly complicated strategy, which takes advantage of the 5'-exonuclease activity of thermostable DNA polymerases (Heid et al. 1996). A short oligonucleotide probe containing two different fluorescent labels is allowed to hybridize downstream from one of the PCR primers. The two dyes are close enough that the emission spectrum of the pair is altered by fluorescence resonance energy transfer. As the primer is extended, the polymerase reaches this probe and degrades it.

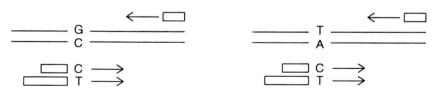

Figure 4.13 Allele-specific PCR used in genetic analysis. Different primer lengths allow the results of amplification of the two different alleles to be distinguished.

A color change is produced because the nucleotide products of the digestion diffuse too far away from each other for energy transfer to be efficient. Thus each chain extension results in the same incremental change in fluorescence. This procedure is called the TaqMan™ assay. A related spectroscopic trick, called molecular beacons, has recently been described in which a hybridization probe is designed as an oligonucleotide hairpin with different fluorescent dyes at its 3'- and 5'-ends (Kramer, 1996). In the hairpin these are close enough for efficient energy transfer. When the probe hybridizes to a longer target to form a duplex, its ends are now separated far apart in space, and the energy transfer is eliminated.

Hairpins can also be used to produce selective PCR amplification. In suppression PCR, long GC-rich adapters are ligated onto the ends of a mixture of target fragments (Diatchenko et al., 1996). When the ligation products are melted, the ends of the resulting single strands can form such stable hairpins that these ends become unaccessible for shorter primers complementary to the adapter sequences. However, if molecules in the target mixture contain a known internal target sequence, this can be used to initiate PCR. Chain extension from a primer complementary to the internal sequence will produce a product with only a single adapter. This will now allow conventional PCR amplification with one internal primer and one adapter primer.

A final PCR variant is called *DNA shuffling* (Stemmer, 1994). Here the goal is to enhance the properties of a target gene product by in vitro recombination. Suppose that a series of mutant genes exist with different properties; the goal is to combine them in an optimal way. The genes are randomly cleaved into fragments, pooled, and the resulting mixture is subjected to PCR amplification using primers flanking the gene. Random assembly of overlapping fragments will lead to products that can be chain extended until full length reassembled genes are produced. These then support exponential PCR amplification. The resulting populations of mutants are cloned and characterized by some kind of screen or selection in order to concentrate those with the desirable properties. This new method appears to be extremely promising. A very interesting alternative method to shuffle DNA segments uses catalytic RNAs (Mikheeva and Jarrell, 1996).

TOTAL GENOME AMPLIFICATION METHODS

A frequently encountered problem in biological research is insufficient amounts of sample. If the sample is a cultured cell or microorganism, the simplest solution is to grow more material. However, many samples of interest cannot be cultured. For example, many differentiated cells cannot be induced to divide without destroying or altering their phenotype. Sperm cells are incapable of division. Most microorganisms cannot be cultured by any known technique—we know of their existence only because we can see their cells or detect aspects of their DNA. Fossil samples and various clinical biopsies are other examples of rare materials with insufficient DNA for convenient analysis. Finally sorted chromosomes (Chapter 2) present the challenge of a very useful resource for which there is always more demand than supply.

In each of these cases mentioned above, one could use a particular set of primers to amplify any given known DNA region of interest. However, once this were done, the rest of the sample would be lost for further analysis. Instead, what would be useful is an amplification method that first samples all of the DNA in the rare material. This can then be stockpiled for future experiments of a variety of types including more specific PCR, when needed. The issue is how to do this in such a way that the stockpile represents a complete, or at least a relatively complete and even sampling of the original sample. The danger of course is that the sample will consist of a set of regions with very different amplification efficiencies with any particular set of primers. After PCR the stockpile will now be a highly distorted version of the original, and future experiments will all be plagued by this distorted view.

One approach to PCR sampling of an entire genome is the method of primer extension preamplification (PEP). This was designed to be used on a single cell or single sperm. The detailed rationale for PEP will become apparent when genetics by single sperm PCR is discussed in Chapter 7. PEP is illustrated schematically in Figure 4.14a. A mixture of all possible 4^n primers of length n is generated by automated oligonucleotide synthesis, using at each step all four nucleotides rather than just a single one. This extremely complex mixture is then used as a primer. Although the concentration of any one primer is vanishingly small, there are always enough primers present that any particular DNA segment has a reasonable chance of amplification. Norman Arnheim and his coworkers have reported reasonable success at using this approach with $n = 15$ (Arnheim and Ehrlich, 1993). They use 50 cycles of amplification and estimate that at least 78% of the genome will be amplified to 30 or more copies by this method.

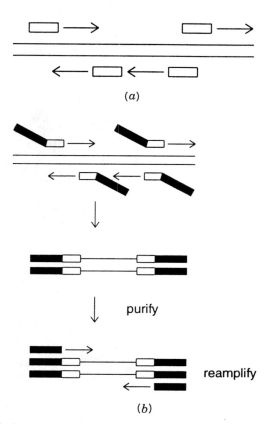

(a)

purify

reamplify

(b)

Figure 4.14 Methods for total genome PCR. *(a)* Primer extension preamplification (PEP). *(b)* Tagged random primer PCR (T-PCR).

When we attempted to use the PEP method, we ran into the difficulty that a large number of template-independent sequences were amplified in the reaction mixture. This is shown in the results of Figure 4.15, which illustrates the pattern of hybridization seen when PEP-amplified *S. pombe* chromosome 1 is hybridized to an arrayed cosmid library providing a fivefold coverage of the *S. pombe* genome. Almost all of the clones are detected with comparable intensities, even though only about 40% of them should contain material from chromosome 1. We reasoned that the complex set of long primers might allow for very significant levels of primer dimers (Fig. 4.3*b*) to be produced, and since the primers represented all possible DNA sequence, their dimers would also represent a broad population of sequences. Thus, when used in hybridization, this mixture should detect almost everything, which, indeed, it seems to do.

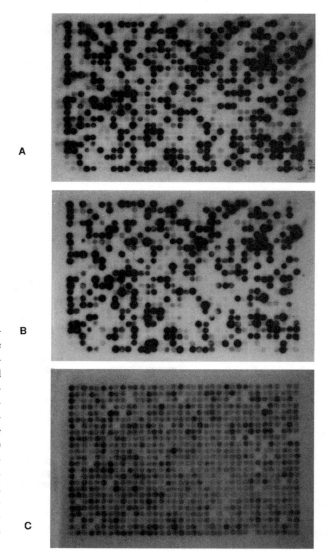

A

B

C

Figure 4.15 Examples of amplification of *S. pombe* chromosome I DNA. *(a)* A schematic of an experiment in which directly labeled chromosome I or amplified chromosome I is used as a hybridization probe against an arrayed library of the entire *S. pombe* genome cloned into cosmids. *(b)* Actual hybridization results for labeled chromosome I *(top)*, chromosome I labeled after T-PCR amplification *(center)*, chromosome I labeled after PEP amplification *(bottom)*. (From Grothues et al., 1993.)

Three variations were introduced to circumvent the primer dimer problem in PEP. Together these constitute an approach we call *T-PCR,* for tagged random primer PCR (Grothues et al., 1993). Our primers consist of all 4^9 nonanucleotides. Their shorter length and smaller complexity should be an advantage compared with the 4^{15} compounds used in ordinary PEP. Each primer was equipped at its 5'-end with a constant 17 base sequence; this is the tag. Thus the actual primers used were

$$GTTTTCCCAGTCACGACN_9$$

where N is a mixture of A, T, G, and C. After a few rounds of PCR, the resulting mixture was fractionated by gel filtration, and material small enough to be primer dimers was discarded. Then the remaining mixture was used as a target for amplification with only the tag sequence as a primer (Fig. 4.14*b*). As shown in Figure 4.15, this yielded reaction products that produced a pattern of hybridization with the *S. pombe* cosmid array almost identical to that seen with directly labeled chromosome 1. Thus we feel that T-PCR offers very good prospects for uniformly sampling a complex DNA sample. In our hands this approach has been successful thus far with as little as 10^{-12} g DNA, which corresponds to less than a single human cell.

Quite a few variations on this approach have been developed by others. One example is degenerate oligonucleotide-primed PCR (DOP–PCR) described by Telenius et al. (1992). Here primers are constructed like

$$AAGTCGCGGCCGCN_6ATG$$

with a six base totally degenerate sequence flanked by a long 5' unique sequence and a specific 3 to 6 base unique sequence. The 3'-sequence serves to select a subset of potential PCR start points. The degenerate sequence acts to stabilize the primer-template complex. The constant 5'-sequence can be used for efficient amplification in subsequent steps just as the tag sequence is used in T-PCR. It is not yet certain how to optimize whole genome PCR methods for particular applications. Issues that must be considered include the overall efficiency of the amplification, the uniformity of the product distribution, and the fraction of the original target that is present in the final amplified product. A recently published DOP–PCR protocol (Cheung and Nelson, 1996) looks particularly promising.

APPLICATION OF PCR TO DETECT MOLECULES OTHER THAN DNA

A natural extension of PCR is its use to detect RNA. Two general approaches for doing this are summarized in Figure 4.16. In one, which is specific for polyadenylated mRNA, an oligo-dT primer is used, with reverse transcriptase, to make a DNA copy of the RNA. Then conventional PCR can be used to amplify that DNA. In the other approach, which is more general, random $(dN)_n$'s are used to prime reverse transcriptase to make initial DNA copies, and then ordinary PCR ensues.

Less obvious is the use of PCR to detect antigens or other non-nucleic acid molecules. We originally demonstrated the feasibility of this approach, which should have a broad

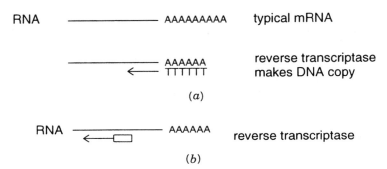

Figure 4.16 Two methods for PCR amplification of RNA. *(a)* Use of reverse transcriptase with an oligo-dT primer. *(b)* Use of reverse transcriptase with short random oligonucleotide primers.

range of applications, and it should be generalizable to almost any class of molecule. The basic principle of what we call *immuno-PCR* (i-PCR) is shown in Figure 4.17. DNA is used as the label to indirectly tag an antibody. Then the DNA is detected by ordinary PCR.

In the test case, shown in Figure 4.17, the antibody is allowed to detect an immobilized antigen and to bind to it in the conventional way (Fig. 4.18). Then the sample is exhaustively washed to remove free antibody. Next a molecule is added that serves to couple DNA to the bound antibody. That molecule is a chimeric protein fusion between the protein streptavidin and two domains of staphylococcal protein A. The chimera was made by conventional genetic engineering methods and expressed as a gene fusion in *E. coli*. After purification the chimeric protein is fully active. Its properties are summarized in Table 4.1. The chimera is a tetramer. It is capable of binding four immunoglobulin G's and four biotins.

After the chimera is bound to the immobilized antibody, any unbound excess material is removed, and now biotinylated DNA is added (Fig. 4.18). In our hands, end-biotinylated linearized pUC19 was used. This was prepared by filling in the ends of a restriction enzyme-digested plasmid with biotinylated dpppU, but it could just as easily have been made by PCR with biotinylated primers. The biotinylated DNA binds to the immobilized chimera.

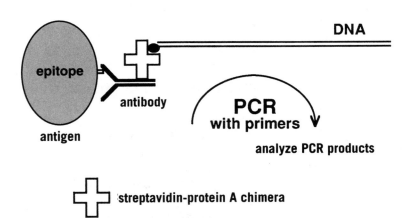

Figure 4.17 Basic scheme for immuno-PCR: Detection of antigens with DNA-labeled antibodies.

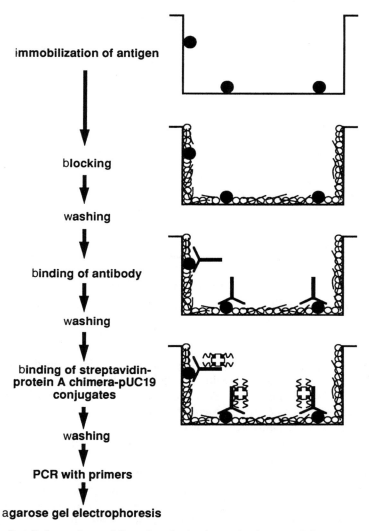

Figure 4.18 Detailed experimental flow chart for implementing immuno-PCR to detect an immobilized antigen.

TABLE 4.1   Streptavidin-Protein A Chimera

Expression vector:	pTSAPA-2
Amino acid residues:	289 per subunit
Subunits:	4 (subunit tetramer)
Molecular mass:	31.4 kDa per subunit
	126 kDa per molecule
Biotin binding:	4 per molecule
	1 per subunit
IgG binding:	4 per molecule
	1 per subunit
	(human IgG)

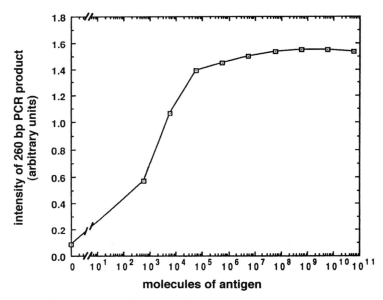

Figure 4.19 Result of detection of tenfold serial dilutions of an antigen, bovine serum albumin, by immuno-PCR. (Taken from Sano et al., 1992.)

Excess unbound DNA is carefully removed by extensive washing. Now the entire sample is subjected to PCR using primers specific for the particular DNA tag. Typical results are shown for a set of serial dilutions of antigen in Figure 4.19. The results indicate that i-PCR promises to be an antigen-detection system of unparalleled sensitivity. Less than 600 molecules of antigen could be easily detected without any effort to optimize the system. This is 10^5 times more sensitive than conventional immunoassays. A major advantage of i-PCR is that the DNA molecule used is purely arbitrary. It can be changed at will to prevent the buildup of laboratory contaminants. It need not correspond (indeed it should not correspond) to any sequences found in the samples. Thus there should be no interference from sample DNA. Finally a number of different DNA labels could be detected simultaneously, which would open the way for multiplex PCR detection of several antigens simultaneously. Such an application has recently been reported (Hendrickson et al., 1995).

DNA AMPLIFICATION WITHOUT THERMAL CYCLING AND OTHER ALTERNATIVES TO PCR

From a practical viewpoint it is difficult to fault PCR. If there is any step that is tedious, it is the need for stringent control at several different temperatures. It would be nice to eliminate this requirement. From a commercial standpoint, existing PCR patents create quite a powerful band of protection around this technology and make the notion of potential competing technologies quite attractive as lucrative business ventures. Taken together, these considerations have fueled a number of attempts to create alternate DNA amplification procedures. Several of these have been shown to be practical. Some appear to be very attractive alternates to PCR for certain applications. None yet have shown the generality or versatility of PCR. The degree of amplification achievable by these methods is quite impressive, but it is still considerably less than that seen with conventional PCR (Table 4.2).

TABLE 4.2   Comparison of Various In Vitro Nucleic Acid Amplification Procedures

Method	Amplified Species	Temperature Used (°C)	Target-specific Probes Needed	Amplification Extent
PCR	Target	50–98 cycle	2 or more	10^{12}
QβR	Probe	37 isothermal	1	10^{9}
LCR	Probe	50–98 cycle	4	10^{5}
3SR	Target	42 isothermal	2	10^{10}
SDA	Target	37 isothermal	4	10^{7}

Source: Adapted from Abramson and Myers (1993).

Isothermal self-sustained sequence replication (3SR) is illustrated in Figure 4.20. In this technique an RNA target is the preferred starting material. DNA targets can always be copied by extending a primer containing a promoter site for an enzyme like T7 RNA polymerase (and then that enzyme is used to generate an RNA copy of the original DNA). The complementary DNA strand of the RNA is synthesized by Avian myeloblastosis virus (AMV) reverse transcriptase (RT) using a primer that simultaneously introduces a promoter of T7 RNA polymerase. AMV RT contains an intrinsic RNase H activity. This activity specifically degrades the RNA strand of an RNA-DNA duplex. Thus, as AMV RT synthesizes the DNA complement, it degrades the RNA template. The result is a single-stranded DNA complement of the original RNA. Now a second primer, specific for the target sequence, is used to prime the RT to synthesize a double-stranded DNA (Fig. 4.20a). When this is completed, the resulting duplex now contains an intact promoter for T7 RNA polymerase so that enzyme can, rapidly, synthesize many RNA copies. These RNAs are the complement of the original RNA target (Fig. 4.20b).

Now, in a cyclical process, the RT makes DNA complements of the RNAs, degrading them in the process by its RNaseH activity. RT then turns the single-stranded DNAs into duplexes. These duplexes in turn serve as templates for T7 RNA polymerase to make many more copies of single-stranded RNA. The key point is that all these reactions can be carried on simultaneously at a constant temperature. A substantial level of amplification is observed, and in principle, many of the same tricks and variations of PCR can be implemented through the 3SR approach. Primer nesting does appear to be more difficult, and it is not clear how well this technique will work in multiplexing.

A method that is similar in spirit but rather different in detail is strand displacement amplification (SDA). This is illustrated in Figure 4.21. It is based on the peculiarities of the restriction endonuclease *Hinc* II which recognizes the hexanucleotide sequence and cleaves it, as shown below:

$$\begin{array}{l} \text{GTTGAC} \qquad \text{GTT} + \text{GAC} \\ \text{CAACTG} \rightarrow \text{CAA} \quad \text{C T G} \end{array}$$

The key feature of this enzyme exploited in SDA is the effect of alpha thio-substituted phosphates on the enzyme. These can be introduced into DNA by the use of alpha-S-dpppA. When this is incorporated into the top strand of the recognition sequence, there is no effect. However, in the bottom strand the thio derivatives inhibit cleavage (Fig. 4.21a). How this peculiarity is used for isothermal amplification is illustrated in Figure 4.21b.

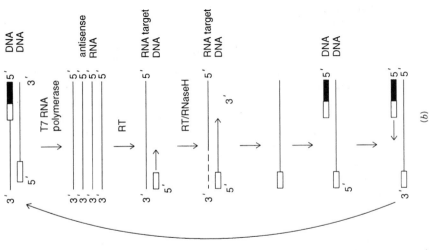

Figure 4.20 Isothermal self-sustained sequence replication (3SR). (a) Synthesis of a double-stranded duplex from a starting RNA, using one transcript-specific primer, TSP, fused to a sequence containing a promoter site (solid bar) for T7 RNA polymerase and a second transcript-specific primer B. (b) Cyclical amplification by transcription and conversion of the resulting RNA molecules to duplex DNAs.

124

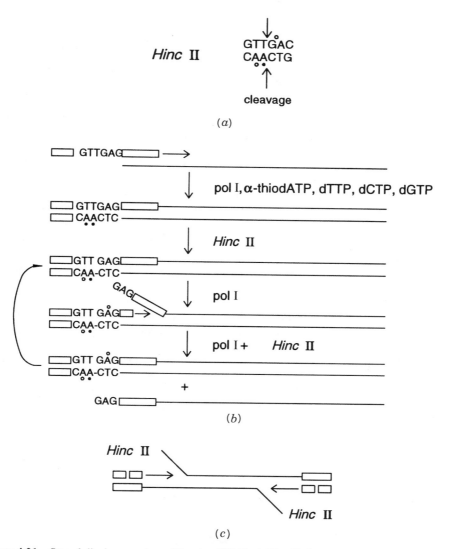

Figure 4.21 Strand displacement amplification (SDA). A *Hinc* II site with alpha thio DNA derivatives that block *(solid circles)* and do not block *(open circles)* cleavage. *(b)* Linear amplification by strand displacement from one *Hinc* II cleavage site. *(c)* Exponential amplification from two *Hinc* II cleavage sites.

A DNA polymerase I mutant with no 5'-exonuclease activity is used. This leads to strand displacement. Consider first the effect of this enzyme on the target-primer complex shown in the figure. The primer has a potential *Hinc* II site overhanging the template. The polymerase extends both the template and the primer, incorporating alpha-S-A. The top strand of the resulting duplex can be cleaved by *Hinc* II; the bottom strand is resistant. This creates a target for the polymerase that can strand displace most of the top strand and continually make copies of it, resulting in linear amplification.

If primers are established with overhanging *Hinc* II sites on both sides of a target duplex, each strand can be made by linear amplification (Fig. 4.21*c*). Now, however, each newly synthesized strand can anneal with the original primer to form new complexes capable of further *Hinc* II cleavage and DNA polymerase-catalyzed amplification. Thus the overall system will show exponential amplification in the presence of excess primers.

In order to use SDA, one must have the desired target sequence flanked by *Hinc* II sites (or sites for any other restriction enzyme that might display similar properties). These sites can be introduced by using primers flanking the target sequence and tagged with additional 5'-sequences containing the desired restriction enzyme cleavage sites. These flanking primers are used in a single cycle of conventional PCR; then SDA initiates spontaneously, and the amplification can be continued isothermally.

Overall, SDA is a very clever procedure that combines a number of tricks in DNA enzymology. It would appear to have some genuinely useful applications. However, SDA as originally described seems unlikely to become a generally used method because the resulting products have alpha-S-A, which is not always desirable, and the primers needed are rather complex and idiosyncratic. Recently variants on this scheme were developed that have fewer restrictions.

Other modes of DNA polymerase-based amplification are still in their infancy, including rolling circle amplification (Fire and Xu, 1995) and protein-primed DNA amplification (Blanco et al., 1994). Alternate schemes for DNA amplification have been developed that avoid the use of DNA polymerase altogether. Foremost among these is the ligase chain reaction (LCR). This is illustrated in Figure 4.22. The target DNA is first denatured. Then two oligonucleotides are annealed to one strand of the target. Unlike PCR, these two sequences must be adjacent in the genome, and they correspond to the same DNA strand.

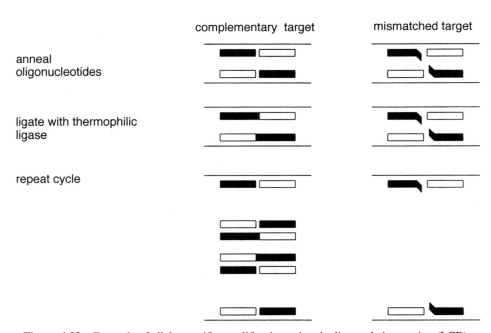

Figure 4.22   Example of allele-specific amplification using the ligase chain reaction (LCR).

If they match the target sequence exactly, DNA ligase will efficiently seal them together provided that a 5'-phosphate is present to form the phosphodiester bond between them. A complementary set of oligonucleotides can be used to form a ligation product directed by the other DNA strand. The overall result is to double the number of DNA strands. Both duplexes are melted, more oligonucleotides anneal, and the process can be continued indefinitely. With continual thermal cycling, the result is exponential amplification of the target. This is most easily detected in automated systems by using a capture tag on one of the oligonucleotides to be ligated and a color-producing tag on the other. This procedure is obviously limited to small DNA target sequences, but it could form a powerful alternative to allele-specific PCR. Hybrid amplification procedures that combine LCR and polymerase extension reactions also appear to be very promising.

The final amplification scheme we will discuss is carried out by the enzyme $Q\beta$ replicase. This occurs strictly at the RNA level. Appropriate RNA targets can be made by subcloning DNA samples into vectors that embed the desired targets within $Q\beta$ sequences and place them all downstream from a T7 RNA polymerase promoter so that an RNA copy can be made to start the $Q\beta$ replication process ($Q\beta$R). A much more general approach is to construct two separate RNA probes that can anneal to adjacent sequences on a target RNA. In the presence of T4 DNA ligase, the two probes will become covalently joined. Neither prone alone is a substrate for $Q\beta$ replicase. However, the ligation product is a substrate and is efficiently amplified (Fig. 4.23).

$Q\beta$ has an unusual mode of replication. No primer is needed. No double-stranded intermediate is formed. The enzyme recognizes specific secondary structure features and sequence elements on the template, and then makes a complementary copy of it. That copy dissociates from the template as it is made, and it folds into its own stable secondary structure which is a complement of that of the template. This structure also can serve as a template for replication. Thus the overall process continually produces both strands as targets, much in the manner of a dance in which the two partners move frenetically but never stay in contact for an extended period.

The usual mode of $Q\beta$ replication is very efficient. It is not uncommon to make 10^7 to 10^8 copies of the original target. One can start from the single molecule level. However, the system is not that easy to manipulate; $Q\beta$ replicase itself is a complex four-subunit enzyme not that commonly available. The procedures needed to prepare the DNA target for $Q\beta$ replication are somewhat elaborate, and there are considerable restrictions on what sorts of RNA insertions can be tolerated by the polymerase. For all these reasons the $Q\beta$ amplification system is most unlikely to replace PCR as a general tool for DNA analysis. It may, however, find unique niches for analyses where the idiosyncrasies of the system do not interfere, and where very high levels of amplification at constant temperature are needed.

FUTURE OF PCR

In this chapter we have illustrated myriad variations and potential applications of PCR. In viewing these, it is important to keep in mind that PCR is a young technique. It is by no means clear that today's versions are the optimal ones or the most easily adaptable ones for the large-scale automation eventually needed for high-throughput genome analysis. Much additional thought needs to be given on how best to format PCR for widespread use and how to eliminate many of the current glitches and irreproducibility inherent in such a high-gain amplification system.

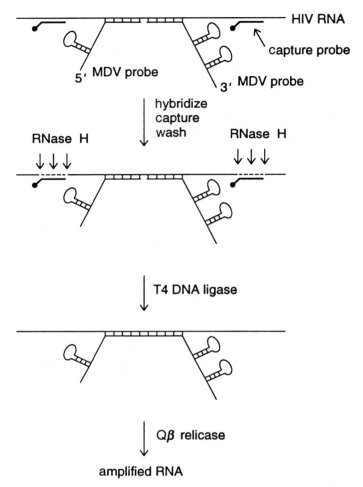

Figure 4.23   Example of the use of Qβ to detect a target, HIV RNA, by ligation. Here two specific RNA probes are designed to be complementary to adjacent target sequences. Each probe alone contains insufficient secondary structure features to support Qβ replicase amplification. Two additional end-biotinylated DNA capture probes are used to allow a streptavidin-based solid phase purification of the probe-target complexes and then release of these complexes by digestion with RNAse H which specifically cleaves RNA-DNA duplexes. The released complexes are treated with T4 DNA ligase, which will work on a pure double-stranded RNA substrate. Ligation produces an RNA that is now recognized and amplified by Qβ replicase. (Adapted from Kramer and Tyagi, 1996.

SOURCES AND ADDITIONAL READINGS

Abramson R. D., and Myers, T. W. 1993. Nucleic acid amplification technologies. *Current Biology* 4:41–47.

Arnheim, N., and Erlich, H. 1992. Polymerase chain reaction strategy. *Annual Review of Biochemistry* 61:131–156.

Barany, F. 1991. Genetic disease detection and DNA amplification using cloned thermostable ligase. *Proceedings of the National Academy of Sciences USA* 88:189–193.

Barnes, W. 1994. PCR Amplification of up to 35-kb DNA with high fidelity and high yield from bacteriophage templates. *Proceedings of the National Academy of Sciences USA* 91:2216–2220.

Blanco, L., Lazaro, J. M., De Vega, M., Bonnin, A., and Salas, M. 1994. Terminal protein-primed DNA amplification. *Proceedings of the National Academy of Sciences USA* 91: 12198–12202.

Bloch, W. 1991. A biochemical perspective of the polymerase chain reaction. *Biochemistry* 30: 2735–2747.

Caetano-Annolés, G. 1996. Scanning of nucleic acids by in vitro amplification: New developments and applications. *Nature Biotechnology* 14: 1668–1674.

Cheng, S., Fockler, C., Barnes, W. M., and Higuchi, R. 1994. Effective amplification of long targets from cloned inserts and human genomic DNA. *Proceedings of the National Academy of Sciences USA* 91: 5695–5699.

Cheung, V. G., and Nelson, S. F. 1996. Whole genome amplification using a degenerate oligonucleotide primer allows hundreds of genotypes to be performed on less than one nanogram of genomic DNA. *Proceedings of the National Academy of Sciences USA* 93: 14676–14679.

Chou, W., Russell, M., Birch, D. E., Raymond, J., and Block, W. 1992. Prevention of pre-PCR mispriming and primer dimerization improves low-copy-number amplifications. *Nucleic Acids Research* 20: 1717–1723.

Cobb, B. D., and Clarkson, J. M. 1994. A Simple procedure for optimising the polymerase chain reaction (PCR) using modified Taguchi methods. *Nucleic Acids Research* 22: 3801–3805.

Diatchenko, L., Lau, Y. F., Campbell, A. P., Chenchik, A., Moqadam, F., Huang, B., Lukyanov, S., Lukyanov, K., Gurskaya, N., and Sverdlov, E. D. 1996. Suppression subtractive hybridization: A method for generating differentially regulated or tissue-specific cDNA probes and libraries. *Proceedings of the National Academy of Sciences USA* 93: 6025–6030.

Erlich, H. A., Gelfand, D., and Sninsky, J. J. 1991. Recent advances in the polymerase chain reaction. *Science* 252: 1643–1651.

Fahy, E., Kwoh, D. Y., and Gingeras, T. R. 1991. Self-sustained sequence replication (3SR): An isothermal transcription-based amplification system alternative to PCR. *PCR Methods and Applications* 1: 25–33.

Fire, A., and Xu, S., 1995. Rolling replication of short DNA circles. *Proceedings of the National Academy of Sciences USA* 92: 4641–4645.

Grothues, D., Cantor, C. R., and Smith, C. L. 1993. PCR amplification of megabase DNA with tagged random primers (T-PCR). *Nucleic Acids Research* 21:1321–1322.

Heid, C. A., Stevens, J., Livak, K. J., and Williams, P. M. 1996. Real time quantitative PCR. *Genome Research* 6: 986–994.

Hendrickson, E. R., Hatfield-Truby, T. M., Joerger, R. D., Majarian, W. R., and Ebersole, R. C. 1995. High sensitivity multianalyte immunoassay using covalent DNA-labeled antibodies and polymerase chain reaction. *Nucleic Acids Research* 23: 522–529.

Lagerström, M., Parik, J., Malmgren, H., Stewart, J., Pettersson, U., and Landegren, U. 1991. Capture PCR: Efficient amplification of DNA fragments adjacent to a known sequence in human and YAC DNA. *PCR Methods and Applications* 1: 111–119.

Landegren, U. 1992. DNA probes and automation. *Current Opinion in Biotechnology* 3: 12–17.

Lizardi, P. M., Guerra, C. E., Lomeli, H., Tussie-Luna, I., and Kramer, F. R. 1988. Exponential amplification of recombinant-RNA hybridization probes. *Biotechnology* 6: 1197–1202.

Mikheeva, S., and Jarrell, K. A. 1996. Use of engineered ribozymes to catalyze chimeric gene assembly. *Proceedings of the National Academy of Sciences USA* 93: 7486–7490.

Nickerson, D. A., Kaiser, R., Lappin, S., Stewart, J., Hood, L., and Landegren, U. 1990. Automated DNA diagnostics using an ELISA-based oligonucleotide ligation assay. *Proceedings of the National Academy of Sciences USA* 87: 8923–8927.

Sano, T., and Cantor, C. R. 1991. A streptavidin-protein A chimera that allows one-step production of a variety of specific antibody conjugates. *Bio/Technology* 9: 1378–1381.

Sano, T., Smith, C. L., and Cantor, C. R. 1992. Immuno-PCR: Very sensitive antigen detection by means of specific antibody-DNA conjugates. *Science* 258: 120–122.

Siebert, P. D., Chenchik, A., Kellogg, D. E., Lukyanov, K. A., and Lukyanov, S. A. 1995. An improved PCR method for walking in uncloned genomic DNA. *Nucleic Acids Research* 23: 1087–1088.

Stemmer, W. P. C. 1994. Rapid evolution of a protein *in vitro* by DNA shuffling. *Nature* 370: 389–391.

Telenius, H., Carter, N. P., Bebb, C. E., Nordenskjöld, M., Ponder, B. A. J., and Tunnacliffe, A. 1992. Degenerate oligonucleotide-primed PCR: General amplification of target DNA by a single degenerate primer. *Genomics* 13: 718–725.

Telenius, H., Pelmear, A., Tunnacliffe, A., Carter, N. P., Behmel, A., Ferguson-Smith, M. A., Nordenskjöld, M., Pfragner, R., and Ponder, B. A. J. 1992. Cytogenetic analysis by chromosome painting using DOP-PCR amplified flow-sorted chromosomes. *Genes, Chromosomes and Cancer* 4: 257–263.

Tyagi, S., and Kramer, F. R. 1996. Molecular beacons: Probes that fluoresce upon hybridization. *Nature Biotechnology* 14: 303–308.

Tyagi, S., Landedren, U., Tazi, M., Lizardi, P. M., and Kramer, F. R. 1996. Extremely sensitive, background-free gene detection using binary probes and Qβ replicase. *Proceedings of the National Academy of Sciences USA* 93: 5395–5400.

Walker, G. T., Little, M., Nadeau, J. G., and Shank, D. D. 1992. Isothermal in vitro amplification of DNA by a restriction enzyme/DNA polymerase system. *Proceedings of the National Academy of Sciences USA* 89: 392–396.

White, T. J. 1996. The future of PCR technology: diversification of technologies and applications. *Trends in Biotechnology* 14: 478–483.

Wittwer, C. T., Herrmann, M. G., Moss, A. A., and Rasmussen, R. P. 1997. Continuous fluorescence monitoring of rapid cycle DNA amplification. *BioTechniques* 22: 130–138.

5 Principles of DNA Electrophoresis

PHYSICAL FRACTIONATION OF DNA

The methods we have described thus far all deal with DNA sequences. PCR allows, in principle, the selective isolation of any short DNA sequence. Hybridization allows, in principle, the sequence-specific capture of almost any DNA strand. However, to analyze the results of these powerful sequence-directed methods, we usually resort to length-dependent fractionations. There are two reasons for this. Separation of DNAs by length were developed before we had much ability to manipulate DNA sequences. Hence the methods were familiar and validated, and it was natural to incorporate them into most protocols for using DNAs. Second, our ability to fractionate DNA by length is actually remarkably good. The fact that DNAs are stiff and very highly charged greatly facilitates these fractionations. In this chapter we briefly review DNA fractionations other than electrophoresis. The bulk of the chapter will be spent on trying to integrate the maze of experimental and modeling observations that constitute our present-day knowledge of the principles that underlie DNA electrophoresis. Practical aspects of DNA separations are treated in Boxes 5.1 and 5.2.

SEPARATION OF DNA IN THE ULTRACENTRIFUGE

Velocity or equilibrium ultracentrifugation of DNA represents the only serious alternative to DNA electrophoresis. For certain applications it is a powerful tool. However, for most applications, the resolution of ultracentrifugation just isn't high enough to compete with electrophoresis. DNA can be separated by size in the ultracentrifuge by zonal sedimentation. Commonly density gradients of small molecules like sucrose are employed to prevent convection caused by gravitational instabilities. Sucrose gradient sedimentation is tedious because ultracentrifuges typically allow only half a dozen samples to be analyzed simultaneously.

One unique and very useful application of the ultracentrifuge to DNA fractionation is the separation of DNA fragments of different base composition by equilibrium density gradient centrifugation. There are slight density differences between A–T and G–C base pairs. These densities match different CsCl concentrations. Thus, when DNA fragments are subjected to centrifugation in an equilibrium gradient of CsCl, different species will show different buoyant densities. That is, they will concentrate in different regions of the CsCl gradient where their density matches that of the bulk fluid. A schematic example is shown in Figure 5.1. Because of these density differences, it is possible to obtain fractions of DNA physically isolated on the basis of their average base composition. The small intrinsic effects of base composition on density can be enhanced by the use of density shift ligands that bind differentially to A + T or G + C rich DNA. This allows much higher-resolution density fractionation of DNA.

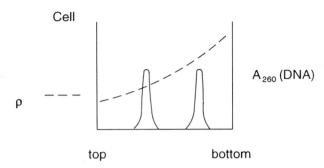

Figure 5.1   Schematic illustration of separation *(from top to bottom of a tube)* of DNAs of different base composition by density gradient equilibrium ultracentrifugation; ρ is density, Absorbance (A_{260}) is proportional to DNA concentration.

The utility of such preparative methods will be illustrated in Chapter 15, when the biological properties of isochores, specific density fractions of DNA, are discussed.

The other aspect of DNA structure that has been conveniently probed by ultracentrifugation is topological properties. Velocity sedimentation can be used to discriminate among linear DNAs, relaxed circles, and supercoiled circles. Equilibrium density gradient separations are also very powerful for this discrimination because, under the right circumstances, supercoils will bind very different amounts of intercalating dyes, like ethidium bromide, than bound by relaxed circles or linear duplexes (see Cantor and Schimmel, 1980, ch. 24). This leads to large density shifts. Thus such fractionations have played a major role in early studies of DNA topology, and they have served as major tools for the separation of plasmid DNAs key in many recombinant DNA experiments. Once again, however, as the need for dealing with very large numbers of samples in parallel grows, these centrifugation methods have to be replaced by others. The capacities of existing centrifuges simply cannot cope with large numbers of DNA samples, and loading and unloading centrifuges is a procedure that is very difficult to automate efficiently. In addition even the best protocols for DNA size or topology separation by ultracentrifugation have far lower resolution than typical electrophoretic methods.

ELECTROPHORETIC SIZE SEPARATIONS OF DNA

In contemporary methods of DNA analysis, electrophoresis dominates size fractionations for all species from oligonucleotides with a few bases up to intact chromosomal DNAs with sizes in excess of 10^7 bp. Rarely does a single method prove to be so powerful across such a broad size range. The size resolution for optimized separations is a single base up to more than 1000 bases for single-stranded DNAs, a few percent in size for double strands up to about 10 kb, and progressively lower resolution as DNA sizes increase until, for species in excess of 5 Mb, only about 10% size resolution is available.

Size is the overwhelming factor that determines the separation power of double-stranded DNA electrophoresis. In general, double-stranded DNA fractionations are carried out in dilute aqueous buffer near neutral pH. Base composition and base sequence have almost no effect on the gross separation patterns. There is a significant effect of base

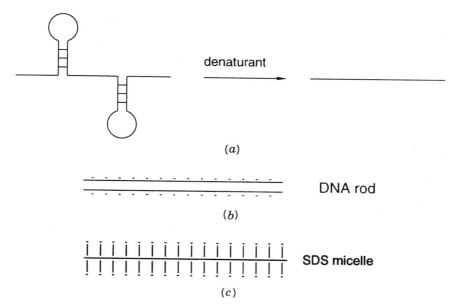

denaturant

(a)

DNA rod

(b)

SDS micelle

(c)

Figure 5.2 Behavior of highly charged molecules in electrophoresis. *(a)* Structure of single-stranded DNA without denaturants, or in the presence of strong denaturants where charge and friction are both proportional to length. *(b)* Structure of short double-stranded DNA, where charge and friction are both proportional to length. *(c)* Proteins, denatured by SDS, form micelles where charge and friction are both proportional to length.

sequence in the special case of sequences that lead to pronounced DNA bending. The most common example are AA's or TT's spaced at intervals comparable to the helical repeat distance. When this occurs, the slight bends at each AA are in phase, and the overall structure can be markedly curved.

Electrophoretic fractionations of single-stranded DNA are mostly carried out in the presence of strong denaturants such as high temperature or high concentrations of urea or formamide. Under these conditions most of the potential secondary structure of the DNA strands is eliminated, and size dominates the electrophoretic separations (Fig. 5.2a). Sometimes, however, single-stranded DNA electrophoresis is carried out under conditions that allow substantial secondary structure formation. In this case it is the extent of folded structure that tends to dominate the separation pattern. We will discuss these cases in more detail in Chapter 13 when we describe the single-stranded–conformational polymorphism (SSCP) method of looking for mutations at the DNA level.

ELECTROPHORESIS WITHOUT GELS

Essentially all DNA electrophoresis is carried out within gel matrices. Crosslinked polyacrylamide is used for short single-stranded DNAs, agarose is used for very large double-stranded DNAs, and certain other specialized gel matrices have been optimized for particular intermediate separations. To understand the key role played by the gel in these separations, it is instructive to consider DNA electrophoresis without gels. The key parameter measured in electrophoresis is the mobility, μ. This is the velocity per unit field. In one dimension,

$$\mu = \frac{v}{E}$$

where the velocity, v, is measured in cm/s and the electric field, **E,** is measured in volts/cm. Double-stranded DNA behaves in electrophoresis as a free-draining coil. This means that each segment of the chain is able to interact with the solvent in a manner essentially independent from any of the others. Under these conditions the frictional coefficient of the molecule, f, felt by the coil as it moves through the fluid is proportional to the length, L, of the coil:

$$f = \alpha_1 L$$

DNA has a constant charge per unit length. There is one negative charge for each phosphate; a significant fraction of this charges is effectively screened by bound counterions, but the net result is still a charge **Z** proportional to L (Fig. 5.2b):

$$\mathbf{Z} = \alpha_2 L$$

The net steady state velocity in electrophoresis is the result of equal, opposite electrostatic forces accelerating the molecule, **ZE,** and frictional forces, $f\mathbf{v}$, retarding the motion. This lets us set $\mathbf{ZE} = f\mathbf{v}$, where **Z** is the net charge on the molecule. It can be rearranged to

$$\mathbf{v} = \frac{\mathbf{ZE}}{f} = \frac{\alpha_2 L\mathbf{E}}{\alpha_1 L} = \frac{\alpha_2\mathbf{E}}{\alpha_1}$$

Thus from simple considerations we are led to the conclusion that the mobility of DNA in electrophoresis should be independent of size. Indeed this prediction was verified by the work of Norman Davidson and his collaborators more than 20 years ago. Electrophoresis of DNA in free solution fails to achieve any size fractionation at all. Why, then is electrophoresis such a powerful tool in fractionating DNA. The answers will all have to lie with the ways in which DNA molecules interact with gels under the influence of electrical fields.

In passing, it is worth noting that the problem of size-independent electrophoretic mobility is not limited to DNA molecules. A common form of protein electrophoresis is the fractionation of proteins denatured by the detergent sodium dodecyl sulphate (SDS). For proteins without disulphide crosslinks, SDS denaturation produces a highly extended protein chain saturated by bound SDS molecules, as shown in Figure 5.2c. This produces a tubular micellar structure that superficially resembles a DNA double helix in size and shape and charge characteristics. SDS-protein micelles have an approximately constant charge per unit length, and they behave as free-draining structures. As a result, like DNA molecules, proteins in SDS show a size-independent electrophoretic mobility in the absence of a gel. The great power of SDS electrophoresis to fractionate proteins also rests in the nature of the interaction of the SDS micelles with the gel. Here, however, the similarities end. The SDS micelles seem to be mostly sieved by the gels; the interaction of DNA with the gels turns out to be much more complex than this.

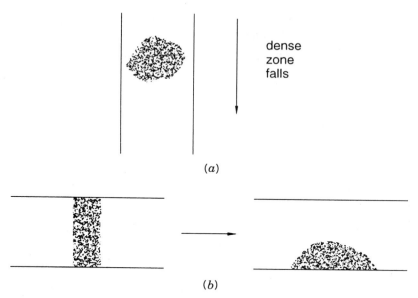

Figure 5.3 Convective instabilities in electrophoresis. *(a)* Vertical electrophoresis. *(b)* Horizontal electrophoresis.

Electrophoresis in free solution is not a common technique, for reasons that actually have little to do with the above considerations. Almost all electrophoresis is done with gels or other support matrices, even when the molecules involved do not behave as free draining. Why use a gel at all under these circumstances? The reason is to prevent convective instabilities. Placement of an electrical field across a conductive solution leads to significant current flow and significant heating. This heating produces nonuniformities in temperature, and convective solvent motion will result from these, much as convection patterns are established if water is heated on the surface of a stove. The presence of bands of dissolved solute molecules leads to regions with local bulk density differences, as illustrated in Figure 5.3. These dense zones are gravitationally unstable. If the electrophoresis is carried out vertically, the dense zone can simply fall through the solution like a droplet of mercury in water. If the electrophoresis is horizontal, any local fluctuations will lead to collapse of the zone as shown in the figure. Gels are one way to minimize the effects of these gravitational and convective instabilities. Another is to use density gradients generated by high concentrations of uncharged molecules, such as sucrose or glycerol. In practice, however, density gradient electrophoresis is a cumbersome and rarely used technique.

MOTIONS OF DNA MOLECULES IN GELS

It is extremely useful to contrast the effect of a gel on two types of macromolecular separation processes: gel filtration or molecule sieve chromatography, and gel electrophoresis. Both of these techniques are illustrated schematically in Figure 5.4. In both cases the gel matrix potentially acts as a sieve. In gel filtration, particles of gel matrix are suspended in a bulk fluid phase. Molecules too large to enter the gel matrix see the gel particles as solid objects.

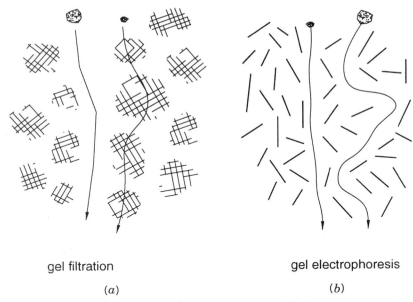

gel filtration

(*a*)

gel electrophoresis

(*b*)

Figure 5.4   Motion *(shown by arrows)* of molecules in gels. *(a)* Gel filtration (molecular sieve chromatography) where large molecules elute more rapidly than small molecules. *(b)* Gel electrophoresis where small molecules migrate more rapidly than large molecules.

As bulk liquid flow occurs, these large molecules move through the fluid interstices between the gel particles. They can find relatively straight paths, and thus their net velocity is rapid. In contrast, smaller molecules can enter the gel matrix. They experience a far larger effective volume as they move through the gel, since they can meander through both the gel matrix and the spaces between the particles. As a result their net motion is considerably retarded. Thus large species elute more rapidly than small species in gel filtration chromatography.

In typical gel electrophoresis the entire sample is a continuous gel matrix. There is no free volume between gel particles. All molecules must move directly through the gel matrix; thus the pore sizes must be large enough to accommodate this. Generally the gel matrices used have a very wide range of pore sizes. As a result relatively small molecules can move through almost all the pores. They experience the gel matrix as an open network, and they can take relatively straight paths under the influence of the electrical field. In contrast, larger molecules can pass through only a restricted subset of the gel pores. To find this subset, they have to take longer paths. Thus, even though their local velocity is the same as that of the small molecules, their net effective motion in the direction of the field is much slower. This means that small molecules will move much faster than large ones in gel electrophoresis.

COMPLEX EFFECTS OF GEL STRUCTURE AND BEHAVIOR

The picture described above is the classical view of molecule motions in gels. It successfully explains the behavior of proteins and small DNAs under conditions where sieving is, in fact, the dominant solute-gel interaction. The details of the gel structure and its interaction with the solute are not considered at all. However, for DNA electrophoresis we now know that the details of the gel structure, its behavior, and its interactions with the solute matter considerably. Unfortunately, our current knowledge about the structure of gels, and the ways macromolecules interact with them, is very slight. For example, experiments have shown that the gels used for typical DNA separations can respond directly to electrical fields. The result is to change the orientation of gel fibers in order to minimize their interaction energy with the applied field. In general, macroscopic measures of gel fiber orientation show that these direct gel-field effects occur at field strengths higher than those typically employed is most electrophoresis.

When DNA is present, applied electrical fields lead to changes in the gel not seen in the absence of DNA. These changes presumably reflect distortion and orientation of the gel caused by motion of the DNA and direct gel-DNA interactions. It is not known if DNA has any indirect effect on the interaction between the gel and the electrical field. What is even less clear is whether the effects of DNA electrophoresis through the gel matrix lead to any irreversible changes in the gel structure. Certainly the overall forces of interaction between a highly charged macromolecule and an obstructive stationary phase could be considerable. We know that gels can usually not be reused more than a few times for electrophoresis without a serious degradation in their performance. This is attributed to some breakdown in necessary gel properties. How much of this is due to electrochemical attack on the gel matrix, and how much to DNA-mediated damage, remains unknown.

The buffer system used to cast the gel, and used in the actual gel electrophoresis itself, matters a great deal. For example, agarose gels cast or run in Tris (trihydroxymethyl-aminomethane) acetate and Tris borate behave very differently. This is believed to reflect the direct interaction of borate ion with cis hydroxyls on the agarose fibers. Finally the specific monomers and crosslinkers used in gels, like polyacrylamide, can have a profound effect on gel running speed, resolution, and the ability of the gel to discriminate unusual DNA structures. For simplicity, we will ignore all of these complications in the discussion that follows.

In general, polyacrylamide-like matrices are used for small, single-stranded DNAs in the size range of 1 to 2 kb. They are also used for small double-stranded DNAs. Agaroses, with much larger pores, are used for larger double-stranded DNAs, ranging from 1 kb to more than 10 Mb in size. These gel systems were not specifically designed to handle nucleic acids, and there is no particular reason to think that they are anywhere optimal for these materials. However, some alternative materials are clearly undesirable for DNA separations because they cannot be made sufficiently free of nucleases or because they demonstrate nonspecific adsorption of DNA molecules.

The ultimate solution for DNA separations may be to use micro-fabricated separation matrices. Here one uses techniques like microlithography to manufacture a custom-designed surface or volume containing specific obstacles to modulate DNA movement and prevent gross convection. Robert Austin at Princeton University has recently demonstrated the electrophoresis of DNA on a microlithographic array. He constructed a regular array of posts spaced at $1\text{-}\mu$ intervals, sticking up from a planer surface. In the presence of an electrical field, DNAs in solution move through these posts (Fig. 5.5). The much larger DNA molecules tend to get hooked on the posts; they eventually are pulled off and net motion through the array continues. This carefully controlled situation appears to mimic a number of key features of DNA electrophoresis in natural gels, as will soon be demonstrated. Whether such a regular array can actually outperform the separation characteristics of natural gels remains to be seen.

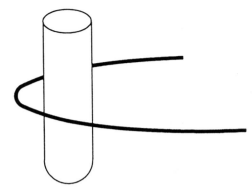

Figure 5.5   Example of DNA behavior seen during electrophoresis of DNA molecules on a microlithographic array of posts.

BIASED REPTATION MODEL OF DNA BEHAVIOR IN GELS

A typical DNA molecule in a polyacrylamide or agarose gel will extend across tens to thousands of discrete pores or channels. The notion of treating this as a simple sieving problem clearly makes no sense. A number of groups including Lumpkin and Zimm, Slater and Noolandi, and Lerman and Frisch, have adapted the models of condensed polymer phases originally developed by Pierre DeGennes to explain DNA gel electrophoresis. DeGennes coined the term *reptation* to explain how a polymer diffuses through a condensed polymer solution or a gel. The gel defines a tube in which a particular DNA molecule slithers (Fig. 5.6a). Diffusion of DNA in the tube is restricted to sliding forward or backward. A net motion in either direction means that the head or tail of the DNA has entered a new channel of the pore network, and the remainder of the molecule has followed.

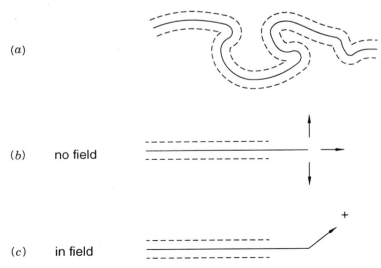

Figure 5.6   Behavior of DNA in gels according to the reptation model. *(a)* The path in the gel defines a tube. *(b)* With no field, motion of the head (or tail) of the DNA occurs by random diffusion. *(c)* With an electrical field, the direction chosen by the head is biased.

This redefines the tube. The model has the DNA behaving much in the same way as a snake in a burrow; hence the term reptation from the word reptile.

In polymer diffusion, when the head of the chain emerges from its tube, it makes a random choice from among the pores available to it (Fig. 5.6b). The elegant simplification offered by the reptation model is that to explain DNA motion, one need only consider what is happening at the head and the tail. The remainder of the complex network of pores, and the detailed configuration of the DNA (except for the overall length of the tube) can be ignored. Indeed this model works very well to explain polymer diffusion in condensed phases. To adapt the reptation model to electrophoresis, it was proposed that the influence of the electrical field biases the choice of pores made by the head of the DNA molecule (Fig. 5.6c). In this biased reptation model, one only needs to consider the detailed effect of the field on the head; the remainder of the chain follows, and its interaction with the gel and the field can be modeled in a simple way that does not require knowledge of the detailed configuration of the DNA or the gel pores.

The biased reptation model makes the following predictions: For relatively small DNAs (small for a particular combination of field strength and gel matrix), diffusion tends to balance out most of the effects of the applied field. The DNA molecule retains a chain configuration that is highly coiled, and the ends have a wide choice of pores. For much larger DNAs, the electrical field dominates over random diffusion. The molecule becomes highly elongated and highly oriented. The ends have a relatively small choice of pores, and there is more frictional interaction between the DNA and the pores. The predicted result of this on electrophoretic behavior is shown in Figure 5.7. Most experiments conform very well to this prediction. At a given field strength, up to some critical DNA length, the mobility drops progressively with DNA size. However, once a point is reached where the DNAs are fully oriented, they all move at the same speed. For typical agarose gel electrophoresis conditions, this plateau occurs at around 10 kb. Fully oriented molecules in gels display a constant charge and friction per unit length; thus their electrophoretic mobility is size independent.

Early attempts to circumvent the problems of DNA orientation and the resultant loss in electrophoretic resolution above a size threshold were not very successful. Lowering the field strength increased the separation range but at the great cost of much longer experimental running times. Lowering the gel concentration increased pore size and allowed larger molecules to be handled, but the resulting gels were extremely difficult to use because of their softness and fragility.

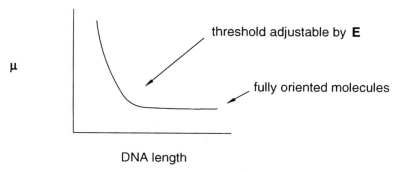

Figure 5.7 Dependence of electrophoretic mobility on DNA length predicted by the biased reptation model and observed under ordinary gel electrophoretic conditions.

PULSED FIELD GEL ELECTROPHORESIS (PFG)

A totally new method for electrophoresis, PFG, was originally conceived as a way to circumvent the limitations on ordinary gel electrophoresis caused by DNA orientation. However, PFG has had a much broader impact because, as attempts to unravel the mechanism of this technique have progressed, they have revealed that the fundamental picture of DNA electrophoresis offered by the biased reptation model is totally inadequate, even for ordinary electrophoresis. The original rationale for PFG is shown in Figure 5.8. Imagine DNA molecules moving in a gel, fully elongated and oriented in response to the field. Suppose that the field direction is suddenly switched. After a while the DNA molecules will find themselves moving in the new field direction and fully oriented once again. However, to achieve this, they have to reorient, and this presumably requires passing through an intermediate state where the molecules are less elongated. Without considering the detailed mechanism of this reorientation (which is still unknown), it seems intuitively reasonable to guess that larger DNA molecules will take longer to reorient than smaller ones. Therefore, if periodically alternating field directions were used, larger DNAs should display slower net motion than smaller ones. This is because they will spend a larger percentage of each field cycle in reorientation, without much productive net motion (Fig. 5.9). Some examples of the power of PFG to separate large DNA molecules are illustrated in Box 5.1.

In retrospect, it is not surprising that PFG enhanced the ability to separate larger DNAs. It is still surprising that the effect is so dramatic. The first PFG experiments demonstrated the ability to resolve, easily, DNA molecules up to 1 Mb in size. Subsequent refinements have pushed these limits a factor of ten or more higher. Thus PFG has expanded the domain of useful electrophoretic separations of DNA by a factor of 100 to 1000. We still are not completely sure why.

There are countless variations on the type of PFG apparatus used, and the details of the electrical field directions employed. With rare exceptions, square wave field pulses have

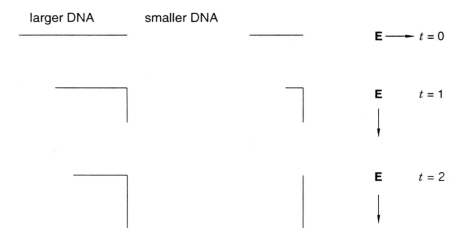

Figure 5.8   Original rationale for the development of pulsed field gel electrophoresis. The notion was that longer molecules would reorient more slowly in the gel in response to a change in electrical field direction.

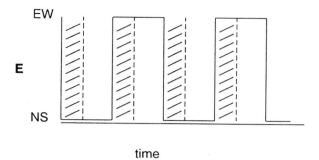

Figure 5.9 Effect of reorientation time on PFG mobility, predicted from very simple consideration of the fraction of each pulse time needed for reorientation *(striped sections).*

been used. These can be characterized by two parameters: the pulse time, which is the period during which the field direction is constant, and the field strength, during that pulse.
In most general modes of PFG, the field strength is always constant. Commonly two
alternate field directions are used; the net motion of the DNA is along the resultant
of these two field directions as shown in Figure 5.10*a*. The usual angle between the two
applied fields is 120° Changes in field direction can be accomplished by switching between pairs (or multiples) of electrodes, by physically rotating the electrodes relative to
the sample, or by physically rotating the sample relative to a fixed pair of electrodes. The
applied fields can be homogeneous, in which case the DNA molecules will move in
straight paths, or inhomogeneous. In the latter case, where field gradients exist, DNA
molecules do not move in straight paths. This makes it more difficult to compute the mobility of a sample or to compare, quantitatively, results on samples in different lanes, that
is, samples with different starting positions in the gel. However, field inhomogeneities,
properly applied, can produce band sharpening because one can create a situation where
the leading edge of a zone of DNA is always moving slightly slower than the trailing
edge.

BOX 5.1
EXAMPLES OF PFG FRACTIONATIONS OF LARGE
DNA MOLECULES

The cohesive ends of bacteriophage lambda DNA were discussed in Box 2.3. These
allow intramolecular circularization, but at high concentration, linear concatenation-
ization is thermodynamically preferred. (For a quantitative discussion, see Cantor and
Schimmel, 1980.) Variants of lambda with different sizes and other viral DNAs with
similar cohesive ends are also known. These samples provide sets of molecules with
known lengths spaced at regular intervals. Such concatemers are the primary size standards used in most PFG work. An example of what a PFG separation of such molecules looks like is shown in Panel A.

(continued)

BOX 5.1 *(Continued)*

Panel A. Separations of Concatemeric Assemblies of Bacteriophage DNAs

An overview of the DNA in a whole microbial genome can be gained by digestion of that genome with a restriction nuclease that has relatively rare recognition sites. This produces a set of discrete fragments that can be displayed by PFG size fractionation. An estimate of the total genome size can be made reliably by adding the sizes of the fragments. Variations in different strains show up readily as evidenced by the example in Panel B.

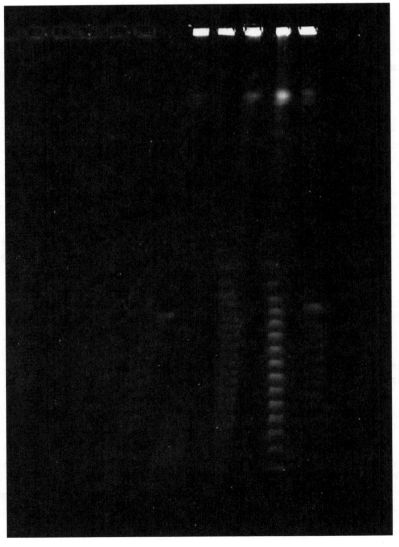

(continued)

BOX 5.1 *(Continued)*

Panel B. Separations of Total Restriction Nuclease Digests of Different Strains of *Escherichia coli*

Microorganisms, like yeasts, that contain linear chromosomal DNAs can be analyzed by PFG to yield a molecular karyotype. Any major rearrangements of these chromosomes are usually revealed by shifts in the size or number of chromosomes in the patterns of chromosomes that hybridize to a particular DNA probe. Since each chromosome is a genetic linkage group (Chapter 6), an initial PFG analysis (e.g., the examples shown in Panel C for yeasts) provides an instant overview of the genetics of an organism and greatly facilitates subsequent gene mapping by pure physical procedures.

(continued)

BOX 5.1 *(Continued)*

Panel C. Separations of Yeast Chromosomes

In part (a), lane 2 is lambda DNA concatemer; lane 3 is some of the smaller *S. cerevisial* chromosomal DNAs. In part (b), lane 2 is *S. pombe* chromosomal DNAs; lane 3 is the largest *S. cerevisial* chromosomal DNAs.

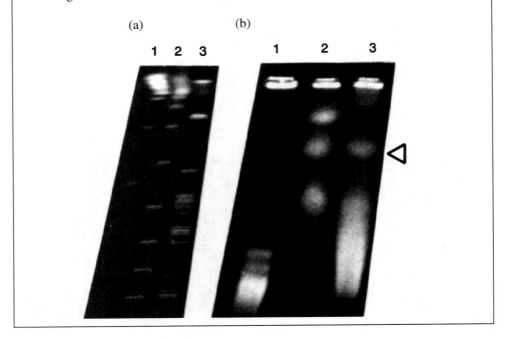

A number of variants of PFG exist where multiple field directions are used. These are generally not employed for routine PFG analyses. One convenient type of PFG apparatus uses multiple point electrodes, each individually adjusted by a computer-controlled power supply. The Poisson equation, $(\nabla)^2 \phi = \rho$, can be used to compute the electrostatic potential, ϕ, in the gel, by recognizing that the free charge, ρ, is zero everywhere in the gel, and

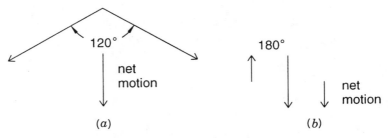

Figure 5.10   Two common experimental arrangements for PFG. *(a)* Field directions that alternate 120(*a*). *(b)* FIGE where field directions alternate by 180(*a*).

using as boundary conditions the voltages set at the electrodes. Then the electrical field, **E,** at each position in the gel can be calculated as $\mathbf{E} = \text{grad } \phi$. This permits a single apparatus can be used for multiple field shapes and directions without the need to physically move or rewire numerous electrodes (Fig. 5.11). A popular variant of this approach uses voltages set at individual electrodes to ensure very homogeneous field shapes and thus very straight lanes. This version is called the contour-clamped homogeneous electrical fields (CHEF).

One version of PFG is basically different. It uses 180° angles between the applied fields (Fig. 5.10*b*). In this technique, called field inversion gel electrophoresis (FIGE), either the length of the forward and backward pulses must be different or their field strengths must be different; otherwise, there will be no net DNA motion at all. FIGE is quite popular because the apparatus needed for it is very simple and because FIGE can achieve very high resolution separations under some conditions. However, the effect of DNA size on FIGE mobility is complex, as we will demonstrate, and it is also rather sensitive to overloading by too high sample concentrations.

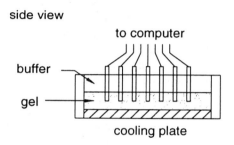

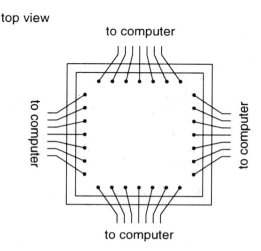

Figure 5.11 Schematic of a contemporary PFG apparatus in which computer-controlled individual electrodes can be used to generate a variety of field shapes in a single apparatus.

MACROSCOPIC BEHAVIOR OF DNA IN PFG

The results of a large number of systematic studies on DNA mobilities in PFG are summarized in Figures 5.12 through 5.14. The major variables explored have been DNA size, the field strength, the angle between the fields, and the pulse time. Other parameters are also known to be important, such as the gel concentration (and the details of the type of agarose used), the buffer type, the ionic strength, and the temperature. These will not be considered further. The effect of angle is relatively slight in the range around 120°. However, regular oscillation of homogeneous electrical fields between two sets of parallel electrodes 90° apart does not produce PFG fractionations. More complex sets of 90° pulses, with varying durations or field strengths, or multiple directions, have been shown to be effective under some circumstances.

The PFG behavior of DNAs up to about a Mb in size with 120° alternating fields is shown in Figure 5.12 as a function of pulse time. A very simple picture suffices to explain these results, but it begs the question of the details of how DNA moves in a gel. At very long pulse times, the process of reorientation should require an insignificant fraction of each pulse period. As a result the DNA moves by essentially ordinary electrophoresis, in a zigzag pattern centered along the average of the two field directions. Because ordinary electrophoresis is dominant, there is no net effect of DNA size on mobility; the DNAs are essentially fully oriented almost all the time. At very short pulse times, the field changes direction much more rapidly than the DNA molecules can reorient. They experience a constant net field which is just the vector sum of the two distinct applied fields. They move in response to this net average field by ordinary electrophoresis. Since they become oriented and elongated, their net mobility is size independent and in fact is the same as their mobility at very long pulse times.

At intermediate pulse times, DNA reorientation processes occupy a significant fraction of each pulse period. This leads to a marked decrease in overall electrophoretic mobility. The pulse time at which mobility is a minimum increases roughly linearly with DNA size. Larger DNAs have a progressively broader response to pulse time. The result is that one can find pulse times that afford very good resolution of particular DNA size classes.

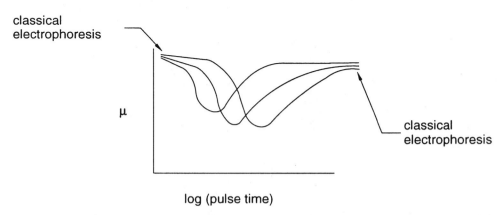

Figure 5.12 PFG mobility as a function of pulse time for DNAs of different size, observed using 120° alternate field directions.

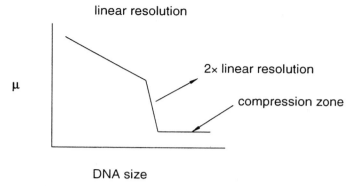

Figure 5.13 Dependence of PFG mobility on DNA size, with 120° alternation in field direction.

Resolution is most easily gauged by plotting mobility as a function of DNA size for a fixed pulse time. This is illustrated in Figure 5.13. Relatively short DNAs have a mobility that decreases linearly with DNA size. Above a sharp size threshold, the slope of this linear decrease doubles. In this size zone the resolution of PFG is particularly high. At even larger DNA sizes, the mobility of DNA becomes size independent. This size range is called the *compression zone*. The DNAs are still migrating in the gel, but there is no size resolution. The compression zone is useful where, for example, one wishes to purify all DNAs above a certain critical size. From the simple picture of PFG described above, the compression zone should consist of DNAs with sizes too big to reorient during the pulse time. However, the simple picture fails to explain the zone with especially high resolution. Indeed no model of PFG has yet explained this.

The dependence of FIGE mobility on DNA size is shown in Figure 5.14. It is dramatically different from the behavior seen in ordinary PFG. With FIGE, the mobility of DNA is not a monotonic function of its size. Very large and very small molecules move at comparable speeds. Molecules with intermediate sizes move slower, and the retardation of the slowest species is quite marked. A simple explanation of FIGE behavior, based on the notions we have explored thus far would say that small DNAs orient rapidly with each ong forward pulse and move efficiently. Large DNAs are never disoriented by the short

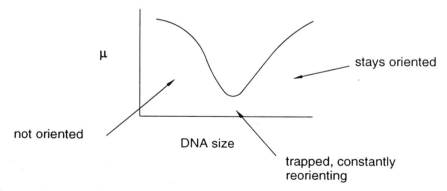

Figure 5.14 DNA mobility as a function of size in typical FIGE using only a single pulse time.

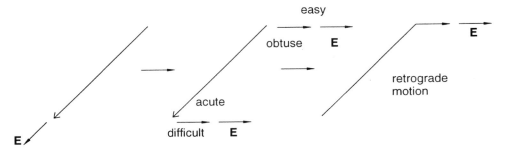

Figure 5.15   The ratchet model for PFG, proposed by Edwin Southern of Oxford University.

backward pulses; thus they remain oriented and move efficiently during the forward pulses. Intermediate size DNAs never achieve a configuration that allows efficient motion for an appreciable fraction of the forward pulse cycle.

The complex dependence of FIGE mobility on DNA size makes it difficult to use FIGE, at any particular set of conditions, to fingerprint a population of different DNA sizes. To circumvent this problem, one can progressively vary either the FIGE pulse times or field strengths during the course of an experiment. The use of such programs produces a superposition of a spectrum of FIGE experiments in which an approximate monotonic decrease in DNA mobility with increasing size is restored.

Early in the development of PFG, a very simple model was put forth by Southern that effectively explained why 90° angles might be ineffective, and why mobility should decrease linearly with DNA size. This ratchet model of DNA motion in gels is shown in Figure 5.15. In the ratchet model the head of the moving DNA is oriented along the applied field direction. When the field direction is switched, the molecule attempts to reorient by a reptation motion. If this were led by the head, it would require that the chain bend through an acute angle. It would seem easier, instead, for the tail to lead the reorientation, since that would require bending the chain through a much less sharp obtuse angle. This leads to retrograde motion until the entire chain has changed orientation. The ratchet model leads directly to a linear dependence of mobility on DNA size because the mobility decreases with the length of the retrograde motion. It also justifies the poor performance of angles 90° and smaller, since these eliminate the need for retrograde motion. However, the ratchet model does not easily account for FIGE, nor for the zone of enhanced resolution in PFG. Finally the ratchet model predicts that molecules greater than a specific size will not move at all, contrary to the observation that the molecules in the compression zone move, albeit slowly.

INADEQUACY OF REPTATION MODELS FOR PFG

Three specific observations of macroscopic DNA behavior are very difficult to reconcile with any type of biased reptation model such as the ratchet or simple reorientation pictures described above. The first of these are measurements of the field strength dependence of PFG mobility, especially for relatively short DNAs. These showed μ proportional to E^2, where E is the field strength in volts/cm. This behavior clearly reflects a complex mechanism, since μ must be an odd function of E to ensure net motion in a particular field direction, while E^2 is an even function that implies no net migration direction.

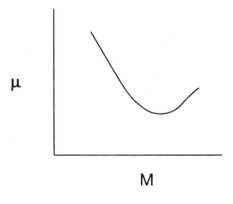

μ

M

Figure 5.16 Behavior of DNA in ordinary electrophoresis at high field strengths in dense gels: Mobility, μ, as a function of molecular weight, M.

The second perplexing observation is DNA trapping. For DNAs with sizes above a particular threshold, at high enough fields no PFG motion could be observed. The threshold size decreased with increasing field strength. This DNA trapping implies more complex DNA-gel interactions than contained in simple reptation models. A final, devastating observation was made in ordinary DNA electrophoresis in high concentration gels at very high field strengths. Here it was observed that, under some conditions, the mobility of DNA was no longer a monotonically decreasing function of DNA size. Instead, at a particular DNA size, a minimum in mobility is observed, just as in FIGE, even though in these experiments a constant, uniform applied electrical field was employed (Figure 5.16).

Three different approaches have been used to examine in detail the nature of DNA motions in gels under the influence of electrical fields. All of these produced unexpected results, totally inconsistent with simple reptation pictures. This was true, even in ordinary electrophoresis with constant fields, or pulsed fields in a single direction. Several different groups reported, almost simultaneously, the detection, by UV fluorescence microscopy, of single DNA molecules undergoing gel electrophoresis. In order to do these experiments, the DNAs were prestained with an intercalating dye like acridine orange or ethidium bromide. These dyes bind at every other base pair and increase the overall length of the DNA by about 50%. Studies of macroscopic DNA electrophoresis, or PFG, with and without bound dyes, indicate that there is little perturbation of the behavior of the DNA by the dyes, except for the predictable consequences of this increase in length.

When observed in the microscope at constant field strength, the motion of DNA molecules was very irregular. In the absence of a field, the molecules were coiled and moderately compact (Figure 5.17). This is expected from random walk models of polymer chain statistics. On application of an electrical field, the molecules elongate, orient, and move parallel to the field. However, the head soon collides with some obstacle in the gel. It stops moving, but the rest of the molecule does not. As a result the elongated chain collapses; the tail may even overtake the head. A very condensed configuration is formed around the obstacle. Eventually one end of the DNA works free and starts to move, pulling some of the chain with it. If the DNA is still attached to the obstacle, both ends may pull free, and the result is a very elongated, tethered structure. Finally the motion of one end dominates; the DNA slips free of the obstacle and starts to run as an elongated aligned structure until it impacts on the next obstacle. Thus the DNA spends most of its time entrapped on obstacles, and the detailed dynamics of how it becomes trapped and freed dominate the overall electrophoretic behavior. The overall motion is very irregular.

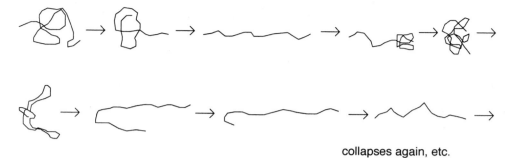

collapses again, etc.

Figure 5.17   DNA behavior in conventional gel electrophoresis as visualized by fluorescence microscopy of individual stained DNA molecules.

There may be DNA sizes where unhooking from obstacles is especially difficult at particular field strengths. This would explain the minimum in ordinary electrophoretic mobility seen macroscopically for certain DNA size ranges at high field strengths in dense gels.

When DNA molecules undergoing PFG are viewed by fluorescence microscopy, additional unexpected behavior is seen (Fig. 5.18). When the field is rotated by 90°, DNA molecules respond to the new direction not by motions of their ends but by herniation at several internal sites. This produces a series of kinks which start to move in the direction of the new field. These kinks grow and compete. Eventually one dominates, and this becomes the leading edge of the moving DNA. At some subsequent point the hairpin structure presumably unravels, and the DNA attains full elongation again. Note that this picture is quite at odds with the reptation model where the DNA is supposed to remain in its tube, except for motions at the ends. It suggests that a tube model is not at all appropriate for DNA in a gel.

The second experimental approach that revealed unexpected complexities of DNA behavior in gels was measurement of bulk electrophoretic orientation by linear dichroism (LD). Here the absorbance of polarized UV light by DNA in gels in the presence of an electrical field was measured. The base pairs of DNA are the dominant absorber of near-UV light (wavelengths around 260 nm). As shown in Figure 5.19, the base pairs preferentially absorb light polarized in the plane of the bases. DNA tends to orient in an electrical field with the helix axis parallel to the field. The LD is defined as

$$LD = A_z - A_y$$

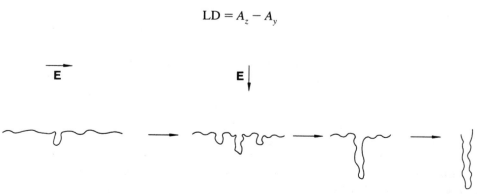

Figure 5.18   Fluorescence microscopic images of DNA in a gel after a 90° rotation of the electrical field direction. Shown from left to right are successive time points.

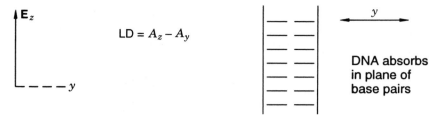

Figure 5.19 Electric dichroism of DNA. An electrical field is applied in the z direction (E_z). Absorption of light polarized in the z and y directions is compared. A_z and A_y for DNA will be negative because the molecule orients along the z axis but preferentially absorbs light in the planes of the base pairs which will be perpendicular to the z axis.

where A_z is the absorbance with polarizers set parallel to the field, and A_y is the absorbance with perpendicular polarizers. This means that the net LD of oriented DNA will be negative, as shown in Figure 5.19. The magnitude of the LD is a measure of the net local orientation of the DNA helix axis.

The LD of DNA in the absence of an electrical field is zero because there is no net orientation. When the LD of DNA in a gel is monitored during the application of a square wave electrical field pulse, very surprising results are seen. What was expected is shown in Figure 5.20a: a monotonic increase in $-$ LD until the orientation saturates; then a monotonic

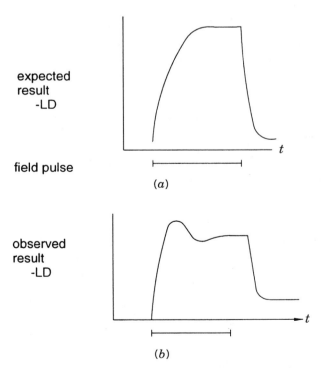

Figure 5.20 Linear dichroism of DNA in a gel, as a result of a single applied electrical field pulse of sufficient intensity to cause a saturating level of orientation. (*a*) Expected result. (*b*) Observed result (from Nordén et al., 1991).

decrease in $-$ LD soon reaching zero, after the field is removed. What is actually observed, under a fairly wide range of conditions, is shown in Figure 5.20b. Upon application of the field, the $-$ LD increases, but it overshoots above the steady state orientation value and then goes through one or more oscillations before saturation occurs. When the field is turned off, most $-$ LD is lost as expected, but a small amount takes a very long time to decay to zero. Furthermore the decay kinetics depend on the original field strength, even though the decay takes place in the absence of the field. This DNA behavior is very complex and difficult to explain by simple models.

The orientation of the agarose molecules that make up the gel itself can be examined by measuring the linear birefringence, which is just n_z-n_y, where n is the refractive index and z and y indicate light polarization axes, as illustrated in Figure 5.19. Birefringence must be used to examine the gel rather than LD because the gel has only very weak near-UV absorbance. The birefringence is dominated by the gel, since it is present at much higher concentrations than the DNA. Without DNA, no field-dependent birefringence is seen at the field strengths used in these experiments. However, in the presence of DNA, the gel shows a rapid orientation after application of an electrical field pulse, and a much slower disorientation after the field is removed. This indicates that the DNA is interacting with the gel and remodeling its shape under the influence of the electrical field. Clearly simplistic explanations will not be adequate to explain DNA electrophoresis in gels.

When LD measurements are applied to monitor the effects of successive field pulses, even more complications emerge. These results are illustrated in Figure 5.21. Two patterns of behavior are seen when pulses are applied spaced by an interval Δt. When Δt is comparable to the pulse time, and the second pulse is either in the same direction as the first or at 90° to the first, no overshoot or orientation oscillations are seen in response to the second pulse. In contrast, when Δt is comparable to the pulse time, but the second pulse is oriented at 180°, or alternatively, when Δt is very long relative to the pulse time, then the second pulse produces an overshoot and oscillations comparable to those produced by the first pulse. These observations have a number of profound implications. They suggest that the relaxed DNA coil is not equivalent to any moving state. The initial response of this relaxed configuration leads to hyperorientation, presumably because of the nature of the way the DNA becomes hooked on gel obstacles. Regaining the original relaxed configuration after a pulse can be very slow; it can take up to 30 minutes for 100 kb DNA, which is much longer than the times inferred for orientation and disorientation from macroscopic observations of PFG mobilities. Apparently inverted pulses act effectively to produce a relaxed-like state. Perhaps under these circumstances the DNA can largely retrace the paths it took when it became entangled with obstacles. Perpendicular pulses do not restore a relaxed configuration. One way to interpret the overshoot seen by LD is to postulate a phased response of the original population of relaxed molecules to the applied field. Once the DNAs have equilibrated in the field, a more complex set of orientations occurs which dephases subsequent responses.

To mimic macroscopic PFG more closely, LD measurements have been performed with continuous alternate pulsing. The time-averaged $-$ LD was measured as a function of pulse time for different angles, as shown in Figure 5.22. It is apparent that relatively good, net local DNA orientation is seen with both long and short pulse times. However, with intermediate pulse times, much poorer net orientation is seen. This picture fits very well with the earliest ideas about a minimum in DNA orientation at interme-

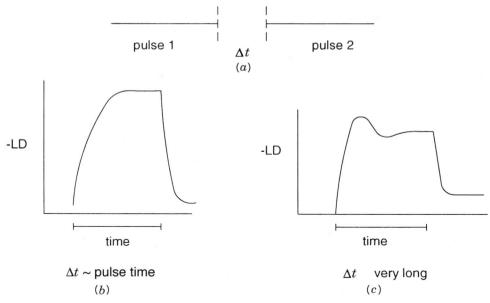

Figure 5.21   Dichroism of DNA in a gel produced by the second of two successive pulses, separated in time by Δt. *(a)* Time course of pulses. The remaining panels show LD in response to the second pulse. *(b)* Results of Δt is comparable to the pulse time with the field direction of the second pulse either parallel to the first pulse or perpendicular to it. *(c)* Results if Δt is very long compared to the pulse time, or if the second pulse is opposite in direction to the first pulse. (Adapted from Nordén et al., 1991.)

diate pulse times. However, the mechanisms that underlie these minima appear to be much more complex than originally envisioned. Note that the minimum in orientation is much more pronounced with 120° pulses than with 90° pulses. This correlates with the much more effective ability of 120° PFG to fractionate different size DNA molecules.

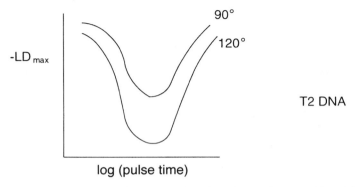

Figure 5.22   Linear dichroism seen for maximally oriented bacteriophage T2 DNA produced by continually alternating electrical field directions. (Adapted from Nordén et al., 1991.)

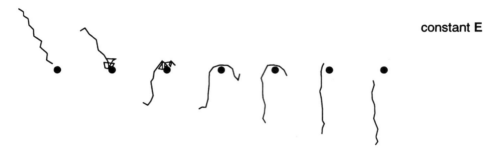

constant **E**

Figure 5.23 Behavior of DNA in gel electrophoresis as simulated by Monte Carlo calculations. Shown from left to right are successive time points. The black dot is an obstacle. The electrical field is vertical.

The third approach that has been used to improve our understanding of DNA electrophoresis in gels is computer simulations using Monte Carlo methods. Here the DNA is typically modeled as a set of charged masses attached by springs. The fluid is modeled by its contribution to the kinetic energy of the DNA through brownian motion. The gel is modeled as a set of rigid obstacles. Most early simulations were done in two dimensions to limit the amount of computer time required. This is unlikely to allow a realistic picture of either the gel or the DNA. However, more recent three-dimensional simulations are at least in qualitative agreement with the earlier two-dimensional results. Undoubtedly the model of the gel used in these simulations is unrealistically crude; the potential used to describe the interaction of the DNA with the gel is hopelessly oversimplified. The DNA model itself is not terribly accurate. Despite all these reservations, the picture of DNA behavior that emerges from the simulations is remarkably close to what is observed by microscopy. An example is shown in Figure 5.23. Collisions with obstacles dominate the motion of the DNA; hyperelongation of hooked structures, until they are released, apparently accounts for the overshoots seen in LD experiments.

Only a few simulations have yet been reported for PFG. These suggest that kinked structures play important roles. It seems logical that kinks should enhance the chances of DNA hooking onto obstacles (Fig. 5.24). Thus conditions that promote kinking might lead to minima in mobility. This notion, which is not yet proved, would at least be consistent with what is generally seen in macroscopic PFG experiments. The speed of most computers, and the efficiency of the algorithms used, has limited most simulations of electrophoresis to relatively short DNA chains at very dilute concentrations. Recently Yaneer Bar-Yam at Boston University has demonstrated orders of magnitude increases in simulation speed by implementing a molecular automaton approach on an intensely parallel computer architecture. This increase in the power of computer simulations should allow a wide variety of experimental conditions to be modeled much more efficiently.

Figure 5.24 Expected behavior of kinked DNA in electrophoresis.

Such increases in computation speed are needed because two additional variations of PFG show considerable promise for enhanced size resolution, but each of these introduces additional complications into both the experiments and any attempts to simulate them.

DNA TRAPPING ELECTROPHORESIS

The technique of DNA trapping electrophoresis was devised by Levi Ulanovsky and Walter Gilbert as an approach to improving the resolution of DNA sequencing gels. These gels examine single-stranded DNA in crosslinked polyacrylamide under the denaturing conditions of 7 M urea and elevated temperatures. Typical behavior of DNA under such conditions is shown in Figure 5.25. A monotonic decrease in mobility with increasing DNA size is seen until at some threshold, usually around 1 kb, a limiting mobility is reached, and all larger molecules move through the gel at the same velocity. This presumably reflects complete orientation, just as we have argued for double-stranded DNA in agarose. To circumvent the loss in resolution at large DNA sizes, a globular protein was attached to one end of the DNA, as shown schematically in Figure 5.26. The actual protein used was streptavidin because of the ease of placing it specifically on the end-biotinylated target DNA. Streptavidin is a 50 kDa tetramer that can have a diameter of 40Å. This is considerably fatter than the 25-Å diameter of the DNA double helix. Note that streptavidin has such a stable tertiary and quaternary structure that it remains as a folded globular tetramer even under the harsh, denaturing conditions of DNA sequencing gel electrophoresis.

The larger size of the streptavidin, and its lack of charge under the electrophoretic conditions used, should ensure that the tagged end exists predominantly as the DNA tail. Periodically the head of the DNA chain will enter gel pores too large for the bulky tail to penetrate. This will lead to enhanced trapping, since the entire DNA will have to back out of the pore in order for net motion to occur. Above a certain DNA size, thermal energies may be insufficient to allow such backtracking, and once trapped, the tethered DNA might remain so indefinitely. A typical experimental result is shown in Figure 5.25.

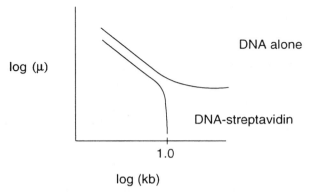

Figure 5.25 DNA trapping electrophoresis. Shown is the dependence of the electrophoretic mobility of single-stranded DNA on size for ordinary DNA and DNA end-labeled with streptavidin (see Fig. 5.26). (Adapted from Ulanovsky et al., 1990.)

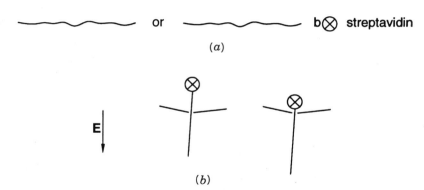

(a)

(b)

Figure 5.26 Effect of a bulky end label on DNA mobility in gels. *(a)* Comparison of DNA with DNA terminally labeled with streptavidin. *(b)* Trapping of DNA in a pore because of the bulkiness of the streptavidin.

Instead of a limiting mobility, tagged DNA shows a zone of super high resolution until, at a key size, the mobility drops to zero. This is all in accord with the simple view illustrated in Figure 5.23.

If the picture just described is accurate, one can make the further prediction that FIGE will assist the escape of the trapped DNA. These results are shown in Figure 5.27. The three curves in this figure show the size dependence of DNA mobility under three different FIGE conditions with DNAs tethered to streptavidin. Very slow FIGE leads to a reappearance of the limiting mobility. Apparently under these conditions all trapped molecules can be rescued. Very rapid FIGE fails to remove any trapping effects.

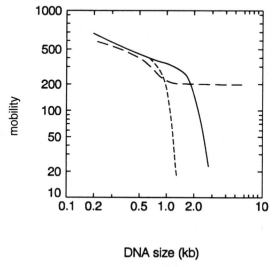

Figure 5.27 DNA trapping electrophoresis under FIGE conditions. Shown is FIGE mobility for three different sets of conditions, Forward/reverse (ms): 800/200 (long dashed line), 80/20 (solid line), 8/2 (short dashed line). In each case the sample was a single-stranded DNA terminally labeled with a streptavidin. (Adapted from Ulanovsky et al., 1990.)

Apparently these pulses are too short to allow the trapped molecules to escape. Intermediate FIGE conditions produce a very impressive increase in the useful separation range of the gel. Overall, DNA trapping electrophoresis appears to be a very interesting idea worth further elaboration and pursuit. It underscores the fact that trapping, and not sieving, appears to be the dominant effect underlying DNA electrophoretic size separations.

SECONDARY PULSED FIELD GEL ELECTROPHORESIS (SPFG)

The history of SPFG is amusing. Tai Yong Zhang, a technician then working with us in the early 1990s, was instructed to perform the experiment shown schematically in Figure 5.28. The notion was to use a series of rapid field alternations to perturb the motion of the head of an oriented DNA chain moving in response to a slowly varying field. The result should be to fold the chain into a zigzag pattern, which ought to affect its subsequent orientation kinetics markedly when the primary field direction is switched. Zhang, whose English was quite imperfect at the time, misunderstood the original instructions and did the experiment shown schematically in Figure 5.29a. He applied periodic, short intense pulses, along the direction of net DNA motion. At about the same time Jan Noolandi's group did a similar experiment, shown in Figure 5.29b. They applied short, intense pulses opposite to the direction of net DNA motion. The results of both experiments, now called SPFG, are quite similar. The overall rate of DNA motion is dramatically increased. In addition, under some SPFG conditions, greatly improved resolution is seen, and larger molecules can be handled than is possible with conventional PFG alone under comparable field strengths and primary pulse times. Some examples are shown in Figure 5.30.

   While the actual mechanism for the success of SPFG is unknown, one suggested mechanism shown in Figure 5.31 seems quite reasonable. DNA in PFG spends most of its time trapped on obstacles. As long as the field direction remains constant, the applied field will tend to keep the DNA trapped, as illustrated directly in trapping electrophoresis.

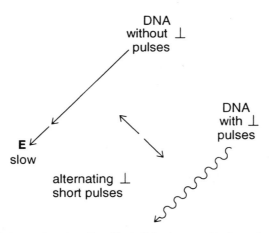

Figure 5.28   Untried experiment on the effect of short perpendicular pulses on PFG mobility.

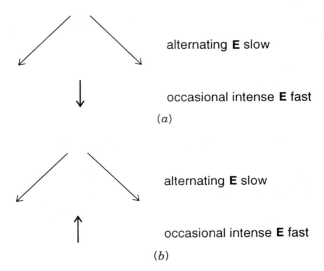

Figure 5.29   Electrical field configurations actually used in SPFG. *(a)* In our own work, *(b)* in the work of Jan Noolandi and collaborators.

Any tangential field, such as applied in either implementation of SPFG, will tend to bias the configuration of the trapped DNA, making it easier to slip off the obstacle in response to the primary field. At present it is difficult to find optimal SPFG conditions because of the large number of experimental variables involved, and the fact that these variables are highly interactive (Table 5.1). Perhaps the vastly improved rates of simulations of DNA electrophoresis will soon be applied to increase our understanding of SPFG.

ENTRY OF DNAs INTO GELS

An additional, surprising aspect of SPFG was revealed when the behavior of molecules as large as (presumably) intact human DNAs was observed. It was found that the conditions required for gel entry were far more stringent than the conditions needed to move DNA, once the DNA was inside the running gel matrix. This may reflect a trivial artifact, that the DNA must be broken in order to enter the matrix, or it could be revealing an intriguing aspect of DNA behavior. Note that DNA samples for PFG or SPFG are all made in situ in agarose, as shown schematically in Figure 5.32. The sample agarose is a low-gelling temperature variety that has different pore sizes than the running gel. In addition the sample gel (0.5%) usually has half the agarose concentration of the running gel (1.0 to 1.1%). However, what is probably more significant is that once the sample cells are lysed, and the chromosomal DNAs freed of bound protein and any other cellular constituents, they find themselves in free solution inside a chamber much bigger than the gel pores. When the electrical field is turned on, the DNA coils in free solution are swept to one side of the chamber where they may encounter the gel in quite a different state than they experience once they have threaded its way into the first series of gel pores.

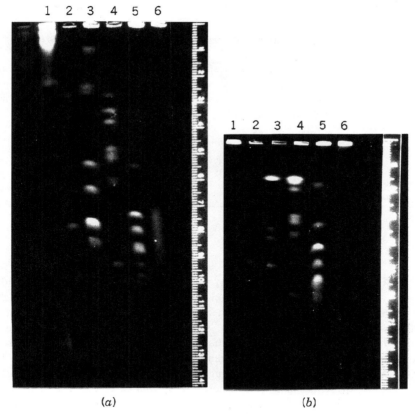

Figure 5.30 Examples of the effect of secondary pulses on PFG behavior. Separation of chromosomal DNA molecules in the size range between 50 kb and 5.8 Mb by SPFG. Samples are (lane 1) *S. pombe,* (lane 2) Pichia 1A which consists of a mixture of *P. scolyti* and *P. mississipiensis,* (lane 3) *P. scolyti,* (lane 4) *P. mississippiensis,* (lane 5) *S. cerevisiae,* and (lane 6) lambda concatemers. Separation on a 1% agarose gel using *(a)* a pulse program as follows: (i) 4800 s, 2.2 volts/cm primary pulses with 1 : 15 s (1 second pulse every 15 seconds), 6 volts/cm secondary pulses for 12 h, (ii) 2400 s, 2.8 volts/cm primary pulses with 1 : 15 s, 6 volts, cm secondary pulses for 12 h, (iii) 240 s, 6 volts/cm primary pulses with 1 : 15 s; 10 volts/cm secondary pulses for 21 h, and (iv) 120 s, 6 volts/cm primary pulses with 1 : 15 s, 10 volts/cm secondary pulses for 10 h. *(b)* The same primary pulse program shown in *(a)* but without secondary pulses. (From Zhang et al., 1991.)

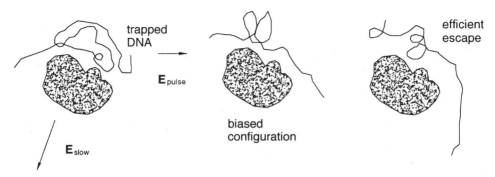

Figure 5.31 Schematic picture of how secondary pulses may accelerate PFG by releasing DNAs from obstacles.

TABLE 5.1 Interactive Parameters in Secondary Pulsed Field Electrophoresis (SPFG)

Electrical Parameters	Other Parameters[a]
Primary field strength	Agarose gel concentration
Primary pulse time	Particular type of agarose (low melting, high
Angle of alternation of primary fields	endoosmosis, etc.)
Secondary field strength	Ionic strength
Direction of secondary field	Temperature
Secondary pulse time	Specific ions present (e.g., acetate versus borate)
Phase between secondary and primary fields	
Homogeneity of primary field	
Homogeneity of secondary field	

[a]Known to affect PFG and thus assumed to affect SPFG as well.

One possible consequence of the way in which large DNA is made and the obstacles it must overcome to enter a gel is shown schematically in Figure 5.33. Vaughan Jones, a mathematical topologist at the University of California, Berkeley, in thinking about large DNA molecules, suggested in 1989 that above a critical DNA size the conformation of these chains would always contain at least one potential knot. Much earlier, Maxim Frank-Kamenetskii, in thinking about DNA cyclization reactions, had very similar thoughts (Frank-Kamenetskii, 1997). Knot formation poses no problems inside cells where topoisomerases exist in abundance that can tie and untie knots at will. Similarly it is no problem for an isolated DNA molecule in solution. However, as DNA enters a gel, any knots it contains are prime targets for entanglement on obstacles. Despite the fact that this has occurred, the remainder of the DNA molecule will tend to be pulled further into the gel by the influence of the electrical field. This will result in tightening the knot, and it may lead to permanent trapping of the DNA at the site of entry into the gel. Secondary pulses could help to alleviate this problem by continually untrapping the knotted DNA from the obstacle until it has, by chance, assumed an unknotted configuration that allows it to fully enter the gel and move unimpeded. This idea has been explored recently by Jean Viovy who has demonstrated, using a clever multidimensional version of PFG, that DNA molecules do become irreversibly trapped when they are forced to run in agarose above a certain field strength (Viovy et al., 1992). Viovy argues from considerations of the forces involved that the effect of secondary pulses may be to prevent knots from tightening rather than to untie them once formed. What is needed is a clever way to test this intriguing but unproved suggestion.

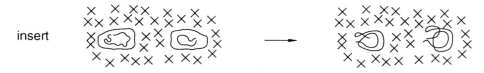

Figure 5.32 Preparation of large DNA molecules by in-gel cell lysis and deproteinization.

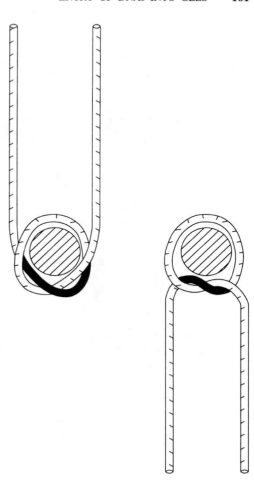

Figure 5.33 How DNA knotting may prevent gel entry. Jean Viovy has obtained evidence that is consistent with the notion that knots are responsible for irreversible trapping of DNA within the gels. But there is still no direct proof for this idea (Viovy et al., 1992).

BOX 5.2
PRACTICAL ASPECTS OF DNA ELECTROPHORESIS

The ideal PFG fractionations would be rapid, high-yield resolution separations and allow larger quantities of sample to be purified to increase the ease and sensitivity of subsequent analyses. This ideal is not achievable because both speed and sample size generally lead to a loss in resolution, for reasons that are not fully understood.

The quality of PFG separations strongly deteriorates as DNA concentration is raised. Samples should be run at the lowest concentrations that allow proper visualization or other necessary analysis of the data or utilization of the separated samples. Examples are presented in Panel A.

(continued)

BOX 5.2 *(Continued)*

Panel A. Effect of DNA Concentration on Apparent Electrophoretic Mobility (Adapted from Doggett et al. 1992)

In ordinary agarose electrophoresis of DNA, most band broadening occurs not by diffusion but by an electrical-field-dependent process called *dispersion.* The factors that produce this dispersion remain to be clarified (Yarmola et al., 1996).

In PFG separations in all generally used apparatus, a trade-off must be made between band sharpness (resolution) and how straight the lanes are. The use of field gradients, an inherent property of the first PFG instrument designs, leads to band sharpening. However, no successful design of a PFG apparatus has been demonstrated that produces straight lanes and still allows gradients to be present to sharpen the bands. Examples of the trade-offs before resolution and straightness are shown in Panel B.

Panel B. Effect of a Gradient in Electrical Field on the Apparent Sharpness of Bands

The basic mechanism by which gradients lead to band sharpening is shown in Panel C:

(continued)

BOX 5.2 *(Continued)*

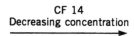

CF 14
Decreasing concentration

Amount loaded   1/16 1/8 1/4 1/2 3/4 /   /   /

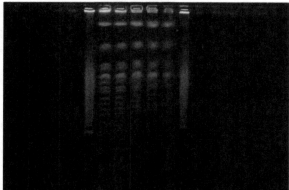

1/16 1/8 1/4 1/2 3/4

Panel C. Rationale of the Effect of Electrical Field Gradients

Because molecules at the back of the zone are moving faster than those at the front, they will catch up until limits on zone thickness are reached that broaden the zone either by dispersion or by concentration-dependent effects.

Band sharpening can also be achieved by other methods. For example, a gradient of increasing gel concentration will mimic the effect of a gradient of decreasing electrical field strengths.

SOURCES AND ADDITIONAL READINGS

Birren, B., and Lai, E., eds. 1993. *Pulsed Field Gel Electrophoresis*. San Diego: Academic Press.

Birren, B., and Lai, E. 1994. Rapid pulsed field separation of DNA molecules up to 250 kb. *Nucleic Acids Research* 22: 5366–5370.

Bustamante, C., Vesenka, J., Tang, C. L., Rees, W., Guthold, M., and Keller, R. 1992. Circular DNA molecules imaged in air by scanning force microscopy. *Biochemistry* 31: 22–26.

Cantor, C. R., and Schimmel, P. R. 1980. *Biophysical Chemistry*. San Francisco: W. H. Freeman, ch. 24.

Carlsson, C., and Larsson, A. 1996. Simulations of the overshoot in the build-up of orientation of long DNA during gel electrophoresis based on a distribution of oscillation times. *Electrophoresis* 17: 1425–1435.

Desruisseaux, C., and Slater, G. W. 1996. Pulsed-field trapping electrophoresis: A computer simulation study. *Electrophoresis* 17: 623–632.

Doggett, N. A., Smith, C. L., and Cantor, C. R. 1992. The effect of DNA concentration on mobility in pulsed-field gel electrophoresis. *Nucleic Acids Research* 20: 859–864.

Duke, T. A. J., Austin, R. H., Cox, E. C., and Chan, S. S. 1996. Pulsed-field electrophoresis in microlithographic arrays. *Electrophoresis* 17: 1075–1079.

Frank-Kamenetskii, M. F. 1997. *Unraveling DNA*. Reading MA: Addison-Wesley.

Gurrieri, S., Smith, S. B., Wells, K. S., Johnson, I. D., and Bustamante, C. 1996. Real-time imaging of the reorientation mechanisms of YOYO-labelled DNA molecules during 90 degrees and 120 degrees pulsed field gel electrophoresis. *Nucleic Acids Research* 24: 4759–4767.

Monaco, A. P., ed. 1995. *Pulsed Field Gel Electrophoresis: A Practical Approach*. New York: Oxford University Press.

Nordén, B., Elvingson, C., Jonsson, M. and Åkerman, B. 1991. Microscopic behavior of DNA during electrophoresis: electrophoretic orientation. *Quarterly Review of Biophysics* 24: 103–164.

Schwartz, D., and Cantor, C. R. 1984. Separation of yeast chromosome-sized DNAs by pulsed field gradient gel electrophoresis. *Cell* 37: 67–75.

Smith, M. A., and Bar-Yam, Y. 1993. Cellular automaton simulation of pulsed field gel electrophoresis. *Electrophoresis* 14: 337–343.

Turmel, C., Brassard, E., Slater, G. W., and Noolandi, J. 1990. Molecular detrapping and band narrowing with high frequency modulation of pulsed field electrophoresis. *Nucleic Acids Research* 18: 569–575.

Ulanovsky, L., Drouin, G., and Gilbert, W. 1990. DNA trapping electrophoresis. *Nature* 343: 190–192.

Viovy, J. L., Miomandre, F., Miquel, M. C., Caron, F., and Sor, F. 1992. Irreversible trapping of DNA during crossed-field gel electrophoresis. *Electrophoresis* 13: 1–6.

Whitcomb, R. W., and Holzwarth, G. 1990. On the movement and alignment of DNA during 120 degrees pulsed-field gel electrophoresis. *Nucleic Acids Research* 18: 6331–6338.

Yarmola, E., Sokoloff, H., and Chrambach, A. 1996. The relative contributions of dispersion and diffusion to band spreading (resolution) in gel electrophoresis. *Electrophoresis* 17: 1416–1419.

Zhang, T.-Y., Smith, C. L., and Cantor, C. R. 1991. Secondary pulsed field gel electrophoresis: a new method for faster separation of larger DNA molecules. *Nucleic Acids Research* 19: 1291–1296.

6 Genetic Analysis

WHY WE NEED GENETICS

Despite the great power of molecular biology to examine the information coded for by DNA, we have to know where on the DNA to look to find information of relevance to particular phenomena. Some genes have very complex phenotypic consequences. We may never have the ability to look at a DNA sequence and infer, directly, that it regulates some aspect of facial features or mathematical reasoning ability. Genetic analysis offers a totally independent approach to determining the location of genes responsible for inherited traits. In this chapter we will explore the power of this approach and also some of its particular limitations. Genetics, in the human, will mostly serve as a low resolution method for localizing genes, but it is the only method available if the only prior information about a gene is some hypothesis about its function, or some presumed phenotype that results from a particular DNA sequence (allele) of that gene.

BASIC STRATEGY FOR GENETIC ANALYSIS IN THE HUMAN: LINKAGE MAPPING

In most organisms, genetics is carried out by breeding specific pairs of parents and examining the characteristics of their offspring. Clearly this approach is not practical in the human. Even leaving aside the tremendous ethical issues such an approach would raise, the small size of our families, and the long lifespan of our species, make genetic experimentation all but impossible. Instead, what must be done is to perform retrospective analyses of inheritance in families. Statistical analysis of the pattern of inheritance is used instead of direct genetic manipulation to test hypotheses about the genetic mechanism underlying a particular trait.

Human beings and other mammals are diploid organisms. They contain two copies of each homologous chromosome. Normally an offspring receives one of its homologs from each parent (Fig. 6.1). This is accomplished by a specialized cell division process, meiosis, which occurs during the formation of sperm and eggs, a process known as gametogenesis. Sperm cells and egg cells (ova) are haploid; both contain only a single one of each of the homologous pairs of the parental chromosomes. When sperm and egg combine in fertilization, this reconstitutes a diploid cell which is a mosaic of halves of the genomes of its parents.

Linkage is the tendency for two observable genetic traits, called *markers,* to be coinherited if they lie near each other on the same chromosome. To be distinguishable genetically, a marker must occur in more than one form (e.g., eye color) in different members of the population. We call these forms alleles. Consider the case shown in Figure 6.1 where two detectable markers, one with alleles A or a and the other with alleles B or b, are being followed in a family. Markers A and B are both inherited from the father.

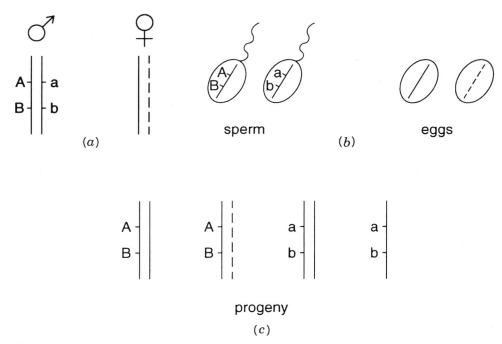

Figure 6.1   Basic scheme for chromosome segregation in the genetics of diploid organisms. *(a)* parental genome. *(b)* Possible gametes. *(c)* Possible offspring (all equally probable).

Markers on different chromosomes would, by chance, pass together to an offspring only 25% of the time, since there is a 50% chance of the offspring receiving each particular homologous chromosome. In contrast, markers on the same chromosome will always be coinherited, so long as the chromosome remains intact. Half of the offspring in this case would receive both allele A and allele B. This simple picture of inheritance is complicated by the process known as *meiotic recombination*. During the formation of sperm and eggs, at one stage in meiosis, homologous chromosomes pair up physically, and the DNA molecules within these chromosomes can form a four-ended structure known as a *Holliday junction*. This structure can be reverted to unpaired DNA duplexes in two different ways. Half the time the result is to produce a DNA molecule that consists of part of each parental DNA, as shown in Figure 6.2. When these rearranged chromosomes are passed to offspring, the result is that part of each parental homolog is inherited. This can separate two markers that were originally linked on one of the parental chromosomes.

Meiotic recombination in the human occurs at least once for each pair of homologous chromosomes.[1] If two markers A and B are 1 Mb apart, there is roughly a 1%

---

[1]It is interesting to speculate why meiotic recombination appears to be obligatory. In the absence of recombination, a chromosome that acquired a dominant lethal mutation would be lost to all future generations. Alleles that accidentally resided on this chromosome would also be lost. The resulting drift toward homozygosity could be harmful from an evolutionary standpoint. This drift is largely prevented by the requirement for meiotic recombination.

meiotic recombination during gametogenesis

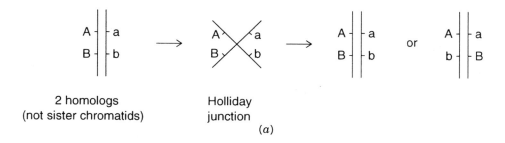

2 homologs
(not sister chromatids)

Holliday
junction

(a)

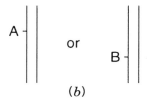

(b)

Figure 6.2 Minimalistic view of meiotic recombination in a single pair of homologous chromosomes. *(a)* Formation and resolution of a Holliday junction. *(b)* Typical offspring after a recombination event.

chance that they will be separated, and an offspring will be produced with A and b not B, and vice versa. Because human families are generally small, it becomes very difficult to detect recombination events for closely linked markers. In other words, the resolution of genetics in the human is quite limited. It would take families with 100 offspring to detect a single recombination event that would place two genes 1 Mb apart. Since such families are unavailable, one must pool data from different families in order to have a large enough population to provide a reasonable chance of seeing a rare recombination event. Such pooling makes the presumption that markers A and B in different families actually correspond to the same genes. For many reasons this is not always the case. For example, if the marker is the elimination of a complex metabolic pathway, it is possible that the phenotype might be similar regardless of which gene for one of the enzymes in the pathway was inactivated. Such multigenic complications may be quite common in human genetics, and they can severely impair our ability to use linkage analysis to find genes.

An additional complication is the restricted and uncontrolled variability of human DNA. In order for a marker to be useful, one must be able to distinguish which of the two parental homologous chromosomes has been inherited. If both chromosomes of a parent contain the same marker A, there is no way to tell from this marker alone which chromosome the offspring received. We say that this marker is uninformative. If both parents have the same set of allele pairs, A and a, as shown in Figure 6.3, the child can have allele pairs A, A or a, a or A, a. Once again we have no way of discerning from this marker alone which chromosome came from which parent. To carry out human genetics effectively, it is necessary to have large numbers of markers which are informative. That is, at any particular region, we would like to be able to distinguish easily among the four homologous chromosomes originally carried by the two parents.

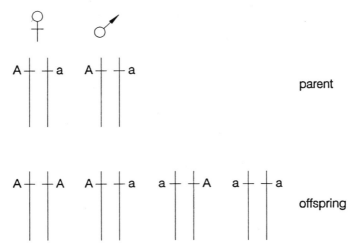

Figure 6.3 Examples of cases where particular combinations of alleles in parents and offspring confound attempts to trace genetic events uniquely.

BOX 6.1
MEIOSIS

The process of meiosis and the recombination events that occur during meiosis, are actually much more complicated than the simple picture given in the text. Figure 6.4 illustrates this process for the relatively simple case of a cell with only two different chromosomes. The parental cell is diploid; it contains two copies of each homologous chromosome. When this cell duplicates its genome, in preparation for cell division, it becomes tetraploid. At this point each chromosome exists as a pair of sister chromatids. During the first meiotic cell division, the homologous chromosomes line up and associate, as shown in Figure 6.4. The resulting structures are pairs of homologous chromosomes, each containing two paired sister chromatids (identical copies of each other). Thus, altogether, four DNA molecules (eight DNA strands!) are involved in close proximity. In order for these molecules to separate, there appears to be an obligatory recombination event that involves at least one chromatid from each pair of the homologous chromosomes. As shown in Figure 6.4, this results in a reciprocal exchange of DNA between two chromatids, while, in the simple case illustrated, the other two chromatids remain unaltered.

The homologous chromosomes segregate to separate daughter cells during the last stages of the first meiotic division. When this is completed, the result is two diploid cells as in ordinary mitotic division. However, what is profoundly different about these cells is that each contains paired homologous chromosomes that now consist of some regions where both copies of the homolog arose by duplication of a single one of the initial homologous pairs. Because of recombination, one member of each chromosome pair at this point contains some material that is a mosaic of the two original parental homologs. In the second meiotic cell division, each the pair of homologous chromosomes is segregated into a separate daughter cell that goes on, eventually, to become a haploid gamete. Thus, as shown in Figure 6.4, some gametes receive chromosomes that are unaltered versions of the original parental material. Others receive chromosomes that have been rearranged by meiotic recombination. The true picture is even more complex than this, and we will return to it later.

BOX 6.1 *(Continued)*

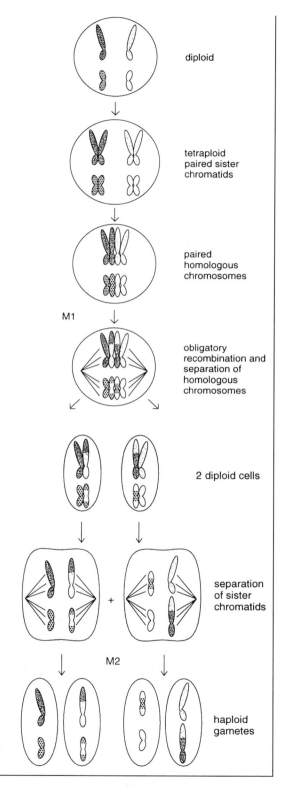

diploid

tetraploid
paired sister
chromatids

paired
homologous
chromosomes

M1

obligatory
recombination and
separation of
homologous
chromosomes

2 diploid cells

separation
of sister
chromatids

M2

haploid
gametes

Figure 6.4 Some of the steps in meiosis, illustrated for a hypothetical cell with only two types of chromosomes.

A GLOSSARY OF GENETIC TERMS

The science of genetics emerged and matured before DNA was known to be the genetic material. Many terms and concepts were coined without a knowledge of their chemical basis. Here we define some of the commonly used genetic terms needed to follow arguments in human genetics, and we provide an indication of their molecular basis. The first few of these notions are illustrated schematically in Figure 6.5.

Gene. A commonly used term for a unit of genetic information sufficient to determine an observable trait. In more contemporary terms, most genes determine the sequence and structure of a particular protein molecule. However, genes may code for untranslated RNAs like tRNA or rRNA, or they may code for more than one different protein when complex mRNA processing occurs.

Locus. A site on a chromosome, that is, site on a DNA molecule.

Allele. A variant at a particular locus, namely a DNA sequence variant (simple or complex) at the particular locus of interest.

Homozygosity. The two homologous chromosomes in a diploid cell have the same allele at the locus of interest.

Heterozygosity. The two homologous chromosomes in a diploid cell have different alleles at the locus of interest.

Independent segregation. The notion that each of the parental homologous chromosomes in a diploid cell has an equal chance of being passed to a particular daughter cell or offspring. Thus, considering only two chromosomes in Figure 6.6, there will be four possible types of gametes (neglecting recombination), and each of these will occur at equal frequencies.

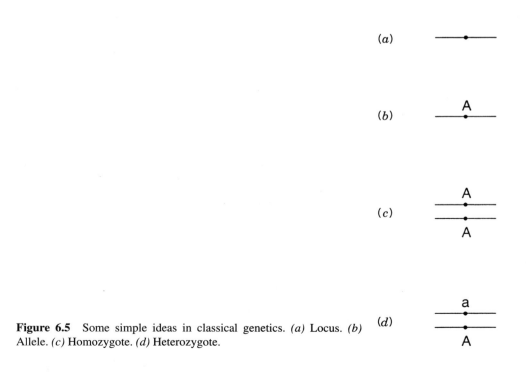

Figure 6.5 Some simple ideas in classical genetics. *(a)* Locus. *(b)* Allele. *(c)* Homozygote. *(d)* Heterozygote.

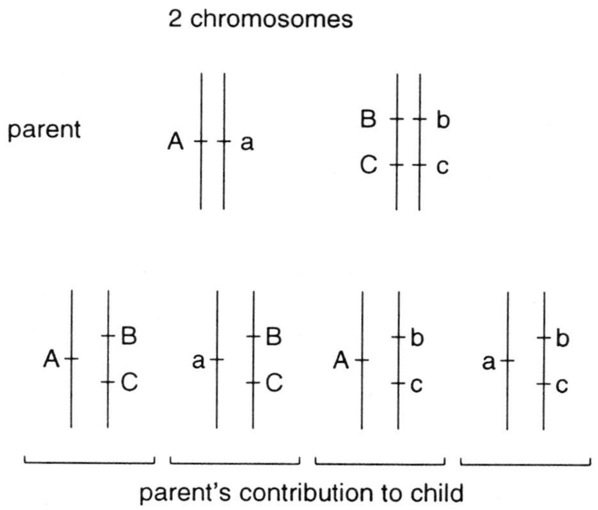

2 chromosomes

Figure 6.6 Independent segregation of alleles on two different chromosomes. Shown are parental genomes and possible gametes arising from these genomes.

Phenotype. What one sees, namely the complete set of observable inherited characteristics of an organism, or the particular set of characteristics controlled by the locus or loci of interest.

Genotype. What one gets, namely the genetic constitution, which is the particular set of alleles inherited by the organism as a whole, or for the particular locus or loci of interest.

Haplotype. The actual pattern of alleles on one chromosome, that is, on a single DNA molecule.

Dominant. An allele that produces an identical phenotype if it is present on one homologous chromosome, or on both. True simple dominant traits are rare in the human; one example may be Huntington's disease where individuals carrying one or two disease alleles appear to have indistinguishable disease phenotypes.

Recessive. An allele that produces a particular phenotype only if it is present on both homologous chromosomes. Obviously dominant and recessive are oversimplifications that ignore molecular mechanisms and the possibility of more than two alleles at a locus. In principle, with three alleles, A, a, and à, there are six possible genotypes: AA, Aa, Aà, aa, aà, and àà, and each of these might yield a distinguishable phenotype if we use precise enough definitions and measurements.

Penetrance. The chance that an allele yields the expected phenotype. This term covers a multitude of sins and oversimplifications. In simple genetic systems the penetrance should normally be 1.0. In outbred species like the human, there can be unknown genetic variants that interact with the gene product at the locus in question and produce different phenotypes in different individuals. Purely at the DNA level we expect all phenotypes to be distinguishable and to show complete penetrance because the phenotype at the level of DNA is identical to the genotype. An exception occurs if the phenotype requires the expression of a particular allele in an environmental context that might not be experienced by all individuals. For example, an allele that determined a serious adult emotional illness might require certain physical or mental stress during the development of the nervous system in order to actually result in phenotypically detectable disease.

Phase. The distribution of particular alleles on homologous chromosomes. Suppose that alleles B and b can occur at one locus, alleles C and c at another. An individual with the phenotype BbCc can have two possible genotypes, as shown in Figure 6.7a. These are arbitrarily called cis and trans in analogy with chemical isomers. The two possible phases in this simple case can be distinguished easily by performing a genetic cross. An individual with phenotype BbCc is mated with an individual like bbcc (or BBCC). As shown in Figure 6.7b, this allows an unambiguous phase determination. In humans, where we cannot control mating, we need to find family members whose particular inheritance pattern allows an unambiguous determination of the phase of markers to be determined. This is not always possible, and one must frequently make unprovable assumptions about recombination (or the lack of it) in the family in order to assess the phase.

Linkage. A statement that alleles at two loci, say B and C, tend to be coinherited. If linkage is complete they will be coinherited 100% of the time. In molecular terms, these alleles occur on loci that lie near to each other on the same DNA molecule.

Linkage group. A set of loci that in general tend to be inherited together. In effect, a linkage group is a chromosome, a set of genes on a single DNA molecule.

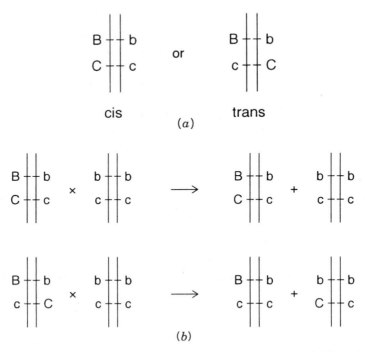

Figure 6.7 Phase of alleles at two loci on the same chromosome. *(a)* Definition of two possible phases, arbitrarily called cis and trans. *(b)* Determination of the phase in a particular individual by performing a genetic cross with an appropriately informative partner.

Unlinked loci. Two loci that show random cosegregation. Ordinarily, if an allele at one locus, A, is inherited, there is a 50% chance that an allele, B, at another unlinked locus will also be inherited. Loci on different (nonhomologous) chromosomes behave as unlinked. So do loci very far apart on the same large chromosome.

Meiotic recombination. A process, already described at the chromosomal level in Box 6.1, that makes the description of genetic inheritance much more complex. However, it is also the major event that allows us to map the location of genes. Simply put, meiotic recombination is the breakage and reunion of DNA molecules prior to their segregation to daughter cells during the process of gamete formation. Because of meiotic recombination, alleles at linked loci may occasionally fail to show co-segregation (Fig. 6.8a). In the human genome, meiotic recombination occurs at a frequency of about 1% for two loci spaced 1 Mb apart. The actual frequency observed between two loci is called Θ. We define a 1% recombination frequency as 1 cM (centiMorgan after the geneticist Morgan). Most human chromosomes are more than 100 cM in length; thus genes at opposite ends of these chromosomes behave as unlinked. Note that recombination is an obligatory event for each pair of homologs in meiosis. This implies that smaller chromosomes may have overall higher average levels of recombination.

Genetic map. An ordered set of loci with relative spacing (and order) determined from measured recombination frequencies. The simplest map construction would be based on pairwise recombination frequencies (Fig. 6.8b). However, more accurate maps can be made by considering the inheritance of multiple loci simultaneously. This allows for multiple crossovers in recombination, multiple DNA breaks and joins, to be considered. Frequently genetic maps are displayed as statements about the relative likelihood of the order shown. For each pair of adjacent loci, what is presented is the relative odds in favor of the order shown, say ABCD, as opposed to an inverted order,

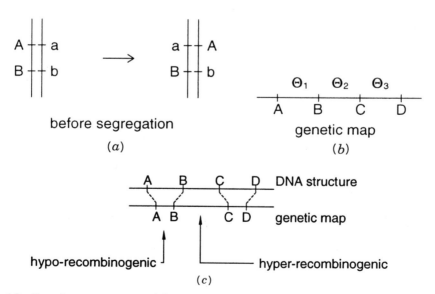

Figure 6.8 Genetic maps constructed from recombination frequencies between pairs of adjacent loci. *(a)* Effect of recombination on two adjacent loci before segregation. *(b)* Genetic map constructed from recombination frequencies. *(c)* Typical relationship between a genetic map and the underlying structure of the DNA, a physical map.

ACBD, given the available recombination data. In practice, the current human genetic map has an average spacing between markers of around 2 cM. The ultimate map we are likely to achieve in the absence of significant improvements in our analytical methods is about a 1 cM map (see Fig. 6.29c).

RELATIONSHIP BETWEEN THE PHYSICAL AND THE GENETIC MAPS

The order of loci must be the same on the genetic map and on its corresponding physical map. Both must ultimately reflect the DNA sequence of the chromosome. However, the genetic map is based on recombination frequencies, while the physical map is based directly on measurements of DNA structure. The meiotic recombination frequency is not uniform along most chromosomes. Regions are observed where recombination is much more frequent than average, hot spots, and where recombination is much less frequent than average, cold spots. Such regions lead to considerable metrical distortion when genetic and physical maps are laid side by side (Fig. 6.8c). We know very little about the properties of recombination hot spots. Attempts to narrow them down to short specific DNA sequence elements have not yet succeeded in mammalian samples. However, we do know that recombination hot spots can, themselves, be inherited alleles. This can lead to serious complications in trying to interpret genetic maps directly in terms of relative physical distances along the DNA.

Two other complications can lead to errors in the construction of genetic maps, and confuse the issue when these maps are compared directly with physical maps. If not properly identified and accounted for, both can lead to inconsistent evidence for gene order. Multiple crossovers can occur between a pair of DNAs strand during meiotic recombination. One schematic example of this is given in Figure 6.9, and the consequences for puta-

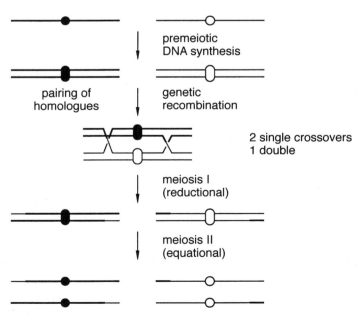

Figure 6.9 A more detailed view of meiotic recombination showing the association between pairs of homologous chromosomes after DNA synthesis, and the possibility of multiple crossovers before segregation.

multiple crossovers

Figure 6.10   Effect of a double crossover on the pattern of inheritance at three loci.

tive gene order are illustrated in Figure 6.10. Obviously multiple crossovers between very close loci are improbable, in general, but they are observed, particularly in regions where recombination hotspots abound. Such a double crossover makes more distant loci appear closer than they should.

The second genetic complication is gene conversion. Here information from one homologous chromosome is copied onto the corresponding region of another homolog. A schematic mechanism is given in Figure 6.11, and the consequences for gene order are shown in Figure 6.12. Gene conversion appears to be relatively frequent in yeast. Evidence for human or mammalian gene conversion is much more spotty, but it is generally believed that this process does play a significant role. Gene conversion can make a nearby marker appear to be far away, as shown in Figure 6.12. The exact biological functions of gene conversion remain to be clarified. It potentially forms a mechanism for very rapid evolution, since it allows a change in one copy of a gene to be spread among identical or nearly identical copies. Gene conversion is also believed to play a role in the sorting out of homologous chromosome pairs prior to crossing over, as described in Box 6.2.

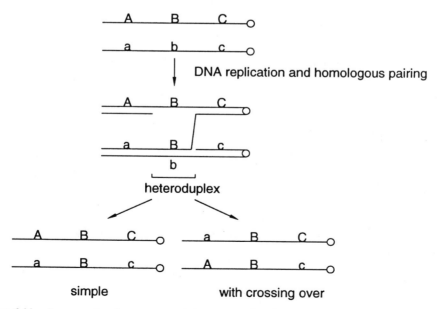

Figure 6.11   An example of gene conversion, an event that destroys the usual 1:1 segregation pattern seen in typical Mendelian inheritance.

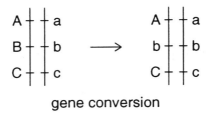

gene conversion

Figure 6.12 Effect of a gene conversion event on the pattern of inheritance at three loci.

BOX 6.2
TWO-STAGE MODEL OF RECOMBINATION

For recombination to occur, homologous chromosomes must pair up with their DNA sequences aligned. Studies in yeast have led to a rather complex model for this process. The complications occur because DNA sequence similarities or identities occur not only between the pairs of homologs but also between different chromosomes as a result of dispersed repeated DNA sequences (see Chapter 14) or dispersed gene families with similar or identical members. The recombination apparatus in bacteria, yeast, and presumably all higher organisms has the ability to catalyze sequence similarity searches. In these, a DNA duplex is nicked, and a single-stranded region is exposed and covered with protein. This is then used to scan duplex DNA by processes we still do not understand very well (see discussion of *rec*A protein in Chapter 14). When a close sequence match is found, the single strand can invade the corresponding duplex and displace its equivalent sequence there. Depending on what happens next, this can result either in a gene conversion event in which information is copied from one homolog to the other, a Holliday structure, which may eventually lead to strand rearrangement, or in simple displacement of the invading strand and restoration of the original DNA molecules.

Figure 6.13 illustrates some of the stages of these processes in a hypothetical example of a cell with two different homologous pairs of chromosomes. In meiosis each chromosome starts as a pair of identical sister chromatids linked at the centromere. Initial strand exchange (shown as dotted lines between the chromosomes) occurs both between homologs and between nonhomologs (Fig. 6.13a). Some gene conversion events may result at this stage. As the system is driven toward increasing amounts of strand exchange (in a manner we do not know), it is clearly much more likely that homologous pairing dominates (Fig. 6.13b). Finally, after suitable alignment is reached (again, judged by mechanisms that we have no current knowledge about), crossing-over events occur (Fig. 6.13c), and the homologs segregate to daughter cells (Fig. 6.13d). Each chromosome at that point consists of a pair of sister chromatids that are no longer identical because of the different gene conversion and crossing-over events they have experienced.

(continued)

BOX 6.2 *(Continued)*

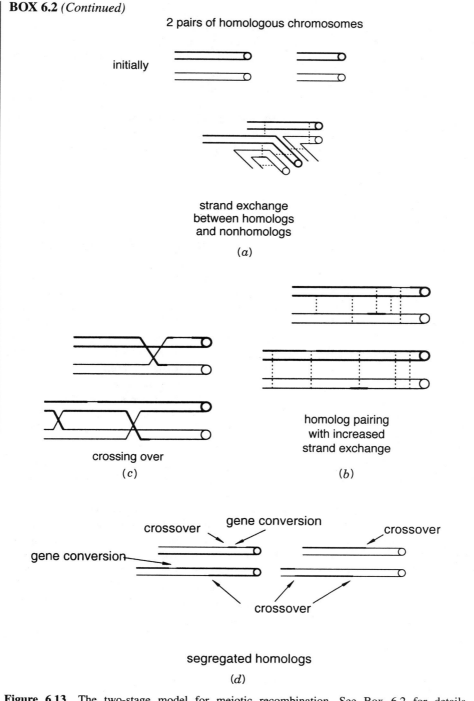

Figure 6.13 The two-stage model for meiotic recombination. See Box 6.2 for details. (Adapted from Roeder, 1992.)

To distinguish single and multiple crossover events, and gene conversion, large numbers of highly informative loci are very helpful. Several features of genetic loci make them particularly powerful for mapping. Ideally one has many different alleles, and these occur at reasonable frequencies in the population. Where this occurs, there is a very good chance that all four parental homologs are distinguishable at the locus because they each carry different alleles. A major thrust in human genetics has been the systematic collection of such highly informative loci. This will be discussed in a later section. It is also helpful to have many offspring in families under study and to have many generations of family members. Because it is so difficult to satisfy these conditions, human genetics is often rendered rather impotent compared with the genetics of more easily manipulatable experimental organisms.

POWER OF MOUSE GENETICS

Mice are among the smallest common mammals. They are relatively inexpensive to maintain, have large numbers of offspring, and mature quickly enough to allow many generations to be examined. However, this alone does not explain why mice form such a powerful genetic system. Several reasons exist that make the mouse the preeminent model for mammalian genetics. First, so much is already known about so many mouse genes, that an extensive genetic map already exists. Second, many highly inbred strains of laboratory mice exist. These tend to be homozygous at most alleles, and thus a description of their phenotype and genotype is relatively simple. Also gene transfer and knockout technology exist for mice. What is particularly useful is that rather distant inbred strains of mice can be interbred to give at least some fertile offspring. This property dramatically simplifies the construction of high-resolution genetic maps. Several different sets of inbred strains can be used. One of the earlier choices was *Mus musculus,* the common laboratory mouse, and *Mus spretus,* a distant cousin.

Figure 6.14 illustrates how crosses between *M. musculus* and *M. spretus* generate very useful genetic information. Because these mice are so different, the F1 offspring of a direct cross tend to be heterozygous at almost any locus examined. It turns out that the males that result from such a cross are sterile, but the females are fertile. When an F1 *spretus* × *musculus* female is bred with an *M. musculus* (a procedure called a *backcross*), in most regions of the genome the progeny either resemble wild type *M. musculus* homozygotes, or F1 *musculus* × *spretus* heterozygotes. However, every time a recombination event occurs, there is a switch between the homozygous pattern and the heterozygous pattern (Fig. 6.14*b*). Given the dense set of genetic markers available in these organisms, the location of many recombination events can be determined in each set of experimental animals. The result is that one develops and refines genetic maps extremely rapidly.

WEAKNESS OF HUMAN GENETICS

Humans, from a genetic standpoint, are a stark contrast with mice. We have a relatively long generation time; it precludes the simultaneous availability of large numbers of generations. Our families are small; most are far too small for effective genetics. Crosses cannot be controlled, and there are no inbred strains, only the occasional result of very limited inbreeding in particular cultures that promote such practices as marriages between

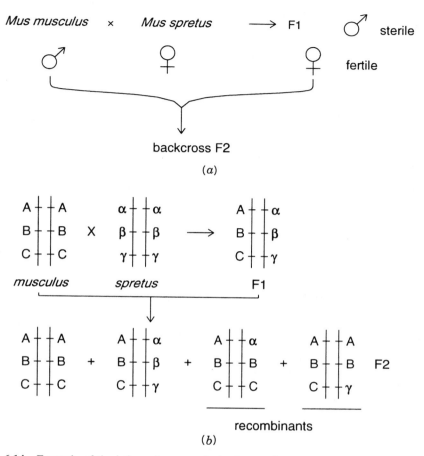

Figure 6.14 Example of the informativeness of a back cross between two distant mouse species *M. musculus* and *M. spretus*. *(a)* Design of the back cross. *(b)* Alleles seen for three loci in the parents and the first (F1) and typical second (F2) generation of offspring, with and without recombination.

cousins. For all of these reasons, the design and execution of prospective genetic experiments is impossible in the human. Instead, one must do retrospective genetic analysis on existing families. Ideally these will consist of family units where at lest three generations are accessible for study. Results from many different families usually must be pooled in order to have enough individuals segregating a trait of interest to allow a statistically significant test of hypotheses about the model for its inheritance. Usually that test is to ask if the trait is linked to any other known trait in the genome. This is a very tedious task, as we will illustrate. However, large numbers of ongoing studies use this approach because it is the only effective way we have to find a human gene location if the only available information we have is a disease phenotype. If there are animal models for the trait in question, or if one has a hint about the functional defect, one can sometimes cut short a search of the entire genome by focusing on candidate genes. However, these genes still must be examined in linkage studies with human markers because the arrangement of genes in humans and model organisms, while similar (Fig. 2.24), is not identical.

There are several additional major weaknesses that compromise the power of human genetics. In many cases our ability to evaluate the phenotype is imprecise. For example, in inherited diseases it is not at all uncommon to have imprecise or even incorrect diagnoses. These result in the equivalent of a recombination event as far as genetics is concerned, and a few such errors can often destroy the chances that a search for a gene will be successful. A second common problem is missing or uncooperative family members. In such cases the family tree, called a *pedigree,* is incomplete, and phase or other information about the inheritance of a disease trait, or a potential nearby marker, is lost. Homozygosity in key individuals is another frequent problem. As illustrated earlier, this makes it impossible to determine which parental homolog in the region of interest has been inherited. As genetic markers become denser and more informative, this problem is becoming less severe, but it is by no means uncommon yet.

The final problem that frequently plagues human genetic studies is mispaternity. This is relatively easy to discover by using the highly informative genetic markers currently available. Usually the true parent is not identified; this results in a missing family member for genetic studies.

LINKAGE ANALYSIS IGNORING RECOMBINATION

The statistical analysis of linked inheritance is the tool used for almost all genetic studies in the human. Here we introduce this approach, and the Bayesian statistics used to provide a quantitative evaluation of the pattern of inheritance, assuming for the moment that recombination does not occur. The result will be a test of whether two loci are linked, but there will be no information about how far away on the genetic map these linked loci are. The treatment in this and the two following sections follows closely a previous exposition by Eric Lander and David Botstein (1986).

Consider the simple family shown in Figure 6.15. This is in fact the simplest case that can be used to illustrate the basic features of linkage analysis. We deal with two loci with two alleles each: A and a, D and d. We assume that phenotypic analysis allows all the possible independent genotypes at these two loci to be distinguished. Thus all individuals can be typed as AA, Aa, or aa and as DD, Dd, or dd. In linkage studies we ask whether particular individuals tend to inherit alleles at the two loci independently or in common. Our simple family has two parents, one heterozygous at the two loci and one homozygous at both loci. There are two offspring; both are heterozygous at both loci. The issue at hand is to assess the statistical significance of these data to reveal whether or not the two loci at linked; that is, whether they are on the same chromosome.

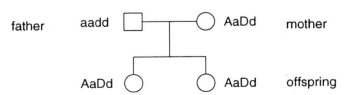

Figure 6.15 A typical family used to test the notion that two genetic loci A/a and D/d might be linked. (Adapted from Lander and Botstein, 1986). Circles show females, squares show males.

Suppose that the two loci are on different chromosomes. We then know the genotypes of the mother and the father unambiguously as shown in Figure 6.16. Independent segregation of alleles on different chromosomes leads to four possible genotypes for the offspring of these parents. A priori the probability of occurrence of each of these phenotypes should be equal, so each should occur with an expected frequency of $\frac{1}{4}$. Thus the a priori probability that the family in question should have both children with the same particular phenotype, AaDd is $\frac{1}{4} \times \frac{1}{4} = \frac{1}{16}$. We need to compare this probability, calculated from the hypothesis that the loci are unlinked, with the probability of seeing the same result if the loci are linked.

If the two loci are linked, they are on the same chromosome. In this case the genome of the homozygous father can be described unambiguously, but there are two possible phases for the mother. These are shown in Figure 6.17a. Unless we have some a priori knowledge about the mother's genotype, we must assume a priori that these two possible phases are equally probable. Thus there is a $\frac{1}{2}$ chance that the mother is cis and $\frac{1}{2}$ that she is trans. Since we are not allowing for the possibility of recombination, if the mother is trans, the probability that she would have an AaDd daughter is zero, since both the A and D alleles have to come from the mother in the family shown in Figure 6.15, and this would be impossible in the trans configuration where the A and D alleles are on different homologs.

If the mother is cis for the two loci, there are two possible genotypes for any offspring. These are shown in Figure 6.17b. In the absence of any intervening factors, the a priori probability of observing these genotypes should be equal. Thus the chance of seeing a child with the genotype AaDd under these circumstances is $\frac{1}{2}$. The chance of a family with two children both of whom are AaDd will be $\frac{1}{2} \times \frac{1}{2} = \frac{1}{4}$. However, since we have no a priori knowledge about the phase of the mother, we must average the chances of seeing the expected offspring across both possible phases. The overall odds of the observed family inheritance pattern is then $\frac{1}{2}(0) + \frac{1}{2}(\frac{1}{4}) = \frac{1}{8}$ if the two loci are linked.

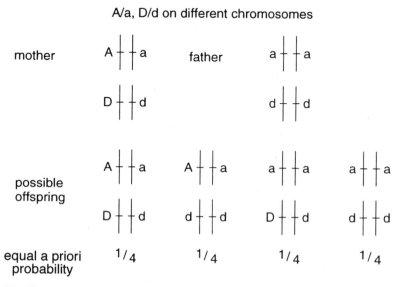

Figure 6.16 Genotypes of the parents and expected offspring if the two loci and unlinked, and there is no recombination.

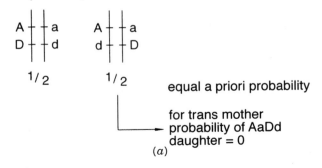

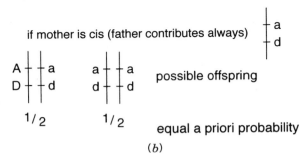

Figure 6.17 Inheritance patterns with linkage, but no recombination. *(a)* Possible maternal geno-types if the two loci are linked. *(b)* Possible offspring if the mother is cis. For the example consid-ered, other offspring genotypes will be seen only if the loci are not linked.

The odds ratio is a test of the likelihood that our hypothesis that linkage exists is cor-rect. This is the ratio of the calculated probabilities with and without linkage:

$$\text{odds ratio} = \frac{P_{\text{linked}}}{P_{\text{unlinked}}} = \frac{\frac{1}{8}}{\frac{1}{16}} = 2$$

In human genetics we frequently know the phase of an individual because this is available from data on other generations or other family members. In this case the odds ratio becomes

$$\text{odds ratio} = \frac{P_{\text{linked}}}{P_{\text{unlinked}}} = \frac{\frac{1}{4}}{\frac{1}{16}} = 4$$

The greater power of linkage analysis with known phase is readily apparent.

With or without known phase, the single family shown in Figure 6.15 provides a small amount of statistical evidence in favor of linkage. To strengthen (or contradict) this evidence, we need to pool data from many such families. The overall odds ratio then becomes

$$\frac{P_{\text{linked}}}{P_{\text{unlinked}}} = \text{odds}_1 \times \text{odds}_2 \times \text{odds}_3 \ldots$$

where the subscripts refer to different families under study. It is convenient mathematically to deal with a sum rather than a product of such data, and this is accomplished by taking the logarithm of both sides of the above equation. Since

$$\log(A \times B \times C \ . \ . \ .) = \log(A) + \log(B) + \log(C) + . \ . \ .$$

the result is

$$\text{LOD} = \log\left(\frac{P_{\text{linked}}}{P_{\text{unlinked}}}\right) = \log(\text{odds}_1) + \log(\text{odds}_2) + \log(\text{odds}_3) \ . \ . \ .$$

This is called a *LOD* (rhymes with cod) *score*. The LOD score is calculated from the data seen with a particular family or set of families. Some feeling for the number of individuals that must be examined for a LOD score to be statistically significant can be captured by the expected LOD score calculated for a particular inheritance model, called an *ELOD*. However, the inheritance model we use has to include the possibility of recombination to be realistic enough to represent actual data.

LINKAGE ANALYSIS WITH RECOMBINATION

Consider a pair of markers at loci that appear to be linked by available data. There are three possible cases to deal with

1. The markers are unlinked, but random segregation gives the appearance of linkage.
2. The markers are really linked.
3. The markers are linked, but recombination has disguised this linkage.

We will deal with two markers as in the previous case. Here, however, it simplifies matters if one of these is a locus where D is an allele of a disease gene that we are trying to find. A is an allele at another locus, and we are interested in testing the hypothesis that in a particular family A and D are linked. The chance that a recombination event occurs between the two loci in each meiosis is an unknown variable Θ. We need to calculate the odds in favor of linkage, for data from a particular family, as a function of Θ. Actual LOD(Θ) calculations are complex. To illustrate the considerations that go into such calculations, we will calculate the expected contribution of a single individual observed to inherit the disease allele D to the overall LOD score. This contribution is called the expected LOD or ELOD(Θ).

We will analyze the case where a parent is AaDd. Usually we are dealing with a relatively rare disease, and the other parent does not have the D allele. We assume for simplicity that the healthy parent also either has the a allele or some other allele that we can distinguish from A and a. We look only at offspring that are detected to carry the disease allele D. If the two loci are unlinked, the offspring inherit pairs of two different chromosomes carrying A or a and D or d at random, as shown in Figure 6.18a. We look only at offspring carrying D; thus there is a 0.5 probability that such an offspring will also inherit A.

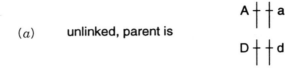

(a)   unlinked, parent is

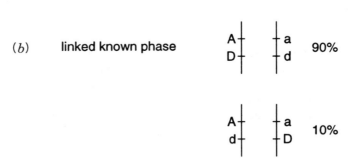

(b)   linked known phase

Figure 6.18   Analysis of the inheritance of a disease allele D and a possible linked allele A. *(a)* Possible parental contributions to an offspring if no linkage occurs. *(b)* Possible parental contributions to an offspring if the loci are linked but the recombination frequency between them is 0.1.

We will consider the case where two loci are linked and the phase in the parent is known to be cis (Fig. 6.18*b*). What we want to calculate is the effect of observing one child of this parent on the odds in favor of linkage of A and D. Suppose that Θ is 0.1. Such a 10% chance of recombination corresponds to an average distance of 10 Mb in the genome. This is near the maximum distance across which linkage is visible in the analysis of only two loci at once. We can calculate the chance of two outcomes:

1. Probability that a child with D inherits AD from the parent is 0.9.
2. Probability that a child with D inherits aD from the parent is 0.1.

The contribution of case 1 to the expected LOD score is

$$\log\left(\frac{0.9}{0.5}\right)$$

which is the ratio of the odds of seeing A and D with linkage versus without linkage. The contribution of case 2 to the expected LOD score is

$$\log\left(\frac{0.1}{0.5}\right)$$

which is the ratio of the odds seeing aD with linkage to the odds of seeing aD without linkage.

Thus the average contribution from one child to the ELOD is the sum of these two cases weighted by their expected frequency. Since recombination across 10 cM occurs only 10% of the time,

$$\text{ELOD}(0.1) = 0.9 \log\left(\frac{0.9}{0.5}\right) + 0.1 \log\left(\frac{0.1}{0.5}\right)$$

$$\text{ELOD}(0.1) = +0.23 - 0.07 = 0.16$$

Thus observation of cosegregation of A and D adds to the probability of linkage, while observation of separation of A and D subtracts from the evidence for linkage.

What we need to do is develop the tools to assess the statistical significance of a particular ELOD score. Since some markers will appear to cosegregate by chance in any study with a relatively small number of affected individuals, there is always a chance of seeing a significantly positive LOD score, simply because of the random fluctuations. A near consensus in human genetics is that an observed LOD of 3.0 or higher is required before the probability of purely accidental linkage can be reduced to the point where few errors are made. For the example just described, the number of individuals segregating D with unambiguous pedigrees that would have to be combined to generate a LOD score of 3.0 can be estimated as 3/0.16 = 18. For common inherited diseases this is not a problem, but for very rare diseases it may be extremely difficult to find 18 genetically informative individuals for a particular marker with an unambiguous diagnosis.

Note that several constraints apply to the linkage analysis described above. One must have access to a parent with known phase between A and D. The marker A to be tested for linkage must have useful heterozygosity. The diagnosis of D must be unambiguous in all the individuals tested. Note that failing to diagnose an individual who is carrying D (a false negative) does not hurt the analysis, since in this case the individual and the parent are not scored. However, misclassifying an individual as carrying D instead of d (a false positive) causes serious problems because it will weaken the evidence about which alleles at other loci are cosegregating with D.

INTERVAL MAPPING

Once a genetic map is available for a region of interest, the process of linkage analysis can be made more powerful by examining several markers simultaneously. We will consider the simplest possible case, illustrated in Figure 6.19a. As in the previous discussion of simple linkage analysis, we will calculate the average contribution of the LOD score from a single, informative individual inheriting a disease allele D. We wish to test a region of the genome containing two linked loci with markers A and B to see if the disease allele D lies between them or is unlinked. (Here we ignore the case that it might be linked to A and B but lie outside them rather than between them.)

Suppose that the loci containing A and B are 20 cM apart. This is a reasonable model for how human genetic maps are used in average regions of the genome. $\Theta_{AB} = 0.2$. First we calculate the possible contributions from a parent carrying D to a child, also carrying D, if there is no linkage between A and B with D (Fig. 6.19b). Since A and B are on the same chromosome, D, if unlinked, they must lie on a different chromosome. Assuming that the parent is heterozygous and informative at all these loci, there are four possible contributions from the parent to the child (Fig. 6.19c).

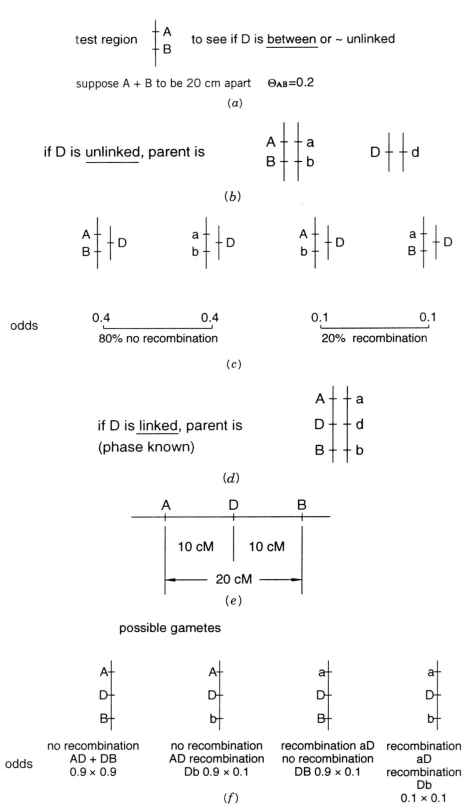

test region to see if D is <u>between</u> or ~ unlinked

suppose A + B to be 20 cm apart Θ_{AB}=0.2

(a)

if D is <u>unlinked</u>, parent is

(b)

odds

0.4   0.4

80% no recombination

0.1   0.1

20% recombination

(c)

if D is <u>linked</u>, parent is
(phase known)

(d)

A   D   B

10 cM | 10 cM

20 cM

(e)

possible gametes

odds

no recombination
AD + DB
0.9 × 0.9

no recombination
AD recombination
Db 0.9 × 0.1

recombination aD
no recombination
DB 0.9 × 0.1

recombination
aD
recombination
Db
0.1 × 0.1

(f)

If no recombination between A and B occurs (80% probability for markers 20 cM apart), the child will either inherit ABD (0.4 odds) or abD (0.4 odds). If recombination between A and B occurs (20% probability), the child will inherit either AbD (0.1 odds) or aBD (0.1 odds).

If D is linked and located between A and B, assuming the phase of the parent is known, the two homologous chromosomes of the parent carry alleles ADB and adb, as shown in Figure 6.19d. In principle, D may lie anywhere between A and B and the actual position of D is a variable that must be included in the calculations. Here we will consider the simple case where D lies midway between A and B. Assuming that the recombination frequency is uniform in this region of the chromosome, we can then place D 10 cM from A and 10 cM from B (Fig. 6.19e). There are four possible sets of alleles that can be passed from this parent to a child who inherits D (Fig. 6.19f). These are as follows:

ADB: resulting from no recombination between A and D, and no recombination between D and B (odds are 0.9×0.9).

ADb: resulting from no recombination between A and D but recombination has occurred between D and B (odds are 0.9×0.1).

aDB: recombination has occurred between A and D, but no recombination has occurred between D abd B (odds are 0.1×0.9).

aDb (a double crossover event): recombination has occurred both between A and D and between D and B (odds are 0.1×0.1).

Thus the same four possible genotypes can arise either with or without linkage. However, the odds of particular genotypes vary considerably in the two cases. For the four possible offspring:

Alleles	ADB	ADb	aDB	aDb
Odds (linked/unlinked)	0.81/0.4	0.09/0.1	0.09/0.1	0.01/0.4

The ELOD for a single statistically representative child can be calculated from these results by realizing that if there is linkage, the probabilities of seeing the four patterns of alleles are 0.81, 0.09, 0.09, and 0.01, respectively. Thus the ELOD is given by

$$
\text{ELOD}(0.2) = 0.81 \log\left(\frac{0.81}{0.4}\right) + 0.09 \log\left(\frac{0.09}{0.1}\right) + 0.09 \log\left(\frac{0.09}{0.1}\right)
$$
$$
+ 0.01 \log\left(\frac{0.01}{0.4}\right)
$$
$$
\text{ELOD}(0.2) = +0.25 - 0.004 - 0.004 - 0.02 = 0.23
$$

Figure 6.19 Interval mapping to test the hypothesis that a disease allele D is located equidistant between two linked markers A and B, separated by 20 cM. *(a)* Map of the test region. *(b)* Parental chromosomes if D is unlinked to A and B. *(c)* Possible chromosomes inherited by an offspring carrying the disease allele in the absence of linkage. *(d)* Parental chromosomes if D lies between A and B. *(e)* Map location assumed for D for the example calculated in the text. *(f)* Possible parental contributions to an offspring inheriting the disease allele.

Note that this ELOD is larger in the case of interval mapping than in the simple case of linkage analysis we considered earlier. The number of informative individuals that would have to be examined to achieve a LOD score of 3 would be 3/0.23 = 14.

FINDING GENES BY GENETIC MAPPING

What is done, in practice, is to repeat the kinds of calculations previously described with all possible values of Θ as a variable using actual genotype data from real families. For simple linkage analysis the sorts of results obtained are shown schematically in Figure 6.20. These yield the expected LOD score as a function of Θ. The critical results are the maximum LOD value and the confidence limits on possible values of Θ. With interval mapping, the results are more complex, but the basic kind of information obtained is similar, as shown by the example in Figure 6.21. For details, see Ott (1991) and Lalouel and White (1966).

In a typical case, no a priori information exists about the putative location of a gene of interest. To have a reasonable chance of finding it, one must test the hypothesis that it lies near (or between) any of about 150 informative markers. This will subdivide the genome into intervals spaced about 20 cM apart. Each marker must be tested with a sufficient number of informative individuals to achieve a LOD score of 3.0 or higher if that particular marker is linked to the gene. With present technology this search is often carried out one marker and one individual at a time. It is easy to estimate that around 150 markers $\times$ 40 to 60 individuals (parents and offspring) must be tested in ideal cases where parental phase is known and markers are very informative. If the analysis is carried out by ordinary Southern blotting (Chapter 3) of DNA bands 6000 to 12,000 gel electrophoresis lanes have to be examined by hybridization to afford a reasonable chance of finding a gene, and this is an ideal case! Schemes have recently been described that can reduce the workload by an order of magnitude through the use of pools of samples (Churchill et al., 1993; see also Chapter 9 for examples of the power of pooling).

If a LOD score of 3.0 or greater is achieved, there is a reasonable chance that the correct location of the disease gene of interest has been found. What is usually done is to celebrate, publish a preliminary report, and fend off overoptimistic members of the press or families segregating the disease of interest who confuse the first sighting of a gene location with the identification of the actual disease gene itself. Knowing the location of a disease gene does provide improved diagnostics for the disease but, initially, only in those families where the phase of the disease allele and nearby markers is known.

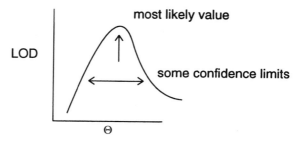

Figure 6.20   LOD score for linkage of two genes, with a particular recombination frequency, that would be seen in a typical set of family inheritance data.

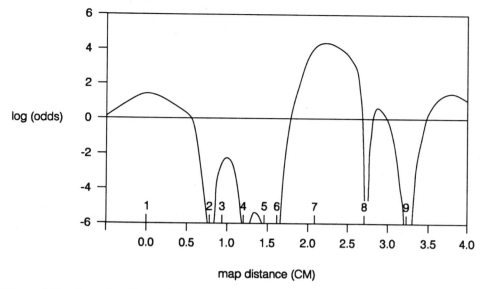

Figure 6.21 Example of interval mapping data. Shown is the expected LOD score [log(odds)] as a function of the possible location of a gene within the interval mapped by two known linked genes. (Adapted from Leppert et al., 1987).

Furthermore, at a 10 cM distance, the amount of recombination between the marker and the disease allele in each meiosis is still 10%, so the accuracy of any genetic testing is quite limited. More accurate approaches are outlined in Box 6.3 and Box 6.4.

BOX 6.3
MULTIPOINT MAPPING

More accurate genetic maps can be constructed by considering all the loci simultaneously rather than just dealing with pairs of loci. In this case what one establishes, primarily, is the order of the loci and the relative odds in favor of that order based on the sum of all the available data. In principle, one can write down all possible genetic maps and calculate the relative likelihood of each being correct in the context of the available data. In practice, it is usually quite tedious to do this. Instead, as shown in Figure 6.22, one usually plots the most likely map, and gives the relative odds that the order of each successive pair of markers is reversed from the true order.

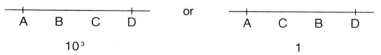

Figure 6.22 Typical map data by multipoint analysis. Shown are the relative odds in favor of two orderings of the markers A, B, C, and D.

The next goals are to strengthen the evidence for linkage and narrow the putative location of the gene. Additional examples of affected individuals can be examined using only the closest known markers. If this increases the LOD score, there is little doubt that the gene location has been correctly identified. Once the interval containing the gene is known, one can look for additional markers in the region of interest. Various methods to find polymorphic markers in selected DNA regions will be described later. These methods are quite powerful so long as the region is actually polymorphic in the population. Note that once the approximate gene location is found, the markers used to refine that location need not be informative in all patients in the sample. What is key is to find particular individuals who demonstrate recombination between the disease gene and nearby markers. Until linkage was established such individuals actually weakened the search because there was no way of knowing a priori that they were recombinants, and thus, as shown in earlier examples, they subtracted from the expected LOD score. Once the gene is known to be nearby, such individuals can be recognized as recombinants and properly scored as shown by the example in Figure 6.23. Just two informative individuals with recombination events defined by their haplotypes (patterns of alleles on a single chromosome) are sufficient to pinpoint the location of the disease gene, barring the unlikely occurrence of a gene conversion or double crossover.

MOVING FROM WEAK LINKAGE CLOSER TO A GENE

Failure to find a linked marker in an initial test does not mean that no marker is linked to the gene. A disease gene must lie somewhere in the genome. A possibility is that the model for inheritance used in the linkage study was wrong. One must consider dominant and recessive inheritance as well more complex cases where multiple alleles or even multiple genes are involved. It is very tempting in cases where the maximum LOD score obtained is less than 3.0 to review individual families contributing to the LOD score and ask if the score can be improved by dropping some of the families. This implicitly challenges the diagnosis in these families or presumes that the disease is heterogeneous—that it is influenced by other factors in addition to the particular gene in question.

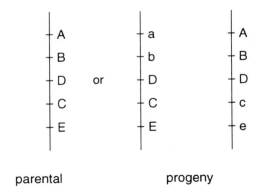

Figure 6.23 Examples of two recombinant genotypes seen from a parent with known phase. Once the disease allele D is known to lie in this region, the genotype of the two recombinants restricts the possible location of the disease gene to between markers b and c.

This is a very dangerous practice, since if one starts with a sufficient number of families, it will almost always to possible to achieve an alluring LOD score by selectively choosing among them. Clearly the appropriate statistical tests must be employed to discount the resulting LOD score against such selective manipulation of the data. The real issue is not whether one can increase a LOD score by dropping a family with a negative contribution. The issue is whether the magnitude of the increase in LOD is sufficient to justify the additional parameterization implicit in dropping this family. A much safer procedure is to collect more families and try additional markers near the ones that have already shown a hint of linkage if not yet compelling evidence for linkage. When this has been done, some LODs of 2.0 eventually have produced the desired gene; others have faded into oblivion.

Eventually genetic linkage studies may narrow down the location of a gene to a 2 cM region. However, in such an interval of the genome, there may be a single gene or more than 80. It is very difficult to use conventional linkage analysis to narrow the location further. The available families are likely to have only a limited number of recombination events in the region of interest because they represent just a few generations, which means any recombinations seen must have occurred recently. A 2 cM localization means that already 50 informative meioses have been found. It is usually not efficient to keep gathering more families at this point, although it is efficient to keep trying to find additional informative markers, since these can narrow down the location of any recombination events.

LINKAGE DISEQUILIBRIUM

In fortunate cases, a variant on linkage analysis can be used to home in on the likely location of a disease gene once it has been assigned to a mapped region of a chromosome. Suppose that most affected individuals have the same disease allele D. This is the case, for example, with the Huntington's disease individuals who live near Lake Maricaibo in Venezuela; it is also the case with individuals affected with sickle cell disease, and with most individuals of northern European descent afflicted with severe cystic fibrosis. In such cases it is possible that the disease is the result of a founder effect: all affected individuals have inherited the same disease allele-carrying chromosome from a common progenitor. (When no evidence for a single disease allele exists, but phenotypic variation in the disease is evident, one can try to subtype the disease by severity, age of onset, particular symptoms, and test the presumption that, for this subtype, a founder effect may exist.)

If a disease allele arose once by mutation on a single chromosome, it will be created in the context of a particular haplotype (Fig. 6.24). The chromosome that first carries the disease will have a particular set of polymorphic markers. It will have a particular genetic background. As this chromosome is passed through many generations of offspring, it

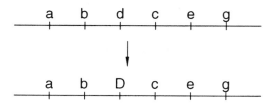

Figure 6.24   Generation of a disease allele by a mutation on a founder haplotype sets the stage for linkage disequilibrium.

will suffer frequent meiotic recombination events. These will tend to blur the memory of the original haplotype of the chromosome; they will average out the original genetic background with the general distribution of markers in the human population. However, those markers very close to the disease gene will tend, more likely than average, to retain the haplotype of the original chromosome because, as the distance to the disease gene shrinks, it becomes less likely that recombination events will have occurred in this particular location.

Humans are an outbred population. Most alleles were established when the species was established, and a sufficient number of generations has passed since then that frequent recombination events have occurred between any pair of neighboring loci resolved on our genetic maps. For this reason the distribution of particular haplotypes in neighboring loci in the population (as opposed to particular families) should be close to random. Consider the case shown in Figure 6.25, for two neighboring loci with two alleles each. Within the population, the frequencies X of the alleles at a particular locus must sum to 1.0.

$$X_a + X_A = 1.0 \quad X_b + X_B = 1.0$$

The frequencies of particular haplotypes, f, should be given by simple binomial statistics:

$$f_{AB} = X_A X_B \quad f_{Ab} = X_A X_b \quad f_{aB} = X_a X_B \quad f_{ab} = X_a X_b$$

Deviations from these results, measured, for example, as f_{AB} observed $- f_{AB}$ calculated, are evidence for linkage disequilibrium, and they indicate that the individuals examined are not a random sample of the population. Note, however, that deviation of allele frequencies from those expected by binomial statistics may have other causes besides genetic linkage. Deviations can reflect improper sampling of the population, or they can reflect actual functional association between specific alleles. The latter process could occur, for example, if the protein products of the two genes in question actually interacted biochemically. (For further discussion see Ott, 1991.)

To search for a gene by linkage disequilibrium, one does not examine families segregating a disease allele D. Instead, one looks across a broad spectrum of the population for unrelated individuals who have the disease allele D. If evidence for linkage disequilibrium is found, it reflects recombinations along the chromosome all the way back in time to the original founder. Since this may extend back hundreds of years, more than ten generations may be involved, and thus the number of recombination events viewed will be much greater than possible with any contemporary family. In the case of linkage disequilibrium, we expect to see the general results shown in Figure 6.26. There will be a gradient of increasing deviation from equilibrium as the neighborhood of the disease gene is reached because of the diminishing likelihood of recombination events occurring in an ever-shrinking region.

Figure 6.25 Possible haplotypes in a two-allele system used to examine whether loci are at equilibrium.

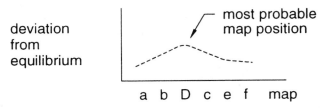

Figure 6.26 Gradient of linkage disequilibrium seen near a disease allele in a case where a founder effect occurred.

COMPLICATIONS IN LINKAGE DISEQUILIBRIUM AND GENETIC MAPS IN GENERAL

The human genome is a potential minefield of uncharted genetic events, hidden re-arrangements, new mutations, and genetic heterogeneity. Failure to see linkage disequilibrium near a gene does not mean that the gene is far away. Two of the most plausible potential complications are the existence of more than one founder or the existence of a significant fraction of alleles in the population that have arisen by new mutations. For example, in the case of dominant lethal diseases (those in which, nominally, the affected individuals have no offspring), one must expect that most disease alleles will be new mutations. Multiple founders can occur in distinct geographical populations, and they can be tested for by subdividing the linkage disequilibrium analysis accordingly. However, our increasingly mobile population, at least in developed countries, will make such analyses increasingly difficult.

Two other reasonable explanations for a failure to see linkage disequilibrium near a disease gene of interest are shown in Figure 6.27. The first of these is the possible presence of recombination hot spots. If the recombination pattern in the region of interest is punctate, then an even gradient of linkage disequilibrium will not be seen. Instead, markers that lie within a pair of hot spots will appear to be in disequilibrium, while those that lie on opposite sides of a hot spot will appear to have equilibrated. The occurrence of any disequilibrium in the region is presumptive evidence that a disease gene is there, since this is the basis for selection of the particular set of individuals to be examined. However, the complex pattern of allele statistics in the region will make it difficult to narrow in on the location of the disease gene.

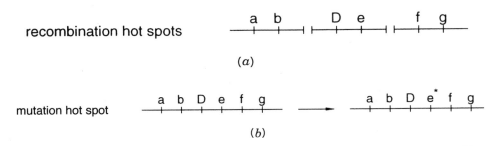

Figure 6.27 complications that can obscure evidence for linkage disequilibrium. *(a)* Recombination hotspots near the disease gene. *(b)* Mutation hot spots near the disease gene.

A second potential source of confusion is the presence of mutation hotspots. These are quite common in the human genome. For example, the sequence CpG is quite mutagenic in those regions of the genome where the C is methylated, as discussed in Chapter 1. When mutation hotspots are present, these alleles appear to have equilibrated with their neighbors, while more distant pairs of alleles may still show deviations from equilibrium. As in the case of recombination hotspots, disequilibrium indicates that one has not sampled the population randomly. This is presumptive evidence for a disease gene nearby, but mutation hotspots weaken the power of the disequilibrium approach to actually focus in on the location of the desired gene.

DISTORTIONS IN THE GENETIC MAP

We have already discussed briefly the occurrence of recombination hot spots and their deleterious effect on attempts to find genes by linkage disequilibrium. Some hot spots are inherited; in the mouse Major Histocompatibility Complex (MHC), a set of genes that regulates immune response, a hot spot allele has been found that raises the local frequency of recombination by a hundredfold. All of the recombination events caused by this hot spot have been mapped within the second intron of the E^b gene, 4.3 kb in size. While we are not sure what has caused this hot spot, the region has been sequenced, and one peculiarity is the occurrence of four sequences with 9/11 bases equal to a consensus sequence TGGAAATCCCC. Such sequences have also been found in regions associated with other recombination hot spots.

The genetic map of the human, and other organisms is not uniform. Recombination is generally higher near the telomers and lower near centromeres. The map is strikingly different in males and females—that is, meiosis in males and females appears to display a very different pattern of recombination hot spots. A typical example is shown for a selected region of human chromosome 1 in Figure 6.28. Note that some regions that have short genetic distances in the female have long distances in the male, and vice versa. Genetic linkage analysis is more powerful in regions where recombination is prevalent because, the more recombinants per Mb, the more finely the genetic data will serve to subdivide the region. In general, genetic maps based on female meioses are considerably longer than those based on male meioses. This is summarized in Table 6.1. A frequent practice is to pool data and show a sex-averaged genetic map. It is not very clear that this is a reasonable thing to do. Instead, it would seem that once a region of interest has been selected, meioses should be chosen from either the female or the male depending on which set produces the most expanded and informative map of the region. At present it does not appear that most workers pay much attention to this.

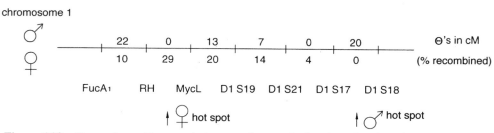

Figure 6.28 Comparison of low-resolution genetic maps in female and male meiosis. Shown is a portion of the map of human chromosome 1.

TABLE 6.1   Genetic and Physical Map Lengths of the Human Chromosomes

Chromosome[a]	Physical Size (Mb)	Genetic Size					
		Sex Averaged	cM/Mb	Female	cM/Mb	Male	cM/Mb
1	263	292.7	1.11	358.2	1.36	220.3	0.84
2	255	277.0	1.09	324.8	1.27	210.6	0.83
3	214	233.0	1.09	269.3	1.26	182.6	0.85
4	203	212.2	1.05	270.7	1.33	157.2	0.77
5	194	197.6	1.02	242.1	1.25	147.2	0.76
6	183	201.1	1.10	265.0	1.45	135.2	0.74
7	171	184.0	1.08	187.2	1.09	178.1	1.04
8	155	166.4	1.07	221.0	1.43	113.1	0.73
9	145	166.5	1.15	194.5	1.34	138.5	0.96
10	144	181.7	1.26	209.7	1.46	146.1	1.01
11	144	156.1	1.08	180.0	1.25	121.9	0.85
12	143	169.1	1.18	211.8	1.48	126.2	0.88
13q	98	117.5	1.20	132.3	1.35	97.2	0.99
14q	93	128.6	1.38	154.4	1.66	103.6	1.11
15q	89	110.2	1.24	131.4	1.48	91.7	1.03
16	98	130.8	1.33	169.1	1.73	98.5	1.01
17	92	128.7	1.40	145.4	1.58	104.0	1.13
18	85	123.8	1.46	151.3	1.78	92.7	1.09
19	67	109.9	1.64	115.0	1.72	98.0	1.46
20	72	96.5	1.34	120.3	1.67	73.3	1.02
21q	39	59.6	1.53	70.6	1.81	46.8	1.20
22q	43	58.1	1.35	74.7	1.74	46.9	1.09
X	164	198.1	1.21	198.1	1.21		

Source: Adapted from Dib et al. (1996).

[a]Only the long arm is shown for five chromosomes in which the short arm consists largely of tandemly repeating ribosomal DNA. The approximate full length of these chromosomes are given in parentheses.

CURRENT STATE OF THE HUMAN GENETIC MAP

Several major efforts to make genetic maps of the human genome have occurred during the past few years, and recent emphasis has been on merging these efforts to forge consensus maps. A few examples of the status of these maps several years ago are given in Figure 6.29a and b. These are sex-averaged maps because they have a larger density of markers. The ideal framework map will have markers spaced uniformly at distances around 3 to 5 cM, since this is the most efficient sort of map to use to try to find additional markers or disease genes with current technology. On this basis the map shown for chromosome 21 is quite mature, while the map for chromosome 20, a much less studied chromosome, is still relatively immature. Chromosome 21 is completely covered (except for the rDNA-containing short arm) with relatively uniform markers; in contrast, chromosome 20 has several regions where the genetic resolution is much worse than 5 cM. The current status of the maps can only be summarized highly schematically, as shown in Figure 6.29c. Here 2335 positions defined by 5264 markers are plotted. On average, this is almost 1 position per cM.

20

147 cM

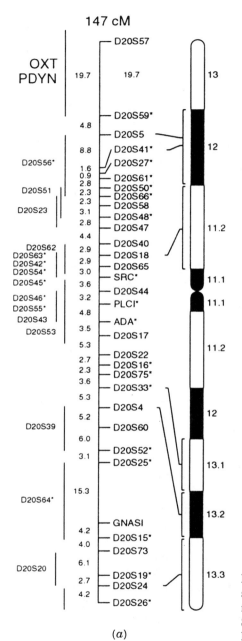

(a)

21

67 cM

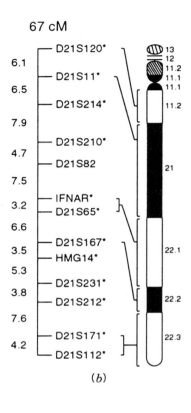

(b)

Figure 6.29 Genetic maps of human chromosomes. *(a,b)* Status of sex-averaged maps of chromosomes 20 and 21 several years ago. *(c)* Schematic summary of current genetic map (from Dib et al., 1996).

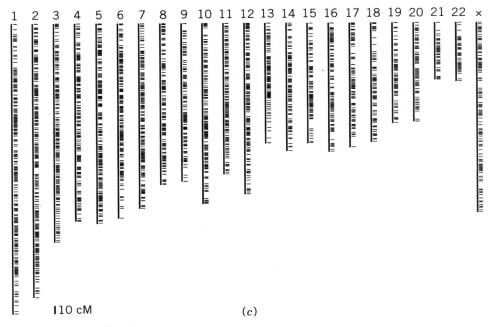

Figure 6.29   *(Continued)*

The physical lengths of the human chromosomes are estimated to range from 50 Mb for chromosome 21, the smallest, to 263 Mb for chromosome 1, the largest. Table 6.1 summarizes these values and compares them with the genetic lengths, seen separately from male and female meioses. Several interesting generalizations emerge from an inspection of Table 6.1. The average recombination frequency per unit length (cM per Mb) varies over a broad range from 0.73 to 1.46 for male meioses and 1.09 to 1.81 for female meioses. Smaller chromosomes tend to have proportionally greater genetic lengths, but this effect is by no means uniform. Recombination along the X chromosome (seen only in females) is markedly suppressed compared with autosomal recombination.

Leaving the details of the genetic map aside, the current version is an extremely useful tool for finding genes on almost every chromosome. As this map is used on a broader range of individuals, we should start to be able to pinpoint potential recombination hot spots and explore whether these are common throughout the human population or whether some or all of them are allelic variants. The present map is already a landmark accomplishment in human biology.

GENETICS IN THE PSEUDOAUTOSOMAL REGION

Meiotic recombination in the male produces a special situation. The male has only one X and one Y chromosome. These must segregate properly to daughter cells. They must pair with each other in meiotic metaphase, and by analogy with the autosomes, one expects that meiotic recombination will be an obligatory event in proper chromosome segregation.

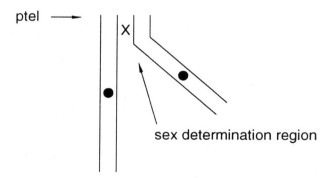

Figure 6.30   Pairing between the short arms of the X and Y chromosomes in male meiosis.

The problem this raises is that the X and Y are very different in size, and much of these two chromosomes appear to share little homology. How then does recombination occur between them? It turns out that there are at least two regions of the Y chromosome that share close enough of the X to allow recombination. These are called *pseudoautosomal regions,* for reasons that will become apparent momentarily.

The major psuedoautosomal region of the Y chromosome is located at the tip of the short arm. It is about 2.5 Mb in length and corresponds closely in DNA sequence with the 2.5 Mb short arm terminus of the X chromosome. These two regions are observed to pair up during meiosis, and recombination presumably must occur in this region (Fig. 6.30). If we imagine an average of 0.5 crossovers per cell division, this is a very high recombination rate indeed compared to a typical autosomal region.

Since the Y chromosome confers a male phenotype, somewhere on this chromosome must lie a gene or genes responsible for male sex determination. We know that this region lies below the pseudoautomal boundary, a place about 2.5 Mb from the short term telomere. Below this boundary, genes appear to be sex linked because, by definition, these genes must not be able to separate away from the sex determination region in meiosis. A genetic map of the pseudoautomal region of the Y chromosome is shown in Figure 6.31. There is a gradient of recombination probability across the region. Near the p telomere, all genes will show 50% recombination with the sex-determining region because of the obligatory recombination event during meiosis. Thus these genes appear to be autosomal, even though they are located on a sex chromosome, because like genes on autosomes they

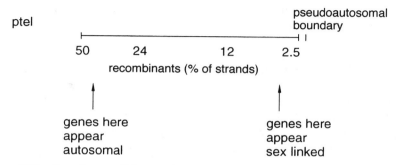

Figure 6.31   Genetic map of the pseudoautosomal region of the X and Y chromosomes.

have a 50% probability of being inherited with each sex. As one nears the pseudoautosomal boundary, the recombination frequency of genes with the sex determination region approaches a more normal value, and these genes appear to be almost completely sex linked.

Recently data have been obtained that indicate that a second significant pseudoautosomal region may lie at the extreme ends of the long arms of the X and Y chromosomes. About 400 kb of DNA in these regions appears to consist of homologous sequences, and a 2% recombination frequency in male meioses between two highly informative loci in these regions has been observed. The significance of DNA pairing and exchange in this region for the overall mechanism of male meiosis is not yet known. It is also of interest to examine whether in female meiosis any or all of the X chromosome regions that are homologous to the pseudoautosomal region of the Y chromosome show anomalous recombination frequencies.

BOX 6.4
MAPPING FUNCTIONS: ACCOUNTING
FOR MULTIPLE RECOMBINATIONS

If two markers are not close, there is a significant chance that multiple DNA crossovers may occur between them in a particular meiosis. What one observes experimentally is the net probability of recombination. A more accurate measure of genetic distance will be the average number of crossovers that has occurred between the two markers. We need to correct for the occurrence of multiple crossovers in order to compute the expected number of crossovers from the observed recombination frequency. This is done by using a mapping function.

The various possible recombination events for zero, one, and two crossovers are illustrated schematically in Figure 6.32. In each case we are interested in correlating the observed number of recombinations between two distant markers and the actual average number of crossovers among the DNA strands. In all the examples discussed below it is important to realize that any crossovers that occur between sister chromatids (identical copies of the parental homologs) have no effect on the final numerical results. The simplest case occurs where there are no crossovers between the markers; clearly in this case there is no recombination between the markers. Next, consider the case where there is a single crossover between two different homologs (Fig. 6.32b). The net result is a 0.5 probability of recombination because half of the sister chromatids will have been involved in the crossover and half will not have been.

When two crossovers occur between the markers, the results are much more complex. Three different cases are illustrated in Figure 6.32c. Two single crossovers can occur, each between a different set of sister chromatids. The net result, shown in the figure, is that all the gametes show recombination between the markers; the recombination frequency is 1.0. There are four discrete ways in which these crossovers can occur. Alternatively, the two crossovers may occur between the same set of sister chromatids. This is a double-crossover event. The net result is no observed recombination between the distant markers. There are four discrete ways in which a double crossover can occur. Note that the net result of the two general cases we have considered thus far is 0.5 recombinant per crossover.

(continued)

BOX 6.4 *(Continued)*

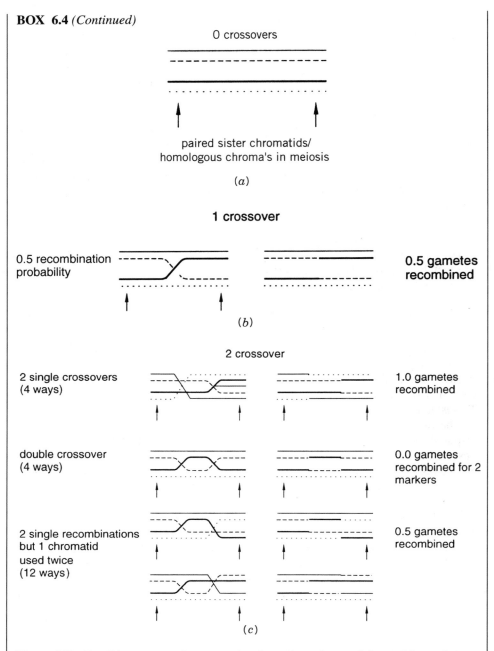

Figure 6.32   Possible crossovers between pairs of genetic markers and the resulting meiotic recombination frequencies that would be observed between two markers (vertical arrows) flanking the region. *(a)* No crossovers. *(b)* One crossover. *(c)* Various ways in which two crossovers can occur.

(continued)

BOX 6.4 *(Continued)*

The final case we need to consider for two crossovers are those in which three of the four paired sister chromatids are involved; two are used once and one is used twice. There are 12 possible ways in which this can occur. Two representative examples are shown in Figure 6.32c. Each results in half of the DNA strands showing a recombination event between distant markers, and half showing no evidence for such an event. Thus, on average, this case yields an observed recombination frequency of 0.5. This is also the average for all the cases we have considered except the case where no crossovers have occurred at all. It turns out that it is possible to generalize this argument to any number of crossovers. The observed recombination frequency, Θ_{obs}, is

$$\Theta_{obs} = 0.5 \, (1 - P_0)$$

where P_0 is the fraction of meioses in which no crossovers occur between a particular pair of markers.

It is a reasonable approximation to suppose that the number of crossovers between two markers will be given by a Poisson distribution, where μ represents the mean number of crossovers that take place in an interval the size of the spacing between the markers. The frequency of n crossovers predicted by the Poisson distribution is

$$P_n = \frac{\mu^n \exp(-\mu)}{n!}$$

and the frequency of zero crossovers is just $P_0 = \exp(-\mu)$. Using this, we can rewrite

$$\Theta_{obs} = 0.5 \, (1 - \exp(-\mu))$$

This can easily be rearranged to give μ as a function of Θ_{obs}:

$$\mu = -\ln(1 - 2\,\Theta_{obs})$$

The parameter μ is the true measure of mapping distance corrected for multiple crossovers. It is the desired mapping function.

WHY GENETICS NEEDS DNA ANALYSIS

In almost all of the preceding discussion, we assumed the ability to determine genotype uniquely from phenotype. We allowed that we could always find heterozygous markers when we needed them, and that there was never any ambiguity in determining the true genotype from the phenotype. This is an ideal situation never quite achieved in practice, but we can come very close to it by the use of DNA sequences as genetic markers.

The simplest DNA marker in common use is a two allele polymorphism—a single inherited base pair difference. This is shown schematically in Figure 6.33. From the DNA sequence it is possible to distinguish the two homozygotes from the heterozygote. Earlier, in Chapter 4, we demonstrated how this can be accomplished using allele-specific PCR.

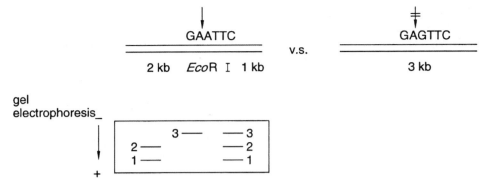

Figure 6.33 Example of a simple RFLP and how it is analyzed by gel electrophoresis and Southern blotting. Such an allele can also be analyzed by PCR or allele-specific PCR, as described in Chapter 4.

An alternative approach, less general but with great historical precedent, is the examination of restriction fragment length polymorphisms (RFLPs). Such a case is illustrated in Figure 6.33. Where a single-base polymorphism adventitiously lies within the sequence recognized and cleaved by a restriction endonuclease, the polymorphic sequence results in a polymorphic cleavage pattern. All three possible genotypes are distinguishable from the pattern of DNA fragments seen in an appropriate double digest. If this is analyzed by Southern hybridization, it is helpful to have a DNA probe that samples both sides of the polymorphic restriction site, since this prevents confusion from other possible polymorphisms in the region of interest.

The difficulty with two-allele systems is that there are many cases where they will not be informative in a family linkage study, since too many of the family members will be homozygotes or other noninformative genotypes. These problems are rendered less serious when more alleles are available. For example, two single-site polymorphisms in a region combine to generate a four-allele system. If these occur at restriction sites, the alleles are both sites cut, one site cut, the other site cut, and no sites cut. Most of the time the resulting DNA lengths will all be distinguishable.

The ideal DNA marker will have a great many easily distinguished alleles. In practice, the most useful markers have turned out to be minisatellites or variable number tandem repeated sequences (VNTRs). An example is a block like $(AAAG)_n$ situated between two single-copy DNA sequences. This is analyzed by PCR from the two single-copy flanking regions or by hybridization using a probe from one of the single-copy regions (Figure 6.34). The alleles correspond to the number of repeats. There are a large number of possible alleles. VNTRs are quite prevalent in the human genome. Many of them have a large

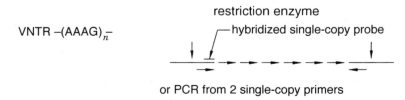

Figure 6.34 Use of PCR to analyze length variations in typical VNTR. In genetics such markers are extremely informative and can be easily found and analyzed.

number of alleles in the actual human population, and thus they are very powerful genetic markers. See, for example, Box 6.5. Most individuals are heterozygous for these alleles, and most parents have different alleles. Thus the particular homologous chromosomes inherited by an offspring can usually be determined unambiguously.

BOX 6.5
A HIGHLY INFORMATIVE POLYMORPHIC MARKER

A particularly elegant example of the power of VNTRs is a probe described by Alec Jeffries. This single-copy probe which detects a nearby minisatellite was originally called MS32 by Jeffries (MS means minisatellite). When its location was mapped on the genome, it was found to lie on chromosome 1 and was assigned by official designation D1S8. Here D refers to the fact that the marker is a DNA sequence, 1 means that it is on chromosome 1, S means that it is a single copy DNA sequence, and 8 means that this was the eighth such probe assigned to chromosome 1. The power of D1S8 is illustrated in Figure 6.35. The minisatellite contains a *Hin*f I cleavage site within each 29 base repeat. In addition some repeats contain an internal *Hae* II cleav-

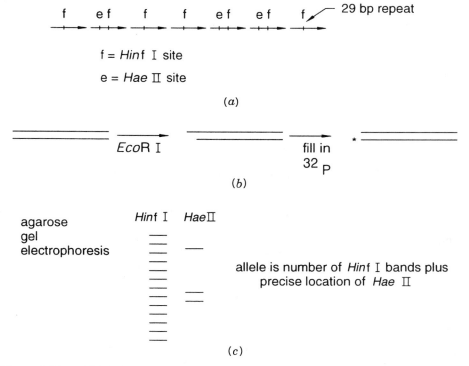

Figure 6.35 A highly informative genetic marker D1S8 that can be used for personal identity determinations. *(a)* Repeating DNA structure of the marker. *(b)* PCR production of probes to analyze the marker. *(c)* Typical results observed when the marker is analyzed after separate, partial *Hae* II and *Hin*f I digestions.

(continued)

BOX 6.5 *(Continued)*

age site. When appropriately chosen PCR primers are used, one can amplify the region containing the repeat and radiolabel just one side of it. Then a partial digestion with the restriction enzyme *Hae* II or *Hin*f I generates a series of DNA bands whose sizes reflect all of the enzyme-cutting sites within the repeat. This reveals not only the number of repeats but also the locations of the specific *Hae* II sites. When this information is combined, it turns out that there are more than 10^{70} possible alleles of this sequence. Almost every member of the human population (exempting identical twins) would be expected to have a different genotype here; thus this probe is an ideal one not only for genetic analysis but also for personal identification.

By comparing the alleles of D1S8 in males and in sperm samples, the mutation rage of this VNTR has been estimated. It is extremely high, about 10^{-3} per meiosis or 10^5 higher than the average rate expected within the human genome. The mechanism believed to be responsible for this very high mutation rate can be inferred from a detailed analysis of the mutant alleles. It turns out that almost all of the mutations arise from interallelic events, as shown in Figure 6.36. These include possible slippage of the DNA during DNA synthesis, and unequal sister chromatid exchange. Only 6% of the observed mutations arise from ordinary meiotic recombination events between homologous chromosomes.

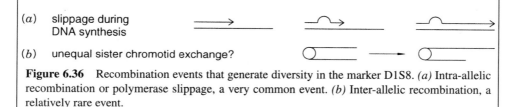

(a) slippage during DNA synthesis

(b) unequal sister chromotid exchange?

Figure 6.36   Recombination events that generate diversity in the marker D1S8. *(a)* Intra-allelic recombination or polymerase slippage, a very common event. *(b)* Inter-allelic recombination, a relatively rare event.

DETECTION OF HOMOZYGOUS REGIONS

Because the human species is highly outbred, homozygous regions are rare. Such regions can be found, however, by traditional methods or by some of the fairly novel methods that will be described in Chapter 13. Homozygous regions are very useful both in the diagnosis of cancer and in certain types of genetic mapping. The significance of homozygous regions in cancer is shown in Figure 6.37. Oncogenes are ordinary cellular genes, or foreign genes that under appropriate circumstances can lead to uncontrolled cell growth, that is, to cancer. Quite a few oncogenes have been found to be recessive alleles, ordinarily silenced in the heterozygous

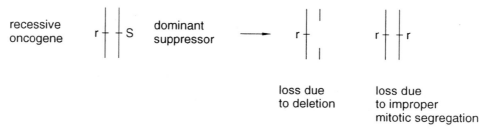

recessive oncogene

dominant suppressor

loss due to deletion

loss due to improper mitotic segregation

Figure 6.37   Generation of homozygous DNA regions in cancer.

state by the presence of a homologous dominant suppressor gene (called a tumor suppressor gene). Consider what happens, when that suppressor is lost either due to mutation, deletion, or improper mitotic segregation so that a daughter cell receives two copies of the homologous chromosome with the recessive oncogene on it. These are all rare events, but once they occur, the resulting cells have a growth advantage; there is a selection in their favor. In the case of a loss due to deletion, the resulting cells will be hemizygous in the region containing the oncogene. Strictly speaking, they will show only single alleles for all polymorphisms in this region, and thus a zone of apparent homozygosity will mark the location of the oncogene. In the case of improper mitotic segregation, the entire chromosome carrying the oncogene may be homozygous, and once this is discovered, it is useful for diagnostic purposes but less useful for finding the oncogene, since an entire chromosome is still a very large target to search.

A second application of homozygous regions is the genetic technique known as homozygosity mapping. This is illustrated in Figure 6.38. It is useful in those relatively rare

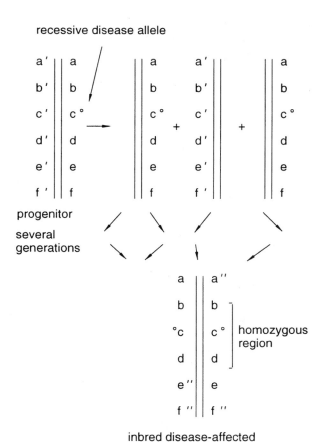

Figure 6.38 Example of homozygosity mapping. Inbreeding within a family can reveal an ancestral disease allele by a pattern of homozygous alleles surrounding it.

BOX 6.6
IMPRINTING

For a small number of genes in the human and mouse, the phenotype caused by an allele depends on the parent who transmitted it. Some alleles are preferentially expressed from the gene coded for by the father's gamete (for example, insulinlike growth factor 2 and small ribonucleoprotein peptide n [Snrpn] in the mouse). Others are expressed preferentially from the mother's gamete (such as insulinlike growth factor 2 receptor and H19 in the mouse). These effects can be seen strikingly when one parent or the other contains a deletion in an imprinted region.

DNA methylation is believed to play the major role in marking which parent of origin will dominate. Generally methylation silences gene expression, so methylation of a region in a particular parent will silence the effect of that parent. The process of imprinting is complicated. Imprinting may not be felt the same in all tissues in which the gene is expressed. The imprint must be established in the gametes; it may be maintained during embryogenesis and in somatic adult tissues, but must be erased in the early germline so that a new imprint can be re-established in the gametes of the individual who is now ready to transmit it to the next generation. There are well-known examples of human diseases that show a parent of origin phenotype arising from differences in methylation. For example, Prader-Willi syndrome is used to describe the disease phenotype when a damaged or deleted human analog of the mouse Snrpn gene is inherited from the father, whereas Angelman syndrome describes the rather different phenotype when the deleted allele is inherited from the mother. For further reading see Razin and Cedar (1994).

---

cases where there is significant inbreeding in the human population. Consider a progenitor carrying a recessive disease allele. Several generations later come two individuals who have inherited this allele from the original progenitor mate and have offspring together. One or more offspring receive two copies of the disease allele, and as a result they are affected by the inherited disease. Because these offspring have inherited the allele from the same original chromosome, they are likely to share not only that allele but also nearby alleles. While the chromosome will have suffered several recombination events during the generations that have ensued between the progenitor and the affected individuals, most of the original chromosome has presumably remained intact. This approach will be useful in finding disease genes that segregate in inbred populations, and it has the advantage that only affected individuals need to be examined to find the zone of homozygosity that marks the approximate location of the disease gene.

A third application of homozygous regions is their frequent occurrence in the rather remarkable genetic phenomenon called imprinting. Here, as described in Box 6.6, it matters from which parent a gene is inherited.

SOURCES AND ADDITIONAL READINGS

Churchill, G. A., Giovannoni, J. J., and Tanksley, S. D. 1993 Pooled-samplings makes high-resolution mapping practical with DNA markers. *Proceedings of the National Academy of Sciences USA* 90: 16–20.

Dib, C., Fauré, S., Fizames, C., Samson, D., et al. 1996. A comprehensive genetic map of the human genome based on 5,264 microsatellites. *Nature* 380: 152–154.

Dietrich, W. F., Miller, J., Steen, R., Merchant, M. A., et al. 1996. A comprehensive genetic map of the mouse genome. *Nature* 380: 149–152.

Donis-Keller, H., Green, P., Helms, C., et al. 1987. A genetic linkage map of the human genome. *Cell* 51: 319–337.

Lalouel, J.-M., and White, R. L. 1996. Analysis of genetic linkage. In Rimoin, D. L., Connor, J. M., and Pyeritz, R. E., eds. *Principles and Practice of Medical Genetics.* New York: Churchill Livingstone, pp. 111–125.

Lander, E., and Botstein, D. 1986. Mapping complex genetic traits in humans: New methods using a complete RFLP linkage map. *Cold Spring Harbor Symposium* 51: 1–15.

Lander, E., and Botstein, D. 1987. Homozygosity mapping: A way to map human recessive traits with the DNA of inbred children. *Science* 236: 1567–1570.

Leppart, M., Dobbs, M., Scambler, P. et al. 1987. The gene for familial polyposis coli maps to the long arm of chromosome 5. *Science* 238: 1411–1413.

Lindahl K. F. 1991. His and hers recombinational hotspots. *Trends in Genetics* 7: 273–276.

Nicholls, R. D., Knoll, J. H. M., Butler, M. G., Karam, S., and Lalande, M. 1989. Genetic imprinting suggested by maternal heterdisomy in non-delection Prader-Willi syndrome. *Nature* 342: 281-286.

Ott, J. 1991. *Analysis of Human Genetic Linkage*, rev. ed. Baltimore: John Hopkins University Press.

Razin, A. and Cedar, H. 1994. DNA methylation and genomic imprinting. *Cell* 77: 473–476.

Roeder, G. S. 1990. Chromosome synapsis and genetic recombination: their roles in meiotic chromosome segregation. *Trends in Genetics* 6: 385–389.

Shuler, G. D., Boguski, M. S., Stewart, E. A., Stein, L. D., et al. 1996. A gene map of the human genome. *Science* 275: 540–546.

Swain, J. L., Stewart, T. A., and Leber, P. 1987. Parental legacy determines methylation and expression of an autosomal transgene: A molecular mechanism for parental imprinting. *Cell* 50: 719-727.

7 Cytogenetics and Pseudogenetics

WHY GENETICS IS INSUFFICIENT

The previous chapter showed the potential power of linkage analysis in locating human disease genes and other inherited traits. However, as demonstrated in the chapter, genetic analysis by current methods is extremely tedious and inefficient. Genetics is the court of last resort when no more direct way to locate a gene of interest is available. Frequently one already has a DNA marker believed to be of interest in the search for a particular gene, or just of interest as a potential tool for higher-resolution genome mapping. In such cases it is almost always useful to pinpoint the approximate location of this marker within the genome as rapidly and efficiently as possible. This is the realm of cytogenetic and pseudogenetic methods. While some of these have potentially very high resolution, most often they are used at relatively low resolution to provide the first evidence for the chromosomal and subchromosomal location of a DNA marker of interest. Unlike genetics, which can work with just a phenotype, most cytogenetics and pseudogenetics are best accomplished by direct hybridization using a cloned DNA probe or by PCR.

SOMATIC CELL GENETICS

In Chapter 2 we described the availability of human-rodent hybrid cells that maintain the full complement of mouse or hamster chromosomes but lose most of their human complement. Such cells provide a resource for assigning the chromosomal location of DNA probes or PCR products. In the simplest case a panel of 24 cell lines can be used, each one containing only a single human chromosome. In practice, it is more efficient to use cell lines that contain pools of human chromosomes. Then simple binary logic (presence or absence of a signal) allows the chromosomal location of a probe to be inferred with a smaller number of hybridization experiments or PCR amplifications. In Chapter 9 some of the principles for constructing pools will be described. However, the general power of this approach is indicated by the example in Figure 7.1. The major disadvantage of using pools is that they are prone to errors if the probe in question is not a true single-copy DNA sequence but, in fact, derives from a gene family with representatives present on more than one human chromosome. In this case the apparent assignment derived from the use of pools may be totally erroneous. A second problem with the use of hybrid cells, in general, is the possibility that a given human DNA probe might cross-hybridize with rodent sequences. For this reason it is preferable to measure the hybridization of the probe not by a dot blot but by a Southern blot after a suitable restriction enzyme digestion. The reason for this is that rodents and humans are sufficiently diverged that although a similar DNA sequence may be present, there is a good chance that it will lie on a different-sized DNA restriction fragment. This may allow the true human-specific hybridization to be distinguished from the rodent cross-hybridization.

Chromosome	Base 3	← Right digit →			← Center digit →			← Left digit →		
		Pool A = 0	Pool B = 1	Pool C = 2	Pool D = 0	Pool E = 1	Pool F = 2	Pool G = 0	Pool H = 1	Pool I = 2
1	001		+		+			+		
2	002			+	+			+		
3	010	+				+		+		
4	011		+			+		+		
5	012			+		+		+		
6	020	+					+	+		
7	021		+				+	+		
8	022			+			+	+		
9	100	+			+				+	
10	101		+		+				+	
11	102			+	+				+	
12	110	+				+			+	
13	111		+			+			+	
14	112			+		+			+	
15	120	+					+		+	
16	121		+				+		+	
17	122			+			+		+	
18	200	+			+					+
19	201		+		+					+
20	202			+	+					+
21	210	+				+				+
22	211		+			+				+
X (23)	212			+		+				+
Y (24)	220	+					+			+

Examples:

samples	pool analysis (left, center, right)	conclusions
mixture 5 + 8	0, 1 and 2, 2	=> 012, 022 = 5 + 8
mixture 10 + 22	1 and 2, 0 and 1, 1	=> 101, 211 = 10 + 22
		or 111, 201 = 13 + 19

Figure 7.1 Assignment of a cloned DNA probe to a chromosome by hybridization against a panel of cell lines containing limited sets of human chromosomes in a rodent background. This hypothetical example is based on a base three pooling (see Chapter 9) which is quite efficient compared with randomly selected pools. If the probe is derived from a single chromosome, it can be uniquely assigned. If it contains material that hybridizes to two or more chromosomes, in some cases all of the chromosomes involved can be identified, but in most cases the results will be ambiguous as shown by the example at the bottom of the figure.

An alternative way to circumvent the problem of rodent cross-hybridization in assigning DNA probes to chromosomes is to use flow-sorted human chromosomes, preferably chromosomes that have been sorted from human-rodent hybrid cells so that the purified human fraction has very little contamination from other human chromosomes (Chapter 2). This is a very powerful approach; its major limitation is the current scarcity of flow-sorted human chromosomal DNA.

SUBCHROMOSOMAL MAPPING PANELS

Frequently it is possible to assemble sets of cell lines containing fragments of only one chromosome of interest. The occurrences of a DNA probe sequence in a subset of these cell lines can be used to assign the probe to a specific region of the chromosome. The accuracy of that assignment is determined by the accuracy with which the particular chromosome fragments contained in each cell line are known. This can vary quite considerably, but in general this approach is quite a powerful and popular one. A simple example is shown in Figure 7.2. Here three cell lines are used to divide a chromosome into three regions. The key aspect of these cell lines are breakpoints that determine which chromosome segments are present and which are absent.

A more complex example of cell lines suitable for regional assignment is given in Figure 7.3. Here the properties of a chromosome 21 mapping panel are illustrated. It is evident that some regions of the chromosome are divided by the use of this panel into very high resolution zones while others are defined much less precisely. In the case shown in Figure 7.3, human cell lines were assembled from available patient material: individuals with translocations or chromosomal deletions. Then these chromosomes were moved by cell fusion into rodent lines to simplify hybridization analyses. Because no systematic method was used to assemble the panel, the resulting breakpoint distribution is very uneven.

If a selectable marker exists near the end of a chromosome, it is possible to assemble a very orderly and convenient mapping panel. Assuming the selection can be applied in a hybrid cell, what is done is to irradiate the cell line lightly with X rays so as to cause an average of less than one break per chromosome, and then grow the cells in culture and allow broken chromosomes to heal. Since the selection applies to only one end of the human chromosome, there will be a tendency to keep chromosomes containing that end and lose pieces without that end. The result is a fine, ordered subdivision of the chromosome, as shown in Figure 7.4. This resource is exceptionally useful for mapping the location of cloned probes because the results are usually unambiguous. The pattern of hybridization should be a step function and the point where positive signal disappears indicates the most distal possible location of the probe.

A few caveats are in order concerning the use of cell lines that derive from broken or rearranged human chromosomes. Broken chromosome are always healed by fusion with other chromosomes (or by acquisition of telomeres). For example, cell lines described as 21q- are missing the long arm of chromosome 21. Frequently such cell lines have other human chromosomes present that have been quietly ignored or forgotten. This is not a problem when a single-copy probe is used, assuming that this probe is already known to

Figure 7.2 A simple example of regional assignment of a DNA sample by hybridization to three cell lines containing incomplete copies of the chromosome of interest. The two break points in the chromosomes are used to divide the chromosome into three intervals.

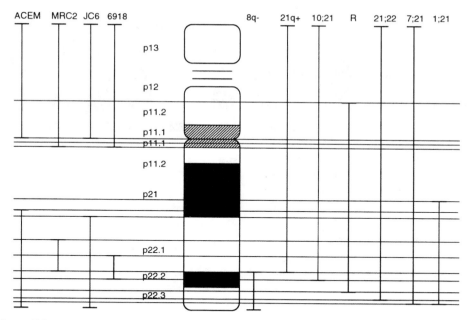

Figure 7.3 A mapping panel of hybrid cell lines used to assign probe locations on human chromosome 21. Vertical lines indicate the human chromosome content of each cell line. Note the unevenness of the resulting intervals (horizontal lines).

be located on the chromosome of interest, and the goal is just narrowing down its subchromosomal location. However, the problem is much more serious if the probe of interest comes from a gene family or contains repeated sequences. The actual chromosome content of a set of cell lines described as chromosome 21 hybrids is illustrated in Figure 7.5. Many of these lines, in fact, contain other human chromosomes. Cross-hybridization to rodent background is always a potentially serious additional complication when working with hybrid cells. For this reason it is best to do Southern blots or PCR analysis rather than dot blots, as we already explained earlier in the chapter.

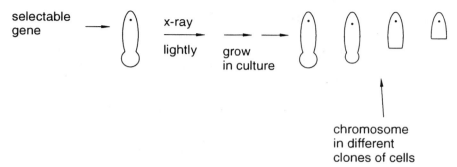

Figure 7.4 A method for selection of broken chromosomes that can be used to generate a much more even mapping panel than the one shown in Figure 7.3.

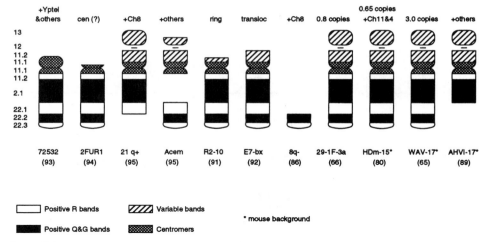

Figure 7.5 Actual chromosome content of some of the cell lines used to produce the mapping panel summarized in Figure 7.3. Note that many contain other human chromosomes besides number 21.

RADIATION HYBRIDS

It is possible to make hybrid cells containing small fragments of human chromosomes. These cell lines represent, essentially, a way of cloning small continuous stretches of human DNA. The method was originally described by S. Goss and H. Harris, and later rediscovered and elaborated by D. Cox and R. Meyers (Cox et al., 1990). The basic technique for producing these cells, called *radiation hybrids,* is shown schematically in Figure 7.6. The starting point is a hybrid cell line containing only a single human chromosome in a Chinese hamster ovary (CHO) cell with a wild type hypoxanthine ribosyl transferase (hprt) gene. This cell line is subjected to a very high X-ray dose, 8000 rads, which is sufficient to break every chromosome into five pieces on average. After irradiation the fragments are allowed to heal, and the resulting cells are fused with an hprt$^-$CHO line, one in which an inactive allele of the hprt gene is present. Cell clones that have received and retained hprt$^+$ DNA are selected and maintained by growth on a particular set of conditions called *HAT medium,* which require an active hprt gene. Some of these clones are recipients of both CHO and human DNA. Note that no human-specific selection has been used to focus the choice of hybrid cells on any particular region of the human chromosome in question. Thus the population of hybrids should represent a random sampling of that chromosome if there is no intrinsic selection method operative.

Selection for an hprt$^+$ phenotype ensures that all of the cell lines maintained will have been recipients of some foreign DNA. Those that have human DNA present can be detected by hybridization with human-specific repeated sequences. Then these lines are allowed to grow in the absence of any selection for human markers. This will lead to random loss of human sequences as time progresses. The cells are continually screened for the presence of human markers until 30% to 60% of the hybrids contain a particular marker of interest. At this point the radiation hybrids contain quite a few discrete fragments of human DNA integrated between hamster DNA segments.

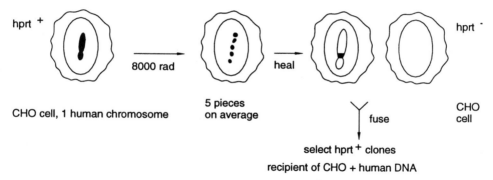

hprt $^+$

8000 rad

heal

hprt $^\cdot$

CHO cell, 1 human chromosome

5 pieces on average

fuse

CHO cell

select hprt $^+$ clones

recipient of CHO + human DNA

Figure 7.6   Steps in the production of a set of radiation hybrid cell lines. Note that there is no selection for any of the human DNA fragments, so they will be lost progressively as the cells are grown.

They are now used to look at the statistics of co-retention of pairs of human DNA markers. The basic idea is illustrated below, for three markers on a chromosome:

————————A —— B————————————C——

Even when the chromosome is fragmented into five pieces, there is a good chance that A and B will still be found on the same DNA piece; it is much more likely that C will be on a different piece. Thus cell lines that have lost A are also likely to have lost B, but they are less likely to have lost C. By establishing the presence or absence of particular markers, we gain information about the relative proximity of the markers because nearby markers tend to be co-retained.

Here we will examine the statistics of co-retention of marker pairs. There is a basic problem with this approach. If only two markers at a time are considered, it turns out that there are more variables than constraints, and approximations have to be used to analyze the results. With more than two simultaneous markers, the mathematics becomes more robust but only at a cost of much greater complexity. For didactic purposes we will analyze the two-marker case; anyone with a serious interest in pursuing this further should be aware that in practice, one must consider four or more markers at once to avoid the additional assumptions used later in this section. The parameters we must consider are illustrated below. The two markers A and B reside either on the same DNA fragment or on separate fragments as shown by the horizontal dashed lines. These fragments will have individual probabilities of retention, as indicated.

————————A————————B————————   retention probability P_{AB}

————————A————   ————B————   retention probabilities P_A, P_B

The probability that a break has occurred between A and B in a particular irradiation treatment is given by the additional parameter ϕ. All four parameters P_A, P_B, P_{AB}, and ϕ have values between 0 and 1.

What is done experimentally is to examine a set of hybrid cell clones and to ask in each whether markers A and B are present. The fraction of them that have both markers present is f_{AB}, the fractions with only A or only B are f_A and f_B, respectively, and the fraction that have retained neither marker is f_0. These four observables can be related to the four parameters by the following equations:

$$f_{AB} = P_{AB}(1 - \phi) + \phi P_A P_B$$

The first term denotes cells that have maintained A and B on an unbroken piece of DNA; the second term considers those cells that have suffered a break between A and B but have subsequently retained both separate DNA pieces.

$$f_A = \phi P_A(1 - P_B)$$
$$f_B = \phi P_B(1 - P_A)$$

The top equation indicates cells that have had a break between A and B and kept only A; the bottom equation indicates cells with a similar break, but these have retained only B.

$$f_0 = (1 - P_{AB})(1 - \phi) + \phi(1 - P_A)(1 - P_B)$$

The first term indicates cells that originally contained A and B on a continuous DNA piece but subsequently lost this piece (probability $1 - P_{AB}$); the second term indicates cells that originally had a break between A and B and subsequently lost both of these DNA pieces.

The four types of cells represent all possibilities;

$$f_{AB} + f_A + f_B + f_0 = 1$$

This means that we have only three independent observables, but there are four independent unknown parameters. The simplest way around this dilemma would be to assume that the probability of retention of individual markers was equal, for example, that $P_A = P_B$. However, available experimental data contradict this. Presumably some markers lie in regions that because of chromosome structural elements like centromeres or telomers, or metabolically active genes, convey a relative advantage or disadvantage for retention of their region. The approximation actually used by Cox and Myers to allow analysis of their data is to assume that P_A and P_B are related to the total amount of A and B retained, respectively:

$$P_A = f_{AB} + f_A$$
$$P_B = f_{AB} + f_B$$

Upon inspection this assumption can be shown to be algebraically false, but in practice, it seems to work well enough to allow sensible analysis of existing experimental data.

The easiest way to use the two remaining independent observables is to perform the sum indicated below:

$$f_A + f_B = \phi P_A(1 - P_B) + \phi P_B(1 - P_A)$$

This can be rearranged to yield a value for ϕ, which is the actual parameter that should be directly related to the inverse distance between the markers A and B:

$$\phi = \frac{f_A + f_B}{P_A + P_B - 2\,P_A\,P_B}$$

Here the parameters P_A and P_B are evaluated from measured values of f_{AB}, f_A, and f_B as described above.

Note that in using radiation hybrids, one can score any DNA markers A and B whether they are just physical pieces of DNA or whether they are inherited traits that can somehow be detected in the hybrid cell lines. However, the resulting radiation hybrid map, based on values of ϕ, is expected to be a physical map and not a genetic map. This is because ϕ is related to the probability of X-ray breakage, which is usually considered to be a property of the DNA itself and not in any way related to meiotic recombination events. In most studies to date, ϕ has been assumed to be relatively uniform. This means that in addition to marker order, estimates for DNA distances between markers were inferred from measured values of ϕ.

An example of a relatively mature radiation hybrid map, for the long arm of human chromosome 21, is shown in Figure 7.7. The units of the map are in centirays (cR), where 1 cR is a distance that corresponds to 1% breakage at the X-radiation dosage used. The density of markers on this map is impressively high. However, the distances in this map are not as uniform as previously thought. As shown in Table 7.1, the radiation hybrid map is considerably elongated at both the telomeric and centromeric edges of the chromosome arm, relative to the true physical distances revealed by a restriction map of the chromosome. It is not known at present whether these discrepancies arose because of the oversimplifications used to construct the map (discussed above) or because, as has been shown in several studies, the probability of breakage or subsequent repair is nonuniform along the DNA. A radiation hybrid map now exists for the entire human genome. It is compared to the genetic map and to physical maps in Table 7.2.

SINGLE-SPERM PCR

A human male will never have a million progeny, but he will have an essentially unlimited number of sperm cells. Each will display the particular meiotic recombination events that generated its haploid set of chromosomes. These cells are easily sampled, but in

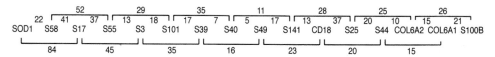

Figure 7.7 A radiation hybrid map of part of human chromosome 21. The units of the map are in centirays (cR). The map shows the order of probes and the distance between each pair. The telomere is to the right, near S100B. Distances for adjacent markers are shown between the markers, and for nearest neighbor pairs above and below the map. (Adapted from Burmeister et al., 1994.)

TABLE 7.1   Comparison of the Human Chromosome 21 *Not* I Genomic Map with the Radiation Hybrid Map

Marker Pairs	Physical Distance, kb[a]	cR[b]	kb/cR[a]
D21S16–D21S11, 1	5300 ± 2400	87	36 ± 16
D21S11, 1–D21S12	7300 ± 2200	140	52 ± 16
D21S12–SOD	6065 ± 2495	66	92 ± 38
SOD–D21S58	1490 ± 480	22	68 ± 22
D21S58–D21S17	2310 ± 1040	39	59 ± 26
D21S17–D21S55	3860 ± 1600	35	110 ± 46
D21S55–D21S39	3700 ± 1350	43	86 ± 31
D21S39–D21S141, 25	3690 ± 660	40	91 ± 16
D21S151, 25–COL6A	2330 ± 540	54	43 ± 10

Source: From Wang and Smith (1994).

[a]The uncertainty shown for each distance is the maximum possible range. The distance given is the mean of that range. The uncertainty shown for each ratio is the maximum range, considering only uncertainty in the placement of markers on the restriction map and ignoring any uncertainty in the radiation hybrid map.

[b]Taken from the data of Cox et al. (1990) and Burmeister et al. (1991). These data were obtained at 8000 rads.

TABLE 7.2   Characteristics of the Human Radiation Hybrid Map

Chromosome	Physical Length (Mb)	Average Relative Metrics	
		cM/cR[a]	kb/cR[a]
1	263	0.20	197
2	255	0.21	225
3	214	0.27	233
4	203	0.28	256
5	194	0.29	272
6	183	0.24	243
7	171	0.23	229
8	155	0.26	271
9	145	0.38	305
10	144	0.26	253
11	144	0.30	270
12	143	0.21	234
13q	98	0.22	179
14q	93	0.34	208
15q	89	0.36	203
16	98	0.43	201
17	92	0.23	147
18	85	0.30	172
19	67	0.20	110
20	72	0.30	191
21q	39	0.31	151
22q	43	0.95	185
X	164	0.31	231
Genome	3154	0.31	208

Source: Adapted from McCarthy (1996).

Note: The data were obtained at 3000 rads. Hence physical distances relative to cR are roughly 8/3 as large as the data shown in Table 7.1.

[a]cM/cR scales the genetic and radiation hybrid maps. kb/cR scales the physical and radiation hybrid maps.

order to analyze specific recombination events, it is necessary to be able to examine the markers *cosegregating in each single sperm cell.* Sperm cannot be biologically amplified except by fertilization with an oocyte, and the impracticality of such experiments in the human is all too apparent. Thus, until in vitro amplification became a reality, it was impossible to think of analyzing the DNA of single sperm. With PCR, however, this picture has changed dramatically.

A series of feasibility tests for single-sperm PCR was performed by Norman Arnheim and co-workers. First they asked whether single diploid human cells could be genotyped by PCR. They mixed cells from donors who had normal hemoglobin, homozygous for the β_A gene with cells from donors who had sickle cell anemia, homozygous for the β_S gene. Single cells were selected by micromanipulation and were tested by PCR amplification of the β locus followed by hybridization of a dot blot with allele-specific oligonucleotide probes. The purpose of the experiment was to determine whether the pure alleles could be detected reliably, or whether there would be cross-contamination between the two types of cells. The results for 37 cells analyzed were 19 β_A, 12 β_S, 6 none, and 0 both. This was very encouraging.

The next test involved the analysis of single sperm from donors with two different LDL receptor alleles. For 80 sperm analyzed, the results were 22 allele 1, 21 allele 2, 1 both alleles, and the remainder neither allele. Thus the efficiency of the single-sperm analysis was less than with diploid cells, but the cross-contamination level was low enough to be easily dealt with. A final test was to look at two nonlinked two-allele systems. One was on chromosome 6 (a_1 or a_2), and the other on chromosome 19 (b_1 or b_2). Four types of gametes were expected in equal ratios. What was actually observed is

a_1b_1	a_1b_2	a_2b_1	a_2b_2
21	18	14	17

This is really quite close to what was expected statistically,

The final step was to do single-sperm linkage studies. In this case one does simultaneous PCR analyses on linked markers. Consider the example shown in Figure 7.8. Nothing needs to be known about the parental haplotypes to start with. PCR analysis automatically provides the parental phase. In the case shown in the figure, most of the sperm measured at loci a and b have either alleles a_1 and b_2 or a_3 and b_1. This indicates that these are the haplotypes of the two homologous chromosomes in the donor. Occasional recombinants are seen with alleles a_1 and b_1 or a_3 and b_2. The frequency at which these are observed is a true measure of the male meiotic recombination frequency between markers a and b. Since large numbers of sperm can be measured, one can determine this frequency very accurately. Perhaps more important, with large numbers of sperm, very rare meiotic recombination events can be detected, and thus very short genetic distances can be measured.

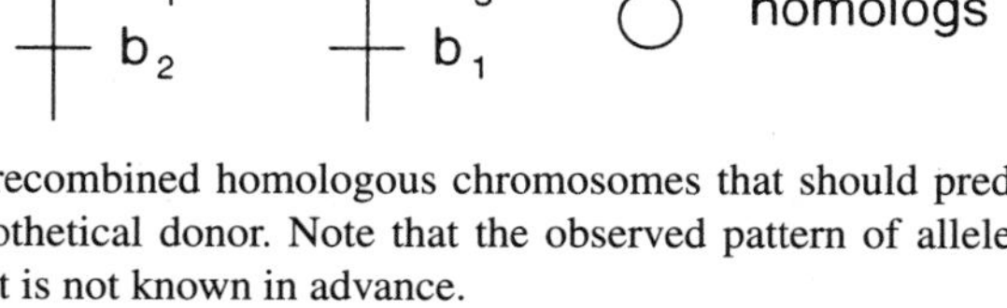

Figure 7.8   Two unrecombined homologous chromosomes that should predominate in the sperm expected from a hypothetical donor. Note that the observed pattern of alleles also determines the phase at these loci if it is not known in advance.

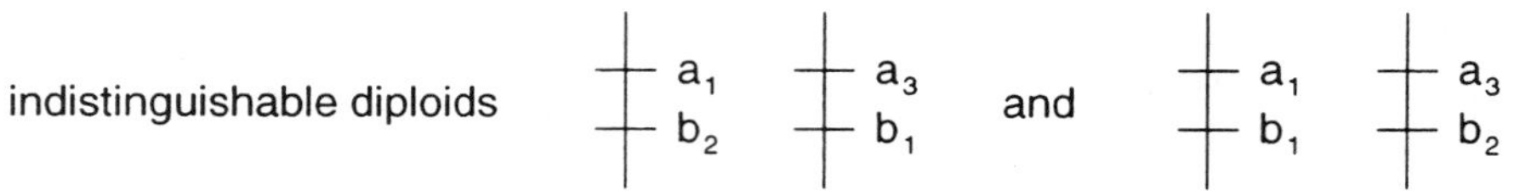

indistinguishable diploids $\dashv$ a$_1$ $\dashv$ a$_3$ and $\dashv$ a$_1$ $\dashv$ a$_3$
$\dashv$ b$_2$ $\dashv$ b$_1$ $\dashv$ b$_1$ $\dashv$ b$_2$

Figure 7.9   The allele pairs present in single DNA molecules will determine the phase of two markers in a pair of homologous chromosomes.

The map that results from single-sperm PCR is a true genetic map because it is a direct measure of male meiotic recombination frequencies. However, it has a number of limitations. Only male meioses can be measured. Only DNA markers capable of efficient and unique PCR amplification can be used. Thus this genetic map cannot be used to locate genes on the basis of their phenotype. Not only must DNA be available, but enough of it must be sequenced to allow for the production of appropriate PCR primers. The phenotype is irrelevant, and in fact it is invisible. Perhaps the greatest limitation of direct single-sperm PCR is that the sperm are destroyed by the PCR reaction. In principle, all markers of interest on a particular sperm cell must be analyzed simultaneously, since one will never be able to recreate that precise sperm cell again. This is not very practical, since simultaneous multi-locus PCR with many sets of primers has proved to be very noisy. One solution is to first do random primed PCR (PEP), tagged random primed (T-PCR), or degenerate oligonucleotide primed (DOP)-PCR, as described in Chapter 4. The sample is saved, and aliquots are used for the subsequent analysis of specific loci, one at a time, in individual, separate ordinary PCR reactions.

A variation on single-sperm PCR is single DNA molecule genetics. Here one starts with single diploid cells, prepares samples of their DNA, and dilutes these samples until most aliquots have either a single DNA molecule or none at all. The rationale behind this tactic is that it allows determination of the phase of an individual without any genetic information about any other relatives. A frequent problem in clinical genetics is that only a single parent is available because the other is uncooperative, aspermic, or dead. Phase determination means distinguishing among the two cases shown in Figure 7.9. It is clear that simultaneous PCR analysis of the alleles present on particular individual DNA molecules will reveal the phase in a straightforward manner. This is a very important step in making genetic mapping more efficient.

IN SITU HYBRIDIZATION

A number of techniques are available to allow the location of DNA sequences within cells or chromosomes to be visualized by hybridization with a labeled specific DNA probe. These are collectively called in situ hybridization. This term actually refers to any experiment in which an optical image of a sample (large or small) is superimposed on an image generated by detecting a specifically labeled nucleic acid component. In the current context we mean superimposing images of chromosomes or DNA molecules in the light microscope with images of the locations of labeled DNA probes. Radioactive labels were originally used, and the resulting probe location was determined by autoradiography superimposed on a photomicrograph of a chromosome. Fluorescent labels have almost totally supplanted radioisotopes in these techniques because of their higher spatial resolution, and because both the chromosome image and the specific hybridization image can be captured on the same film, eliminating the need for a separate autoradiographic development step. Short-lived radioisotopes like ^{32}P have decay tracks that are too long for microscopic images, and considerable resolution is lost by imprecision in the location of the origin of a track. Longer-lived isotopes like ^{3}H or ^{14}C would have shorter tracks, but these would be less efficient to detect, and more seriously, one would have to wait unrealistically long periods of time before an image could be detected. The technique in widespread use today depends on several cycles of stoichiometric amplification such as streptavidin biotin amplification (see Chapter 3) to increase the sensitivity of fluorescent detection. It is called FISH, for fluorescent in situ hybridization.

Metaphase chromosomes, until recently, were the predominant samples used for DNA mapping by in situ hybridization. A schematic illustration is given in Figure 7.10. In typical protocols, cell division is stopped, and cells are arrested in metaphase by adding drugs such as colchicine; chromosomes are isolated, dropped from a specified height onto a microscope slide, fixed (partially denatured and covalently crosslinked), and aged. If this seems like quite a bit of magic, it is. The goal is to strike a proper balance between maintaining sufficient chromosome morphology to allow each chromosome to be recognized and measured, but disrupting enough chromosome structure to expose DNA sequences that can hybridize with the labeled DNA probe. Under such circumstances the hybridization reactions are usually very inefficient. As a result a probe with relatively high DNA complexity must be used to provide sufficient illumination of the target site. Typically one starts with probes containing a total of 10^4 to 10^5 base pairs of DNA from the site of interest, and these probes are broken into small pieces prior to annealing to facilitate the hybridization.

Recently improved protocols and better fluorescence microscopes have allowed the use of smaller DNA probes for in situ hybridization. For instance, one recent procedure called PRINS (primed in situ hybridization) can use very small probes. In this method the probes hybridized to long target DNAs are used as primers for a DNA polymerase extentions. During the DNA polymerase extentions, modified bases containing biotin or digoxygenin are incorporated. This allows subsequent signal amplification by methods described in Chapter 3. Thus even short cDNAs can be used.

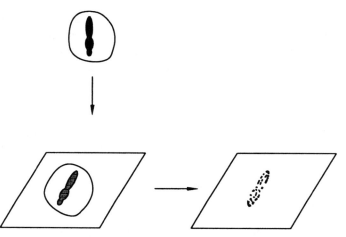

Figure 7.10   Schematic illustration of the preparation of metaphase chromosomes for in situ hybridization. See text for explanation.

In practice, it is usually much easier to first use the cDNA to find a corresponding cosmid clone, and then use that as a probe. In conventional FISH, 40 kb clones are used as probes. Complex probes like cosmids or YACs always have repeated human DNA sequences on them. It is necessary to eliminate the effect of these sequences; otherwise, the probe will hybridize all over the genome. An effective way to eliminate the complications caused by repeats is to fragment the probe into small pieces and prehybridize these with a great excess of $C_0t = 1$ DNA (see Chapter 3). A typical metaphase FISH result with a single-copy DNA probe is shown schematically in Figure 7.11. In an actual color image a bright pair of yellow fluorescent spots would be seen on a single red chromosome. The yellow comes from fluorescein conjugated to the DNA probe via streptavidin and biotin. The red comes from a DNA stain like DAPI used to mark the entire chromosome. The pair of yellow dots result from hybridization with each of the paired sister chromatids. In the simplest case one can determine the approximate chromosomal location by measuring the relative distance of the pair of spots from the ends of the chromosome. The identity of the particular chromosome is revealed in most cases by its size and the position of its centromeric constriction.

The interpretation of FISH results requires quantitative analysis of the image in the fluorescent microscope. It is not efficient to do this by photography and then processing of the image. Instead, direct on-line imaging methods are used. One possibility is to equip a standard fluorescence microscope with a charge couple device array (CCD) camera that can acquire and process the image as discrete pixels of information. The alternate approach is to use a scanning microscope like a confocal laser microscope that records the image's intensity as a function of position. In either case it is important to realize that the number of bits of information in a single microscope image is considerable; only a small amount of it actually finds its way into the final analyzed probe position. Either a large amount of mass storage must be devoted to archiving FISH images, or a procedure must be developed to allow some arbitration or reanalysis of any discrepancies in the data after the original raw images have been discarded.

Many enhancements have been described that allow FISH to provide more accurate chromosomal locations than the simple straightforward approach illustrated in Figure 7.11. Several examples are given in Figure 7.12. Chromosome banding provides a much more accurate way of identifying individual chromosomes and subchromosomal regions than simple measurements of size and centromere position. Each chromosome in the microscope is an individual—the amount of stretching and the nature of any distortion can vary considerably. Clearly, by superimposing banding on the emission from single-copy labeled probes, one not only provides a unique chromosome identifier but also local markers at the band locations that allow more accurate positioning of the single-copy probe.

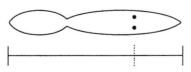

Figure 7.11 Appearance of a typical metaphase chromosome in FISH when a single fluorescein-labeled DNA probe is used along with a counterstain that lightly labels all DNA. Shown below is the coordinate system used to assign the map location of the probe.

map by relative position

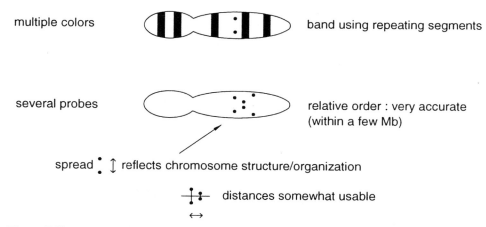

multiple colors   band using repeating segments

several probes   relative order : very accurate
                                   (within a few Mb)

spread ⁝ ↕ reflects chromosome structure/organization

distances somewhat usable

Figure 7.12 *Some of the refinements possible in metaphase FISH. (a) Simultaneous use of a single-copy probe and a repeated DNA probe like the human Alu sequence which replicates the banding pattern seen in ordinary Giemsa staining. (b) Use of three different single-copy DNA probes.*

Standard DNA banding stains are incompatible with the fluorescent procedures used for visualizing single-copy DNA. Fortunately it turns out that the major repeated DNA sequences in the human have preferential locations within the traditional Giemsa bands. Thus *Alu* repeats prefer light G bands, while L1 repeats prefer dark G bands. One can use two different colored fluorescent probes simultaneously, one with a single-copy sequence, the other with a cloned repeat, and the results, like those shown in Figure 7.12, represent a major improvement.

The accuracy gained by FISH over conventional radioactively labeled in situ hybridization is illustrated in Figure 7.13. The further increase in accuracy when FISH is used on top of a banded chromosome stain is also indicated. It is clear that the new procedures completely change the nature of the technique from a rather crude method of chromosome location to a highly precise mapping tool. It is possible to improve the resolution of FISH mapping even further by the simultaneous use of multiple single-copy DNA probes. For example, as shown in Figure 7.12, when metaphase chromosomes are hybridized with three nearby DNA segments, each labeled with a different color fluorophore, a distinct pattern of six ordered dots is seen. Each color is a pair of dots on the two sister chromatids. The order of the colors gives the order of the probes down to a resolution limit of about 1 to 2 Mb. The spread of the pair of dots relative to the long axis of the chromatids reflects details of chromosome structure and organization that we do not understand well today. It is a reproducible pattern, but our lack of knowledge about the detailed arrangement of packing of DNA in chromosomes compromises current abilities to turn this information into quantitative distance estimates between the probes. This is frustrating, but fortunately there is an easy solution, as described below. Ultimately, as we understand more about chromosome structure, and as high-resolution physical maps of DNA become available, FISH on metaphase chromosomes will undoubtedly turn out to be a rich source of information about the higher-order packing of chromatin within condensed chromosomes.

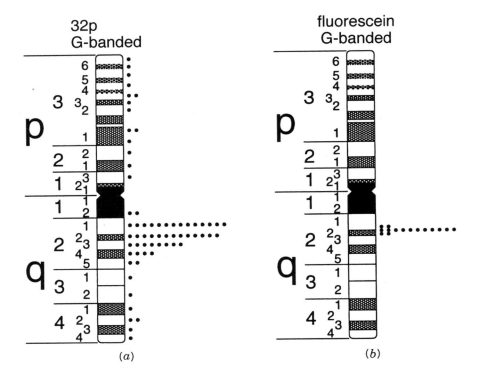

(a)

(b)

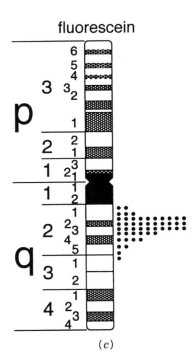

(c)

Figure 7.13 Examples of the accuracy of various in situ hybridization mapping techniques. In each case the probe is the blast1 gene. Dots show the apparent position of the probe on individual chromosomes. (a) ^{32}P-labeled probe on Giemsa-banded chromosomes. (b) Fluorescein-labeled probe on fluorescently banded chromosomes. (c) Fluorescein-labeled probe on unbanded chromosomes. (Adapted from Lawrence et al., 1990.)

Today metaphase FISH provides an extremely effective way to assign a large number of DNA probes into bins along a chromosome of interest. An example from work on chromosome 11 is shown in Figure 7.14. Note, in this case, that the mapped, cloned probes are not distributed evenly along the chromosome; they tend to cluster very much in several of the light Giemsa bands. This is a commonly observed cloning bias, and it severely complicates some of the approaches, used to construct ordered libraries of clones, which will be described in Chapters 8 and 9.

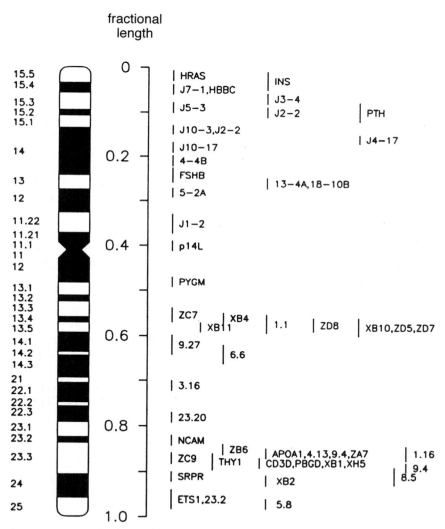

Figure 7.14   Regional assignment of a set of chromosome 11 cosmids by FISH. Note the regional biases in the distribution of clones. (Adapted from Lichter et al., 1990b.)

HIGH-RESOLUTION FISH

To increase the resolution of FISH over what is achievable with metaphase chromosomes, it is necessary to use more extended chromosome preparations. In most cases this results in sufficient disruption of chromosome morphology that no apriori assignment of a single probes to particular chromosomes or regions is possible. Instead, what is done is to determine the relative order and distances among a set of closely spaced DNA probes by simultaneous multicolor FISH, as we have already described for metaphase chromosomes. The different approaches reflect mostly variations in the particular preparation of extended chromosomes used.

Several different methods exist for systematic preparation of partially decondensed chromosomes. One approach is to catch cells at the pro-metaphase stage, before chromosomes have become completely condensed. A second approach to fertilize a hamster oocyte with a human sperm. The result of this attempted interspecies cross is called a humster. It does not develop, but within the hamster nucleus the human chromosomes, which are highly condensed in the sperm head, become partially decondensed; in this state they are very convenient targets for hybridization. A third approach is to use conditions that lead to premature chromosome condensation.

The most extreme versions of FISH use highly extended DNA samples. One way to accomplish this is to look at interphase nuclei. Under these conditions the chromatin is mostly in the form of 30 nm fibers (Chapter 2). One can make a crude estimate of the length of such fiber expected for a given length of DNA as follows. The volume of 10 bp of DNA double helix is given by $\pi r^2 d$ where r is the radius and d is the pitch. Evaluating these as roughly 10 Å and 34 Å, respectively, yields 10^4 Å per 10 bp or 10^3 Å per bp. The volume of a micron of 30-nm filament is $\pi r^2 d$, where r is 150 Å and d is 1 $\mu = 10^4$ Å. This is 8×10^8 Å^3, and roughly half of it is DNA. Thus one predicts that a micron of 30-nm filament will contain on average about 0.4 Mb of DNA. This estimate is not in bad agreement with what is observed experimentally (Box 7.1). Since the resolution of the fluorescence microscope is about 0.25 μ, the ultimate resolution of FISH based on interphase chromatin should be around 0.1 Mb.

Even higher resolution is possible if the DNA is extended further in methods called *fiber FISH*. One way to do this is the technique known as a Weigant halo. Here, as shown schematically in Figure 7.15, nuclei are prepared and then treated so that the DNA is deproteinized and exploded from the nucleus. Since naked DNA is about 3 Å per base pair, such samples should show an extension of 3×10^3 Å **per kb, which is 0.3** μ per kb. Thus the ultimate resolution of FISH under these circumstances could approach 1000 base pairs. To take advantage of the high resolution afforded by extended DNA samples, one must use multicolored probes to distinguish their order, and then estimate the distance between the different colors. The probes themselves will occupy a significant distance along the length of the DNA, as shown in Figure 17.16.

Interphase chromatin, or naked DNA, in contrast to metaphase chromosomes, is not a unique structure, and it is not rigid (Fig. 7.17a). To estimate the true distance between markers from the apparent separation of two probes in the microscope, one must correct for the fact that the DNA between the markers is not straight. The problem becomes ever more severe as the distance between the markers increases (Fig. 7.17b). No single molecule measurement will suffice because there is no way of knowing what the unseen DNA configuration is. Instead it is necessary to average the results over observations on many molecules, using a model for the expected chain configuration of the DNA.

Figure 7.15 A Weigant halo, which is produced when naked DNA is allowed to explode out of a nucleus with portions of the nuclear matrix still intact.

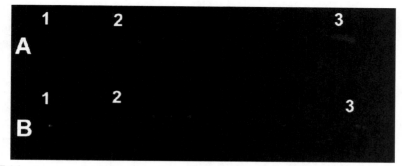

Figure 7.16 An example of the sorts of results obtainable with current procedures for interphase FISH. (From Heiskanen et al., 1996.) Figure also appears in color insert.

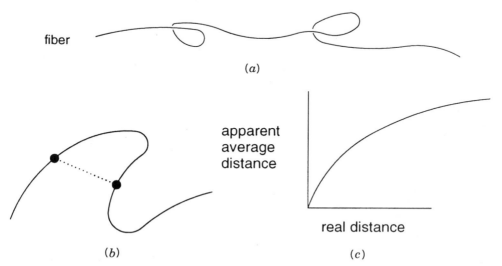

Figure 7.17 Typical configuration of a chromatin fiber in interphase FISH. (*a*) Appearance of one region of a fiber. (*b*) Apparent versus real distance between two loci. (*c*) Expected dependence of the apparent distance on the true distance for a random walk model.

BOX 7.1
QUANTITATIVE HIGH-RESOLUTION FISH

The quantitative analysis of distances in high-resolution FISH has been pioneered by two groups headed by Barb Trask, currently at the University of Washington, and Jeanne Lawrence, at the University of Massachusetts in Amherst. Others have learned their methods and begun to practice them. A few representative analyses are shown in Figures 7.18 and 7.19. The distribution of measured distances between two fixed markers in individual samples of interphase chromatin varies over quite a wide range, as shown in Figure 7.18a. However, when these measurements are averaged and plotted as a function of known distance along the DNA, for relatively short distances a reasonably straight plot is observed, but for unknown reasons, in this study, it does not pass through the origin (Fig. 7.18b). In another study, data were analyzed two ways. The cumulative probability of molecular distances was plotted as a function of measured distance, and the resulting curves appeared to be well fit by a random walk model (Fig. 7.19a). Alternatively, the square of the measured distance was an approxi-

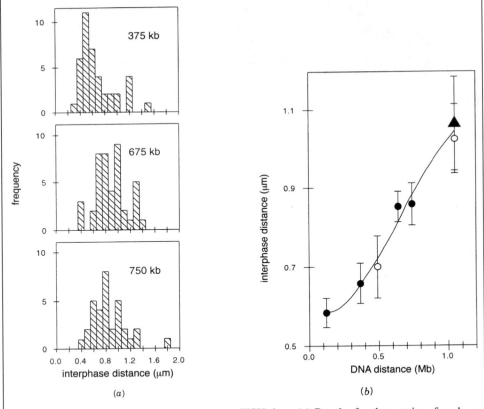

(a) (b)

Figure 7.18 Some representative interphase FISH data. (*a*) Results for three pairs of probes: Histograms indicate the number of molecules seen with each apparent size. (*c*) Apparent DNA distance as a function of the true distance, for a larger set of probes. (Adapted from Lawrence et al., 1990.)

(continued)

BOX 7.1 (*Continued*)

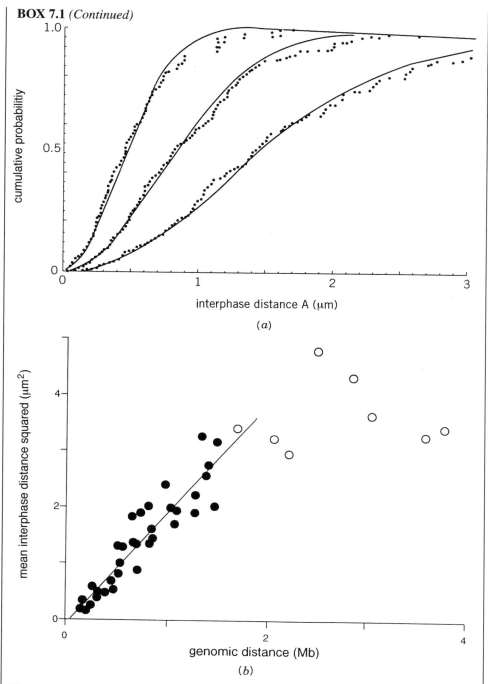

Figure 7.19 Additional examples of interphase FISH data. (*a*) Distribution of apparent distances seen for three probes. What is plotted is the fraction of molecules with an observed distance less than or equal to a particular value, as a function of that value. The solid lines are the best fit of the data to a random walk model. (*b*) Plot of the square of the apparent distance as a function of the true distance for a large set of probes. It is clear that the random walk model works quite well for distances up to about 2 Mb. (Adapted from Van den Engh et al., 1992.)

(continued)

BOX 7.1 *(Continued)*

mately linear function of the true distance, out to about 1.5 Mb (Fig. 7.19*b*). Thus these results support the use of a random walk model. The difficulty with the available results to date is that they do not agree on the scaling. Three studies are summarized below:

INVESTIGATOR	DISTANCES MEASURED	REAL
Lawrence	1 μ	1 Mb
Trask	1 μ	0.5 Mb
Skare	1 μ	1.2 Mb or more

It is not clear if these considerable differences are due to differences in the methods used to prepare the interphase chromatin or differences in the properties of interphase chromatin in different regions of the genome. Note that the scaling, at worst, is within a factor of three of what was estimated in the text from a very crude model for the structure of the 30-nm filament.

If this is taken to be a random walk, then the measured distance between markers should be proportional to the square root of the real distance between them (Fig. 7.17*c; see also* Box 7.1).

An example of the utility of interphase FISH is shown in Figure 7.20. Here interphase in situ hybridization was used to estimate the size of a gap that existed in the macrorestriction map of a region of human chromosome 4 near the tip of the short arm and now known to contain the gene for Huntington's disease. The gap was present because no clones or probes could be found between markers E4 and A252. In all other segments in the region, the accord between distances on the macrorestriction map and distances inferred from interphase in situ hybridization (using a scaling of 0.5 Mb per μ) is quite good. This has allowed the conclusion that the gap in the physical map would have to be small to maintain this consistency.

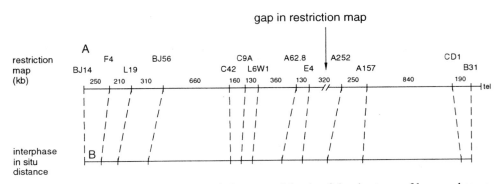

Figure 7.20 Comparison of a macrorestriction map of the tip of the short arm of human chromosome 4 with interphase FISH results for the same set of probes. (Adapted from Van den Engh et al., 1992.)

CHROMOSOME PAINTING

For most chromosomes a dense set of mapped markers now exists. An even larger set of clones is available and assigned to a particular chromosome but not yet mapped. These clones can be used to examine the state of the entire chromosome, either in metaphase or in less condensed states. This practice is called *chromosome painting*. Metaphase chromosome painting is a useful tool to see if a chromosome is intact or has rearranged in some way. A novel application of this is illustrated in Figure 7.21. Here the high degree of homology between chimp and human DNA sequences was used to examine the relationship between human chromosome 2 and its equivalent in the chimp. Probes from human chromosome 2 were used first on the human, where they indicated even and exhaustive coverage of the chromosome. Next the same set of probes was used on the chimp. Here two smaller chimp chromosomes were painted, numbers 12 and 13. Each of these is acrocentric, while human chromosome 2 is metacentric. These results make it clear that human chromosome 2 must have arisen by a Robertsonian (centromeric) fusion of the two smaller chimp chromosomes.

When interphase chromatin is painted, one can observe the cellular location of particular segments of DNA. The complexity of interphase chromatin makes it difficult to view more than small DNA regions simultaneously. This technique is still in its infancy, but it is already clear that it has the potential to provide an enormous amount of information of how DNA is organized in a functioning nucleus (Chandley et al., 1996, Seong et al., 1994).

A recently developed variation of chromosome painting shows considerable promise both in facilitating the search for gene involved in diseases like cancer and in assisting the development of improved clinical diagnostic tests for chromosomal disorders. In this technique, called *comparative genome hybridization* (CGH), total DNA from two samples to be compared is labeled using different specificity tags. Actual procedures use nick translation: For one sample a biotinylated dNTP is used; for the other a digoxigenin-labeled dNTP is used (see Chapter 3). These two samples are then allowed to hybridize simultaneously to a metaphase chromosome spread. The results are visualized by two-

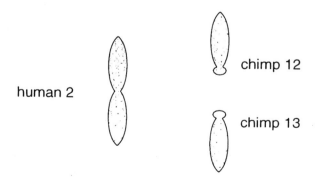

Figure 7.21 An example of metaphase chromosome painting, in which a human chromosome 2 probe is hybridized to the full set of human chromosomes, and separately to the full set of chimpanzee chromosomes. Shown are just those resulting chromosomes from both species that show significant hybridization.

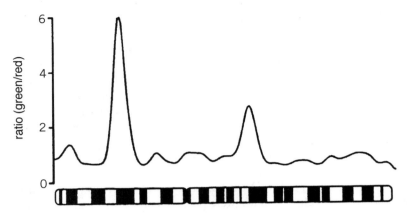

Figure 7.22 An example of comparative genome hybridization. Shown is the green to red color ratio seen along human chromosome 2 after competitive hybridization with DNA from two cell lines. The red cell line is normal human DNA. Three regions of DNA amplification are apparent in the green labeled cell line. The most prominent of these corresponds to the oncogene N-*myc* known to be amplified in the cell line used. (From Kallioniemi et al., 1992.)

color detection using fluorescein-labeled streptavidin or avidin to detect the biotin and rhodamine-labeled antidigoxigenin to detect the digoxigenin. The ratio of the two colors should indicate the relative amounts of the two probes hybridized. This should be the same if the initial probes have equal concentrations everywhere in the genome. However, if there are regions in one target sample that are amplified or deleted, the observed color ratio will shift. The results can be dramatic as shown by the example in Figure 7.22. The color ratio shifts provide an indication of the relative amounts of each amplification or deletion, and they also allow the locations of all such variations between two samples to be mapped in a single experiment.

CHROMOSOME MICRODISSECTION

Frequently the search for a gene has focused down to a small region of a chromosome. The immediate task at hand is to obtain additional DNA probes for this region. These are needed to improve the local genetic map and also to assist the construction of physical maps. They may also be useful for screening cDNA libraries if there is any hint of the preferred tissue for expression of the gene of interest. The question is how to focus on a small region of the genome efficiently. One approach has been microdissection of that region from metaphase chromosomes and microcloning of the DNA that results. Two approaches to microdissection are shown schematically in Figure 7.23. In one, a fine glass needle is used to scratch out the desired chromosome region and transfer the material to a site where it can be collected. In the other approach, a laser is used to ablate all of the genome except for the small chromosome region of interest. In either case, the problem in early versions of this method was to develop efficient cloning schemes that could start from the very small amounts of DNA that could be collected by microdissection.

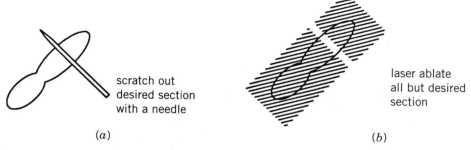

scratch out
desired section
with a needle

laser ablate
all but desired
section

(*a*)   (*b*)

Figure 7.23 Two examples of chromosome microdissection. (*a*) A fine needle is used to scratch a chromosome and transfer a section of it to another site where it can be further manipulated. (*b*) A laser is used to destroy the DNA in all but a preselected segment of a chromosome.

Hundreds of microdissection products would be combined and placed in liquid micro-drops suspended in oil (Fig. 7.24). Restriction enzymes, vectors, and other components needed for cloning were delivered by micromanipulators. The result was a technical *tour de force* that did deliver clones from the desired region but usually in very small numbers.

With the development of genome-wide PCR amplification methods (Chapter 4), the need for microcloning dissected chromosome samples is eliminated. Using these methods, it is relatively easy to amplify the dissected chromosome material by PCR until there is enough DNA to be handled by more conventional methods. Note that the microdissection method has several important potential uses beyond the hunt for specific disease genes. Most DNA libraries in any type of vector are biased, and some regions are under-represented or not represented at all. Microdissection offers an attractive way to compensate for cloning biases, especially when they are severe. The alternate approach, in common use today, is to combine clones from different types of libraries in the hope that biases will compensate for each other. This can be a very effective strategy, but it increases substantially the number of DNA samples that must be handled. Another potential use for microdissection will be to pull clones from a particular individual who may have a region of special interest, such as a suspected rearrangement. This will not always be an effective strategy, but it may well be necessary in cases where simpler approaches fail to yield a definitive picture of what has happened.

liquid microdrop
suspended in oil

Figure 7.24 Microcloning of DNA in a tiny droplet containing microdissected chromosomes.

SOURCES AND ADDITIONAL READINGS

Boyle, A. L., Feltquite, D. M., Dracopoli, N. C., Housman, D. E., and Ward, D. C. 1992. Rapid physical mapping of cloned DNA on banded mouse chromosomes by fluorescence *in situ* hybridization. *Genomics* 12: 106–115.

Burmeister, M., Kim, S., Price, E. R., de Lange, T., Tantravahi, U., Meyers, R. M., and Cox, D. 1991. A map of the distal region of the long arm of human chromosome 21 constructed by radiation hybrid mapping and pulsed field gel electrophoresis. *Genomics* 9: 19–30.

Chandley, A. C., Speed, R. M., and Leitch, A. R. 1996. Different distributions of homologous chromosomes in adult human Sertoli cells and in lymphocytes signify nuclear differentiation. *Journal of Cell Science* 109: 773–776.

Cherif, D., Julier, C., Delattre, O., Derré, J., Lathrop, G. M., and Berger, R. 1990. Simultaneous localization of cosmids and chromosome R-banding by fluorescence microscopy: Application to regional mapping of human chromosome 11. *Proceedings of the National Academy of Sciences USA* 87: 6639–6643.

Cox, D. R., Burmeister, M., Price, E. R., Kim, S., and Meyers, R. 1990. Radiation hybrid mapping: A somatic cell genetic method for constructing high-resolution maps of mammalian cell chromosomes. *Science* 250: 245–251.

Ellis, N., and Goodfellow, P. N. 1989. The mammalian pseudoautosomal region. *Trends in Genetics* 5: 406–410.

Heiskanen M., Kallioniemi, O., and Palotie, A. 1996. Fiber-FISH: Experiences and a refined protocol. *Genetic Analysis (Biomolecular Engineering)* 12: 179–184.

Jeffreys, A. J., Neumann, R., and Wilson, V. 1990. Repeat unit sequence variation in minisatellites: A novel source of DNA polymorphism for studying variation and mutation by single molecule analysis. *Cell* 60: 473–485.

Kallioniemi, A., Kallioniemi, O. P., Sudar, D., Rutovitz, D., Gray, J. W., Waldman, F., and Pinkel, D. 1992. Comparative genomic hybridization for molecular cytogenetic analysis of solid tumors. *Science* 258: 818–821.

Lawrence, J., Singer, R. H., and McNeil, J. A. 1990. Interphase and metaphase resolution of different distances within the human dystrophin gene. *Science* 249: 928–932.

Li, H., Cui, A., and Arnheim, N. 1990. Direct electrophoresis detection of the allelic state of single DNA molecules in human sperm by using the polymerase chain reaction. *Proceedings of the National Academy of Sciences USA* 87: 4580–4584.

Li, H., Gyllensten, U. B., Cui, X., Saiki, R. K., Erlich, H. A., and Arnheim, N. 1988. Amplification and analysis of DNA sequences in single human sperm and diploid cells. *Nature* 335: 414–417.

Lichter, P., Ledbetter, S. A., Ledbetter, D. H., and Ward, D. C. 1990. Fluorescence in situ hybridization with *Alu* and L1 polymerase chain reaction probes for rapid characterization of human chromosomes in hybrid cell lines. *Proceedings of the National Academy of Sciences USA* 87: 6634–6638.

Lichter, P., Tang, C. C., Call, K., Hermanson, G., Evans, G. A., Housman, D., and Ward, D. C. 1990. High resolution mapping of human chromosome 11 by in situ hybridization with cosmid clones. *Science* 247: 64–69.

McCarthy, L. C. 1996. Whole genome radiation hybrid mapping. *Trends in Genetics* 12: 491–493.

Nilsson, M., Krejci, K., Koch, J., Kwiatkowski, M., Gustavsson, P., and Landegren, U. 1997. Padlock probes reveal single nucleotide differences, parent of origin and in situ distribution of centromeric sequences in human chromosomes 13 and 21. *Nature Genetics* 16: 252–255.

Ruano, G., Kidd, K. K., and Stephens, J. C. 1990. Haplotype of multiple polymorphisms resolved by enzymatic amplification of single DNA molecules. *Proceedings of the National Academy of Sciences USA* 87: 6296–6300.

Seong, D. C., Song, M. Y., Henske, E. P., Zimmerman, S. O., Champlin, R. E., Deisseroth, A. B., and Siciliano, M. J. 1994. Analysis of interphase cells for the Philadelphia translocation using painting probe made by Inter-*Alu*-polymerase chain reaction from a radiation hybrid. *Blood* 83: 2268–2273.

Silva, A. J., and White, R. 1988. Inheritance of allelic blueprints for methylation patterns. *Cell* 54: 145–152.

Stewart, E. A., McKusick, K. B., Aggarwal, A., et al. 1997. An STS-based radiation hybrid map of the human genome. *Genome Research* 7: 422–433.

Strong, S. J., Ohta, Y., Litman, G. W., and Amemiya, C. T. 1997. Marked improvement of PAC and BAC cloning is achieved using electroelution of pulsed-field gel-separated partial digests of genomic DNA. *Nucleic Acids Research* 25: 3959–3961.

Van de Engh, G., Sachs, R., and Trask, B. J. 1992. Estimating genomic distance from DNA sequence location in cell nuclei by a random walk model. *Science* 257: 1410–1412.

Wang, D., and Smith, C. L. 1994. Large-scale structure conservation along the entire long arm of human chromosome 21. *Genomics* 20: 441–451.

Wu, B-L., Milunsky, A., Nelson, D., Schmeckpeper. B, Porta, G., Schlessinger, D., and Skare, J. 1993. High-resolution mapping of probes near the X-linked lymphoproliferative disease (XLP) locus. *Genomics* 17: 163–170.

Yu, J., Hartz, J., Xu, Y., Gemmill, R. M., Korenberg, J. R., Patterson, D., and Kao, F.-T. 1992. Isolation, characterization, and regional mapping of microclones from a human chromosome 21 microdissection library. *American Journal of Human Genetics* 51: 263–272.

8 Physical Mapping

WHY HIGH-RESOLUTION PHYSICAL MAPS ARE NEEDED

Physical maps are needed because ordinary human genetic maps are not detailed enough to allow the DNA that corresponds to particular genes to be isolated efficiently. Physical maps are also needed as the source for the DNA samples that can serve as the actual substrate for large-scale DNA sequencing projects. Genetic linkage mapping provides a set of ordered markers. In experimental organisms this set can be almost as dense as one wishes. In humans one is much more limited because of the inability to control breeding and produce large numbers of offspring. Distances that emerge from human genetic mapping efforts are vague because of the uneven distribution of meiotic recombination events across the genome. Cytogenetic mapping, until recently, was low resolution; it is improving now, and if it were simpler to automate, it could well provide a method that would supplant most others. Unfortunately, current conventional approaches to cytogenetic mapping seem difficult to automate, and they are slower than many other approaches that do not involve image analysis.

It is worth noting here that in some other organisms, cytogenetics is more powerful than in the human. For example, in Drosophila the presence of polytene salivary gland chromosomes provides a tool of remarkable power and simplicity. The greater extension and higher-resolution imaging of polytene chromosomes allows bands to be seen at a typical resolution of 50 kb (Fig. 8.1). This is more than 20 times the resolution in the best human metaphase FISH. Furthermore the large number of DNA copies in the metaphase salivary chromosomes means that microdissection and cloning are much more powerful here than they are in the human. It would be nice if there were a convenient way to place large continuous segments of human DNA into Drosophila and then use the power of the cytogenetics in this simple organism to map the human DNA inserts.

Radiation hybrids offer, in principle, a way to measure distances between markers accurately. However, in practice, the relative distances in radiation hybrid maps appear to be distorted (Chapter 7). In addition there are a considerable number of unknown features of these cells; for example, the types of unseen DNA rearrangements that may be present need to be characterized. Thus it is not yet clear that the use of radiation hybrids alone can produce a complete accurate map or yield a set of DNA samples worth subcloning and characterizing at higher resolution. Instead, currently other methods must be used to accomplish the three major goals in physical mapping:

1. Provide an ordered set of all of the DNA of a chromosome (or genome).
2. Provide accurate distances between a dense set of DNA markers.
3. Provide a set of DNA samples from which direct DNA sequencing of the chromosome or genome is possible. This is sometimes called a *sequence-ready map*.

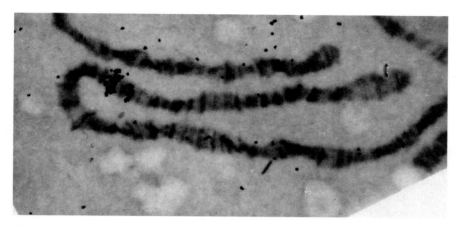

Figure 8.1 Banding seen in polytene Drosophila salivary gland chromosomes. Shown is part of chromosome 3 of for which a radiolabeled probe to the gene for heat-shock protein hsp70 has been hybridized. A strong and weak site are seen. (Figure kindly provided by Mary Lou Pardue.)

Most workers in the field focus their efforts on particular chromosomes or chromosome regions. Usually a variety of different methods are brought to bear on the problems of isolating clones and probes and using these to order each other and ultimately produce finished maps. However, some research efforts have produced complete genomic maps with a restricted number of approaches. Such samples are turned over to communities interested in particular regions or chromosomes that finish a high-resolution map of the regions of interest.

There are basically two kinds of physical maps commonly used in genome studies: restriction maps and ordered clone banks or ordered libraries. We will discuss the basic methodologies used in both approaches separately, and then, in Chapter 9, we will show how the two methods can be merged in more powerful second generation strategies. First we outline the relative advantages and disadvantages of each type of map.

RESTRICTION MAPS

A typical restriction map is shown in Figure 8.2. It consists of an ordered set of DNA fragments that can be generated from a chromosome by cleavage with restriction enzymes individually or in pairs. Distances along the map are known as precisely as the lengths of the DNA fragments generated by the enzymes can be measured. In practice, the lengths are measured by electrophoresis in virtually all currently used methods; they are accurate to a single base pair up to DNAs around around 1 kb in sizes. Lengths can be measured with better than 1% accuracy for fragments up to 10 kb, and with a low percent of accuracy for fragments up to 1 Mb in size. Above this length, measurements today are still fairly qualitative, and it is always best to try to subdivide a target into pieces less than 1 Mb before any quantitative claims are made about its true total size.

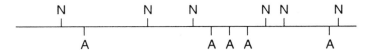

Figure 8.2   Typical section of a restriction map generated by digestion of genomic or cloned DNA with two enzymes with different recognition sites A, and N.

In an ideal restriction map each DNA fragment is pinned to markers on other maps. Note that if this is done with randomly chosen probes, the locations of these probes within each DNA fragment are generally unknown (Fig. 8.3). Probes that correspond to the ends of the DNA fragments are more useful, when they are available, because their position on the restriction map is known precisely. Originally probes consisted of unsequenced DNA segments. However, the power of PCR has increasingly favored the use of sequenced DNA segments.

A major advantage of a restriction map is that accurate lengths are known between sets of reference points, even at very early stages in the construction of the map. A second advantage is that most restriction mapping can be carried out using a top-down strategy that preserves an overview of the target and that reaches a nearly complete map relatively quickly. A third advantage of restriction mapping is that one is working with genomic DNA fragments rather than cloned DNA. Thus all of the potential artifacts that can arise from cloning procedures are avoided. Filling in the last few small pieces is always a chore in restriction mapping, but the overall map is a useful tool long before this is accomplished, and experience has shown that restriction maps can be accurately and completely constructed in reasonably short time periods.

In top-down mapping one successively divides a chromosome target into finer regions and orders these (Fig. 8.4). Usually a chromosome is selected by choosing a hybrid cell in which it is the only material of interest. There have been some concerns about the use of hybrid cells as the source of DNA for mapping projects. In a typical hybrid cell there is no compelling reason for the structure of most of the human DNA to remain intact. The biological selection that is used to retain the human chromosome is actually applicable to only a single gene on it. However, the available results, at least for chromosome 21, indicate that there are no significant differences in the order of DNA markers in a selected set of hybrid and human cell lines (see Fig. 8.49). In a way this is not surprising; even in a human cell line most of the human genome is silent, and if loss or rearrangement of DNA were facile under these circumstances, it should have been observed. Of course, for model organisms with small genomes, there is no need to resort to a hybrid cell at all—their genomes can be studied intact, or the chromosomes can be purified in bulk by PFG.

In a typical restriction mapping effort, any preexisting genetic map information can be used as a framework for constructing the physical map. Alternatively, the chromosome of interest can be divided into regions by cytogenetic methods or low-resolution FISH.

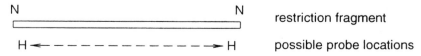

Figure 8.3   Ambiguity in the location of a hybridization probe on a DNA fragment.

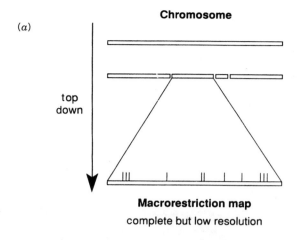

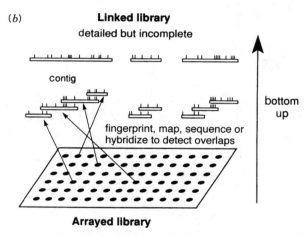

Figure 8.4 Schematic illustration of methods used in physical mapping. (*a*) Top-down strategy used in restriction mapping. (*b*) Bottom-up strategy used in making an ordered library.

Large DNA fragments are produced by cutting the chromosome with restriction enzymes with very rare recognition sites. The fragments are separated by size and assigned to regions by hybridization with genetically or cytogenetically mapped DNA probes. Then the fragments are assembled into contiguous blocks, by methods that will be described later in this chapter. The result at this point is called a *macrorestriction map*. The fragments may average 1 Mb in size. For a simple genome this means that only 25 fragments will have to be ordered. This is relatively straightforward. For an intact human genome, the corresponding number is 3600 fragments. This is an unthinkable task unless the fragments are first assorted into individual chromosomes.

If a finer map is desired, it can be constructed by taking the ordered fragments one at a time, and dissecting these with more frequently cutting restriction nucleases. An advantage of this reductionist mapping approach is that the finer maps can be made only in those regions where there is sufficient interest to justify this much more arduous task.

The major disadvantage of most restriction mapping efforts is that they do not produce the DNA in a convenient, immortal form where it can be distributed or sequenced by available methods. One could try to clone the large DNA fragments that compose the macrorestriction map, and there has been some progress in developing the vectors and techniques needed to do this. One could also use PCR to generate segments of these large fragments (see Chapter 14). For a small genome, most of the macrorestriction fragments it contains can usually be separated and purified by a single PFG fractionation. An example is shown in Figure 8.5. In cases like this, one does really possess the

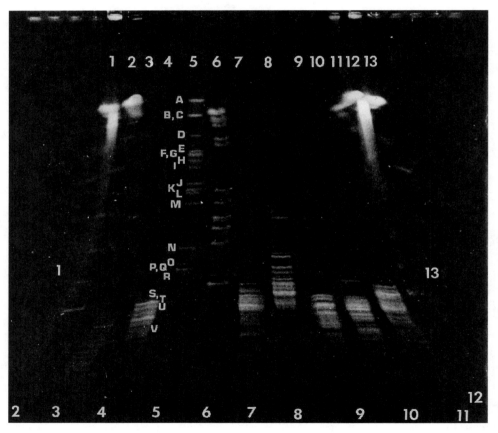

Figure 8.5 Example of the fractionation of a restriction enzyme digest of an entire small genome by PFG. *Above: Not* I digest of the 4.6 Mb *E. coli* genome shown in lane 5; *Sfi* I digest in lane 6. Other lanes shown enzymes that cut too frequently to be useful for mapping. (Adapted from Smith et al., 1987.) *Left:* Structure of the ethidium cation used to stain the DNA fragments.

genome, but it is not in a form where it is easy to handle by most existing techniques. For a large genome, PFG does not have sufficient resolution to resolve individual macrorestriction fragments. If one starts instead with a hybrid cell, containing only a single human chromosome or chromosome fragment, most of the human macrorestriction fragments will be separable from one another. But they will still be contaminated, each by many other background fragments from the rodent host.

ORDERED LIBRARIES

Most genomic libraries are made by partial digestion with a relatively frequently cutting restriction enzymes, size selection of the fragments to provide a fairly uniform set of DNA inserts, and then cloning these into a vector appropriate for the size range of interest. Because of the method by which they were produced, the cloned fragments are a nearly random set of DNA pieces. Within the library, a given small DNA region will be present on many different clones. These extend to varying degrees on both sides of the particular region (Fig. 8.6). Because the clones contain overlapping regions of the genome, it is possible to detect these overlaps by various fingerprinting methods that examine patterns of sequence on particular clones. The random nature of the cloned fragments means that many more clones exist than the minimum set necessary to cover the genome. In practice, the redundancy of the library is usually set at five- to tenfold in order to ensure that almost all regions of the genome will have been sampled at least once (as discussed in Chapter 2). From this vast library the goal is to assemble and to order the minimum set of clones that covers the genome in one contiguous block. This set is called the *tiling path*.

Clone libraries have usually been ordered by a bottom-up approach. Here individual clones are initially selected from the library at random. Usually the library is handled as an array of samples so that each clone has a unique location on a set of microtitre plates, and the chances of accidentally confusing two different clones can be minimized. The clone is fingerprinted, by hybridization, by restriction mapping, or by determining bits of DNA sequence. Eventually clones appear that share some or all of the same fingerprint pattern (Fig. 8.7). These are clearly overlapping, if not identical, and they are assembled into overlapping sets, called *contigs,* which is short for contiguous blocks. There are several obvious advantages to this approach. Since the DNA is handled as clones, the map is built up of immortal samples that are easily distributed and that are potentially suitable for direct sequencing. The maps are usually fairly high resolution when small clones are used, and some forms of fingerprinting provide very useful internal information about each clone.

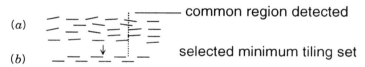

Figure 8.6 Example of a dense library of clones. (*a*) The large number of clones insures that a given DNA probe or region (vertical dashed line) will occur on quite a few different clones in the library. (*b*) The minimum tiling set is the smallest number of clones that can be selected to span the entire sample of DNA.

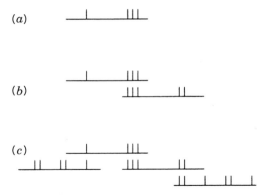

Figure 8.7 Example of a bottom-up fingerprinting strategy to order a dense set of clones. (*a*) A clone is selected at random and fingerprinted. (*b*) Two clones that share an overlapping fingerprint pattern are assembled into a contig. (*c*) Longer contigs are assembled as more overlapping clones are found.

There are a number of disadvantages to bottom-up mapping. While this process is easy to automate, no overview of the chromosome or genome is provided by the fingerprinting and contig building. Additional experiments have to be done to place contigs on a lower-resolution framework map, and most approaches do not necessarily allow the orientation of each contig to be determined easily. A more serious limitation of pure bottom-up mapping strategies is that they do not reach completion. Even if the original library is a perfectly even representation of the genome, there is a statistical problem associated with the random clone picking used. After a while, most new clones that are picked will fall into contigs that are already saturated. No new information will be gained from the characterization of these new clones. As the map proceeds, a diminishingly smaller fraction of new clones will add any additional useful information. This problem becomes much more serious if the original library is an uneven sample of the chromosome or genome. The problem can be alleviated somewhat if mapped contigs are used to screen new clones prior to selection to try to discard those that cannot yield new information. An additional final problem with bottom-up maps is shown in Figure 8.8. Usually the overlap distance between two adjacent members of a contig is not known with much precision. Therefore distances on a typical bottom-up map are not well defined.

The number of samples or clones that must be handled in top-down or bottom-up mapping projects can be daunting. This number also scales linearly with the resolution desired. To gain some perspective on the problem, consider the task of mapping a 150-Mb chromosome.

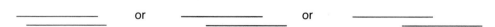

Figure 8.8 Ambiguity in the degree of clone overlap resulting from most fingerprinting or clone ordering methods.

This is the average size of a human chromosome. The numbers of samples needed are shown below:

RESOLUTION	RESTRICTION MAP	ORDERED LIBRARY ($5\times$ REDUNDANCY)
1 Mb	150 fragments	750 clones
0.1 Mb	1500 fragments	7500 clones
0.01 Mb	15,000 fragments	75,000 clones

With existing methods the current convenient range for constructing physical maps of entire human chromosomes allows a resolution of somewhere between 50 kb (for the most arduous bottom-up approaches attempted) to 1 Mb for much easier restriction mapping or large insert clone contig building.

The resolution desired in a map will determine the sorts of clones that are conveniently used in bottom-up approaches. Among the possibilities currently available are

Bacteriophage lambda	10 kb inserts
Cosmids	40 kb inserts
P1 clones	80 to 100 kb inserts
Bacterial artificial chromosomes (BACs)	100 to 400 kb inserts
Yeast artificial chromosomes (YACs)	100 to 1300 kb inserts

The first three types of clones can be easily grown in large numbers of copies per cell. This greatly simplifies DNA preparation, hybridization, and other analytical procedures. The last two types of clones are usually handled as single copies per host cell, although some methods exist for amplifying them. Thus they are more difficult to work with, individually, but their larger insert size makes low-resolution mapping much more rapid and efficient.

RESTRICTION NUCLEASE GENOMIC DIGESTS

Generating DNA fragments of the desired size range is critical for producing useful libraries and genomic restriction maps. If genomes were statistically random collections of the four bases, simple binomial statistics would allow us to estimate the average fragment length that would result from a given restriction nuclease recognition site in a total digest. For a genome where each of the four bases is equally represented, the probability of occurrence of a particular site of size n is 4^{-n}; therefore the average fragment length generated by that enzyme will be 4^n. In practice, this becomes

SITE SIZE (N)	AVERAGE FRAGMENT LENGTH (kb)
4	1
6	4
8	64
10	1000

This tabulation indicates that enzymes with four or six base sites are convenient for the construction of partial digest small insert libraries. Enzymes with sites ten bases long would be the preferred choice for low-resolution macrorestriction mapping, but such enzymes are unknown. Enzymes with eight-base sites would be most useful for currently achievable large-insert cloning. More accurate schemes for predicting cutting frequencies are discussed in Box 8.1.

A list of the enzymes currently available that have relatively rare cutting sites in mammalian genomes is given in Table 8.1. Unfortunately, there are only a few known restriction enzymes with eight-base specificity, and most of these have sites that are not well-predicted by random statistics. A few enzymes are known with larger recognition sequences. In most cases there is not yet convincing evidence that the available preparations of these enzymes have low enough contamination with random nonspecific nucleases to allow them to be used for a complete digest to generate a discrete nonoverlapping set of large DNA fragments. Several enzymes exist that have sites so rare that they will not occur at all in natural genomes. To use these enzymes, one must employ strategies in which the sites are introduced into the genome, their location is determined, and then cutting at the site is used to generate fragments containing segments of the region where the site was inserted. Such strategies are potentially quite powerful, but they are still in their infancy.

The unfortunate conclusion, from experimental studies on the pattern of DNA fragment lengths generated by genomic restriction nuclease digestion, is that most currently available 8-base specific enzymes are not useful for generating fragments suitable for macrorestriction mapping. Either the average fragment length is too short, or the digestion is not complete, leading to an overly complicated set of reaction products. However, mammalian genomes are very poorly modeled by binomial statistics, and thus some enzymes, which statistically might be thought to be useless because they would generate fragments that are too small, in fact generate fragments in useful size ranges. As a specific example, consider the first two enzymes known with eight-base recognition sequences:

Sfi I	GGCCN^NNNNGGCC
Not I	GC^GGCCGC

Here the symbol N indicates that any of the four bases can occupy this site; the caret (^) indicates the cleavage site on the strand shown; there is a corresponding site on the second strand.

Human DNA is A–T rich; like that of other mammals it contains approximately 60% A+T. When this is factored into the predictions (Box 8.1), on the basis of base composition, alone, these enzymes would be expected to cut human DNA every 100 to 300 kb. In fact *Sfi* I digestion of the human genome yields DNA fragments that predominantly range in size from 100 to 300 kb. In contrast, *Not* I generates DNA fragments that average closer to 1 Mb in size. The size range generated by *Sfi* I would make it potentially useful for some applications, but the cleavage specificity of *Sfi* I leads to some unfortunate complications. This enzyme cuts within an unspecified sequence. Thus the fragments it generates cannot be directly cloned in a straightforward manner because the three base overhang generated by *Sfi* I is a mixture of 64 different sequences. Another problem, introduced by the location of the *Sfi* I cutting site, is that different sequences turn out to be cut at very different rates. This makes it difficult to achieve total digests efficiently, and as described later, it also makes it very difficult to use *Sfi* I in mapping strategies that depend on the analysis of partial digests.

BOX 8.1
PREDICTION OF RESTRICTION ENZYME-CUTTING FREQUENCIES

An accurate prediction of cutting frequencies requires an accurate statistical estimation of the probability of occurrence of the enzyme recognition site. To accomplish this, one must take into account two factors: First, the sample of interest, that is, the human genome, is unlikely to have a base composition that is precisely 50% G + C, 50% A + T. Second, the frequencies of particular dinucleotide sequences often vary quite substantially from that predicted by simple binomial statistics based on the base composition. A rigorous treatment should also take the mosaicism into account (see Chapters 2 and 15).

Base composition effects alone can be considered, for double-stranded DNA with a single variable: $X_{G + C} = 1 - X_{A + T}$, where X is a mole fraction. Then the expected frequency of occurrence of a site with n G's or C's and m A's or T's is just

$$\left(\frac{1}{2}\right)^{n+m}(X_{G+C})^n(1 - X_{G+C})^m$$

To take base sequence into account, Markov statistics can be used. In Markov chain statistics the probability of the nth event in a series can be influenced by the specific outcome of the prior events such as the $(n - 1)$th and $(n - 2)$th events. Thus Markov chain statistics can take into account known frequency information about the occurrences of sequences of events. In other words, this kind of statistics is ideally suited for the analysis of sequences.

Suppose that the frequencies of particular dinucleotide sequences are known for the sample of interest. There are 16 possible dinucleotide sequences: The frequency $X_{A,C}$, for example, indicates the fraction of dinucleotides that has a AC sequence on one strand base paired to a complementary GT on the other. The sum of these frequencies is 1. Only 10 of the 16 dinucleotide sequences are distinct unless we consider the two strands separately. On each strand, we can relate dinucleotide and mononucleotide frequencies by four equations:

$$X_{A,C} + X_{A,A} + X_{A,T} + X_{A,G} = X_A$$

since base A must always be followed by some other base (the X's indicate mole fraction). The expected frequency of occurrence of a particular sequence string, based on these nearest-neighbor frequencies is just

$$X_{12} \prod_{i=2}^{n} \frac{X_{i,i+1}}{X_i}$$

where the product is taken successively over the mole fractions $X_{i,i+1}$, respectively, and X_i of all successive dinucleotides $i, i + 1$, and mononucleotides i in the sequence of interest. Predictions done in this way are usually more accurate than predictions based solely on the base composition. Where methylation occurs that can block the cutting of certain restriction nucleases, the issue becomes much more complex, as discussed in the text.

TABLE 8.1 Restriction Enzymes Useful for Genomic Mapping

Enzyme[a]	Recognition Site (5'–3')[b]	Source[a,c]
Enzymes with Extended Recognition Sites		
I-*Sce* I	TAGGGATAA/CAGGGTAAT	B
VDE	TATSYATGYYGGTGY/	O
	GGRGAARKMGKKAAWGAAWG	
I-*Ceu* I	TAACTATAACGGTCCTA/AGGTAGCGA	N
I-*Tli* I	GGTTCTTTATGCGGACAC/TGACGGCTTTATG	N
I-*Ppo* I	CTCTCTTAA/GGTAGC	P
Enzymes with >6-bp Recognition Site		
Pac I	TTAAT/TAA	N
Pme I	GTTT/AAAC	N
Swa I	ATTT/AAAT	B
*Sse*83888t1	CCTGCA/GC	T
Enzymes with >6-bp Recognition that Cut in CpG Islands		
Rsr II (*Csp**)	CG/GWCCG	N, P*
SgrA I	CR/CCGGYG	B
Not I	GC/GGCCGC	B, N, S, P
Srf I	GCCC/GGGC	P, S
Fse	GGCCGGCC	B
Sfi I[d]	GGCCNNNN/NGGCC	B, N, P
Asc I	GG/CGCGCC	N
Enzymes that cut in CpG Islands: Fragments Average >200 kb		
Mlu I	A/CGCGT	B, N, P, S
Sal I	G/TCGAC	N
Nru I	TCG/CGA	N
Bss HII	G/CGCGC	
Sac II	CCGC/GG	N
Eag I (*EcI* XI*, *Xma* III)	C/GGCCG	B*, N
Enzymes that cut in CpG Islands: Fragments Average <200 kb		
Nar I	GG/CGCC	N, P
Sma I	CCC/GGG	N
Xho I	C/TCGAG	N
Pvu I	CGAT/CG	N
Apa I	GGGCC/C	N
Enzymes with TAG in their Recognition Sequence		
Avr II (*Bln* I*)	C/CTAGG	N, T*
Nhe I	G/CTAGC	N, P, S
Xba I	T/CTAGA	B, N, P, S
Spe I	A/CTAGT	B, N, P, S, T
Nhe I	G/CTAGC	P
Dra I	TTT/AAA	B, P, S
Ssp I	AAT/ATT	P

[a] Asterisk indicates preferred enzyme and source.

[b] R, A, or G; Y, C, or T; M, A, or C; K, G, or T; S, G, or C; W, A, or T; N, A, or C or G or T.

[c] B, Boehringer Mannheim; N, New England Biolabs; O, not commercially available; P, Promega; S, Stratagene; T, Takara.

[d] Two sites are needed in order for cleavage to occur at both of them.

Note that the recognition site for *Sfi* I contains no CpG's, while the site recognized by *Not* I contains 2 CpG's. In Chapter 1 we showed that the frequency of occurrence of this dinucleotide sequence is reduced in mammalian DNA to about 1/4 of the level expected statistically. On this basis *Not* I can be expected to behave like a ten-base specific enzyme rather than an eight-base specific enzyme; as a result the expected fragment sizes are predicted to lie around 1 Mb, in agreement with experimental results. However, this is an oversimplified argument for it ignores the effect of DNA methylation. Overall, in the human genome about 80% of the CpG sequences are methylated to 5-meC, and *Not* I (and most other restriction nucleases) is unable to cleave at sites that are methylated. If we factor this effect into the calculation, we can now predict that for random methylation only about 1/25 of the *Not* I sites have no ^mC and thus are cleavable. This yields an average DNA fragment size of 25 Mb, making *Not* I all but useless for conventional macrorestriction mapping. Fortunately the distribution of CpG methylation is not random. In practice, it appears that about 90% of the *Not* I sites in the human genome are not methylated, and so they are accessible to cleavage by the enzyme. This explains why *Not* I can generate fragments that average about 1 Mb in size.

HTF ISLANDS

The peculiar distribution of methylated CpG, which has such a dramatic effect on the frequency of *Not* I cutting sites, is a reflection of a more general statistical unevenness in mammalian genomes. This was first discovered when genomic digestions were carried out with a much less specific restriction enzyme, *Hpa* II. A typical genomic digest generated by this enzyme that recognizes the sequence CCGG is shown in Figure 8.9. A statistically random genome would be expected to give a roughly Gaussian distribution of fragment sizes. Instead, the two-phase distribution observed in practice is striking. *Hpa* II is inhibited by DNA methylation. It cannot cut the sequence C^{m}CGG. To as good approximation, the fragment sizes shown in Figure 8.9 can be fit by assuming that the genome is divided into regions where no methylation occurs and regions where most of the CpG's are methylated. The large fragments in the latter regions are what were expected from an *Hpa* II digest. The small fragments were unexpected. They were named *Hpa* II tiny fragments (HTFs), and the regions that contain them have been called *HTF islands.* As we have learned more about the properties of these regions, many researchers have preferred to call them *CpG islands,* but for others, the original term has stuck.

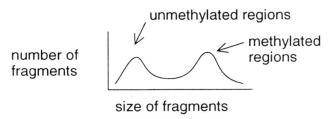

Figure 8.9 Distribution of the sizes of the DNA fragments generated by a complete *Hpa* II digest of a typical mammalian genome.

HTF islands have a number of very interesting properties. They tend to be located near genes, most often at the 5′-edge of genes. They are very rich in G+C. The frequency of CpG is as expected from binomial statistics—that is, there is no suppression of CpG in these regions and no elevation of TpG (which results from ^{m}CpG mutagenesis as described in Chapter 1.) In HTF islands the CpG sequences are unmethylated. These results are self-consistent: If there is no methylation in these regions, there should be no progressive loss of CpG by mutation, and thus the frequency of this sequence should just reflect the local C+G content. More than 90% of the known *Not* I sites appear to be located in HTF islands; this is understandable. To produce a cleavable *Not* I site requires two nearby unmethylated CpG's. This is an event most unlikely to occur in the bulk of the genome where CpG's are both very rare and methylated.

Many HTF islands have been studied by sequencing the DNA flanking *Not* I sites. This is relatively easy to do because, as we will discuss later in the chapter, there are straightforward ways to clone DNA that contains a cutting site for this enzyme (or almost any other enzyme for that matter). Two representative human DNA sequences flanking *Not* I sites on chromosome 21 are described in Figure 8.10, which shows the local base composition as a function of distance from the *Not* I site. The first of these examples is extraordinarily G + C rich throughout a 600–800 base region. The second example shows a transition from an HTF island to more ordinary genomic DNA; the *Not* I site is near the edge of the island. The distribution of CpG, GpC, and TpG sequences in the first of these examples is plotted as a function of position in Figure 8.11. It is evident that GpC's and GpC's are extraordinarily prevalent; more significantly, their prevalence is roughly equivalent, showing the lack of any significant CpG suppression.

ORDERING RESTRICTION FRAGMENTS

The simplest way to view top-down mapping is by projecting a low-resolution map with ill-defined distances onto a higher-resolution map with more carefully defined distances. Suppose that a genetic or cytogenetic map already exists for a chromosome of interest, and this is actually the case today for almost all regions of the human genome, at least at 2 Mb resolution (Chapter 6). Each genetically mapped or cytogenetically mapped DNA marker can be radiolabeled and used as a hybridization probe to identify the corresponding large DNA fragments that it resides on, as shown in Figure 8.12. If DNA fragments can be generated that are comparable in size to the density of available DNA markers, then most of the construction of a restriction map would be accomplished with a relatively small number of direct DNA hybridizations. In reality, for almost all regions of the genome, the probes available today are not dense enough to order a complete set of restriction fragments. Instead one must utilize procedures that allow the restriction map to be extended outward from the position of known fragments into neighboring regions. A second problem is that the DNA fragment sizes generated by total digestion with available rare-cutting enzymes are quite diverse. Thus, although the average fragment size seen with *Not* I is about 1 Mb, many fragments are 3 Mb in size, and many are 0.2 Mb or smaller. Available DNA probes are very likely to be found that recognize the 3-Mb fragments; it is much less likely that any preexisting probes will correspond to the 0.2-Mb fragments at typical probe densities of 1 per 1 to 2 Mb.

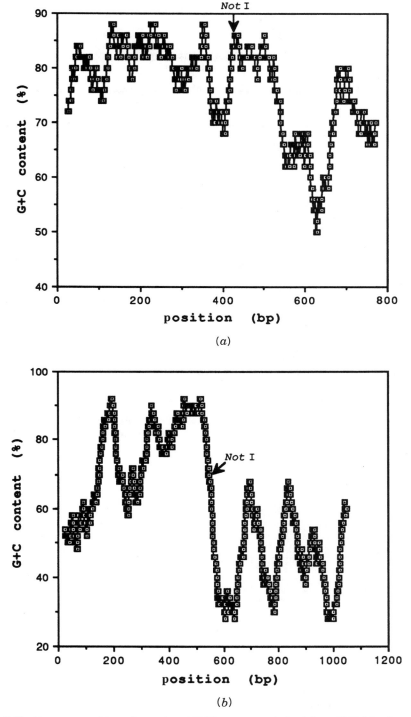

Figure 8.10 Base composition of two cloned DNA segments derived from HTF islands on human chromosome 21. Plotted is the local average base composition as a function of the position within the clone. A *Not* I site that allowed the selective cloning of these DNA pieces is also indicated. (*a*) Clone centered in an HTF island. (*b*) Clone at one edge of an HTF island. (Adapted from Zhu et al., 1993.)

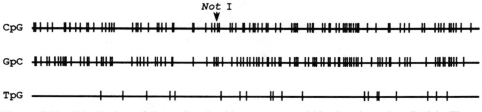

Figure 8.11 Distribution of three dinucleotide sequences within the clone described in Figure 8.10*a*.

To carry out macrorestriction mapping projects efficiently, a number of needs must be met that allow one to circumvent the general problems raised above:

1. Probes must be isolated that are not preferentially located on large restriction fragments.
2. DNA must be cut less frequently to generate large fragments in any desired area of the genome.
3. Isolated probes must correspond to fragments of interest, those fragments where probes are needed to complete a map.
4. Neighboring fragments must be unequivocally identified.
5. Any tiny fragments generated by chance, by enzymes that on average yield very large fragments, must not be ignored.

Methods now exist that deal with all of these problems reasonably efficiently. Most will be described in this chapter; a few will be deferred to Chapter 14 where we deal with some of the specialized methods that have been developed to manipulate particular DNA sequences.

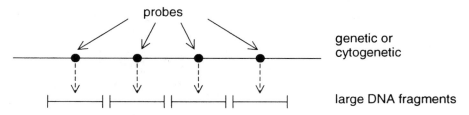

Figure 8.12 The genetic or cytogenetic map provides a set of anchor points to place selected large DNA fragments in order.

IDENTIFYING THE DNA FRAGMENTS GENERATED BY A RARE-CUTTING RESTRICTION ENZYME

The restriction digestion pattern generated by any enzyme that yields pieces large enough to be useful in genomic or chromosomal restriction mapping must be analyzed by PFG. Since this separation technique has a limited window of high resolution, under any fixed experimental conditions it is usually necessary to fractionate the digest by PFG using a range of three or four different pulse times: 30-second pulses for fragments 0 to 200 kb, 60-second pulses for fragments 200 to 1000 kb, 1000-second pulses at lower field strengths for 1 to 3 Mb pieces, and 3600-second pulses or secondary PFG for fragments larger than this.

The entire distribution of DNA fragments can be visualized by staining the PFG fractionated material in the gel with ethidium bromide. This dye binds in between every other base pair (total stoichiometry 0.5 dye per base pair), and shows a more than 25-fold fluorescence enhancement when bound. Ethidium bromide appears to have no significant specificity for any particular base compositions or base sequences. It is sensitive enough for all routine use. Recently a number of other dyes have been reported that may offer the promise of higher-sensitivity DNA detection than ethidium bromide. However, the interactions of these dyes with very large DNA molecules have not yet been fully characterized.

If the genome under analysis is a relatively simple one, ethidium staining allows all the pieces to be visualized, as we showed earlier in Figure 8.5. Since the amount of ethidium bound is proportional to the size of the DNA fragment, the resulting distribution of fluorescence intensity in a result like Figure 8.5 gives the weight average distribution of DNA. Small fragments are very hard to see. In general, a monotonic increase in fragment intensity is expected as fragment size increases. Deviations from this pattern indicate multiplets: size fractions that contain two or more unresolved DNA pieces, or heterogeneity; DNA fragments present in substoichiometric amounts because they arise from a contaminant in the sample, from a restriction site that has been only partially cleaved, or from DNA partially degraded (by nuclease or even by electrophoresis itself). These complications aside, a quantitative analysis of the pattern of ethidium staining will usually allow an accurate analysis of the number of DNA fragments in the genome. When these sizes are summed, an estimate for the total genome size is produced. Indeed, when ethidium staining is carried out very carefully, the stained intensity of DNA bands is not just a monotonic function of their size; it is very close to a linear function of their size up to a few Mb. Thus from the relative staining intensity alone it is sometimes possible to make reliable estimates of DNA sizes, even when proper size standards, for some reason, are not present or useable.

With complex genomes only a smear of ethidium staining intensity is generally seen. At the highest obtainable PFG resolution, the entire human and mouse genome actually show a very discrete and reproducible banding pattern in ethidium-stained gels of DNA digested with *Not* I or other rare-cutting restriction enzymes (Fig. 8.13). The significance of this pattern has never been explained—but it does allow DNA from different species to be identified. In those unusual cases where a major repeating sequence has a rare-cutting site in it, a bright DNA band will sometimes be seen above the background ethidium smear.

To develop a restriction map of a chromosome from a complex genome, like the human genome, it is best to start with that chromosome in a hybrid cell. Even better, for the larger mammalian chromosomes, are hybrid cells that contain only a 10 to 50 Mb fragment of the chromosome of interest. There are two major advantages for starting with such a hybrid cell. First, the chromosome of interest will almost certainly have a unique genotype. Even if multiple copies are present, these are likely to be identical. In effect the sample is homozygous. This eliminates any confusion that would otherwise result if two different polymorphic structures of a region were merged into the same map.

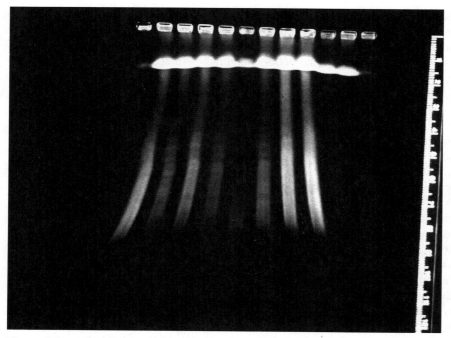

Figure 8.13 Ethidum-stained PFG fractionation of *Not* I digested human and mouse DNA from different cell lines. Fragments resolved in this gel range in size from 50 kb to 1 Mb. The distinct pattern of banding seen depends on the particular species and enzyme used.

An example of such a useful cell line is WAV17 which has two to three identical copies of human chromosome 21 in a mouse background.

The second advantage of using a hybrid cell is that it is possible to view most or all of the human component above the background of DNA from the rodent. This is accomplished by hybridizing a blot of a PFG-fractionated restriction enzyme digest of the cellular DNA with various human-specific repeating DNA sequences. For example, the most common human-specific repeat is the *Alu* sequence. This will be described in much greater detail in Chapter 14. Here, however, it is sufficient to note that this sequence occurs on average about once every 3 kb in some regions of the human genome and about once every 10 kb in others. Probes for *Alu* exist that show little significant cross-hybridization with rodent DNA. These can then be used as a human-specific stain to detect the presence of human restriction fragments in a mouse or hamster cell. An example of such an analysis is shown in Figure 8.14.

The human *Alu* sequence should be seen on almost every large fragment of human DNA if our genome were well fit by a statistically random model. This can be shown by using a simple Poisson model for the distribution of *Alu*'s. From the Poisson distribution (Chapter 6) we can estimate that the probability of an *Alu* not occurring on a fragment of interest will be

$$P(\text{no } Alu) = e^{(-\text{average fragment size}/Alu \text{ spacing})}$$

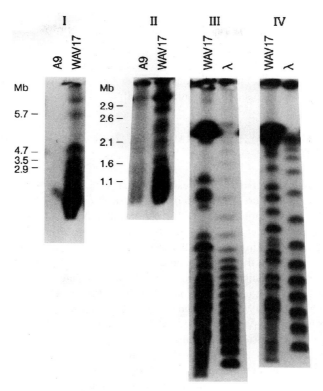

Figure 8.14 PFG fractionation of *Not* I-digested DNA from a mouse cell line that contains chromosome 21 as the only human component. After electrophoresis the gel was blotted, and the blot was hybridized with the highly repeated human-specific *Alu* sequence. Shown in some of the panels are size standards used to estimate the length of particular human DNA fragments. (Taken from Sainz et al., 1992.)

The smallest fragments of much interest in human macrorestriction mapping are around 50 kb; the least *Alu*-dense regions of the genome have a 10-kb spacing between *Alu*'s; thus in these regions the chance that a 50-kb fragment has no *Alu* is exp(-5). The chance, at random, that a 1 Mb fragment should lack an *Alu* is infinitesimal. We have actually characterized the distribution of *Alu*'s on more than 50 *Not* I fragments of human chromosome 21. In practice, all but one contain *Alu* as detected by hybridization experiments. The sole exception, however, is a 2.3 Mb fragment. The probability of seeing such an event at random is e^{-230} assuming it derives from an *Alu*-poor region of the chromosome. Clearly we will have to refine our statistical picture of human DNA sequences quite considerably as more cases like this surface.

When human DNA fragments are detected by hybridization with human-specific repeated sequences the resulting distribution of intensity should still be generally proportional to fragment size. Note, however, that considerable variations around this mean will occur because repeated sequences tend to cluster and because small fragments with small numbers of repeats on average will show typical small number fluctuations. One way to generate labeling intensity that is independent of fragment size is to label the ends.

This can be done with enzymes like polynucleotide kinase that place a phosphate on the 5′-end of a DNA chain, or it can be done by filling in any inset 3′-ends of duplex DNA with DNA polymerase and radiolabeled dppN's. This procedure is readily applied within agarose gels, and it has been useful in the analysis of small genomes. For mammalian genomes the procedure is not useful because there is no specific way to label just the ends of the human DNA fragments in a rodent hybrid cell.

MAPPING IN CASES WHERE FRAGMENT LENGTHS CAN BE MEASURED DIRECTLY

The classical approach to constructing restriction maps of small DNAs like plasmids and viruses is illustrated in Figure 8.15. Two or more restriction enzymes are used separately and in double digests to fragment the DNA of interest. The sizes of all pieces seen in an ethidium-stained gel are measured. Usually the pattern of sizes allows alignment of the different cutting sites in a single map. This procedure is clearly not a rigorous one, and it breaks down severely once the maps become complex, or when many similar-sized fragments are involved. In principle, each fragment could be isolated by electrophoretic fractionation, radiolabeled, and used as a probe. This would allow all overlapping fragments from digests with other enzymes to be unambiguously identified. In practice, however, it is usually easier to employ partial digestion strategies with end-labeled probes, as we will describe later for macrorestriction mapping.

With mammalian DNAs, multiple restriction enzyme digest mapping is much less effective. Although one can determine the size of the fragments generated by each enzyme, by using repeated sequence hybridizations, there is an annoying tendency for many of the restriction enzymes that yield large DNA fragments to cut in the same regions. This is because most such enzymes prefer HTF islands. Thus the usefulness of double digestion in most regions of mammalian genomes is far less than illustrated by the example in Figure 8.15. A more serious problem is that with large numbers of fragments in single-enzyme digests, double digests become hopelessly complicated to analyze. With mammalian DNA, in contrast to DNA from simple genomes, one cannot easily access each purified fragment because it is contaminated by other human or rodent fragments. PCR can help circumvent this problem, as we will demonstrate in Chapter 14. In general, though, one must rely on hybridization with single-copy probes in order to simplify the pattern of DNA fragments to the point where it can be analyzed. The example in Figure 8.15 shows clearly that fragments from the end of large DNA pieces are particularly useful hybridization probes. The figure also indicates that with only a limited set of probes from the region, double digests are frequently impossible to analyze because many of the fragments in these digests will not correspond to any of the available probes. This discussion should make it clear that new strategies had to be developed to simplify the construction of macrorestriction maps of segments of complex genomes.

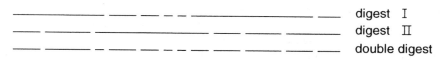

Figure 8.15 Schematic illustration of the double digestion procedure used to assemble simple restriction maps.

GENERATION OF LARGER DNA FRAGMENT SIZES

Some of the problems illustrated by the example in Figure 8.15 would be alleviated if there were a systematic way to generate large DNA fragments in a region of interest. One general approach for doing this will be discussed here; others, like the RARE method, are deferred to Chapter 14.

It has already been mentioned that most restriction enzymes are inhibited by DNA methylation. One can take advantage of this to increase the specificity of certain restriction enzymes by methylating a subset of their cutting sites. Methylation can also be used as a general way to promote partial digests by using a methylase that recognizes all the cutting sites but does not allow the reaction to go to completion. Here, however, we deal with methylation reactions that are carried to completion. Consider the DNA sequence shown below. It contains a *Not* I site flanked by additional G–C pairs shown in boldface:

$$\textbf{C}\text{GCGGCCGC}\textbf{G} \quad \rightarrow \quad {}^m\text{CGCGGC}{}^m\text{CGCG}$$

$$\textbf{G}\text{CGCCGGCG}\textbf{C} \qquad \text{GCG}{}^m\text{CCGGCG}{}^m\text{C}$$

Roughly one-quarter of all *Not* I sites will have an extra C at their 5′-end; an additional quarter will have an extra G at their 3′-end as shown in the above example. These extra residues generate a recognition site for the *Fnu*D II methylase that converts CGCG to ^{m}CGCG. Since some of the methylation is within the *Not* I cutting site, this inhibits any subsequent attempts to cleave the site with *Not* I. Thus by methylation one can inactivate about half of all the *Not* I cleavage sites and double the average fragment size generated by this enzyme. Many variations on this theme exist.

Methylation also plays a key role in a whole set of potential schemes for site-selective cleavage of DNA employing the unusual restriction endonuclease *Dpn* I. This enzyme recognizes the sequence GATC, but it requires that the A be methylated in order for cleavage to occur. The preferred substrate is

$$\text{G}{}^m\text{ATC}$$

$$\text{CT}{}^m\text{AG}$$

A major complication is that the monomethyl derivative is also cut, but much more slowly. Hence, in the schemes described below, it is essential to expose the DNA substrate to *Dpn* I for the minimum time needed to cleave at the desired sites before the background of undesired additional cleavages becomes overwhelming.

Dpn I is converted to an infrequently cutting enzyme by starting with unmethylated target DNA, which *Dpn* I cannot cleave at all, and selectively introducing methyl groups by treatment with methylases with recognition sequences that overlap part of the *Dpn* I site. The simplest example, shown below, employs the *Taq* I methylase. This enzyme recognizes the sequence TCGA and converts it to TCG^{m}A. If two such sequences lie adjacent to each other in a genome, the result, once both are methylated is

$$\text{TCG}{}^m\text{ATCG}{}^m\text{A}$$

$${}^m\text{AGCT}{}^m\text{AGCT}$$

Comparison of this sequence with the *Dpn* I recognition site shown above indicates that a *Dpn* I cleavage site has been generated. An eight-base sequence is required to specify this site by the procedure we have used. Since it contains two CpG's, it will be a very rare site in mammalian genomes. Thirty to 40 variations on this theme exist, some generating sites as large as 16 base pairs. Some of these sites are predicted to occur less than once, on average, in most genomes. However, one difficulty with these schemes, is the lack of availability of many of the necessary methylases in sufficiently nuclease-free form to generate the very large DNA fragments specified by such large recognition sites. At present the exploitation of *Dpn* I remains an extremely attractive method that is not yet generally practiced because of some of these unsolved experimental limitations.

LINKING CLONES

In this and the next few sections we discuss several of the methods that allow the order of restriction fragments to be determined, de novo, without access to other mapping information. By far the most robust of these, in many respects, is the use of specialized clones called *linking clones.* These clones contain the same rare-cutting sites that were used to produce macrorestriction fragments from the sample to be mapped. As shown in Figure 8.16, because these clones have an internal rare-cutting site, they must overlap two adjacent large DNA fragments. (The sequence properties of two authentic human linking clones were illustrated in Figs. 8.10 and 8.11.) When used as hybridization probes, the linking clones should identify two fragments that are adjacent in the genome. This will occur unless the rare-cutting site is so close to one edge of the linking clone that not enough single-copy material remains beyond the site to hybridize effectively. To eliminate this possibility, it is useful to work with more than one set of linking clones, constructed in such a way as to have different distributions of DNA flanking the rare-cutting site. A convenient way to do this is to start with separate total digest libraries made with different restriction nucleases such as *Eco*R I and *Hin*d III. In principle, there is sufficient information in linking clone hybridizations that if a complete linking library were available, its use in hybridizations would produce data that would allow the entire set of restriction fragments to be ordered.

Two of the methods that have been used to prepare libraries of linking clones are summarized in Figure 8.17. In each case one starts with a small insert genomic library contained into a circular vector that is lacking any sites for the restriction enzyme of interest. If such a library does not preexist, it is easily constructed by subcloning an existing library into such a vector. DNA from the entire library is purified and digested with the rare-cutting restriction nuclease. Only those clones containing a genomic insert that includes a site for this enzyme will be linearized; the rest remain circular. Two different methods can be used to select out the clones that have been linearized. One approach is to use trapping electrophoresis: ordinary or PFG electrophoresis at high fields such that linear molecules migrate into the gel effectively while circles remain trapped in the sample

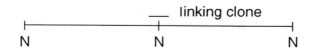

Figure 8.16   A linking clone spans two adjacent large DNA fragments.

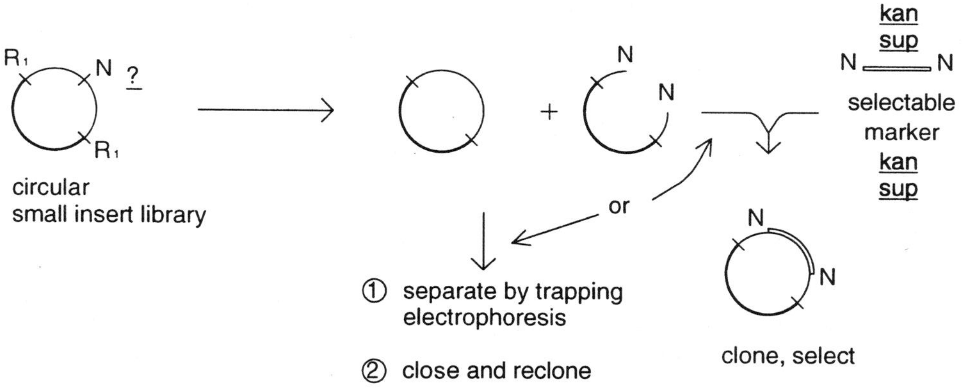

Figure 8.17 Two procedures used to make libraries of linking clones. Both require starting with a library of the desired chromosome (or other target) in a circular vector.

well or sample plug. In this way the linking clones are selectively captured. They can be religated at low concentration to avoid the production of chimeras, and then re-introduced into a bacterial host.

The alternative approach, after the rare-site-containing clones have been linearized, is to ligate a selectable marker into the rare-cutting site. Both suppressor tRNA genes and kanamycin resistance genes have been successfully used for this purpose. The library is then used to transform a bacterial host in the presence of the selection. The only clones that should survive are the linking clones. Both of the methods described work reasonably well. However, the procedures are not perfect, and it is important, after a linking library has been made, to screen out artifactual clones. These may include such annoyances as chimeras of rare cutter fragments, clones that contain a rare cutter site that is not actually cut in the parental genome because it is methylated, and clones containing small individual rare cutter fragments. One way to screen for useful clones is to use them as probes in Southern hybridizations with genomic DNA singly and doubly digested with the restriction enzyme used for constructing the library such as *Eco*R I or *Hin*d III, and the rare-cutting enzyme. Proper linking clones will show single bands or no bands in the absence of rare cutter cleavage, and double bands once the rare site has been cut.

Once linking clones are available, they have many other useful applications besides the ordering of macrorestriction fragments. Linking clones serve as effective bridges between two different kinds of restriction nuclease digests (or other types of DNA samples) because they come from a known position, the ends, of large DNA fragments. As a second example, one can cut linking clones at their internal rare cutter site and isolate the two fragments. The resulting samples are called *half-linking clones*. They are useful for the analysis of the internal structure of large DNA fragments, as shown by the schematic experiment illustrated in Figure 8.18. Here a *Not* I half-linking clone is used as an indirect end label by hybridizing it to a blot of a PFG-fractionated partial *Eco*R I digest of a *Not* I fragment. The sizes of the DNA pieces seen in the partial digest provide the location of the internal *Eco*R I sites. This type of analysis was originally developed for smaller DNA pieces by Smith and Birnstiel. It is a very powerful approach. Half-linking clones are also very useful for the analysis of partial digests with enzymes that cut infrequently, as we will demonstrate later.

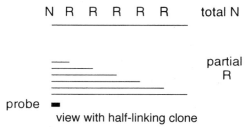

view with half-linking clone

Figure 8.18   Use of a half-linking clone as an indirect end label to reveal the pattern of internal restriction sites in a partial digest. This approach to restriction mapping, called the Smith-Birnstiel method after the researchers who first described it, remains the most powerful way to accumulate large amounts of restriction map data rapidly.

Another potential use of linking clones will be to order complete digest libraries of cloned large DNA fragments, once we are able to produce such libraries. The difficulty today is that too large a percentage of the fragments generated from mammalian DNA by an enzyme like *Not* I exceed the capacities of current large insert cloning vectors. However, an example of the potential power of this approach can be seen for an enzyme like *Eag* I that recognizes the sequence CCGCCG, which is the internal 6 base pairs of the *Not* I recognition sequence. *Eag* I cuts genomic DNA into fragments that average 200 to 300 kb in size. These fragments can be cloned into conventional YAC vectors once the vectors are equipped with the proper cloning sites. The result of probing such a library with *Eag* I linking clones is shown in Figure 8.19. In principle, there is enough information in the two sets of clones, linking clones and YACs, to completely order both sets of samples, and the method has two advantages. The materials used are only a tiny bit larger than the minimum possible tiling path for any ordered library, and distances along this path are known with great precision. In practice, there is no example of a library that has been fully ordered in this way.

One limitation with currently used linking clones must be realized. These clones tend to come from very G + C rich regions because this is where the sites of most rare-cutting enzymes are preferentially located. These regions make subsequent PCR analyses fairly difficult because, in current PCR protocols, very G + C rich primers and templates do not work especially well. Secondary structure in the single strands of these materials is presumably the major cause of the problems, but a generally effective solution to these problems is not yet in hand.

Figure 8.19   Linking clones and large DNA fragments generated by the same enzyme could be assembled, in principle, into an ordered library with almost the minimum possible tiling length.

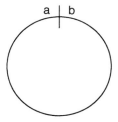

Figure 8.20 Basic notion of a jumping clone: Two discontinuous pieces of the genome (*a*) and (*b*), but related by some map or fragment information, are assembled into the same clone.

JUMPING LIBRARIES

Jumping clones offer a way of dealing with two discontinuous pieces of a chromosome. The basic notion of a jumping clone is shown in Figure 8.20. It is an ordinary small insert clone except that two distinct and distant pieces of the original DNA target are brought together and fused in the process of cloning. A simple case of a jumping clone would be one that contained the two ends of a large DNA fragment, but all of the internal DNA of the fragment had been excised out. The way in which such clones can be made is shown in Figure 8.21.

A genomic digest with a rare-cutting enzyme is diluted to very low concentration and ligated in the presence of a vector containing a selectable marker. The goal is to cyclize the large fragments around vector sequences. At sufficiently low concentrations of target, it becomes very unlikely that any intermolecular ligation of target fragments will occur. One can use excess vector, provided that it is dephosphorylated so that vector–vector ligation is not possible. When successful, ligations at low concentrations can produce very large DNA circles. These are then digested with a restriction enzyme that has much more frequent sites in the target DNA but no sites in the vector. This can be arranged by suitable design of the vector.

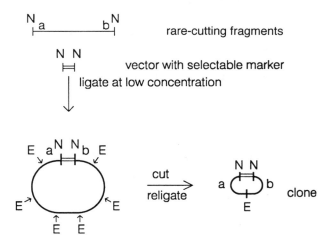

Figure 8.21 Method for producing a library of jumping clones derived from the ends of large DNA fragments.

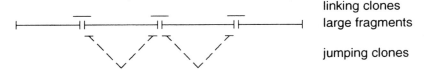

Figure 8.22 Potential utility of jumping clones that span large DNA fragments.

The result is a mixture of relatively small DNA pieces; only those near the junction of the two ends of the original large DNA fragments contain vector sequences. A second ligation is now carried out, again at very low concentration. This circularizes all of the DNAs in the sample. When these are reintroduced into *E. coli,* under conditions where selectable markers on the vector are required for growth, a jumping library results.

There are potentially powerful uses for jumping libraries that span adjacent sites of infrequently cutting restriction enzymes. These are illustrated in Figure 8.22. Note that each of the jumping clones will overlap two linking clones. If one systematically examines which linking clones share DNA sequences with jumping clones, and vice versa, the result will be to order all the linking clones and all the jumping clones. This could be done by hybridization, PCR, or by direct DNA sequencing near the rare-cutting sites. In the latter case DNA sequence comparisons will reveal clone overlaps. Such an approach, called *sequence-tagged rare restriction sites* (STARs), becomes increasingly attractive as the ease and throughput of automated DNA sequencing improves.

A more general form of jumping library is shown schematically in Figure 8.23. Here a genome is partially digested by a relatively frequent cutting restriction nuclease. The resulting, very complex, mixture is fractionated by PFG, and a very narrow size range of material is selected. If this is centered about 500 kb, then the sample contains all contiguous blocks of 500-kb DNA in the genome. This material is used to make a jumping library in the same general manner as described above, by circularizing it around a vector at low concentration, excising internal material, and recircularizing. The resulting library is a size-selected jumping library, consisting, in principle, of all discontinuous sets of short DNA sequences spaced 500 kb apart in the genome. The major disadvantage of this library is that it is very complex. However, it is also very useful, as shown in Figure 8.24a. Suppose that one has a marker in a region of interest, and one would like another marker spaced approximately 500 kb away. The original marker is used to screen a 500-

(*a*) 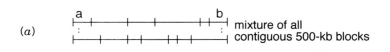 mixture of all contiguous 500-kb blocks

(*b*)   all short sequences spaced 900 kb in genome

Figure 8.23 DNA preparation (*a*) used to generate a more general jumping library (*b*) consisting of a very dense sampling of genomic fragments of a fixed length.

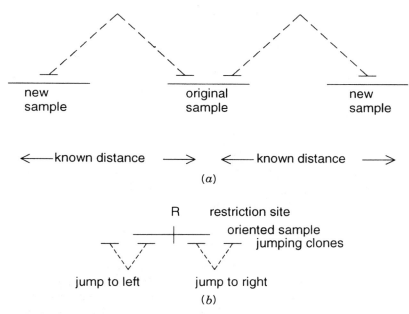

Figure 8.24 Example of the use of a general jumping library to move rapidly from one site in a genome to distantly spaced sites. (*a*) Half-jumping clones provide new probes, at known distances from the starting probe. (*b*) Information about the direction of the jump can be preserved if the map orientation of the original probe is known.

kb jumping library. This will result in the identification of clones that can be cut into half-jumping probes that flank the original marker by 500 kb on either side. It is possible to tell which jump has occurred in which direction if the original marker is oriented with respect to some other markers in the genome and if it contains a convenient internal restriction site (Fig. 8.24*b*). Selection of jumping clones by using portions of the original marker will let information about its orientation be preserved after the jump. The major limitation in the use of this otherwise very powerful approach is that it is relatively hard to make long jumps because it is difficult to ligate such large DNA circles efficiently.

PARTIAL DIGESTION

In many regions of the human genome, it is difficult to find any restriction enzyme that consistently gives DNA fragments larger than 200 kb. In most regions, genomic restriction fragments are rarely more than a few Mb in size. This is quite inefficient for low-resolution mapping, since PFG is capable of resolving DNA fragments up to 5 or 7 Mb in size. To take advantage of the very large size range of PFG, it is often extremely useful to do partial digests with enzymes that cut the genome infrequently. However, there is a basic problem in trying to analyze the result of such a digest to generate a unique map of the region. This problem is illustrated in Figure 8.25. If a single probe is used to examine the digest by hybridization after PFG fractionation, the probe will detect a mixture of DNA fragments that extends from its original location in both directions. It is not straightforward to analyze this mixture of products and deduce the map that generated it.

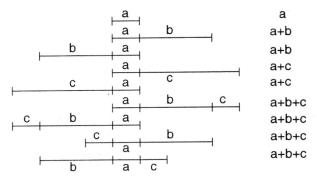

Figure 8.25   Ambiguity in interpreting the fragment pattern seen in a partial restriction nuclease digest, when hybridized with only a single probe.

If the enzyme cut all sites with equal kinetics, one could probably use likelihood or maximum entropy arguments to select the most likely fragment orderings consistent with a given digestion pattern. The difficulty is that many restriction enzymes appear to have preferred cutting sites. In the case of *Sfi* I these are generated by the peculiarities of the interrupted recognition sequence. In the case of *Not* I and many other enzymes, hot spots for cutting appear to occur in regions where a number of sites are clustered very closely in the genome (Fig. 8.26). Since the possibility of a hot spot in a particular region is hard to rule out, a priori, one plausible candidate for a restriction map that fits the partial digestion data is a null map in which all partially cleaved sites lie to one side of the probe and a fully cut hot spot lies to the other side (Fig. 8.27). While such a map is statistically implausible in a random genome, it becomes quite reasonable once the possibility of hot spots is allowed.

There are three special cases of partial digest patterns that can be analyzed without the complications we have just raised above. The first of these occurs when the region is flanked by a site that is cut in every molecule under examination. There are two ways to generate this site (which is, in effect, a known hot spot). Suppose that the probe used to analyze the digest is very close to the end of a chromosome (Fig. 8.28). Then the null map is the correct map. Using a telomeric probe is equivalent to the Smith-Birnstiel mapping approach we discussed earlier for smaller DNA fragments. It is worth noting that even if the probe is not at the very end of the mapped region, any prior knowledge that it is close to the end will greatly simplify the analysis of partial digest data. A second case of such a site occurs if we can integrate into a chromosome the recognition sequence of a very rare cutting enzyme, such that this will be the only site of its kind in the region of interest.

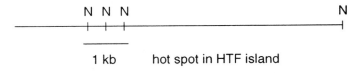

Figure 8.26   Example of a restriction enzyme hot spot that can confound attempts to use partial digest mapping methods.

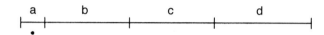

Figure 8.27 Null ordering of fragments seen in a partial digest.

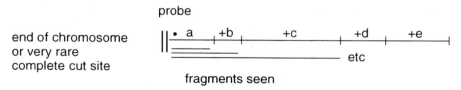

Figure 8.28 Assembling a restriction map from partial digest data when a totally digested site (or a chromosome end) is known to be present in the region.

Then we cleave at the very rare site completely and partially digest with an enzyme that has somewhat more frequent cleavage sites. The power of this approach as a rapid mapping method is considerable. However, it is dependent on the ability to drive the very rare cleavage to completion. As we have indicated before, in many of the existing schemes for very rare cleavage, total digestion cannot be guaranteed. Thierry and Dujon in Paris have recently shown that sites for the nuclease I-*Sce* I (Table 8.1) can be inserted at convenient densities in the yeast genome and that the enzyme cuts completely enough to make the strategy we have just described very effective.

A second case occurs that allows analysis of partial digestion data if the region of interest is next to a very large DNA fragment. Then, as shown in Figure 8.29, all DNA fragments seen in the partial digest that have a size less than the very large fragment can be ordered so long as the digest is probed by hybridization with a probe that is located on the first small piece next to the very large fragment. Again, as in the case of telomeric probes, relaxing this constraint a bit still enables considerable map information to be inferred from the digest.

The third case where partial digests can be analyzed occurs when one has two probes known, a priori, to lie on adjacent DNA fragments. This is precisely the case in hand, for example, when two half-linking clones are used as probes of a partial digest made by using the same rare site present on the clones. In this case, as shown in Figure 8.30, those fragment sizes seen with one probe and not the other must extend in the direction of that probe. Bands seen with both probes are generally not that informative. Thus linking clones play a key role in the efficient analysis of partial digests.

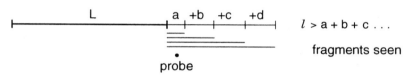

Figure 8.29 Assembling a restriction map from partial digest data adjacent to a very large restriction fragment.

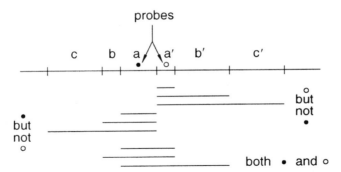

Figure 8.30 Assembling a restriction map when two hybridization probes are available and are known to lie on adjacent DNA fragments. Such probes are available, for example, as the halves of a linking clone.

The power of the approaches that employ partial digests is that a probe in one location can be used to reach out and obtain map data across considerable distances, where no other probes may yet exist. In order to carry out these experiments, however, two requirements must be met. First, reliable length standards and good high-resolution PFG fractionations are essential. The data afforded by a partial digest is all inherent in the lengths of the DNA bands seen. For example, if one used a probe known to reside on a 400-kb band, and found in a partial 1000-kb band, this is evidence that a 600-kb band neighbors the 400-kb band. One could then try to isolate specific probes from the 600-kb region to try to prove this assignment. However, if the 1000-kb band was mis-sized, and it was really only 900 kb, when one went to find probes in the 600-kb region of a gel, this would be the wrong region. The second constraint for effective partial digest analysis is very sensitive hybridization protocols. The yields of DNA pieces in partials can easily be only 1 to 10%. Detecting these requires hybridizations that are 10 to 100 times as sensitive as those needed for ordinary single-copy DNA targets. Autoradiographic exposures of one to two weeks are not uncommon in the analysis of partial digests with infrequently cutting enzymes.

EXPLOITING DNA POLYMORPHISMS TO ASSIST MAPPING

Suppose that two different DNA probes detect a *Not* I fragment 800 kb long. How can we tell if they lie on the same fragment, or if it just a coincidence and they derive from two different fragments with sizes too similar to resolve? One approach is to cut the *Not* I digest with a second relatively rare-cutting enzyme. If the band seen by one probe shortens, and the band seen by the other does not, we know two different fragments are involved. If both bands shorten after the second digestion, the result is ambiguous, unless two different size bands are seen by the two probes and the sum of their sizes is greater than the size of the band originally seen in the *Not* I digest.

A more reliable approach is to try the two different probes on DNA isolated from a series of different cell lines. In practice, eight cell lines with very different characteristics usually suffices. When this is done with a single DNA probe, usually one or more of the lines shows a significant size difference or cutting difference from the others (Fig. 8.31). There are many potential origins for this polymorphism. Mammalian genomes are rampant with tandem repeats, and these differ in size substantially from individual to individual.

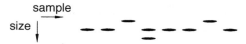

Figure 8.31 Example of the polymorphism in large restriction fragments seen with a single-copy DNA probe when a number of different cell lines are compared.

Most rare-cutting sites are methylation sensitive, and especially in tissue culture cell lines, a quite heterogeneous pattern of methylation frequently occurs. There are also frequent genetic polymorphisms at rare-cutting enzyme sites: These RFLPs arise because the sites contain CpGs that are potential mutation hotspots. For the purposes of map construction, the source of the polymorphism is almost irrelevant. The basic idea is that if two different probes share the same pattern of polymorphisms across a series of cell lines, whatever its cause, they almost certainly must derive from the same DNA fragment. One caveat to this statement must be noted. Overloading a sample used for PFG analysis will slow down all bands. If one cell line is overloaded, it will look like the apparent size of a particular *Not* I band has increased, but in practice, this is an artifact. To avoid this problem, it is best to work at very low DNA concentrations, especially when PFG is used to analyze polymorphisms. Further details are given in Chapter 5.

Polymorphism patterns can also help to link up adjacent fragments. Consider the example shown in Figure 8.32. Here one has probes for two DNA fragments that happen to share a common polymorphic site. In a typical case the site is cut in some cell lines, partially cut in others, and not cut at all in others. The pattern of bands seen by the two

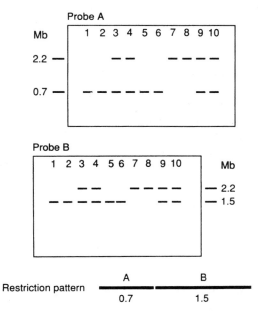

Figure 8.32 Effect of a polymorphic restriction site on the patterns of DNA fragments seen with two probes that flank this site (from Oliva et al., 1991). Hybridization of DNA from different cell lines (lanes 1–10) with two putatively linked probes (A and B) leads to the detection of different and common fragment sizes but identical polymorphism patterns.

probes will be correlated across a set of cell lines in a precise way. Furthermore one should see a band in some cell lines that is the sum of the two fragment lengths seen in others. This provides evidence that the two probes are, in fact, seeing neighboring bands. More complex cases exist where there are multiple polymorphisms in a small region. These become very hard to analyze. In fact such regions may not be that rare; there is some evidence from a few of the restriction maps that have been made thus far that polymorphism, at the level of macrorestriction maps is a patchy phenomenon with regions relatively homogeneous across a set of cell lines alternating with regions that are much more variable. Methylation differences are probably at the heart of this phenomenon, but whether it has any biological significance or is just an artifact of in vitro tissue culture conditions is not yet known.

PLACING SMALL FRAGMENTS ON MAPS

In most macrorestriction maps that have been constructed to date, occasional very small restriction fragments have been missed. These fragments add little value to the map, and they inevitably show up in subsequent work when a region is mapped more finely or converted to an ordered clone bank. However, from a purely aesthetic standpoint, it is undesirable to leave gaps in a map. The real question is how to fill these gaps with minimum effort. One approach that has been reasonably successful, is to end label DNA fragments from a complete digest with an infrequently cutting enzyme and use PFG with short short pulses or conventional electrophoresis to detect the presence of any small fragments. Once these are identified, a simple way to place them on the restriction map is to use them as a probe in hybridization against a partial digest. As shown by Figure 8.33, the small fragment itself is invisible in such a digest. However, when it is fused to fragments on either side, the sizes of these pieces will be detected in the partial digest, and in almost all cases these sizes will be sufficiently characteristic to place the small fragment uniquely on the map.

An alternative approach to identifying small fragments is to use PCR. Splints are ligated onto the sticky ends generated by the restriction enzyme digestion. Primers specific for the splints are then used for PCR amplification. Given the limited size range of typical PCR amplifications, no macrorestriction fragments will be amplified. The PCR product will consist of only small restriction fragments if any of these were generated by the digest. These can be fractionated by size and individually used to probe a partial digest, as described above, in order to determine the position of the small fragments on the macrorestriction map.

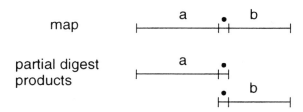

Figure 8.33 Placing a small DNA fragment on a macrorestriction map.

REACHING THE ENDS OF THE PHYSICAL MAP: CLONING TELOMERES

Ten years ago there was a tremendous flurry of activity in cloning mammalian telomere-associated sequences. This was motivated by several factors. The simple sequence repeat at the very end of mammalian chromosomes had been determined, but little was known about what kinds of DNA sequences lay immediately proximal to this. Such sequences should be of considerable biological interest, since they ought to play a role in chromosome function, and they might help determine the identity of different chromosomes. Clones near telomeres will be extremely useful probes for partial digests, as we have already described. Finally telomeres are, by definition, the ends of all linear chromosome maps, and until they are anchored to the remainder of a genetic or physical map, one cannot say that that map is truly complete.

True telomeres would not be expected to be present in plasmid, bacteriophage, or cosmid libraries, and in fact it has been all but impossible to find them there. There are at least two explanations for this. First, the structure of the simple sequence telomere repeat contains a hairpin rather than a normal duplex or single-stranded overhang (Chapter 2). This hairpin is quite stable under the conditions used for ordinary DNA cloning. It will not ligate under these conditions, and thus one would expect that the telomeres would fail to be cloned. Second, simple tandem repeats are not very stable in typical cloning vectors; they are easily lost by recombination; the 10- to 30-kb simple human telomeric repeat would almost certainly not be stable in typical bacterial cloning strains. The sequences distal to the simple repeat are themselves rich in medium-size tandem repeats, and whole blocks of such sequences are repeated. Thus these regions are also likely to be quite unstable in *E. coli*. For all these reasons it is not surprising that early attempts to find telomeres in conventional libraries failed.

To circumvent the problems discussed above and find mammalian telomeric DNA, about six research groups simultaneously developed procedures for selectively cloning these sequences in the yeast, *S. cerevisiae*. The standard cloning vector used in yeast is a yeast artificial chromosome (YAC; see Box 8.2). We and others have reasoned that because telomeres are such key elements in chromosome function, and because the properties of telomeres are so well conserved through most species, there was a chance that a mammalian telomere can function in yeast even if its structure is not identical to the yeast telomere and associated sequences. To test this idea, we used half YACs as cloning vectors. To survive in yeast as a stable species, the half YAC would have to become ligated to another telomere-containing DNA fragment.

We did not expect that the process of cloning telomeres would be very efficient. To enhance the chances for a human telomere to be successfully cloned, we developed the procedure shown in Figure 8.34. Total genomic human DNA was digested with the restriction enzyme *Eco*R I. The reaction mixture was diluted, and ligase was added. All normal *Eco*R I DNA fragments have two sticky ends generated by the original restriction enzyme digest. These are capable of ligation, and at the reduced concentrations the primary ligation products become intramolecular circles (just as in the case of the jumping library construction). The only fragments that do not lose their sticky ends by ligation are telomeres, which have only one sticky end and cannot circularize, and very short *Eco*R I fragments, which cannot bend enough to form a circle. By this ligation procedure most of the nontelomeric restriction fragments can be selectively eliminated; in practice, about a 10^4 enrichment for telomeric restriction fragments is produced.

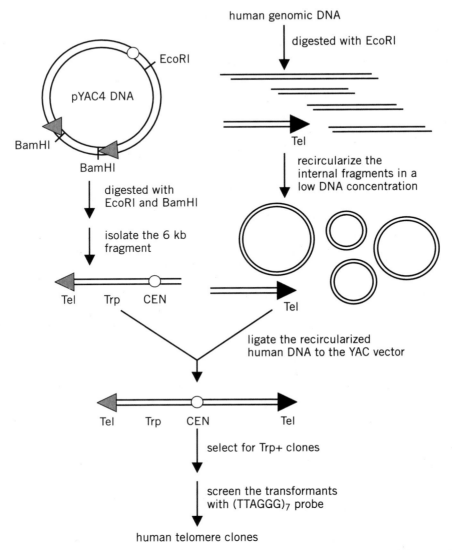

Figure 8.34   Strategy used to pre-enrich human DNA for telomeric fragments and then to clone these fragments. (From Cheng et al., 1989.)

The resulting mixture was then ligated to an *Eco*R I-digested YAC vector. DNA was transfected into yeast, and colonies that grew in the presence of the selectable marker carried on the half YAC arm were screened for the presence of human repeated sequences and human telomere simple repeats. The first clone found in this way proved to be an authentic human telomere. Its structure is shown in Figure 8.35. The human origin of the clone is guaranteed by the presence of a human *Alu* repeat, and a short stretch of human-specific telomere simple sequence. Most of the simple sequence that originally had to be present at the authentic human telomere has been lost, and some of it has been replaced

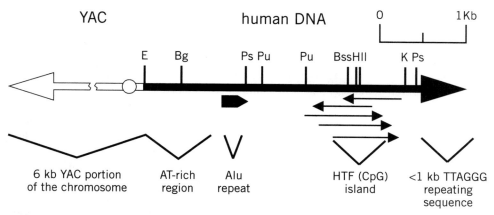

Figure 8.35 Structure of a cloned human telomere, as isolated in yeast. Clone yHT1 is a chimeric minichromosome containing human DNA (filled arrow) and yeast DNA (open arrow). Arrowheads represent regions of telomeric simple repeats. The open circle represents a yeast centromere. Locations of *Eco*R I(E), *Bgl* II (Bg), Pst I (Ps), *Pvu* II (Pv), and *Kpn* I (K) restriction sites are shown above the map. The position of an *Alu* repetitive element is indicated by a solid arrowhead below the map, which points in the direction of the poly(dA) stretch. (From Cheng et al., 1989.)

BOX 8.2
CLONING LARGE DNA FRAGMENTS AS
ARTIFICIAL CHROMOSOMES

Both yeast and bacterial large fragment cloning systems are developed. Yeast artificial chromosomes (YACs) are constructed using small plasmids that are grown in *E. coli*, as shown in Figure 8.36. The YAC vectors have telomere sequences at each of their ends, a centromere sequence, and usually a DNA replication origin and selectable marker in each arm of the vector. Large fragments are cloned into a multicloning site occurring between the two arms of the YAC. Initially both arms of the YAC were cloned into the same plasmid. This meant that the plasmid was cleaved both at the cloning site and at a site between the converging telomeric sequences. Subsequently the vector arms were divided between two plasmids. The vector arms are ligated to high molecular DNA that has been partially digested with a restriction enzyme like *Eco*R I. The recombinant DNAs are introduced into *S. cerevisiae* cells chemically treated so that they could take up DNA. YACs up to about 1 Mb have been created.

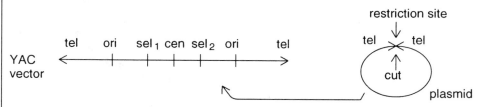

Figure 8.36 General design of yeast artificial chromosome (YAC) cloning vectors and how they are propagated in bacteria: tel, telomere, ori, replication origin, cen, centromere, sel1 and sel2, selectable markers.

(continued)

BOX 8.2 *(Continued)*

Even though transformation frequencies were very low (hundreds per mg of input DNA), eventually most of the human genome was cloned in this manner. The YAC libraries have facilitated the cloning of many human genes. YACs are not as easy to manipulate as bacterial clones, and many are chimeric, containing DNA from more than one region of the human genome, or contain interstitial deletions.

The large insert bacterial cloning systems are based on well-characterized autonomously replicating extrachromosomal DNA elements of *E. coli*. P1 artificial chromosomes (PACs) use vector sequences for P1 bacteriophage (Box 2.3), whereas bacterial artificial chromosomes (BACs) are based on the fertility factor (F-factor, Figure 8.37). P1 bacteriophage replicates as an extrachromosomal low-copy plasmid and as a high-copy lytic phage. Induction of the lytic replicon in PACs ensures that large amounts of recombinant PAC DNA can be synthesized. Although both the P1 bacteriophage and F-factors genomes are 100 kb in size, only a small portion of these genome codes for essential functions and can be deleted for cloning. Up to about 2.5 Mb of *E. coli* DNA has been known to be stably maintained in F-plasmids. Initially PACs were developed so that efficient transfection systems could be used to introduce recombinant DNA into cells. This limited the recombinant DNA size to ~100 kb so that it could be packaged into bacteriophage particles. Later recombinant PAC DNAs up to about 250 kb were introduced into cells using electroporation. Unlike YACs which are linear DNA molecular, PACs and BACs must be circular to be stable in *E. coli*. The efficiency of circulation in vitro by DNA ligase decreases with the size of the molecule. Hence BAC and PAC systems were developed that took advantage of a P1 coded site-specific recombination enzyme, namely cre recombinance. This enzyme promotes recombination between two loxP sites. Cre-promoted recombination circularizes a linear DNA fragment containing loxP sites at both ends. Genes specifying proteins such as the cre enzyme are moved to the host chromosome to minimize the BAC or PAC vector sequences and to allow for independent gene expression. Bacterial-based BAC or PAC cloning systems are easier to manipulate than YAC systems, appear to contain much fewer rearrangements and allow for easy manipulation of clone DNA.

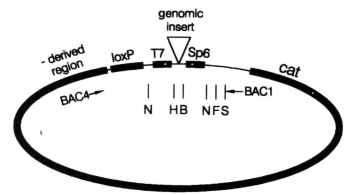

Figure 8.37   A typical bacterial artificial chromosome (BAC) cloning vector. It contains a loxP sequence to promote circularization, and bacteriophage T7 and Sp6 promoters to allow strand-specific transcription of the cloned insert. Also shown inside the circle are restriction enzyme cleavage sites and useful PCR primers.

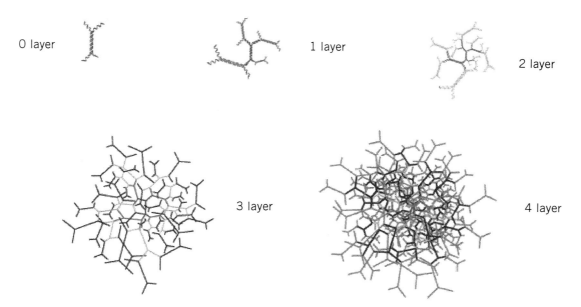

Figure 3.27 Dendrimer layer growth.

Figure 7.16 An example of the sorts of results obtainable with current procedures for interphase FISH. (From Heiskanen et al., 1996.)

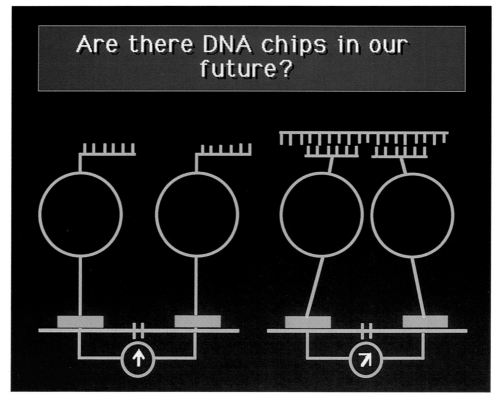

Figure 12.24 Possible future direct reading oligonucleotide hybridization chip.

by yeast telomeric simple sequence repeats. This indicates that the human simple sequence is unstable in yeast but that some feature of it is eventually recognized by the yeast telomerase, which then converts the end of the clone into a yeastlike telomere.

The particular clone that we first isolated turned out to be useful in mapping a few Mb of DNA at the long-arm telomeres of human chromosomes 4 and 21. However, it could only be used in a hybrid cell line because no where on this clone could any single-copy human DNA be found. Since this early work, others have systematically cloned and characterized single-copy probes from a number of human telomeres. These findings promise to be useful also for clinical diagnostics, since DNA rearrangements at telomeres are not uncommon in diseases like cancer.

OPTICAL MAPPING

Recently an optical method for restriction mapping has been described that could speed the process considerably (Kai et al., 1995; Samad et al., 1995). This method has successfully been applied to bacteriphage and YAC clones and to natural yeast chromosomes. Recently it has been semi-automated and extended to even larger DNAs. In optical mapping DNA molecules are elongated by gentle flow as they are fixed by capture onto poly-L-lysine derivatized glass surfaces. The restriction enzyme cleavage is used to fragment the fixed molecules. A small portion of the stretched chain relaxes at each cleavage site. This leaves a gap that is visible by fluorescence microscopy after staining the DNA samples. The contour length of each fragment seen indicates its size. However, a very significant feature of this method is that the pattern of organization of the fragments is maintained by the initial fixation, and thus the order of the fragments is immediately known. Because this is a single-molecule method, it can deal effectively with any sample heterogeneity. If sufficient numbers of molecules are examined, a complete map of each class of species in the sample should be revealed.

BOTTOM-UP LIBRARY ORDERING

The conventional approach to constructing an ordered library of clones is to fingerprint a dense set of samples and look for clones that share overlapping properties. A key variable in designing such strategies is the minimum fraction of the clones that must be in common in order for their overlap to be detectable. The smaller the overlap required, the fewer clones needed to produce an ordered set, and the faster the process proceeds (Fig. 8.38).

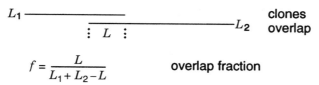

Figure 8.38 The degree of overlap, f, of two clones of length $L1$ and $L2$ will determine the resolution of the fingerprinting procedure needed to identify them.

Typical methods usually require a five- to tenfold redundant set of clones to ensure that there is a good chance of sampling the entire target at sufficient density to allow overlap detection. A simplistic estimate based on this redundancy would indicate that overlaps between clones should be perfectly scored at the 80 to 90% overlap level and perhaps half detected at the 40 to 50% overlap level. The key thing is to avoid false positives: predicting an overlap where none exists in fact. This is a serious error because it will result in the assignment of clones to incorrect regions of the target. A number of different methods have been used to detect overlaps in past studies. In all of these a key requirement is to have some sort of statistical way of determining the most likely set of overlaps in cases where there are ambiguities or potential inconsistencies. One very effective approach for evaluating overlap data is summarized in Box 8.3.

The earliest clone fingerprinting methods used restriction fragment sizes seen in single and double digests with 6-base specific enzymes like *Eco*R I and *Hind* III. This approach was first used to order bacteriophage lambda clones of *S. cerevisiae*. It was actually very difficult because of inaccuracies in sizing DNA fragments in the 1 to 10 kb range, particularly when results obtained on different gels electrophoresed on different days had to be compared. More accurate sizing is possible with smaller DNA fragments, but the number of such fragments that result from digestion of a bacteriophage or larger clone with a 4-base specific restriction enzyme is too large to allow all of them to be separated cleanly. One way around this problem is to use an end-labeling strategy as shown in Figure 8.39. This was first developed to order a cosmid library from the nematode *C. elegans*. First the clone is digested with a 6-base specific enzyme. The ends of the resulting fragments clone are labeled. Then a second digest is done with a 4-base specific enzyme. This results in DNA pieces that can be analyzed with single-base resolution of DNA sequencing gels, but their number is restricted to a manageable set.

An alternative approach for clone fingerprinting by restriction enzyme digestion is illustrated in Figure 8.40. This procedure was first used by Kohara in constructing an *E. coli* library in bacteriophage lambda clones. Indirect end labeling from probes in the vector sequence was used to determine the positions of restriction sites seen in separate partial digests, each generated with one of eight different restriction enzymes. The key advantage of this approach was that all eight digests were analyzed in a single gel electrophoresis. Under these circumstances, even though the DNA size information might be

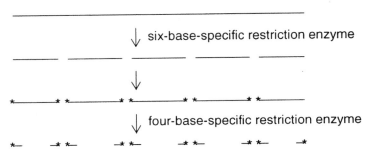

Figure 8.39   End-labeling procedure (asterisk) to produce a set of discrete DNA sizes which serve as a fingerprint.

BOX 8.3
STATISTICAL EVALUATION OF CLONE OVERLAPS

A very effective method that can be used for clone analysis is an adaptation of the procedure originally developed by Branscomb et al. (1996) for contig building by restriction fragment fingerprinting. However, the method can also be applied to hybridization based fingerprinting like the *S. pombe* example discussed in Chapter 9. Clones are considered in pairs; each pair either overlaps (O), with a fraction of overlap f, defined in Figure 8.38, or does not overlap (N). Each piece of fingerprinting data available is tested for its support of the hypothesis that two clones overlap with a fraction f. The data can be a concordant or a discordant hybridization result with a particular probe or the presence or absence of particular size restriction fragments or restriction site order. Let q represent the fraction of all of the clones that do not have a particular restriction fragment size or order (or alternatively that do not hybridize to a certain probe). Then $1 - q$ is the fraction that would be expected to show that fragment size (or hybridize to a certain probe) due to chance alone. We consider six cases in all: $P(+ +)$ means that both clones are detected, $P(+ -)$ means that only one clone is detected (scored positive), and $P(- -)$ means that neither clone is detected. Simple statistical considerations have been used to develop the following equations:

$$P(+ +, N) = (1 - q)^2$$

$$P(+ -, N) = 2q(1 - q)$$

$$P(- -, N) = q^2$$

$$P(+ +, O) = 1 - P(+ -, O) - P(- -, O)$$

$$P(+ -, O) = 2(1 - qq^{-f})q$$

$$P(- -, O) = q^2q^{-f}$$

The first three equations are obvious because they derive from simple binomial statistics for two uncorrelated clones. The remaining terms are not rigorous, but they do model the statistics of coincident detection of partially overlapping clones approximately. For example, the sixth equation, if the overlap, f, is zero, becomes equal to the third equation. If the overlap is 1, $P(- -, O) = q$ which is reasonable, since in effect one now is dealing with only a single distinct clone. The fifth equation, in the limit $f = 0$, is equal to the second equation. In the limit $f = 1$, the fifth equation is zero which is the desired behavior because identical clones cannot be expected to show discordant behavior. The fourth equation, $P(+ +, O)$ also observes the correct limits. These equations have been used successfully in the construction of ordered libraries by fingerprinting and by hybridization. A more exact treatment must take into account the relative lengths of the probes and the clones in hybridization assays. For example, if two clones partially overlap, a probe larger than the clones has a higher chance of detecting both than a probe smaller than the clones.

(continued)

BOX 8.3 *(Continued)*

To use the above equations one must evaluate $P(- -, O)$ and $P(+ -, O)$ before $P(+ +, O)$ can be calculated. When these results are considered for the entire set of restriction digests or hybridization results seen, we can compute the likelihood, L, of any particular overlap value, f, for each pair of clones as

$$L(f) = \prod_n \frac{P(X_n, O)}{P(X_n, N)}$$

where the outcome for the comparison of the two clones with the nth test is x_n, and the product is taken over the entire set of tests. The calculation is very time intensive because all pairs of probes must be considered separately for all probes and all overlap values. The procedure can be simplified by using results with longer clones to arrange the shorter clones into bins (see Fig. 9.4). Then overlap calculations need to be carried out only within bins and between the edges of neighboring bins.

imprecise, the relative order of the different restriction sites would be known with great accuracy. This order information was the major source of data used to fingerprint the clones.

Some of these approaches can be improved upon by using the power of current automated four-color DNA sequence readers (see Chapter 10). An example is shown in Figure 8.41. Here restriction fragments are filled in by end labeling in order to reduce the number of fragments seen to a manageable level, as just described. In this case, however, instead of just labeling with a single color (radioactive phosphate), a restriction site is filled in with a mixture of four different colored dpppNs. By chosing a restriction enzyme that cuts outside of its recognition sequence, as shown in the figure, one develops a very informative fingerprint of the end of the restriction fragments labeled. Now, instead of knowing that two clones share a common length restriction fragment, one learns that they share this length, plus they share a particular terminal sequence that will only occur, by chance, in 1/16 of the sites cleaved by the particular enzyme used. Thus the chances of distinguishing true overlaps, from accidental similarities, become greatly enhanced.

The example just described shows the power of obtaining "color" information about a clone; that is, more information beyond just fragment sizes. Another way to do this is shown schematically in Figure 8.42. Here each clone is digested with several restriction enzymes, the fragments are separated by electrophoresis, and the resulting gel is blotted and hybridized with several different interspersed repeating sequence probes. These probes provide a signature that goes beyond pure size measurements and indicates those sizes that contain particular repeats. This adds considerable information to each analysis, and it makes it much less likely that coincident similarities will be scored as false positives. The kind of clarity with which this approach allows overlaps to be viewed is shown by the example in Figure 8.43.

Several other powerful variations of clone fingerprinting have been described, and some of these are now being tested intensively. If repeats can be nulled out, then the use of clones or sets of clones as hybridization probes against other clones or sets of clones becomes an effective fingerprinting method. This approach will be described in detail in

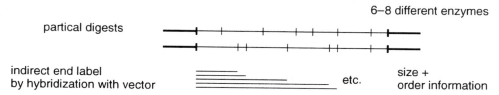

Figure 8.40 Fingerprinting clones by Smith-Birnstiel restriction mapping. An indirect end label is used to probe the pattern of fragments seen in a set of different, separate partial digests.

the next chapter. An extreme version is to use individual or mixtures of arbitrary, short oligonucleotides as hybridization probes to fingerprint individual clones. In the limit of this approach, one would actually determine the DNA sequence of all of the clones by repeated hybridization experiments. The powers and limitations of oligonucleotide hybridization fingerprinting will be discussed in Chapter 12. Here it is sufficient to note that this approach has worked well in the construction of ordered libraries.

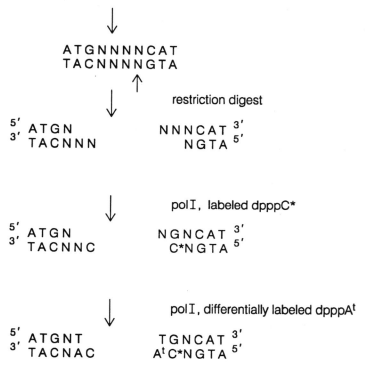

Figure 8.41 Use of restriction enzymes with imperfectly defined cutting sites to label different restriction sites with characteristic colors. This greatly increases the informativeness of the fingerprint pattern generated by the sizes of these fragments.

total restriction digest

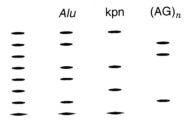

Figure 8.42 Fingerprinting a clone by hybridization with different repeated DNA sequences. See Chapter 14 for a description of these sequences.

hybridization pattern

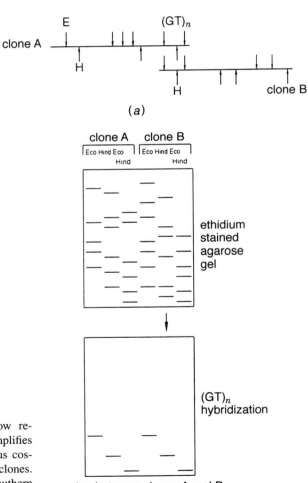

Figure 8.43 An example of how repeated sequence hybridization simplifies the identification of two contiguous cosmid clones. (*a*) Two overlapping clones. (*b*) Restriction fragments and Southern blot. (Adapted from Stallings et al., 1990.)

MEASUREMENTS OF PROGRESS IN BUILDING ORDERED LIBRARIES

The process of assembling contigs by fingerprinting clones can be treated in relatively straightforward mathematical ways. One makes the key assumption that the genome is being sampled uniformly, and sets, as a parameter, the degree of overlap between two clones necessary to constitute positive evidence that they are contiguous. Lander and Waterman have modeled this process of clone ordering. The sorts of results they obtained are shown in Figure 8.44. It is assumed that clones are fingerprinted one at a time, and the number of clones assembled into contigs of two or more clones is plotted as a function of the number of clones fingerprinted. At early times in the project, there are almost no contigs because the odds of picking overlapping clones, chosen at random, are small. Eventually overlaps start to build up, but most contigs contain just two clones. These begin to coalesce into larger contigs as the genome is sampled deeper and deeper. However, the effectiveness of the contig building begins to saturate long before all clones are assembled into a single contig. This saturation is partly determined by the lack of completeness of the library; if any regions are not represented at all, contigs cannot be built across them. The saturation is also a function of the effectiveness of overlap detection; due to chance, some clones that actually are contiguous may not overlap enough to be counted as a positive score. Several ongoing programs in contig building have been evaluated by the Lander-Waterman approach. Actual progress on these projects is in remarkably good agreement with predictions.

Eventually the pure bottom-up approach must be abandoned if a complete ordered library is desired. The point at which a switch in strategy is profitable is said to be somewhere between 60% and 90% coverage, when almost all progress in typical bottom-up mapping stops. The early stages of bottom-up mapping are very efficient. DNA preparations, fingerprinting, and data analysis have all been completely automated for some of the schemes we have described. Contig assembly is also done by computer software. Once the saturation point is reached, a typical project will still have hundreds or thousands of separate contigs. The challenge is to close the gaps between them in an efficient way. Several different approaches are useful at this stage. The contigs can be ordered by FISH localization of individual clone representatives from each contig. Once one knows that two contigs are very close to each other, frequently overlap data that were marginal

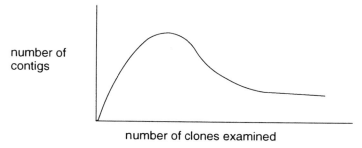

number of contigs

number of clones examined

Figure 8.44 Progress in a pure bottom-up clone-ordering strategy, as calculated from the Lander-Waterman model. Plotted is the number of contigs as a function of randomly chosen clones examined.

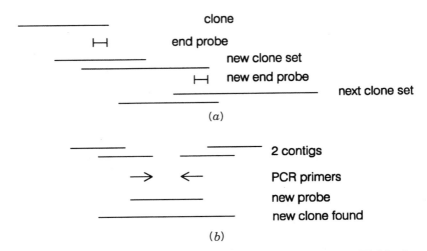

Figure 8.45   Two strategies for finishing the construction of contig maps. (*a*) Walking by probing existing or new libraries of clones with the ends of existing contigs. (*b*) Attempting to PCR across the gaps between two contigs suspected of being adjacent on the map (generally hinted at from other data such as FISH results).

before can now be used to fuse the contigs. The easiest way to fill major gaps where they are suspected is to switch to another library. Here regional assignment of clones from that library (or microdissection, Chapter 7) can help to focus on clones most likely to lie in regions where contiguity is not yet established.

A generally useful endgame strategy is to use existing contigs to screen a library of clones and subtract out those that have already been found. This greatly improves the odds of finding new, useful clones, once additional random picking from the remainder is reinitiated. Perhaps the single most useful method, once a dense set of contigs exists, is walking (Fig. 8.45*a*). Here one takes clones from the ends of existing tiling path contigs and uses them to screen libraries. Both the original library and totally new libraries can be used. The goal is to identify new clones that allow the contig to be extended. It is often particularly useful to change from one type of library to another in the walking process. Frequently a gap will exist because the sequence within it is not cloneable, say in cosmids, but it may be easily cloneable in YACs, and vice versa. Multiplex walking methods have been described that allow the simultaneous walking from many contig ends.

A final useful endgame strategy is to sequence the ends of contigs. Sequence information is much more robust than any other kind of fingerprinting. Even if two clones overlap by as few as 15 base pairs, sequence information can determine that they actually overlap. Sequence information at the ends of contigs can also be used to design PCR primers that face outward from the contigs (Fig. 8.45*b*). These primers can be used to test systematically whether two contigs suspected of being located near enough to each other are actually within a few kb apart. This technique turns out to be extremely powerful, in practice, because in actual projects, thus far, many of the hardest to close gaps turn out to be very small, and PCR can be carried out across them.

SURVEY OF RESTRICTION MAP AND ORDERED LIBRARY CONSTRUCTION

Complete macrorestriction maps have been produced for a number of prokaryotic genomes, some simpler eukaryotic genomes, and sections of complex genomes. The first of these maps, a *Not* I map of *E. coli,* is shown in Figure 8.46. The most complex of all these maps, that for human chromosome 21q, is shown in Figure 8.47. A number of features of this map are of interest. Note that small *Not* I fragments and large *Not* I fragments tend to cluster. This must eflect wide oscillations in the density of HTF islands along the chromosome, since *Not* I sites occur almost exclusive in these islands.

The *Not* I map of human chromosome 21 was actually executed, not in a single cell line but in a set of eight cell lines. Polymorphisms among these lines were helpful in establishing the map as described earlier in the chapter. The full pattern of polymorphisms is illustrated in Figure 8.48. While the extent of polymorphism is considerable, almost all of it is consistent with varying degrees of methylation in the cell lines studied. There is little or no compelling evidence for major shifts in the lengths of DNA between existing *Not* I sites. Most important, there is no evidence that any significant amounts of DNA have been rearranged or lost in these cell lines.

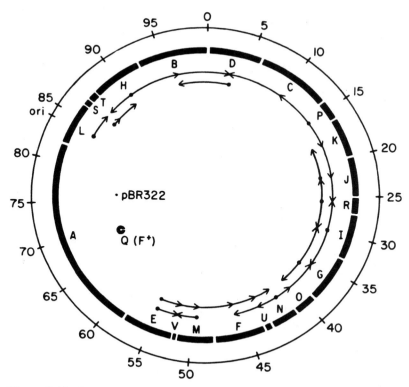

Figure 8.46   *Not* I restriction map of *E. coli.* (Adapted from Smith et al., 1987.)

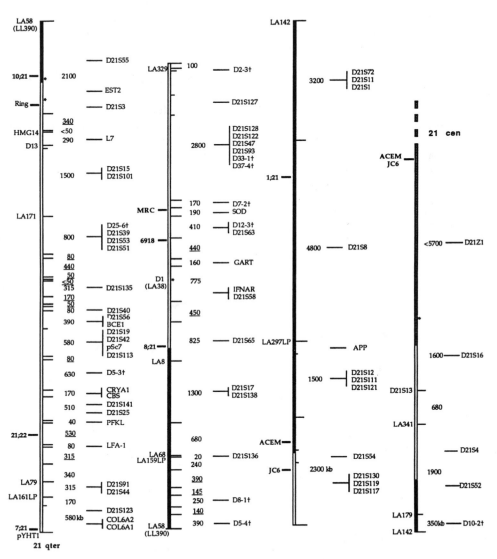

Figure 8.47 *Not* I restriction map of the long arm of human chromosome 21. (Taken from Wang and Smith, 1994.)

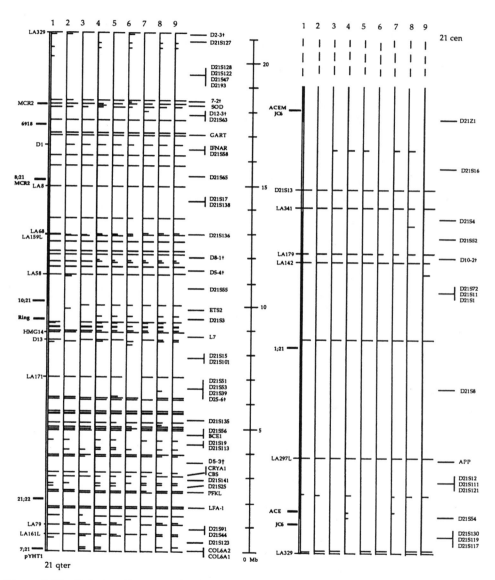

Figure 8.48 Polymorphisms seen in the *Not* I map of human chromosome 21q in nine different cell lines (lanes 1 to 9). (Taken from Wang and Smith, 1994.)

A number of successful projects have been reported that have produced complete, or almost complete ordered clone libraries. The first of these was the ordered bacteriophage lambda library covering the *E. coli* genome. Other model organisms now mapped include the yeasts *S. cerevisiae,* and *S. pombe,* and the nematode *C. elegans.* Extensive map data also exists for Drosophila and for the human genome. A relatively complete YAC map covering the informative part of the Y chromosome has been reported, and a complete YAC maps exist that cover most human chromosomes. Extensive cosmid ordering projects on chromosomes 16 and 19 are virtually complete. Gaps not covered in cosmids are mostly covered in YACs or BACs.

An example of some of the data used to construct the chromosome YAC 21 map is shown in Figure 8.49. It is apparent that at the present stage some of the overlap evidence would be strengthened by interpolating results from additional clones, and some YACs used show evidence of rearrangements that are potential sources of error. Indeed, when the YAC contig for chromosome 21 is compared with the *Not* I restriction map, several YACs appear to be assigned to the wrong locations on the chromosome (Fig. 8.50). This is almost certainly partly the result of YAC chimeras which can seriously confuse clone ordering (see Chapter 9). Other discrepancies appear to result from the use of several probes with confused identities. Nevertheless, a remarkable amount of information and a goodly number of useful clones are now available for this chromosome.

A complete YAC map and three complete cosmid maps are available for the yeast *S. pombe.* The tiling path YACs from this map are shown in Figure 8.51, alongside the *Not* I restriction map of this organism and a sketch of the genetic map. This view, which presents a very simple looking map, hides the complex process that actually went into the construction of the map. Figure 8.52 illustrates the actual YAC clones

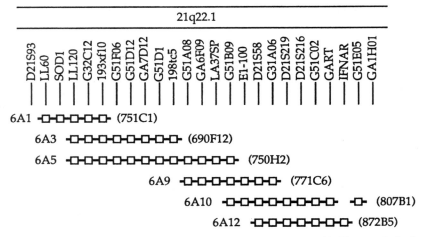

Figure 8.49   A contiguous section of YACs from human chromosome 21. The contig is about 2.3 Mb long; 18 probes (STSs) were needed to assemble it. Note that several of the YACs appear to have internal deletions.

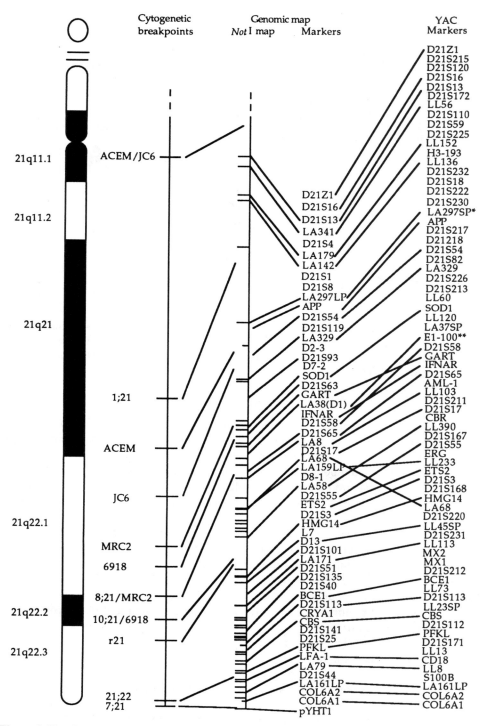

Figure 8.50 Comparison of marker order in the *Not* I restriction map of human chromosome 21 and the chromosome 21 YAC contig map. (Taken from Wang and Smith, 1994.)

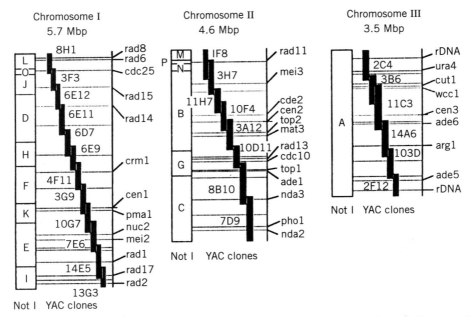

Figure 8.51 Three maps of the fission yeast *S. pombe*. Plotted are the *Not* I restriction map, the 26-clone tiling set of a YAC contig map, and markers from the genetic map. Dotted lines indicate genetic markers and cosmids, which were hybridized to *Not* I digests of *S. pombe* and to YACs. (Taken from Maier et al., 1992.)

and probes studied along the way to map completion, and the selection of a simple tiling set. The large number of samples required, even for a simple organism, can barely be displayed as a legible figure. This should make it clear that any mapping project, with contemporary technology, is not to be undertaken lightly. The cosmid maps of *S. pombe* are even more complex and hard to display visually. Some details about the procedures that were used to construct one of these maps will be given in Chapter 9.

An issue that still leads to considerable debate is when to end a mapping project. How important is it to close the last gap, that is, to confirm the relative order within a contig to beyond any doubt? The simplest way to deal with this question is to recall the purpose of maps. We need them to access the genome, both for biological studies and for eventual DNA sequencing. A map that is 70% complete has seen only the beginning of the effort required to make a fully finished map—but it already provides access to 70% of the chromosome. A 90% map is frankly, for most purposes, almost as useful as a fully completely map, unless one is so unfortunate as to need clones or sequence data in some of the regions that are still in small fragments or contigs. In general, the usefulness of mapping projects grows very rapidly in the early stages and then begins to increase much more slowly as the maps near completion. It is important to consider this in deciding how much effort should be devoted to fitting in the last contig, as opposed to breaking out into new, uncharted territory on another chromosome or in another genome.

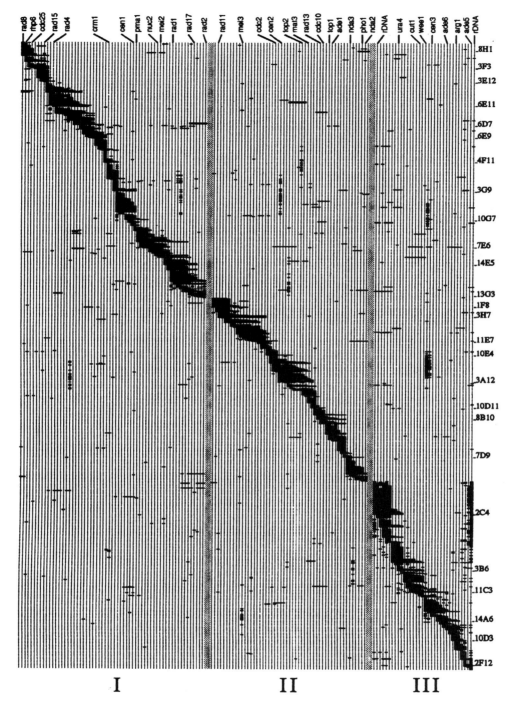

Figure 8.52 Actual sets of YACs and probes needed to generate the YAC tiling set in Figure 8.51. YAC clones are shown on the vertical axis, where a subset of 26 clones spanning the entire genome is indicated. Probes are drawn on the horizontal axis; some of the genetic markers used are identified. Vertical gray bars separate the three chromosomes. Positive signal outside the contructed contigs indicates the locations of repeats. (Taken from Maier et al., 1992.)

SOURCES AND ADDITIONAL READINGS

Branscomb, E., Slezak, T., Pae, R., Galas, D., Carrano, A. V., and Watermann, M. 1990. Optimizing restriction fragment fingerprinting methods for ordering large genomic libraries. *Genomics* 8: 351–366.

Cai, W., Aburatani, H., Stanton, V. P., Housman, D. E., Wang, Y. K., and Schwartz, D. C. 1995. Ordered restriction endonuclease maps of yeast artificial chromosomes created by optical mapping on surfaces. *Proceedings of the National Academy of Sciences USA* 92: 5164–5168.

Cheng, J. F., Smith, C. L., and Cantor, C. R. 1989. Isolation and characterization of a human telomere. *Nucleic Acids Research* 17: 6109.

Huang, M. E., Chuat, J. C., Thierry, A., Dujon, B., and Galibert, F. 1994. Construction of a cosmid contig and of an *Eco*R I restriction map of yeast chromosome X. *DNA Sequence* 4: 293–300.

Ioannou, P. A., Amemiya, C. T., Garnes, J., Kroisel, P. M., Shizya, H., Chen, H., Batzer, M. A., and de Jong, P. J. 1994. A new bacteriophage p1-derived vector for the propagation of large human DNA fragments. *Nature Genetics* 6: 84–89.

Jasin, M. 1996. Genetic manipulation of genomes with rare-cutting endonucleases. *Trends in Genetics* 12: 224–228.

Lennon, G. G., and Lehrach, H. 1991. Hybridization analyses of arrayed cDNA libraries. *Trends in Genetics* 7: 314–317.

Maier, E., Hoheisel, J. D., McCarthy, L., Mott, R., Grigoriev, A. V., Monaco, A. P., Larin, Z., and Lehrach, H. 1992. Compiete coverage of the Schizosaccharomyces pombe genome in yeast artificial chromosomes. *Nature Genetics* 1: 273–277.

Oliva, R., Lawrance, S. K., Wue, T., and Smith, C. L. 1991. Chromosomes: Molecular studies. In Dulbecco, R., et al., eds., *Encyclopedia of Human Biology*. San Diego: Academic Press, pp. 475–488.

Palazzolo, M. J., Sawyer, S. A., Martin, C. H., Smoller, D. A., and Hartl, D. L. 1991. Optimized strategies for sequence-tagged-site selection in genome mapping. *Proceedings of the National Academy of Sciences USA* 88: 8034–8038.

Sainz, J., Pevny, L., Wu, Y., Cantor, C. R., and Smith, C. L. 1992. Distribution of interspersed repeats (*Alu* and *Kpn*) on *Not* I restriction fragments of human chromosome 21. *Proceedings of the National Academy of Sciences USA* 89: 1080–1085.

Samad, A., Huff, E. J., Cai, W., and Schwartz, D. 1995. Optical mapping: A novel, single-molecule approach to genomic analysis. *Genome Research* 5: 1–4.

Smith, C. L., Econome, J. G., Schutt, A., Klco, S., and Cantor, C. R. 1987. A physical map of the *Escherichia coli* K12 genome. *Science* 236: 1448–1453.

Stallings, R. L., Torney, D. C., Hildebrand, C. E., Longmire, J. L., Deaven, L. L., Jett, J. H., Doggett, N. A., and Moyzis, R. K. 1990. Physical mapping of human chromosomes by repetitive sequence fingerprinting. *Proceedings of the National Academy of Sciences USA* 87: 6218–6222.

Sternberg, N. 1992. Cloning high molecular weight DNA fragments by bacteriophage p1 system. *Genetic Analysis: Techniques and Applications* 7: 126–132.

Stubbs, L. 1992. Long-range walking techniques in positional cloning strategies. *Mammalian Genome* 3: 127–142.

Wang, D., Fang, H., Cantor, C. R., and Smith, C. L. 1992. A contiguous *Not* I restriction map of band q22.3 of human chromosome 21. *Proceedings of the National Academy of Sciences USA* 89: 3222–3226.

Wang, D., and Smith, C. L. 1994. Large scale structure conservation along the entire long arm of human chromosome 21. *Genomics* 20: 441–451.

Zhu, Y., Cantor, C. R., and Smith, C. L. 1993. DNA sequence analysis of human chromosome 21 *Not* I linking clones. *Genomics* 18: 199–205.

9 Enhanced Methods for Physical Mapping

WHY BETTER MAPPING METHODS ARE NEEDED

In Chapter 8 we described the original top-down and bottom-up approaches that have led to the construction of a fair number of macrorestriction maps and ordered libraries. These methods are quite laborious, and it would be difficult to replicate them on very large numbers of mammalian genomes. New methods will be needed that are much more powerful if we are ever to be able to explore the full range of evolutionary diversity and the full range of human diversity by genome analysis. It seems fairly clear that in the future we will want the ability to go into any individual genome and obtain samples suitable for sequencing of large contiguous blocks of DNA. At least from the present perspective, this could require prior mapping studies to prepare the samples needed for subsequent sequencing. The key will be to develop approaches that allow a rapid focus on a particular region of interest and then a rapid collection and ordering of samples suitable for direct sequencing. It would be especially desirable if methods could eventually be developed that focus directly on map differences between individuals or species. Many future studies will doubtlessly be interested only in differences between two otherwise fairly homologous samples. Today techniques for effective differential mapping are unknown, and we will largely focus instead on methods for making direct mapping approaches much more efficient.

LARGER YEAST ARTIFICIAL CHROMOSOMES (YACs)

YACs are a major tool currently used for making ordered libraries of large insert clones. The basic design and generation of YACs was described in Chapter 8. A major issue with YACs has been the size of the DNA insert. The first YAC libraries made had average insert sizes of 200 to 300 kb. This is a vast improvement over cosmid clones when used in schemes for rapid walking. However, since the first libraries, the sizes of YACs have continued to grow steadily. At least two improvements in YAC design have assisted this. Early on, in YAC development it was noted that mammalian DNA was not always rich in sequences that could serve serendipitously as yeast replication origins (autonomously replicating sequences). The original YAC vectors had only a single origin in one arm of the YAC vector. Requiring that this origin replicate the entire chromosome places potentially severe kinetic constraints on the viability of the chromosome. This problem can be alleviated considerably by building authentic YAC origins into both vector arms.

A second technique for increasing the size of YAC inserts has been to size-fractionate the DNA to be cloned both before and after ligation of YAC vector arms. The ligation is normally carried out in a melted agarose sample. Under these conditions Mb DNA is quite susceptible to shear breakage, which increases as the square of the length of the DNA. Any DNA that is fragmented by shear breakage or nuclease contamination during the ligation procedure, and also any contaminating vector arms, will be eliminated by the second size-fractionation. This is important because, otherwise, large numbers of vector arms will contaminate the true YACs. Since these carry the selectable markers, and can recombine with yeast chromosomal DNA, they lead to a high background of useless clones. Several groups have reported the construction of YAC libraries with average insert sizes of 500 to 700 kb and even larger. However, the greatest success has been seen with a continuing effort at Genethon to make larger and larger YACs. This has resulted in a series of libraries with average insert sizes of 700 kb, 1.1 Mb, 1.3 Mb, and 1.4 Mb. The largest insert libraries resulted from an extensive effort at Genethon. These had average insert sizes in excess of 1 Mb. The protocols for producing these megaYACs do not seem to be reproducible at this stage. Instead, by having the same team concentrate on the repeated construction of YAC libraries, the quality of these libraries appeared to improve on average, for unknown reasons as the team gained more experience. All of these libraries are made from PFG-fractionated *Eco*R I partial digests of genomic DNA. Usually DNA is transformed into yeast by electroporation.

The major problem with megaYACs (and most YAC libraries for that matter) are rearranged clones. These include chimeric clones, clones with deletions, and clones with insertions of yeast DNA. These are illustrated in Figure 9.1. Deletions and yeast insertions make it difficult to use the YACs directly as DNA sources for finer mapping or sequencing. However, such clones are still useful for the kinds of mapping strategies we will describe later in this chapter. Chimeric clones are more of a problem because they can lead to serious errors in mapping if they are not detected. The chimeric clones appear to contain two or more disconnected genomic regions. In some YAC libraries more than 50% of the clones are chimeras.

There are two potential origins for the occurrence of chimeras. Some may arise during ligation, especially if the insert DNA is at too high a concentration relative to the amount of YAC arms present. Co-ligation can be reduced substantially by more complex cloning strategies than used in early YAC library construction (Wada et al., 1994).

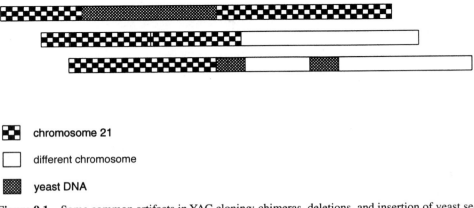

   chromosome 21

   different chromosome

   yeast DNA

Figure 9.1 Some common artifacts in YAC cloning: chimeras, deletions, and insertion of yeast sequences.

For example, suppose that genomic DNA is partially digested with *Mbo* I in agarose, which produces fragments ending in

$$5'\text{———}$$

$$3'\text{———CTAG}$$

The single-stranded ends of the resulting sample are then partially filled in treatment with Klenow fragment DNA polymerase and dpppA and dpppG only. The resulting genomic DNA fragments will then have ends like

$$5'\text{———GA}$$

$$3'\text{———CTAG}$$

and thus they cannot be ligated to each other. In parallel, the YAC cloning vector is digested to completion with *Bam*H I to generate telomeres (see Box 8.1) and then digested with *Sal* I to yield fragments that end in

$$5'\text{———G}$$

$$3'\text{———CAGCT}$$

The ends of these are then partially filled in by treatment with Klenow DNA polymerase in the presence of only dpppT and dpppC. This yields fragments ending in

$$5'\text{———GTC}$$

$$3'\text{———CAGCT}$$

Now the vector arms produced cannot ligate to each other, but they are still capable of ligating to the genomic DNA fragments prepared as described above.

A second source of chimeras will arise from recombination. In preparing yeast for DNA transformation, usually a small fraction of the cells is rendered competent to pick up DNA, and it is not at all uncommon for these cells to pick up several YACs. Usually the YACs will separate in subsequent cell divisions. Occasionally a cell will stably maintain two different YACs. However, since mitotic recombination is very prevalent in yeast, two different YACs can recombine at shared DNA sequences, and as a result two chimeric daughters are produced. If each of these retains a centromere, they will usually segregate to separate daughter cells, each of which will now maintain a different single chimeric YAC (Fig. 9.2). Evidence for such recombination between two YACs has been obtained in at least one case where the two original inserts corresponded to DNA of known sequence, and thus the site of the recombination event could be identified. Alternatively, a dicentric YAC and an acentric YAC could be produced by a recombination event. In this case the latter clone will be lost, and the former may break unless one of the centromeres becomes inactivated.

Human DNA is likely to be a very favorable recombination target in yeast because of the large amount of interspersed repeated DNA sequences. This can also lead to instabilities within a single YAC which may lose part of its insert by an intramolecular recombination event. Just how prevalent these rearrangements are in particular libraries is not

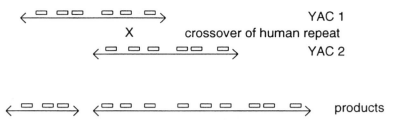

Figure 9.2   Recombination between human repeated sequences (shown as boxes) as a mechanism for the production of chimeric YACs.

known, but there are reasons to think that these are serious problems. The yeast strains used for almost all current YAC library construction have not had their recombination functions disabled. When a recombination deficient yeast host was used, a dramatic decrease in the fraction of chimeras was observed (Ling et al., 1993). Additional indirect evidence that recombination and not co-cloning is the major cause of chimeric YACs comes from observations on libraries made from hybrid cell lines. In most of these cases, the amount of chimerism is very low. This presumably arises because the human DNA is much more dilute in these samples, and human-rodent recombination is much less efficient because most repeated sequences are not well conserved between the two species.

When YACs are prepared directly from DNA obtained from flow-sorted chromosomes, the frequency of chimeras is also quite low. This is partly a result of the low DNA concentrations, which will diminish coligation and recombination events. However, the decreased chimera frequency also is likely to reflect the fact that in these samples all of the DNA preparation and manipulation was carried out in agarose. Agarose will reduce the number of DNA fragments with broken, unligatable ends, which are highly recombinogenic in yeast.

HOW FAR CAN YACs GO?

Larger insert clones facilitate mapping projects in several ways. The number of clones one needs to order to fill the minimum tiling path is reduced. This greatly simplifies the process of clone ordering. If one has a fingerprinting method that requires a certain absolute amount DNA for demonstrating an overlap, the larger the clone the smaller a fraction of the clone this amount represents. Finally larger clones can easily be used to order clones one-half to one-third their size. Thus, as we will describe later, by having a tiered set of samples, the whole process of ordering them is greatly facilitated. The limit of the tier is determined by the largest stable and reliable clones that are available.

The chromosomes of *S. cerevisiae* range in size from 250 kb to 2.4 Mb. Thus the largest current YACs are in the midsize range of yeast chromosomes. It is not clear whether there is a size limit to yeast chromosomes. One issue already mentioned is the frequency of replication origins in the insert DNA. Early studies with artificial chromosomes in yeast indicated that stability, measured as retention over many generations of growth, actually increased sharply as a function of size, but these studies were not carried out up to the size range of current megaYACs. At some point the amount of foreign DNA that any organism can tolerate becomes limited by its competition for binding of key cellular enzymes or regulatory proteins. Where this limit occurs in yeast is unknown.

Some hints that *S. cerevisiae* can tolerate really large amounts of foreign DNA are available from studies with amplifiable YACs. The basic scheme behind such cloning vectors is shown in Figure 9.3*a*. Yeast centromeres can be inactivated if transcription occurs across them. To take advantage of this effect, a YAC vector arm has been designed with a strong, regulatable promoter extending toward a centromere. This vector is used to transform DNA into yeast in the ordinary way. The yeast is allowed to grow in the presence of selectable markers on the vector arms, first in the absence of transcription from the regulatable promoter. Then the promoter is activated, and growth is continued in the presence of selection. What happens is that with centromeres inactivated, the YACs segregate unevenly into daughter cells (Fig. 9.3*b*). Those daughters that receive many copies of the YACs have a selective advantage; those that receive very few copies are killed by the selection. The result is to increase, progressively, the average number of YACs per viable cell. This process continues up to the point where there are 10 to 20 copies of each YAC per cell. At this point the YAC DNA is 20 to 40% of the entire yeast DNA. This technique appears to be very promising, for it produces cells that are much easier to screen by hybridization than ordinary single-copy YACs. However, it has not yet been applied to libraries of megaYACs.

One potential improvement over standard YAC cloning methods might come from the use of *S. pombe* rather than *S. cerevisiae*. The former yeast has about the same genome size as the latter. However, *S. pombe* has only three chromosomes that range in size from 3.6 to 5.8 Mb. Based solely on this observation, it seems reasonable to speculate that *S. pombe* ought to be able to accommodate large YACs if there were a way to get them into the cell. One potential complication is that the centromeres of *S. pombe* and *S. cerevisiae* are very different in size. *S. cerevisiae* has a functional centromere that covers only a few hundred base pairs.

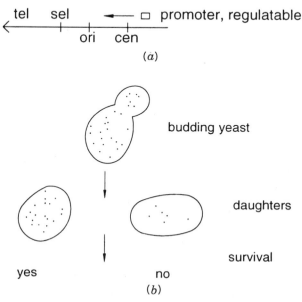

Figure 9.3 Amplifiable YACs. (*a*) Vector used that allows regulation of centromere function. (*b*) Random segregation in mitosis leads to selective survival of cells with large numbers of YACs (dots).

In contrast, while the irreducible minimum centromere of *S. pombe* is unknown, past experience suggests that it could well be in excess of 100 kb. This would make the construction of cloning vectors for use in *S. pombe* very cumbersome.

An alternative cloning system for large DNA where considerable recent progress has been made uses bacterial artificial chromosomes or BACs (see Box 8.2). This system employs single-copy *E. coli* F-plasmids as vectors for DNA inserts. Natural F-factors can be a megabase in size. Thus BACs ought to have capacities in this size range if the resulting DNAs can be transfected into *E. coli* efficiently. The first BACs were rather small with inserts mostly in the 100- to 200-kb size range. However, recently larger BACs, with sizes from greater than 300 kb, have been reported. It remains to be seen how much this range can be enhanced by further modification and protocol optimization. BACs have the intrinsic advantage that the background *E. coli* DNA is only a third that of background yeast DNA. It is also relatively easy to purify the BAC DNA away from the host genomic DNA. Powerful bacterial genetic procedures can be used to manipulate the BAC sequences in vivo, and it is fair to say that comparable procedures can be used to manipulate YACs within their host cells. A key feature for both systems is that we now know the entire DNA sequence of both *E. coli* and *S. cerevisiae*. Procedures are likely to be developed and in place soon for direct DNA sequencing of large insert clones in one or both of these organisms. By knowing all of the host DNA sequence ahead of time, it will be possible to design sequencing or PCR primers in an intelligent, directed way. Any accidental host sequence that results from primer errors or homology will be immediately apparent once the putative clone sequence is compared with that of its host genome. Another bacterial large DNA cloning system that is in widespread use is the P1 artificial chromosome (PACs), described in Box 2.3.

VECTOR OBSOLESCENCE

Based on past experience with ordinary recombinant DNA procedures over the past decade, the highly desirable vector of today is an inefficient, undesirable vector tomorrow. There is no way we can predict what the optimal vectors will be like five years from now; what bells and whistles they must have to facilitate the rapid mapping and sequencing procedures that will then be in use. To demonstrate how cloudy our crystal ball is in this respect, within five years the development of rapid methods to screen clones for possible functions is very likely. Just what form these screens will take, and what requirements they will impose on cloning vectors, are entirely unknown.

Imagine that tomorrow a vast improvement in some cloning vector has been achieved. All of the current map data and samples do not use this vector. How will we transfer the enormous number of samples used in genomic mapping from yesterday's obsolete vectors to the new ones? Certainly it will not be efficient to do this clone by clone. New strategies are badly needed that allow flexibility in the handling of samples to allow mass recloning or rescreening of entire ordered or partially ordered libraries to retain useful order information but equip the clones with the newly desired features. It is fair to say that today, while the problem is recognized, creative solutions to it are still lacking. We will either need clever selection, very effective automation for large numbers of separate samples, or very effective multiplexing that will allow many samples to be handled together and then sorted out in some very simple way afterward.

One consequence of the virtual certainty of vector obsolescence is that it is desirable to minimize the numbers of samples archived for storage and subsequent redistribution. Instead, it seems more efficient to develop procedures that will allow desired clones to be pulled easily from whatever new libraries are made. The advantage of PCR-based approaches is obvious in this regard. These approaches require storing only DNA sequence information that allows primers to be made whenever they are needed to assay for a given sequence in a sensitive way. Whatever library a desired DNA sequence is in, it should then be possible to find it in a relatively quick and inexpensive way, by PCR assays on pools of clones or hybridization assays on arrays of clones from that library, as we will describe later in this chapter. In this way no large numbers of samples need be stored for long time periods nor for mass distribution. Only pools of samples will have to be archived.

HYBRID MAPPING STRATEGIES: CROSS-CONNECTIONS BETWEEN LIBRARIES

In any physical mapping project, sooner or later there is the need to handle a number of different types of samples. These include cell lines, radiation hybrids, and large restriction fragments for regional assignments and gaining an overview, as well as various clone libraries such as megaYACs, YACs, P1, and cosmid clones that actually form the eventual basis for DNA sequencing. Past projects have tended to concentrate on ordering at most a few of these samples across the chromosome, and then they resorted to using other types of samples in selected regions where these were needed to address particular problems. Based on these experiences, it now seems evident that much of the labor in handling all of these types of samples is preparing a dense set of labeled DNA probes or PCR primers (e.g., STSs *vide infra*) that are needed for most fingerprinting or mapping activities.

Once one has such probes or primers, an attractive scheme for ordering them is shown in Figure 9.4. The labeled probes or primers are used to interrogate, simultaneously, all samples that are of potential interest for the chromosome of question. If a dense enough set of probes or primers exists, and if the clone libraries are highly redundant, we will show that the result of the interrogation should be to order all the samples of interest in parallel. Ordering by any of the methods currently at our disposal involves finding overlap information. The larger the number of different samples used in the overlap procedure, the more likely one or more will cover the key region needed to form an informative overlap. This approach is neither top down or bottom up; in most respects it combines the best features of the two extreme views of map construction. They key issue is how to implement such a strategy in an efficient way. There are three basic issues: how to handle the probes, how to handle the samples, and how to do the interrogation.

Tens of thousands of DNA samples are involved in most large-scale mapping efforts. These cannot be handled routinely as individual liquid DNA preparations. One viable approach is to make dense arrays of DNA spots on filters. As an example consider the filter shown schematically in Figure 9.5. It consists of 2×10^4 individual cosmids. This must be prepared by an automated arraying device (Box 9.1). The key fact is that once the samples are prepared, the device can make as many copies of the filter as needed with relatively little additional effort. If each cosmid has an average of 4×10^4 base pairs of DNA, the array represents a total of 8×10^8 base pairs. This is a fivefold redundant coverage of

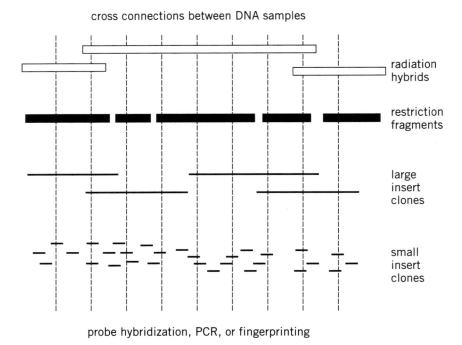

Figure 9.4 Mapping by making cross-connections between a set of different DNA samples. The key variables are the density of available probes and the density or coverage of available samples.

an average, 150 Mb, human chromosome. It is sufficient for most mapping exercises that one can contemplate. For example, hybridization of a labeled probe to the filter will identify any cosmids that contain corresponding DNA sequences. All the cosmids can be examined in a single experiment, and if the signal to noise in the hybridization is good, the resulting data should be fairly unequivocal. We assume that any repeated DNA sequences in the probe will be competed out by methods described earlier in the book.

The alternative approach to making arrays is to make pools of samples. This has the potential advantage that the DNA is handled in homogeneous solution, which facilitates some screening procedures like PCR. It also has the advantage that a relatively small number of pools can replace a very large number of individual samples or spots on an array. Procedures for constructing these pools intelligently will be a major theme of this chapter. However, regardless of how they are made, a disadvantage of pools is that a single interrogation will not usually identify unique clone targets identified by a probe. Instead, one usually has to perform several successive probings or PCRs in order to determine which elements of particular pools were responsible for positive signals generated by the probe.

Hundreds to thousands of probes or primers are used in a large-scale mapping effort. The complexity of these samples is almost as great as that of the clones themselves. In the past most probes or primer pairs have been handled individually. Probes have consisted of small DNA clones, cosmids, YACs, large DNA fragments, or radiation hybrids. It is necessary to compete out any repeated DNA in these probes to prevent a background level of hybridization that would be totally unacceptable. PCR tricks abound that can be used to

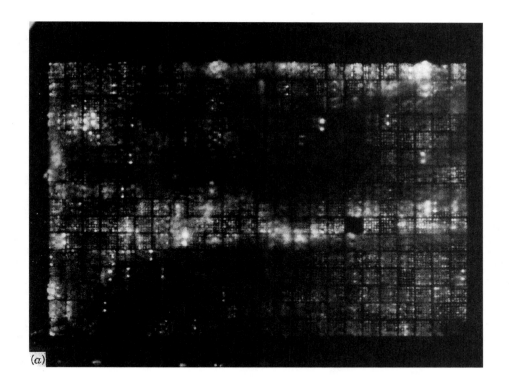

Figure 9.5 An example of a dense sample array. Cosmid clones from a chromosome 19-specific library were (*a*) arrayed by 36-fold compression from 384-well microtitre plates and (*b*) probed by hybridization with a randomly primed pool of five cosmids. (Unpublished work of A. Copeland, J. Pesavento, R. Mariella, and D. Masquelier of LLNL. Photographs kindly provided by Elbert Branscomb.)

BOX 9.1
AUTOMATED MANIPULATION OF SAMPLE ARRAYS

A considerable background of experience exists in automated handling of liquid samples contained in microtitre plates. While it is by no means clear that this is the optimal format for mapping and sequencing automation, the availability of laboratory robots already capable of manipulating these plates has resulted in most workers adopting this format. A typical microtitre plate contains 96 wells in an 8 by 12 format, which is about 3×5 cm in size (Fig. 9.6 bottom). Each well holds about 100 μl (liquid) of sample. Higher-density plates have recently become available: An 18 by 24 sample plate, the same size as the standard plate with proportionally smaller sample wells, seems easily adapted to current instruments and protocols (Fig. 9.6 top). A fourfold higher-density 36 by 48 sample plate has been made, but it is not yet clear that many existing robots have sufficient mechanical accuracy and existing detection systems sufficient sensitivity to allow this plate to be adopted immediately.

Figure 9.6 Typical microtitre plates: 384-well plate (top), 96-well plate (bottom).

(continued)

BOX 9.1 *(Continued)*

Liquid-handling robots such as the Beckman Biomec, the Hewlett-Packard Orca chemical robot, among others, can use microtitre plates singly or several at a time. More complex sets of plates can be handled by storage and retrieval from vertical racks called microtitre plate hotels. A number of custom-made robots have also been built to handle specialized aspects of microtitre plate manipulation efficiently, such as plate duplication, custom sample pooling, and array making from plates. All of these instruments share a number of common design features. Plate wells can be filled or sampled singly with individual pipetting devices, addressable in the *x-y* plane. Rows and columns can be sampled or fed by multiple-headed pipetors. Entire plates can be filled in a single step by 96-head pipetors. This is done, for instance, when all the wells must be filled with the same sample medium for cell growth or the same buffer solution for PCR.

Most standard biochemical and microbiological manipulations can be carried out in the wells of the microtitre plates. A sterile atmosphere can be provided in various ways to allow colony inoculation, growth, and monitoring by absorbance. Temperature control allows incubation or PCR. Solid state DNA preparations such as agarose plug preparations of large DNA, or immobilized magnetic microbead-based preparations of plasmid DNA or PCR samples are all easily adapted to a microtitre plate format. The result is that hundreds to thousands of samples can be prepared at once. Standard liquid phase preparations of DNA are much more difficult to automate in the microtitre plate format because they usually require centrifugation. While centrifuges have been built that handle microtitre plates, loading and unloading them is tedious.

For microbiological samples, automated colony pickers have been built that start with a conventional array of clones or plaques in an ordinary petri dish (or a rectangular dish for more accurate mechanical positioning), optically detect the colonies, pick them one at a time by poking with a sharp object, and transfer them to a rectilinear array in microtitre plates. The rate-limiting step in most automated handling of bacteria or yeast colonies appears to be sterilization of the sharp object used for picking, which must be done after each step to avoid cross-contamination. With liquid handling, cross-contamination can also be a problem in many applications. Here one has the choice of extensive rinsing of the pipet tips between each sample, which is time-consuming, or the use of disposable pipet tips, which is very costly. As we gain more experience with this type of automation, more clever designs are sure to emerge that improve the throughput by parallelizing some of the steps. A typical example would be to have multiple sample tips or pipetors so that some are being sterilized or rinsed off line while others are being used. At present, most of the robots that have been developed to aid mapping and sequencing are effective but often painfully slow.

Dense sample arrays on filters are usually made by using offset printing to compress microtitre plate arrays. An example is shown in Figure 9.7. Here a 96-pin tool is used to sample a 96-well microtitre plate of DNA samples, and stamp its image onto a filter, for subsequent hybridization. Because the pin tips are much smaller than the microtitre plate wells, it is possible to intersperse the impressions of many plates to place all of these within the same area originally occupied by a single plate. With presently available robots, usually the maximum compression attained is 16-fold (a 4×4 array of 96 well images). This leads to a 3×5 cm filter area with about 1600 samples. Thus a 10×10 cm filter can hold about 10^4 samples, more than enough for most current

(continued)

BOX 9.1 (*Continued*)

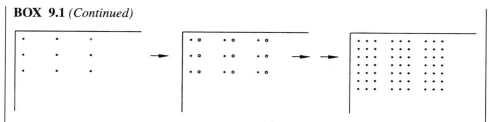

Figure 9.7   Making a dense array of samples by offset spotting of a more dilute array.

applications. Most dense arrays that have been made thus far for mapping projects are random. That is, no information is available about any of the DNA samples at the time the array is made. As the map develops, the array becomes much more informative, but the x, y indexes of each sample have only historical significance.

Once a map has been completed, it is convenient to reconfigure the array of samples so that they are placed in the actual order dictated by the map. While this is not absolutely necessary, it does allow for visual inspection of a hybridization to be instantly interpretable in many cases. There is no difficulty in instructing a robot to reconfigure an array. This procedure needs to be done only once, however slow the process, and then replicas of the new configuration can be made rapidly. The great motivation for achieving more compressed arrays is sample storage. Many clones and most DNA samples are stored at low temperature. Large libraries of samples can rapidly saturate all available temperature-controlled laboratory storage space, especially if a large number of replicas is made for subsequent distribution.

select out human material from hybrid cell lines or to reduce the complexity of a probe to desirable levels. Some of these techniques were described in Chapter 4; others will be dealt with in Chapter 14. It is worth noting that randomly chosen oligonucleotides often make very effective probes for fingerprinting. For example, a randomly chosen 10-mer should detect 1 out of every 25 cosmid clones. A key issue that we will address later in this chapter is whether it is possible to increase the efficiency of handling probes by pooling them instead of using them individually.

SCREENING BY PCR VERSUS HYBRIDIZATION

A key variable in contemporary mapping efforts is whether the connections between probes and samples are made by hybridization or by PCR. The two techniques have compensating disadvantages and advantages, and the choice of which to use will depend on the nature of the samples available and the complexity of the mapping project. We will consider these differences at the level of the target, the probe, and the organization of the project.

The sensitivity of PCR at detecting small amounts of target is unquestionably greater than hybridization. The advantage of being able to use smaller samples is that with presently available methods, it is much easier to automate the preparation of relatively small DNA samples. It is also cheaper to make such samples. When pools of targets are used, the greater sensitivity of PCR allows more complex pools with larger numbers of samples to be used. What matters for detection is the concentration of the particular target

DNA that will be detected by one method or the other. With PCR this concentration can be almost arbitrarily low, so long as there are not contaminants that will give an unacceptable PCR background.

Hybridization has the advantage when a large number of physically discrete samples (or pools) must be examined simultaneously. We have already described the use of dense arrays of samples on filters to process in parallel large numbers of targets in hybridization against a single probe. These arrays can easily contain 10^4 samples. In comparison, typical PCR reactions must be done in individually isolated liquid samples. Microtitre plates handle around 10^2 of these. Devices have been built that can do thousands of simultaneous PCRs, but these are large; with objects on such a scale one could easily handle 10^5 samples at once by hybridization.

At the level of the probe, PCR is more demanding than hybridization. Hybridization probes can be made by PCR or by radiolabeling any DNA sample available. Typical PCR analyses require a knowledge of the DNA sequence of the target. In contrast, hybridization requires only a sample of the DNA that corresponds to one element of the target. PCR primers require custom synthesis, which is still expensive and time-consuming although recent progress in automated synthesis has lowered the unit cost of these materials considerably. PCR with large numbers of different primers is not very convenient because in most current protocols the PCR reactions must be done individually or at most in small pools. Pooling of probes (or using very complex probes) is a powerful way to speed up mapping, as we will illustrate later in this chapter. However, in PCR, primers are difficult to pool. The reason is that with increasing numbers of PCR primers, the possible set of reactions rises as the square of the number of primers: Each primer, in principle, could amplify with any other primer if a suitable target were present. Thus the expected background will increase as the square of the number of primers. The desired signal will increase only linearly with the number of primers. Clearly this rapidly becomes a losing proposition. In contrast, it is relatively easier to use many hybridization probes simultaneously, since here the background will increase only linearly with the number of probes.

Both PCR and hybridization schemes lend themselves to large scale organized projects, but the implications and mechanics are very different. With hybridization, filter replicas of an array can be sent out to a set of distant users. However, the power of the array increases, the more about it one knows. Therefore, for an array to have optimal impact, it is highly desirable that all results of probe hybridizations against it be compiled in one centrally accessible location. In practice, this lends itself to schemes in place in Europe where the array hybridizations are actually done in central locations, and data are compiled there. Someone who wishes to interrogate an array with a probe, mails in that probe to a central site.

With PCR screening, the key information is the sequences of the DNA primers. This information can easily be compiled and stored on a centrally accessible database. Users simply have to access this database, and either make the primers needed or obtain them from others who have already made them. This allows a large number of individual laboratories to use the map, as it develops, and to participate in the mapping without any kind of elaborate distribution scheme for samples and without centralized experimental facilities. The PCR-based screening approach has thus far been more popular in the United States.

In the long run it would be nice to have a hybrid approach that blends the advantages of both PCR and hybridization. One way to think about this would be to use small primers or repeated sequence primers to reduce the complexity of complex probes by sampling either useful portions of pools of probes or large insert clones such as YACs. This would greatly increase the efficiency of PCR for handling complex pools and also decrease the cost of making large numbers of custom, specific PCR primers. What is still needed for an optimal method is spatially resolved PCR. The idea is to have a method that would allow one to probe a filter array of samples directly by PCR without having to set up a separate PCR reaction for each sample. However, in order for this to be efficient, the PCR products from each element of the array have to be kept from mixing. While some progress at developing in situ PCR has been reported, it is not yet clear that this methodology is generally applicable for mapping. One nice approach is to do PCR inside permeabilized cells, and then the PCR products are retained inside the cells. However, this procedure, thus far, cannot be carried out to high levels of amplification (Teo and Shaunak, 1995).

TIERED SETS OF SAMPLES

In cross-connecting different sets of DNA samples and libraries, it is helpful to have dense sets of targets that span a range of sizes in an orderly way. The rationale behind this statement is illustrated in Figure 9.8. The figure shows an attempt to connect two sets of total digest fragments or clones by cross-hybridization or some other kind of complementary fingerprinting. When large fragment *J* is used as a probe, it detects three smaller fragments *a, b,* and *c* but does not indicate their order. When the smaller fragments *a, b,* and *c* are used as probes, *b* detects only *J*, which means that *b* is the central small fragment; *a* and *b* each detect additional flanking larger fragments, so they must be external. A generalization of this argument indicates that it is more efficient if one has access to a progression of samples where average sizes diminish roughly by factors of three. Less sharp size decreases will mean an unnecessarily large number of different sets of samples. Larger size decreases will mean that ordering of each smaller set will be too ambiguous.

The same kinds of arguments are applicable in more complex cases where overlapping sets of fragments or clones exist. Consider the example shown in Figure 9.9. The objective is to subdivide a 500 kb YAC into cosmid clones so that these can be used as more convenient sources of DNA for finding polymorphisms or for sequencing. The traditional approach to this problem would have been to subclone the purified YAC DNA into cosmids. However, this involves a tremendous amount of work. The more modern approach is to start with a YAC contig flanking the clone of interest. This contig will automatically arise in the context of producing an ordered YAC library. With a fivefold redundant array of YACs, the contig will have typically five members in this region. Therefore, if these members are used separately for hybridization or fingerprinting, they will divide the region into intervals that average about 100 kb in size. Each interval will serve as a bin to assign cosmids to the region. Since the intervals are less than three times the cosmid in-

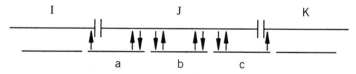

Figure 9.8   Clone ordering when tiered sets of clones are available.

sert size, the ordering should be reasonably effective. However, the key point in the over-all strategy is that one need not make the YAC contig in advance. Cross-connecting the various sets of samples in parallel will eventually provide all of the information needed to order all of them.

The example shown in Figure 9.9 is a somewhat difficult case because the size step taken, from 500 kb clones to 40 kb clones, is a big one. It would be better to have one or two tiers of samples in between; say 250 kb YACs and 100 kb P1 clones. This will compensate quite a bit for the inevitable irregularities in the distribution and coverage of each of the libraries in particular regions. An analogy that may not be totally farfetched is that the intermediate tiers of samples can help strengthen the ordering process in the same manner as intermediate levels in a neural net can enhance its performance (see Chapter 15).

YAC and cosmid libraries involve large numbers of clones, and it would be very ineffi-cient to handle such samples one at a time. We have already shown that cosmids can be handled very efficiently as dense arrays of DNAs on filters. YACs are less easily handled in this manner. The amount of specific sample DNA is much less, since typical YACs are single-copy clones in a 13-Mb genome background, while cosmids are multicopy in a 4.8-Mb background. Thus hybridization screening of arrays of YAC clones has not always been very successful. It would greatly help if there were an effective way to purify the YAC DNA away from the yeast genomic DNA. Automated procedures already have been developed to purify total DNA from many YAC-bearing strains at once (Box 9.1). Now what is needed is a simple automatable method for plucking the YACs out of this mixture. An alternative to purification would be YAC amplification. As described earlier in this chapter, this is possible but not yet widely used, and the amount of amplification is still only modest. Perhaps the most effective method currently available for increasing our sensitivity of working with YAC DNA is PCR. In Chapter 14 we will illustrate how human-specific PCR, based on repeating sequence primers like *Alu*'s, can be used to see just the human DNA in a complex sample. Almost every YAC is expected to contain mul-tiple *Alu* repeats. It is possible to do hundreds of PCR reactions simultaneously with com-monly available thermal cyclers, and ten to perhaps a hundred times larger number of samples is manageable with equipment that has been designed and built specially for this purpose.

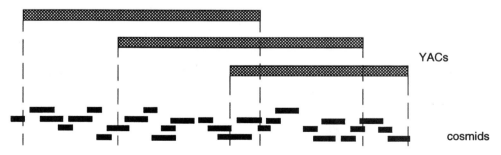

hybridization pattern arranges cosmids in (ordered) bins

Figure 9.9 Division of a YAC into a contig of cosmids by taking advantage of other YACs known to be contiguous.

SIMPLE POOLING STRATEGIES FOR FINDING A CLONE OF INTEREST

A major problem in genome analysis is to find in a library the clones that correspond to a probe of interest. This is typically the situation one faces in trying to find clones in a region marked by a specific probe that is suspected to be near a gene of interest. With cosmid clones, one screens a filter array, as illustrated earlier. With YAC clones, filter arrays have not worked well in many hands, and instead, PCR is used. But with today's methods it is inconvenient and expensive to analyze the YACs in a large library, individually, by PCR. For example, a single coverage 300-kb human genomic YAC library is 10^4 clones. Fivefold coverage would require 5×10^4 clones. This number of PCR reactions is still daunting. However, if the objective is to find a small number of clones in the library that overlap a single probe, there are more efficient schemes for searching the library. These involve pooling clones and doing PCR reactions on the pools instead of on individual clones.

One of the simplest and most straightforward YAC pooling schemes involves three tiers of samples (Fig. 9.10). The YAC library is normally distributed in 96-well microtitre plates (Fig. 9.6a). Thus 100 plates would be required to hold 10^4 clones. Pools are made from the clones on each plate. This can be done by sampling them individually or by using multiple headed pipetors or other tools as described in Box 9.1. Each plate pool contains 96 samples. The plate pools are combined ten at a time to make super pools. First, ten super pools are screened by PCR. This takes 10 PCR reactions (or 50 if a fivefold redundant library is used). Each positive superpool is then screened by subsequent, separate PCR reactions of each of the ten plate pools it contains. In turn, each plate pool that shows a positive PCR is divided into 12 column pools (of 9 YACs each) and, separately, 9 row pools (of 12 YACs each), and a separate PCR analysis is done on each of these samples. In most cases this should result in a unique row-column positive combination that serves to identify the single positive YAC that has been responsible for the entire tier of PCR amplifications. Each positive clone found in this manner will require around 41 PCR

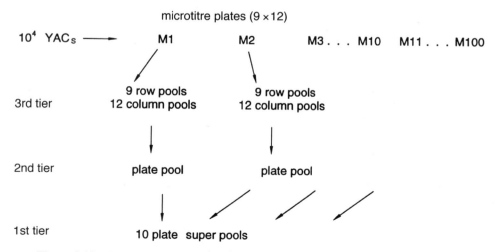

Figure 9.10   Three-tier pooling strategy for finding a clone of interest in a YAC library.

reactions. This is a vast improvement over the 10^4 reactions required if YACs are examined one at a time. For a fivefold redundant library, if the super pools are kept the same size, then the total number of PCRs needed will be 81 per positive clone. This is still quite reasonable.

SEQUENCE-SPECIFIC TAGS

The power of the simple pooling approach just described had a strong effect on early strategies developed for genome mapping. In reality the power is greatest when individual clones are sought, and it diminishes considerably when the goal is to order a whole library. However, the attractiveness of PCR-based screening has led to much consideration about the kinds of PCR primers that would be suitable for genome analysis. Since PCR ordinarily requires known sequences, the issue really becomes what kinds of DNA sequences should be used for finding genes or for genome mapping. A number of different types of approaches are currently being used; these are described below.

  STS. This is short for a sequence tagged site. The original notion was that any arbitrary bit of known DNA sequence could be used as a probe if it successfully generated useful PCR primers. One early notion was to take all of the existing polymorphic genetic probes and retro-fit these as STSs by determining a partial DNA sequence, and then developing useful PCR primers based on this sequence. This approach was never executed on a large scale, which is probably good in retrospect because recently developed genetic markers are far more useful for mapping than earlier ones, and the process for finding these automatically includes the development of unique sequence tags suitable for PCR.

  STAR. This stands for sequence tagged rare restriction site. The great utility of such probes has already been described in considering strategies for efficient restriction mapping or ordering total digest libraries. The appeal of STARs is that they allow precise placement of probes on physical maps, even at rather early stages in map construction. The disadvantage of STARs, mentioned earlier, is that many cleavage sites for rare restriction nucleases turn out to be very G + C rich, and thus PCR in these regions is more difficult to perform. For any kind of probe, clone ordering is most efficient if the probes come from the very ends of DNA fragments or clone inserts. As shown in Figure 9.11, this allows PCR to be done by using primers in the vector arms in addition to primers in the insert. Thus the number of unique primers that must be made for each probe is halved. STARs in total digest libraries naturally come from the ends of DNA fragments, and thus they promote efficient mapping strategies.

Figure 9.11 Use of vector primers in PCR to amplify the ends of clones like YACs.

STP or STRP. These abbreviations refer to sequence tagged polymorphism or polymorphic sequence tag. A very simple notion is involved here. If the PCR tag is a polymorphic sequence, then the genetic and physical maps can be directly aligned at the position of this tag. This allows the genetic and physical maps to be built in parallel. A possible limitation here is that some of the most useful genetic probes are tandemly repeating sequences, and a certain subset of these, usually very simple repeats like $(AC)_n$, tend to give extra unwanted amplification products in typical PCR protocols. However, it seems possible to find slightly more complex repeats, like $(AAAC)_n$, that are equally useful as genetic probes but show fewer PCR artifacts.

EST. This stands for expressed sequence tag. It could really refer to any piece of coding DNA sequence for which PCR primers have been established. However, in practice, EST almost always refers to a segment of the DNA sequence of a cDNA. These samples are usually obtained by starting with an existing cDNA library, choosing clones at random, sequencing as much of them as can be done in a single pass, and then using this sequence information to place the clone on a physical map (through somatic cell genetics or FISH). There are many advantages to such sequences as probes. One knows that a gene is involved. Therefore the region of the chromosome is of potential interest. The bit of DNA sequence obtained may be interesting: It may match something already in the database or be interpretable in some way (see Chapter 15). In general, the kinds of PCR artifacts observed with STPs, STRPs, and STARs are much less likely to occur with ESTs.

Despite their considerable appeal there are a number of potential problems in dealing with ESTs as mapping reagents. As shown in Figure 9.12, cDNAs are discontinuous samples of genomic DNA. They will typically span many exons. This can be very confusing. If the EST crosses a large intron, the probe will show PCR amplification, but genomic DNA or a YAC clone will not. A common strategy for EST production uses largely untranslated DNA sequence at the 3′-ends of the cDNA clones. It is relatively easy to clone these regions, and they are more polymorphic than the internal coding region. Furthermore cDNAs from gene families will tend to have rather different 3′-untranslated regions, and thus one will avoid some of the problems otherwise encountered with multiple positive PCR reactions from members of a gene family. These 3′-end sequences will also tend to contain only a single exon in front of the untranslated region. However, all of these advantages carry a price: The 3′-end sequence is less interesting and interpretable than the remainder of the cDNA.

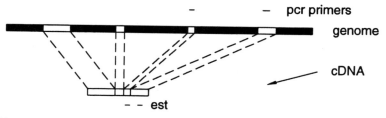

Figure 9.12 A potential problem with ESTs as mapping reagents is that an EST can cross one or more introns.

Eugene Sverdlov and his colleagues in Moscow have developed an efficient procedure for preparing chromosome-specific cDNA libraries. Their three-step procedure is an extension of simpler procedures that were tried earlier by others. This procedure uses an initial *Alu*-primed PCR reaction to make a cDNA copy of the hnRNA produced in a hybrid cell containing just the chromosome of interest. The resulting DNA is equipped with an oligo-G tail, and then a first round PCR is carried out using an oligo-C containing primer and an *Alu* primer. Then a second round of PCR is done with a nested *Alu* primer. The PCR primers are also designed so that the first round introduces one restriction site and the second round another. The resulting products are then directionally cloned into a vector requiring both sites. In studies to date, Sverdlov and his coworkers have found that this scheme produces a diverse set of highly enriched human cDNAs. Because these come from *Alu*s in hnRNA, they will contain introns, and this gives them potential advantages as mapping probes when compared with conventional cDNAs. As with the 3′-cDNAs discussed above, cDNA from hnRNAs will be more effective than ordinary cDNAs in dealing with gene families and in avoiding cross-hybridization with conserved exonic sequences in rodent-human cell hybrids.

POOLING IN MAPPING STRATEGIES

All of the above methods for screening individual samples in a complex library are fine for finding genes. However, they are very inefficient in ordering a whole library. The schematic result of a successful screen for a gene-containing clone in an arrayed library is shown in Figure 9.13a. A single positive clone is detected, presumably containing the sample of interest. However, from the viewpoint of information retrieval, this result is very weak. A sample array is potentially an extremely informative source of information about the order of the samples it contains. A single positive hybridization extracts the minimum possible amount of information from the array and requires a time-consuming

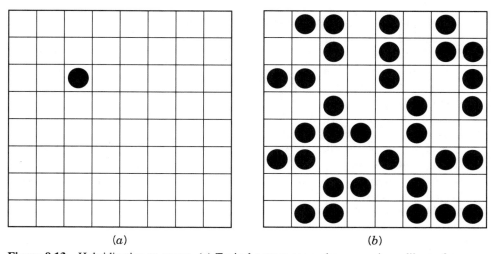

(a)   (b)

Figure 9.13 Hybridization to arrays. (*a*) Typical pattern seen when screening a library for a particular gene. (*b*) Ideal pattern of hybridization in interrogating a dense sample array.

experiment. From the viewpoint of information theory, a much better designed interrogation of the array would produce the result shown in Figure 9.13*b*. In this ideal black-and-white case, roughly half the clones would be scored positive and half negative if we are considering only positive and negative hybridization results. When the amount of signal can be quantitated, much more information is potentially available from a single test of the array through all the gray levels seen for each individual clone.

The practical issue is how to take advantage of the power of arrays, or pools of samples, in a maximally efficient way so that all of the clones are ordered accurately with a minimum number of experiments. The answer will depend critically on the nature of the errors involved in interrogating arrays or pools. We start by considering fairly ideal cases. A detailed analysis of the effects of errors in real applications is still not available, but the power of these approaches is so great that it appears that a reasonable error rate can be tolerated.

The goal is to design probe or sample pools and arrays that allow roughly half of the targets to be scored positive in each hybridization or PCR. There are a number of different ways to try to accomplish this. One general approach is to increase the amount of different DNA sequences in the probes or the targets. The other general approach is to increase the fraction of potential target DNA that will be complementary to probe sequences. In principle, both approaches can be combined. A very simple strategy is to use complex probes. For example, purified large DNA fragments can be used as hybridization probes. DNA from hybrid cell lines or radiation hybrids can be used as probes. In some or all of these cases, it is helpful to employ human-specific PCR amplification so that the probe contains sufficient concentrations of the sequences that are actually serving to hybridize with specific samples in the target.

The logic behind the use of large DNA probes is that they automatically contain continuity information, and sets of target clones detected in such a hybridization should lie in the same region of the genome. It is far more efficient, for example, to assign an array of clones to chromosome regions by hybridizing with DNA purified from those regions, than it is to assign the regional location of clones one at a time by hybridizing to a panel of cell lines. The key technical advances that makes these new strategies possible are PCR amplification of desired sequences and suppression hybridization of undesired repeated sequences.

An alternative to large DNA probes is simple sequence probes. Oligonucleotides of lengths 10 to 12 will hybridize with a significant fraction of arrayed cosmids or YACs. Alternatively, one could use simple sequence PCR primers, as we discussed in Chapter 4. There is no reason why probe molecules must be used individually. Instead, one could make pools of probes and use this pool directly in hybridization. By constructing different sets of pools, it is possible, after the fact, to sort out which members of which pools were responsible for positive hybridization. This turns out to be more efficient in principle. Any sequences can be used in pools. One approach is to select arbitrary, nonoverlapping sets of single-copy sequences. Another approach is to build up pools from mixtures of short oligonucleotides.

Another basic strategy for increasing the efficiency of mapping is to use pools of samples. This is a necessary part of PCR screening methods, but there is no reason why it also could not be used for hybridization analyses. We will describe some of the principles that are involved in pooling strategies in the next few sections. These principles apply equally whether pools of probes or pools of samples are involved. Basically it would seem that one should be able to combine simultaneously sample pooling and probe pooling to increase the throughput of experiments even further. However, the most efficient ways to do this have not yet been worked out.

A caveat for all pooling approaches concerns the threshold between background noise or cross-hybridization and true positive hybridization signals. Theoretical, noise-free strategies are not likely to be viable with real biological samples. Instead of striving for the most ambitious and efficient possible pooling strategy, it is prudent to use a more overdetermined approach and sacrifice a bit of efficiency for a margin of safety. It is important to keep in mind that a pooling approach does not need to be perfect. Once potential clones of interest are found or mapped, additional experiments can always be done to confirm their location. The key idea is to get the map approximately right with a minimum number of experiments. Then the actual work involved in confirming the pooling is finite and can be done regionally by groups interested in particular locales.

PROBE POOLING IN *S. POMBE* MAPPING

An example of the power of pooling is illustrated by results obtained in ordering a cosmid library of the yeast *S. pombe*. This organism was chosen for a model mapping project because a low-resolution restriction map was already available and because of the interest in this organism as a potential target for genomic sequencing. An arrayed cosmid library was available. This was first screened by hybridization with each of the three *S. pombe* chromosomes purified by PFG. Typical results are shown in Figure 9.14. It is readily apparent that most cosmids fall clearly on one chromosome by their hybridization. However there are significant variations in the amount of DNA present in each spot of the array. Thus a considerable amount of numerical manipulation of the data is required to correct for DNA sample variation, and for differences in the signals seen in successive hybridizations. When this is done, it is possible to assign, uniquely, more than 85% of the clones to a single chromosome based on only three hybridizations.

The next step is to make a regional assignment of the clones. Here purified restriction fragments are labeled and used as hybridization probes with the cosmid array. An example is shown in Figure 9.15. Note that it is inefficient to use only a single restriction fragment at a time. For example, once one knows the chromosomal location of the cosmids, one can mix restriction fragments from different chromosomes and use them simultaneously with little chance of introducing errors. After a small number of such experiments, one has most of the cosmids assigned to a well-defined region.

To fingerprint the cosmid array further, and begin to link up cosmid contigs, arbitrary mixtures of single-copy DNA probes were used. These were generated by the method shown in Figure 9.16. A FIGE separation of a total restriction enzyme digest of *S. pombe* genomic DNA was sliced into fractions. Because of the very high resolution of FIGE in the size range of interest, these slices should essentially contain nonoverlapping DNA sequences. Each slice was then used as a hybridization probe. As an additional fingerprinting tool, mixtures of any available *S. pombe* cloned single-copy sequences were made and used as probes. For all of these data to be analyzed, it is essential to consider the quantitative hybridization signal, and correct it both for background and for differences in the amount of DNA in each clone, the day-to-day variations in labeling, and overall hybridization efficiency.

The hybridization profile of each of the cosmid clones with the various probes used is an indication of where the clone is located. A likelihood analysis was developed to match

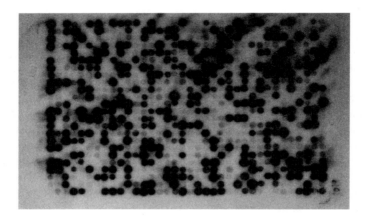

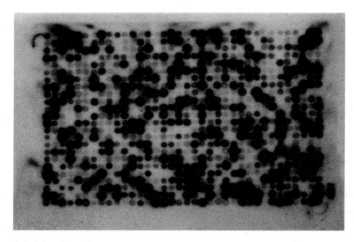

Figure 9.14   Hybridization of a cosmid array of *S. pombe* clones with intact *S. pombe* chromosomes I (*a*) and II (*b*).

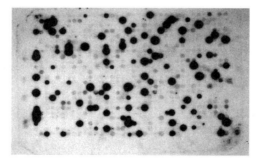

S. pombe Sfi I fragments K and L

Figure 9.15   Hybridization of the same cosmid array shown in Figure 9.14 with two large restriction fragments purified from the *S. pombe* genome.

high resolution separation of small
restriction fragments by FIGE

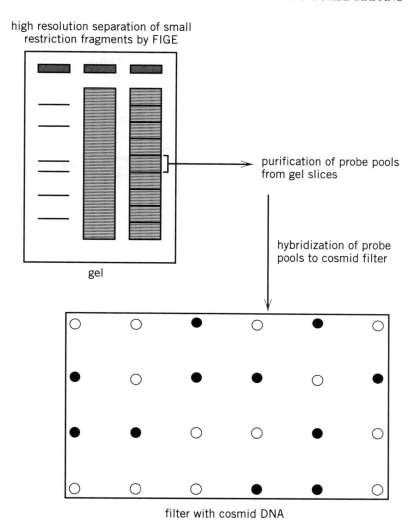

purification of probe pools
from gel slices

hybridization of probe
pools to cosmid filter

gel

filter with cosmid DNA

Figure 9.16   Preparation of nonoverlapping pools of probes by high-resolution FIGE separation of restriction enzyme-digested *S. pombe* genomic DNA.

up clones with similar hybridization profiles that indicate possible overlaps (Box 8.3). The basic logic behind this method is shown in Figure 9.17. Regardless of how pools of probes are made, actually overlapping clones will tend to show a concordant pattern of hybridization when many probe pools are examined. The likelihoods that reflect the concordancy of the patterns were then used in a series of different clone-ordering algorithms, including those developed at LLNL, LANL, a cluster analysis, a simulated annealing analysis, and the method sketched in Box 9.2. In general, we found that the different methods gave consistent results. Where inconsistencies were seen, these could often be resolved or rationalized by manual inspection of the data. Figure 9.18 shows the hybridization profile of three clones. This example was chosen because the interpretation of the profile is straightforward. The clones share hybridization with the same chromosome, the same restriction fragments, and most of the same complex probe mixtures. Thus they must be located nearby, and in fact they form a contig. The clones show some hybridization differences when very simple probe mixtures are used, since they are not identical clones. An example of the kinds of contigs that emerge from such an analysis is shown in Figure 9.19. These are clearly large and redundant contigs of the sort one hopes to see in a robust map.

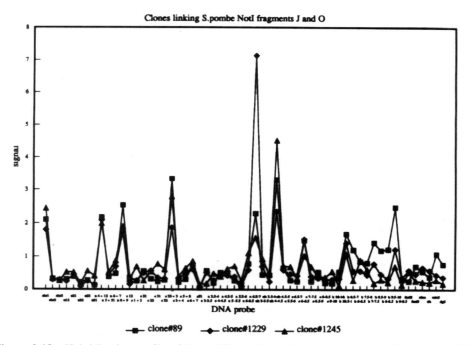

only cosmids marked by arrow are detected with both probe sets a and b

Figure 9.17   Even when complex pools of probes are used (e.g., pools *a* and *b*) overlapping clones will still tend to show a concordant pattern of hybridization.

Figure 9.18   Hybridization profile of three different *S. pombe* cosmid clones, with a series of 63 different pools of probes.

Figure 9.19   *S. pombe* cosmid map constructed from the patterns of probe hybridization to the *S. pombe* cosmid array. Cosmids are indicated by horizontal lines along the maps. Letters shown above the map and fragment names (e.g., SHNF = *Sfi* I fragment H and *Not* I fragment F). Gaps in contigs are indicated by a vertical bar at the right end. Positions of LTRs, 5S rDNAs, and markers are indicated by *, #, and ×, respectively.

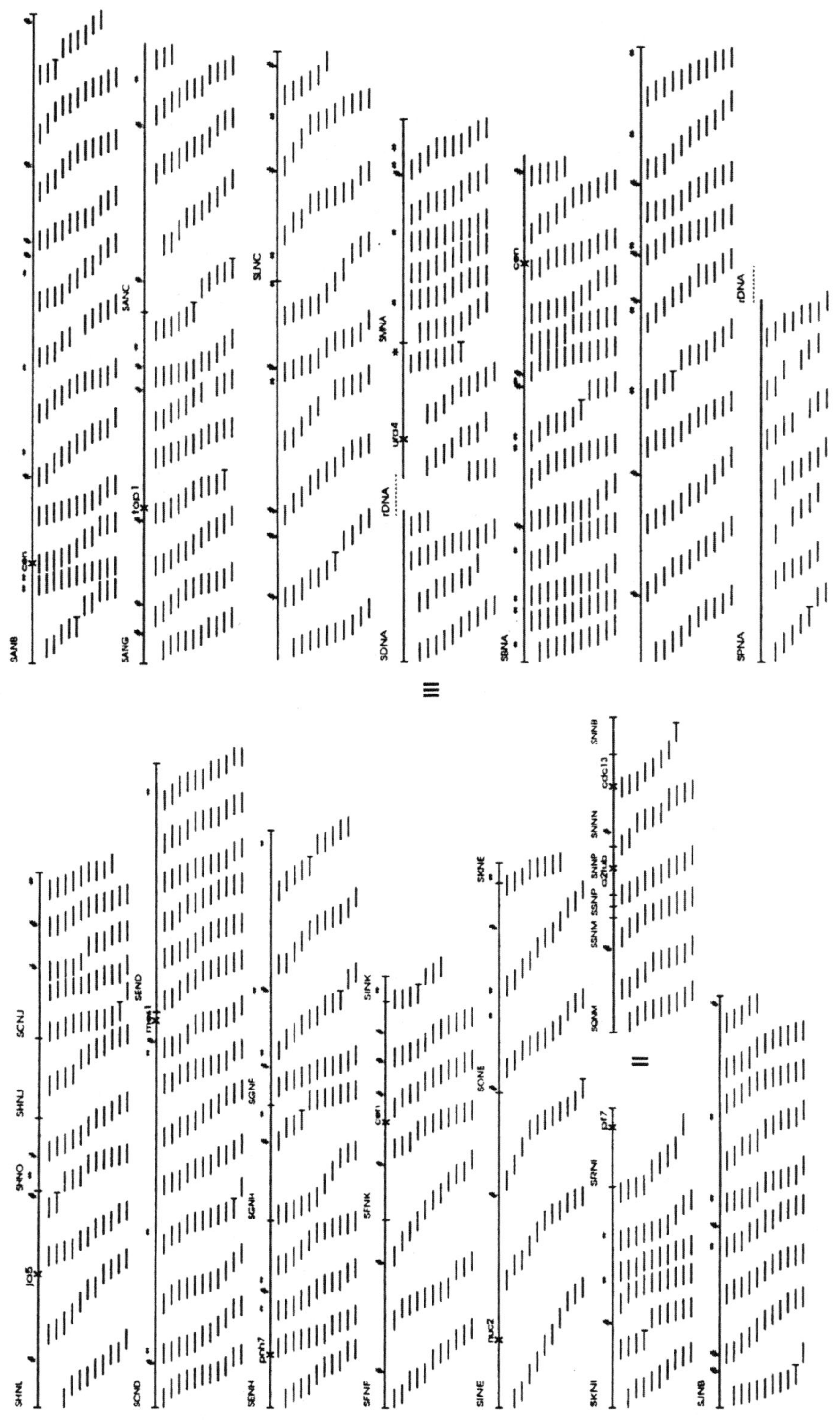

BOX 9.2
CONSTRUCTION OF AN ORDERED CLONE MAP

Once a set of likelihood estimates has been obtained for clone overlap, the goal is to assemble a map of these clones that represents the most probable order given the available overlap data. This is a computationally intensive task. A number of different algorithms have been developed for the purpose. They all examine the consistency of particular clone orders with the likelihood results. The available methods all appear to be less than perfectly rigorous, since they all deal with data only on clone pairs and not on higher clusters. Nevertheless, these methods are fairly successful at establishing good ordered clone maps.

Figure 9.20 shows the principle behind a method we used in the construction of a cosmid map of *S. pombe*. The objective in this limited example is to test the evidence in favor of particular schemes for ordering three clones A, B, and C. Various possible arrangements of these clone are written as the columns and rows of a matrix, each element of this matrix is represented by the maximum likelihood estimate that the clones *i* and *j* overlap by a fraction *f*: $L_{ij}(f)$. For each possible arrangement of clones, we calculated the weight of the matrix, W_m defined as

$$W_m = \sum_{j>i} \sum_i |i - j| L_{ij}(f)$$

The result for a simple case is shown in the figure. The true map will have an arrangement of clones that gives a minimum weight or very close to this. The method is particularly effective in penalizing arrangements of clones where good evidence for overlap exists, and yet the clones are postulated to be nonoverlapping in the final assembled map.

	A	B	C
A	x	L_{AB}	L_{AC}
B		x	L_{BC}
C			x

$W_m = L_{AB} + 2 L_{AC} + L_{BC}$

	A	C	B
A	x	L_{AC}	L_{AB}
C		x	L_{BC}
B			x

$W_m = L_{AC} + 2 L_{AB} + L_{BC}$

Figure 9.20 Example of an algorithm used to construct ordered clone maps from likelihood data. Details are given in Box 9.2.

The methods developed on *S. pombe* allowed a cosmid map to be completed to about the 98% stage in 1.5 person years of effort. Most of this effort was method or algorithm development, and we estimate that to repeat this process on a similar size genome would take only 3 person-months of effort. This is in stark contrast with earlier mapping approaches. Strict bottom-up fingerprinting methods, scaled to the size of the *S. pombe* genome, would require around 8 to 10 person-years of effort. Thus by the use of pooling and complex probes, we have gained more than an order of magnitude in mapping speed. The issue that remains to be tested is how well this kind of approach will do with mammalian samples where the effects of repeated DNA sequences will have to be eliminated. However, between the use of competition hybridization, which has been so successful in FISH, and selective PCR, we can be reasonably optimistic that probe pooling will be a generally applicable method.

FALSE POSITIVES WITH SIMPLE POOLING SCHEMES

Row and column pools are very natural ideas for speeding the analysis of an array of samples. It is easy to implement these kinds of pools with simple tools and robots. However, they lead to a significant level of false positives when the density of positive samples in an array becomes large. Here we illustrate this problem in detail, since it will be an even more serious problem in more complex pooling schemes. Consider the simple example shown in Figure 9.21. Here two clones in the array hybridize with a probe (or show PCR amplification with the probe primers, in the case of YAC screening). If row and column pools are used for the analysis, rather than individual clones, four potentially positive clones are identified by the combination of two positive rows and two positive columns. Two are true positives; two are false positives. To decide among them, each isolated clone can be checked individually. In the case shown, only four additional hybridizations or PCR reactions would have to be done. This is a small addition to the number of tests required for screening the rows and columns. However, as the number of true positives grows linearly, the number of false positives grows quadratically. It soon becomes hopelessly inefficient to screen them all individually. An alternative approach, which is much more efficient, in the limit of high numbers of positive samples, is to construct alternate pools. For example, in the case shown in Figure 9.21, if one also included pools made along the diagonals, most true and false positives could be distinguished.

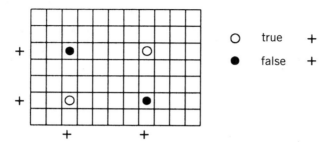

Figure 9.21 An example of how false positives are generated when a pooled array is probed. Here row and sample pooling will reveal four apparent positive clones whenever only two real positives occur (unless the two true positives happen to fall on the same row or column).

MORE GENERAL POOLING SCHEMES

A branch of mathematics called *sampling theory* is well developed and instructs us how to design pools effectively. In the most general case there are two significant variables: the number of dimensions used for the array and the pools, and the number of alternate pool configurations employed. In the case described in Figure 9.21, the array is two dimensional, and the pools are one dimensional. Rows and columns represent one configuration of the array. Diagonals, in essence, represent another configuration of the array because they would be rows and columns if the elements of the array were placed in a different order. Here we want to generalize these ideas. It is most important to realize that the dimensionality of an array or a pool is a mathematical statement about how we chose to describe it. It is not a statement about how the array is actually composed in space. An example is shown by the pooling scheme illustrated in Figure 9.22. Here plate pools are used in conjunction with vertical pools, made by combining each sample at a fixed *x-y* location on all the plates. The arrangement of plates appears to be three dimensional; the plate pools are two dimensional, but the vertical pools are only one dimensional.

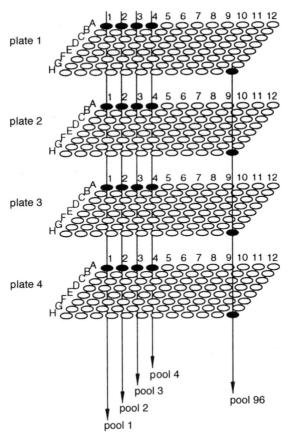

Figure 9.22 A pooling scheme that appears, superficially, to be three dimensional, but, in practice, is only two dimensional.

We consider first an N element library, and assume that we have sufficient robotics to sample it in any way we wish. A two-dimensional square array is constructed by assigning to each element a location:

$$a_{ij} \qquad \text{where } i = 1, N^{1/2}; j = 1, N^{1/2}$$

If we pool rows and columns, each of these one-dimensional pools has $N^{1/2}$ components, and there are $2N^{1/2}$ different pools if we just consider rows and columns. The actual construction of a pool consists of setting one index constant, say $i = 3$, and then combining all samples that share that index.

The same N element library can be treated as a three-dimensional cubic array. This is constructed, by analogy, with the two-dimensional array by assigning to each element a location:

$$a_{ijk} \qquad \text{where } i = 1, N^{1/3}; j = 1, N^{1/3}; k = 1, N^{1/3}$$

If we pool two-dimensional surfaces of this array, each of these pools has $N^{2/3}$ components. There are $3N^{1/3}$ different pools if we consider three orthogonal sets of planes. The actual process of constructing these pools consists of setting one index constant, say $j = 2$, and then combining all samples that share this index.

We can easily generalize the pooling scheme to higher dimensions. These become hard to depict visually, but there is no difficulty at handling them mathematically. To make a four-dimensional array of the library, we assign each of the N clones an index:

$$a_{ijkl} \qquad \text{where } i = 1, N^{1/4}; j = 1, N^{1/4}; k = 1, N^{1/4}; l = 1, N^{1/4}$$

This array is what is actually called a hypercube. We can make cubic pools of samples from the array by setting one index constant, say $k = 4$, and then combining all samples that share this index. The result is $4N^{1/4}$ different pools, each with $N^{3/4}$ elements. The pools actually correspond to four orthogonal sets of sample cubes.

The process can be extended to five and higher dimensions as needed. Note that the result of increasing the dimensionality is to decrease, steadily, the number of different pools needed, at a cost of increasing the size of each pool. Therefore the usefulness of higher-dimension pooling will depend very much on experimental sensitivity. Can the true positives b distinguished among an increasingly higher level of background noise as the complexity of the pools grows?

The highest dimension possible for a pooling scheme is called a binary sieve. It is definitely the most efficient way to find a rare event in a complex sample so long as one is dealing with perfect data. In the examples discussed above, note that the range over which each sample index runs keeps dropping steadily, from $i = 1, N^{1/2}$ to $i = 1, N^{1/4}$ as the dimension of the array is increased from two to four. The most extreme case possible would allow each index only a single value; in this case there is really no array, the samples are just being numbered. A pooling scheme one step short of this extreme would be to allow each index to run over just two numbers. If we kept strictly to the above analogy we would say $i = 1, 2$; however, it is more useful to let the indices run from 0 to 1. Then we assign to a particular clone an index like a_{101100}. This is just a binary number (the equivalent decimal is 44 in this case). Pools are constructed, just as before, by selecting all clones with a fixed index, like a_{ijk0mn}.

With a binary sieve we are nominally restricted to libraries where N is a power of 2: $N = 2^q$. Then the array is constructed by indexing

$$a_{ijklmn} \ldots \qquad \text{where } i, j, k, l, m, n, \ldots = 0, 1$$

Each of the pools made by setting one index to a fixed value will have $N/2$ elements. The number of pools will be $q = \log(N)/\log(2)$. This implicitly assumes that one scores each clone for 1 or 0 at each index, but not both. The notion in a pure binary sieve is that if the clone is not in one pool ($k = 1$), it must certainly be in the other ($k = 0$), and so there is no need to test them both. With real samples, one would almost certainly want to test both pools to avoid what are usually rather frequent false negatives. The size of each pool is enormous—it contains half of the clones in the library. However, the number of pools is very small. It cannot be further reduced without including additional sorts of information about the samples, such as intensity, color, or other measurable characteristics.

The binary sieve is constructed by numbering the samples with binary numbers, namely integers in base two. One can back off from the extreme example of the binary sieve by using indices in other bases. For example, with base three indices, the array is constructed as $a_{ijkl} \ldots$, where $i, j, k, l, \ldots = 0, 1, 2$. This results in a larger number of pools, each less complex than the pools used in the binary sieve. It is clear that to construct the actual pools used in binary sieves and related schemes would be quite complex if one had to do it by hand. However, it is relatively easy to instruct an x-y robot to sample in the required manner. Specialized tools could probably be utilized to make the pooling process more rapid.

A numerical example will be helpful, here. Consider a library where $N = 2^{14} = 16,384$ clones. For a two-dimensional array, we need 2^7 by 2^7 or 128 by 128 clones. The row and column pools will have 128 elements each. There are 256 pools needed. In contrast, the binary array requires only 14 pools (28 if we want to protect against false negatives). Each pool, however, will have 8192 clones in it! Constructing these pools is not conceivable unless the procedure is fully automated.

As the dimensionality of pooling increases, there are trade-offs between the reduced number of pools and the increased complexity of the pools. The advantage of reduced pool number is obvious: Fewer PCR reactions or hybridizations will have to be done. A disadvantage of increased pool complexity, beyond background problems that we have already discussed, is the increasing number of false positives when the array has a high density of positive targets. For example, with a two-dimensional array, two positive clones a_{36} and a_{24} imply that false positives will appear at a_{34} and a_{26}. In three dimensions, two positive clones a_{826} and a_{534} will generate false positives at a_{824}, a_{836}, a_{834}, a_{536}, a_{524}, and a_{526}. The number of false positives is smaller if some share a common orthogonal plane. In general for two positive clones there will be up to $2^n - 2$ false positives in an n-dimensional pool. By the time a binary sieve is reached, the false positives become totally impossible to handle. Thus the binary sieve will be useful only for finding very rare needles in very large haystacks. For realistic screening of libraries, much lower dimensionality pooling is needed.

The actual optimum dimension pooling scheme to use will depend on the number of clones in the array, the redundancy of the library (which will increase the rate of false positives) and the number of false positives that one is willing to tolerate, and then rescreen for individually or with an alternate array configuration. Figure 9.23 gives some

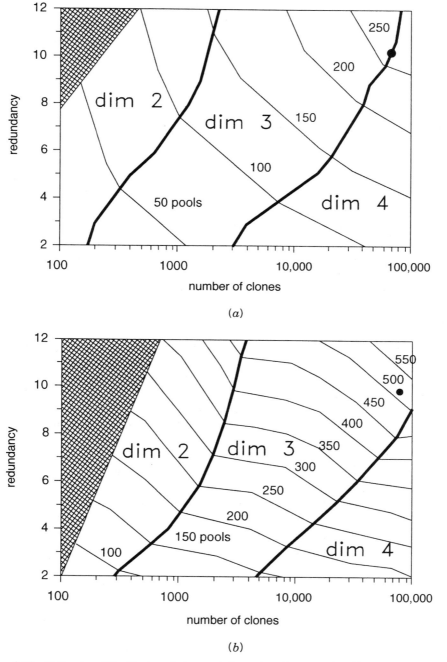

(a)

(b)

Figure 9.23 Estimates of the ideal pool sizes and dimensions for screening libraries with different numbers of clones and different extents of redundancy. The actual characteristics of the CEPH YAC library are indicated by a dot. (a) Where one false positive is tolerable for each true positive. (b) Where only 0.01 false positive is tolerated per screen. Note that the shaded area represents situations where pooling strategies are not useful. (Taken from Barillot et al., 1991.)

calculated results for two realistic cases. These results cover cases where one false positive per true positive is acceptable, and where only 0.01 false positive per screen will be seen. In general, the results indicate that for small numbers of clones, pooling is not efficient. The larger the library, the higher is the dimensionality of effective pools; however, the higher the redundancy, the worse is the problem of false positives, and the lower is the optimum dimensionality.

ALTERNATE ARRAY CONFIGURATIONS

There is a generally applicable strategy to distinguish between true and false positives in pools of arrayed clones. This strategy is applicable even where the density of true positives becomes very high. The basic principle behind the strategy is illustrated in Figure 9.24 for a two-dimensional array with two positive clones. Suppose that one has two versions of this array in which the positive clones happen to be at different locations, but the relationship between the two versions is known. That is, the identity of each clone at each position on both versions of the array is known. When row and column pools of the array are tested, each configuration gives two positive rows and two positive columns, resulting in four potentially positive clones in each case. However, when the actual identity of the putative positive clones is examined, it turns out that the same true positives will occur in both configurations, but usually the false positives will be different in the two configurations. Thus they can be eliminated.

The use of multiple configurations of the array is rather foolish and inefficient for small arrays and small numbers of false positives. However, this process becomes very efficient for large arrays with large numbers of false positives. Procedures exist called *transformation matrices* or *Latin squares* that each show how to reconfigure an original two-dimensional array into informative alternates. It is not obvious if efficient procedures are known for arrays in higher dimensions. It is also not clear that any general reconfiguration procedure is optimal, since the best scheme at a given point may depend on the prior results. Suffice it to say that the use of several alternate configurations appears to be a very powerful tool. This is illustrated by the example shown in Figure 9.25. Here five positive clones are correctly identified using row and column pools from three configurations of a 7 × 7 array. This requires testing a total of 42 pools, which is only slightly less than the 49 tests needed if the clones were examined individually. Again, however, the efficiency of the approach grows tremendously as the array size increases. Additional discussion of quantitative aspects of pooling strategies can be found in Bruno et al., (1995).

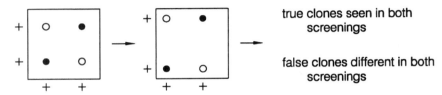

Figure 9.24   A simple example of how two different configurations of an array can be used to resolve the ambiguities caused by false positives.

1st configuration

$$
\begin{bmatrix}
c_{0,0}^+ & c_{0,1}^+ & c_{0,2}^+ & c_{0,3}^+ & c_{0,4}^+ & c_{0,5}^+ & c_{0,6}^+ \\
c_{1,0}^+ & c_{1,1}^+ & c_{1,2}^+ & c_{1,3}^+ & c_{1,4}^+ & c_{1,5}^+ & c_{1,6}^+ \\
c_{2,0}^+ & c_{2,1}^+ & c_{2,2}^+ & c_{2,3}^+ & c_{2,4}^+ & c_{2,5}^+ & c_{2,6}^+ \\
c_{3,0}^+ & c_{3,1}^+ & c_{3,2}^+ & c_{3,3}^+ & c_{3,4}^+ & c_{3,5}^+ & c_{3,6}^+ \\
c_{4,0}^+ & c_{4,1}^+ & c_{4,2}^+ & c_{4,3}^+ & c_{4,4}^+ & c_{4,5}^+ & c_{4,6}^+ \\
c_{5,0} & c_{5,1} & c_{5,2} & c_{5,3} & c_{5,4} & c_{5,5} & c_{5,6} \\
c_{6,0} & c_{6,1} & c_{6,2} & c_{6,3} & c_{6,4} & c_{6,5} & c_{6,6}
\end{bmatrix}
$$

$\downarrow A$

2nd configuration

$$
\begin{bmatrix}
c_{0,0}^+ & c_{6,2} & c_{5,4} & c_{4,6}^- & c_{3,1} & c_{2,3}^+ & c_{1,5}^+ \\
c_{1,6} & c_{0,1} & c_{6,3} & c_{5,5} & c_{4,0} & c_{3,2} & c_{2,4} \\
c_{2,5}^+ & c_{1,0} & c_{0,2} & c_{6,4}^- & c_{5,6} & c_{4,1}^+ & c_{3,3}^+ \\
c_{3,4} & c_{2,6} & c_{1,1} & c_{0,3} & c_{6,5} & c_{5,0} & c_{4,2} \\
c_{4,3}^+ & c_{3,5} & c_{2,0} & c_{1,2}^+ & c_{0,4} & c_{6,6}^- & c_{5,1}^- \\
c_{5,2} & c_{4,4} & c_{2,6} & c_{2,1} & c_{1,3} & c_{0,5} & c_{6,0} \\
c_{6,1} & c_{5,3} & c_{4,5} & c_{3,0} & c_{2,2} & c_{1,4} & c_{0,6}
\end{bmatrix}
$$

$\downarrow A^2$

3rd configuration

$$
\begin{bmatrix}
c_{0,0}^+ & c_{4,5}^- & c_{1,3}^- & c_{5,1} & c_{2,6} & c_{6,4} & c_{3,2} \\
c_{2,4} & c_{6,2} & c_{3,0} & c_{0,5} & c_{4,3} & c_{1,1} & c_{5,0} \\
c_{4,1}^+ & c_{1,6}^- & c_{5,4}^- & c_{2,2} & c_{6,0} & c_{3,5} & c_{0,3} \\
c_{6,5}^- & c_{3,3}^+ & c_{0,1}^- & c_{4,6} & c_{1,4} & c_{5,2} & c_{2,0} \\
c_{1,2}^+ & c_{5,0}^- & c_{2,5}^+ & c_{0,3} & c_{3,1} & c_{0,6} & c_{4,4} \\
c_{3,0} & c_{0,4} & c_{4,2} & c_{1,0} & c_{5,5} & c_{2,3} & c_{6,1} \\
c_{5,3} & c_{2,1} & c_{6,6} & c_{3,4} & c_{0,2} & c_{4,0} & c_{1,5}
\end{bmatrix}
$$

Figure 9.25 A more complex example of the use of multiple array configurations to distinguish true and false positives. In this case three configurations allowed the detection of five true positives (larger font characters at positions 0, 0; 1, 2; 2, 5; 3, 3; and 4, 1 in the first configuration) in a set of 49 clones. The +'s in the first configuration indicate positive rows and columns. In the second and third configuration the +'s indicate clones that could be positive; the −'s indicate those that are excluded by the results of the first and second configurations, respectively. (Taken from Barillot et al., 1991.)

INNER PRODUCT MAPPING

The enormous attractiveness of large sample arrays as genome mapping tools has been made evident. What a pity it is that with most contemporary methods there is no way to make these arrays systematically. Suppose that we had an arrayed library from a sample, and we wished to construct the equivalent array from a closely related sample (Fig. 9.26). The samples could be libraries from two different people, or from two closely related species. Constructing an array from the second sample that is parallel to the first sample is really making a map of the second sample, as efficiently as possible. We could do this one clone at a time, by testing each clone from the second library against the array of the first library. We could do this more rapidly by pooling clones from the second library. But ultimately we would have to systematically place each of the clones in the second library into its proper place in an array. This is an extremely tedious process. What we need, in the future, is a way of using the first array as a tool to order simultaneously all of the clones from the second array. In principle, it should be possible to use hybridization to do this; we just have to develop a strategy that works efficiently in practice.

One very attractive strategy for cross-correlating different sets of samples has recently been developed by Mark Perlin at Carnegie Mellon. The method has been validated by the construction of a YAC map for human chromosome 11. The basic idea behind the procedure, called *inner product mapping* (IPM) is shown in Figure 9.27. Radiation hybrids (RHs; see Chapter 7) can be analyzed relatively easily by PCR using STSs of known order on a chromosome. Because each RH encompasses a relatively large amount of human DNA, relatively few STS measurements suffice to produce a good RH map. RHs in turn provide an excellent source of DNA to identify YACs in corresponding regions. In practice, what is done is inter-*Alu* PCR (Chapter 14) both on each separate YAC and each

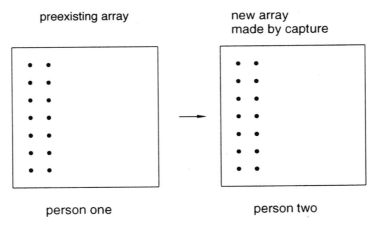

Figure 9.26   Unsolved problem of efficiently mapping one dense array onto another.

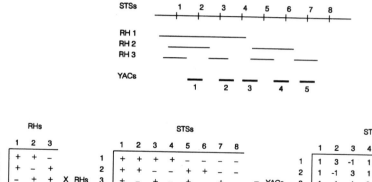

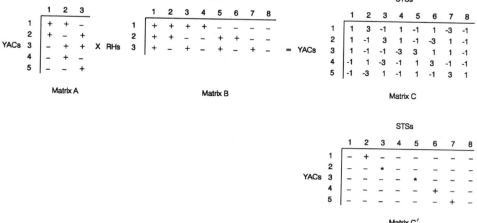

Figure 9.27 Inner product mapping (IPM). (See text for details.)

RH. An array is made of the YAC PCR products, and this is probed successively by hybridization with each RH PCR product. This assigns the YACs to RHs.

In previous STS mapping strategies, PCR with STS primers had to be used directly to analyze YACs in YAC pools. This is relatively inefficient for reasons that have been described earlier in this chapter. Instead, in IPM, the STS-YAC correspondences are built mathematically as shown in Figure 9.27. The inner product C of two matrices A and B is computed as

$$C_{ij} = \sum_k A_{ik}B_{kj}$$

The YAC versus RH hybridization results can be scored as positive $(+1)$ or negative (-1) as shown for matrix A. The RH versus STS PCR results can also be scored as positive $(+1)$ or negative (-1) as shown in matrix B. The inner product matrix C is computed element by element in a very simple fashion. It is a matrix that describes comparisons between STS's and YACs that reflects the separate RH results with each. The best estimate for each YAC-STS direct correspondence is the largest (most positive) element in each row of matrix C. This is shown as + or * in the simplified matrix C' which indicates only the largest elements of C. The + symbols in C' indicate a YAC (column) that actually contains the indicated STS (row). The * symbols in *C'* indicate a YAC does not actually contain the STS (i.e., it would be scored negative in a direct PCR test) but must be located near this STS for the RH data to be self-consistent. This simple example indicates that IPM has more mapping power than direct STS interrogation of YACs even though the latter process would involve far more work.

The IPM mapping project of human chromosome 11 used 73 RHs, 1319 YACs, and 240 STSs to construct a YAC map. A total of 241 RH hybridizations of YAC clone arrays were required, and 240 STS interrogations of the 73 RHs were done with duplicate PCRs.

SLICED PFG FRACTIONATIONS AS NATURAL POOLS OF SAMPLES

Previous considerations make clear that working with pools of probes or samples is often a big advantage. Sometimes this is also unavoidable when a region is too unstable or too toxic to clone in available vector systems. Cloning a region, even if it is stable, may also be too time-consuming or costly if one needs to examine the region in a large number of different samples. This would be the case where, for example, a region expanded in a set of different tumor samples is to be characterized. One way to circumvent the problem of subcloning a region is to find one or more slices of a PFG-fractionated restriction digest that contains the region. With enzymes like *Not* I that have rare recognition sites, generally 1 to 2 Mb regions will reside on at most a few fragments. Only a few probes from the region will suffice to identify these fragments. PFG separation conditions can then be optimized to produce these fragments in separation domains where size resolution is optimum. The resulting slice of separation gel will then contain, typically, about 2% of the total genome. For a 600-kb human DNA fragment, this slice will consist of 100 fragments, only one of which is the fragment of interest. This is probably too dilute to permit any kind of direct isolation or purification. But the slice can serve as an efficient sample for PCR amplifications that try to assign additional STSs or ESTs to the region. If the slice is examined from a digest of a single chromosome hybrid, it will contain only 1 or 2 human DNA fragments. Then, as shown in Chapter 14, PCR amplification based on human-specific repeating sequences can be used to produce numerous human-specific DNA probes from the region of interest.

RESTRICTION LANDMARK GENOME SCANNING

An alternative method for systematically generating a dense array of samples from a genome has been developed. This method is called *restriction landmark genome scanning* (RGLS). It was originally conceived as a way of facilitating the construction of genetic maps by finding large numbers of useful polymorphic sequences. Thus it is a method set up to reveal differences between two genomes, and as such it fits the spirit of the kind of differential analysis that needs to be developed. RLGS, as currently practiced, is based, however, on genomic DNA rather than on cloned DNA. The basic idea behind RLGS is illustrated in Figure 9.28. A genome is digested with a rare-cutting restriction nuclease like *Not* I, and the ends of the fragments are labeled. This generates about 6000 labeled sites because there are about 3000 *Not* I sites in the genome, and each fragment will be

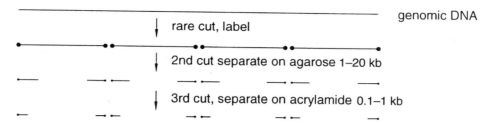

Figure 9.28 Steps in the preparation of DNA samples for restriction landmark genome scanning (RLGS).

labeled on both ends. The sample is then digested with a second restriction enzyme, one that cuts more frequently, say at a six base site. The resulting fragments are then fractionated by agarose gel electrophoresis in the size range of 1 to 20 kb. The agarose lane is excised and digested in situ with a third, more frequently cutting enzyme, one that recognizes a four-base sequence. The resulting small DNA fragments are now separated in a second electrophoretic dimension on polyacrylamide, which fractionates in the 0.1 to 1 kb size range. The result is a systematic pattern of thousands of spots, as shown in Figure 9.29a. Each spot reveals the distance between the original *Not* I site and the nearest site

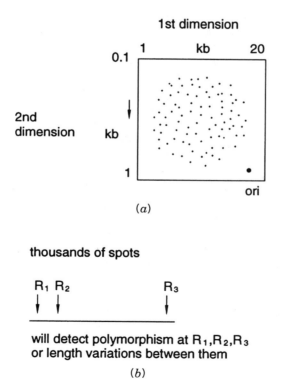

Figure 9.29 Example of the results seen by RLGS. (*a*) Two-dimensional electrophoretic separation of DNA fragments. (*b*) Sites where polymorphisms can be detected on a typical fragment.

for the second and third enzymes (Fig. 9.29*b*). Any polymorphisms in these distances, caused either by altered restriction sites or by DNA insertions or deletions, will appear as displaced spots in the two-dimensional fractionation. The method appears to be very powerful because so many spots can be resolved, and the patterns, at least for mouse DNA where the method was developed, are very reproducible.

PROGNOSIS FOR THE FUTURE OF GENOME MAPPING

In the human, mouse, and other species officially sanctioned as part of the human genome project, genetic mapping is proceeding rapidly and effectively. Indeed the rate of progress appears to be better than originally projected. Dense sets of polymorphic genetic markers have been generated. These have served well to order megaYACs. Finer maps are still needed because of the difficulties in handling megaYACs and the need to break them down into smaller samples for subsequent manipulations. These finer maps, however, will be more easy to construct by using preexisting megaYAC contigs, just as the preexisting *S. pombe* restriction map allowed the design of efficient strategies to order an *S. pombe* cosmid library. BAC, PAC, or cosmid maps are still needed for current direct DNA sequencing technology. Direct sequencing from YACs or from genomic DNA is possible, as we will describe in Chapter 10, but it is not yet reliable enough to be routinely used in large-scale sequencing projects. As the genome project concentrates on DNA sequencing, the notion of a sequence-ready map has become important. Such a map consists of samples ready for DNA sequencing. Detailed order information on these samples could be known in advance, or it could be obtained in the process of DNA sequencing. See Chapter 11 for further discussion.

For species other than those already intensively studied, the best strategies will depend on the kinds of samples that are available. If radiation hybrids and mega clone libraries are made, these will obviously be valuable resources. If dense genetic maps can be made, the probes from these will order the megaYACs. If a genetic map is not feasible, FISH provides a readily accessible alternative. In other cases it may be possible to purify the chromosomes or fragments efficiently by flow sorting, improved microdissection, or other tricks to be described in Chapter 14.

The notion of having to make a map of a person for diagnostic purposes is still awesomely difficult. Mapping methods are complex and make major demands on both instrumentation and skilled personnel. New approaches will be needed before diagnostic mapping can be considered at all realistic. The use of radioactive ^{32}P pervades most current mapping methods, and this is surely something to be avoided in a technique proposed for widescale clinical use. One area that may impact heavily on the prospects for diagnostic, mapping is the development of improved, sensitive, nonradioactive detection techniques. These will be described as we deal with DNA sequencing methods because it is here where these methods have first been used or tested.

SOURCES AND ADDITIONAL READINGS

Allshire, R. C. 1995. Elements of chromosome structure and function in fission yeast. *Seminars in Cell Biology* 6: 55–64.

Ashworth, L. K., Hartman, M.-A., Burgin, M., Devlin, L., Carrano, A. V., and Batzer, M. A. 1995. Assembly of high-resolution bacterial artificial chromosome, P1-derived artificial chromosome, and cosmid contigs. *Analytical Biochemistry* 224: 565–571.

Barillot, E., Lacroix, B., and Cohen, D. 1991. Theoretical analysis of library screening using a *N*-dimensional pooling strategy. *Nucleic Acids Research* 19: 6241–6347.

Bruno, W. J., Knill, E., Balding, D. S., Bruce, D. C., Doggett, N. A., Sawhill, W. W., Stallings, R. L., Whittaker, C. C., and Torney, D. C. 1995. Efficient pooling designs for library screening. *Genomics* 26: 21–30.

Green, E. D., Riethman, H. C., Dutchik, J. E., and Olson, M. V. 1991. Detection and characterization of chimeric yeast artificial-chromosome clones. *Genomics* 11: 658–669.

Grigoriev, A., Mott, R., and Lehrach, H. 1994. An algorithm to detect chimeric clones and random noise in genomic mapping. *Genomics* 22: 482–486.

Grothues, D., Cantor, C. R., and Smith, C. L. 1994. Top-down construction of an ordered *Schizosaccharomyces pombe* cosmid library. *Proceedings of the National Academy of Sciences USA* 91: 4461–4465.

Hatada, I., Hayashizaki, Y., Hirotsune, S., Komatsubara, H., and Mukai, T. 1991. A genomic scanning method for higher organisms using restriction sitas landmarks. *Proceedings of the National Academy of Sciences USA* 88: 9523–9527.

Ling, L., Ma, N. S.-F., Smith, D. R., Miller, D. D., and Moir, D. T. 1993. Reduced occurrence of chimeric YACs in recombination-deficient hosts. *Nucleic Acids Research* 21: 6045–6046.

Mejia, J. E., and Monaco, A. P. 1997. Retrofitting vectors for *Escherichia coli*-based artificial chromosomes (PACs and BACs) with markers for transfection studies. *Genome Research* 7: 179–186.

Perlin, M., and Chakravarti, A. 1993. Efficient construction of high-resolution physical maps from yeast artificial chromosomes using radiation hybrids: Inner product mapping. *Genomics* 18: 283–289.

Perlin, M., Duggan, D., Davis, K., Farr, J., Findler, R., Higgins, M., Nowak, N., Evans, G., Qin, S., Zhang, J., Shows, T., James, M., and Richard, C. W. III. 1995. Rapid construction of integrated maps using inner product mapping: YAC coverage of human chromosome 11. *Genomics* 28: 315–327.

Smith, D. R., Smyth, A. P., and Moir, D. T. 1990. Amplification of large artificial chromosomes. *Proceedings of the National Academy of Sciences USA* 87: 8242–8246.

Teo, I., and Shaunak, S. 1995. Polymerase chain reaction in situ: An appraisal of an emerging technique. *Histochemical Journal* 27: 647–659.

Wada, M., Abe, K., Okumura, K., Taguchi, H., Kohno, K., Imamoto, F., Schlessinger, D., and Kuwano, M. 1994. Chimeric YACs were generated at unreduced rates in conditions that suppress coligation. *Nucleic Acids Research* 22: 1651–1654.

Whittaker, C. C., Mundt, M. O., Faber, V., Balding, D. J., Dougherty, R. L., Stallings, R. L., White, S. W., and Torney, D. C. 1993. Computations for mapping genomes with clones. *International Journal of Genome Research* 1: 195–226.

10 DNA Sequencing: Current Tactics

WHY DETERMINE DNA SEQUENCE

A complete DNA sequence of a representative human genome is the major goal of the human genome project. Complete DNA sequences of other genomes are also sought. Why do we want or need this information? All descriptions of the organization of a genome, at lower resolution than the sequence, appear to offer little insight into genome function. Sometimes genes with common or related functions are clustered. This is particularly true in bacteria where the clustering allows polycistronic messages to ensure even production of a set of interactive gene products. However, in higher cells, related genes are not necessarily close together. For example, in humans, genes for alpha and beta globin chains are located on different chromosomes, even though it is desirable to produce their products in equal amounts because they associate to form a heterotetramer, (alpha)$_2$(beta)$_2$. The major purpose served by low-resolution maps is that they help us find things in the genome. We usually want to find genes in order to study or characterize their function. It is only at the level of the DNA sequence where we have any chance of drawing direct inferences about the function of a gene from its structure. Admittedly, our ability to do this today is still rather limited, as will be demonstrated in Chapter 15. However, from the rate of progress in our ability to interpret DNA sequences de novo in terms of plausible gene function, we can be reasonably optimistic that by the time the human genome is completely sequenced, coding regions will be identifiable with almost perfect accuracy, and most new genes will carry in their sequence immediately recognizable clues about function.

A second reason to have the DNA sequence of genomes is that it gives us direct access to the DNA molecules of these genomes via PCR. Using the sequence, it will almost always be possible to design primers that will amplify a small DNA target of interest, or to provide a probe that will uniquely allow effective screening of a library for a larger segment of DNA containing the region of interest. The key point is that once DNA sequence is available, clones do not have do be stored and distributed. DNA sequences also often allow us to search for similar genes in related organisms (or even more distant organisms) more efficiently than by using DNA probes of unknown sequence. For example, to find a mouse gene comparable to a human gene, one can try to use the human gene as a hybridization probe at reduced stringency (lower temperature, higher salt) against a mouse library or use the human gene to design PCR primers for probing the mouse genome. But, if one had both the relevant human and mouse DNA sequences available, a comparison among these might reveal consensus regions that are more highly conserved than average and thus better suited for hybridization or PCR to find corresponding genes in other species. This becomes increasingly important when searching for homologs of very distantly related proteins.

A continual debate in the human genome project is whether to determine the DNA sequence of the junk: DNA that as far as we can tell is noncoding. Sydney Brenner was quick to point out early in the project that this DNA is rightly called junk and not garbage because, like junk, this DNA has been retained, while garbage is discarded. Today, admittedly, we cannot interpret much from noncoding DNA sequences. But this does not mean they are nonfunctional. The fact that they remain in the genome argues for function, at least at the level of evolution. However, there are surely also functions for these sequences at the level of gene regulation, chromosome function, and perhaps properties we know nothing about today. The junk is certainly worth sequencing, but it will be best to do this later in the genome project when the cost of DNA sequencing has diminished. An analogy can be made between the genome project and the exploration of a new continent. At the time the interior of North America was first explored, a major target was river valleys because they were accessible and because they were commercially valuable. No one willingly spent much time in deserts or arctic slopes. However, most of our oil deposits are located far from river valleys, and if we had not pushed exploration of the continent to completion, we would never have found very valuable resources. It is probably this way also with the genome; when we finally make our way through the junk, systematically, there will be some unexpectedly valuable finds. We may not know enough today to realize they were valuable, even if we could find them.

DESIGN OF DNA SEQUENCING PROJECTS

The first DNA sequence was determined in 1970 by Ray Wu at Cornell University. It consisted of the 12-base single-stranded overhang at each end of bacteriophage lambda DNA. The samples needed were readily in hand. Two investigators worked on the project for three years. Data handling and analysis did not present any unexpected or formidable problems. The major chore was developing techniques for actually determining the order of the bases. The method employed, selective addition of subsets of the four dpppN's, still has many attractive features, and we will revisit it several times in this and the next chapter.

Today, the complete DNA sequencing of 50-kb DNA targets, the size of the entire bacteriophage lambda, is a common task in specialized high-throughput sequencing laboratories. However, such projects are not yet routine in most laboratories that do DNA sequencing. The sequencing of targets 3 to 90 times larger has been accomplished in quite a few cases. Sequencing of continuous Mb blocks of human DNA is now becoming commonplace in quite a few research groups. These projects, even 50-kb projects, pose obstacles that were inconceivable at the dawn of DNA sequencing.

It is useful to divide discussion about DNA sequencing projects into tactics and strategy. Tactics is how the order of the bases on a single DNA sample is read and confirmed. Strategy, as illustrated in Figure 10.1, has a number of components. Presumably the target is selected in a rational manner, given the amount of effort that is actually required to complete a sequencing project. The upstream strategy is concerned with how the target is reduced to DNA samples suitable for application of the particular tactics selected. The tactics are then used, piece by piece, in as efficient and automated a way as possible. Then the downstream strategy consists in assembling the data into contiguous blocks of DNA sequence, filling any gaps, and correcting the inevitable errors that creep into all DNA sequence data.

Several caveats must be noted when thinking about DNA sequencing projects. Both the ideal tactics and strategy may depend on the types of targets. Effective strategies may

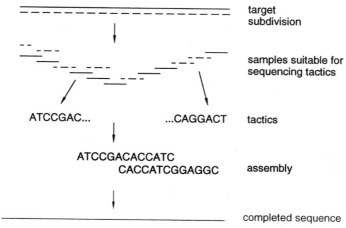

target
subdivision

samples suitable for
sequencing tactics

ATCCGAC...   ...CAGGACT   tactics

ATCCGACACCATC
      CACCATCGGAGGC   assembly

completed sequence

Figure 10.1   Design of a typical DNA sequencing project.

combine several types of targets and several types of tactics. The key variable to judge efficiency and cost is the throughput: the number of base pairs of DNA sequence generated per day for each individual working in the laboratory. With current methods, except at the largest and most efficient genome sequencing groups, personnel costs are the completely dominant expense; chemicals, enzymes, and instrument depreciation all pale in comparison with salaries. In a few very automated and experienced centers, reagents and supplies are now the dominant costs.

Three terms are useful in evaluating sequencing progress. Raw DNA sequence is the direct data read from an experimental curve or photograph with local error correction done, for example, a manual override to correct an ambiguous call by sequence reading software. Finished sequence is the assembled DNA sequence for the entire target, with error corrections made by comparing redundant samples. In general, the complete DNA sequence is read separately from both DNA strands. This is a major contributor to finding and correcting some of the most common kinds of errors. Sequencing redundancy is the ratio of the number of raw base pairs of sequence acquired to the number of base pairs of finished sequence determined. It is usually at least 2, because of the need just cited to examine both strands. In general, the redundancy is dependent on the strategy used, and it has often been as high as 10 in many of the relatively large DNA sequencing projects that have been accomplished to date.

LADDER SEQUENCING TACTICS

Virtually all current de novo DNA sequencing methods are based on the ability to fractionate single-stranded DNA by gel electrophoresis in the presence of a denaturant with single base resolution. Information about the location of particular bases in the sequence is converted into a specific DNA fragment size. Then these fragments are separated and analyzed. The gels used are either polyacrylamide or variants on this matrix like Long Ranger™. The denaturant is usually 7 M urea. Its presence is required to eliminate most of the secondary structure that individual DNA strands can achieve by intramolecular base pairing, where this is allowed by the DNA sequence. It is possible, under ideal cases, to maintain single base resolution up to sizes of 1 kb. Some success has been reported with ever larger sizes by the use of gel-filled capillaries. The use of denaturing gels is an unfortunate aspect of current DNA sequencing. Since urea solutions are not stable to long-term storage, the gels must be cast within a few days of their use, and it is difficult to reuse most gels more than several times without a serious decrease in performance. In the two decades since Wu's first DNA sequencing, the ladder methods we will describe have produced more than 1,000 Mb of DNA sequence deposited in databases, and perhaps an equal amount or more that has not been published or deposited.

Two rather different approaches have been used to generate DNA sizes based on DNA sequence. We will describe how they are carried out starting with a single-stranded DNA template. Slightly more complex procedures are required if the original template is double stranded. The first of these methods, developed by Allan Maxam and Walter Gilbert, is shown in Figure 10.2. The ends of the DNA are distinguished by specifically labeling one of them. Usually this is done directly, and covalently, with a kinase that places a radiolabeled phosphate at the 5'-terminus of the template. There are other ways to label the 5'-end or 3'-end directly, and it is also possible to label either end indirectly, by hybridization with an appropriate complementary sequence. This requires that the end sequence be known; it usually is known, since the DNA template is cloned into a vector of known flanking sequence.

In Maxam-Gilbert sequencing base-specific or base-selective partial chemical cleavage is used to fragment the DNA. This is carried out under conditions where there is an average of only one cut per template molecule with each cleavage scheme employed. Thus a very broad range of fragment sizes is produced that reflects the entire sequence of the template. Four separate chemical fragmentation reactions are carried out; each one favors cleavage after a specific base. The fragments are fractionated, and the sizes of the labeled pieces are measured, usually in four parallel electrophoretic lanes. The DNA sequence can be read directly off the gel as indicated by the example in Figure 10.3. The pattern of bands seen is often called a ladder for reasons obvious from the figure. Note that in the Maxam-Gilbert approach, there are additional fragments produced that are not detected because they are not labeled, but they are present in the sample. For some alternate schemes of detecting DNA fragments for sequencing, like mass spectrometry, these additional pieces are undesirable.

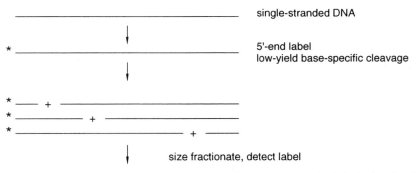

Figure 10.2 Maxam-Gilbert sequence technique: Preparation of end-labeled, size-fractionated DNA sample.

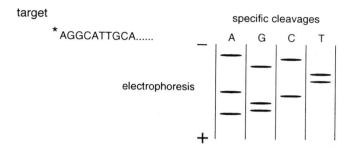

Figure 10.3   Typical Maxam-Gilbert sequencing ladder and its interpretation.

The second general approach to DNA fragmentation for ladder sequencing was developed by Frederick Sanger (Fig. 10.4). This is the approach in widespread use today, for a variety of reasons, including the ability to avoid the use of toxic chemicals and the ease of adapting it to four-color fluorescent detection. One starts with a single-stranded template. A primer is annealed to this template, near the 3′-end of the DNA to be sequenced. The primer must be long enough so that it binds only to one unique place on the template. This primer must correspond to known DNA sequence, either in the target or, more commonly, in the flanking vector sequence. A DNA polymerase is used to extend the primer in a sequence-specific manner along the template. However, the sequence extension is halted, in a base specific manner, by allowing the occasional uptake of chain terminators: dpppN analogs that cannot be further extended by the enzyme. Almost all current DNA sequencing uses dideoxy-pppN's as terminators. As shown in Figure 10.5, these derivatives lack the 3′ OH needed to form the next phosphodiester bond. Four separate chain

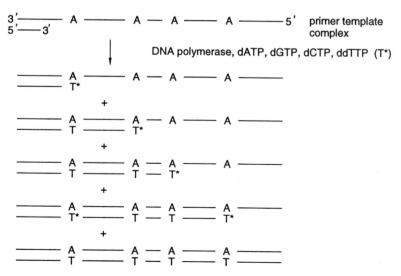

Figure 10.4   Sanger sequencing technique: Preparation of an end-labeled, size-fractionated DNA sample. The actual sequencing ladder will be virtually identical to that seen with the Maxam-Gilbert method.

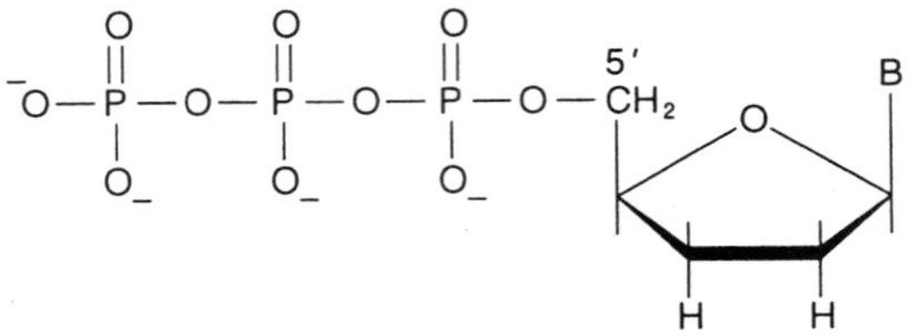

Figure 10.5   Structure of a dideoxynucleoside triphosphate terminator.

extension reactions are carried out—each one with a different terminator. Label can be introduced in several different ways: through the primer, the terminator, or internal dpppN's. The resulting mixture of DNA fragments is melted off the template and analyzed by gel electrophoresis.

ISSUES IN LADDER SEQUENCING

The major goal is to maximize sequencing throughput. A second, significant goal is to minimize the number of sequencing errors. An important element of these goals is to be able to read the longest possible sequencing ladders, accurately. There are two significant variables in this. The resolution of the gel electrophoresis will determine how far the sequencing data can be read, if there are data to be read at all. Ultimately there are trade-offs between how fast the gel can be run, which also affects the throughput, how well certain artifacts can be eliminated, and how much sample must be applied. The more sample we have, the easier is the detection but, in general, the lower is the resolution. Large double-stranded DNAs show negligible diffusion during gel electrophoresis as described earlier in Chapter 5 (Yarmola et al., 1996). This is not the case for the smaller single-stranded DNAs used in sequencing where diffusion is a significant cause of band broadening. This motivates the use of higher fields where shorter running times can be achieved, hence minimizing the effects of diffusion. However, higher fields lead to greater joule heating. This increases the effects of thermal inhomogeneities which also lead to band broadening. The issues are complex because field strength also influences the shape of DNA in a gel and thus affects its diffusion coefficient. Other factors that affect band shape and thus resolution are the volume in which the sample is loaded, the volume sampled by the detector, and any inhomogeneities in gel concentration. For a thorough discussion of the effects of these variables, see Luckey et al. (1993), and Luckey and Smith (1993).

To a good approximation, the velocity, v, of DNA in denaturing acrylamide gel electrophoresis is proportional to $1/L$, where L is the length of the molecule. In automated fluorescent detection (or the bottom wiper shown later in Fig. 10.10), the sample is examined at a constant distance from the starting point, D. The time it takes a fragment of a particular length to reach this distance is proportional to $D/v = DL$. Hence the spacing between two bands of length L and $L - 1$ is $DL - D(L - 1) = D$. Thus the band spacing is independent of size, but it can be increased, more or less at will by using longer and longer running gels.

The second determinant of how far a ladder can be read is the uniformity of the sample fragment yield. It is important to realize that the larger the target is, the smaller the yield of each piece even if the distribution of fragments is absolutely uniform. Thus, with perfect cleavage, sequencing a 100-base piece of DNA will require only 10% the amount of sample that a 1-kb target requires. Put another way, for constant amounts of DNA sample loaded, the detection sensitivity will have to increase in proportion to the length of the DNA target. The relative yield of particular DNA fragments is affected by the choice of DNA polymerase, the nature of the terminators and primers used, the actual DNA template, and the reaction conditions. Much optimization has been required to produce reproducible runs of DNA sequence data that extend longer than 500 bases.

It is also important to realize that throughput is really the product of the number of lanes per gel and the speed of the electrophoresis. Speed can be controlled by the electrical field applied. In fact higher fields appear to improve electrophoretic performance. What limits the speed, once efficient cooling is provided to keep the running temperature of the gel constant, is the sensitivity of the detection scheme, if it is done on line. With off-line detection, the sensitivity is still important, not for speed, but for determining the number of lanes that can be used. The smaller the width of each lane, the more lanes one can place on a single gel but the smaller the amount of DNA one can actually load into each lane.

A major factor that affects the quality of DNA sequence data is the quality of the template DNA. When fluorescent labeling is used, great care must be taken not to introduce fluorescent contaminants into the DNA sample. A number of automated methods for DNA preparation routinely yield DNA suitable for sequencing. These methods are convenient because they are so standardized. A laboratory that tries to sequence DNA from many different types of sources will frequently encounter difficulties.

In early DNA sequencing, ^{32}P was the label of choice, introduced from γ [^{32}P] pppA via kinasing of the primer for Sanger sequencing or the strand to be cleaved for Maxam-Gilbert sequencing. This isotope has a short half-life which results in very high experimental sensitivity. However, ^{32}P also has a relatively high energy beta particle, which causes an artifactual broadening of the thin fragment bands on DNA sequencing gels. Instead of ^{32}P one can use the radioisotope ^{35}S, as γ thio-pppA. This still has a short half-life, but the decay is softer, leading to sharper bands. At first, DNA sequence data were obtained by using X-ray film in autoradiography to make an image of the sequencing gel. This can be read by hand, which is still done by some, perhaps with the help of devices and software to expedite transferring the data into a computer file. Alternatively, the film can be scanned and digitized by a device like a charge-couple device (CCD) camera. This then allows most of the data to be processed by image analysis software, with human intervention needed in difficult places. The accuracy of using film and some of the existing software does not appear to be as good as the fluorescent systems we will describe later.

A new approach to recording data from radioactive decay is the use of imaging plates. These consist of individual pixels that record local decays. After the plate is exposed, it is read out by laser excitation in a raster pattern (scanning successive lines, in the same manner as a TV camera or screen), and the resulting data are transferred into a computer file. A great advantage of imaging plates over film is that their response is a linear function of dose over more than five orders of magnitude in intensity, and most important, they are linear down to the lowest detectable doses. In contrast, film shows a dead zone at very low doses, and it easily saturates at high doses. Imaging plates are reusable, and for the heavy user, the great savings in film that result eventually compensate for the high costs of the imaging plates and the instrument needed to read them out. Although it is possible, in principle, to use several different radioisotopes simultaneously, as is common in liquid scintillation counting, and thus achieve multicolor labeling and detection, in practice, this is rarely done with radioactive DNA sequencing data.

In most contemporary DNA sequencing, radioisotopes have been replaced by fluorescent labels. These can be used on the primers, the terminators, or internally. It may seem surprising that fluorescent detection can be competitive with radioisotopes. However, one can gain enormous amounts of sensitivity in fluorescence by sequential excitation and emission from the same fluorophore until it undergoes some chemical side reaction and becomes bleached. This makes up in large part for the difference in energy between a beta particle and the fluorescent photon. The major determinant of sensitivity in fluorescence detection is, then, not really signal; it is background. Scrupulous care must be taken to avoid the use of reagents, solvents, plastics, glove powder, and detergents that have fluorescent contaminants.

Four different colored fluorescent dyes are used in several of the most common DNA sequencing detection schemes. One dye is used for each base-specific primer extension. The ideal set of dyes would have very similar chemical structures so that their presence would affect the electrophoretic mobility of labeled DNA fragments in identical ways. They would also have emission spectra as distinct as possible, and they would all be excitable by the same wavelength so that a single excitation source would suffice for all four dyes. The dyes would also allow similar very high sensitivity detection so that signal intensities from the four different cleavage reactions would be comparable. Inevitably with currently available dyes there are compromises. For example, a set of nearly identical dye-labeled chain terminators was produced for DNA sequencing that led to very good electrophoretic properties, but the emission spectra of these compounds were too similar for the kind of accuracy needed in reading long sequence ladders. Subsequently a more well-resolved set of fluorescent terminators that are substrates for Sequenase, the most popular enzyme used in Sanger sequencing became commercially available. These have the advantage that all four terminators can be used simultaneously in a single sequencing reaction.

All currently used dyes for four-color DNA sequencing are excited in the UV/visible wavelength range. The limits of this range and the typical widths of emission spectra of high quantum yield dyes make it rather difficult to detect more than four colors simultaneously. The infrared (IR) spectrum is much broader, and work is in progress trying to develop DNA sequencing dyes in this range. If the lower sensitivity of IR detection can be tolerated, such dyes would offer two advantages. The laser sources needed to excite them are inexpensive, and at least eight different colors would be obtainable. This could be used to double the throughput of four-color sequencing, or it could be used to include a known standard in every sequencing lane to improve the accuracy of automatic sequence calling. Recently IR-excited dyes have begun to make an impact on automated DNA sequencing. Multiple IR colors are presumably soon on the horizon.

A significant improvement in fluorescent dyes for automated sequencing is the use of energy transfer methods (Glazer and Mathies, 1997). Primers contain a pair of fluorescent dyes (Fig. 10.6). One dye is common to all four primers. This is optimized to absorb the exciting laser dyes. The second dye is different in each primer, and it is close enough in each case that fluorescence resonance energy transfer is 100% efficient. Thus all the excitation energy migrates to the second dye where it is subsequently emitted. The second dyes are chosen so that they have as different emission spectra as possible to maximize the ability to accurately discriminate the four different colors.

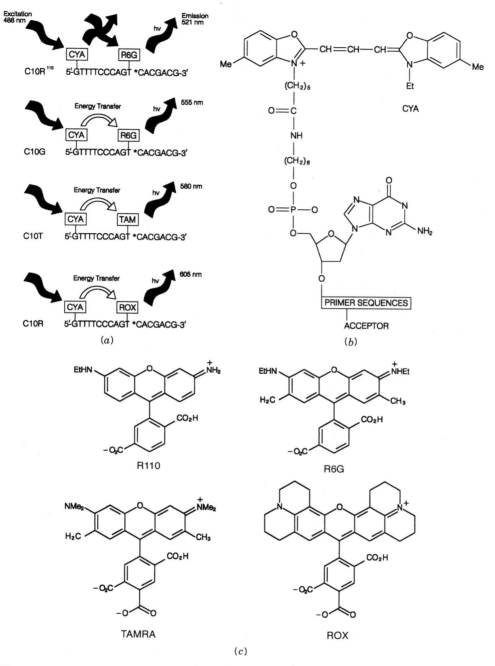

Figure 10.6 Energy transfer primers (provided by Richard Mathies). (*a*) Schematic design of a set of four primers. (*b*) Structure of the donor dye. (*c*) Structure of four different acceptor fluorescent dyes that can be detected simultaneously in DNA sequencing.

An alternative to fluorescent labels is chemiluminescence. This has the great advantage that no exciting light is needed. Thus the sensitivity can be extremely high, since there is no contamination from scattering of the exciting light used in fluorescence, or the effects of fluorescent impurities. Today, chemiluminescent detection schemes exist that can readily be used in DNA sequencing. They have a few disadvantages. Only one color is currently available, and once the chemiluminescence has been read, it is difficult to use the gel or filter again. While this is not often a problem in most forms of DNA sequencing, it is a problem in most mapping applications where the same filter replica of a gel is frequently probed many times in succession. Nevertheless, the sensitivity of chemiluminescence makes it attractive for some mapping applications. The advantages of four-color fluorescence are also beginning to be felt in some aspects of genome mapping. An example was given in Chapter 8.

CURRENT FLUORESCENT DNA SEQUENCING

There are two basically distinct implementations of fluorescent detected DNA sequence determination. These are the current commonly available state-of-the-art tools used today in most large-scale DNA sequencing projects. They each can produce more than 10^4 to 10^5 bp of raw DNA sequence per laboratory worker per day. Most allow 400 to 800 bases of data to be read per lane; most of the lanes give readable data when proper DNA preparation methods are used. The detection schemes used in the two approaches are illustrated in Figure 10.7. Both are on-line gel readers. These two schemes have a number of serious trade-offs. In the Applied Biosystems (ABI) instrument, based on original developments by Leroy Hood and Lloyd Smith, four different colored dyes are used to analyze a mixture of four different samples in a single gel lane (Fig. 10.7a). This allows four times more samples to be loaded per gel, if the width of the lanes is kept constant. The use of four colors in a single lane avoids the problem of compensating for any differences in the mobility of fragments in adjacent lanes—that is, there is no lane registration problem. In order to do the four-color analysis, a laser perpendicular to the gel is used to excite one lane at a time, and the signal is detected through a rotating four-color wheel to separate the emission from the four different dyes. Thus the effective power of the laser is the time shared among the lanes and the colors. With 20 lanes, the actual time-averaged illumination available is, at most, 1/80 the laser intensity.

In the alternative implementation, embodied in the Pharmacia automated laser fluorescence (ALF) instrument, only a single fluorescent dye is used (Fig. 10.7b). The dye originally selected was fluorescein because it is the most sensitive available for the particular laser exciting wavelength used. In a newer version of the instrument, a different laser and an infrared emitting dye, Cy5, are used. The key feature of the ALF is that the laser excitation is in the plane of the gel, through all the sample lanes simultaneously. This design, which is based on an instrument originally developed by Wilhelm Ansorge, is possible because at the concentrations of label used for DNA sequencing the samples are optically thin. This means that the amount of light absorbed at each lane is an insignificant fraction of the original laser intensity, so all lanes receive, effectively, equal excitation. The emission from all the lanes is recorded simultaneously by an array of detectors, one for each lane. While these could be made four-color detectors, in principle, the cost and complexity is not warranted. Instead the ALF reads data from four closely spaced lanes, one for each base-specific fragmentation. Thus the number of lanes needed for one sample in the

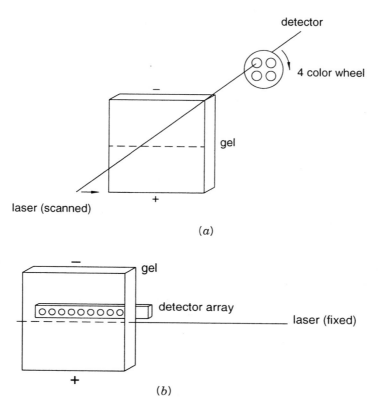

Figure 10.7   Schematic operation of current commercially available automated DNA sequencing gel readers. (*a*) ABI four-color instrument. (*b*) Pharmacia one-color instrument.

ALF is four times the number used in the ABI instrument. The use of a single color simplifies the construction of labeled primers, since only a single primer is needed for the four sequencing reactions, whereas with the ABI approach, if fluorescent primers are used, four different ones must be made—one for each color. The basic advantage of the ALF is the higher signal to noise for a given laser strength, since the full intensity of the exciting beam can be used to illuminate each sample continuously. This higher signal to noise allows faster running and, in principle, smaller lanes.

At present the advantages of the two different fluorescent approaches really depend on the application for which they are intended. If massive amounts of sample throughput, on relatively small DNA fragments, is most important, the ABI has an edge because of the larger number of samples that can be run per gel. If longer fragments are important, or if sample amounts are limited, or if the raw data must be scrutinized as in mutation detection (see Chapter 13), the ALF has the edge because of its greater sensitivity.

The remaining status of most state-of-the-art DNA sequencing is easy to summarize. The gels are still made manually. Attempts to manufacture and distribute precast gels have been a dismal failure. Samples can be loaded manually; however, semiautomatic methods are widely available, like multiple-headed microsyringes. Fully automated gel-loading robots will be commonplace soon. The sequencing chemistry done can be done in a fully automated form for almost all choices of templates and tactics (described later in this chapter). Several different robotic systems that perform the sequencing reactions in microtitreplate wells are now available (see Chapter 9). Three basic choices are available for the template. The sample to be sequenced can be cloned into the single-stranded DNA bacteriophage M13, and then a single M13 sequencing primer can be used for all templates. Alternatively, PCR can be used to prepare the template by a process called cycle sequencing in which one strand is differentially labeled or synthesized during cycles of linear amplification and terminators are introduced to allow the sequence to be read. The third general approach is to sequence directly from genomic DNA. Here the primer is used directly on double-stranded plasmid, bacteriophage, cosmid, or even bacterial DNA after that sample has been melted.

VARIATIONS IN CONTEMPORARY DNA SEQUENCING TACTICS

A tremendous amount of energy and cleverness has gone into attempts to improve and optimize current DNA sequencing technology. Here we cover some recent developments near the cutting edge of conventional DNA sequencing. The first improved the throughput by increasing the running speed. Major improvements in DNA sequencing rates are achievable by the use of thin gel samples. These can be either gel-filled capillaries or thin gel slabs. The advantage of a thin gel is that heat dissipation is more effective. The sample has a more even temperature distribution, and probably more important, higher-field strengths can be used that can increase the running speed by up to an order of magnitude. This increases sample throughput, and it diminishes any residual effects of diffusion. However, greater detection sensitivity is needed to process the fluorescence as the samples whip by the detector. Special gel materials are also available that allow faster running. For example, the Long Ranger™ gel speeds up the electrophoresis; it also appears to have better resolution than standard polyacrylamide when longer than conventional running gels are used. Perhaps most significant, these gels can be reloaded and rerun several times before their performance starts to deteriorate.

The major improvement in recent DNA sequencing chemistry has been the use of engineered polymerases like the modified form of bacteriophage T7 DNA polymerase, called Sequenase (version 2). This genetically modified form of the enzyme has improved processivity, which means it makes more even sequencing ladders. The behavior of the enzyme is further enhanced by using Mn^{2+} ions instead of Mg^{2+} (Fig. 10.8). With Sequenase, the limiting factor in long DNA sequencing reads is at the level of the electrophoresis and not the sequencing chemistry. Different genetically engineered polymerases have been optimized for cycle sequencing. These include Amplitaq FS and Thermosequenase.

In most DNA sequencing the position of a band is used as the sole source of information. The intensity of the band is ignored. With the very even sequencing ladders provided by the use of Sequenase, and the high signal-to-noise ratio of the ALF-type systems, it is possible to use intensity as a base specific label. A typical result is shown in Figure 10.9. Here different amounts of fluorescent and nonlabeled primers were used for the four different base-specific terminations. All samples were combined into a single lane. The result is equivalent to four-color sequencing in terms of throughput and eliminating the need to register adjacent lanes. However, it is not clear how resistant the intensity-labeling process is to various potential errors. Thus this method has not seen widespread use.

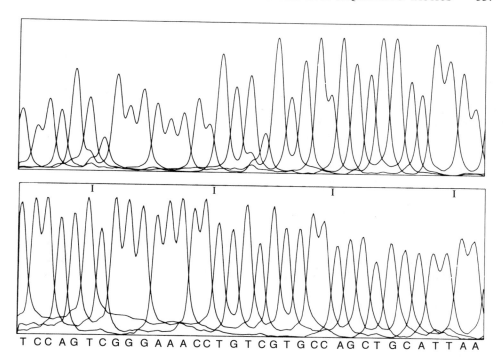

T C C A G T C G G G A A A C C T G T C G T G C C A G C T G C A T T A A

Figure 10.8 Example of how engineered polymerases (*bottom*) provide more even DNA sequencing ladders than natural polymerases (*top*). Provided by Wilhelm Ansorge.

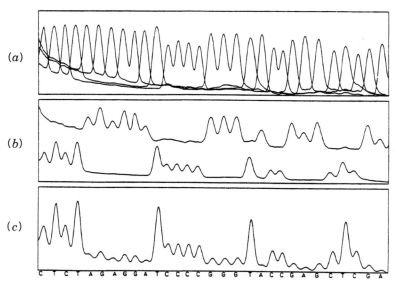

(a)

(b)

(c)

C T C T A G A G G A T C C C C G G G T A C C G A G C T C G A

Figure 10.9 Example of the use of bases labeled with different amounts of a single dye to get information about all four bases in a single lane with only one color. (*a*) Normal sequencing (four separate lanes). (*b*) Two bases at a time (two lanes). (*c*) All four bases in one lane. Taken from Ansorge et al. (1990).

Three other DNA sequencing tricks are worth describing briefly. One solution to the problems of fast on-line analysis is off-line analysis. A very clever way to do this is the bottom wiper. As shown in Figure 10.10, this consists of a short sequencing gel atop a moving membrane. As the electrophoresis proceeds, samples are automatically eluted and transferred to the membrane. The resulting blot can then be analyzed off line in any way one wishes. The unique aspect of this approach is that the spacing between adjacent bands is not only a function of the electrophoresis, it can also be manipulated by how fast the membrane moves. Thus samples that are too close to each other to be well resolved by a given detector, even though the bands are resolved in the electrophoresis, can be separated and analyzed individually.

The second method, developed by Barbara Shaw, is a novel variation on Sanger sequencing chemistry. Here the use of dideoxynucleoside triphosphate terminators is avoided. Instead, trace amounts of boron derivatives of the normal deoxynucleoside triphosphates are added to the cocktail of substrates used for DNA polymerase. Compounds like $5'$-α-[P-borano]-triphosphates are well tolerated by most DNA polymerases. These compounds have a BH_3 group replacing an oxygen on the alpha phosphate (nearest the sugar). Thus they lead to incorporation of boron-substituted phosphates into the DNA chain. The polymerization process is efficient enough that the derivatives can be introduced as part of PCR amplification. These derivatives are resistant to exonuclease III cleavage. Thus, when the PCR product is digested with exonuclease III, a ladder of fragments is produced that terminates at the first location at which a boron derivative is encountered.

A final trick that appears to be extremely promising is internal labeling. Here a primer is used adjacent to some $3'$ known flanking sequence. Label is introduced into the primer by selective extension in the absence of one of the four ordinary dpppN's (essentially the original Ray Wu sequencing strategy) selected from the known sequence. Either fluorescein-labeled or IR-dye-labeled dU or dA can be used (Fig. 10.11). The latter appears to be superior. There are several advantages in this approach. The fluorophores introduced are internal, which protects them from any exonuclease degradation; there is no background from a great excess of labeled primer, and there is no need to synthesize fluorescent primers. An example of the success in using internal labeling is shown in Figure 10.12. Admittedly this is an extraordinarily good sequencing result. However, recent experience at the EMBL where this technique was developed indicates that even in a teaching setting, the use of internal labeling routinely proved highly accurate raw sequencing

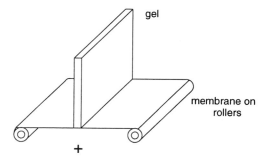

Figure 10.10   Schematic illustration of the bottom wiper used to transfer DNA sequencing ladders to a membrane.

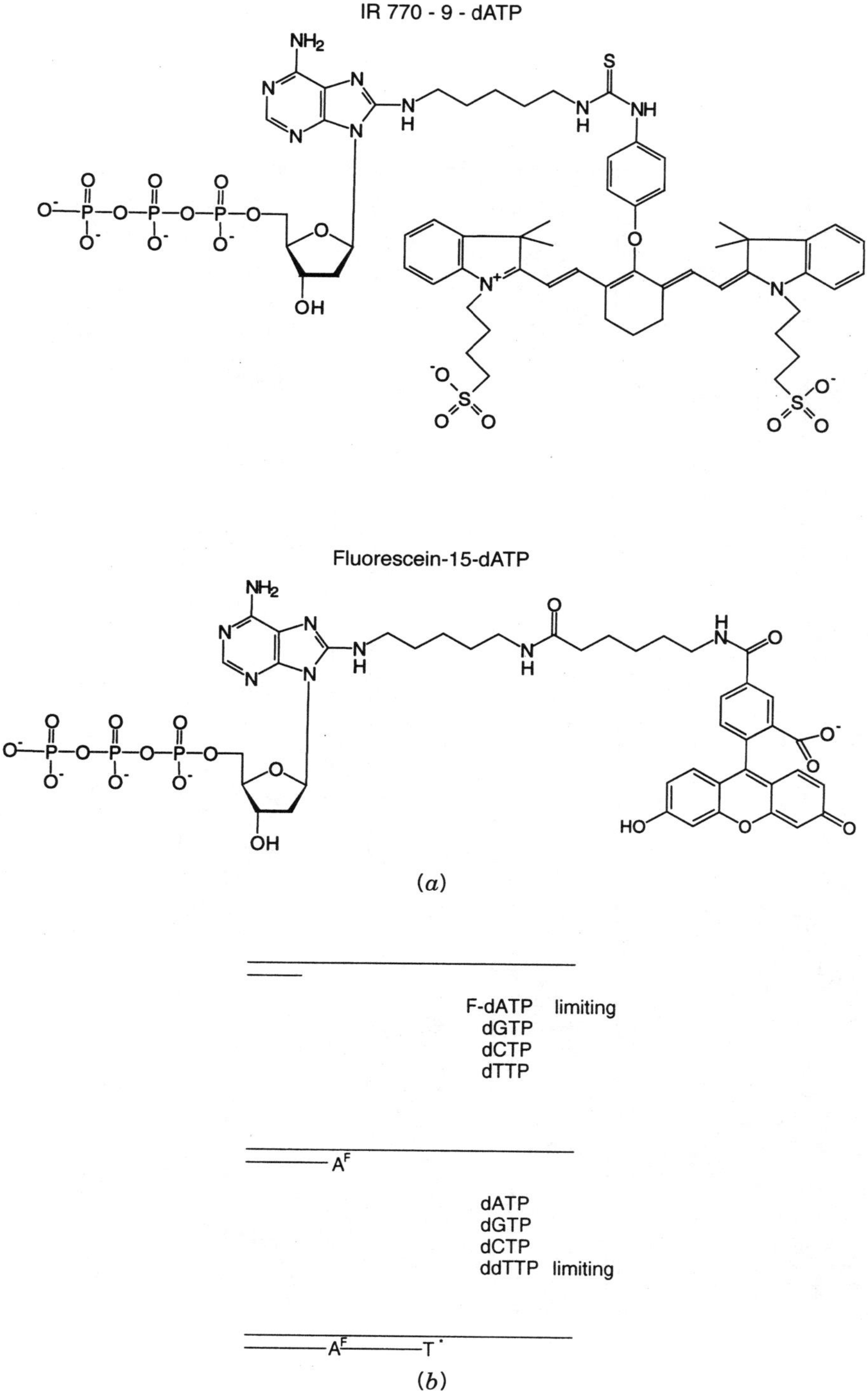

IR 770 - 9 - dATP

Fluorescein-15-dATP

(*a*)

F-dATP   limiting
dGTP
dCTP
dTTP

———— A^F

dATP
dGTP
dCTP
ddTTP   limiting

———— A^F ———— T$^{\cdot}$

(*b*)

Figure 10.11 Fluorescent dyes used for internal labeling. (*a*) Dye structures. (*b*) Typical procedure for internal labeling.

Figure 10.12 Results of a very good but not exceptional DNA sequencing ladder obtained with internal labeling. Taken from Grothues et al. (1993).

340

data with typical reads of 400 to 500 bases. Since the required fluorescent materials are readily available, there is to reason why they should not be incorporated into all one-color fluorescent DNA sequencing methods.

ERRORS IN DNA SEQUENCING

A major factor in automated DNA sequencing is continued improvements in the software used to analyze the data. From manual sequence calling, the state of the art has progressed to automated sequence reading, regardless of whether radioactive, chemiluminescent, or fluorescent sequencing data are recorded. Most software allows manual editing and overriding. Ideal software indicates where bases are known with great accuracy and where there are ambiguities. One recent report cited a 1% automatic calling accuracy in 350-base fluorescent sequencing runs, with significant deterioration beyond this point to 17% error in 500 base runs (Koop et al., 1993). A more recent study (Table 10.1) is more optimistic and demonstrates that manual editing can improve, sometimes substantially, on the accuracy of automated calling software (Naeve et al., 1995). The best current automatic software is claimed to read 500 base runs with less than 1% error. It is very important to realize that most of the issues that have been addressed thus far at the software level are how to deal with cases where adjacent fragments are only partially resolved. The resolution of successive bands in DNA sequencing gel electrophoresis gradually deteriorates as the size of the bands increases. Larger fragments spend more time in the gel and have correspondingly more time to disperse. They also are increasingly subject to electrical orientation (Chapter 5) which leads to a loss in size-dependent electrophoretic mobility. For these reasons one usually tries to sequence both strands of a target, and this provides the most accurate sequence possible at both ends of the target.

Many of the errors and ambiguities that occur in DNA sequencing data are systematic. This encourages the use of clever computer algorithms or artificial intelligence approaches to refine, even more, automated sequence calling. Since the average separation between adjacent fragment lengths varies very gradually in DNA electrophoresis, from the width of a band, one usually knows how many bases it represents even if these are not well resolved, as in the case of a run of the same base at very large fragment sizes. However, when the results are examined more closely, it is apparent that the band spacing is not perfect; it varies slightly depending on the identity of the last base in the chain, and perhaps the one before (Fig. 10.13a). Band intensities are also a function of the local sequence, and they are markedly affected by the particular DNA polymerase used. Secondary structure in the single-stranded DNA is not totally eliminated by the denaturing conditions used (Fig. 10.13b). This can be partially compensated for by using base analogs like 7-deazaG instead of G, since these form weaker secondary structures. When DNA strands form hairpins under the conditions of sequencing gel electrophoresis, they migrate faster than expected. The result is called a compression; part of the sequence may be missed or may just be impossible to read (Fig. 10.14). Frequently compressions occur on one strand but not on the complementary strand; this is one of the major reasons to sequence both strands. At first, the strand dependence of compressions may seem puzzling. After all, whatever intrastrand base pairs that can be formed by one strand can also be formed by its complement. However, two additional complications arise. First is that any strand with a high local density of G residues can form unusual helical structures with G–G pairing. These are very stable; fortunately the complementary strand will be C-rich

TABLE 10.1 Accuracy of Automated DNA Sequencing as a Function of the Distance from the Primer

Method	Maximum Correct (%)						Median Correct (%)						Minimum Correct (%)					
	1–100	101–200	201–300	301–400	401–500	501–600	1–100	101–200	201–300	301–400	401–500	501–600	1–100	101–200	201–300	301–400	401–500	501–600
Dye-primer, unedited	98	100	100	100	99	96	93	100	100	99	92	44	38	99	96	71	14	9
Dye-primer, edited	96	100	100	100	99	65	93	100	199	99	98	61	38	100	100	98	69	0
Dye-terminator, unedited	100	100	100	100	100	75	97	100	99	98	83	27	51	92	88	36	0	0
Dye-terminator, edited	100	100	100	100	100	84	97	100	100	100	84	15	21	94	95	39	0	0

Source: Adapted from Naeve et al. (1995).

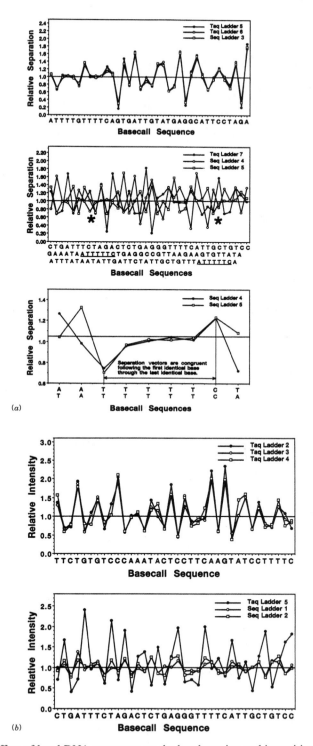

Figure 10.13 Effect of local DNA sequences on the band spacing and intensities seen in sequencing data. Taken from Tibetts et al. (1996). (*a*) Relative band spacings are intrinsic to the sequence. (*b*) Relative band intensities depend on the polymerase.

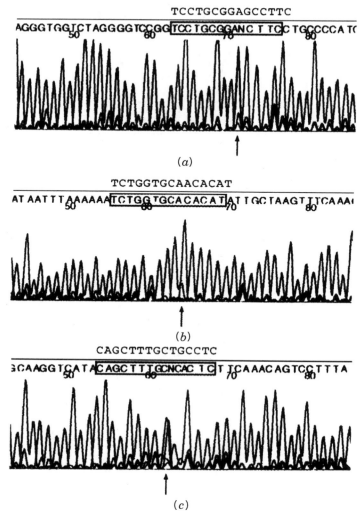

Figure 10.14   Example of compression artifacts in DNA sequencing caused by stable secondary structures in the single-stranded sample. Shown with the sequence data are the automatic software call, and the true sequence is shown above. Taken from Yamakawa et al., (1996)

instead of G-rich. The use of 7-deazaG presumably eliminates such structures. The second, and more serious complication is the sequence of the loop of the hairpin. It turns out that certain loop sequences promote particularly stable hairpin formation. An example is GCGAAAGC, which forms a hairpin with only two base pairs, but its melting temperature is 76°C in 0.15 M salt (Hirao et al., 1990). A recent study surveyed a large number of natural and synthetic sequences that led to compressions (Yamakama et al., 1996). Remarkably all but 2% of these were formed by only two types of sequence motifs. About a third, which showed up on both strands were hairpins with a G+C-rich stem (≥ 3 bp) connected by a 3–4 base loop. The remaining two-thirds occurred on only 1

Figure 10.15 Example of background in DNA sequencing ladders generated by mispriming.

strand which carried the consensus sequence YGN1–2AR, where Y and R are complementary. Note that this motif is contained in the extraordinarily stable hairpin just described above. Now that the prevalence of this motif in compressions is understood, it can be used to correct the misread sequence as shown by the example in Figure 10.14.

Other sources of error in DNA sequencing are caused by mispriming. Not uncommonly, there will be a secondary site on the template where the primer can bind and be extended by the polymerase (Fig. 10.15). This adds a bit of low-level, specific noise to the primary sequencing data. Another source of ambiguity arises when the sample is heterozygous or a mixture. The basic point is that most of these errors can be partially or even totally corrected if the software is clever enough to search out these possibilities. As more raw sequence data are obtained, and ultimately corrected into finished sequence, it should be possible to go back to the raw data and refine the algorithms used to process it. In short, the automated analysis of DNA sequencing data ought to be able to improve itself continually with time. The ideal software, which does not yet exist, would actually give use the probability of each of the four bases occurring at a given position. At a given site the result might be

$$A = 0.01$$
$$G = 0.98$$
$$C = 0.00$$
$$T = 0.01$$

This would be the best data to feed back into artificial intelligence approaches to refine the software further. A nice step in this direction is Phil Green's phred algorithm which automatically calls sequences and assigns a quality score, q, to each base,

$$q = -10 \log p$$

where p is the estimated error probability for that base. Hence phred scores of 30 or better indicate sequences that are likely to be perfect.

AUTOMATED DNA SEQUENCING CHEMISTRY

The one remaining area we need to describe, where great progress has been made is the automated preparation of DNA for sequencing. The most success appears to be seen with solid state DNA preparations. These were developed by Mathias Uhlen, and a recent in-

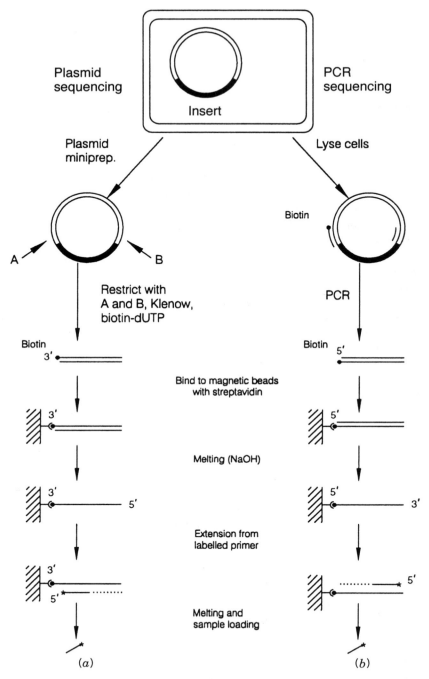

Figure 10.16   Two solid state DNA sequencing schemes. (*a*) From a plasmid DNA miniprep. (*b*) From DNA prepared by PCR. Provided by Mathias Uhlen. See Holtman et al. (1989).

teresting modification has been accomplished by Ulf Landegren. In both methods the idea is to capture DNA onto a solid surface via streptavidin-biotin technology, and then do at least one strand of the DNA sequencing on that surface. Two schemes developed by Uhlen are illustrated in Figure 10.16. They are pretty much self-explanatory. The Uhlen implementation of these solid state preparations uses magnetic microbeads containing immobilized streptavidin. The DNA is biotinylated either by filling in a restriction site with a biotinylated base analog or by using a biotinylated PCR primer. Once the duplex DNA is captured, the nontethered strand is removed by alkali. An essential aspect of the procedure is that the streptavidin-biotin link is resistant to the harsh alkali treatment needed to melt the DNA. Sequencing chemistry is then carried out on the immobilized DNA strand that remains. If desired, the strand released into solution can also be subsequently captured in a different way and sequenced. The great advantage of this approach is the ease with which it can be automated and the very clean DNA preparations that are provided because of the efficient sample washing possible in this format.

Multiple samples can be manipulated with a permanent magnet in a microtitre plate format as shown in Figure 10.17. The alternative implementation, also using immobilized

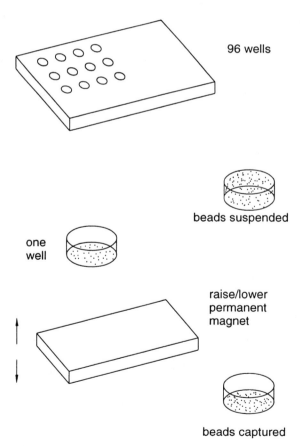

96 wells

beads suspended

one well

raise/lower permanent magnet

beads captured

Figure 10.17 Microtitre plate magnetic separator used by Uhlen for automated solid state DNA sequencing.

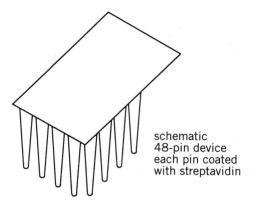

schematic
48-pin device
each pin coated
with streptavidin

Figure 10.18 Microtitre plate multipin device used by Landegren for automated DNA sequencing and related automated DNA manipulations.

streptavidin, employs a 48-pin device instead of magnets (Fig. 10.18). Here the immobilized DNA is captured on the ends of plastic pins which have been loaded with strepta-vidin-conjugated microbeads. A very high density of strepavidin can be generated in this way. It seems clear that by combining magnetic beads and plastic pins, one may be able to automate even more elaborate protocols easily. Recently Landegren reported a very clever variation of this scheme in which the DNA sequencing chemistry is carried out on a plastic comb of the type used to cast the sample slots in a sequencing gel. The teeth of the comb contained immobilized streptavidin beads. Once the chemistry was completed, the contents of the entire comb were loaded onto a DNA sequencing gel by inserting the comb into a gel with wells containing formamide. This solvent disrupts the binding between streptavidin and biotin, denatures the DNA, and releases the DNA samples into the gel. Apparently the formamide has no serious deleterious consequences on the subsequent electrophoresis. Thus, in a very simple way, the problem of automated gel loading has effectively been solved.

FUTURE IMPROVEMENTS IN LADDER SEQUENCING

Using all the power of current technology, the very best sequencing laboratories can generate more than 10^5 bp of raw DNA sequence per day per worker. Lanes read to 600 and 700 bases are common. The entire process is fully automated from colony picking to DNA sample preparation to gel loading and running, to the raw sequence analysis and entry into a database. Only gel casting and sequence editing are still manual.

A number of different approaches are being tested to see if the throughput of ladder sequencing can be further improved. Here we will describe some or the more promising or more novel attempts. The basic issues are how to extend a ladder to longer sizes, how to perform the fractionation more rapidly, how to increase the number of different samples that can be handled simultaneously, and how to read the data more rapidly. A number of the approaches share the feature that they use very thin samples. The advantages of such gels for increasing speed were described earlier. A disadvantage is that thin gels mean lower amounts of sample, and this requires greater detection sensitivity.

As detectors are improved, it is to be expected that larger numbers of samples will be loaded on each gel by using closer spaced and narrower lanes. One limitation with the current ALF system is its single color detection; yet the high sensitivity afforded by having the laser in the plane of the gel is a clear advantage. In principle, one could use multiple lasers in the gel, at different positions, and each could be accompanied by a suitable detector array. Ansorge has developed such an instrument, which clearly will have higher throughput since each lane will then be available for multicolor sequencing.

One method of diminishing sample size while retaining sensitive detection is to use a fluorescent microscope as the detector. In Chapter 7 the power of confocal scanning laser microscopy for FISH was described. This microscope also makes an excellent detector for direct scanning of fluorescent-labeled DNA samples in gels. The advantage of the confocal microscope is that it gathers emission very efficiently from a very narrow vertical slice through the sample. Light that emanates from above or below this plane is not imaged. Thus the confocal microscope can detect fluorescence from inside a capillary or thin slab without background due to scattering from the interface between the capillary and the gel, or the interface between the capillary and the external surroundings. This is a major improvement. One consequence is that the capillaries are scanned off line, in order to take full advantage of the scanning speed of the microscope. Dense bundles of capillaries can be made, loaded in parallel with multiple headed syringes, run in parallel, and scanned together (Fig. 10.19). The increase throughput ultimately achievable with this approach may be considerable. A potential limitation is the difficulty in making gel-filled capillaries. This will be alleviated somewhat as it becomes possible to use liquid (noncrosslinked) gels instead of solid (crosslinked) gels. This is because solid gels must be polymerized within the capillary, while liquid gels can just be poured into the capillary. The alternative to capillaries is to use large thin gel slabs. This simplifies the optics needed for on-line detection of the DNA. In anticipation of the considerable demands placed on detector systems by fast running, thin gels, a number of alternative new detectors are being developed as possible readout devices for fluorescence-based DNA sequencing.

APPROACHES TO DNA SEQUENCING BY MASS SPECTROMETRY

A separate approach to improving ladder sequencing is to change the way in which the labeled DNA fragments are detected. Here considerable attention has been given to mass spectrometry. There are actually three ways in which mass spectrometry might be used, in principle, to assist DNA sequencing. In the simplest case the mass spectrometer is used as a detector for a mass label attached to the DNA strand in lieu of a fluorescent label. Alternatively, the mass of the DNA molecule itself can be measured. In this case the mass spectrometer replaces the need for gel electrophoresis; it separates the DNA molecules and detects their sizes. The most ambitious and difficult potential use of mass spectrometry would involve a fragmentation analysis of the DNA and the determination of all of the resulting species. In this way the mass spectrometer would replace all of the chemistry and electrophoresis steps in conventional ladder sequencing. We are a long way from accomplishing this. In this section we will discuss each of these potential applications of mass spectrometry to enhance DNA sequencing.

Mass spectroscopy is almost as sensitive a detector as fluorescence, with some instruments having sensitivities of the order of thousands of atoms or molecules, and special

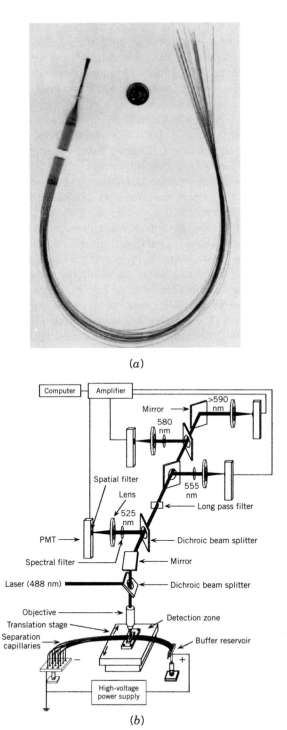

Figure 10.19 Apparatus for DNA sequencing by capillary electrophoresis. (a) An array of gel-filled capillaries used for DNA sequencing. (b) On-line detection by confocal scanning fluorescence microscopy. Figure provided by Richard Mathies. Figure also appears in color insert.

techniques such as ion cyclotron resonance mass spectrometry having even greater sensitivity. However, the principal potential advantages of mass spectra over fluorescence is that isotopic labeling leads to much less of a perturbation of electrophoretic properties than fluorescent labeling, and the number of easily used isotopic labels far exceeds the number of fluorophores that could be used simultaneously. Mass spectrometers actually measure the ratio of mass to charge; the best instruments have a mass to charge resolution of better than 1 part in a million. Thus asking a mass spectrometer to distinguish between, say, two isotopes like ^{34}S and ^{36}S is not very demanding if these isotopes reside in small molecules.

One basic strategy in using mass spectrometry as a DNA sequencing detector simply replaces the fluorophore with a stable isotope. Two approaches have been explored. In one case four different stable isotopes of sulfur would be used as a 5′ label incorporated, for example, as thiophosphate. In the other case a metal chelate is attached at the 5′-end of the primer, and different stable metal isotopes are used. Some of the possibilities are shown in Table 10.1. Since many of the divalent ions in the table have very similar chemistry, chelates can be built that, in principle, would bind many different elements. Thus, when all the isotopes are considered, there is the possibility of doing analyses with more than 30 different colors. Whether sulfur or metal isotopes are used, the sample must be vaporized and the DNA destroyed so that the only mass detected is that of a small molecule or single atom containing the isotope. With sulfur labeling, one possible role for mass spectrometry is as an on-line detector for capillary electrophoresis. DNA fragments are eluted from the capillary into a chamber where the sample is burned, and the resulting SO_2 is ionized and detected.

With metal labeling, a much more complex process is used to analyze the sample by mass spectrometry. This is a technique called resonance ionization spectroscopy (RIS), and it is illustrated in Figure 10.20. Here mass spectrometry would serve to analyze a filter blot, or a thin gel, directly, off line. In RIS just the top few microns of a sample are examined. Either a strong laser beam or an ion beam is used to vaporize the surface of the sample, creating a mixture of atoms and ions. The beam scans the surface in a raster pattern. Any ions produced are pulled away by a strong electric field. Then a set of lasers is used to ionize a particular element of interest; in our case this is the metal atom used as the label. Because ionization energies are higher than the energy in any single laser photon, two or more lasers must be used in tandem to pump the atom up to its ionization state. Then it is detected by mass spectrometry. The same set of lasers can be used to

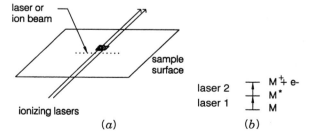

Figure 10.20 Resonance ionization mass spectrometry (RIS). (*a*) Schematic design of the instrument used to scan a surface. (*b*) Three electronic states used for the resonance ionization of metal atoms.

excite all of the different stable isotopes of a particular element; however different lasers may be required when different elements are to be analyzed.

An example of RIS as applied to the reading of a DNA sequencing gel is shown in Figure 10.21. The method clearly works; however, it would be helpful to have higher signal to noise. Actually RIS is an extremely sensitive method, with detection limits of the order of a few hundreds to a few thousands of atoms. Very little background should be expected from most of the particular isotopes listed in Table 10.2, since many of these are not common in nature, and in fact most of the elements involved, with the notable exception of iron and zinc, are not common in biological materials. The problem is that gel electrophoresis is a bulk fractionation; very few of the DNA molecules separated actually lie in the thin surface layer that can be scanned. Similarly typical blotting membranes are also not really surfaces; DNA molecules penetrate into them for quite a considerable distance.

To assist mass spectrometric analysis of DNA, it would clearly be helpful to have simple, reproducible ways of introducing large numbers of metal atoms into a DNA molecule and firmly anchoring them there. One approach to this is to synthesize base analogs that have metal chelates attached to them, in a way that does not interfere with their ability to

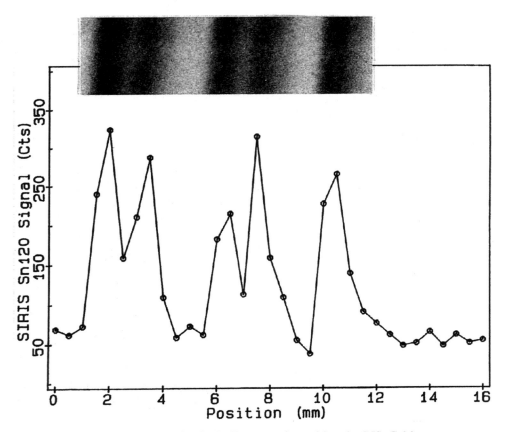

Figure 10.21   Example of analysis of a DNA sequencing gel lane by RIS. Gel image appears at top; RIS signal below. Provided by Bruce Jacobson. See Jacobson et al. (1990).

TABLE 10.2   Stable Metal Isotopes Bound by Metallothionein

$_{26}$Fe	^{54}Fe	$_{50}$Sn	^{112}Sn
	^{56}Fe		^{114}Sn
	^{57}Fe		^{115}Sn
	^{58}Fe		^{116}Sn
			^{117}Sn
$_{27}$Co	^{56}Co		^{118}Sn
			^{119}Sn
$_{28}$Ni	^{58}Ni		^{120}Sn
	^{60}Ni		^{122}Sn
	^{61}Ni		^{124}Sn
	^{62}Ni		
	^{64}Ni	$_{79}$Au	^{197}Au
$_{29}$Cu	^{63}Cu	$_{80}$Hg	^{196}Hg
	^{65}Cu		^{198}Hg
			^{199}Hg
$_{30}$Zn	^{64}Zn		^{200}Hg
	^{66}Zn		^{201}Hg
	^{67}Zn		^{202}Hg
	^{68}Zn		^{204}Hg
	^{70}Zn		
		$_{82}$Pb	^{204}Pb
$_{47}$Ag	^{107}Ag		^{206}Pb
	^{109}Ag		^{207}Pb
			^{208}Pb
$_{48}$Cd	^{106}Cd		
	^{108}Cd	$_{83}$Bi	^{209}Bi
	^{110}Cd		
	^{111}Cd		
	^{112}Cd		
	^{113}Cd		
	^{114}Cd		
	^{116}Cd		

Total 50 species

hybridize to complementary sequences. An example is shown in Figure 10.22. An alternative approach is to adapt the streptavidin-biotin system to place metals wherever in a DNA one places a biotin. This can be done by using the chimeric fusion protein shown in Figure 10.23. This fusion combines the streptavidin moiety with metallothionein, a small cysteine-rich protein that can bind 8 divalent metals or up to 12 heavy univalent metals. The list of different elements known to bind tightly to metallothionein is quite extensive. All of the isotopes in Table 10.2 are from elements that bind to metallothionein. The fusion protein is a tetramer because its quaternary structure is dominated by the extremely stable streptavidin tetramer. Thus there are four metallothioneins in the complex, and each retains its full metal binding ability. As a result, when this fusion protein is used to label biotinylated DNA, one can place 28 to 48 metals at the site of each biotin. The use of this fusion protein should provide a very substantial increase in the sensitivity of RIS for DNA detection.

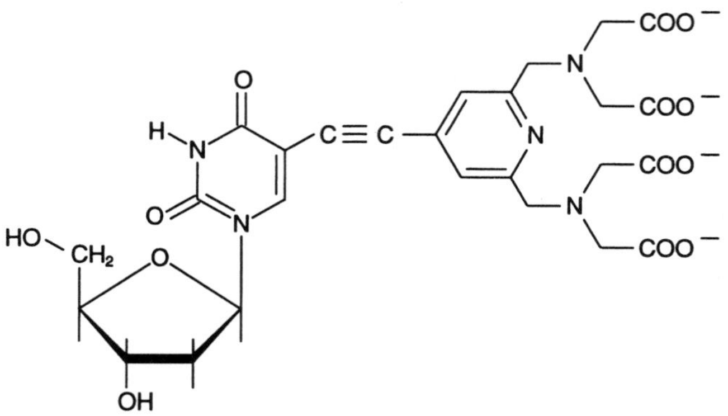

Figure 10.22 A metal chelate derivative of a DNA base suitable as an RIS label.

While mass spectrometry has great potential to detect metal labels in biological systems, a drawback of the method is that current RIS instrumentation is quite costly. Another limitation is that RIS destroys the surface of the sample, so it may be difficult to read each gel or blot more than once. Alternative schemes for the use of metals as labels in DNA sequencing exist. One is described in Box 10.1

The second way to use mass spectrometry to analyze DNA sequences ladders is to attempt to place the DNA molecules that constitute the sequencing ladder into the vapor phase and detect their masses. In essence this is DNA electrophoresis in vacuum. A key requirement of the approach is to minimize fragmentation once the molecules have been placed in vacuum, since all of the desired fragmentation needed to read the sequence has already been carried out through prior Sanger chemistry; any additional fragmentation is confusing and leads to a loss in experimental sensitivity. Two methods show great promise for placing macromolecules into the gas phase. In one, called electrospray (ES), a fine mist of macromolecular solution is sprayed into a vacuum chamber; the solvent evaporates and is pumped away. The macromolecule retains an intrinsic charge and can be accelerated by electrical fields, and its mass subsequently measured. In the second approach, matrix-assisted laser desorption and ionization (MALDI), the macromolecule is suspended in a matrix that can be disintegrated by light absorption. After excitation with a pulsed laser, the macromolecule finds itself suspended in the vapor phase; it can be accelerated if it has a charge. These two procedures appear to work very well for proteins. They also work very well on oligonucleotides, and a few examples of high-resolution mass spectra have been reported for compounds with as many as 100 nucleotides.

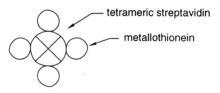

Figure 10.23 Structure of a streptavidin-metallothionein chimeric protein capable of bringing 28 to 48 metal atoms to a single biotin-labeled site in a DNA.

BOX 10.1
MULTIPHOTON DETECTION (MPD)

In contrast to RIS, a much simpler, nondestructive detector has been developed by Andy Druckier that has the capacity to analyze many different metal isotopes simultaneously, but these must be short lived radioisotopes. The principle used in this exquisitely sensitive detector is shown in Figure 10. 24. It is based on certain special radioisotopes that emit three particles in close succession. First a positron and an electron are emitted simultaneously, at 180-degree angles. Shortly thereafter a gamma ray is emitted. The gamma energy depends on the particular isotope in question. Hence many different isotopes can be discriminated by the same detector. The electron and positron events are used to gate the detector. In this way the background produced by gamma rays from the environment is extremely low. The potential sensitivity of this device is just a few atoms. It is a little early to tell just how well suited it is for routine applications in DNA analysis. Prototype many-element detector arrays have been built that examine not just what kind of decay event occurred but also where it occurred. Such position sensitive detectors are in routine use in high-energy physics. The advantage of detector arrays is that an entire gel sample can be analyzed in parallel rather than having to be scanned in a raster fashion. This leads to an enormous increase in sensitivity.

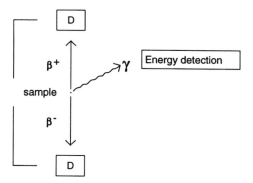

Figure 10.24   MPD: Ultra sensitive detection of many different radioisotopes by triple coincidence counting.

However, there appear to be a number of technical difficulties that must be resolved if ES or MALDI mass spectrometry are to become a generally used tool for DNA sequencing. One complication is that a number of different charged species are typically generated. This is a particularly severe problem in ES, but it is also seen in MALDI. The multiplicity of species is an advantage in some respects, since one knows that charges must be integral, and thus the species can serve as an internal calibration. However, each species leads to a different charge/mass peak in the spectrum which makes the overall spectrum complex. The most commonly used mass analyzer for DNA work has been time of flight (TOF). This is a particularly simple instrument well adapted for MALDI work—one simply measures the time lag between the initial laser excitation pulse and the time DNA samples reach the detector after a linear flight. TOF is an enormously rapid measurement, but it does not have as high resolution or sensitivity as some other methods. Today, a typical good TOF mass resolution on an oligonucleotide would be 1/1000. This is sufficient to sequence short targets as shown by the example in Figure 10.25, but DNAs longer than 60 nucleotides typically show far worse resolution that does not allow DNA sequencing. A more serious problem with current MALDI-TOF mass spectrometry is that starting samples of about 100 fmol are required. This is larger than needed with capillary electrophoretic DNA sequencing apparatuses which should become widely available within the next few years.

While current MALDI-TOF mass spectral sequence reads are short, they are extraordinarily accurate. An example is shown in Figure 10.26, where a compression makes a section of sequence impossible to call when the analysis is performed electrophoretically. In

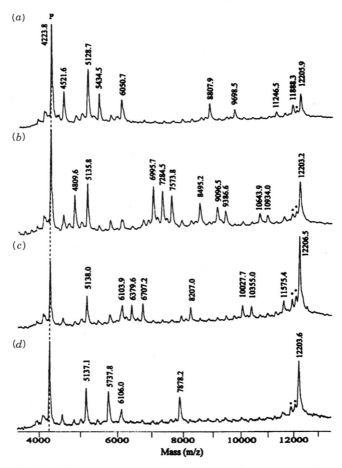

Figure 10.25 MALDI-TOF mass spectra of sequencing ladders generated from an immobilized 39-base template strand d(TCT GGC CTG GTG CAG GGC CTA TTG TAG TGA CGT ACA). P indicates the primer d(TGT ACG TCA CAA CT). The peaks resulting from depurination are labeled by an asterisk. (*a*) A-reaction, (*b*) C-reaction, (*c*) G-reaction, and (*d*) T-reaction. MALDI-TOF MS measurements were taken on a reflectron TOF MS. From Köster et al. (1996).

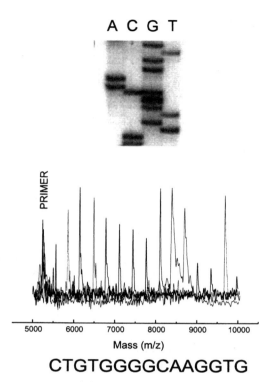

Figure 10.26 Sequencing a compression region in the beta globin gene by gel electrophoresis (*top*) and MALDI-TOF mass spectrometry (*bottom*). Note in the gel data a poorly resolved set of G's topped by what appears to be a C, G heterozygote. The true sequence, GGGGC, is obvious in the mass spectrum. Provided by Andi Braun. Figure also appears in color insert.

MS, however, all that is measured is the ratio of mass to charge, so any secondary structure effects are invisible. Thus the resulting sequence ladder is unambiguous.

A second mass analyzer that has been used for analysis of DNA is Fourier transform ion cyclotron resonance (FT-ICR) mass spectrometry. Here ions are placed in a stable circular orbit, confined by a magnetic field. Since they continuously circulate, one can repeatedly detect the DNA, and thus very high experimental sensitivity is potentially achievable. The other great advantage of FT-ICR instruments is that they have extraordinary resolution, approaching $1/10^6$ in some cases. This produces extremely complex spectra because of the effects of stable isotopes like ^{13}C. However, since the mass differences caused by such isotopes are known in advance, the bands they create can be used to assist spectral assignment and calibration. FT-ICR can be used for direct examination of Sanger ladders, or it can be used for more complex strategies as described below. The major disadvantages of FT-ICR are the cost and complexity of current instrumentation and the greater complexity of the spectra. An example of a FT-ICR spectrum of an oligonucleotide is shown in Figure 10.27.

The third strategy in mass spectrometric sequencing is the traditional one in this field. A sample is placed into the vapor phase and then fragmentation is stimulated either by collision with other atoms or molecules or by laser irradiation. Most fragmentation under these conditions appears to occur by successive release of mononucleotides from either

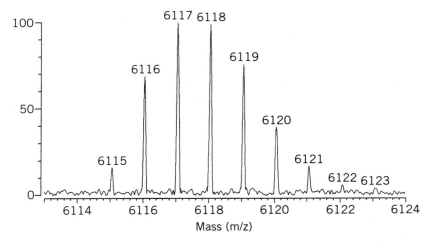

Figure 10.27   An example of FT-ICR mass spectrometry of an oligonucleotide. Here the complex set of bands seen for the parent ion of a single compound are molecules with different numbers of ^{13}C atoms. Provided by Kai Tang.

end of the DNA chain. While this approach is feasible and has yielded the sequence of one 50-base DNA fragment, the resulting spectra are so complex that at least at present this approach does not appear likely to become a routine tool for DNA analysis.

One great potential power of mass spectroscopy for direct analysis of DNA ladders is speed. In principle, at most a few seconds of analysis would be needed to collect sufficient data to analyze one ladder. This is much faster than any current extrapolation of electrophoresis rates. The major ultimate challenge may be finding a way to introduce samples into a mass spectrometer fast enough to keep up with its sequencing speed.

RATE-LIMITING STEPS IN CURRENT DNA SEQUENCING

Today, laboratories that can produce 10^4 to 10^5 raw base pairs of DNA sequence per person per day are virtually unanimous in their conclusion that at this rate, analysis of the data is the rate-limiting step. So long as one has a good supply of samples worth sequencing, the rate of data acquisition with current automated equipment is not a barrier. A few laboratories have equipment that operates at ten times this rate. This has not yet been used in a steady production mode for sequence determination, but once it enters this mode, there is every reason to believe that data analysis will continue to be rate limiting. Some of the potential enhancements we have described for ladder sequencing promise almost certainly to yield an additional factor of 10 improvement in throughput over the next few years, and a factor of 100, eventually, is not inconceivable. This will surely exacerbate the current problems of data analysis.

The rapid rate of acquisition of DNA sequence data makes it critical that we improve our methods of designing large-scale DNA sequencing projects and develop improved abilities for on-line analysis of the data. This analysis includes error correction, assembly of raw sequence into finished sequence, and interpreting the significance of the sequencing data. In the next chapter we will deal with the first two issues as part of our consideration of the strategies for large-scale DNA sequencing. The issue of interpreting DNA sequence is deferred until Chapter 15.

SOURCES AND ADDITIONAL READINGS

Ansorge, W., Zimmermann, J., Schwager, C., Stegemann, J., Erfle, H., and Voss, H. 1990. One label, one tube, Sanger DNA sequencing in one and two lanes on a gel. *Nucleic Acids Research* 18: 3419–3420.

Carrilho, E., Ruiz-Martinez, M. C., Berka, J., Smirov, I., Goetzinger, W., Miller, A. W., Brady, D., and Karger, B. L. 1996. Rapid DNA sequencing of more than 1000 bases per run by capillary electrophoresis using replaceable linear polyacrylamide solutions. *Analytical Chemistry* 68: 3305–3313.

Ewing, B., Hillier, L., Wendl, M. C., and Green, P. 1998. Base-calling of automated sequencer traces using Phred. I. Accuracy assessment. *Genome Research* 8: 175–185.

Ewing, B., and Green, P. 1998. Base-calling of automated sequencer traces using Phred. II. Error possibilities. *Genome Research* 8: 186–194.

Glazer, A. N., and Mathies, R. A. 1997. Energy-transfer fluorescent reagents for DNA analyses. *Current Opinion in Biotechnology* 8: 94–102.

Grothues, D., Voss, H., Stegemann, J., Wiemann, S., Sensen, C., Zimmerman, J., Schwager, C., Erfle, H., Rupp, T., and Ansorge, W. 1993. Separation of up to 1000 bases on a modified A.L.F. DNA sequencer. *Nucleic Acids Research* 21: 6042–6044.

H. S. Rye, S. Yue, D. E. Wemmer, M. A. Quesada, R. P. Haugland, R. A. Mathies, and A. N. Glazer. Stable Fluorescent Complexes of Double-Stranded DNA with Bis-Intercalating Asymmetric Cyanine Dyes: Properties and Applications. *Nucleic Acids Research* 20, 3803-3812 (1992).

Hultman, T., Stahl, S., Hornes, E., and Uhlen, M. 1989. Direct solid phase sequencing of genomic and plasmid DNA using magnetic beads as solid support. *Nucleic Acids Research* 19: 4937–4936.

Jacobson, K. B., Arlinghaus, H. F., Schmitt, H. W., Sacherleben, R. A., et al. 1990. An approach to the use of stable isotopes for DNA sequencing. *Genomics* 8: 1–9.

Kalman, L. V., Abramson, R. D., and Gelfand, D. H. 1995. Thermostable DNA polymerases with altered discrimination properties. *Genome Science and Technology* 1: 42.

Koster, H., Tang, K., Fu, D.-J., Braun, A., van den Boom, D., Smith, C. L., Cotter, R. J., and Cantor, C. R. 1996. A strategy for rapid and efficient DNA sequencing by mass spectrometry. *Nature Biotechnology* 14: 1123–1128.

Kustichka, A. J., Marchbanks, M., Brumley, R. L., Drossman, H., and Smith, L. M. 1992. High speed automated DNA sequencing in ultrathin slab gels. *Bio/Technology* 10: 78–81.

Kwiatkowski, M., Samiotaki, M., Lamminmaki, U., Mukkala, V.-M., and Landegren U. 1994. Solid-phase synthesis of chelate-labelled oligonucleotides: Application in triple-color ligase-mediated gene analysis. *Nucleic Acids Research* 22: 2604–2611.

Luckey, J. A., Norris, T. A., and Smith, L. M. 1993. Analysis of resolution in DNA sequencing by capillary gel electrophoresis. *Journal of Physical Chemistry* 97: 3067–3075.

Luckey, J. A., and Smith, L. M. 1993. Optimization of electric field strength for DNA sequencing in capillary gel electrophoresis. *Analytical Chemistry* 65: 2841–2850.

Naeve, C. W., Buck, G. A., Niece, R. L., Pon, R. T., Robertson, M., and Smith, A. J. 1995. Accuracy of automated DNA sequencing: A multi-laboratory comparison of sequencing results. *BioTechniques* 19: 448–453.

Parker, L. T., Deng, Q., Zakeri, H., Carlson, C., Nickerson, D. A., and Kwok, P. Y. 1995. Peak height variations in automated sequencing of PCR products using Taq dye-terminator chemistry. *BioTechniques* 19: 116–121.

Porter, K. W., Briley, J. D., and Shaw, B. R. 1997. Direct PCR sequencing with boronated nucleotides. *Nucleic Acids Research* 25: 1611–1617.

Stegemann, J., Schwager, C., Erfle, H., Hewitt, N., Voss, H., Zimmermann, J., and Ansorge, W. 1991. High speed on-line DNA sequencing on ultathin slab gels. *Nucleic Acids Research* 19: 675–676.

Tabor, S., and Richardson, C. C. 1995. A single residue in DNA polymerases of the *Escherichia coli* DNA polymerase I family is critical for distinguishing between deoxy- and dideoxyribonucleotides. *Proceedings of the National Academy of Sciences USA* 92: 6339–6343.

Yamakawa, H., Nakajima, D., and Ohara, O. 1996. Identification of sequence motifs causing band compressions on human cDNA sequencing. *DNA Research* 3: 81–86.

Yarmola, E., Sokoloff, H., and Chrambach, A. 1996. The relative contributions of dispersion and diffusion to band spreading (resolution) in gel electrophoresis. *Electrophoresis* 17: 1416–1419.

11 Strategies for Large-Scale DNA Sequencing

WHY STRATEGIES ARE NEEDED

Consider the task of determining the sequence of a 40-kb cosmid insert. Suppose that each sequence read generates 400 base pairs of raw DNA sequence. At the minimum two-fold redundancy needed to assemble a completed sequence, 200 distinct DNA samples will have to be prepared, sequenced, and analyzed. At a more typical redundancy of ten, 1000 samples will have to be processed. The chain of events that takes the cosmid, converts it into these samples, and provides whatever sequencing primers are needed is the upstream part of a DNA sequencing strategy. The collation of the data, their analysis, and their insertion into databases is the downstream part of the strategy. It is obvious that this number of samples and the amount of data involved cannot be handled manually, and the project cannot be managed in an unsystematic fashion. A key part of the strategy will be developing the protocols needed to name samples generated along the way, retaining an audit of their history, and developing procedures that minimize lost samples, unnecessary duplication of samples or sequence data, and prevent samples from becoming mixed up.

There are several well-established, basic strategies for sequencing DNA targets on the scale of cosmids. It is still not yet totally clear how these can best be adopted for Mb scale sequencing, but it is daunting to realize that this is an increase in scale by a factor of 50, which is not something to be undertaken lightly. The basic strategies range from shotgun, which is a purely random approach, to highly directed walking methods. In this chapter the principles behind these strategies will be illustrated, and some of their relative advantages and disadvantages will be described. In addition we will discuss the relative merits of sequencing genomes and sequencing just the genes, by the use of cDNA libraries. Current approaches for both of these objectives will be described.

SHOTGUN DNA SEQUENCING

This is the method that has been used, up to now, for virtually all large-scale DNA sequencing projects. Typically the genomic (or cosmid) DNA is broken by random shear. The DNA fragments are trimmed to make sure they have blunt ends, and as such they are subcloned into an M13 sequencing vector. By using shear breakage, one can generate fragments of reasonably uniform size, and all regions of the target should be equally represented. At first, randomly selected clones are used for sequencing. When this becomes inefficient, because the same clones start to appear often, the set of clones already sequenced can be pooled and hybridized back to the library to exclude clones that come from regions that have already been over sampled (a similar procedure is described later in the chapter for cDNA sequencing; see Fig. 11.17). The use of a random set of small targets ensures that overlapping sequence data will be obtained, across the entire target, and that both strands will be sampled. An obvious complication that this causes is that one does not know a priori to which strand a given segment of sequence corresponds. This has to be sorted out by software as part of problem of assembling the shotgun clones into finished sequence.

Shotgun sequencing has a number of distinct advantages. It works, it uses only technology that is fully developed and tested, and this technology is all fairly easy to automate. Shotgun sequencing is very easy at the start. As redundancy builds, it catches some of the errors. The number of primers needed is very small. If one-color sequencing or dye terminator sequencing is used, two primers suffice for the entire project if both sides of each clone are examined; only one primer is needed if a single side of each clone is examined. With four-color dye primer sequencing the number of primers needed are eight and four, respectively.

The disadvantages of shotgun sequencing fall into three categories. The redundancy is very high, definitely higher than more orderly methods, and so, in principle, optimized shotgun sequencing will be less efficient than optimized directed methods. The assembly of shotgun sequence data is computationally very intensive. Algorithms exist that do the job effectively, such as Phil Green's phrap program, but these use much computer time. The reason is that one must try, in assembly, to match each new sequence obtained with all of the other sequences and their complements in order to compensate for the lack of prior knowledge of where in the target and which strand the sequence derives from. Repeats can confuse the assembly process. A simple example is shown in Figure 11.1. Suppose that two sequence fragments have been identified with a nearly identical repeat. One possibility is that the two form a contig; they derive from two adjacent clones that overlap at the repeat, and the nonidentity is due to sequencing errors. The other possibility is that the two sequences are not a true contig, and they should not be assembled. The software must keep track of all these ambiguities and try to resolve them, by appropriate statistical tests, as the amount of data accumulates.

Closure is a considerable problem in most shotgun sequencing. Because of the statistics of random picking and the occurrence of uncloneable sequences, even at high redundancy, and even if hybridization is used to help select new clones, there will be gaps. These have to be filled by an alternate strategy; since the gaps are usually small, PCR-based sequencing across them is usually quite effective. A typical shotgun approach to se-

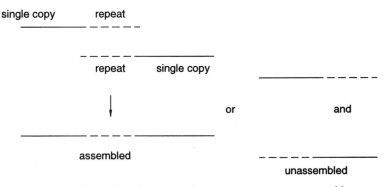

Figure 11.1   Effect of a repeated sequence on sequence assembly.

quencing an entire cosmid will require a total of 400 kb of sequence data (redundancy of 10) and 1000 four-color sequencing gel lanes. If 32 samples are analyzed at once, and three runs can be done per day, this will require 10 working days with a single automated instrument. In practice, around 20% sample failure can be anticipated, and most laboratories can only manage two runs a day. Therefore a more conservative time estimate is that a typical laboratory, with one automated instrument at efficient production speed, will require about 18 working days to sequence a cosmid. This is a dramatic advance from the rates of data acquisition in the very first DNA sequencing projects. However, it is not yet a speed at which the entire human genome can be done efficiently. At 8×10^4 cosmid equivalents, the human genome would require 4×10^3 laboratory years of effort at this rate, even assuming seven-day work weeks. It is not an inconceivable task, but the prospects will be much more attractive once an additional factor of 10 in throughput can be achieved. This is basically achievable by the optimization of existing gel-based methods described in Chapter 10.

DIRECTED SEQUENCING WITH WALKING PRIMERS

This strategy is the opposite of shotgun sequencing, in that nothing is left to chance. The basic idea is illustrated in Figure 11.2. A primer is selected in the vector arm, adjacent to the genomic insert. This is used to read as far along the sequence as one can in a single pass. Then the end of the newly determined sequence is used to synthesize another primer, and this in turn serves to generate the next patch of sequence data. One can walk in from both ends of the clone, and in practice, this will allow both strands to be examined.

The immediate, obvious advantage of primer walking is that only about a twofold redundancy is required. There are also no assembly or closure problems. With good sensitivity in the sequence reading, it is possible to perform walking primer sequencing directly on bacteriophage or cosmid DNA. This eliminates almost all of the steps involved in subcloning and DNA sample preparation. These advantages are all significant, and would make directed priming the method of choice for large-scale DNA projects if a number of the current disadvantages in this approach could be overcome.

The major, obvious, disadvantage in primer walking is that a great deal of DNA synthesis is required. If 20-base-long primers are used, to guarantee that the sequence selected is unique in the genome, one must synthesize 5% of the total sequence for each

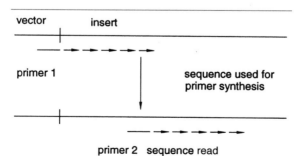

Figure 11.2 Basic design of a walking primer sequencing strategy.

strand, or 10% of the overall sequence. By accepting a small failure rate, the size of the primer can be reduced. For example, in 45 kb of DNA sequence, one can estimate by Poisson statistics that the probability of a primer sequence not occurring elsewhere, and only at the site of interest is

Primer length	10	11	12
Unique	0.918	0.979	0.998

Thus primers less than 12 bases long are not going to be reliable. With 12 mers, one still has to synthesize 6% of the total sequence to examine both strands. This is a great deal of effort, and considerable expense is involved with the current technology for automated oligonucleotide synthesis.

The problem is that existing automated oligonucleotide synthesizers make a minimum of a thousand times the amount of primer actually needed for a sequence determination. The major cost is in the chemicals used for the synthesis. Since the primers are effectively unique, they can be used only once, and therefore all of the cost involved in their synthesis is essentially wasted. Methods are badly needed to scale down existing automated DNA synthesis. The first multiplex synthesizers built could only make 4 to 10 samples at a time; this is far less than would be needed for large-scale DNA sequencing by primer walking. Recently instruments have been described that can make 100 primers at a time in amounts 10% or less of what is now done conventionally. However, such instruments are not yet widely available. In the near future primers will be made thousands at a time by in situ synthesis in an array format (see Chapter 12).

A second major problem with directed sequencing is that each sample must be done serially. This is illustrated in Figure 11.3. Until the first patch of sequence is determined, one does not know which primer to synthesize for the next patch of sequence. What this means is that it is terribly inefficient to sequence only a single sample at once by primer walking. Instead, one must work with at least as many samples as there are DNA sequencing lanes available, and ideally one would be processing two to three times this amount so that synthesis is never a bottleneck. This means that the whole project has to be orchestrated carefully so that an efficient flow of samples occurs. Directed sequencing is not appropriate for a single laboratory interested in only one target. Its attractiveness is for a laboratory that is interested in many different targets simultaneously, or that works on a system that can be divided into many targets that can be handled in parallel.

At present, a sensible compromise is to start a sequencing project with the shotgun strategy. This will generate a fairly large number of contigs. Then at some point one switches to a directed strategy to close all of the gaps in the sequence. It ought to be possible to make this switch at an earlier stage than has been done in the past as walking becomes more efficient and can cover larger regions. In such a hybrid scheme the initial

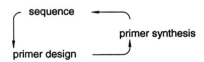

Figure 11.3 Successive cycles of primer synthesis and sequence determination lead to a potential bottleneck in directed sequencing strategies.

shotgunning serves to accumulate large numbers of samples, which are then extended by parallel directed sequencing.

Three practical considerations enter the selection of particular primer sequences. These can be dealt with by existing software packages that scan the sequence and suggest potentially good primers. The primer must be as specific as possible for the template. Ideally a sequence is chosen where slight mismatches elsewhere on the template are not possible. The primer sequence must occur once and only once in the template. Finally the primer should not be capable of forming stable secondary structure with itself—intramolecularly, as a hairpin, or intermolecularly—as a dimer. The former category is particularly troublesome, since, as illustrated in Chapter 10, there are occasional DNA sequences that form unexpectedly stable hairpins, and we do not yet understand these well enough to generalize and be able to identify and reject them a priori.

PRIMING WITH MIXTURES OF SHORT OLIGONUCLEOTIDES

Several groups have recently developed schemes to circumvent the difficulties required by custom synthesis of large numbers of unique DNA sequencing primers. The relative merits of these different schemes are still not fully evaluated, but on the surface, they seem quite powerful. The first notion, put forth by William Studier, was to select, arbitrarily, a subset of the 4^9 possible 9-mers and make all of these compounds in advance. The chances that a particular 9-mer would be present in a 45-kb cosmid are given below, calculated from the Poisson distribution $P(n)$, where n is the expected number of occurrences:

$$P(O) = 0.709 \quad P(1) = 0.244 \quad P(> 1) = 0.047$$

Therefore, with no prior knowledge about the sequence, a randomly selected 9-mer primer will give a readable sequence about 20% of the time. If it does, as shown in Figure 11.4, its complement will also give readable sequence. Thus the original subset of 4^9 compounds should be constructed of 9-mers and their complements.

One advantage of this approach is that no sequence information about the sample is needed initially. The same set of primers will be used over and over again. As sequence data accumulate, some will contain sites that allow priming from members of the premade 9-mer subset. Actual simulations indicate that the whole process could be rather efficient. It is also parallelizable in an unlimited fashion because one can carry as many different targets along as one has sequencing gel lanes available. An obvious and serious

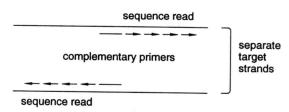

Figure 11.4 Use of two complementary 9-base primers allows sequence reading in two directions.

disadvantage of this method is the high fraction of wasted reactions. There are possible ways to improve this, although they have not yet been tried in practice. For example one could pre-screen target-oligonucleotide pairs by hybridization, or by DNA polymerase-catalyzed total incorporation of dpppN's. This would allow those combinations in which the primer is not present in the target to be eliminated before the time-consuming steps of running and analyzing a sequencing reaction are undertaken.

At about the same time that Studier proposed and began to test his scheme, an alternative solution to this problem was put forth by Wacslaw Szybalski. His original scheme was to make all possible 6-mers using current automated DNA synthesis techniques. With current methods this is quite a manageable task. Keeping track of 4096 similar compounds is feasible with microtitre plate handling robots, although this is at the high end of the current capacity of commercial equipment. To prime DNA polymerase, three adjacent 6-mers are used in tandem, as illustrated in Figure 11.5. They are bound to the target, covalently linked together with DNA ligase, and then DNA polymerase and its substrates are added. The principle behind this approach is that end stacking (see Chapter 12) will favor adjacent binding of the 6-mers, even though there will be competing sites elsewhere on the template. Ligation will help ensure a preference for sequence-specific, adjacent 6-mers because ligase requires this for its own function. The beauty of this approach is that the same 6-mer set will be used ad infinitum. Otherwise, the basic scheme is ordinary directed sequencing by primer walking. After each step the new sequence must be scanned to determine which combination of 6-mers should be selected as the optimal choice for the next primer construction.

The early attempts to carry out tandem 6-mer priming seemed discouraging. The method appeared to be noisy and unreliable. Too many of the lanes gave unreadable sequence. This prompted several variants to be developed, as described below. More recently, however, conditions have been found which make the ligation-mediated 6-mer priming very robust and dependable. This approach, or one of the variants below, is worth considering seriously for some of the next generation large-scale DNA sequencing projects. The key issues that need to be resolved are its generality and the ultimate cost savings it affords against any possible loss in throughput. Smaller-scale, highly multiplexed synthesis of longer primers will surely be developed. These larger primers are always

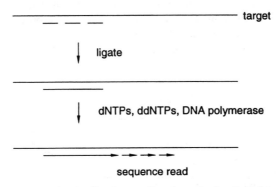

Figure 11.5 Directed sequencing by ligation-mediated synthesis of a primer, directed by the very template that constitutes the sequencing target.

likely to perform slightly better than ligated mixtures of smaller compounds because the possibility of using the 6-mers in unintended orders, or using only two of the three, is avoided. There is enough experience with longer primers to know how to design them to minimize chances of failure. Similar experience must be gained with 6-mers before an accurate evaluation of their relative effectiveness can be made.

Two more recent attempts to use 6-base oligonucleotide mixtures as primers avoid the ligation step. This saves time, and it also saves any potential complexities arising from ligase side reactions. Studier's approach is to add the 6-mers to DNA in the presence of a nearly saturating amount of single-stranded DNA binding protein (ssb). This protein binds to DNA single strands selectively and cooperatively. Its effect on the binding of sets of small complementary short oligonucleotides is shown schematically in Figure 11.6. The key point is the high cooperativity of ssb to single-stranded DNA. What this does is to make configurations in which 6-mers are bound at separate sites less favorable, energetically, than configurations in which they are stacked together. In essence, ssb cooperativity is translated into 6-mer DNA binding cooperativity. This cuts down on false starts from individual or mispaired 6-mers. An unresolved issue is how much mispairing still can go on under the conditions used. We have faced before the problem in evaluating this; there are many possible 6-mers, and many more possible sets of 6-mer mixtures, so only a minute fraction of these possibilities has yet been tried experimentally. One simplification is the use of primers with 5-mC and 2-aminoA, instead of C and A, respectively. This allows five base primers to be substituted for six base primers, reducing the complexity of the overall set of compounds needed to only 1024.

An alternative approach to the problem has been developed by Levy Ulanovsky. He has shown that conditions can be found where it is possible to eliminate both the ligase and the ssb and still see highly reliable priming. Depending on the particular sequence, some five base primers can be substituted for six base primers. The effectiveness of the approach appears to derive from the intrinsic preference of DNA polymerase for long primers, so a stacked set is preferentially utilized even though individual members of the set may bind to the DNA template elsewhere alone or in pairs. It remains to be seen, as each of these approaches is developed further, which will prove to be the most efficient and accurate. Note that in any directed priming scheme, for a 40-kb cosmid, one must sequence 80 kb of DNA. If a net of 400 new bases of high-quality DNA sequence can be completed per run, it will take 200 steps to complete the walk. As a worst-case analysis, one may have to search the 3'-terminal 100 bases of a new sequence to find a target suitable for priming. Around 500 base pairs of sequence must actually be accumulated per run to achieve this efficiency. Thus a directed priming cosmid sequencing project is still a considerable amount of work.

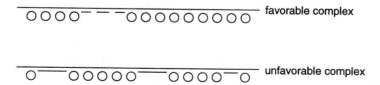

Figure 11.6 Use of single-stranded binding protein (ssb) to improve the effectiveness of tandem, short sequencing primers.

ORDERED SUBDIVISION OF DNA TARGETS

There are a number of strategies for DNA sequencing that lie between the extreme randomness of the shotgun approach and the totally planned order of primer walking. These schemes all have in common a set of cloning or physical purification steps that divides the DNA target into samples that can be sequenced and that also provides information about their location. In this way a large number of samples can be prepared for sequencing in parallel, which circumvents some of the potential log jams that are undesirable features of primer walking. At the same time from the known locations one can pick clones to sequence in an intelligent, directed way and thus cut down substantially on the high redundancy of shotgun sequencing. One must keep in mind, however, that the cloning or separation steps needed for ordered subdivision of the target are themselves quite labor intensive and not yet easily automated. Thus there is no current consensus on which is the best approach. One strategy used recently to sequence a number of small bacterial genomes is the creation of a dense (highly redundant) set of clones in a sequence-ready vector—that is, a vector like a cosmid which allows direct sequencing without additional subcloning. The sequence at both ends of each of the clones is determined by using primers from known vector sequence. As more primary DNA sequence data become known and available in databases, the chances that one or both ends of each clone will match previously determined sequence increase considerably. Such database matches, as well as any matches among the set of clones themselves, accomplish two purposes at once. They order the clones, thus obviating much or all of the need for prior mapping efforts. In addition they serve as staging points for the construction of additional primers for directed sequencing. Extensions of the use of sequence-ready clones are discussed later in the chapter.

Today, there is no proven efficient strategy to subdivide a mega-YAC for DNA sequencing. Some strategies that have proved effective for subdividing smaller clones are described in the next few sections. A megabase YAC is, at an absolute minimum, going to correspond to 25 cosmids. However, there is no reason to think that this is the optimal way to subdivide it. Even if direct genomic sequencing from YACs proves feasible, one will want to have a way of knowing, at least approximately, where in the YAC a given primer or clone is located, to avoid becoming hopelessly confused by interspersed repeating sequences. However, for large YACs there is an additional problem. As described elsewhere, these clones are plagued by rearrangements and unwanted insertions of yeast genomic DNA. For sequencing, the former problem may be tolerable, but not the latter. To use mega-YACs for direct sequencing, we would first need to develop methods to remove or ignore contaminating yeast genomic DNA.

TRANSPOSON-MEDIATED DNA SEQUENCING

The structure of a typical transposon is shown in Figure 11.7. It has terminal repeated sequences, a selectable marker, and codes for a transposase that allows the transposon to hop from one DNA molecule (or location) to another. The ideal transposon for DNA sequencing will have recognition sites for a rare restriction enzyme, like Not I, located at both ends, and it will have different, known, unique DNA sequences close to each end. A scheme for using transposons to assist the DNA sequencing of clones in *E. coli* is shown in Figure 11.8. The clone to be sequenced is transformed into a transposon-carrying *E.*

unique seqence

Figure 11.7 Structure of a typical transposon useful for DNA sequencing. N is a Not I site; a and b are unique transposon sequences. Arrow heads show the terminal inverted repeats of the transposon.

coli. It is present in an original vector containing a single infrequent cutting site (which can be the same as the ones on the transposon) and two unique known DNA sequences near to the vector-insert boundary. The transposon may jump into the cloned DNA. If it does so, this appears to occur approximately at random. The nonselectivity of transposon hopping is a key assumption in the potential effectiveness of this strategy. From results that have been obtained thus far, the randomness of transposon insertions seems quite sufficient to make this approach viable.

DNA from cells in which transposon jumping has potentially occurred can be transferred to a new host cell that does not have a resident transposon, and recovered by selection. The transfer can be done by mating or by transformation. Since the transposon carries a selectable marker, and the vector which originally contained the clone carries a second selectable marker, there will be no difficulty in capturing the desired transposon insertions. The goal now is to find a large set of different insertions and map them within the original cloned insert as efficiently as possible. The infrequent cutting sites placed into the transposon and the vector greatly facilitate this analysis. DNA is prepared from all of the putative transposon insertions, cut with the infrequent cutter, and analyzed by ordinary agarose gel electrophoresis, or PFG, depending on the size of the clones. Each insert should give two fragments. Their sizes give the position of the insert. There is an ambiguity about which fragment is on the left and which is on the right. This can be resolved if the separated DNAs are blotted and hybridized with one of the two unique sequences originally placed in the vector (labeled 1 and 2 in Fig. 11.8).

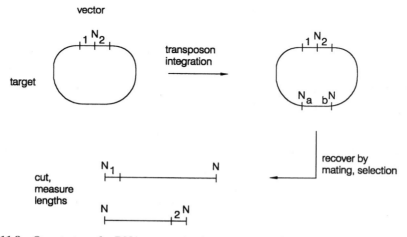

Figure 11.8 One strategy for DNA sequencing by transposon hopping. a, b, and N are as defined in Figure 11.7; 1 and 2 are vector unique sequences.

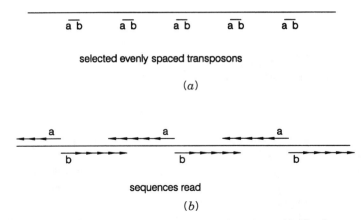

Figure 11.9   DNA sequencing from a set of transposon insertions. (*a*) Chosing a set of properly spaced transposons. (*b*) Use of transposon sequences as primers.

Now the position of the transposon insertion is known. One can select samples that extend through the target at intervals than can be spanned by sequencing, as shown in Figure 11.9. To actually determine the sequence, one can use the two unique sequences within the transposon as primers. These will always allow sequence to be obtained in both directions extending out from the site of the transposon insertion (Fig. 11.9). One does not a priori know the orientation of the two sequences primed from the transposon relative to the initial clone. However, this can be determined, if necessary, by reprobing the infrequent-cutter digest fingerprint of the clone with one of the transposon unique sequences. In practice, though, if a sufficiently dense set of transposon inserts is selected, there will be sufficient sequence overlap to assemble the sequence and confirm the original map of the transposon insertion sites.

DELTA RESTRICTION CLONING

This is a relatively straightforward procedure that allows DNA sequences from vector arms to be used multiple times as sequencing primers. It is a particularly efficient strategy for sampling the DNA sequence of a target in numerous places and for subsequent, parallel, directed primer sequencing. Delta restriction cloning requires that the target of interest be contained near a multicloning site in the vector, as shown in Figure 11.10. The various enzyme recognition sequences in that site are used, singly, to digest the clone, and all of the resulting digests are analyzed by gel electrophoresis. One looks for enzymes that cut the clone into two or more fragments. These enzymes will have cutting sites within the insert. From the size of the fragments generated, one can determine approximately where in the insert the internal cleavage site is located. Those cuts that occur near regions of the insert where sequence is desired can be converted to targets for sequencing by diluting the cut clones, religating them, and transforming them back into *E. coli*. The effect of the cleavage of religation is to drop out from the clone all of the DNA between the multicloning site and the internal restriction site. This brings the vector arm containing

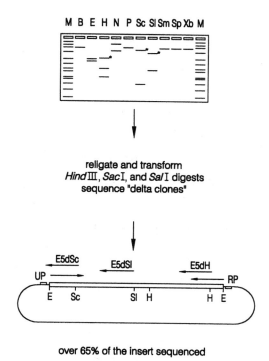

Figure 11.10   Delta restriction subcloning. Top panel shows restriction digests of the target plasmid. Bottom panel shows sequences read from delta subclones. Adapted from Ansorge et al. (1996).

the multicloning site adjacent to what was originally an internal segment of the insert and allows vector sequence to be used as a primer to obtain this internal sequence. In test cases about two-thirds of a 2- to 3-kb insert can be sequenced by testing 10 enzymes that cut within the polylinker. Only a few of these will need to be used for actual subcloning. The problem with this approach is that one is at the mercy of an unknown distribution of restriction sites, and at present, it is not clear how the whole process could be automated to the point where human intervention becomes unnecessary.

NESTED DELETIONS

This is a more systematic variant of the type of delta restriction cloning approach just described. Here a clone is systematically truncated from one or both ends by the use of exonucleases. The original procedure, developed by Stephen Henikoff, is illustrated in Figure 11.11. A DNA target is cut with two different restriction nucleases. One yields a 3'-overhang; the other yields a 5'-overhang. The enzyme *E. coli* exonuclease III degrades a 3'-overhang very inefficiently, while it degrades the 3'-strand in a 5'-overhang very efficiently. The result is degradation from only a single end of the DNA. After exonuclease treatment, the ends of the shortened insert must be trimmed to produce cloneable blunt ends. The DNA target is then taken up in a new vector and sequenced using primers from

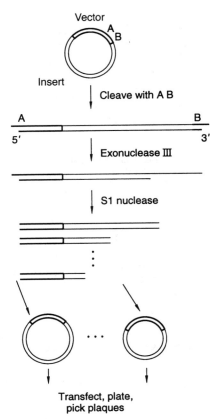

Figure 11.11 Preparation of nested deletion clones. *A* and *B* are restriction enzyme cleavage sites that give the overhanging ends indicated. Adapted from Henikoff (1984).

that new vector. In principle, this process ought to be quite efficient. In practice, while this proposed strategy has been known for many years, it does not seem to have found many adherents.

A variation on the original exonuclease III procedure for generating nested deletions has been described by Chuan Li and Philip Tucker. In this method, termed exoquence DNA sequencing, a DNA fragment with different overhangs at its ends is produced and one end is selectively degraded with exonuclease III. At various time points the reaction is stopped, and the resulting template-primer complexes are treated separately with one of several different restriction enzymes and then subjected to Sanger sequencing reactions, as shown in Figure 11.12. The final DNA sequencing ladders are examined directly by gel electrophoresis. Thus no cloning is required. If the restriction enzymes are chosen well, and the times at which the reactions are stopped are spaced sufficiently closely, sufficient sequence data will be revealed to generate overlaps that will allow the reconstruction of contiguous sequence. This is an attractive method in principle; it remains to be seen whether it will prove more generally appealing than the original nested deletion cloning approach.

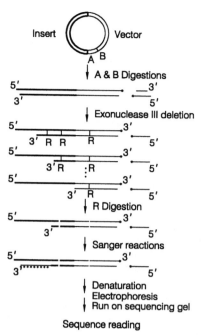

Figure 11.12 Strategy for exoquence DNA sequencing. Shown is a relatively simple case; there are more elaborate cases if the restriction enzymes used cut more frequently. *A* and *B* are restriction enzyme sites as in Figure 11.11; *R* is an additional restriction enyzme cleavage site. Taken from Li and Tucker (1993).

PRIMER JUMPING

This strategy has been discussed quite a bit. However, there are yet no reported examples of its implementation. The basic notion is outlined in Figure 11.13. It is similar in many ways to delta subcloning, but it differs in a number of significant features. PCR is used, rather than subcloning. A very specific set of restriction enzymes is used: one rare cutter which can have any cleavage pattern and an additional pair of restriction enzymes consisting of an eight base cutter and a four or six base cutter; but they have to produce the same set of complementary single-stranded ends. Examples are *Not* I (GC/GGCCGC) and *Sse*8387 I (CCTGCA/GG) for the eight cutters, and *Ene* I (Y/GGCCR) and *Nsi* I (ATGCA/T) or *Pst* I (CTGCA/G), respectively, as more frequent cutters. In principle, the approach shown in Figure 11.13 ought to be applicable to much larger DNA than delta subcloning, based on the past success at making reasonably large jumping libraries (Chapter 8).

For primer jumping the insert is cloned next to a vector fragment containing any desired infrequent cleavage site between two known unique DNA sequences, shown as *a* and *b* in Figure 11.13. The vector fragment is constructed so that it contains no cleavage sites for the 4-base or 6-base cutting enzyme between the unique sequences and the insert, but it does contain a second infrequent cleavage site, *N* in Figure 11.13, as close to the upstream unique sequence as possible. Ideally the vector will be one arm of a

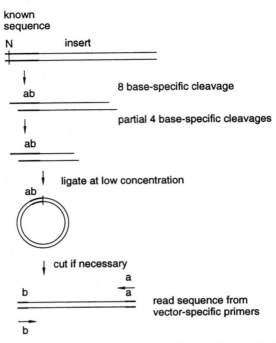

Figure 11.13   Primer jumping, an untested but potentially attractive method for directed DNA sequencing. Restriction enzyme site N must produce an end that is complementary to the end generated by the four-base cutter. Sites a and b can be anything so long as they are not present in the target, but there must a site for infrequent cleavage between them.

YAC, and the other arm could be treated in a similar fashion. The clone is cut at the distal rare cleavage to completion and then partially digested with the frequently cutting enzyme. The resulting fragments are separated by length, and the separation gel is sliced into pieces. The resulting DNA fragments are diluted to very low concentration and ligated. This will produce DNA circles in which the vector sequence, including segments a and b in Figure 11.13, is now located next to each site in the partial digest which was cleaved by the frequently cleaving enzyme. Thus the known sequence can now be used for starting a primer walk. The approximate position of the walk within the large clone will be known from the size of the fragment. With the 800-bp to 1-kb sequence reads now being achieved under good circumstances, it is conceivable that one would be able to sequence from the cleaved site up to the next equivalent restriction site without the need to make additional primers, in most cases. If this were the case, one could do a directed walking strategy on a large DNA target using only two primers—one for each vector arm.

In a similar vein to primer jumping, if single-sided PCR ever works well enough (Chapter 4), these methods could be used for directed cycle sequencing by the approaches just described.

PRIMER MULTIPLEXING

This is a potentially very powerful strategy for large-scale DNA sequencing. It was developed by George Church and has been elaborated, independently, by Raymond Gesteland. There are a number of features that set multiplexing aside from many other approaches. A major peculiarity of the method is that it does not scale down efficiently, so it is best suited for fairly massive projects, typically several hundred kb of DNA sequence or more. The basic scheme for primer multiplexing is shown in Figure 11.14. In the particular case shown, a multiplexing of 40 is used. Forty different vectors are constructed; each has a unique 20-base sequence on each side of the cloning site. The DNA target of interest is shotgun cloned, separately, into all 40 vectors. This produces 40 different libraries. Pools are constructed by selecting one clone from each of the libraries and mixing them. These 40-clone pools are the samples on which DNA sequencing is performed. The pools are subjected to standard DNA sequencing chemistry to generate a mixture of 40 different ladders, but no radioactivity or other label is introduced into the DNA at this stage. The mixture is fractionated by polyacrylamide electrophoresis and blotted onto a membrane. A particularly convenient way to do this is by the bottom wiper described in Chapter 10. The blotted DNA is crosslinked onto the filter by UV irradiation to attach it very stably. This is a key step, since the filters will be reused many times.

To read the DNA sequence from each pool of clones, the filter is hybridized with a probe corresponding to one of the 40 unique 20-base sequences. By this indirect end-labeling method (introduced in Chapter 8), only one of the 40 clones in the sequence ladder is visible. The probe is removed from the filter by washing, and then the hybridization and washing are repeated successively for each of the other unique sequence primers. By this multiplexing approach, most stages of the project are streamlined by a factor of 40.

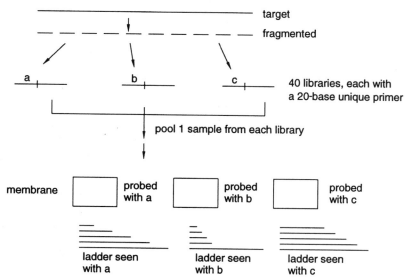

Figure 11.14 Basic scheme used for primer multiplexing: *a, b, c,* and so on, represent unique vector primer sequences.

The exceptions are the hybridization, autoradiography, or other color detection, and washing. Thus great care must be taken to automate these steps in the most efficient way. Recently a fairly successful demonstration of the efficiency that can be achieved with primer multiplexing, combined with transposon jumping, was reported by Robert Weiss and Raymond Gesteland.

MULTIPLEX GENOMIC WALKING

A different approach to multiplex sequencing has been suggested by Walter Gilbert. This is designed to be used for the sequencing of entire small genomes like *E. coli* where direct genomic DNA sequencing is feasible. The method is illustrated in Figure 11.15a. The great appeal of this method is that absolutely no cloning is required. The total genome is digested separately with a set of different restriction enzymes. The products of this digestion are loaded onto polyacrylamide gels in adjacent lanes and fractionated. A highly labeled probe with an arbitrary sequence is selected (with a length chosen to occur on aver-

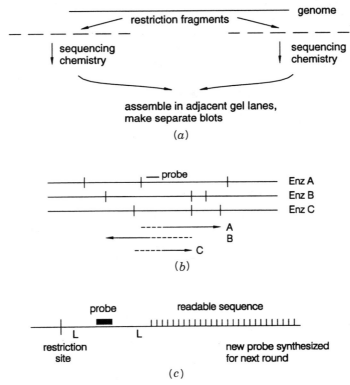

Figure 11.15 Multiplex genomic walking. (*a*) Basic outline of the experiment. (*b*) Restriction map in a typical region, and resulting segments of sequence, *A, B, C* revealed by hybridization with one specific probe. (*c*) Sections of readable and unreadable sequence on a particular restriction fragment. The probe is located *L* bases from the end of the fragment.

age once per genome) and used to hybridize with a blot of the separated fragments (Fig. 11.15*b*). In most of the lanes this probe will give a readable sequence. Suppose that the probe lies 60 bp upstream from a given restriction site. The first 60 bases of sequence will be unreadable because data will extend in both directions (Fig. 11.15*c*). However, longer regions of the ladder will be interpretable, since they must lie in the direction away from the nearby restriction site. In general, one will expect to get a number of usable reads in both directions from the probe, just by the fortuitous occurrence of useful restriction sites. These reads are assembled into a segment of DNA sequence. Next probes are designed from the most distal regions of the segment, and these are used to continue the genomic walk.

In principle, multiplex genomic walking is a very elegant and spartan approach to DNA sequencing. One has a choice at any time whether to use additional arbitrary probes, and so increase the number of parallel sequencing thrusts, or whether to focus on directed walking. Thus one has a method with some of the advantages of both random and directed strategies. A potential weakness is the relatively high fraction of failed lanes that will occur unless the probe has a single binding site in the genome. Another problem is the technical demands that genomic sequencing makes. It is also not obvious how easy this strategy will be to automate. It does work, but the overall efficiency needs to be established before the method can be compared quantitatively with others.

GLOBAL STRATEGIES

A basic issue that has confronted the human genome project since its conception is not how to sequence but what to sequence. From a purely biological standpoint, the most interesting sequencing targets are genes. The choice of genes depends on the sorts of biological questions one is interested in. An evolutionary biologist may want to sequence one homologous gene in a wide variety of organisms. Cell biologists or physiologists may want to focus on a set of functionally related genes or gene families within just a few organisms. However, from the point of view of whole genome studies, the purpose of sequencing is really to find genes and make them available for subsequent biological studies. This puts a very different tilt on the issues that affect the choice of sequencing targets.

For simple gene-rich organisms like bacteria and yeasts, there is little doubt that complete genomic sequencing is desired and worth doing even with existing DNA sequencing technology. Indeed sequencing projects have been completed on many bacteria including *H. influenzae, Mycoplasma genitalium, Mycoplasma pneumoniae, Methanococcus jannaschii, Synechocystis* strain pcc6803, and *Escherichia coli,* and one yeast, *S. cerevisiae* (see Chapter 15). Additional projects are well underway with a number of other microorganisms, including the bacterium *Mycobacterium tuberculosis* and the yeast, *S. pombe. E. coli* is an obvious choice as the focus of much of our fundamental studies in prokaryotic molecular biology. Mycoplasmas represent the smallest known free-living genomes. *Mycobacterium tuberculans* is important because of the current medical crisis with drug-resistant tuberculosis. The two yeasts account for most of our current knowledge and technical power in fungal genetics. They are also very different from each other, so much will be learned from comparisons between them. The real issue that will have to be faced in the future is at what stage in DNA sequencing technology is it desirable and affordable to sequence the genomes of many other simple organisms?

There are a number of more advanced organisms that appear to have relatively high coding percentages of DNA. These include a simple plant, *Arabidopsis thaliana,* a much more economically important plant, rice, the fruitfly, *Drosophila melanogaster,* and the nematode, *Caenorhabditis elegans.* There are strong arguments in favor of obtaining complete DNA sequences on these organisms rapidly. They all are systems where a great deal of past genetics has been done, and a great deal of ongoing interest in biological studies remains. Certain primitive fishes may also have small genomes as does the puffer fish. Here the argument in favor of sequencing is that it will be relatively easy to find most of the genes. However, these organisms are currently pretty much in a biological vacuum.

For more complex, gene-dilute organisms, the selection of sequencing targets is, not surprisingly, also more complex. Here there is little debate that *Homo sapiens* and the mouse, *Mus musculus,* are the obvious first choice. It is much less clear what should come after this. Do we target other primates because they will be most useful in understanding the very large fraction of human genes that are believed to be central nervous system specific? Do we examine genomes of organisms that have long been the focus of physiological studies like rats, dogs, and cats. Or do we aim for a much broader representation of evolutionary diversity? Alternatively, how important should the commercial value of potential genome targets be? Cows, horses, pine trees, maize, and salmon have a much more important economic role than *Arabidopsis* or *C. elegans.* These questions are interesting to ponder, but they really do not require answers at the present time. If sufficiently inexpensive DNA sequencing methods are developed in the future, we will want to sequence every genome of biological interest. For the present, technology pretty much limits us to a few choices.

With most complex organisms, only a few percent of the genome is known to be coding sequence. The function of the rest, which we earlier termed junk, is unknown, today. With limited resources, and relatively slow sequencing technology, most involved groups are choosing to focus on selectively sequencing genes from human or other sources. There are two ways to go about this. One approach is to find a gene-rich region in a genome and sequence it completely. Regions that have been selected include the T-cell receptor loci, immunoglobulin gene families, and the major histocompatibility complex. All of these regions are of intense interest in understanding the function of the immune system. Another region of interest is the Huntington's disease region because it is very gene rich, and in the process of finding the particular gene responsible for the disease a large set of cloned DNA samples from this region has become available.

An alternative to genomic sequencing in a gene-rich region is to sequence cDNAs, DNA copies of expressed mRNAs. These are relatively easy to produce, and many cDNA libraries are available. Each represents the pattern of gene expression of the particular tissue or sample from which the original mRNA was obtained. In sequencing a cDNA, one knows one is dealing with an expressed gene, therefore a functional gene. This is a considerable advantage over genomic sequencing where one has no knowledge a priori that a particular gene found at the DNA level is actually ever used by the organism. With cDNA sequencing, one is always examining genes or nearby flanking sequences. This is another great advantage over genomic sequencing where, even in the best of cases, most of the sequence will not be coding. However, there are some potential difficulties with projects to examine massive numbers of cDNA sequences, as we will demonstrate.

SEQUENCE-READY LIBRARIES

Today, the notion of sequencing an entire human chromosome from left to right telomere is being considered seriously at a number of Genome Centers. In some cases the plans are based on a preexisting minimum tiling set of clones. Here, as long as the set is complete and exists in a vector like a cosmid or a BAC that allows direct sequencing, the strategy is predetermined. The clones are selected and sequenced one by one by whatever method is deemed optimal at the time for 50- to 150-kb clones.

Suppose, however, that, with sequencing as the eventual goal, one wishes to create an optimal library to facilitate subsequent sequencing of any particular region deemed interesting. There are two basically similar strategies for achieving this objective. If a dense ordered library already exists in an appropriate vector, one can sequence the ends of all of the clones in a relatively easy and cost-effective manner. Since vector priming can be used, the goal is to read into the cloned insert as far as possible in a single pass of raw DNA sequencing. If this is done for all the clones, the result is a sampling of the genomic sequence (Smith et al., 1994). For example, suppose that the initial library is 20-fold redundant 50-kb cosmids. A cosmid end on average would occur every 1.25 kb. A 700-base sequence read at each end would generate a total of 28 kb of sequence. When realistic failure rates and some inevitable overlaps are considered, the result would still be roughly half the total sequence. This is sufficiently dense that almost any cDNA sequence from the region would be represented in some of the available genomic DNA sequence. Thus all sequenced cDNAs could be mapped by software sequence comparisons without the need for any additional experiments. The average spacing between sequenced genomic regions would be short enough so that PCR primers could be designed to close any of the gaps by cycle sequencing.

For many targets, however, there is no existing clone map. The effort to create one de novo is considerable, even by the enhanced methods described in Chapter 9. For this reason, as automated DNA sequencing becomes more and more efficient, strategies that avoid the construction of a map altogether become attractive. One recent proposal for such a scheme also relies on the sequencing of the ends of the clones (Venter et al., 1996). Consider, for example, an ordered tenfold redundant BAC library of the human genome. With 150-kb inserts, 200,000 BACs are required. If each of these is sequenced for 500 bp from both ends, the resulting data set will contain 400,000 sequence reads encompassing 200 Mb of DNA. On average, the density of DNA sequence is a 500-bp block every 7.5 kb. Once created, such a resource would serve two functions. Many cDNAs would still match up with a segment of BAC sequence, and they could serve to correlate the BAC library with other existing genome resources and information. The utility of the BACs in this regard could be improved if, for example, they were created so that their ends had a bias to occur in coding sequence. However, even in the absence of cDNA information, the BACs will serve as a starting point for the genomic sequencing of any region of interest. One could choose any BAC that corresponds to the region of interest and sequence it completely. Then, by inspection, the BACs in the library that overlapped least with the first sequenced BAC could be picked out and used for the next round of sequencing. The process would continue until the region of interest were completed. In this way the sequencing project itself would create the minimum tiling set of BACs needed for the region.

SEQUENCING cDNA LIBRARIES

Usually cDNA libraries are made by a scheme like that shown in Figure 11.16. To prepare high-quality cDNAs, it is important to start with a population of intact mRNAs. This
is not always easy; mRNAs are very susceptible to cleavage by endogenous cellular ri-
bonucleases, and some tissues or samples are very rich in these enzymes. Most eukaryotic
mRNAs have several hundred bases of A at their 3′-end. This poly A tail can be used to
capture these mRNAs and remove contaminating rRNA, tRNA, and other small cytoplas-
mic and nuclear RNAs. Unfortunately, one also loses that fraction of mRNAs that lack a
poly A tail. An oligo-T primer can then be used with reverse transcriptase to make a DNA
copy of the mRNA strand. Alternatively, random primers can be used to copy the
mRNAs, or specific primers can be used if one is searching for a particular mRNA or
class of mRNAs. There are two general methods to convert the resulting RNA-DNA duplexes into cDNAs. Left to their own devices, some reverse transcriptases will, once the
RNA strand is displaced or degraded, continue synthesis, after making a hairpin, until
they have copied the entire DNA strand of the duplex. As shown in Figure 11.16*a,* S1 nuclease can then be used to cleave the hairpin and generate a cloneable end. Unfortunately,
the S1 nuclease treatment can also destroy some of the ends of the cDNA. An alternative
procedure is to use RNase H to nick the RNA strand of the duplex. The resulting nicks
can serve as primer for DNA polymerases like *E. coli* DNA polymerase I. This eventually
leads to a complete DNA copy except for a few nicks which can be sealed by DNA lig-

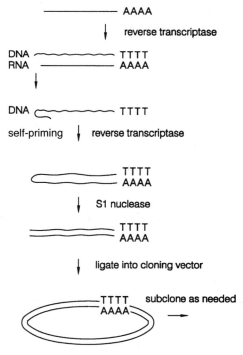

Figure 11.16   Approaches to the construction of cDNA libraries: Use of S1 nuclease to generate
clonable inserts.

ase. This procedure is generally preferred over the S1 nuclease method because it tends to produce longer, more intact cDNAs.

An unfortunate fact about many cDNA clones is that they are biased toward the 3′-end of the original message because of the poly A capture and the oligo-T priming used to prepare them. The true 5′-end is often missing and needs to be found in other clones or in the genome. Some attempts have been made to take advantage of the specialized cap structure at the 5′-end of eukaryotic mRNAs to purify intact molecules. One possibility is to try to produce high-affinity monoclonal antibodies specific for this cap structure. More effective has been the use of an enzyme called tobacco pyrophosphatase. This cleaves off the cap to leave an ordinary 5′-phosphate-terminated DNA strand that then can serve as a substrate in a ligation reaction, which can be used to add a known sequence. This known sequence will serve as the staging site for subsequent PCR amplification. Several different Japanese groups have recently perfected such strategies to the point where 5′-end-containing cDNA libraries can now be made quite reliably.

DEALING WITH UNEVEN cDNA DISTRIBUTION

With relatively rare exceptions like rDNAs, genes in the genome are in approximately a 1:1 ratio. In contrast, the relative amount of mRNAs present in a typical cell extends over a range of more than 10^5. This leads to very serious biases in most cDNA libraries. These will tend to be overrepresented with a relatively small numbers of different high-frequency clones. In addition existing cloning methods will tend to bias the libraries toward short mRNAs. If one attempts to sequence cDNAs at random from a library, in most cases the high copy number clones will be re-sequenced over and over again, while most rare mRNAs will never be sampled. It is important to stress that the problems of random selection and library biases seriously interfere with genomic DNA sequencing projects, even though one is starting with an almost uniform sample of the genome. With cDNAs these problems are much more serious and must be dealt with directly and forcefully.

One simple approach to systematic sequencing of cDNA libraries is shown in Figure 11.17. One starts with an arrayed library. A small number of clones, say 100, are selected and sequenced. All of the sequenced clones are pooled, labeled, and hybridized back to

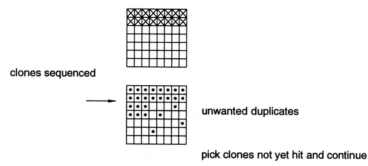

clones sequenced

unwanted duplicates

pick clones not yet hit and continue

Figure 11.17 Basic scheme for sequencing an arrayed cDNA library, and periodically screening the library to detect repeats of clones that have already been sequenced. The schematic array shown has only 56 clones; a real array would have tens to hundreds of times more.

the library. Any clones that are detected by this hybridization are not included in the next set of 100 to be sequenced. By continuing in this way, most duplication can be avoided. Unfortunately, there will also be a tendency to discard cDNAs from gene families, so many of the members of these families will be underrepresented or missed. As an alternative to handling the clones as arrays, one can carry out a solution hybridization of the entire cDNA library with an excess of sequenced clones, discard all the samples that hybridize, and regrow the remainder. This effectively replaces a screen by a physical selection process.

A more complex approach to compensating for uneven cDNA distribution is to try to equalize or normalize the library. The distribution of mRNAs in a typical cell is shown in Table 11.1. Roughly speaking there are three classes of messages: a few very common species, then approximately equal total amounts of species 20 times more rare, and species another factor of 20 rarer still. The goal is to try to even out these differences. The approach used is to anneal the library to itself and remove all the double-stranded species that are formed. We will do this by allowing the reannealing to occur at a very high C_0t: Typically $C_0t = 250$ is used. From Chapter 3, we can write for the fraction of single strand remaining in a hybridization:

$$f_s = \frac{1}{1 + n\, C_0t\, k_2/2N} = \frac{1}{1 + C_0t/C_0t^{1/2}} = \frac{C_0t^{1/2}}{C_0t^{1/2} + C_0t}$$

where all the quantities in this equation have been defined in Chapter 3. When $C_0t \gg C_0t^{1/2}$, we can approximate this result as

$$f_s = \frac{C_0t^{1/2}}{C_0t}$$

Note that the $C_0t^{1/2}$ value for a given sequence depends on the ratio of genome size, N, and the number of times the sequence is represented, n. For a cDNA library, N is the total complexity of the DNA sequences represented in the library, and n is the number of times a given sequence is represented. Thus, for highly frequent cDNAs, N/n will be small so that the $C_0t^{1/2}$ will be small, and these species will renature relatively more rapidly. Note that the amount of a particular cDNA remaining after extensive annealing will be proportional to its original abundance n and to its hybridization rate, which will scale as N/n. Thus, at very long times in the reaction, a relatively even distribution of cDNAs should be produced. We can evaluate the expected results for an attempt to normalize the typical cell mRNAs shown in Table 11.1. This is given in Table 11.2.

TABLE 11.1   Distribution of mRNA in a Typical Cell

Species	Percent	Number of Species	Relative Frequency	$C_0t^{1/2}$
Common	10	10	330	0.08
Medium	45	1000	15	1.7
Rare	45	15,000	1	25.

TABLE 11.2   Effect of Self-Annealing a cDNA Library

Class	f_s at $C_0t = 250$	Initial Frequency	Equalized Frequency
Common	3×10^{-4}	330	9.9×10^{-2}
Medium	7×10^{-3}	15	1.0×10^{-1}
Rare	9×10^{-2}	1	9.0×10^{-2}

The predictions in Table 11.2 look very encouraging. However, a serious potential problem is that one has to discard most of the library in order to achieve this result. PCR or efficient recloning must be used to recover the cDNA clones which have not self-annealed.

An actual scheme for efficient cDNA normalization is shown in Figure 11.18. This has been developed by Bento Soares, Argiris Efstratiadis, and their collaborators. It is designed to avoid the preferential loss of long cDNA clones during the self-annealing, and also to avoid the loss of cDNAs from closely related gene families. Long clones would be

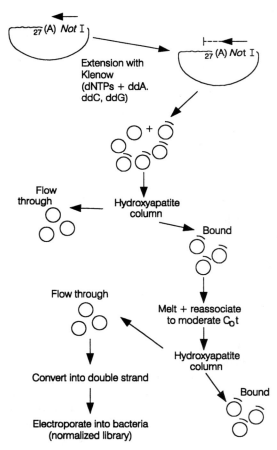

Figure 11.18   A relatively elaborate scheme for cDNA normalization that attempts to prevent a bias against shorter cDNAs and the loss of cDNAs from gene families. Adapted from Soares et al. (1994).

TABLE 11.3 cDNA Library Normalization

Probes	Original Library	Normalized (HAP-FT)	HAP-Bound
Human C_0t 1 DNA	10%	2%	2.6%
Elongation factor 1a	4.6%	0.04%	3.7%
a-Tubulin	3.7%–4.4%	0.045%	6%
b-Tubulin	2.9%	0.4%	0.85%
Mitochondrial 16S rRNA	1.3%	1%	
Myelin basic protein	1%	0.09%	
g-Actin	0.35%	0.1%	1.3%
Hsp 89	0.4%	0.05%	<0.14%
CH13-cDNA#8	0.009%	0.035%	

Source: Data adapted from Soares et al. (1994).

lost if entire cDNA sequences were used for hybridization because, as described in Chapter 3, their rate constants for duplex formation (k_2's) are larger and because there are more places to nucleate potential duplex. The trick used in the scheme of Figure 11.18 is to start with single-stranded cDNA clones and produce a short duplex at the 3'-end of each cDNA clone by primer extension in the presence of chain terminators. By focusing on this region, one will ensure that the new strands synthesized preferentially come from 3' noncoding flanking regions where even closely related genes have significant divergence, since the sequences are not translated and presumably have little function. Any cDNAs that have not successfully templated the synthesis of a short duplex are discarded by chromatography on hydroxyapatite, which specifically binds only duplexes, under the conditions used. These duplex-containing clones are then eluted, melted, and allowed to self-anneal to high C_0t. Now any clones with duplexes are removed, and the clones that have remained as single strands represent the normalized library. These are then amplified and sequenced.

Some actual results using the scheme of Figure 11.18 are given in Table 11.3. It is apparent that the equalization is far from perfect. However, it represents a major improvement over nonnormalized libraries, and materials made in this way are currently being used extensively for cDNA sequencing. Two additional schemes for cDNA normalization are described in Box 11.1. It is not clear at the present time just which schemes will ultimately be widely adopted.

LARGE-SCALE cDNA SEQUENCING

In the past three years at least five separate efforts have been made to collect massive amounts of cDNA sequence. One of these is a collaboration between the Institute for Genome Research (TIGR) and Human Genome Sciences, Inc. At least initially, this effort took an anatomical approach. Libraries of cDNAs from as many different major tissues as possible were collected, and large numbers of clones from each of these were sequenced. The second approach was orchestrated by Incyte Pharmaceuticals, Inc. Here the emphasis was on cell physiology. Sets of cDNA libraries were collected from pairs of cells in known, related physiological states, such as activated or unactivated macrophages. A fixed number of cDNAs, 5000 in the earliest studies, was randomly selected for each of the cell pairs and sequenced. In this way information was obtained about the frequencies of common cDNAs in addition to the sequence information from all classes of cDNAs.

BOX 11.1
ALTERNATE SCHEMES FOR NORMALIZATION OF cDNA LIBRARIES

Two different schemes for the production of normalized cDNA libraries have been described. The first, proposed by Sherman Weissman and coworkers, is shown schematically in Figure 11.19. First PCR is used to amplify cloned cDNAs. Then, as in the Soares and Efstratiadis scheme described in the text, hydroxylapatite fractionation is used to deplete a reaction mixture of double-stranded products. Next a nested set of PCR primers is used to amplify the single-stranded material that survives hydroxyapatite. Finally this material is cloned to make the normalized library. A survey of typical results is given in Table 11.4.

The scheme developed by Michio Oishi is quite different (Fig. 11.20). Here cDNA immobilized on microbeads is annealed to a vast excess of mRNA from the same source. Under these conditions the kinetics of hybridization become pseudo–first order as described in Chapter 3. The highly overrepresented components in the mRNA will actually deplete the corresponding cDNAs below the level of normalization. The resulting cDNA library will be enriched for rare cDNA sequences. A survey of typical results is given in Table 11.5.

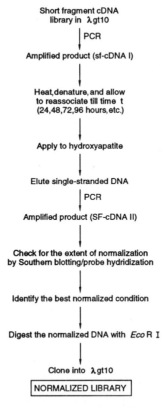

Figure 11.19 A relatively simple scheme for cDNA normalization. Adapted from Sankhavaram et al. (1991).

(continued)

BOX 11.1 *(Continued)*

TABLE 11.4   Effect of Normalization on a cDNA Library

	Number of Clones Identified per 100,000 Plaques		
Probe	STH	NSTH I	NSTH II
R-DNA	30,000	94	12
Blur-8	800	450	360
γ-actin	110	37	NT
HLA-H	104	80	10
CD4	28	37	12
CD8	15	55	12
Oct-1	9	NT	8
β-globin	7	NT	10
c-myc	5	NT	11
TCR	5	NT	8
TNF-α	5	NT	6
α-fodrin	3	NT	9

Source: From Patanjali et al. (1991).

Note: cDNAs present at various levels of abundance in STH library become almost identically abundant in the normalized (NSTH) libraries. Increased reassociation times, as indicated by the increased C_0t value, render better normalized libraries. NT, not tested. For NSTH I the C_0t value was 41.7 mol-s/liter, and for NSTH II the C_0t value was 59.0 mol-s/liter.

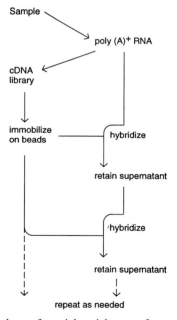

Figure 11.20   A scheme for the preferential enrichment of rare cDNAs. Adapted from Sasaki et al. (1994a).

(continued)

BOX 11.1 (*Continued*)

TABLE 11.5   Change of the Proportion of cDNA Clones Before and After Self-Hybridization

Probe	Input[a]	Before	Percentage	After	Percentage	After/Before
Rabbit β-globin	1	111/10,500	1.067	5/55,000	0.009	0.086
φX174*Hae* III 0.6 kb	0.01	2/30,000	0.0067	8/35,000	0.023	3.43
φX174*Hae* III 0.9 kb	0.01	1/30,000	0.0033	6/30,000	0.02	6
neo[r]	0.0001	0/250,000	<0.0004	2/250,000[b]	0.0008	>2
β-actin		54/10,000	0.54	2/10,000	0.02	0.037
IL-4		0/320,000	<0.0003	3/320,000[b]	0.0009	>3
IL-2		0/320,000	<0.0003	0/320,000	<0.0003	—

Source: Adapted from Sasaki et al. (1994).

[a]Percent of total RNA (w/w).

[b]The positive clones were confirmed by sequencing approximately 300 bp of the inserts.

The frequency information, when pairs of cells are compared, is often quite interesting, and it suggests potential functional roles for a number of newly discovered genes in the libraries. Incyte has termed such comparisons transcript imaging. A third large-scale cDNA sequencing effort is the Image consortium involving several academic or government laboratories and Merck, Inc. Here normalized libraries are serving as the source of clones for sequencing, and the goal is to collect at least one representative cDNA sequence from all human genes.

At present, the sequencing of human cDNAs is being carried out in large laboratory efforts like the three just described as well as many smaller, more focused efforts. Within each laboratory the amount of duplication seen thus far has been relatively small. Thus the early course of the cDNA sequencing strategy appears to be very effective. At what point it will peter out into a morass of duplicate clones is unknown. It is also really unclear what fraction of the total amount of genes will actually be found first through their cDNAs. The tissues used for the majority of these studies are those where large numbers of different genes are expected to be active. These include early embryos, hytidaform moles, which are differentiated but disordered tumors with many different tissue types, liver, and a number of parts of the brain. Whether many specialized tissues will have to be looked at to get genes expressed only in these tissues, or whether there is a broad enough low-level synthesis of almost any mRNA in one or more of the common tissues to let all genes be found there, is an issue that has not yet been answered. One way to try to extend the cDNA approach to find all of the human genes is described in Box 11.2

BOX 11.2
PREPARATION AND USE OF hncDNAs

A major purpose of making ordered libraries is to assist the finding and mapping of genes. Eugene Sverdlov and his colleagues in Moscow have developed an efficient procedure for preparing chromosome-specific hncDNA libraries. Their method is an elaboration of the scheme originally described by Corbo et al. (1990). The procedure is outlined in Figure 11.21. It uses an initial Alu-primed PCR reaction to make an hncDNA copy of the hnRNA produced in a hybrid cell containing just the chromosome of interest. (See Chapter 14 for details about the Alu repeat and Alu-specific PCR primers.) The resulting DNA is equipped with an oligo-G tail, and then a first round PCR is carried out using an oligo-C containing primer and an Alu primer. Then a second round of PCR is done with a nested Alu primer. The PCR primers are also designed so that the first round introduces one restriction site and the second round another. The resulting products are then directionally cloned into a vector requiring both restriction enzyme cleavage sites. In studies to date, Sverdlov and his coworkers have found that this scheme produces a diverse set of highly enriched human cDNAs. Since these come from Alu's in hnRNA, they will contain introns, but they can be used to locate genes on the chromosome equally well if not better than conventional cDNAs.

Note that the hncDNA clones as produced by the Sverdlov method actually contain substantial amounts of intronic regions. This means that they will be more effective than the ordinary cDNAs in dealing with gene families and in avoiding cross-hybridization with conserved exonic sequences in rodent-human cell hybrids.

(continued)

BOX 11.2 *(Continued)*

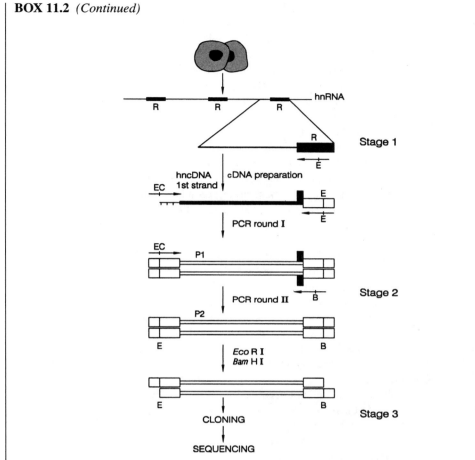

Figure 11.21 Steps involved in making an hncDNA library. Interspersed repeat elements in hnRNA are represented in the upper line by solid boxes (*R*). Arrows indicate primers. Vertical lines crossing the arrows symbolise the primers with sites for *Eco*R I and *Bam*H I restriction endonucleases (*E*) or (*B*). EC is 5′GAGAATTC(C)203′. The open boxes with similar symbols represent sequences corresponding to primers that are included in PCR products P-1 and P-2. Provided by Eugene Sverdlov.

WHAT IS MEANT BY A COMPLETE GENOME SEQUENCE?

If the strictest definition is used for a complete genome sequence, namely every base on every chromosome in a cell has been identified, then it is probably safe to say that we will never accomplish this. This is not to say that the task couldn't be accomplished in principle; it could be, but for several reasons it is a foolish task, at least for the human genome. The human genome is quite variable. This will be discussed in more detail later. Suffice it to say here that there are millions of differences in DNA sequence between the set of two homologous chromosomes in a diploid cell. Unless one could separate these into separate, cloned libraries, inevitable confusion will develop as to which homolog one is on. Hybrid cells make these separations for us, and libraries made from chromosomes of hybrid cells are major candidates for eventual large-scale sequencing. However, because of the history behind the construction of such hybrids, we rarely have separate clones of two homologous chromosomes from a single individual. Even more important, most different single chromosome hybrid cell lines have been made from different individuals. So the real answer to the often asked question "Who will be sequenced in the human genome project" is that the sequence will inevitably represent a mosaic of many individuals. This is probably quite appropriate, given the global nature and implications of the project.

There are other bars to total genome sequencing, or even total sequencing of a given chromosome. We have indicated many times already that closure in mapping is a difficult task; closure in large-scale sequencing projects will also be extremely difficult. For whatever reason, there are bound to be a few regions in any chromosome that cannot be cloned by any of our existing methods, or that may not be approachable even by PCR or genomic sequencing. Sequences with very peculiar secondary structures or sequences toxic to the enzymes or cells we must rely on could lead to this kind of problem. The issue of how much effort should be devoted to a few missing stretches has not yet been forced upon us, but placed in any kind of reasonable perspective, it cannot have high priority relative to more productive use of large-scale sequencing facilities.

Finally, some regions of chromosomes are either very variable or dull, at least at the level of fine details. Examples are long simple sequence or tandemly repeating sequence repeats. Human centromeres appear to have millions of base pairs of such repeats. Other heterochromatic regions are occasionally seen on certain chromosomes. Some of these regions show quite significant size variation within the human population. For example, a case is known of an apparently healthy individual with one copy of chromosome 21 that is 50% longer than average. Surely we will not select these extra long variants for initial mapping or sequencing projects. However, the key point is that extensive, large-scale sequencing of simple repeats does not seem to be justified at the present time by any hints that this large amount of sequencing data will be interesting or interpretable. Furthermore our current methods are actually incapable of dealing with such regions of the genome. Thus we will almost certainly have to claim completeness, missing the centromeres and certain other unmanageable genome regions.

SEQUENCING THE FIFTH BASE

When we have sequenced all of the cloned DNAs from each human single chromosome library, we will not have the complete DNA sequence of an individual, for the reasons cited above. Some of the troublesome regions are almost certainly not in our libraries or, if they are present, they probably represent badly rearranged remnants of what was actually in the genome. If we ever want to look at such sequences in their native state, we may have to sequence them directly from the genome. For the reasons cited above, this may not be a terribly interesting or useful thing to do. However, there is a tremendously important additional reason to perfect methods for direct genomic mammalian sequencing. This is to look at the fifth base, ^{m}C, which is lost in all common current cloning systems. It is also lost in PCR amplification. Therefore, to find the location of ^{m}C in mammalian or other higher eukaryotic DNA sequences, it is necessary to immortalize the positions of these residues before any amplification.

PCR can be used to determine DNA sequence directly from genomic samples with mammalian complexity by a ligation technique that is shown in Figure 11.22a. The ge-

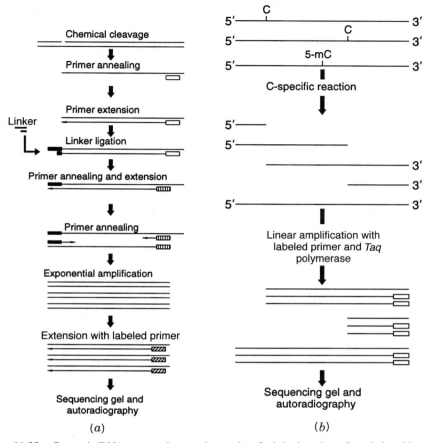

Figure 11.22   Genomic DNA sequencing can be used to find the location of methylated bases. (*a*) Basic scheme for PCR-mediated genomic Maxam-Gilbert DNA sequencing. (*b*) A methylated base prevents normal Maxam-Gilbert cleavage and thus alters the appearance of the DNA sequencing ladder. Adapted from Riggs et al. (1991).

nomic DNA is fragmented chemically as in the Maxam-Gilbert procedure. A sequence-specific primer is annealed to the DNA after this cleavage and extended up to the site of the cleavage. This creates a ligatable end. A splint is ligated onto all the genomic DNA pieces. Then primers are used to extend the DNA past the unique region of the splint, as we described earlier for single-sided PCR reactions in Chapter 4. Finally two primers are used for exponential amplification. The sizes of the amplified products reveal where the original cleavages were in the DNA. If all bases in the genome are accessible to the specific cleavage reactions used, then a perfectly normal sequencing ladder should result.

To find the location of ^{m}C in genomic sequence, one takes advantage of the fact that this base renders DNA resistant to cleavage by the normal C-specific reaction used in the Maxam-Gilbert method. The result, is that the locations of the methylated C's drop out of the sequencing ladder (Fig. 11.22*b*). If the target sequence is already completely known except for the sites of methylation, this is all the information one needs. If not, one can always repeat the sequencing on a cloned sample to confirm the location of the ^{m}C's by their new appearance as C's. This method is quite powerful; it is providing interesting insights into the patterns of DNA methylation in cell differentiation and in X-chromosome inactivation.

SOURCES AND ADDITIONAL READINGS

Ansorge, W., Voos, H., and Zimmermann, J., eds. 1995. *DNA Sequencing: Automated and Advanced Approaches.* New York: Wiley.

Azhikina, T., Veselovskaya, S., Myasnikov, V., Potapov, V., Ermolayeva, O., and Sverdlov, E. 1993. Strings of contiguous modified pentanucleotides with increased DNA-binded affinity can be used for DNA sequencing by primer walking. *Proceedings of the National Academy of Sciences USA* 90: 11460–11462.

Beskin, A. D., Sonkin-Zevin, D., Sobolev, I. A., and Ulanovsky, L. E. 1995. On the mechanism of the modular primer effect. *Nucleic Acids Research* 23: 2881–2885.

Borodin, A., Kopatnzev, E., Wagner, L., Volik, S., Ermolaeva, O., Lebedev, Y., Monastyrskaya, G., Kunz, J., Grzeschik, K.-H., and Sverdlov, E. 1995. An arrayed library enriched in hncDNA corresponding to transcribed sequences of human chromosome 19: Preparation and analysis. *Genetic Analysis* 12: 23–31.

Church, G. M., and Kieffer-Higgins, S. 1988. Multiplex DNA sequencing. *Science* 240: 185–188.

Gordon, D., Abajian, C., and Green, P. 1998. Consed: A graphical tool for sequence finishing. *Genome Research* 8: 195–202.

Henikoff, S. 1984. Unidirectional digestion with exonuclease III creates targeted breakpoints for DNA sequencing. *Gene* 28: 351–359.

Hillier, L., et al. 1996. Generation and analysis of 280,000 human expressed sequence tags. *Genome Research* 6: 807–828.

Hunkapiller, T., Kaiser, R. J., Koop, B. F., and Hood, L. 1991. Large-scale and automated DNA sequence determination. *Science* 254: 59–67.

Kieleczawa, J., Dunn, J. J., and Studier, W. F. 1992. DNA sequencing by primer walking with strings of contiguous hexamers. *Science* 258: 1787–1791.

Kotler, L., Sobolev, I., and Ulanovsky, L. 1994. DNA sequencing: Modular primers for automated walking. *BioTechniques* 17: 554–558.

Kozak, M. 1996. Interpreting cDNA sequences: Some insights from studies on translation. *Mammalian Genome* 7: 563–574.

Li, C., and Tucker, P.W. 1993. Exoquence DNA sequencing. *Nucleic Acids Research* 21: 1239–1244.

Makatowski, W., Zhang, J., and Boguski, M. S. 1996. Comparative analysis of 1196 Orthologous mouse and human full-length mRNA and protein sequences. *Genome Research* 6: 846–857.

McCombie, W. R., and Kieleczawa, J. 1994. Automated DNA sequencing using 4-color fluorescent detection of reactions primed with hexamer strings. *BioTechniques* 17: 574–579.

Ohara, O., Dorff, R. L., and Gilbert, W. 1989. Direct genomic sequencing of bacterial DNA: The pyruvate kinase I gene of *Escherichia coli. Proceedings of the National Academy of Sciences USA* 86: 6883–6887.

Patanjali, S. R., Parimod, S., and Weissman, S. M. 1991. Construction of a uniform-abundance (normalized) cDNA library. *Proceedings of the National Academy of Sciences* 88: 1943–1947.

Raja, M. C., Zevin-Sonkin, D., Shwartburd, J., Rozovskaya, T. A., Sobolev, I. A., Chertkov, O., Ramanathan, V., Lvovsky, L., and Ulanovksy, U. 1997. DNA sequencing using differential expression with nucleotide subsets (DENS). *Nucleic Acids Research* 25: 800–805.

Riggs, A., Saluz, H. P., Wiebauer, K., and Wallace, A. 1991. Studying DNA modifications and DNA-protein interactions in vivo. *Trends in Genetics* 7: 207–211.

Saluz, H. P., Wiebauer, K., and Wallace, A. 1991. Studying DNA modifications and DNA-protein interactions in vivo. *Trends in Genetics* 7: 207–211.

Sasaki, Y. F., Ayusawa, D., and Oishi, M. 1994a. Construction of a normalized cDNA library by introduction of a semi-solid mRNA-cDNA hybridization system. *Nucleic Acids Research* 22: 987–992.

Sasaki, Y. F., Iwasaki, T., Kobayashi, H., Tsuji, S., Ayusawa, D., and Oishi, M. 1994b. Construction of an equalized cDNA library from human brain by semi-solid self-hybridization system. *DNA Research* 1: 91–96.

Schuler, G.D., et al. 1996. A gene map of the human genome. *Science* 274: 540–546.

Smith, M. W., Holmsen, A. L., Wei, Y. H., Peterson, M., and Evans, G. A. 1994. Genomic sequence sampling: A strategy for high resolution sequence-based physical mapping of complex genomes. *Nature Genetics* 7: 40–47.

Soares, M. B., Bonaldo, M. D. F., Jelene, P., Su, L., Lawton, L., and Efstratiadis, A. 1994. Construction and characterization of a normalized cDNA library. *Proceedings of the National Academy of Sciences USA* 91: 9228–9232.

Strausbaugh, L. D., Bourke, M. T., Sommer, M. T., Coon, M. E., and Berg, C. M. 1990. Probe mapping to facilitate transposon-based DNA sequencing. *Proceedings of the National Academy of Sciences USA* 87: 5213–6217.

Studier, F. W. 1989. A strategy for high-volume sequencing of cosmid DNAs: Random and directed priming with a library of oligonucleotides. *Proceedings of the National Academy of Sciences USA* 86: 6917–6921.

Szybalski, W. 1990. Proposal for sequencing DNA using ligation of hexamers to generate sequential elongation primers (SPEL-6). *Gene* 90: 177–178.

Venter, J. C., Smith, H. O., and Hood L. 1996. A new strategy for genome sequencing. *Nature* 381: 364–366.

12 Future DNA Sequencing without Length Fractionation

WHY TRY TO AVOID LENGTH FRACTIONATIONS?

All but one of the methods we have described for DNA sequencing in Chapter 10 involved a gel electrophoretic fractionation of DNA. The exception used mass spectroscopy instead of electrophoresis, but a length fractionation was still needed because all of the information about base location had been translated into fragment sizes prior to analysis. There are several motivations to try to get away from this basic paradigm and develop DNA sequencing methods that do not depend on size fractionation. First, size fractionation, except by mass spectroscopy, is really quite a slow and indirect method of reading sequence data. Second, fractionations are intrinsically hard to parallelize. Third, it is not obvious how fractionation methods could ever be used to look efficiently just at sequence differences, and most of the long-term future of DNA sequencing applications probably lies in this key area of differential sequencing. This is true not only in the potential use of DNA sequencing for human diagnostics, but also for evolutionary applications, for population genetic applications, and for ecological screening.

For all of these reasons there is considerable current interest in trying to develop entirely new approaches to DNA sequencing. Many of the techniques that will be mentioned in this chapter are going to be discussed only briefly. While they are probably capable of maturing into methods that can read DNA sequences, they are unlikely to do this soon enough, or ultimately efficiently enough, to be of much use for large-scale DNA sequence processing. However at least one of the second-generation methods treated in this chapter, sequencing by hybridization (SBH), does appear to offer a significant chance of making an impact on the current human genome project, and an even better chance of making a major impact on future DNA sequencing in clinical diagnostics.

SINGLE-MOLECULE SEQUENCING

A number of different potential DNA sequencing methods require that data be obtained from one single molecule at a time. They include handling DNAs or their reaction products in flow systems, or observing DNAs by microscopy. One approach that has been investigated by Richard Keller, would use an exonuclease to degrade a tethered DNA continually, and detect individual nucleotides as they are cleaved by the enzyme and liberated into a flowing stream. This method exploits the power of flow cytometry for very sensitive detection of a fluorescent target whose location is known rather precisely. A schematic illustration of this approach to single-molecule sequencing is given in Figure 12.1.

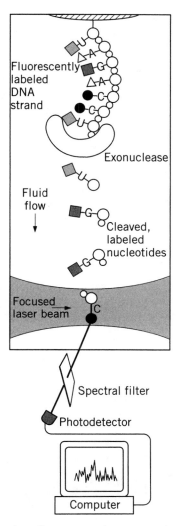

Figure 12.1 Schematic illustration of one approach to sequencing single DNA molecules in solution. Provided by Richard Keller.

To see all four bases in a single molecule, each would have to be labeled with a different fluorophore. There have been suggestions to use the intrinsic fluorescence of each base directly, but this fluorescence is weak, so it would require very sophisticated and expensive detection methods to be practical at the single-molecule limit. The challenge is to make fluorescent base analogs that are acceptable substrates for DNA polymerase. Substitution at every single nucleotide position must be accomplished. This requirement has been met with one and two bases, which is an impressive accomplishment. It remains to be seen if it can be met with all four bases. The ideal exonuclease would liberate bases in a kinetically smooth and rapid rate. It would be processive so that a single enzyme would suffice to degrade the DNA molecule. Otherwise, with the arrangement shown in Figure 12.1, there would be pauses during which no product would be appearing, and one would constantly have to replenish the supply of enzyme as molecules fall off and are lost to the flowing stream. The properties of actual exonucleases, for the most part, are not this ideal, but they are rapid, and some are reasonably processive.

It is possible in a flowing stream to detect single fluorophores of the sort used to label nucleic acid bases, which is like that which would be used with a tethered DNA. Some typical results are illustrated in Figure 12.2. The issue that remains unresolved is the chances of missing a base (a false negative) and the chances of seeing a noise fluctuation, by scattering from a microscopic dust particle, or whatever, that imitates a base when none is present (a false positive). It is interesting to examine what the consequences would be if, instead of one molecule, many were tethered together in the flowing stream, and exonucleases were allowed to process them all. If the digestion could be kept in synchrony, the signal-to-noise problem in detection would be alleviated considerably. However, there is no way to keep the reactions in synchrony. The best one can do is to find a way to start the exonucleases synchronously and use the most processive enzymes available. Even in this case, however, there is an inevitable dephasing of the reaction, as it proceeds, because of the basic stochastic (noisy) nature of chemical reactions. Given

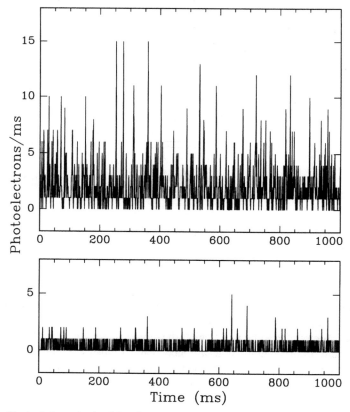

Figure 12.2   Typical data obtained in pilot experiments to test the scheme shown in Figure 12.1. Top panel: A dilute concentration of a labeled nucleotide is allowed to flow past the detector. Bottom panel: A control with no labeled nucleotides.

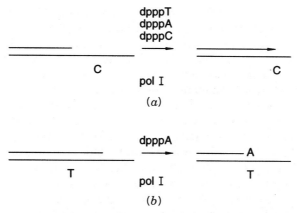

Figure 12.3 Plus-minus DNA sequencing. (*a*) Extension until a template coding for a missing dpppN is encountered. (*b*) Degradation of a chain until a specific nucleotide is reached.

some microscopic rate constant for a reaction, at the level of individual molecules the actual reaction times vary in such a way that a distribution of rates actually contributes the average value seen macroscopically. The reaction zone broadens like a random walk as the reaction proceeds down the chain. Its width soon becomes more than several bases, even under ideal kinetic circumstances.

One way to circumvent the statistical problems with sequencing by exonucleases would be to find a method to stop the reaction at fixed points and then allow it to restart. This, in essence, is what is done with DNA polymerase in plus-minus sequencing, the very earliest method used by Ray Wu. If one dpppN is left out, the reaction proceeds up to the place where this base is demanded by the template, and it stalls there (Fig. 12.3*a*). Adding the missing dpppN then allows the reaction to continue. If a DNA polymerase with a 3′-editing exonuclease activity is used, a similar result can be achieved by having only one dpppN present. In this case the enzyme degrades the 3′-end of a DNA chain, until it reaches a place where the dpppN present is called for by the template (Fig. 12.3*b*). As long as a sufficient supply of dpppN remains, the enzyme will stall at this position. These are useful tricks; they work well for sequencing, and there is no reason why they could not be incorporated into strategies for sequencing a single molecule or small numbers of molecules. However, the major potential advantage of the original scheme proposed by Keller is speed, and steps that require changing substrates numerous times are likely to slow down the whole process considerably.

SEQUENCING BY HIGH-RESOLUTION MICROSCOPY

One of the earliest attempts at the development of alternate DNA sequencing methods was Michael Beer's strategy for determining nucleic acid sequence by electron microscopy. Beer's plan was to label individual bases with particular electron-dense heavy metal clusters and then image these. Two problems made this approach unworkable. First, the nucleic acids were labeled by covalent modification after DNA synthesis. This leads to less than perfect stoichiometry of the desired product, and it undoubtedly also leads to some unwanted side reactions with other bases. The second problem is that sample damage in the conventional electron microscope is considerable; this makes it very difficult to achieve accurate enough images to read the DNA sequence as the spacing between metal-tagged bases, since the structure moves around in response to the electron beam of the molecule. This problem of molecular perturbation by microscopes remains with us today as the greatest obstacle to using high-resolution microscopy for DNA sequencing.

Currently a new generation of ultramicroscopes has reopened the issue of whether DNA could be sequenced, directly or indirectly, by looking at it. The new instruments are scanning tip microscopes; the best studied of these are the scanning tunneling microscope (STM) and the atomic force microscope (AFM). Both of these instruments read the surfaces of samples in much the same way that a blind person reads braille. The surface is scanned with a sharp tip in a raster pattern, as shown schematically in Figure 12.4. In AFM what leads to the image is the force between the tip and the sample. Van der Waals forces will attract the tip to the surface at long distances and repel the tip at short distances. What is usually done is to have a feedback loop via a piezoelectric device. This can be used to place a force on the tip to keep its vertical position constant, and the voltage needed to accomplish this is measured. Alternatively, one can apply a constant force, and measure the vertical displacement of the tip, for example, by bouncing a laser beam off the tip and seeing where it is reflected to. In STM an electrical potential is maintained between the tip and the surface. This leads to a current flow from the tip to the surface at short distances. The current is dependent on the distance between the tip and the surface, and the electrical properties of the surface. In practice, one can adjust the position of the tip to maintain a constant current, and measure the tip position, or keep the vertical height of the tip constant and measure the current.

For AFM or STM to be successful, very flat surfaces are required. With hard samples on such surfaces, atomic resolution is routinely observed, and even subatomic resolution has been reported, where information about the distribution of electron density within the sample is uncovered. DNA is not a hard sample, and it does not easily adhere to most very flat surfaces. These difficulties have produced many frustrations in early attempts to image DNA by AFM or STM. In retrospect, most or all of the spectacular early pictures of DNA have been artifacts, perhaps caused by helixlike imperfections in the underlying surfaces. The best that can be said is that images that looked like DNA were rare and far between, and not generally reproducible. One problem that soon became quite apparent is that the forces used in these early attempts were sufficient in most, if not all, cases to knock the DNA molecules off the surface being imaged.

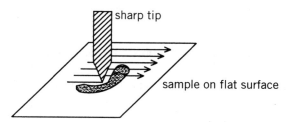

Figure 12.4   Operating principle of a typical scanning tip microscope. In STM the electrical current between the tip and the surface is measured. In AFM the repulse force between the tip and the surface is measured.

More recent attempts to image DNA with scanning tip microscopes, particular with AFM, have been more successful, at least in the sense that dense arrays of molecules can be seen reproducibly. This is accomplished by using surfaces to which DNA adheres better, like freshly cleaved mica, instead of the graphite or silicon surfaces used earlier. Sharper tips give high enough resolution to be able to measure the lengths of the molecules reliably. The current images are, however, a long way from the resolution needed to read the sequence directly by looking at the bases. A number of severe obstacles will have to be overcome if this is ever to be done. First, the current images are mostly of double-stranded DNA. This is understandable since it is a much more regular structure, much more amenable to detection and quantitation in the microscope. However, in the double strand, only short bits of sequence are readable from the outside, as one is forced to do in AFM. This will lead to a difficult, but apparently not insurmountable, sequence reconstruction problem where data from many molecules will have to be combined to synthesize the final sequence. A second problem is that the DNA molecules could still be distorted quite a bit as the tip of the microscope moves over them. This may or may not be alleviated by newer microscope designs that would allow lower forces to be used.

A third problem with AFMs is that the image seen is a convolution of the shape of the tip and the shape of the molecule, as shown in Figure 12.5. Thus, unless very sharp tips can be made, or tips of known shape, it can be difficult with a soft, deformable molecule to deconvolute the image and see the true shape of the DNA. One approach to circumvent many of these difficulties would be to label the DNA with base-specific reagents that are more distinctive either in AFM, where larger, specific shapes could be used, or in STM, where labels with different electrical properties might serve. As a test of this, and to make sure DNA imaging was now reliable in the AFM, proteins were attached to the ends of DNAs before AFM imaging. Two examples of the sorts of images seen are shown in Figure 12.6. The protein used, purely because it was available and of a size that made it easy to distinguish from DNA, was a chimera between streptavidin and a fragment of staphylococcal protein A, which was already introduced in Chapter 4, where it was used for immunoPCR. The DNA was biotinylated, either on one or both ends. Two different lengths of DNAs were used, and since streptavidin is tetrameric, the resulting images show a progression of structures from DNA monomers up to trimers. Because of the nature of these structures, the proper measured lengths of the DNAs within them, and the expected height difference between the protein label at the ends or vertices of the DNA and the DNA itself, one can be very confident that these are true images of DNA and protein. However, the resolution is still far too low to allow sequencing.

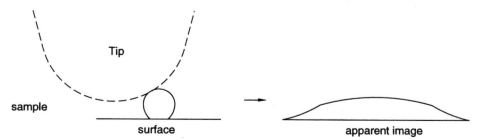

Figure 12.5 In scanning tip microscopy, what is actually measured is a convolution of the shape of the object and the shape of the tip.

Biotinylated DNA + Streptavidin

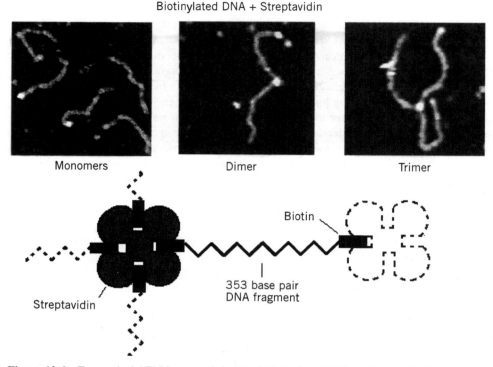

Figure 12.6   Two typical AFM images of short-end biotinylated DNA molecules labeled at one or both ends by a chimeric protein fusion of streptavidin and staphylococcal protein A. Since the streptavidin is tetrameric, one can see figures representing more than one DNA molecule bound to the same streptavidin. Reproduced from Murray et al. (1993).

It should be possible to use progressively smaller labels and to increase the resulting resolution. Whether this will lead to direct AFM DNA sequencing soon is anyone's guess. If it does, a real advantage is that one could sequence a wide variety of different molecules in a single experiment without the need to clone or fractionate. The labeling would almost certainly be introduced by PCR using analogs of the four bases. This will be much more accurate than the original chemical modification methods used for electron microscopy. However, the resulting images, as elegant as they may look some day, might have to be analyzed as images to extract the DNA sequence data. By current methods this could become a serious bottleneck. What is still needed is a way to direct the tip of the microscope so that it tracks just over the DNA molecule of interest, rather than scanning a grid that is mostly background. If this can somehow be achieved, the problem of image analysis ought to become much simpler, and the rate of image acquisition also ought to be increased considerably.

STEPWISE ENZYMATIC SEQUENCING

A major success story in the history of protein sequencing was the development of stepwise chemical degradation. Amino acid residues are removed one at a time from one end of the polypeptide chain and their identity is determined successively. Automated

Edmond degradation currently provides our main source of direct protein sequence data. The yield in each step is the critical variable, since it determines how far from the original end the sequence can be read. Comparable chemical approaches for DNA or RNA sequencing have not been terribly successful. Recently, however, several stepwise enzymatic sequencing approaches have been suggested. As individual processes, they do not at first glance seem all that attractive. However, they have the potential to be implemented in massively parallel configurations, which, if successful, could ultimately provide very high throughput. These schemes are distinct from the single molecule methods described earlier in that any desired number of target molecules of each type can be employed. Thus detection sensitivity is not an issue here.

One strategy, developed by Mathias Uhlen, is to divide the sample into four separate wells, each containing a DNA polymerase without a 3′-proofreading activity. To each well one of the four dpppN's is added. Chain extension will occur only in one well with the concomitant release of pyrophosphate (Fig. 12.7). This product can be detected with great sensitivity; for example, it can be enzymatically converted to ATP, and that can be measured using luciferase, to generate a chemiluminescent signal. The amount of light emitted is proportional to the amount of ATP made. Thus one can quantitate the amount of dpppN incorporated and determine, within limits, how many units the chain was extended by. Sample from the well that was successfully extended is then divided into four new wells, and the process is repeated. Actually three wells would suffice, since one knows that the next base is not the same as the one or ones just added, but it is probably good to have the fourth as an internal control. One obvious complication with the scheme is that the sample keeps getting divided, so one has to either start with a large amount of it or have a sensitive enough assay that only a small aliquot can be removed and assayed. A variation on this basic approach adds dpppNs in a cyclical order. This avoids the problem of sample subdivision. It appears to have considerable promise. The method is called pyrosequencing.

A second strategy is similar in spirit but uses dideoxy pppN's. This is shown in Figure 12.8. In four separate wells containing target DNA and DNA polymerase is added one of the ddpppN's carrying a label. Alternatively, one could use a single well and a mixture of four different fluorescently labeled ddpppN's. Only one of the ddpppN's becomes incorporated. From the location of the well, or the color, the identity of the base just added is known. The base just added is now removed by treating with the 3′-editing exonuclease activity of DNA polymerase I in the presence of all of the dpppN's except the one just

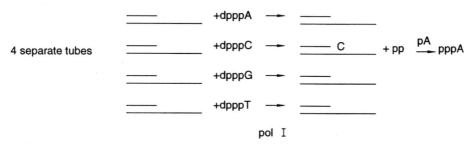

Figure 12.7 One scheme for stepwise enzymatic DNA sequencing. Here, when a particular base is added, pyrophosphatase is used to synthesize ATP from the pyrophosphate (pp) released, and the ATP in turn is used to generate a chemiluminescent signal by serving as a substrate for the enzyme, luciferase.

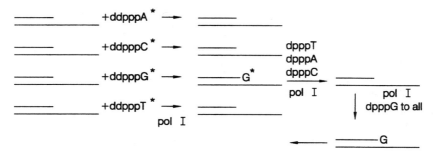

Figure 12.8   A second scheme for stepwise enzymatic DNA sequencing. This is similar in spirit to plus-minus DNA sequencing illustrated in Figure 12.3. It uses fluorescent terminators, ddpppN*, like those employed in conventional automated Sanger sequencing.

added. Next the labeled ddpN just removed is replaced by an unlabeled dpppN by using DNA polymerase with only this particular dpppN present. Then the process is repeated. This scheme avoids the sample division or aliquoting problem of the previous strategy.

To detect a run of the same base, one will have to be able to vary the scheme. What will happen in this case is that the exonuclease treatment will degrade the chain back to the location of the first base in the run. To determine the length of the run, one possible approach is to add a labeled analog of the particular dpppN involved in the run, in the presence of DNA polymerase, and detect the amount of synthesis by quantitating the incorporation of label. Then the entire block of labeled dpN's has to be removed, replaced by unlabeled dpN's, and next base after the block can now be determined. This is an unfortunate complication. However, in principle, the entire scheme could be set up in a microtitre plate format and run in a very parallel way. As in the first scheme the whole process would be best carried out in a solid state sequencing format so that the DNA could be purified away from small molecules and enzymes easily and efficiently after each step.

A third strategy, has not been tested to our knowledge, because it depends on finding a dpppNx derivative that has two special properties. Like a ddpppN the derivative must not be extendable by DNA polymerase. However, there must be a way to change the dpNx after incorporation into the DNA chain so that now it is extendable. The scheme then, as shown in Figure 12.9, is to add dpppNx to the target in four separate tubes. Whichever

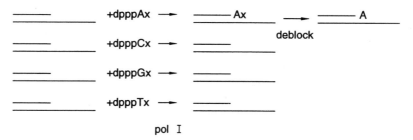

Figure 12.9   A third, as yet unrealized scheme for stepwise enzymatic DNA sequencing. Here the key ingredient would be a 3'-blocked pppNx that could be deblocked after each single base elongation step. Incorporation could be detected by pyrophosphate release as in Figure 12.7 or by the use of a fluorescent blocking agent, x.

one extends could be determined by a pyrophosphatase assay as in the first scheme. Then the incorporated dpNx is converted to dpN, and the process continued. One candidate, in principle, for the desired compound is dpppNp, which, after incorporation of the dpNp into DNA could be converted to dpN by alkaline phosphatase. This latter step certainly works; however, it is uncertain how well DNA polymerase will utilize compounds like dpppNp's. Apparently quite a few other reversible dpppN derivatives have been tried as DNA polymerase substrates without much success thus far. This is a pity because the scheme has real appeal.

In deciding how, eventually, to implement any of the above schemes in an automated fashion, one needs to consider an interesting trade-off between the time to sequence one sample and the number of samples that can be handled simultaneously. Instead of trying all four single base additions simultaneously, they could be tried one at a time, in a cyclical pattern, say A, then G, then C, then T, as in pyrosequencing. With an immobilized DNA sample, the target is simply moved from one set of reagents, after washing, to another, and the point where positive incorporation occurs is recorded. The advantage of this is that the logistics and design of the system become much simpler, particularly for the cases where pyrophosphate release is measured. It will take four times as long to complete a given length of DNA sequence, but one could handle precisely four times as many sequences simultaneously.

DNA SEQUENCING BY HYBRIDIZATION (SBH)

We will devote the rest of this chapter to a number of approaches for determining the sequence of DNA by hybridization. In all of these approaches one uses relatively short oligonucleotides as probes to determine whether the target contains the precise complementary sequences. Essentially SBH reads a word at a time rather than a letter at a time. Intuitively this is quite efficient; it is after all the way written language is usually read. For reasons that will become readily apparent, attempts to perform SBH have usually focused on oligonucleotides with 6 to 10 bases. The conception of SBH appears to have had at least four independent origins. Many groups are now working to develop an efficient practical scheme to implement SBH.

There seems to be a consensus that SBH will eventually work well for some high-throughput DNA sequencing applications, like sequence comparison, sequence checking, and clinical diagnostic sequencing. In all of these cases one is not trying to determine large tracts of sequence de novo; instead, the targets of interest are mostly small differences between a known or expected sequence, and what has actually been found. There is also agreement that SBH will work for determining partial sequences, for fingerprinting, and for mapping. However, SBH may not work for direct complete de novo sequencing unless some of the enhancements or variations that have been proposed to circumvent a number of problems turn out to work in practice.

The two critical features of SBH are illustrated in Figures 12.10 and 12.11. As we demonstrated in Chapter 3, the stability of a perfectly matched duplex is greater than an end mismatched duplex, and much greater than a duplex with an internal mismatch (Fig. 12.10). Thus a key step in SBH is finding conditions where there is excellent discrimination between perfect matches and mismatches. An immediate problem is that for a sequence of length n, there is only one perfect match, there are six possible end mismatches (each base on each end of the target can be any one of the three noncomplementary bases), and there are $3(n - 2)$ possible internal mismatches. Unless the discrimination is

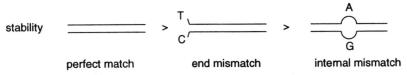

stability   perfect match   >   end mismatch   >   internal mismatch

Figure 12.10   Effect of mismatches on the stability of short DNA duplexes.

TRUESATISFACTIONISNEVER
ESATIS
TISFAC
FACTIO

Figure 12.11   Reading a text by reconstruction from overlapping *n*-tuple words.

very strong, there will be an inevitable background problem where a specific signal is diluted by a large number of weak mismatches.

The second key aspect of SBH is that the sequence can be read by overlapping words, as shown in Figure 12.11. In principle, with perfect data one would not need to try all the words to reconstruct the sequence. The problem of reconstructing a sequence from all *n*-tuple subsegments is highly overdetermined, except for the complications that we will discuss below. Simulations show that reconstructing DNA sequences from oligonucleotide hybridization fingerprints is very robust and very error resistant. Even significant levels of insertions and deletions can be tolerated without badly degrading the final sequence. A key element of SBH that is easily forgotten is that negatives as well as positives are extremely informative. Knowing that a specific oligonucleotide like AACTG-GAC does not exist anywhere in the target provides a constraint that can sometimes be quite useful in assembling the data from words that are found to be present.

BRANCH POINT AMBIGUITIES

The major theoretical limitation with simple direct implementations of SBH is the inability to determine sequences uniquely if repeating sequences are present. There are two kinds of repeats: tandemly repeated sequences and interspersed repeats. The presence of a tandemly repeated sequence can be detected, but it is difficult to determine the number of repeated sequences present. When the length of the monomer repeat sequence length is longer than the SBH words length, then the number of copies of a tandemly repeated sequence can only be determined when a hybridization signal is quantitated and not simply scored positively. These problems are relatively easily dealt with by conventional sequencing or PCR assays, since the unique sequence flanking the simple sequence or tandem repeat will generally be known.

The more serious problem caused by repeats is called a branch point ambiguity (Fig. 12.12). If the SBH word length used is *n*, these ambiguities arise whenever there is an exact recurrence of any sequence with length $n - 1$ in the target. What happens, as shown in Figure 12.12, is that the data produced by the complete pattern of *n*-tuple words can be assembled in two different ways. In general, there is no way to distinguish between the alternative assemblies. In principle, if one could read the sequence out to the very end of the SBH target, the particular ambiguity shown in Figure 12.12 could be resolved.

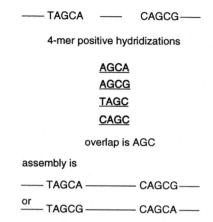

Figure 12.12 A tandem repeating sequence results in a branch point ambiguity: Two different reconstructions are possible from the pattern on *n*-tuple words detected.

However, there is no guarantee that this will be possible. The moment that there are more than one recurrences, the ambiguities become almost intractable. For example, the case shown in Figure 12.13, in which a sequence recurs three times, cannot be resolved even if one could read all the way to the ends of the SBH target.

The probability of any particular sequence 8 long recurring in a target of 200 bp is very low: A rough estimate gives $192/4^8 = 3 \times 10^{-3}$. However the probability that the target will have one or more recurrence once all possible recurrences are considered is actually quite high. An analogy is asking what is the probability that two people in a room have the same birthday. If you specify a particular date, or pick a particular person, the odds of a match are very low. However, if you allow all possible pairings to be considered, the odds are quite high that in a room with 30 people, two will share the same birthday. So sequence recurrences are a serious problem. They limit the length of sequence that can be directly and unambiguously read. In general, the chances of recurrences diminish as the length of the word used increases.

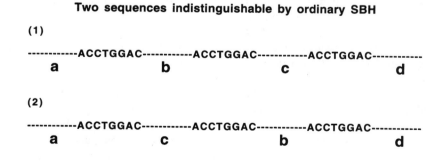

Figure 12.13 A more serious branch point ambiguity that leads to uncertainty in the arrangement of two blocks of single-copy DNA sequence.

There are 4^n possible words of length n, for a four-letter alphabet. Therefore to sequence by hybridization could require examining as many as 65,536 possible 8-mers or 262,144 possible 9-mers. Making complete sets of compounds larger than this and controlling their quality is likely to be challenging with present or currently extrapolated technology. It turns out that an estimate of the average sequence length that can be read before a branch point ambiguity arises is given approximately by the square root of the number of words used. When the words are DNA sequences, this is $4^{n/2}$. For 8-mers, the average length of sequence determined between branch points will be 256. This is quite an acceptable size scale. It seems like a losing proposition to increase the word size much beyond 8, unless technical considerations in the hybridizations demand this. Reducing the word length below 8 will lead to an unacceptably high frequency of branch points, unless some specific additional strategy is introduced to resolve these ambiguities.

Branch point ambiguities have one additional implication that must be dealt with in all attempts to implement a successful SBH strategy. The number of branch points present in a target sequence will grow rapidly as the total length of the target increases. Thus one must subdivide the target into relatively short DNA fragments in order to have a reasonable chance of sequencing each fragment unambiguously. This is a relatively undesirable feature of SBH. However, even if branch points could be resolved some other way, short targets are probably still mandated in order to diminish complications that may arise from intramolecular secondary or tertiary structure in the target. More will be said about this later.

SBH USING OLIGONUCLEOTIDE CHIPS

It is obvious that SBH cannot be practical if one is forced to look at hybridizations between a single oligonucleotide and a single target one at a time. If 8-mers were used, 65,536 different experiments would have to be done to determine a sequence that on average would be a DNA fragment less than 256 bases in length. The major appeal of SBH is that it seems readily adaptable to highly parallel implementations. There are two very different approaches that are being explored for this. The first is to hybridize a single-labeled sample to an array of all of the possible oligonucleotides it may contain. This is sometimes called format I SBH. The ideal array would be very small to minimize the amount of sample that was needed. Hence it is conventional to call the array a chip, by analogy with a semiconductor chip. A schematic illustration of an SBH experiment using such a chip is shown in Figure 12.14. A real chip would probably contain all 65,536 possible 8-mers, probably each present several times to allow signal averaging and control for reproducibility. The location of each particular oligonucleotide would be known. The actual patterns of oligonucleotides would probably be rather particular, a consequence of whatever systematic method is used to produce them. The chip surface itself could be silicon, or glass, or plastic. The key aspect is that the oligonucleotides must be covalently attached to it, and the surface must not interfere with the hybridization. The surface must not show significant amounts of nonspecific adsorption of the target, and it must not hinder, sterically or electrostatically, the approach of the target to the bound oligonucleotides. The ideal surface will also assist, or at least not interfere with, whatever detection system is ultimately used to quantitate the amount of hybridization that has occurred.

Several approaches are being tested to see how to fabricate efficiently a usable chip containing 65,536 8-mers. One basic strategy is to premake all of the compounds in the

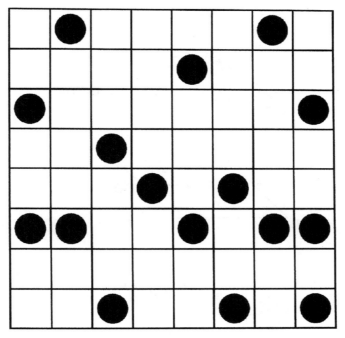

Figure 12.14 An example of the expected hybridization pattern of a labeled target exposed to an oligonucleotide chip. The actual chip might contain 65,000 or more probes.

array separately, and then develop a parallelized automated device to spot or spray the compounds onto the right locations on the chip. The disadvantage of this approach is that the rate of manufacture of each chip could be fairly slow. The major advantage of this approach is that the oligonucleotides only have to be made once, and their individual sequence and purity can be checked. There is no consensus at the present time what the optimal way would be to manufacture chips given samples of all the 65,536 8-mers. A key variable is how they will be attached to the chip surface. A long enough spacer must be used to keep the 8-mers well above the surface. Otherwise, the surface is likely to pose a steric restriction for the much bulkier target DNA.

The alternate strategies involve synthesizing the array on the chip. One potential general way to do this is photolithography, a technique that has been very powerful in the construction of semiconductor chips. It has been used quite successfully by Steven Fodor and others at Affymetrix, Inc. to make dense arrays of peptides, and more recently to make dense arrays of oligonucleotides. The basic requirement is that nucleotide derivatives are needed that are blocked from extending, say because the 3′ OH is esterified. The block used, however, can be removed by photolysis. A mask is used to allow selective illumination of only those chains that require extension by a particular base in this position (Fig. 12.15). Thus the light activates just a subset of the oligomers on the chip. The chain extension reaction is carried out in the dark. Then, in turn, three other masks are used to complete one cycle of synthesis. The key requirement in this approach is that the photoreaction must proceed at virtually 100% yield. Otherwise, the desired sequences will not be made in sufficient purity. This is a very difficult demand to satisfy with photochemical reactions. Instead, one can still use the principles of masks but just do more standard solid

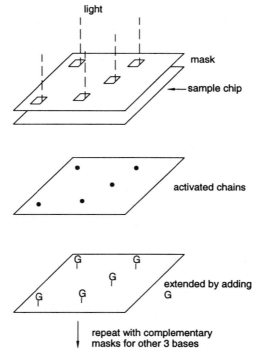

Figure 12.15   Construction of an oligonucleotide chip by in situ synthesis using photolithography techniques.

state oligonucleotide synthesis by spraying liquid reagents through the masks. The great power of the lithographic approach is that one can make any array desired—that is, any compound or compounds can be put in any positions on the array.

A different synthetic approach, with more limited versatility, is shown in Figure 12.16. This is the conception of Edwin Southern. It is actually a very simple lithographic approach that makes one particular array configuration efficiently. Figure 12.16 shows the steps in synthesis by stripes that would be needed to generate all possible tetranucleotides. The configuration that results is similar to the way the genetic code is ordinarily written down. Southern actually uses a glass plate as his chip. The reagents needed for the synthesis are pumped through channels between two glass plates as shown in Figure 12.17. It may be hard to miniaturize this design sufficiently to make a really small chip, but the plates made by Southern in this manner were the first dense oligonucleotide arrays actually being tested in real sequencing experiments.

An alternative approach being used to make arrays of 8-mers involves the use of a thin gel rather than a surface. This has the advantage that the sample thickness potentially allows larger amounts of oligonucleotide to be localized. In this approach, developed by Andrei Mirzabekov and coworkers, a glass plate was covered with a 50-micron thin gel. Pre-made oligonucleotides were deposited on the gel in 1 mm spots. The major effect of using a gel rather than a surface is that one has to be concerned about the local concentration of sample during washing steps. On a surface, solvent exchange is quite rapid, so the concentration of free sample can be reduced to zero quickly, and no back reactions of re-

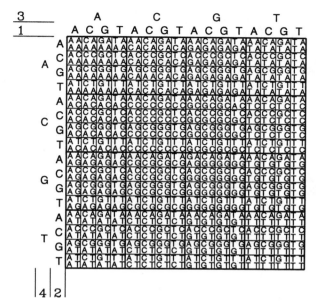

Figure 12.16 Pattern of stepwise DNA synthesis used in Southern's procedure for in situ synthesis of an oligonucleotide chip. Four successive synthetic steps are indicated. Within each square of the array, the sequence of the tetranucleotide synthesized is read left to right from the 3' to 5' direction starting from the upper row and continuing with the lower row.

leased material with the chip need be considered. With a gel, if target is released, it will take quite a while to leave the gel, and during this period there is a significant chance of back reaction with the chip if conditions permit duplex formation. This has both advantages and disadvantages, as we will illustrate later.

Regardless of the method of synthesis, the key technical issue that must be overcome is how the oligonucleotides are anchored at their position in the array. Mirzabekov uses direct chemical coupling to an oxidized ribonucleotide placed at the 3'-end of the 8-mer (Fig. 12.18). This is time-honored nucleic acid chemistry, but it does offer some risk of changing the stability of the resulting duplex because of the altered chemical structure at the sugar. The approach used by Southern is to attach a long hydrophilic linker arm to the glass surface (Fig. 12.19). This arm has a free primary hydroxyl group that can be used to

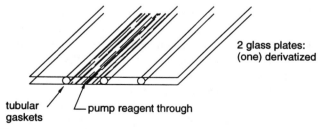

Figure 12.17 Glass plates separated by rubber dams are used to direct the reagents in each step of the procedure illustrated in Figure 12.16.

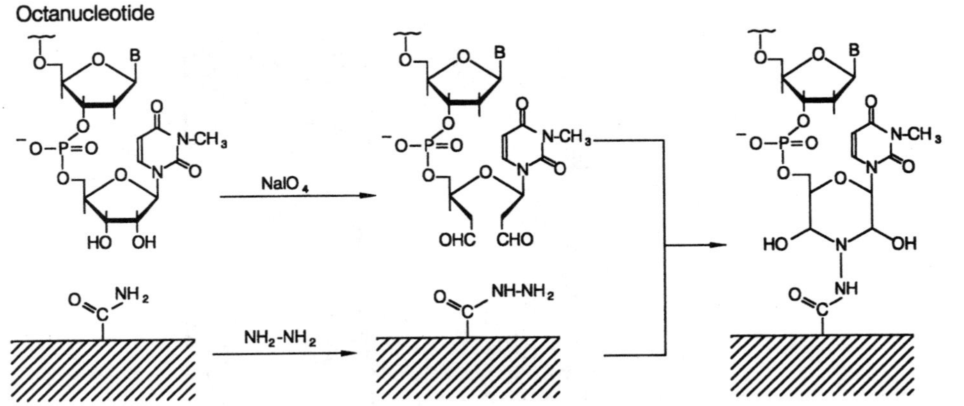

Octanucleotide

Figure 12.18 Method of attachment of oligonucleotides to polyacrylamide gels used by Mirzabekov and coworkers. From Khrapko et al. (1991).

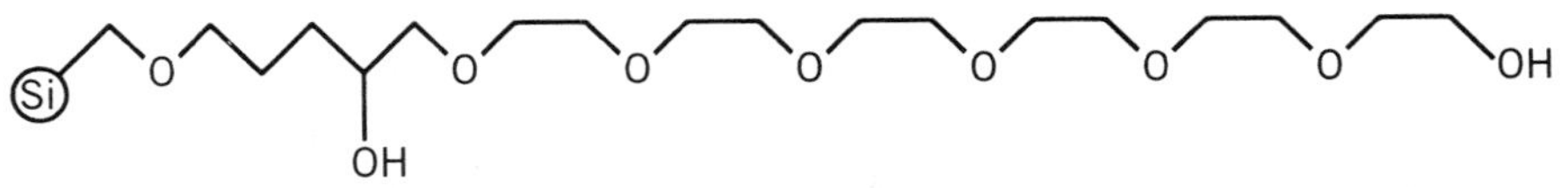

Figure 12.19 Method of attachment of oligonucleotides to glass plates used by Southern et al. (1992).

initiate the synthesis of the first nucleotide of the 8-mer in standard DNA synthesis protocols. It acts chemically exactly like the 3′ OH of a nucleoside in coupling to an activated phosphate of the next nucleotide.

SEQUENCING BY HYBRIDIZATION TO SAMPLE CHIPS

The second general SBH approach is to make a large, dense array of samples and probe it by hybridization with one labeled oligonucleotide at a time. This is sometimes called format II SBH. In this format, while it takes a long time to complete the sequence, one is actually sequencing a large number of samples simultaneously. A schematic illustration of this approach is shown in Figure 12.20. It looks deceptively similar to the use of oligonucleotide chips, but everything is reversed. The array might, for example, correspond to an entire cDNA library, perhaps 2×10^4 clones in all. Because SBH can only use relatively short samples, each cDNA might have to be broken down into fragments. It is not immediately obvious how to do this with large numbers of clones at once. One possibility is to subclone two different restriction enzyme digests of the cDNA inserts. This scrambles up connectivity information in the original clones; however, in most cases that information would be easily restored by the sequencing process itself, or by rehybridization of any ambiguous fragments back to the original, intact clones. If each 1.5-kb average cDNA clone yielded six fragments in each of the two digests, one would want to array 3.6×10^5 subclones in order to maintain the redundancy of coverage of the original library. This would constitute the sample chip.

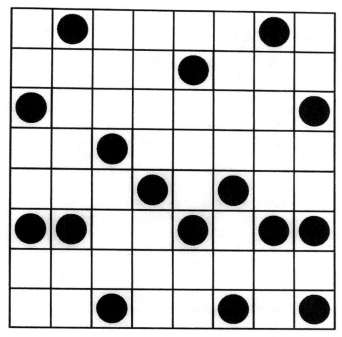

Figure 12.20 An example of the expected hybridization pattern of a labeled oligonucleotide to a sample chip. The actual chip might contain 20,000 or more samples.

Since the individual components are available in any desired quantity, one could, in principle, make as many copies of the array as could conveniently be handled simultaneously. In practice, it does not seem at all unreasonable to suppose that 100 copies of the array could be processed in parallel. It is envisaged that the sample chips be made by using the robotic $x-y$ tables common in the semiconductor industry (Fig. 12.21). These are very accurate and fast. It has been estimated that a sample density of 2×10^4 per 10 to 20

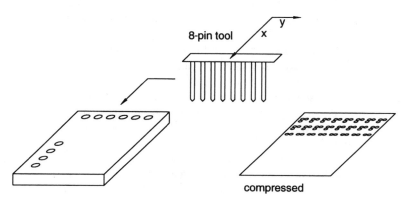

Figure 12.21 How a robotic $x-y$ table can be used in offset printing to construct of dense arrays of samples or probes (at right) starting from more expanded arrays (at left).

cm^2 is quite practical. Thus the entire array of subclones could be contained in 180 to 360 cm^2, which is about 30% to 60% the size of a typical 8.5 × 11 inch piece of paper.

If octanucleotides are used as hybridization probes to this large array, only a small fraction of the samples will show a positive signal. For one 250-bp subclone, the odds of containing any particular 8-mer are $243/4^8 = 3.8 \times 10^{-3}$; the possibility of a positive hybridization is less than 0.4%. When this is multiplied by the number of subclones in the array, there should be an average of 1.4×10^3 positive subclones per hybridization. Since each yields eight bases of DNA sequence data, the rate of sequence acquisition per single hybridization is 1.1×10^4 bp. If 100 chips can really be managed simultaneously, and if three hybridizations can be done per day, the overall throughput is 3.3×10^6. This is quite an impressive rate, and many of the variables used to estimate it are probably conservative.

A major feature of the use of sample chips in sequencing projects is that the approach does not scale down conveniently. Sample chips are only useful if entire libraries are to be sequenced as a unit. Such a method makes good sense for cDNAs and the genomes of model organisms. If all 65,536 8-mers must be used, at the rates we estimated above of 300 hybridizations per day, it will take more than half a year to complete the sequencing. Scaling down would not reduce the time of the effort at all; it would just reduce the amount of sequence data ultimately obtained. In practice, one does not have to use all 65,536 compounds to determine the sequence. Because of the considerable redundancy in the method, one ought to be able to use just a fraction of all 8-mers. The exact fraction will depend on error rates in the hybridization, how branch points will be resolved, and what kind of sequencing accuracy one desires. In some of the enhanced SBH schemes that will be described later, it has been estimated that one might be able to operate close to a redundancy of one rather than eight. However, this remains to be demonstrated in practice.

EARLY EXPERIENCES WITH SBH

A major difficulty in testing the potential of SBH and evaluating the merits of different SBH strategies or particular variations on conditions, sample attachment, and so on, is that the method does not scale down. A particular problem occurs in the use of oligonucleotide chips. It is difficult to vary parameters using the set of all 65,536 8-mers. Indeed, no one has yet actually made this set of compounds. Instead, several more limited tests of SBH have been carried out.

Southern has used the scheme shown in Figure 12.16 to make a chip containing four copies of all possible octapurine sequences (A and G only). An example of some data obtained with this chip is shown in Figure 12.22. In the actual example used, the labeled target DNA was a specific sequence of 24 pyrimidines (C and T). This contains 17 different 8-tuples, and so 17 positive hybridization spots would be expected. The actual results in Figure 12.22 are much more complex than this. Two problems need to be dealt with, which illustrate some of the basic issues in trying to implement SBH on a large scale. First the amount of oligonucleotide at each position in the array differs. More important, the strength of hybridization to different sequences varies quite a bit. Duplexes rich in G+C will be more stable, under most ordinary hybridization conditions than duplexes rich in A+T. There are ways to compensate for this, as we will illustrate later, and one of these was actually used with the samples in Figure 12.22. But the compensation is not

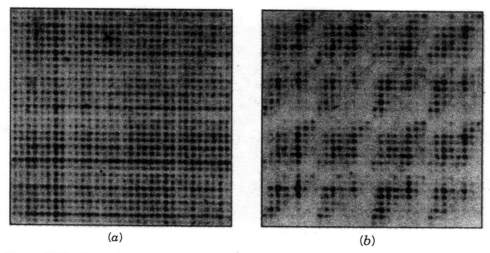

(a) (b)

Figure 12.22 Properties of an octapurine chip which contains four replicas of all 256 octapurines. (*a*) Hybridization pattern with an equimolar mixture of all octapyrimidines to show variation in the amount of attached purine. (*b*) Pattern of hybridization seen with a labeled 24-base target. From Southern et al. (1992).

perfect, and so there is a variability in the signal intensity that needs to be evaluated before a hybridization is scored as positive.

The second basic problem is cross-hybridization with single mismatches. This is a serious problem under the conditions used in Figure 12.22. From the results of these and other experiments, it has been estimated that most of the sample is not hybridized to the correct matches but instead forms a background halo of hybridization with numerous mismatches. Methods are being developed to correct for all these problems and make the best estimates of the right sequence in cases like Figure 12.22. It is too early to judge the effectiveness of these methods. A final potential problem with the test case used by Southern is that homopurine sequences can form triplexes with two antiparallel pyrimidine complements. These triplexes are quite stable, as we will illustrate in Chapter 14. It is conceivable that triplexes could have formed under the conditions used to test the oligopurine arrays, and since they would lead to systematic errors, one could go back and look for them.

Hans Lehrach has shown that oligonucleotide hybridization works well in fingerprinting samples for mapping (Chapter 9). These experiments provide some insight into the potential use of sample arrays for sequencing. Lehrach hybridizes single oligomers, or small pools of compounds, with large arrays of clones. This has successfully led to finished maps, so the sequence specificity under the conditions used must be reasonably good. However, since the mapping systems can tolerate considerable error, this is not a robust test of whether this approach will actually give usable sequence. What greatly expedites these experiments is that for fingerprinting, any oligonucleotide is as good as any other, so a large set of synthetic compounds is not needed to test the basic strategy.

Using the same approach, with a few immobilized samples, Radoje Drmanac and Radomir Crkvenjakov successfully completed two short pilot sequencing projects by SBH. In the first case, the 100-base sequence was known in advance, as was

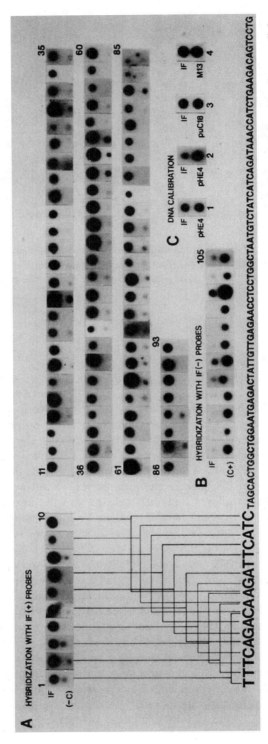

Figure 12.23 Patterns of hybridization seen when an immobilized target sequence and a control sequence are probed successively with adjacent radiolabeled octanucleotides chosen as complements to the target sequence. Taken from Strezoska et al. (1991).

a comparable negative control sequence. Oligonucleotides were selected on the basis of the known sequence; others were added to serve as negative controls. The results are fairly convincing. As shown in Figure 12.23, the discrimination between true positives and negatives is quite good in most of the individual hybridizations. Of course the obvious criticism of this experiment is that with a sequence known in advance, the test is not a truly objective one.

To address these concerns, Drmanac and Crkvenjakov performed a second pilot test of SBH on three closely related unknown sequences containing a total of 343 bases. The design of the test was based on an uninvolved third party who analyzed these sequences and designed a set of oligonucleotides in which only about half corresponded to the sequences in the target samples. In addition the challenge was to determine all three unknown samples and not generate erroneous composites of them by errors in reconstruction. The test was a total success—all three unknown sequences were correctly determined. However, one caveat needs to be considered. Because all 65,536 8-mers were not provided, this automatically supplies enormous amounts of information about the true sequence. Any compound omitted from the set provided is automatically a true negative. Just this information alone restricts the possible sequences tremendously, even before a single experiment has been done. Thus, while the experimental results that have been achieved are impressive, it cannot yet be said that a definitive test of SBH for de novo DNA sequencing has been done. Indeed, in defense of all who work in this field, it will probably not be possible to test the methods definitively until the gamble is taken to make, directly on chips or in bulk for distribution, all of the 65,536 8-mers.

DATA ACQUISITION AND ANALYSIS

Three different methods have been used thus far to detect hybridization in pilot SBH experiments. In each case quantitative data are needed so that positive signals can be discriminated as clearly as possible from background. Southern used image plate analyzers to examine radioisotope decay for the results shown in Figure 12.22. Others have used autoradiograms quantitated with a CCD camera. These approaches were discussed in Chapter 9. Fluorescent probes have been used by Fodor and by Mirzabekov. Here a CCD camera can be used in conjunction with a fluorescence microscope to record quantitative signals. Alternatively, a confocal scanning fluorescence microscope can be used. Other approaches such as mass spectrometry (see Chapter 11) are under development. The very notion of an oligonucleotide or sample chip raises the expectation that it should be possible to find a way to read out the amount of hybridization by a direct electronic method. Kenneth Beattie and Mitchell Eggers have developed one approach to this by detecting the mass of bound sample as it changes the local impedance on a silicon surface. In principle, one ought to be able to enhance such detection by providing the DNA probes or targets with attachments that generate more dramatic effects through altered conductivity, as a source of electrons or holes, or through magnetic properties. Perhaps the ultimate notion, as shown by the purely hypothetical example in Figure 12.24, would be to use the stability of the duplex formed in hybridization to directly manipulate elements of a nanoscale chip and thus lead to a detectable electrical signal.

However the data are obtained, current methods for analyzing data are already quite advanced. While it is difficult to convince people to synthesize 65,536 compounds before a method has proved itself, it is much easier to ask people to simulate the results of these

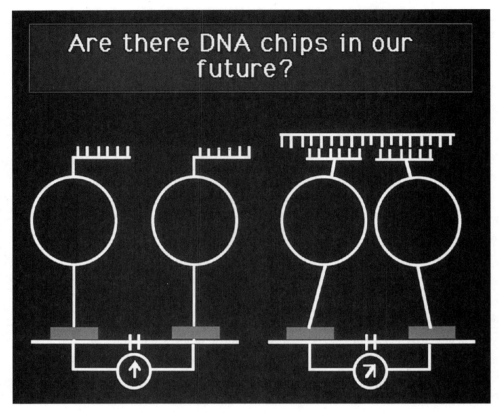

Figure 12.24 Possible future direct reading oligonucleotide hybridization chip. Figure also appears in color insert.

experiments and design software to reconstruct sequences from imperfect *n*-tuple word content. We have already indicated that these simulations are very encouraging, and they suggest that SBH will be a very powerful method, especially if the branch point ambiguities can somehow be dealt with. Two different proposals to handling branch points have been discussed. In the first, shown in Figure 12.25, one takes advantage of the fact that it should be possible to make a sample that consists of a dense set of small overlapping

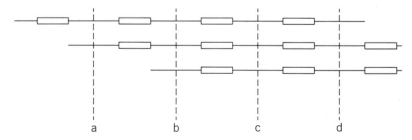

Figure 12.25 Overcoming branch point ambiguities by the simultaneous analysis of clones from a dense overlapping library. Recurrent sequences are shown as hollow bars. Unique hybridization probes are indicated by *a, b, c*. Known clone order implies that *b,* and not *c,* follows *a*.

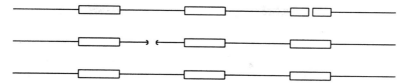

Figure 12.26 Overcoming branch point ambiguities by the use of several homologous but not identical DNA sequence targets.

clones. This is what is done for ordinary shotgun ladder sequencing, except that for SBH the clones would probably have to be even smaller. In these clones unique sequences will lie outside and between the repeats that cause branching ambiguities. Matching up these unique sequences not only places the clones in the proper order, it also resolves the ambiguous internal arrangement of sequences on a clone with three repeats, since the order is determined by the identity of these sequences on the flanking clones. This looks like a powerful approach, but it requires a great deal of experimental redundancy with little overall gain.

A second strategy for resolving branch point ambiguities is shown in Figure 12.26. Here the notion is to determine the DNA sequence of several similar but not identical samples. Because of sequence variations among the samples, exact recurrences in one sample will not necessarily be exact in all the others. Any imperfections in the repeats will break the branch point ambiguities in all of the samples because they can be aligned by homology. In principle, one could use different individuals of the same species and take advantage of natural sequence polymorphism. However, simulations show that the most effective application of this approach would use samples that have about 10% divergence on average. In practice, this may mean that it would be more useful to compare three to five similar species, like human and chimp, rather than compare individuals within a species. Here, as in the previous method, the cost of resolving branch point ambiguities is a considerable increase in the number of samples that have to be examined. However, the additional information that will be obtained will be highly interesting if the species are well chosen.

OBSTACLES TO SUCCESSFUL SBH

The base composition dependence of the melting temperature, T_m poses a very serious challenge to simple and effective implementations of SBH. If a temperature is chosen that allows effective discrimination between perfect matches and mismatches in G+C-rich compounds, many A+T-rich sequences may not form enough duplex to be detected. Alternatively, if one chooses a low enough temperature to stabilize the weakest A+T-rich duplexes, there will not be enough discrimination against mispairing in G+C-rich compounds, and many false positives will result. There are many possible ways to circumvent this problem; quite a few of them are being tested, but no generally acceptable solution has yet been demonstrated in practice.

Ed Southern has been experimenting with the use of high concentrations of tetramethylammonium salts (TMA) instead of more usual low to moderate ionic strength NaCl solutions. These salts have the undesirable feature of slowing down the kinetics of hybridization, but this can be compensated for, if necessary, by adding other agents that speed up hybridizations, such as dextrans which increase the effective concentration of nucleic acids. It has been known for a long time that TMA at the proper concentration can almost equalize the T_m of polynucleotides that are pure A+T and those that are pure G+C. However, when Southern tried TMA in oligonucleotide hybridization, he found that while the T_m's of compounds with extreme base compositions were equalized, a very large effect of DNA sequence on T_m of compounds with intermediate base compositions emerged. Unless this turns out to be an idiosyncracy caused by the use of pure homopurine sequences, it probably means that TMA will have to be abandoned.

An alternative way to even out base composition effects is to use base analogs (Fig. 12.27). One can substitute 2,6-diamino purine for A (an analog that makes three hydrogen bonds with T) and 5-bromoU for T (an analog that has increased vertical stacking energy). This will raise the relative stability of A+T-rich sequences considerably. The base analog 7-deaza G can be used instead of G to lower the stability of G+C-rich sequences. Many more analogs exist that could be tested. The problem is that one really wants to test their effect across the full spectrum of 65,536 8-mers, and there is simply no way to do this efficiently until we have developed much more effective ways to make oligonucleotide chips. Such devices not only provide a way to do SBH, they provide a source of samples that allow the accumulation of massive amounts of duplex T_m data. In model experiments Southern was able to characterize the T_m's of all of the 256 possible homopurine-homopyrimidine 8-mer duplexes under a wide set of experimental conditions. This single set of experiments undoubtedly provided more T_m data than a decade of previous work by several different laboratories.

An alternative approach for compensating for T_m differences has been demonstrated by Mirzabekov. This takes advantage of the fact that chips made of thin gels can rebind significant amounts of released sample at low temperatures. The rate of this rebinding will depend on the concentration of oligonucleotide, since renaturation shows second-order kinetics or pseudo–first-order kinetics (Chapter 3). To reveal these kinetic effects, one first hybridizes a sample to the immobilized probe and then allows a fraction of the duplexes to dissociate with a washing step. By adjusting the relative concentrations of different compounds, one can bring their T_m's very close to the same value. An example is shown in Figure 12.28. These results are very impressive. However, the two samples involved had to be used at a 300-fold concentration difference to achieve them. It is not immediately obvious that this can be done, in general, without leading to serious complications in the detection system used to monitor the hybridization. One will need a system with a very wide dynamic range. It will also be a major effort to try to equalize the melting properties of not just two compounds but 65,536.

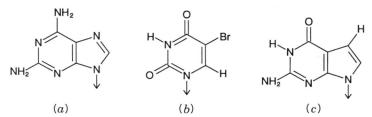

(a)   (b)   (c)

Figure 12.27   Base analogs useful in decreasing the differences in stability between A–T-rich and G–C-rich sequences: (a) 2-Aminoadenine. (b) 5-Bromouracil. (c) 7-Deazaguanine.

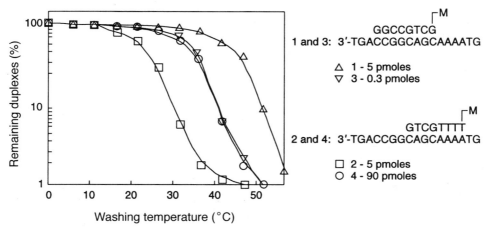

Figure 12.28   Adjusting the concentration of different oligonucleotides can compensate for difference in their melting temperatures. Adapted from Mirzabekov et al.

Instead of attempting to compensate for effects of sequence on the stability of duplexes, one can just measure the hybridization across a range of temperatures. This does not increase the number of samples needed. Instead, one would effectively be recording a melting profile for each sample with the entire set of oligonucleotides. This would increase the experimental time by a factor of ten or more, which is tolerable. In the long run, once extensive data on the thermal stability of each of the 8-mer complexes are known, it may be possible to use a much simpler approach. The set of compounds could be split into groups, each studied at a different optimal temperature. In principle, this could still involve a single chip, except that different regions would be kept at different temperatures. The manufacture of such a split chip would require custom placement of each compound, so simple masking strategies like that employed by Southern are unlikely to suffice. However, this is really not a serious additional manufacturing problem. Ultimately a combination of split chips, base analogs, and special solvents may all be needed for the most effective SBH throughput.

Secondary structure in the target is another potential complication in SBH. This is probably easily circumvented in the sample chip strategy. Here the target could be attached at random but frequent places to the surface under denaturing conditions. This would not be expected to interfere with oligonucleotide hybridization very much. It should effectively remove all but the most stable short sample hairpins (Fig. 12.29). The problem of secondary structure is likely to be more serious when oligonucleotide chips are used. The effect of such structures will be to cause a gap in the readable sequence. This is a serious problem, but since the gaps will be small, they can be filled rather easily by PCR-based cycle sequencing, using the sequence flanking the gaps to design appropriate primers. Thus the real issue is how frequent will such gaps be. If one occurs on each target sample, it will be best to forget SBH and just do the entire project by standard cycle sequencing. Presumably conditions will be found where the problem of secondary structure can be reduced to a much lower level. One way to do this would be to place base analogs in the sample that destabilize intramolecular base pairing more than intermolecular base pairing. These might, for example, be bulky groups where one could be tolerated in the groove of a duplex when the target binds to the probe, but two cannot be tolerated,

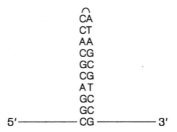

Figure 12.29   An example of a hairpin that is too stable to be detected in SBH.

if the target tries to pair with itself. There is undoubtedly room for much development here and much clever chemistry. A second approach would be to use probes with uncharged backbones. Then low ionic strength conditions can be used to suppress target secondary structure without affecting target-probe interactions. One example of such compounds is polypeptide nucleic acids (PNAs; see Chapter 14). Another example is phosphotriesters in which the oxygen that is normally charged in natural nucleic acids is esterified with an alkyl group. However, this creates an addition optically active center at each phosphorous, which leads to severe stereochemical complexities unless optically pure phosphotriesters are available.

The effects of secondary structure or unusual DNA structures are significant but not yet known in any great depth. In Chapter 2 we discussed the peculiar features of a centromere-associated repeat where the single strands may have a more stable secondary structure than the duplex. In Chapter 10 we illustrated the abnormally stable hairpin formed by a particular short DNA sequence. Whether these cases are representative of 1% of all the DNA sequences, or more or less, is simply unknown at the present time. About the only way we will be able to uncover such idiosyncratic behavior, understand it, and learn to deal with it, is to make large oligonucleotide arrays and start to study them. Unfortunately, this appears to be one of those cases in science where a timid approach is likely to be misleading. At some point we will have to dive in.

SBH IN COMPARATIVE DNA SEQUENCING

Some of the difficulties just described with full de novo SBH approaches have led some experts to doubt that SBH will ever mature into a widespread user-friendly method. For this reason much effort has been concentrated on developing SBH for comparative (or differential) DNA sequencing where one assumes that a reference sequence is known and the objective is to compare it with another sample and look for any potential differences. Comparative sequencing is needed in checking existing sequence data for errors. It is the type of sequencing required for horizontal studies in which many members of a population are examined. This is needed in genetic map construction, genetic diagnostics, the search for disease genes, in mutation detection, and for more biological objectives including ecology, evolution, and profiling gene expression. Some of these applications are discussed in Chapters 13 and 14.

When SBH is considered in the context of sequence comparisons, two problems of the method for de novo sequencing are immediately resolved. It is not necessary to have a probe array consisting of all possible 4^n oligonucleotodes of length n. Instead the array can be customized to look for the desired target and simple sequence variations of that target. Second, since a reference sequence is known, issues of branch point ambiguities are virtually always resolvable by use of the information in that sequence. A particularly powerful version of SBH for comparative sequence has been developed by Affymetrix, Inc. Here a probe array is made that corresponds to all possible strings of length n contained in the original sequence (for a target with L base pairs, $L - n + 1$ substrings are required). For each substring four variants are made corresponding to the expected sequence at the middle position of the substring and all three possible single-base variants there. Thus the array of probes will have $4(L - n + 1)$ elements. This is quite manageable with current photolithographic syntheses for targets in the range of 10 kb.

In actual practice this approach was tested on 16.6-kb human mitochondrial DNA using arrays containing up to 130,000 elements, each of which is a 15- to 25-base probe. (Chee et al., 1996). For convenience these nested targets are arranged, serially, horizontally in the array as shown schematically in Figure 12.30a, with the four possible variants for each central eighth base located vertically. The target is randomly sheared into short fragments (but longer than the length of the probes). A perfectly matched target will hybridize strongly to one member of each vertical set of four probes. A target with a single mismatch will show strong hybridization only to one particular probe in which the central base variant matches the sequence perfectly. For all possible flanking probes, there will be one or two internal mismatches between that target and the probe; hence hybridization will be weak or undetectable. A sample of the actual data seen using this approach is shown in Figure 12.30b. It is impressive. In practice, in most cases a two-color competitive hybridization is used. This allows a sample of the normal sequence (in one color) to be compared with a potential variant (in another color) with most differences in sequence-dependent hybridization efficiency nulled out.

OLIGONUCLEOTIDE STACKING HYBRIDIZATION

There are a number of ways that could potentially increase the length of sequence that can be read with a fixed length oligonucleotide. This is one major way to improve the efficiency of SBH, since the longer the effective word length, the higher the sequencing throughput and also the smaller the number of branch point ambiguities. One approach, specifically designed by Mirzabekov to help resolve branch point ambiguities, is shown in Figure 12.31. It is based on the fact that once a duplex has been formed by hybridization of the target with an 8-mer, it becomes thermodynamically quite favorable to bind a second oligomer immediately adjacent to the 8-mer. The extra thermodynamic stabilization comes from the stacking between the two adjacent duplexes. This same principle was discussed earlier in schemes for directed primer walking (Chapter 11). In practice, Mirzabekov uses pools of ninety 5-mers, chosen specifically to try to resolve known branch points. A test of this approach, with a single perfectly matched 5-mer or various mismatches, is shown in Figure 12.31. It is apparent that the discrimination power of oligonucleotide stacking hybridization is considerable.

Some calculated T_m's for perfect and mismatched duplexes are given in Table 12.1. These are based on average base compositions. The calculations were performed using the equations given in Chapter 3. In the case of oligonucleotide stacking, it is assumed that the first duplex is fully formed under the conditions where the second oligomer is being tested; in practice, this may not always be the case. It is, however, approximately true for the conditions used for the experiments shown in Figure 12.32. The calculations reveal a number of interesting features about stacking hybridization. Note that the binding

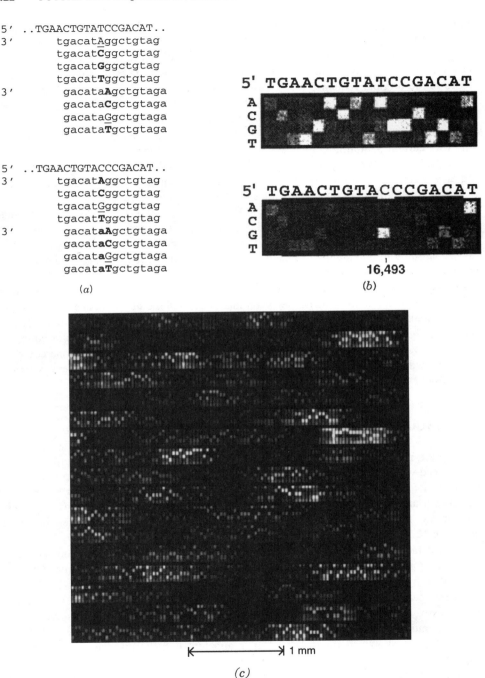

```
5'  ..TGAACTGTATCCGACAT..
3'      tgacatAggctgtag
        tgacatCggctgtag
        tgacatGggctgtag
        tgacatTggctgtag
3'      gacataAgctgtaga
        gacataCgctgtaga
        gacataGgctgtaga
        gacataTgctgtaga


5'  ..TGAACTGTACCCGACAT..
3'      tgacatAggctgtag
        tgacatCggctgtag
        tgacatGggctgtag
        tgacatTggctgtag
3'      gacataAgctgtaga
        gacataCgctgtaga
        gacataGgctgtaga
        gacataTgctgtaga
```

(a)

5' TGAACTGTATCCGACAT

A
C
G
T

5' TGAACTGTACCCGACAT

A
C
G
T

16,493

(b)

1 mm

(c)

Figure 12.30 Use of SBH for comparative hybridization. (*a*) Schematic layout of 15 base probes (*b*) Example of actual data probing for differences in human mitochondrial DNA. Top panel shows hybridization with the same sequence as used to design the array. Bottom panel shows hybridization with a sequence with a single T to C transition in position 16,493. (*c*) Example of hybridization to a full array. Panels (*b*) and (*c*) from Chee et al. (1996).

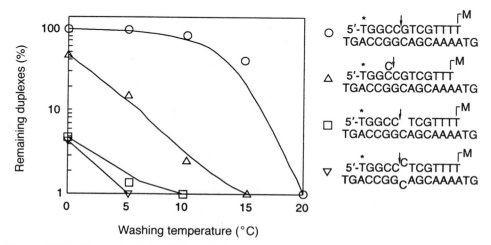

Figure 12.31   Basic strategy in oligonucleotide stacking hybridization

TABLE 12.1   Calculated Thermodynamic Stabilities of Some Ordinary Oligonucleotide Complexes and Other Complexes Involved in Stacking Hybridization

	Energetics of Stacking Hybridization			
Structure[a]	$n =$ 8	7	6	5
▬▬▬▬▬▪▪▪▪▪▪▪▪▪▪▪▪▬▬▬▬	38	33	25	15
▬▬▬▬▬▬▪▪▪▪▪▪▪▪▪▬▬▬▬▬	33	25	15	3
▬▬▬▬▪▪▪▪▪▪◆▪▪▪▪▬▬▬▬	25	15	3	−14
▪▪▪▪▪▪▪▪▪▪▪▪│▪▪▪▪▪▪▪▪▪▪▪▪▪	51	46	40	31
▪▪▪▪▪▪▪▪▪▪▪│▪▪▪▪▪▪▪▪▪▪▪	46	40	31	21
▪▪▪▪▪▪▪▪▪▪│▪▪▪▪▪▪▪▪▪▪▪	40	31	21	11

Note: Calculated T_m (°C, average base composition).

[a] Structures consist of a long target and a probe of length n. The top three samples are ordinary hybridization; the bottom three are stacking hybridization.

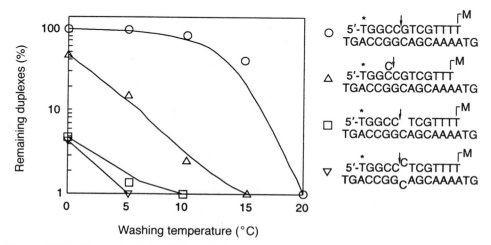

Figure 12.32   Example of the ability of oligonucleotide stacking hybridization to discriminate against mismatches. Taken from Mirzabekov et al.

of a second oligomer next to a preformed duplex provides an extra stability equal to about two base pairs. More interesting still is the fact that mispairing seems to have a larger consequence on stacking hybridization than it does on ordinary hybridization. This is consistent with the very large effects seen in Figure 12.32 for certain types of mispairing. Other types of mispairing are less destabilizing, but there may be a way to eliminate these, as we will discuss, momentarily. In standard hybridization sequencing, a terminal mismatch is the least destabilizing event, and thus it leads to the greatest source of ambiguity or background. For an octanucleotide complex, an average terminal mismatch leads to a 6 °C lowering in T_m. For stacking hybridization, a terminal mismatch on the side away from the preexisting duplex is the least destabilizing event. For a pentamer, this leads to a drop in T_m of 10 °C. These considerations predict that the discrimination power of stacking hybridization in favor of perfect duplexes might be greater than ordinary SBH. They encourage attempts to modify the notion of stacking hybridization so that it becomes a general, stand-alone method for DNA sequencing.

OTHER APPROACHES FOR ENHANCING SBH

Once an oligonucleotide has formed a duplex with the target, it ought to be possible to use enzymatic steps to proofread the accuracy of the hybridization and to read further DNA sequence information from the target. For example, the 3′-end of the oligonucleotide could serve as a primer for DNA polymerase to extend. What is needed is a sufficiently stable primer-template complex to allow the polymerase to function at a suitable temperature. An issue that needs to be explored is whether 8-mers are sufficient for this purpose. There are also potential background problems that will need to be addressed. This general approach has been used quite successfully with longer primers and DNA polymerase extension to detect specific alleles adjacent to the primer in a method called genetic bit analysis (Nikiforov et al., 1994). An alternative method for proofreading and extending a sequence read could use DNA ligase to attach a stacked oligonucleotide next to an initial duplex. This would have the potential advantage that ligase requires proper base pairing and might increase the discrimination of the stacking hybridization. In both cases, and in other schemes that can be contemplated, the label is introduced as a result of the enzymatic reaction. This eliminates much of the current background in SBH that arises from imperfect hybridization products. Some specific examples of how these procedures can be implemented in practice will be described in the next section.

A second, general way to enhance the power of SBH is to use gapped oligonucleotides. Two examples of this are shown below:

$$AGCN_4GAC \quad AGCI_4GAC$$

The first case uses a mixture of 256 possible 10-mers that share the same six external bases. The second uses a single 10-mer, but its four central bases are inosine (I) which can base pair with A, C, or T. In both of these cases the stability of the duplex is increased because it has more base pairs: one can read six bases of sequence but with the stability of a decanucleotide duplex. However, of even greater significance is the fact that the effective reach of the oligonucleotide is increased. Branch point ambiguities are less serious with these gapped molecules than with ungapped oligonucleotides with the same number of well-defined bases. William Baines has simulated SBH experiments with gapped

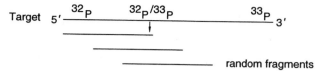

Figure 12.33   Resolving branch point ambiguities by using positional information derived from a gradient of two labels.

probes of various types, and these simulations indicate that this approach really improves the efficiency of SBH.

A third general way to enhance the power of SBH is to use oligonucleotide probes with degenerate ends like

$$N_2AGCTTAAGN_2$$

The advantage of this approach is that any mismatches at the ends of the internal 8-base probe sequence are converted to internal mismatches in the actual 12-base probe used.

Another way to enhance the power of SBH is to use the same kinds of pooling and multicolor detection schemes that we discussed in Chapters 9 and 11 for fast physical mapping and enhanced ladder DNA sequencing. There is every reason to use groups of oligomers simultaneously in hybridization to sample arrays, or groups of samples simultaneously, to oligonucleotide chips. Simulations are needed to help design the most effective strategies to do this. However, very simple arguments show that a considerable increase in throughput ought to be achievable. Earlier we calculated that less than 0.4% of the probes or targets score positive in a single hybridization. Performing 16 hybridizations in binary pools will therefore not entail much risk of ambiguities. Doing this in a single color would result in up to a fourfold increase in throughput. Multiple colors could be used to increase the throughput much more.

Alternatively, multiple colors might be used to help resolve branch point ambiguities. Suppose that one had a way of labeling a target with two colors, such that the ratio of these colors depended on the location of the target within a much larger clone. One way to think about doing this is placing the label in the target by a single cycle of primer extension, varying the relative concentrations of two different labeled dpppN's during the extension. When fragments of this target are hybridized to a oligonucleotide chip, the ratio of the labels will tell, roughly, where in the sequence the particular oligonucleotide is located (Fig. 12.33).

POSITIONAL SEQUENCING BY HYBRIDIZATION (PSBH)

Here we describe a scheme that was developed and refined in our own laboratories as an alternate form of SBH. It is called positional sequencing by hybridization (PSBH). It has a number of potential advantages over conventional SBH but also presents its own set of different obstacles that must be overcome to make the total scheme a practical reality. PSBH relies totally on stacking hybridization. It uses an array of probes constructed as follows, where X_n refers to a single specific DNA sequence of length n, and Y_n is the complement of that sequence:

$$\begin{matrix} 5' & X_nN_m & 3' \\ 3' & Y_n & 5' \end{matrix} \text{ or } \begin{matrix} 5' & X_n & 3' \\ 3' & Y_nN_m & 5' \end{matrix}$$

These probes all share a common duplex stem next to a single-stranded overhang. The details of the duplex sequence are unimportant here. Each element of the probe array will have a different specific overhang. Thus there are 4^m possible probes of each type. These probes, which actually resemble PCR splints, are designed to read a segment of target sequence by stacking hybridization. As shown in Figure 12.34, the 5'-overhang probe allows the 5'-end of a target DNA sequence to be read; the 3'-overhang probe will read the 3'-end of a target.

The basic scheme shown in Figure 12.34, can be improved and elaborated by adding to it most of the enhancements described in the previous section. It seems particularly well suited for incorporating many of these enhancements because the duplex stem of the probe can be made long enough to be totally stable under any of the conditions needed for enzymology. For example, it is possible to use DNA ligase to attach the target to the probe covalently, after hybridization (Fig. 12.35). This has several advantages. Any mispaired probe-target complexes are unlikely to be ligated. Any probes that have hybridized to some internal position in the target (like two of the cases shown in Fig. 12.32) will certainly be unable to ligate. All of the nonligated products can be washed away under conditions where the ligated duplex is completely stable. Thus excellent discrimination between perfectly matched targets and single-base mismatches can be achieved (Table 12.2).

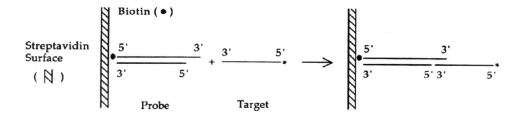

Basic scheme for positional SBH

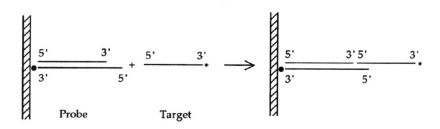

Alternate scheme

Figure 12.34   Basic scheme for positional SBH to read the sequence at the end of a DNA target.

Ligation of target DNA with probe

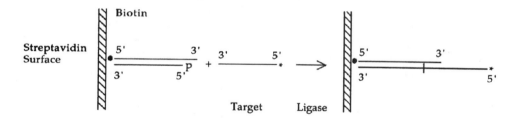

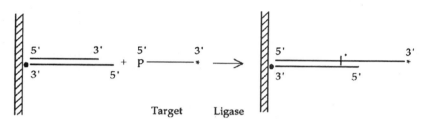

Figure 12.35 Use of DNA ligase to enhance the specificity of positional SBH. Note that since the target is ligated to the constant portion of the DNA probe, the ligation product can be melted off and replaced with a fresh constant portion. Thus a sample chip designed with this type of probes is reusable.

TABLE 12.2   Single-Stranded Target (3′-TCGAGAACCTTGGCT-5′) Annealed and Ligated to Duplexes With 5-base Overhangs with Different Mismatches

Probe[a]	Ligation Efficiency (%)	Discrimination Factor
3′-CTACTAGGCTGCGTAGTC-5′		
5′-b-GATGATCCGACGATCAGCTC-3′	17	
5′-b-GATGATCCGACGCATCAGCT**T**-3′	1	17
5′-b-GATGATCCGACGCATCAGCT**A**-3′	0.5	34
5′-b-GATGATCCGACGCATCAGC**C**C-3′	0.2	85
5′-b-GATGATCCGACGCATCAG**T**TC-3′	0.4	42
5′-b-GATGATCCGACGCATCA**A**CTC-3′	0.1	170

Source: Adapted from Broude et al. (1994).

[a]Each probe contained a constant 18-base duplex region formed by annealing the sequences shown with 3′-CTACTAGGCTGCGTAGTC-5′. Mismatches are shown in boldface.

Once the target has been ligated to the probe, it can serve as a substrate for the acquisition of additional DNA sequence data. For example, as shown in Figure 12.36, the 3'-end of the probe can be used as a primer to read the next base of the target by extension with a single, labeled terminator. Alternatively, any of the single nucleotide addition methods described at the beginning of this chapter can now be used on each immobilized target molecule as in Genetic Bit Analysis (Nikoforov et al., 1994). It would also be possible to do plus/minus sequencing on each immobilized target if one had sufficient quantitation with four colors to tell the amounts of each base incorporated. The basic idea is that the probe array can serve to localize a large number of different target molecules, simultaneously, and determine a bit of their sequence. Most probes will capture only a single target, and each of these complexes can then be sequenced in parallel. This should combine some of the best features of ladder and hybridization sequencing. It should produce sequence reads on each target molecule that are long enough to resolve all the common branch point ambiguities, except for those caused by true interspersed repeating sequences.

A major limitation in the PSBH approach we have described thus far is that it only reads the sequence at one end of the target. This would seem to limit its application to relatively short targets. However, one can circumvent this problem, in principle, by making a nested set of targets, as shown in Figure 12.37. One has to be careful in choosing the strategy for constructing these samples, since the ends of the DNAs must still be able to be ligated. Thus dideoxy terminators could be used, but they would have to be replaced by ordinary nucleotides with a single step of plus/minus sequencing, as we described for single-base addition early in the chapter. Alternatively, chemical cleavage could be used, as described when genomic DNA sequencing was used to locate ^mC's (Chapter 11). The third approach is to use exonuclease digestion to make the nested set. With these nested samples it should be possible to use PSBH to read the entire sequence of a target, limited only by the ability to resolve branch point ambiguities.

A major potential advantage of PSBH over SBH is that stacking hybridization would allow the use of 5-mer or 6-mer overlaps instead of the 8-mer or 9-mer probes required in

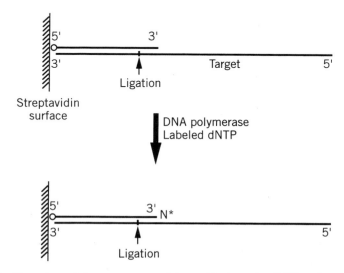

Figure 12.36 Extension of the sequence read by a chip by using DNA polymerase. Note that more sequence would be read but the chip would not be reusable.

Preparation of a nested set of DNAs

Figure 12.37 One way to prepare a nested set of DNA samples so that the entire sequence of a target could be read by positional SBH.

ordinary hybridization. This would decrease the size of the sample array needed by a factor of 64. Thus, for 5-mers, an array of only 1024 elements would be needed for unidirectional reading; twice this number is needed for bidirectional reading. However, this advantage will be offset by the increased frequency of branch point ambiguities unless there is some way to resolve them. A potential solution is afforded by the positional labeling scheme discussed in the previous section. A particularly simple way to mark the location of a branch point ambiguity is to combine a fixed end label and an internal label, as shown in Figure 12.38. The amount of end label would be the same on every target. The amount of internal label would vary depending on the length of the target, and thus on the position of the variable end of the target. The ratio between the internal label and the end label would provide the approximate length of the target. This strategy has not yet been tested in practice, but it seems fairly attractive because the reagents needed for two-color end and internal labeling are readily available (Chapter 10).

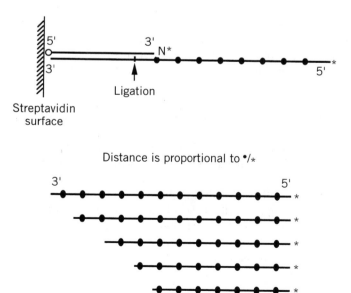

Figure 12.38 Determination of the approximate position of a target sequence by combining an end label with an internal label to provide an estimate of the length of the target.

TABLE 12.3   Single-Stranded Targets Ligated to Duplex Probes With the Indicated 5-Base Overhangs With Different A+T Contents

Probe Overhang[a] (5′ → 3′)	A + T Content	Ligation Efficiency (%)	Discrimination Factor
Match GGCCC	0	30	
Mismatch GGCC**T**		3	10
Match AGCCC	1	36	
Mismatch AGC**T**C		2	18
Match AGCTC	2	17	
Mismatch AGCT**T**		1	17
Match AGATC	3	24	
Mismatch AGAT**T**		1	24
Match ATATC	4	17	
Mismatch ATAT**T**		1	17
Match ATATT	5	31	
Mismatch ATAT**C**		2	16

Source: Adapted from Broude et al. (1994).

[a]Only the variable overhang portion of the probe sequence is shown. Mismatches are shown in boldface.

The major challenge for PSBH, as for SBH, is to build real arrays of probes and use them to test the fraction of sequences that actually perform according to expectations. Base composition and base sequence dependence on the effectiveness of hybridization is probably the greatest obstacle to successful implementation of these methods. The use of enzymatic steps, where feasible, may simplify these problems, since the enzymes do, after all, manage to work with a wide variety of DNA sequences in vivo. In fact initial results with the ligation scheme shown in Figure 12.35 indicate that the relative amount and specificity of the ligation are remarkably insensitive to base composition (Table 12.3). If further PSBH experiments reveal more significant base composition effects, one potential trick to compensate for this would be to allow the adjacent duplex to vary. Thus for an A+T rich overhang, one could use a G+C rich stacking duplex, and vice versa. This will surely not solve all potential problems, but it may be a good place to begin.

COMBINATION OF SBH WITH OTHER SEQUENCING METHODS

The PSBH scheme described in the previous section was initially conceived as a de novo sequencing method. However, it may serve better as a sample preparation method for other forms of rapid DNA sequencing. In essence, PSBH is a sequence-specific DNA capture method. A set of all 1024 PSBH probes can serve as a generic capture device to sort out a set of DNA samples on the basis of their 3′-terminal sequences. Such samples can be prepared either by a set of PCR reactions (with individually selected 5-base tags if necessary) or by digestion of a target with restriction enzymes like *Mwo* I that cut outside their recognition sequence as shown below:

GCNNNNN/NNGC

CGNN/NNNNNCG

The resulting captured set of samples is spatially resolved and can now be subjected to Sanger extension reactions to generate a set of sequence ladders. The appeal of this method is that a set of samples can be processed all at once in a single tube without any need for prior fractionation. Capture has been shown to be efficient with mixtures of up to 25 samples (Broude et al., 1994), and high-quality sequencing ladders have been prepared from mixtures of up to eight samples (Fu et al., 1995). The real promise of this approach probably lies in the preparation of samples for MALDI MS DNA sequencing (Chapter 10) where very large numbers of relatively short samples will need to be processed.

SOURCES AND ADDITIONAL READINGS

Broude, N. E., Sano, T., Smith, C. L., and Cantor, C. R. 1994. Enhanced DNA sequencing by hybridization. *Proceedings of the National Academy of Sciences USA* 91: 3072–3076.

Chee, M., Yang, R., Hubbell, E., Berno, A., Huang, X. C., Stern, D., Winkler, J., Lockhart, D. J., Morris, M. S., and Fodor, S. P. A. 1996. Accessing genetic information with high-density DNA arrays. *Science* 274: 610–614.

Chetverin, A., and Kramer, F. R. 1994. Oligonucleotide array: New concepts and possibilities. *Bio/Technology* 12: 1093–1099.

Dubiley, S. et al. 1997. Fractionation, phosphorylation and ligation on oligonucleotide microchips to enhance sequencing by hybridization. *Nucleic Acids Research* 25: 2259–2265.

Fu, D-J., Broude, N. E., Koster, H., Smith, C. L., and Cantor, C. R. 1995. Efficient preparation of short DNA sequence ladders potentially suitable for MALDI-TOF DNA sequencing. *Genetic Analysis* 12: 137–142.

Fu, D.-J., Broude, N. E., Koster, H., Smith, C. L., and Cantor, C. R. 1995. A DNA sequencing strategy that requires only five bases of known terminal sequence for priming. *Proceedings of the National Academy of Sciences USA* 92: 10162–10166.

Guo, Z., Liu, Q., and Smith, L. M. 1997. Enhanced discrimination of single nucleotide polymorphisms by artificial mismatch hybridization. *Nature Biotechnology* 15: 331–335.

Jurinke, C., van den Boom, D., Jacob, A., Tang, K., Worl, R., and Koster, H. 1996. Analysis of ligase chain reaction products via matrix-assisted laser desorption/ionization time-of-flight mass spectrometry. *Analytical Biochemistry* 237: 174–181.

Khrapko, K. R., Lysov. Y. P., Khorlin, A. A., Ivanov, I. B., Yershov, G. M., Vasilenko, S. K., Florentiev, V. L., and Mirzabekov, A. D. 1991. A method for DNA sequencing by hybridization with oligonucleotide matrix. *Journal of DNA Sequencing and Mapping* 1: 375–368.

Lane, M. J., Paner, T., Kashin, I., Faldasz, B. D., Li, B., Gallo, F. J., and Benight, A. S. 1997. The thermodynamic advantage of DNA oligonucleotide "stacking hybridization" reactions: Energetics of a DNA nick. *Nucleic Acids Research* 25: 611–616.

Li, Y., Tang, K., Little, D.P., Koster, H., Hunter, R. L., and McIver, R. T. Jr. 1996. High-resolution MALDI fourier transform mass spectrometry of oligonucleotides. *Analytical Chemistry* 68: 2090–2096.

Livshits, M. A., Florentiev, M. L., and Mirzabekov, A. D. 1994. Dissociation of duplexes formed by hybridization of DNA with gel-immobilized oligonucleotides. *Journal of Biomolecular Structure Dynamics* 11: 783–795.

Lysov, Y. P. et al. 1994. DNA sequencing by hybridization to oligonucleotide matrix. Calculation of continuous stacking hybridization efficiency. *Journal of Biomolecular Structure Dynamics* 11: 797–812.

Maskos, U., and Southern, E. M. 1992. Parallel analysis of oligodeoxyribonucleotide (oligonu-cleotide) interactions. I. Analysis of factors influencing oligonucleotide duplex formation. *Nucleic Acids Research* 20: 1675–1678.

Maskos, U., and Southern, E. M. 1992. Oligonucleotide hybridisations on glass supports: A novel linker for oligonucleotide synthesis and hybridisation properties of oligonucleotides synthesised in situ. *Nucleic Acids Research* 20: 1679–1684.

Milosavljevic, A. et al. 1996. DNA sequence recognition by hybridization to short oligomers: Experimental verification of the method on the *E. coli* genome. *Genomics* 37: 77–86.

Murray, M. N., Hansma, H. G., Bezanilla, M., Sano, T., Ogletree, D. F., Kolbe, W., Smith, C. L., Cantor, C. R., Spengler, S., Hansma, P. K., and Salmeron, M. 1993. Atomic force microscopy of biochemically tagged DNA. *Proceedings of the National Academy of Sciences USA* 90: 3811–3814.

Nikoforov, T. T., Rendle, R. B., Goelet, P., Rogers, Y.-H., Kotewicz, M. L., Anderson, S., Trainor, G. L., and Knapp, M. R. 1994. Genetic bit analysis: A solid phase method for typing single nu-cleotide polymorphisms. *Nucleic Acids Research* 22: 4167–4175.

Pease, A. C., Solas, D., Sullivan, E., Cronin, M. T., Holmes, C. P., and Fodor, S. P. A. 1994. Light-generated oligonucleotide arrays for rapid DNA sequence analysis. *Proceedings of the National Academy of Sciences USA* 91: 5022–5026.

Ronaghi, M., Uhlen, M., and Nyren, P. (1998). Real-time pyrophosphate detection for DNA se-quencing. *Science* 281: 363–365.

Shalon, D., Smith, S. J., and Brown, P. O. 1996. A DNA microarray system for analyzing complex DNA samples using two-color fluorescent probe hybridization. *Genome Methods* 6: 639–645.

Southern, E. M., Maskos, U., and Elder, J. K. 1992. Analyzing and comparing nucleic acid se-quences by hybridization to arrays of oligonucleotides: Evaluation using experimental models. *Genomics* 13: 1008–1017.

Southern, E. M., Green-Case, S. C., Elder, J. K., Johnson, M., Mir, K. U., Wang, L., and Williams, J. C. 1994. Arrays of complementary oligonucleotides for analysing the hybridisation behavior of nucleic acids. *Nucleic Acids Research* 22: 1368–1373.

Strezoska, Z., Paunesku, T., Radosavljevic, D., Labat, I., Drmanac, R., and Crkvenjakov, R. 1991. DNA sequencing by hybridization: 100 bases read by a non-gel method. *Proceedings of the National Academy of Sciences USA* 88: 10089–10093.

13 Finding Genes and Mutations

DETECTION OF ALTERED DNA SEQUENCES

Genomic DNA maps and sequences are a means to an end. The end is to use this information to understand biological phenomena. At the heart of most applications of mapping and sequencing is the search for altered DNA sequences. These may be sequences involved in an interesting phenotypic trait, an inherited disease, or a noninherited genetic disease due to a DNA change in somatic (nongermline) cells. The way in which maps and sequences can be used to identify altered DNA sequences very much depends on the context of that alteration. Here we will briefly survey the range of applications of maps and sequences, and then we will cover a few examples in considerable depth. However, the emphasis of much of this chapter will be the development of more efficient methods to find any sequence differences between two DNA samples.

Some DNA differences are inherited. There are three levels at which we characterize inherited DNA differences. DNA maps and sequences greatly assist the finding of genes responsible for inherited diseases or other inherited traits. Once a disease gene has been identified, we attempt to develop DNA-based tests for the clinical diagnosis of disease risk. The success of these tests will depend on the complexity of the disease and normal alleles. Even before a disease gene has been identified, DNA-based analyses of linked markers can sometimes offer considerably enhanced presymptomatic or prenatal diagnosis, or carrier screening. Finally DNA tests, in principle, provide a way for us to look for new germline mutations, either at the level of sperm (and ova in principle, but not very easily in practice) or anytime after the creation of an embryo. These mutations are referred to, respectively, as gametic mutations and genetic mutations. The distinction is a subtle one. Any mutations that destroy the ability of a gamete to function will not be inheritable because this gamete will produce no progeny.

Some DNA differences are important at the level of organism function, but they do not affect the germ cells, so they are not passed to the offspring. Examples in normal development occur frequently in the immune system. Both the immunoglobulin genes and the T-cell receptor genes rearrange in lymphocytes, and they also have a high degree of point mutagenesis in certain critical regions. These processes are used to generate the enormous repertoire of immune diversity needed to allow the immune system to detect and combat a wide variety of foreign substances. It has been speculated that DNA rearrangements might also occur in other normal somatic tissues, like the brain, but thus far, evidence for any such functionally significant rearrangements is not convincing. DNA changes in abnormal development appear to be commonplace. Most cancer cells contain DNA rearrangements that somehow interfere with the normal control of cell division. As the resulting cells multiply and spread, they frequently accumulate many additional DNA alterations. Other somatic DNA differences occur when chromosomes segregate incorrectly during mitosis.

A final example where DNA sequence information plays an important role in clinical diagnosis is in infectious disease. For example, strain variations of viruses and bacteria can be of critical importance in predicting their pathogenicity. Examples include virulent versus nonvirulent forms of bacteria like *Mycobacterium tuberculosis,* and various drug-resistant strains of HIV, the virus that causes AIDs. Other examples are quite common in parasitic protozoa, since these organisms, like HIV, use rapid DNA sequence variation as a way of escaping the full surveillance of the immune system of the host. Thus DNA sequence analysis is important in understanding the biology of *Plasmodium falciparum,* the organism that causes malaria, *Trypanosoma brucii* and *Trypanosoma cruzii,* which cause sleeping sickness, and many other organisms that pose significant public health hazards.

In this chapter we will describe the sorts of DNA analyses that can be done to detect genomic changes with present technology, and we will try to extrapolate to see what improvements will be likely in the future.

FINDING GENES

The approach used to find genes based on their location on the genetic map has been called reverse genetics, but a more accurate term is positional cloning. The basic strategy is to use the genetic map to approximate the position of the gene (Fig. 13.1). Then a physical map of the region is constructed if it is not already available. The physical map should provide a number of potential sequence candidates for the gene of interest. It also helps to find additional useful polymorphic markers that narrow the location of the desired gene further. Ultimately one is reduced to a search for a particular set of DNA sequence differences that correlates with a phenotype known to be directed by an allele of the gene. In contrast to positional cloning, genes can sometimes be found by functional cloning. Here an altered biochemical function is traced to an altered protein. This is sequenced, and the resulting string of amino acids is scanned to find regions that allow relatively nondegenerate potential DNA coding sequences to be synthesized and used as hybridization probes to screen genomic or cDNA libraries.

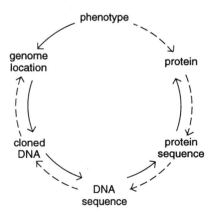

Figure 13.1 Contrasting stages in strategies to find genes by positional cloning (solid line) and by functional cloning (dashed line). Adapted from a slide displayed by Hilger Ropers.

In the past few years there have been many dramatic successes in human positional cloning. Among these are the genes responsible for Duchenne muscular dystrophy, cystic fibrosis, some forms of familial Alzheimers disease, myotonic dystrophy, familial colon cancer, two forms of familial breast cancer, HD, one form of neurofibromatosis, and several genes involved in fragile X-linked mental retardation. Some of the ways in which the genetic map has helped locate and clone these disease genes were discussed in Chapter 6. Here we review, briefly, some of the aspects of this process, with the particular goal of showing where DNA sequencing plays a useful or necessary role. In most gene searches, thus far there have been unexpected benefits in that interesting biological or genetic mechanisms became apparent as correlations became possible between genotype and phenotype. These serendipitous findings may have occurred because so few human disease genes were known previously. However, it is still possible that many additional basic biological surprises remain to be uncovered as much larger numbers of human disease genes are identified. In Box 13.2 we will illustrate one of the most novel disease mechanisms seen in several of these diseases which is caused by unstable repeating trinucleotide sequences.

A successful genetic linkage study within a limited set of families is just the first step in the search for a gene. It reveals that there is a specific single gene involved in a particular disease or phenotype, and it provides the approximate location of that gene in the genome. However, genetic studies in the human can rarely locate a gene to better than 1 to 2 cM. In typical regions of the human genome, this corresponds to 1 to 2 Mb; the problem is that such regions will usually contain 30 to 60 genes. To narrow the search, it is usually necessary to isolate the DNA of the region (Fig. 13.2). Until the advent of YACs and other large insert cloning systems, this was a very time-consuming and costly process. It frequently consisted of parallel attempts at chromosome microdissection and microcloning and attempts at cosmid walking or jumping from the nearest markers flanking the region of interest. Now these steps can usually be carried out much more efficiently by using the larger size YACs and mega-YACs that span most of the human genome. While these have some limitations, discussed in Chapter 9, the DNA of most regions is available just by a telephone call to the nearest YAC repository.

With the DNA of a particular region in hand, one can search for additional polymorphic markers fairly efficiently. For example, simple tandem repeating sequences can be selected by hybridization screening or sequence-specific purification methods (see Chapter 14). These new markers can be used to refine the genetic map in the region. However, a more effective use of nearby markers, as we discussed in Chapter 6, is to pinpoint the location of any recombinants in the region. This is illustrated in Figure 13.3.

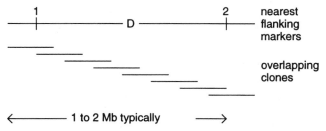

Figure 13.2 Information and samples usually available at the start of the end game in the search for a disease gene.

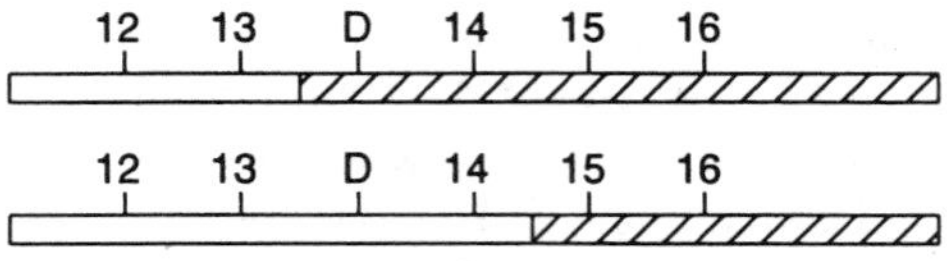

Figure 13.3 The nearest recombination breaking points flank the true location of the gene. Hatched and hollow bars indicate chromosome segments inherited from different parental homologs. D must lie to the right of marker 13 and to the left of marker 15.

Before the gene of interest was successfully linked to markers, any recombinants were damaging, since they subtracted from the statistical power of the linkage tests. Now, however, once the locale of the gene is established beyond doubt, the recombinants are a very valuable resource, and it is often very profitable to search for additional recombinants. As shown in Figure 13.3, the gene location can be narrowed down to a position between the nearest set of available recombinants.

In an ideal case the gene of interest is large, and it occupies a considerable portion of the region. Then frequently a disease allele can be found that contains a large enough size polymorphism to be spotted by PFG analysis of DNA hybridized with available probes in the region. The polymorphism may arise from an insertion, a deletion, or a translocation. Such an association of a disease phenotype with one or more large-scale rearrangements almost always rapidly pinpoints the location of the gene because finer and finer physical mapping can rapidly be employed to position the actual disrupted gene relative to the precise sites of DNA rearrangements. An example of this approach was the search for the gene for Duchenne muscular dystrophy where roughly half of the disease alleles are large deletions in the DNA of the region.

In typical cases one is not lucky enough to spot the gene of interest by using low-resolution mapping approaches. Then it is usually safest to take a number of different approaches simultaneously. This is especially true if, as in many cases of interest, the search for the gene is a competitive one. The genetic approach useful at this point is linkage disequilibrium. This was described in detail in Chapter 6. To reiterate, briefly, if there is a founder effect, that is, if most individuals carrying a disease allele have descended from a common progenitor, they will tend to have similar alleles at other polymorphic sites in the region. This is true even though the individuals have no apparent familial relationships. The closer one is to the gene, the greater the tendency of all individuals with the disease to share a common haplotype. This gradient of genetic similarity allows one to narrow down the location of the gene, but there are many potential pitfalls, as described in Chapter 6.

A second useful approach is to search for individuals who display multiply genetic disorders including the particular disease of interest. Such individuals can frequently be found, and they will often be carriers of microscopic DNA deletions. As shown in Figure 13.4, one can use these individuals to narrow the location of the gene. Low-resolution physical maps of each individual can often reveal the size and position of the deletions. Pooling data from several individuals with different deletions will indicate the boundaries on the possible location of the gene of interest. The process is easiest in cases like X-linked disease, since here, in males, there is only one copy of the region of interest. In somatic disease, there will be two copies, and the altered chromosome will have to be distinguished and analyzed in the presence of the normal one. This general approach can be very productive because after one gene is found, the genes for the additional inherited disorders must lie nearby, and it will be much easier to find them.

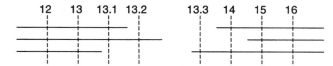

Figure 13.4 The nearest available chromosome breaking points, frequently seen in patients with multiple inherited disorders, flank the true location of the gene. Horizontal lines show markers present in three individuals with a common genetic disorder. The disease gene must lie between markers 13.2 and 13.3.

A third parallel approach is to map and characterize the transcripts coded for by the region. This can be done by using available DNA probes in hybridization against Northern blots (electrophoretically fractionated mRNAs) or against cDNA libraries. If the disease is believed to be predominantly localized in particular tissues, this approach can be very effective, since one can compare mRNAs or cDNAs from tissues believed to be significant sites of expression of the gene of interest with other samples where this gene is not likely to be expressed. With cystic fibrosis, for example, hypotheses about gene expression in sweat glands and in the pancreas were very helpful in narrowing the location of the gene in this way. Alternatively, genes in the target region may already be known as a large number of expressed sequence tags (ESTs) from known tissues are being added to the EST database at GenBank daily and are being mapped to chromosomal regions. Note, however, that considerable pitfalls exist with this approach, since hypotheses about the sites of expression can easily be wrong, and, even if they are correct, the gene of interest may be expressed at too low a level to be seen as mRNA or represented in a typical cDNA library.

DNA of the region can be used in a number of different ways to help find the location of the genes in the region, even where no prior hypotheses about sites of likely expression exist. YACs have been used as hybridization probes to directly isolate corresponding mRNAs or cDNAs, a technique sometimes referred to as fishing (Lovett, 1994). Techniques, such as exon trapping, have been developed to allow specific subcloning of potentially coding DNA sequences from a region (see Box 13.1). Another frequently effective strategy is to look for regions of DNA that are conserved in different mammals or even more distant species. Genes are far more likely to be conserved than noncoding regions. However, this approach is not guaranteed because there is no reason to expect that every gene will be conserved or even exist among a set of species tested. Even genome scanning by direct sequencing has revealed the location of genes.

In some types of disease, other strategies become useful. For example, in dominant lethal disease, most if not all affected individuals are new mutations. These will most likely occur in regions of DNA with high intrinsic mutations rates. While we still have much to learn about how to identify such regions, at least one class of unstable DNA sequence has emerged in recent years that appears to play a major role in human disease. Tandemly repeated DNA sequences have intrinsically high mutation rates because of the possibilities for polymerase stuttering or unequal sister chromatid exchange, as described in Chapter 7. Repeats like $(GAG)_n$ occur in coding regions; $(GAA)_n$ and $(GCC)_n$ occur outside of coding regions. These can shrink or grow rapidly in successive generations and lead to disease phenotypes. Examples of this were first seen in myotonic dystrophy, fragile X-linked mental retardation, and Kennedy's disease (see Box 13.2). A systematic search is now underway to map the locations of these and other trinucleotide repeats, since they may well underlie the cause of additional human diseases. The repeats appear to be fairly widespread as shown by the examples already found (see Table 13.1).

BOX 13.1
EXON TRAPPING METHODS

Exon trapping methods are schemes for selective cloning and screening of coding
DNA sequences. Several different approaches have been described (Duyk et al., 1990;
Buckler et al., 1991; Hamaguchi et al., 1992). Here we will illustrate only the last of
these because it seems to be relatively simple and efficient. The vector used for this
exon trapping scheme is shown in Figure 13.5a. It contains intron 10 of the p53 gene,
which includes a long pyrimidine tract (which appears to prevent exon skipping), and
consensus sequences for the 5'- and 3'-splicing sites (AG/GTGAGT and AG, respectively), and the branch site (TACTCAC) used in an intermediate step in RNA splicing.
The intron contains a *Bgl* II cleavage site used for cloning genomic DNA. Surrounding
the intron are two short p53 exons, flanked by SV40 promoters known to be transcrip-
tionally active in COS-7 cells. Reverse transcriptase is used to make a cDNA copy of
any transcripts, and then PCR with two nested sets of primers is used to detect any
transcripts containing the two p53 exons. When the vector alone is transfected into
COS-7 cells, only a 72-bp transcript is seen. Cloned inserts containing other complete
exons will produce longer transcripts after transfection. In practice, fragments from 90
to 900 bp are screened for because most exons are shorter than 500 bp. These new
fragments will arise by two splicing events as shown in Figure 13.5b. For an example
of recent results using exon trapping, see Chen et al. (1996).

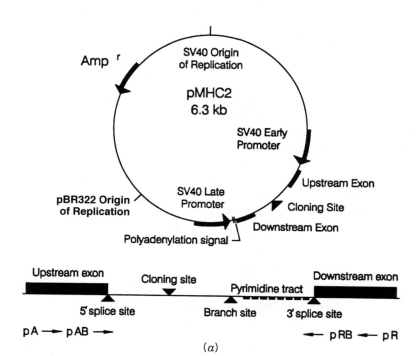

(a)

(continued)

BOX 13.1 *(Continued)*

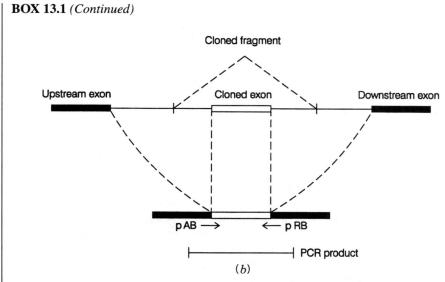

Figure 13.5 Exon trapping to clone expressed DNA sequences. (*a*) Vector and procedures used. Adapted from Hamaguchi et al. (1992). pA, pAB, pRB, and pR are primers used for nested PCR. (*b*) Schematic of the PCR product expected from a cloned exon.

TABLE 13.1   Trinucleotide Repeats in Human Genes

Gene or Encoded Protein	Copy Number[a]	Location
znf6 (zinc finger transcription factor)	8, 3, 3	5' Untranslated region
CENP-B (centromere autoantigen)	5	5' Untranslated region
c-*cbl* (proto-oncogene)	11	5' Untranslated region
Small subunit of calcium-activated neutral protease	10, 6	Coding region (N-terminal)
CAMIII (calmodulin)	6	5' Untranslated region
BCR (breaking point cluster region)	7	5' Untranslated region
Ferritin H chain	5	5' Untranslated region
Transcription elongation factor SII	7	5' Untranslated region
Early growth response 2 protein	5	Coding region (central)
Androgen receptor	17	Coding region (central)
FMR-1 (fragile X disease)	6–60	Not certain yet
(AGC)$_n$ androgen receptor (Kennedy's disease)	13–30	Coding region (central)
DM-1 myotonic dystrophy	5–27	3' Untranslated region
IT 15 Huntington's disease	11–34	Coding region (N-terminal)

Source: Updated, from Sutherland and Richards (1995).

[a]In normal individuals

BOX 13.2
DISEASES CAUSED BY ALTERED TRINUCLEOTIDE REPEATS

Fragile sites on chromosomes have been recognized, cytogenetically, for a long time. When cells are growth under metabolically impaired conditions, some chromosomes, in metaphase, show defects. These give the superficial appearance that the chromosome is broken at a specific locus, as shown in Figure 13.6. Actually it is most unlikely that a real break has occurred; instead, the chromatin has failed to condense normally. A particular fragile spot on the long arm of the human X chromosome, called fraXq27, shows a genetic association with mental retardation. About 60% of the chromosomes in individuals with this syndrome show fragile sites; the incidence in apparently normal individuals is only 1%. Fragile X-linked mental retardation is actually the second most common cause of inherited mental retardation. It occurs in 1 in 2000 males and 4 in 10,000 females. Earlier genetic studies of fragile X syndrome showed a number of very peculiar features that were inexplicable by any simple classical genetic mechanisms.

Now that the molecular genetics of the fragile X has been revealed, and similar events have been seen in many other diseases, including Kennedy's disease, myotonic dystrophy, and Huntington's disease, we can rationalize many of the unusual genetic features of these diseases. A number of fundamental issues, however, remain unresolved. The basic molecular genetic mechanism common to all four diseases and many others is illustrated in Figure 13.7 (Sutherland and Richards, 1995). In each case, near or in the gene, a repeating trinucleotide sequence occurs. Like other variable number tandem repeats (VNTRs), this sequence is polymorphic in the population. Normal individuals are observed to have relatively short repeats: 6 to 60 copies in fragile X syndrome, 13 to 30 in Kennedy's disease, 5 to 27 copies in myotonic dystrophy, and 11 to 34 in Huntington's. Individuals affected with the disease have much larger repeats: more than 200 copies in fragile X, more than 39 in Kennedy's, more than 100 in myotonic dystrophy, and more than 42 in Huntington's.

The case studied in most detail thus far is the fragile X syndrome, and this will be the focus of our attention here. Individuals who are carriers for fragile X, that is, individuals whose offspring or subsequent descendants display the fragile X phenotype, have repeats larger than the 60 copies, which represents the maximum in the normal population, but smaller than 200, the lower bound of individuals with discernable disease phenotypes. This progressive growth in the size of the repeat, from normal to car-

Figure 13.6 Appearance of a fragile X chromosome in the light microscope.

(continued)

BOX 13.1 *(Continued)*

FINDING GENES **441**

BOX 13.2 *(Continued)*

HD - Huntington's Disease

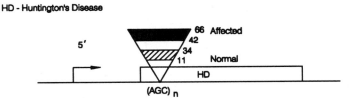

FMR-1 - Fragile X Syndrome

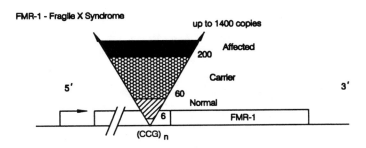

AR - Kennedy's Disease

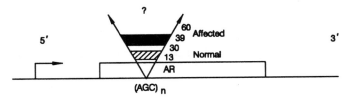

DM-1 - Myotonic Dystrophy

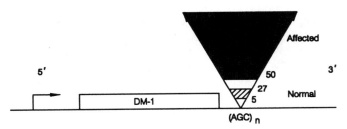

Figure 13.7 Summary of the VNTR expansions seen in four inherited diseases. Shown are repeat sizes to normal alleles and disease-causing alleles. Adapted from Richards and Sutherland (1992).

(continued)

BOX 13.2 *(Continued)*

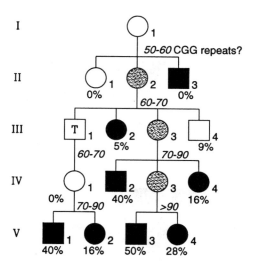

Figure 13.8 Typical fragile X pedigree showing anticipation and a nontransmitting carrier male (T). Black symbols denote mentally retarded individuals; gray symbols are carrier females. Arabic number %'s are the risk of mental retardation based on the general statistics for pedigrees of this type; italic numbers are the copies of CGG repeats present in particular individuals. Adapted from Fu et al. (1991).

rier to affected, explains some of the unusual genetic features of the disease that so puzzled early investigators. Figure 13.8 shows a typical fragile X pedigree. It reveals two nonclassical genetic effects. The individual labeled T in the figure is called a nontransmitting male. He is unaffected, and yet he is an obligate carrier because two of his grandchildren developed the disease. One of these is a male who must have received the disease-carrying chromosome from his mother. The second, more general feature of the pedigree in Figure 13.8 is called anticipation. As the generations proceed, a higher percentage of all the offspring develop the disease phenotype. This is because the number of copies of the repeated sequence in the carriers keeps increasing, until the repeat explodes into the full-blown disease allele. Note that in this pedigree, as is usual, the affected males do not have any offspring. This is because the disease is untreatable and severely disabling. It is effectively a genetic lethal, and thus can only exist at the high frequency observed because the rate of acquisition of new mutations must be high.

Something about the gradual increase in the size of the repeat must eventually trigger a molecular mechanism that leads to a much greater further expansion. Thus above some critical size the sequence is genetically unstable. Figure 13.9 shows two alternate mechanisms that have been proposed to account for this instability. In the first of these, it is postulated that somewhere else in the genome, there is a sequence that normally has no effect on the trinucleotide repeat. However, in a founder chromosome (one that will lead to the carrier state and eventually produce the disease) a mutation occurs in this sequence. This acts, either in cis or in trans, to destabilize the repeat, which then

(continued)

BOX 13.2 *(Continued)*

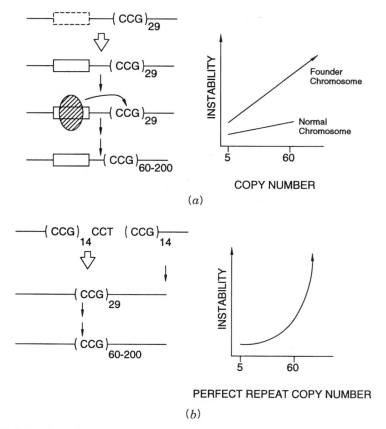

(a)

(b)

Figure 13.9 Two possible mechanisms for the generation of a chromosome with an unstable trinucleotide repeat. (*a*) A mutation affects a site outside the repeat that then acts in cis (directly) or trans (by attracting some other component such as a protein) to destablize the repeat. (*b*) A mutation changes an imperfect repeat into a perfect one, which is assumed then to be intrinsically unstable. Adapted from Richards and Sutherland (1992).

grows larger until, above some critical size, a new mechanism (some abnormality in chromatin packaging, recombination, or replication) leads to the explosive increase in repeat size that produces the disease. There is no a priori reason in this mechanism why the initial mutation rate might be so high.

The second mechanism shown in Figure 13.9 postulates that in a normal chromosome, the repeat is imperfect. This somehow intereferes with processes that lead to expansion in the size of the repeat. If a mutation occurs that makes the sequence a perfect trinucleotide repeat, a founder chromosome is created that is progressively more and more unstable as the repeat size grows. This second mechanism also explains what is seen experimentally, and it may offer clues as to why the observed mutation rate is so high, since an imperfect repeat could be converted to a perfect repeat by a number

(continued)

BOX 13.2 *(Continued)*

of processes including unequal sister chromatid exchange or recombination, or gene conversion (Chapter 6).

Whatever processes lead to the progressive increase in the size of the repeat, the available data clearly show that the degree of instability is strongly affected by the size of the repeat. Figure 13.10 shows the distribution of repeat sizes in the normal population (where a few specific sizes predominant) and in carriers where a much broader range of larger repeat sizes is seen). When the repeat size of parent and offspring is compared, there is a clear correlation between the size of the parental allele and the increase in size seen in the allele of the offspring. These results are shown in Figure 13.11 for offspring that are still carriers and not affected. However, the effects are much more dramatic when affected are also included. This is illustrated in Table 13.2. (It is impractical to include the affected in Fig. 13.11 because some fragile X disease alleles have more than 1000 copies of the trinucleotide repeat, which makes them difficult to display on the same scale.)

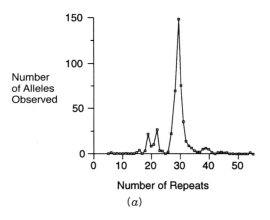

(a)

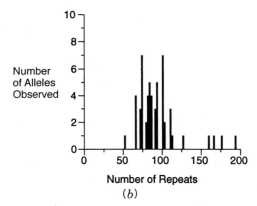

(b)

Figure 13.10   Distribution of CGG repeat lengths in normal individuals (*a*) and premutation carriers (*b*). Adapted from Fu et al. (1991).

(continued)

BOX 13.2 *(Continued)*

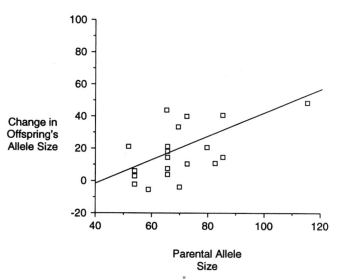

Figure 13.11 Effect of the CGG repeat length in the fragile X gene of an individual on the length of the corresponding repeat in an offspring of that individual. Adapted from Fu et al. (1991).

The data in Table 13.2 also demonstrate that as the size of the carrier allele grows, the probability that it will convert during the next female transmission to the full disease state increases steadily until, above carrier alleles of 90 repeats, the offspring are always affected. In contrast, the expansion of the allele in male transmission is slight, explaining the existence of nontransmitting males shown in Figure 13.8. Thus some difference between male and female transmission must underlie the basic mechanism of expansion of this allele. Repeat expansion in myotonic dystropy, like the fragile X syndrome, requires female transmission. In contrast, Huntington's disease is more severe if it is inherited as a result of male transmission. Since we know that meiotic recombination in the male and female are significantly different (Chapter 6), it is tempting to speculate that meiotic recombination hot spots may somehow be involved in the generation of these disease alleles. However, in most fragile X affected males and females, a mosaic pattern of triplet repeat sizes is seen. In adult tissues, and even in a fetus as young as 21 days, this mosaic pattern is mitotically stable. The inescapable conclusion from these findings is that expansion of the repeat occurs after fertilization, at some early stage in embryogenesis. Perhaps some form of imprinting (methylation pattern) in the X chromosome derived from the mother plays a key role in the subsequent expansion of the repeat to form the disease allele. The mosaicism of the fragile X syndrome is also responsible for the failure of male fragile X patients to transmit the disease to their offspring. (Rare as these offspring may be, a few cases are known.) When sperm from these patients are examined, none contain the full expanded fragile X disease allele. Either this allele interferes with some key process in spermatogenesis,

(continued)

BOX 13.2 *(Continued)*

TABLE 13.2   Comparison of Offspring's Fragile X Alleles with Premutation Alleles in Parents

	Parental Allele	Child's Allele
Male transmissions	66	70, 83
	86	100
	116[a]	163
Female transmissions	52	73
	54[b]	58, 60, 57, 58, 52
	59[b]	54
	66	73, 86, f
	66	60, 73, 110
	70	66, 103
	70	f, f
	73	f, f, f
	73	f, 170[c]
	73	113, f
	77	f, f, f
	80	f
	80	f, f
	80	100
	83	f, f
	83	f, f
	83	f
	83	93, f, f, f
	86	f, f, f
	86	126/193[a]
	90	f, f
	90	f, f
	93	f, f
	93	f
	93	f, f
	93	f, f
	93	f
	100	f, f
	100	f
	100	f
	110	f, f
	113	f

Source: Adapted from Fu et al. (1991).

Note: Numbers indicate number of CGG repeats found in each individual. Those children marked with f received a full mutation.

[a] These individuals are mosaic, and the most prominent allele(s) are indicated.

[b] These transmissions are in a family (CEPH 1408) with no history of fragile X syndrome; all others are fragile X chromosomes segregating in fragile X disease pedigrees.

[c] This allele was measured by Southern blot analysis, and the number of repeats estimated.

(continued)

BOX 13.2 *(Continued)*

or the germ cells all derive from an early embryonic presursor cell that cannot tolerate the full-blown fragile X mutation.

The actual mechanism by which the expanded fragile X allele leads to altered expression of the fragile X gene is a bit mysterious, since the expanded repeat lies outside the coding region. However, a tantalizing hit about this mechanism is provided by the observation that in chromosomes with highly expanded triplet repeats, a nearby HTF island is hypermethylated. Furthermore severity of the illness is modulated by the degree of methylation. Under ordinary circumstances this will be associated with inactivation of gene expression. What is less obvious is the cause and effect relationship between allele expansion and the hypermethylation. For instance, some T-males have expanded repeats but not hypermethylation of the nearby CpG island. A variety of recent studies indicate that the CTG and CGG repeating sequences adopt unusual structures as duplexes, while the separated strands can form stable hairpins that can interfere with replication and lead to a number of other anamolous properties (Wells, 1996). In fact the characteristics of long repeats are so unusual that it is difficult to carry out conventional PCR amplifications of such sequences. Methods have been developed to exploit these unusual characteristics in an attempt to make the detection of new expanded repeats more efficient (Schalling et al., 1993, Broude et al., 1997).

Suppose that all of the approaches described above fail in the search for a particular disease gene. Such a case, for a long time, was Huntington's disease. This was one of the first inherited diseases identified by linkage to polymorphic markers. However, for almost a decade after the first evidence of linkage, and after extensive characterization of the DNA of the region, no specific gene had emerged as a strong contender for the cause of the disease. Huntington's is a case where the disease phenotype is premature death of certain cells in the brain. Thus genes with expression specific to the brain were plausible candidates, but many more indirect mechanisms could not be ruled out. Eventually the gene for Huntington's was identified because a coding sequence isolated from the region by exon trapping (Box 13.1) was found to contain a polymorphic $(CAG)_n$ sequence. However, in less fortunate cases, unless particular biological hypotheses emerge that can be tested, one is left with the particularly unpalatable alternative of sequencing the DNA of the region in a set of affected and unaffected individuals. The disease allele should be one that is common to all affected and not common to others. The problem is that in a Mb region of DNA, there will be several thousand polymorphisms. Only one of these is the disease allele, itself. The remainder are presumably harmless, normal, silent variations in the population.

A sensible strategy would focus on sequencing coding regions first, in the hope that the disease allele is directly expressed as an altered protein sequence. However, there is plenty of precedent for disease alleles that lie outside the coding region and exert their influence by altering RNA splicing or mRNA expression levels. Thus in the worst case one might have to resort, ultimately, to sequencing the entire genomic DNA of the region of interest in a set of individuals. This would not be impossible, even with existing methods, but it would be extremely costly and inefficient. The prospects of such tasks, strongly encourage the development of methods for finding and then sequencing just the differences between pairs of DNA samples. (See Chapter 14.)

A few guidelines exist that can help identify particular polymorphisms that might be responsible for producing disease alleles. DNA sequence changes that disrupt protein sequence are likely to lead to significant disease phenotypes. Prime suspects are frame shifts that will cause massive changes in protein sequence, and usually premature chain termination when a stop codon in the new coding region is reached. Insertions, deletions, and nonconservative amino acid changes are also likely suspects for causing disease. For example, the protein collagen consists mostly of a extensive pro-pro-gly repeat, and the existence of gly every third residue is essential for proper formation of the tertiary structure of the collagen fiber. Thus any mutations of the glycine are likely to have serious consequences, and it has been observed in many inherited collagen disorders studied thus far that mutations of these glycines are involved.

It is important not to trust any guidelines about disease alleles too strictly. It is too easy to get fooled. For example, many disease alleles in familial Alzheimer's disease turn out to be very conservative amino acid changes like isoleucine to valine. Note that disease alleles that are truly catastrophic will usually not be found except in carriers for a recessive disease. The homozygous affected individuals is unlikely to survive beyond early embryogenesis and will usually not be detected because the only result will be an early, usually unnoticed, spontaneous abortion.

In several years the human genome project should provide the DNA sequence of all human genes, including those responsible for all forms of inherited disease. Once all these sequence data are available, the process of linking specific DNA sequence alterations to particular diseases should become much simpler. In some cases comparison of DNA sequence in the human to sequences that have already been studied functionally, in the human or in other organisms, will provide direct clues to possible function (see Chapter 15). In many cases the DNA sequence will allow the construction of useful hybridization probes or PCR primers to examine the pattern of gene expression at the mRNA level. This will help find disease genes where the result of a DNA alteration is a change in mRNA expression at any level, including transcription efficiency or tissue distribution, splicing, or mRNA stability.

In some cases mRNA levels or other characteristics could remain essentially unaltered, even though there was a disruption in the nature of the protein product produced, or the level or cellular location of this product. In such cases the DNA sequence can be used to design peptide antigens that will elicit the production of high-affinity monoclonal antibodies specific for the protein of interest. Such powerful antigenic peptides are called immunodominant epitopes. Their location can often be predicted from the sequence, and the peptides can be synthesized automatically and used for immunization. The resulting antibodies are excellent reagents for examining the level and location of the particular protein in specific cells or tissues. Thus, all in all, a fairly powerful arsenal of approaches is accessible for finding genes with potential functions, once DNA sequence information is available.

DIAGNOSTICS AT THE DNA LEVEL

We have discussed the role that DNA physical mapping and sequencing can play in finding genes involved in human diseases. Once such genes are found (or in more limited cases, once the approximate location of a disease gene is known), it is extremely desirable to develop diagnostic tests for the presymptomatic disease state or for the carrier state.

There are a number of distinct advantages to the direct examination of DNA for diagnostics. Any cell of the body can be used to detect an inherited alteration in DNA. Thus there is no need to sample what may be a relatively inaccessible target tissue where the disease effects will be most pronounced. Since all cells will have the same alteration in DNA, whether or not this produces a disease phenotype in these particular cells, any cells can be investigated. In practice, it is most convenient to use easily accessed samples like blood or epithelia sloughed off in saliva.

In principle, diagnosis at the DNA level will be less sensitive than many tests at the RNA, protein, or metabolic level. The DNA is present in only a few copies per cell; most RNAs and proteins or small molecules are much more frequent, especially if a cell in which the gene or its products are metabolically active is selected. However, for genomic analysis, it turns out that the sensitivity of DNA testing is more than sufficient. This is especially true in most cases because PCR amplification can be used, and thus the samples of cells required can be quite modest in size. The problem is quite different when tests are applied to tumors or infectious diseases. Here only a small fraction of the target cells may have altered DNA sequences present. This demands much higher sensitivity, and sometimes DNA tests may not be sensitive enough for the desired analysis, unless potentially affected cells or infectious agents are purified prior to PCR analysis. However, in some cases sensitive PCR methods have been able to detect low levels of circulating tumor cells (Nawroz et al., 1996).

The perfect DNA test will discriminate precisely between all normal alleles in the population and all disease alleles. How close we can come to this ideal is very different for different inherited diseases. Sickle cell anemia represents an example of the clearest case for DNA analysis. Here the physical diagnostic criteria are clear enough that a single disease allele in the population is responsible for the bulk of the phenotype we call sickle cell disease. The actual change in the gene of HbA, the normal beta chain of hemoglobin, to HbS, with a single altered amino acid, occurs in a restriction site for two enzymes. Thus the disease allele is an RFLP, and it can be analyzed this way to distinguish HbA normals, HbA HbS carriers (sickle cell trait), and HbS individuals affected with sickle cell disease.

HbA	CCTGAGGAG	*Bsu*36 I	CCTNAGG
HbS	CCTGTGGAG	Dde I	CTNAG

Alternatively, allele-specific PCR, as illustrated in Chapter 4, can be used to make the analysis. There are other hemoglobin disorders, collectively called hemoglobinopathies, which produce anemia or other impairments in oxygen transport. Many of these can be successfully analyzed prenatally or in the carrier state by DNA diagnostics. A long catalog of specific abnormalities with particular clinical presentations is now known. Most are quite rare.

Cystic fibrosis is an example of a more complex case for DNA diagnostics. This relatively common disease is undoubtedly far more representative of what will be found for the majority of human inherited disorders, and it raises a number of severe complications. In cystic fibrosis, one particular disease allele, a three-base deletion, resulting in the deletion of a single, key amino acid in the protein chain, phenylalanine 508, accounts for most of the observed disease cases. In populations of northern European extraction, this allele occurs in about 70% of all affected individuals. (Note the implicit complication caused by the heterogeneity of the partially outbred human population. The actual statistics of disease occurrence and alleles responsible are highly dependent on the particular subpopulation of origin of the individuals in question.) In this same group another 10 alleles account for about 20% more of the disease state. Hundreds of additional alleles must be considered, however, to account for all cystic fibrosis cases that have been examined thus far at the DNA level. Ironically some of these produce disease that is phenotypically so different from a classical cystic fibrosis presentation that until DNA analysis of this gene defect became prevalent, these cases were not even recognized as being cystic fibrosis at all. We can extrapolate and conclude that for a large gene like cystic fibrosis, the number of potential disease alleles in our population will be so large that it may never be possible to identify all of them a priori. This will certainly be the case if some are the result of new mutations.

The complexity of disease alleles at the cystic fibrosis locus leads to serious problems when prenatal or postnatal diagnosis for this disease is considered. Suppose that two nonaffected potential parents are tested for the major allele. If neither is positive, it is fairly likely their offspring will be unaffected by cystic fibrosis. A negative result in a test for the next ten most likely alleles would certainly all but rule out the risk of a child with cystic fibrosis. If, instead, both parents test positively for the major allele, or for other frequent disease alleles, the risk of their producing an affected child is one in four (assuming that the homologous allele in each is normal; otherwise, they would be affected individuals). In this case, prenatal diagnosis of a fetus conceived by this couple would be strongly indicated.

A third, not uncommon, potential outcome of genetic tests of the parents leads to a more complicated situation. Suppose that one parent tests positive for the major cystic fibrosis allele (or other common alleles) while the second parent tests negative. There is still a small but significant chance that this second parent is a carrier for a rare disease allele. Thus a prospective child from this couple has a significant risk for cystic fibrosis. This risk can be partially assessed by performing prenatal diagnosis. If the fetus shows an absence of all the major cystic fibrosis alleles, it is very likely that the child is safe from cystic fibrosis. However, if the fetus has inherited the detectable allele from the first parent, one may really want to scrutinize the second parent for minor cystic fibrosis alleles. We need to develop much more cost-effective and accurate ways to do this.

The above example illustrates a major benefit of genetic testing at the DNA level. In a family known to be segregating a disease like cystic fibrosis, the result of a fairly harmless test will, in most cases, rule out any significant chance of a child affected with the disease. Thus the family is spared a great deal of unnecessary anxiety, and the child is spared the trauma of more invasive diagnostic tests. A particularly compelling case for such analysis which is not yet practical but will probably be doable soon is neurofibromatosis. This is a very unpleasant and disabling disease (the elephant man syndrome). The usual earliest clinical presentation of the disease is the presence, at birth, of occasional cafe au lait (light brown pigmentation) spots. Many less severe syndromes can also result in such spots. Hence the real power of genetic tests, once available, will be to rule out the likelihood of a terrible disease in all but a small percentage of those who display these symptoms.

Genetic analysis at the DNA level becomes much more complex for dominant lethal diseases like neurofibromatosis. The problem, discussed earlier, is that most if not all of the alleles responsible for such diseases have to be new mutations. Unless there is a restricted pattern of mutation hot spots in the gene responsible for the disease, one is faced with the difficult prospect of having to search the entire gene for all possible DNA changes. When the enormous size of some genes is considered, the task in some cases could be truly formidable. The gene responsible for Duchenne muscular dystrophy has more than 2.5 million bases; the coding region is more than 15 kb, and it is divided into more than 50 exons. The coding regions for an early onset BRCA2, an early onset breast cancer gene, cover 11,000 base pairs with 26 exons spread over 100 kb.

DNA analyses are also potentially applicable to somatic mutations such as the DNA alterations seen in most kinds of cancer. Some of the types of alterations seen, which are believed to be primary events in the ultimate progression that leads to tumor formation, are shown in Figure 13.12. Some tumor genes are dominant lethals. Once a single copy of the gene responsible for cancer, an oncogene, is affected by a mutation in a somatic cell, uncontrolled growth, or the accumulation of further mutations which lead, in turn, to uncontrolled growth, is triggered (Fig. 13.12a). In other cases the alleles affecting oncogenes are recessive. A single somatic mutation does not immediately lead to cancer. Instead, a second DNA rearrangement must occur to expose the effect of the first gene (Fig. 13.12b). The requirement for a second hit is clear if one imagines the function of the recessive oncogene to be suppression of a gene or a process that otherwise would lead to uncontrolled growth. Loss of one homolog still leaves the other normal allele intact and functional. However, if the normal allele is lost by a second event, which could be local DNA damage, a deletion, or mitotic nondisjunction, the resulting cell is now triggered for

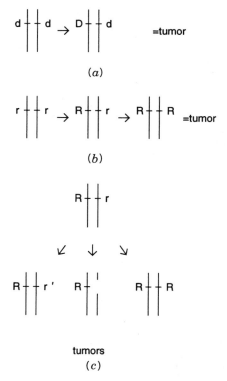

Figure 13.12 DNA changes seen in various types of cancer. (a) A dominant oncogene. (b) A recessive oncogene. Reduction to homozygosity must occur by a second mutation. (c) In familial cancers only one new mutation is required.

uncontrolled growth. This cell now has a selective advantage, and hence its progeny develop into a large population, a tumor.

Many cancers are familial: They have inherited components. Examples include a subset of the individuals affected with several common cancers like colon and breast cancer, as well as many rarer forms like retinoblastoma, a tumor of the eye, and neurofibromatosis, which we have already discussed. In inherited retinoblastoma, the child receives one defective recessive oncogene from the parents; both eyes are highly susceptible to an additional event that damages this gene. As a result bilateral tumors are common. In contrast, with the purely somatic form of the disease, two independent damage events are needed. It is very unlikely for these to occur in both eyes; hence this form of retinoblastoma is usually unilateral.

Mutations in the DNA from many individuals affected by familial (inherited) and spontaneous (somatic) forms of several kinds of cancer have been investigated. The results for colon cancer somatic mutations seen in the adenomous polyposis coli (ApC) gene are summarized in Table 13.13. It is evident, as expected, for such new mutations, that there are a very large number of different disease alleles. This will make it difficult to use DNA analysis to find all possible disease cases unless the entire gene can be scanned. Mutation sites for somatic and familial colon cancer are summarized in Figure 13.13. It is apparent that no significant hot spots are seen for the somatic form. The familial form does display a hot spot area, but the actual mutations here are still widely distributed. Interestingly, regions of the gene that predominant in the familial form are different from those that are common in the somatic form. Thus the challenges of DNA analysis for this fairly common and deadly cancer are truly formidable.

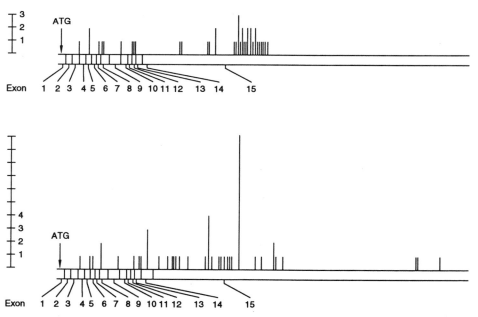

Figure 13.13 Locations of mutations in the APC gene, responsible for many human colon cancers. Shown are numbers of individuals found with (a) somatic mutations and (b) inherited mutations. Adapted from Miyoshi et al. (1992).

TABLE 13.3 Forty-Three Somatic Mutations of Colorectal Tumors in the APC Gene

C113(F)	142	a*a*tag/CTG → atag/CTG	Splice acceptor
C31, C124(F)	213	CGA → *T*GA	Arg → Stop
C24	279	aatttttag/GGT → agtttttag/GGT	Splice acceptor
C47	298	CA*CTC* → CTC	AC deletion
C108(F)	302	CGA → *T*GA	Arg → Stop
C135	438	CAA/gtaa → CAA/g*c*aa	Splice donor
C33	516	AAG/gt → AAG/*t*t	Splice donor
C28	534	AAA → *T*AA	Lys → Stop
C10	540	TTA → TT*A*T	A insertion
C37	906	T*C*TG → TTG	C deletion
A128(F)	911	GAA → G*G*A	Glu → Gly
C23	1068	*TCAA*GGA → GGA	TCAA deletion
C11, C15	1114	CGA → *T*GA	Arg → Stop
C20	1286	GAA → *T*AA	Glu → Stop
A53	1287	ATA → A*A*TA	A insertion
C27	1293	*ACACAGGAAGCAGATTCT*	31 bp deletion
		*GCTAATACCCTGC*AAA → AAA	
C7, C21	1309	G*AAAGA*T → GAT	AAAGA deletion
C14	1309	GAA → *T*AA	Glu → Stop
A41	1313	ACT → *G*CT	Thr → Ala
C31, C42	1315	TCA → T*A*A	Ser → Stop
A44	1338	CAG → T*A*G	Gln → Stop
C22	1353	*GAATTTTCTTC* → TTC	8 bp deletion
A56	1356	TCA → T*G*A	Ser → Stop
C4, C27	1367	CAG → T*A*G	Gln → Stop
C10	1398	*AG*TCG → TCG	AG deletion
C19	1398	AG*T*C → AGC	T deletion
A43	1411	A*G*TG → ATG	G deletion
C16	1420	CC*C*A → CCA	C deletion
C40, A52(F)	1429	GAA → *T*AA	Gln → Stop
C29	1439	C*C*TC → CTC	C deletion
C37	1446	*GCTCAAACCAA*GC → GGC	10 bp deletion
A50(F)	1448	T*T*AT → TAT	T deletion
A49(F)	1465	*AG*TGG → TGG	AG deletion
C23	1490	CA*T*T → CTT	A deletion
C12	1492	GC*C*A → GCA	C deletion
A41	1493	*ACAGAAAGTAC*TCC → TCC	11 bp deletion
C3	1513	GAG → *T*GAG	T insertion

Source: Adapted from Miyoshi et al. (1992).

If we had the ability to do very large-scale DNA sequencing, genetic diseases could be diagnosed by this technique with great power but still not without difficulties. We would still have to develop effective ways to distinguish, for newly found alleles, whether they were just harmless polymorphisms or true disease-causing alleles. While some guidelines for how this might be done were presented earlier in the chapter, it will be hard to do this in general without considerable information about the function of the protein product of the gene. An example of the complex spectrum of spontaneous mutations seen in the human factor IX gene responsible for hemophilia B is shown in Table 13.4. These results

TABLE 13.4   Summary of Sequence Change in 260 Consecutive Cases of Hemophilia B

	Number	Percentage
1. Number with sequence changes in the eight regions of likely functional significance	249	96
2. Of those with sequence changes, number of independent mutations[a]	182	73
3. Of independent mutations, number with a second sequence change	6	3
4. Type of independent mutation:		
Transitions at CpG	48	26
Transitions at non-CpG	65	36
Transversions at CpG	8	4
Transversions at non-CpG	35	19
Small deletions and insertions ($\leq$50 bp)	15	8
Large deletions (>50 bp)	10	6
Large insertions	1	0.6
5. Location of independent mutations:[b]		
Promoter	1	0.5
Coding sequence[c]	163	86
Splice junctions	12	6
Intron sequences away from splice junctions	1	0.5
Poly A region	0	—
Unlocalized (total gene deletions)	5	3
Unknown[b]	8	4
6. Functional consequences of observed independent mutations		
Protein with amino acid substitutions	114	63
Garbled protein (truncated, frameshifted or partial or full deletion of amino acids)	54	30
Abnormal splicing	13	7
Decrease expression	1	0.6

Source: Adapted from Sommer et al. (1992).

[a]Recurrent mutations were judged independent if the haplotypes differed. In a few cases recurrent mutations with the same haplotype were judged independent because the origin of mutation was determined. In four patients recurrent mutations were judged independent because the races of the individuals were different.
[b]Assumes that 11 patients with unknown mutations have the same frequency of independent mutations as patients in which the mutation could be defined (73%). Thus eight independent mutations should be unknown.
[c]Includes partial gene deletions that affect the coding region.

clearly indicate that no single test at the DNA level, even complete DNA sequencing, would be capable of 100% certain diagnoses. It is difficult to measure heterozygous positions with typical gel-based sequencing, and haplotypes of compound heterozygotes cannot be determined at all.

The difficulty of DNA analysis of cancers includes all of these problems, but it is confounded by the heterogeneity of typical tumor tissue. For very early onset diagnosis of somatic cancer, one would need to distinguish, potentially, any altered nucleotide, in a complex gene, present in only a minute fraction of the cells in a sample. It is not easy to see how this could be accomplished by directly DNA sequencing alone. In all these cases, the task is much simpler if only a finite number of specific sequence alleles are correlated with the disease. Then one can set up specific assays for the alleles, and some such assays seem capable of dealing with the sorts of heterogeneous samples encountered in tumors. The emergence of specific cancer or disease-associated alleles can also be tested for in easily assayable fluids. For instance, the fingerprint of prostate tumor cells has been detected in urine, and those of head and neck tumors have been detected in saliva and blood.

ANALYSIS OF DNA SEQUENCE DIFFERENCES

The examples described in the previous section clearly indicate the need to be able to analyze a stretch of DNA sequence to look for abnormalities that might be as small as single base pair. If the DNA target is just a few thousand base pairs, this search can be done by direct sequencing. If the target is much larger, direct sequencing with current methods is impractical. In this section we explore the present status of methods that can detect changes in DNA as small as single base pairs, with less effort than would be required for total DNA sequencing. Many of these methods resort to the formation of DNA heteroduplexes to facilitate the screening for differences between a test sequence and a standard.

Figure 13.14a shows, schematically the three possible genotypes that must be distinguished in making a genetic diagnosis. In general, an individual tested could be normal (dd), heterozygous for the disease allele (dD), or homozygous for the disease allele (DD). If the disease is dominant, one usually expects the affected to be a heterozygote. If one is testing for a carrier status, the test is really looking to see if the individual in question is a heterozygote. In either of these cases, a normal DNA duplex is present that can serve as an internal control for the possible presence of an altered duplex. DNA from the region of interest can be prepared directly from genomic material by PCR, assuming enough known sequence exists to design suitable primers. If this DNA is melted and the separated strands are allowed to reanneal, four distinct products will be formed (Fig. 13.14b). Two of these are the perfectly paired normal and abnormal duplexes. The remaining two are heteroduplexes composed of one normal and one abnormal strand. Any DNA sequence differences between these species will lead to imperfections in the duplex because one or more base pairs will be mismatched.

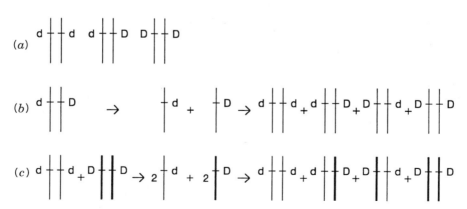

Figure 13.14 Detection of a disease allele as a heteroduplex. (a) Three genotypes that must be distinguished in disease diagnosis. (b) Heteroduplex formation by melting and reannealing DNA from a recessive carrier or a dominant heterozygote. (c) Heteroduplex formation by mixing DNA from a homozygous recessive with DNA from a normal individual.

If the disease in question is recessive, the usual question asked is whether the particular individual being tested is homozygous for the disease allele or heterozygous. Thus a minimal test would be to proceed exactly as described above, except that now, presence of heteroduplex indicates that the individual is a carrier, unaffected for the disease. However, this test would be ambiguous because absence of heteroduplex would imply that either the individual in question is a homozygous normal or a homozygous affected. An additional test is required to resolve this ambiguity. This second test is designed as shown in Figure 13.14c. DNA from the individual to be tested is mixed with a standard DNA sample from a person previously shown to be normal, not a carrier. If the test individual carries the disease, four DNA products will be formed; two of these will be heteroduplexes. If the test individual is normal, no heteroduplex products will be produced. Thus the tests shown in Figure 13.14 reduce the problem of detecting DNA sequence alterations to the problem of detecting heteroduplex DNA.

We will shortly describe a wide variety of techniques that have varying success in distinguishing between DNA heteroduplexes and homoduplexes. One caveat in using these methods for genetic testing must be noted. Successful heteroduplex detection will find any DNA alterations, whether or not these are disease alleles or harmless polymorphisms. Thus what heteroduplex detection does is indicate the presence of a DNA sequence variant. Once this is found, DNA sequencing will frequently be needed to examine the characteristics of the particular variant discovered. Thus the strategy used is to apply a very simple test that can scan large DNA regions to see if any sequence variations exist. If none are found, and the test is reliable, one need go no further. If differences are discovered, then usually a more robust test will need to be applied, but the screening will have narrowed down the DNA target to a much smaller region. For example, PCR can be used to examine the exons of a complex gene one at a time. If a sequence variation is discovered in a single exon, at worst one would have to sequence the DNA of that exon to complete the diagnosis.

HETERODUPLEX DETECTION

The difficulty in designing schemes to detect heteroduplexes is that many possibilities can arise, even from a single altered base pair. This is shown in Figure 13.15. Any mixture of two DNAs with a single base pair difference produces two different heteroduplexes with a single base mismatch. In all, there are eight possible single base mismatches: A–C, A–A, A–G, C–C, C–T, T–T, T–G, G–G, and an acceptable test would have to be able to detect them all. The ideal test would not only detect them, but it would also reveal which exact mismatches were present. A potential complication is that each heteroduplex occurs

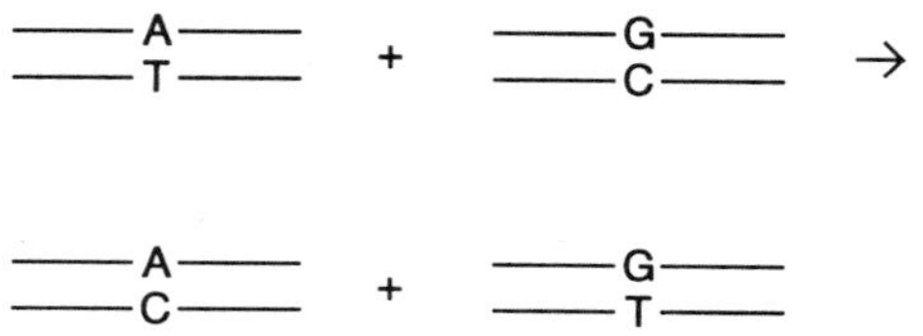

Figure 13.15 A single site mutation will serve to generate the formation of two different heteroduplexes.

within the context of a specific DNA sequence, and the identity of the neighboring base pairs could easily modulate the properties of the heteroduplex. Not much is known about this at present.

In practice, the formation of heteroduplexes from a diploid sample will produce a pair of mismatches. For single base pair differences there are only four possibilities, and each gives a different and discrete set of heteroduplexes. Thus a test that detected a specific half of the possible heteroduplexes would suffice:

MUTATION	HETERODUPLEXES
A–T to T–A	A–A and T–T
G–C to C–G	G–G and C–C
A–T to G–C	G–T and A–C
A–T to C–G	A–G and C–T

Thus, for example, a method that could identify a mispaired T or G but not A or C would suffice to spot the presence of a heteroduplex, but it would probably not have enough resolving power to identify the exact heteroduplex present.

The single-base, mismatched heteroduplexes just illustrated are actually the most difficult case to detect by the methods currently available. Larger mismatches or heteroduplexes arising from insertions or deletions lead to much larger perturbations in the DNA double helical structure, and these are easier to reveal by physical and chemical or enzymatic methods. The principle complication is that the number of possible heteroduplexes becomes rather larger. There are four possible single base insertions or deletions, but these are likely to be susceptible to complexities caused by the local sequence context. For example, in the sequence shown below, two altered mismatched structures compete with each other.

$$
\begin{array}{cccccc}
\text{A–T} & & \text{A–T} & & \text{A–T} & \text{A–T} \\
\text{G–C} & \longrightarrow & \text{G–C}\ \text{C} & \longleftrightarrow & \text{G–C} & \text{G–C} \\
\text{G–C} & & \text{T–A} & & \text{T–A}\ \text{C} & \longleftarrow\ \text{G–C} \\
\text{T–A} & & & & & \text{T–A}
\end{array}
$$

The genetic consequences of the two different structures shown above are identical; however, the presence of two alternate heteroduplex structures could complicate the analysis. Much more work needs to be done to characterize the properties of such structures in more detail. This will have to be done before the overall accuracy of any proposed method of heteroduplex detection can be validated for clinical use.

At least six different basic methods for detecting heteroduplex DNA have been described. All of these tests work well in some cases. However, none have yet been proved to be generally applicable to all possible heteroduplexes. A key issue in these tests is how large a DNA target can be examined directly. The larger the target, the fewer fragments will be needed to cover a whole gene. However, if targets are too large, they may have such a high probability of containing a phenotypically silent polymorphism that the advantage of the test as a primary screen will be lost, since many fragments will test positive. The first four tests, illustrated in Figure 13.16, all have potentially similar characteristics, and all will work, in principle, on very large DNA targets. A straightforward and direct approach is to use single-strand-specific DNases like S1 nuclease to cleave at the

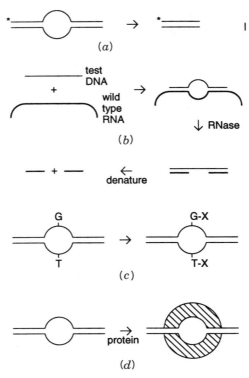

Figure 13.16   Four different methods for direct detection of heteroduplex DNA. (*a*) Nicking with S1 nuclease. (*b*) Trimming and nicking an RNA-DNA hybrid with ribonuclease. (*c*) Reaction of un-paired bases with organic molecules like carbodiimides. (*d*) Binding of *mut*S protein to the site of the mismatch.

mismatched bases in a heteroduplex (Fig. 13.16*a*). Then the resulting shortened DNA fragment could be detected by a Southern blot or by PCR. The difficulty with this approach is that no known combination of enzyme and reaction conditions allows reliable cleavage at single base mismatches. Larger mispaired targets are needed before cleavage becomes efficient and selective. Other nucleases that look very promising for such studies include bacteriophage T4 endonuclease VII.

RNase nicking can be used as an alternative to DNase nicking. Here a single strand of the DNA to be tested is annealed with RNA made from a DNA sample representing the normal allele (Fig. 13.16*b*). This is easily accomplished, for example, by subcloning that allele downstream from a strong in vitro promoter like that for T7 RNA polymerase. One can produce either internally labeled RNA, and look for two shorter fragments as a sign of fragmentation, or end-labeled RNA, and look for one shorter fragment. In this latter case it is necessary to have the test DNA extend beyond the labeled end of the RNA; otherwise, the RNase will remove the label. Some workers swear by the reliability and sensitivity of the RNase approach, but many others have apparently been unable to use it successfully.

Chemical methods can be used instead of enzymes to mark or to cleave at the site of a mismatch. A particularly effective approach has been the use of water-soluble carbodiimides which can react with mismatched T or G (Fig. 13.16*c*). These compounds are available radiolabeled, so one can detect the presence of a mismatch by the incorporation of radioisotope. Alternatively, there are monoclonal antibodies available that are specific for the carbodiimide reaction product. This is a very powerful analytical tool, since it allows physical fractionation of any heteroduplexes, which are thus purified and concentrated if needed for subsequent analysis. The carbodiimide approach does have two distinct disadvantages. It cannot detect all possible mismatches, and it has been reported to be a difficult test to master in the laboratory.

The fourth test based on protein recognition of heteroduplexes is rather different because it does not involve enzymatic cleavage. In *E. coli,* a protein is made by the *mut*S gene that recognizes mismatches and binds to them (Fig. 13.16*d*). This is an early step in the excision and repair of mismatched bases. The binding of *mut*S protein can be detected in a number of different ways. DNA, once bound by *mut*S, will stick to nitrocellulose filters, while free DNA passes through. Alternatively, *mut*S fusions to other proteins involved in color-generating reactions have been made, and monoclonal antibodies against *mut*S are also available. Thus a variety of different methods to exist to detect *mut*S-DNA complexes. An attractive feature of *mut*S, like carbodiimides, is that it allows the selective isolation of intact heteroduplexes. The disadvantage shared by both systems is that not all mismatches are detected. For example, *mut*S fails to recognize a C–C mismatch, and some others. The full extent of the advantages and limitations of analysis of mismatches with *mut*S has not yet been described. However, this general kind of approach is attractive because it mimics a natural biological mechanism, and proteins with properties analogous to *mut*S, but perhaps with even broader mismatch recognition, may well exist in other organisms.

Two additional methods for detection of heteroduplexes are based on the altered electrophoretic properties of these structures. The simplest of these is direct separation of heteroduplexes from homoduplexes of the same length by using specialized gels. A very effective method uses a modified polyacrylamide called MDE, a term that stands for mutation detection electrophoresis. This gel has a somewhat hydrophobic character which alters the mobility of heteroduplexes selectively. Some move faster; most move slower. In good cases a single mismatched base in a 900 base pair duplex is sufficient to give an easily detectable mobility shift. An example of the improved ability of MDE compared with ordinary gel media to resolve heteroplexes from homoduplexes is shown in Figure 13.17. A more complex example of the use of MDE is illustrated in Figure 13.18, where it is clear that different heteroduplexes within the same basic DNA fragments show different mobilities. Thus, in principle, the method offers some promise of revealing, not just that a heteroduplex is present but additional information about its characteristics. The appeal of this method is that it is very easy to perform, since aside from the special properties of the gel, the electrophoretic procedures used are quite ordinary. It is also easy to analyze many samples in parallel. However, the full generality of the method has yet to be proved. For example, it would be good to know how the nature of the mismatch and its location within the duplex affect the ability to detect it. Already we know that deletions lead to large mobility shifts, and DNA molecules with more than one heteroduplex region show very complex behavior. In general, a heteroduplex combination of a deletion and a separate, distant single-base mismatch leads to much larger mobility shifts than expected from the effects seen with the two mutations separately. The reason for this synergistic behavior is not currently understood.

A second electrophoretic method was originally developed by Leonard Lerman and his collaborators, and several variations on this general theme now exist. The original method

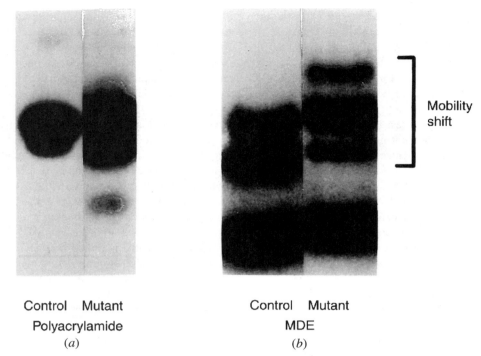

Control   Mutant   Control   Mutant

Polyacrylamide   MDE

(a)   (b)

Figure 13.17   An example of direct detection of heteroduplex DNA by electrophoresis. (a) On polyacrylamide. (b) On MDE gel. Provided by Avitech, Inc.

was called denaturing gradient gel electrophoresis (DGGE). The basic idea behind the method is illustrated in Figure 13.19. DGGE appears to be a general method capable of detecting any mismatch in a DNA sequence that is not at the very end. An internal mismatch in a duplex leads to a substantial decrease in the thermodynamic stability; this is manifested by a drop in the T_m of the duplex (Fig. 13.19a). To test for mismatches, DNA fragments are electrophoresed in a gel through a gradient of increasing denaturant like urea, or a gradient of increasing temperature. At some critical point the section of the duplex containing the destabilizing mismatch reaches conditions above its local T_m, and it

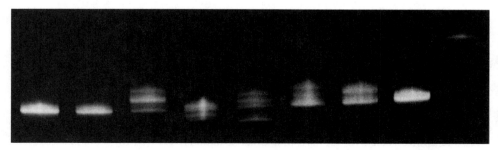

Figure 13.18   Many different single mismatches can be distinguished by electrophoresis on MDE. Each lane is a different cystic fibrosis disease allele except for the left lane which is a normal allele, and shows no heteroduplex formation. Provided by Avitech, Inc.

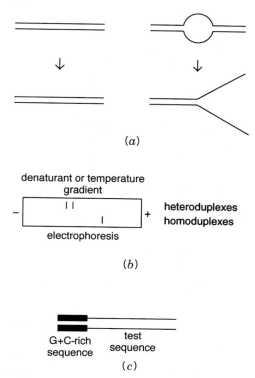

Figure 13.19 Denaturing gradient gel electrophoresis (DGGE) detection of heteroduplexes. (*a*) Destabilization of a duplex by mismatching. (*b*) Electrophoresis in a temperature or denaturant gradient. (*c*) Use of a GC clamp to prevent strand separation in DGGE.

melts. The resulting Y-shaped structure (or one with a large internal loop) has little or no electrophoretic mobility, and so the molecule is trapped in the gel near or at the site where it melted (Fig. 13.19*b*). Thus from a single experiment one can determine not only whether heteroduplexes are present but also what kinds of mismatches were present, since most give characteristic and different shifts in T_m.

It is important that the mismatch does not destabilize the duplex so much that the entire structure melts into separate strands. These would still be free to migrate in the gel. To prevent this unwanted effect, a G+C-rich sequence is usually placed at one end of the duplex to be tested. This is easily done by using PCR primers with an extra overhanging G+C-rich sequence. The use of this so-called GC clamp (Fig. 13.19*c*) prevents complete melting under typical DGGE conditions. A major advantage of DGGE is its generality. Another advantage is that rather complex samples can be analyzed by running a gel composed of a restriction enzyme digest of a clone and then blotting it with particular probes. A major disadvantage of DGGE is that specialized apparatus is needed for this method.

A third electrophoretic method we will describe for detection of altered DNA sequences is also based on electrophoresis, and it takes advantage of the effects of altered DNA sequence on thermodynamic stability. However, in detail, this method is actually quite different from DGGE; it does not involve heteroduplexes. Called single-strand conformational polymorphism (SSCP), the method is based on nondenaturing gel electrophoresis of melted and rapidly cooled samples. Under these conditions individual

DNA strands fold back on themselves to form whatever combination of stems and loops (and other secondary structures, e.g., pseudoknots) that the particular DNA sequence allows. Changes in even a single base can substantially alter the spectrum of secondary structures formed. RNA can be used instead of DNA to enhance the stability of secondary structure formation, and this apparently increases the fraction of heteroduplexes that can be detected. The use of MDE gels also enhances the resolution of SSCP. The results in a typical SSCP analysis are complex. However, SSCP is serving as a very easy and sensitive method to detect a reasonable fraction of all possible heteroduplexes. Two variations of SSCP have been described by Steven Sommer that increase the probability of detecting any mutation in the target. The first of these, called dideoxy fingerprinting (ddf), is a hybrid between Sanger sequencing and SSCP (Sarkar et al., 1992). A Sanger ladder is produced with a single dideoxy pppN, and it is analyzed on a native polyacrylamide sequencing gel. In the second method, restriction endonuclease fingerprinting, the nucleases and the products from these digests are pooled and analyzed together by SSCP (Liu and Sommer, 1995).

A fourth electrophoretic method takes advantage of the power of automated fluorescent DNA sequences. A mutation shows up as an unexpected peak. The sensitivity of the method, called orphan peak analysis (Hattori et al., 1993), is sufficient to allow multiple samples to be probed in each gel lane.

The final method we will mention for detecting altered DNA sequences is based on recent findings by Sergio Pena and coworkers (1994). In using short random DNA primers (RAPD; see Chapter 4) for PCR analysis of human DNA, they noted that the pattern of amplified bands seen was exquisitely sensitive to DNA sequence variations in the neighborhood of the primers such that virtually all alleles tested led to a different, distinct pattern of amplified DNA lengths.

In current practice, faced with a gene to search for mutants, the simplest and most general existing method is probably to do both SSCP and MDE-heteroduplex analysis. The real difficulty that remains is the size of many genes of interest. A typical 3- to 5-kb gene could be scanned in 5 to 10 pieces by selective PCR. In order to do this, however, one has to have available mRNA from the individual to be tested. This may not always be available. A further caveat is that some altered mRNAs may be selectively degraded in a cell if they are not functional. Thus a mutation may make itself invisible at the RNA level. The alternative is do the analysis at the DNA level. Here one can use PCR to look only at the exons. However, the difficulty is that some genes have 30 or more exons, and some exons can be very small. Given the large size of typical introns, each exon will have to be analyzed by a separate PCR reaction; thus the overall test becomes quite complicated.

DIAGNOSIS OF INFECTIOUS DISEASE

DNA sequence is proving to be quite useful in the diagnosis of the presence of infectious organisms. Usually a small bit of DNA sequence will suffice to indicate the presence of a virus, bacterium, protozoan, or fungus. Different species can be identified definitively, once their characteristic DNA sequences are known. The major problem in using DNA analysis for detection of infectious agents is sensitivity; this is the same problem we encountered earlier in examining the prospects of DNA diagnosis for cancer. There may be only a few copies of the DNA (or RNA, for some viruses) genome per organism. The number of infected cells (or the number of organisms free in the blood stream or other body fluids may be very small. PCR is very helpful in amplifying whatever DNA is present, but the real problem is distinguishing a true positive signal from the frequent, unwanted background caused by PCR side reactions. Several strategies have been used to enhance the DNA analysis of infection. If the type of target cell is known, one can often use immobilized monoclonal antibodies against this particular cell type to purify it away from the rest of the sample. This will substantially reduce subsequent PCR background.

Another approach is to screen for the ribosomal RNA of the infectious agent. This is applicable to all agents except viruses, since they do not have ribosomes. Some regions of rRNA vary sufficiently to allow a wide variety of organisms to be distinguished easily. However, the major advantage of looking at rRNA directly is that there are typically 10^4 to 10^5 copies per cell versus only a few copies of most DNA sequences. An alternative, undoubtedly worth exploring for protozoa, will be to use repeating DNA sequences specific to a given species. This is unlikely to work for most simpler organisms because they have very few repeats.

The potential utility of DNA analysis in infectious disease is staggering. We will consider in detail the case of HIV, the virus that causes AIDS. One difficulty in the clinical management of this disease is that the virus has a very high mutation rate. Thus most people have different viral mixtures, and some components of these mixtures are resistant to particular drugs, because of mutations within the HIV reverse transcriptase or protease genes. A brute force approach recently described, which appeared to have some success (although the generality of this success is now disputed), is to treat with a mixture of several different drugs simultaneously. However, each of the drugs has potentially serious side reactions. Furthermore, by treatment with all of the effective agents at once, there is a real possibility of selecting for a viral variant resistant to them all. The complete DNA sequence of the virus is known, and the sequence of many drug-resistant variants has also been determined. Thus we know which sites in the viral reverse transcriptase and protease are likely to mutate and confer resistance to particular drugs.

By direct PCR cycle sequencing of blood samples from AIDs patients undergoing drug therapy, Mathias Uhlen and his coworkers have been able to monitor the course of the disease with a precision not before obtainable (Wahlberg et al., 1992). The single-color, four-lane fluorescent sequencing used by Uhlen is sufficiently quantitative that it not only can distinguish pure viruses, it also can analyze the composition of mixtures of viruses as seen as apparent fractional populations of particular bases at given sequence locations (Fig. 13.20). When the analysis shows that the population of a particular drug-resistant variant is beginning to climb, the physician is alerted to alter the therapy by switching to a different drug. After a while, it is typical to see a relapse in the viral population back to the original major strain, and since this is sensitive to the original drug used, one can switch back to that drug to help control the infection. As this kind of precise diagnostics becomes more affordable and more readily available, it could have a major impact on the practice of medicine.

DETECTION OF NEW MUTATIONS

By definition, a new mutation can occur at any site within a target DNA sequence. To detect new mutations by current methods is very difficult, and it is much more demanding than most of the problems we have discussed earlier in this chapter. New mutations must be detected in the analysis of autosomal dominant lethal diseases, as we have discussed

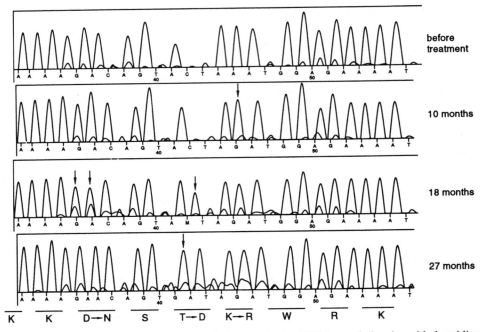

Figure 13.20 DNA sequence analysis of changes in the HIV-1 population in azidothymidine-treated AIDs patients. Shown is the raw sequence data A's (dotted dashed line) and G's (solid line) for a portion of the reverse transcriptase gene before treatment and at various times after treatment. Corresponding amino acid changes are shown below. Adapted from Wahlberg et al. (1992).

before. It is also necessary to detect new mutations if we are to be able to estimate the intrinsic, basal human mutation rate, and how this may be influenced by exposure to various agents in our environment including radioactivity, sunlight, exposure to various chemicals, diet, and various types of radiation such as emissions from electrical power lines, microwave ovens, and color televisions. These types of environmental damage raise serious issues of liability and responsibility which can only be properly assessed if we can monitor, directly and quantitatively, their effect on our genes. Hence the motivation to be able to monitor human new mutations is very high. The sensitivity of different animal species to many of these environmental agents is known to be quite variable. Thus, unfortunately, here we have a case where humans must be studied directly.

There are actually three different types of mutation rates that have to be considered in judging the relative effects of various environmental agents. The three are illustrated in Figure 13.21. Genetic mutations are the type of events we have been discussing throughout most of this text. A new genetic mutation means a change in the genomic DNA of a child resulting in the presence of a sequence that could not have been inherited from either parent. One obvious way that this can come about is mis-paternity. Clearly this trivial explanation must be ruled out for any putative new mutation. Fortunately the power of current DNA personal identity testing makes such screening quite easy and accurate. The second type of mutation one must consider is a gametic mutation. Here one can look in a sperm cell for a DNA sequence not present in the father's genomic DNA. Alternatively, one can compare single sperm, by methods described in Chapter 7, and look for the oc-

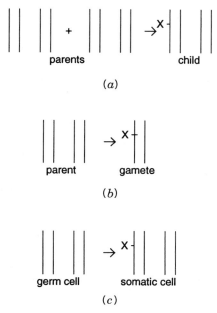

Figure 13.21 Three different types of new mutations that one would like to be able to detect. (*a*) Genetic mutations are inherited by the offspring. (*b*) Gametic mutations are present in the gametes but are lethal, so no offspring are produced. (*c*) Somatic mutations are present in a subset of cells and are not passed to offspring.

currence of sequences that could not have arisen by simple meiotic recombination or gene conversion. The third form of mutation we must consider is at the somatic level. Here we need to look for DNA sequences that are present in some cells but not others. A key element of the problem is that effects of some environmental agents may be different for each type of mutagenesis. For example, an agent that was lethal to gametes would not show up as a genetic mutagen, and yet it would be a source of considerable damage if a significant decrease in fertility resulted. Specific types of cells may be particularly susceptible to certain agents. A few generalities can be made—it seems that rapidly dividing cells are more sensitive to agents that damage DNA; highly transcribed genes seem to be more easily mutagenized. However, in general, our current knowledge of these effects is very slight.

The major difficulty in studying new mutations is that, unless they are accompanied by an easily scored phenotype, the target is a small needle in a very large haystack. The basal, spontaneous mutation rate in the human is estimated to be 10^{-8} per base per meiosis. This means only 30 new mutations per haploid genome. What fraction of these are point mutations or more complex DNA rearrangements is unknown at the present time, although a guess that about half are in each category is probably reasonable. Since the location of the new mutants is unknown a priori, the magnitude of the search required to find them is staggering. Environmental agents will raise the rate of mutations above the basal level. What little evidence we have for typical agents of concern indicates that the increases in mutation caused by typical exposure levels are small. Thus, to quantitate these, we will either need a very good strategy or extraordinary sensitivity.

Two different basic scenarios must be considered that can arise and demand a search for new mutations. In the first of these one may have a small number of individuals exposed to a local, and perhaps very high, dose of toxic agent. An example would be a chemical waste spill. This is a very difficult situation to analyze. One choice is to examine a large percentage of the genomes of the exposed individuals in order to have sufficient sensitivity to see an effect. This is not practical with existing methods. The alternative approach would be to look at potential hot spots for action of the particular agent if enough is known for us to be able to identify such hot spots. Cases where this may eventually be possible are agents that cause very specific kinds of tumors such as acetyl-aminofluorene, an aromatic hydrocarbon that is a selective hepatic carcinogen.

The second scenario is a more favorable case, at least from the limited point of view of DNA analysis. Here one wants to look for mutations in a large number of people at risk (or a large number of sperm at risk) by exposure to a particular agent. An example may be to screen for the genetic effects of depletion of the ozone in the earth's atmosphere. Here a sensible approach is to design assays around regions that are easily tested in large numbers of different samples. Then one can apply these assays to a large population of individuals (or sperm). A risk in this approach is that the region selected may not be representative of the genome as a whole. Some portions of the genome are known to be mutation hot spots, such as VNTRs, and the mitochondrial D loop (origin of replication) because there are no genes there. Many other regions are likely to be identified as we learn more about both the sequence of the human genome and the molecular mechanisms of mutagenesis.

Two types of easy assays can be imagined. In the first strategy, one needs to find a region of DNA that is homozygous in a male, or in both parents. Then a mixture of DNA from parent and child (or sperm) is melted and reannealed. Any heterozygotes are purified away from perfect DNA duplexes. These heterozygotes must represent new mutations. This can now be tested more conclusively by sequence comparisons of the DNAs of interest. The important feature is that a physical purification step is used to examine a very complex mixture of DNA species simultaneously and select just a small fraction of it for subsequent analysis. In this way one begins to approach the ability to handle the amounts of DNA needed to see effects in the 10^{-8} range.

An alternative approach which is potentially very powerful but is only applicable to very particular regions of the genome is shown in Figure 13.22. Here PCR is used to assay for mutations in a restriction enzyme recognition site. Amplification will occur only when the site is not intact. By starting with a set of DNA sequences that contain restriction sites in between PCR primers, only DNA that contains mutations will be amplified.

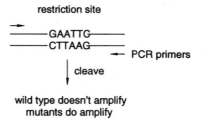

Figure 13.22 A potentially very sensitive PCR method for detecting a mutation in the site recognized by a restriction endonuclease.

This is a most effective way to purify the DNA of interest. However, the difficulty with this approach is that in each sample one only looks at the small numbers of bases that make up a single restriction site. For the assay to be effective at very low mutation frequencies, it will probably be necessary to perform the initial PCR from a large mixture of different sequences, and this makes things potentially complex and noisy. Nevertheless, this sort of approach is methodologically quite attractive, and perhaps variants can be conceived that will be even more sensitive.

SOURCES AND ADDITIONAL READINGS

Birkenkamp, K., and Kemper, B. 1995. In vitro processing of heteroduplex loops and mismatches by endonuclease VII. *DNA Research* 2: 9–14.

Broude, N. E., Chandra, A., and Smith, C. L. 1997. Differential display of genome subsets containing specific interspersed repeats. *Proceedings of the National Academy of Sciences USA* 94: 4548–4553.

Buckler, A. J., Chang, D. D., Graw, S. L., Brook, J. D., Haber, D. A., Sharp, P. A., and Houseman, D. E. 1991. Exon amplification: A strategy to isolate mammalian genes based on RNA splicing. *Proceedings of the National Academy of Sciences USA* 88: 4005–4009.

Chen, X., and Kwok, P.-Y. 1997. Template-directed dye-terminator incorporation (TDI) assay: A homogeneous DNA diagnostic method based on fluorescence resonance energy transfer. *Nucleic Acids Research* 25: 347–353.

Chi, N.-W., and Kolodner, R. D. 1994. Purification and characterization of MSH1, a yeast mitochondrial protein that binds to DNA mismatches. *Journal of Biological Chemistry* 269: 29984–29992.

Cotton, R. G. H. 1997. Slowly but surely towards better scanning for mutations. *Trends in Genetics* 13:43–46.

Duyao, M., et al. 1993. Trinucleotide repeat length instability and age of onset in Huntington's disease. *Nature Genetics* 4: 387–392.

Duyk, G. M., Kim, S., Myers, R. M., and Cox, D. R. 1990. Exon trapping: A genetic screen to identify candidate transcribed sequences in cloned mammalian genomic DNA. *Proceedings of the National Academy of Sciences USA* 87: 8995–8999.

Fu, Y.-H., Kuhl, D. P. A., Pizzuti, A., Pieretti, M., Sutcliffe, J. S., Richards, S., Verkerk, A. J. M. H., Holden, J. J. A., Fenwick Jr., R. G., Warren, S. T., Oostra, B. A. Nelson, D. L., and Caskey, C. T. 1991. Variation of CGG repeat at the fragile X site results in genetic instability: Resolution of the Sherman paradox. *Cell* 67: 1047–1058.

Ganguly, A., and Prockop, D. J. 1990. Detection of single-base mutations by reaction of DNA heterduplexes with a water-soluble carbodimide followed by primer extension: Application to products from the polymerase chain reaction. *Nucleic Acids Research* 18: 3933–3939.

Hacia, J. G., Brody, L. C., Chee, M. S., Fodor, S. P. A., and Collins, F. S. 1996. Detection of heterozygous mutations in BRCA1 using high density oligonucleotide arrays and two-colour fluorescence analysis. *Nature Genetics* 14: 447.

Hamaguchi, M., Sakamoto, H., Tsuruta, H., Sasaki, H., Muto, T., Sugimura, T., and Terada, M. 1992. Establishment of a highly sensitive and specific exon-trapping system. *Proceedings of the National Academy of Sciences USA* 89: 9779–9783.

Hattori, M., Shibata, A., Yoshioka, K., and Sakaki, Y. 1993. Orphan peak analysis: A novel method for detection of point mutations using an automated fluorescence DNA sequencer. *Genomics* 15:415–417.

Ketterling, R. P., Vielhaber, E., and Sommer, S. S. 1994. The rates of G:C → T:A → C:G transversions at CpG dinucleotides in the human factor IX gene. *American Journal of Human Genetics* 54: 831–835.

Khrapko, K., Hanekamp, J. S., Thilly, W. G., Belenkii, A., Foret, F., and Karger, B. L. 1994. Constant denaturant capillary electrophoresis (CDCE): A high resolution approach to mutational analysis. *Nucleic Acids Research* 22: 364–369.

Knight, S. J. L., Flannery, A. V., Hirst, M. C., Campbell, L., Christodoulou, Z., Phelps, S. R., Pointon, J., Middleton-Price, H. R., Barnicoat, A., Pembrey, M. E., Holland, J., Oostra, B. A., Bobrow, M., and Davies, K. E. 1993. Trinucleotide repeat amplification and hypermethylation of a CpG island in *FRAXE* mental retardation. *Cell* 74: 127–134.

Lishanski, A., Ostrander, E. A., and Rine, J. 1994. Mutation detection by mismatch binding protein, MutS, in amplified DNA: Application to the cystic fibrosis gene. *Proceedings of the National Academy of Sciences USA* 91: 2674–2678.

Liu, Q., and Sommer, S. S. 1995. Restriction endonuclease fingerprinting (REF): A sensitive method for screening mutations in long, contigous segments of DNA. *BioTechniques* 18: 470–477.

Lovett, M. 1994. Fishing for complements: Finding genes by direct selection. *Trends in Genetics* 10: 352–357.

Miyoshi, Y., Nagase, H., Ando, H., Horii, A., Ichii, S., Nakatsuru, S., Aoki, T., Miki, Y., Mori, T., and Nakamura, Y. 1992. Somatic mutations of the APC gene in colorectal tumours: Mutation cluster in the APC gene. *Human Molecular Genetics* 1: 229–233.

Nawroz, H., Koch, W., Anker, P., Stroun, M., and Sidransky, D. 1996. Microsatellite alterations in serum DNA of head and neck cancer patients. *Nature Medicine* 2: 1035–1037.

Oliveira, R. P., Broude, N. E., Macedo, A. M., Cantor, C. R., Smith, C. L., and Pena, S. D. 1998. Probing the genetic population structure of *Trypanosoma cruzi* with polymorphic microsatellites. *Proceedings of the National Academy of Sciences USA* 95: 3776–3780.

Pena, S. D. J., Barreto, G., Vago, A. R., De Marco, L., Reinach, F. C., Dias Neto, E., and Simpson, A. J. G. 1994. Sequence-specific "gene signatures" can be obtained by PCR with single specific primers at low stringency. *Proceedings of the National Academy of Sciences USA; Biochemistry* 91: 1946–1949.

Petruska, J., Arnheim, N., and Goodman, M. F. 1996. Stability of intrastrand hairpin structures formed by the CAG/CTG class of DNA triplet repeats associated with neurological diseases. *Nucleic Acids Research* 24: 1992–1998.

Richards, R. I., and Sutherland, G. R. 1992. Dynamic mutations: a new class of mutations causing human disease. *Cell* 70: 709–712.

Sarkar, G., Yoon, H.-S., and Sommer, S. S. 1992. Dideoxy fingerprinting (ddF): A rapid and efficient screen for the presence of mutations. *Genomics* 13: 441–443.

Schalling, M., Hudson, T. J., Buetow, K. H., and Housman, D. E. 1993. Direct detection of novel expanded trinucleotide repeats in the human genome. *Nature Genetics* 4: 135–138.

Smith, J., and Modrich, P. 1996. Mutation detection with MutH, MutL, and MutS mismatch repair proteins. *Proceedings of the National Academy of Sciences USA* 93: 4374–4379.

Snell, R. G., MacMillan, J. C., Cheadle, J. P., Fenton, I., Lazarou, L. P., Davies, P., MacDonald, M. E., Gusella, J. F., Harper, P. S., and Shaw, D. J. 1993. Relationship between trinucleotide repeat expansion and phenotypic variation in Huntington's disease. *Nature Genetics* 4: 393–397.

Sommer, S. S. 1992. Assessing the underlying pattern of human germline mutations: Lessons from the factor IX gene. *FASEB Journal* 6: 2767–2774.

Soto, D., and Sukumar, S. 1992. Improved detection of mutations in the p53 gene in human tumors as single-stranded conformation polymorphs and double-stranded heteroduplex DNA. *PCR Methods and Applications* 2: 96–98.

Sutherland, G., and Richards, R. I. 1995. Simple tandem DNA repeats and human genetic disease. *Proceedings of the National Academy of Sciences USA* 92: 3636–3641.

Wahlberg, J., Albert, J., Lundeberg, J., Cox, S., Wahren, B., and Uhlen, M. 1992. Dynamic changes in HIV-1 quasispecies from azidothymidine (AZT)-treated patients. *FASEB Journal* 6: 2843–2847.

Wells, R. D. 1996. Molecular basis of genetic instability of triplet repeats. *Journal of Biological Chemistry* 271: 2875–2878.

Youil, R., Kemper, B. W., and Cotton, R. G. H. 1995. Screening for mutations by enzyme mismatch cleavage with T4 endonuclease VII. *Proceedings of the National Academy of Sciences USA* 92: 87–91.

Yu, A., Dill, J., Wirth, S. S., Huang, G., Lee, V. H., Haworth, I. S., and Mitas, M. 1995. The trinucleotide repeat sequence d(GTC)$_{15}$ adopts a hairpin conformation. *Nucleic Acids Research* 23: 2706–2714.

14 Sequence-Specific Manipulation of DNA

EXPLOITING THE SPECIFICITY OF BASE-BASE RECOGNITION

In this chapter various methods will be described that take advantage of the specific recognition of DNA sequences to allow analytical or preparative procedures to be carried out on a selected fraction of a complex DNA sample. For example, one can design chemical or enzymatic schemes to cut at extremely specific DNA sites, to purify specific DNA sequences, or to isolate selected classes of DNA sequences. Methods have been developed that allow the presence of repeated DNA sequences in genomes to be used as powerful analytical tools instead of serving as roadblocks for mapping and DNA sequencing. Other methods have been developed that allow the isolation of DNAs that recognize specific ligands. Finally a large number of programs are underway to explore the direct use of DNA or RNA sequences as potential drugs.

In almost all of the objectives just outlined, a fundamental strategic decision must be made at the outset. If the DNA target of interest can be melted without introducing unwanted complications, then the single-stranded DNA sequence can be read, directly, and the full power of PCR can usually be brought to bear to assist in the manipulation of the DNA target. PCR has been well described in Chapter 4, and there is no need to re-introduce the principles here. In those cases where it is not safe or desirable to melt the DNA, alternative methods are needed. Such cases include working with very large DNA, which will break if melted, and working in vivo. Here a very attractive approach is to use DNA triplexes that are capable of recognizing the specific sequence of selected portions of an intact duplex DNA. Triplexes may not have been encountered by some readers before, and so their basic properties will be described before their utility is demonstrated.

STRUCTURE OF TRIPLE-STRANDED DNA

Unanticipated formation of triple-stranded DNA helices was a scourge of early experiments with model DNA polymers. Most of the first available synthetic DNAs were homopolymers like poly dA and poly dT. Contamination of samples or buffers with magnesium ion was rampant. DNAs love to form triplexes under these conditions, if the sequence permits it. Many homopolymeric or simple repeating sequences can form triple-stranded complexes consisting of two purine-rich and one pyrimidine-rich strand or one purine-rich and two pyrimidine-rich strands, depending on the conditions. This is true for DNAs, RNAs, or DNA-RNA mixtures. Eventually conditions were found where the unwanted formation of these triplexes could be suppressed. The whole issue was forgotten and lay dormant for more than a decade. Triplexes were rediscovered, under much more

470

Figure 14.1   Appearance of S1 nuclease hypersensitive sites upstream from the start of transcription of some genes.

interesting circumstances when a decade ago investigators began to explore the chromatin structure surrounding active genes.

The key observation that led to a renaissance of interest in triplexes is a phenomenon called S1 hypersensitivity. S1 nuclease is an enzyme that cleaves single-stranded DNA specifically, usually at slightly acidic pH. It will not cleave double strands; it will not even cleave a single-base mismatch efficiently, although it will cut at larger mismatches. Investigators were using various nucleases to examine the accessibility of DNA segments near or in genes as a function of the potential for gene expression in particular tissues. Unexpectedly, many genes showed occasional sites where S1 could nick one of the DNA strands, upstream from the start of transcription, quite efficiently (Fig. 14.1). The phenomenon was termed S1 hypersensitivity. Its implication was that some unusual structure must exist in the region, rendering the normal duplex DNA susceptible to attack. To identify the sequences responsible for S1 hypersensitivity, upstream sequences were cloned and tested for S1 sensitivity. Fortunately they were initially tested within the plasmids used for cloning. These plasmids were highly supercoiled, and S1 hypersensitive sites were found and rapidly localized to complex homopurine stretches like the example shown in Figure 14.2. The S1 nicks were found to lie predominantly on the purine-rich strand. It was soon realized that the S1 hypersensitivity, under the conditions used, required a supercoiled target. The effect was lost when the plasmid was linearized, even by cuts far away from the purine block.

The problem that remained was to identify the nature of the altered DNA structure responsible for S1 hypersensitivity. The dependence of cleavage on a high degree of superhelicity implied that the sites must, overall, be unwound relative to the normal B DNA duplex. Obvious possibilities were melted loops, left-hand helix formation, or cruciform extrusion (formation of an intramolecular junction of four duplexes like the Holliday structure illustrated in Chapter 1). None of these, however, were consistent with the particular DNA sequences that formed the S1 hypersensitive sites, and none could explain why only the purine strand suffered extensive nicking. The key observation that resolved this dilemma was made by Maxim Frank-Kamenetskii, then working in Moscow. He noted that there was a direct correlation between the amount of supercoiling needed to reveal the S1 hypersensitivity and the pH used for the S1 treatment. A quantitative analysis of this effect indicated that both the amount of unwinding that occurred when the S1 hypersensitive site was created, and the number of protons that had to be bound during this

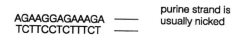

Figure 14.2   DNA sequence of a typical S1 hypersensitive site.

Figure 14.3   Formation of a DNA triplex by disproportionation of two homopurine-homopyrimidine duplexes.

process could be explained by a simple model, which involved the formation of a specific intramolecular pyrimidine-purine-pyrimidine (YRY) triple helix.

It is easiest to examine intermolecular triplex formation before considering the ways in which such structures might be formed intramolecularly at the S1 hypersensitive site. Figure 14.3 illustrates a disproportionation reaction between two duplexes that results in a triplex and a free single strand. This is precisely the sort of reaction that occurred so frequently in early studies with DNA homopolymers and led to the presence of unwanted DNA triplexes. If such a reaction is assayed by S1 sensitivity, disproportionation by the appearance of a single-stranded polypurine would be detected. The corresponding possible intramolecular reactions are illustrated in Figure 14.4. Here a block of homopyrimidine sequence folds back on itself (spaced by a short hairpin) to make an intramolecular triplex; the remaining homopurine stretch, not involved in the triplex, is left as a large single-stranded loop. It is this loop that is the target for the S1 nuclease. The net topological effect is an unwinding of roughly half of the homopurine-homopyrimidine duplex stretch. This is consistent with what is seen experimentally. In order to form base triplets between two pyrimidines and one purine, a T:A−T complex can form directly, but a CH$^+$:G−C complex requires protonation of the N$_3$ of one C, as shown in Figure 14.5.

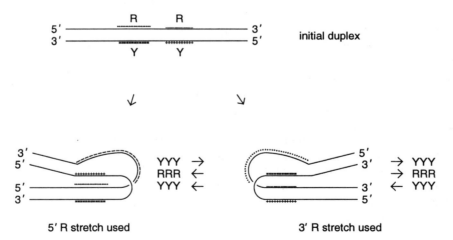

Figure 14.4   Two intramolecular routes for formation of triplexes from a long homopurine-homopyrimidine duplex. The structure shown on the right is the one consistant with a large body of available chemical modification data. Y and R refer, respectively, to homopyrimidine and homopurine tracts.

CH + : G-C   T : A-T

(a)

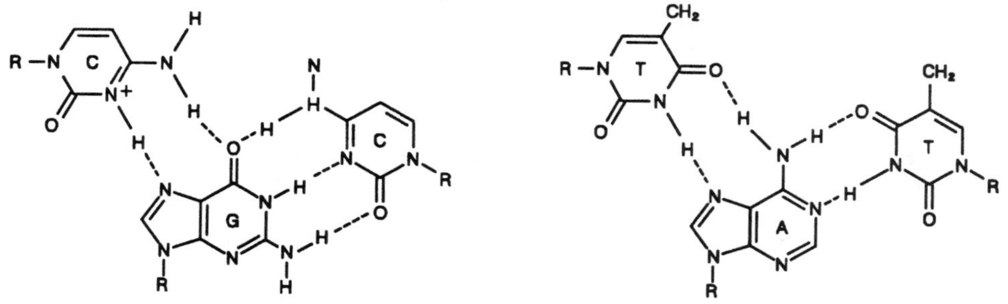

(b)

Figure 14.5 Acid-stabilized triplex base pairing schemes. A dash (-) indicates the normal Watson-Crick base pairing scheme while a colon (:) indicates the base pairs which involve the third strand. *(a)* As written schematically. *(b)* Actual proposed structures.

There are two possible isomeric models consistent with the unwinding and pH dependence of formation of the YRY triple strand. In both the two pyrimidine strands run antiparallel to each other; the Watson–Crick pyrimidine strand is antiparallel to the purine strand; the triplex pyrimidine strand is parallel to the purine strand. The specific structural models proposed require that the pyrimidine sequences have mirror symmetry. This can be tested by manipulating particular DNA sequences, and it turns out to be valid. In addition the two models in Figure 14.4 can be evaluated by looking at the pattern of accessibility of the S1 hypersensitive structure to various agents that chemically modify DNA in a structure-dependent manner. These studies reveal that the correct model for the S1 hypersensitive structure is the one shown on the right in Figure 14.4, where the 3′ segment of the purine stretch is the one incorporated into the triplex. The reason why this structure predominates is not known.

With the principles of triplex formation in S1 hypersensitive sites understood, a number of research groups began to explore the properties of simple linear triplexes more systematically. The need for superhelical density to drive the formation of triplex can be avoided simply by working at a low enough pH. In practice, pH 5 to 6 suffices for most sequences capable of forming triplexes at all. A surprise was the remarkable stability of triplexes. They can survive electrophoresis, even with lengths as small as 12. Proof that the third strand lies in the major groove of the Watson–Crick duplex, and that the third, pyrimidine, strand is antiparallel to the Watson–Crick pyridine strand, was obtained by an elegant series of experiments in which agents were attached to the ends of the third strand that were capable of chemically nicking bases on the duplex. The specific pattern of nicks provided a detailed picture of the structure of the complex (Fig. 14.6).

A second type of DNA triplex, stable at pH 7 was soon rediscovered. This purine-purine-pyrimidine (RRY) was precisely the form known two decades before, stabilized

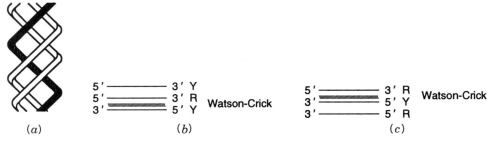

$$
\begin{array}{ll}
5'\text{———————}3'\ Y & \\
5'\text{═══════}3'\ R & \text{Watson-Crick} \\
3'\text{~~~~~~~~}5'\ Y &
\end{array}
\qquad\qquad
\begin{array}{ll}
5'\text{———————}3'\ R & \\
3'\text{~~~~~~~~}5'\ Y & \text{Watson-Crick} \\
3'\text{———————}5'\ R &
\end{array}
$$

(a)   (b)   (c)

Figure 14.6   Structure of DNA triplexes determined from chemical modification experiments. *(a)* The third strand lies in the major groove of the duplex helix. *(b)* Strand directions in structures with two pyrimidine strands and one purine strand. *(c)* Strand directions in structures with two purine strands and one pyrimidine strand. Shading in *(b)* and *(c)* indicates the Watson-Crick duplex.

by Mg^{2+} or other polyvalent cations. This structure could also lead to an S1 hypersensitive site in supercoiled plasmids. However, here the pyrimidine-rich strand was nicked by the enzyme instead of the purine-rich strand. Some DNA sequences can actually form both types of triplexes, depending on the conditions. The structure of the S1 hypersensitive sites favored by divalent ions is shown in Figure 14.7. This particular isomer is the one consistent with the observed pattern of modification with various chemical agents that react with DNA covalently. Three types of base triples can be accommodated in this structure: G:G–C, A:A–T, and T:A–T. Their patterns of hydrogen bonding are shown in Figure 14.8. The two non–Watson–Crick base-paired strands in these complexes are antiparallel; this is supported by studies on particular DNA sequences. As in the type of triplex described earlier, the third strand, in this case an additional purine or an additional pyrimidine strand, lies in the major groove of the Watson–Crick duplex (Fig. 14.6). Studies using circular oligonucleotides can help confirm assignments about the direction of strands in triplexes (Kool, 1995).

More complex triple helices can also be made. An example is shown in Figure 14.9. Here all three strands must contain blocks of alternating homopurine and homopyrimidine sequences. The third strand lies down in the major groove of the Watson–Crick duplex, and alternate blocks made triplexes with two pyrimidine and one purine strand and triplexes with one pyrimidine and two purine strands. As our knowledge of triplex structures increases, and as base analogs are tested, it will undoubtedly be possible to design a wealth of triplexes in which a third strand can be used to recognize a wide variety of DNA duplex sequences. Based on experience to date, these triplexes are likely to be quite stable. One strong caveat to using them in various biological applications must be noted. The kinetics of triplex formation and dissociation are very slow, much slower than the rates of corresponding processes in duplexes.

$$
\begin{array}{l}
5'\text{——————}\quad RRR\ \rightarrow \\
3'\text{——————}\quad YYY\ \leftarrow \\
3'\text{——————}\quad RRR\ \leftarrow \\
5'\qquad\qquad\text{— exposed Y strand}
\end{array}
$$

Figure 14.7   Intramolecular triplex structure formed at neutral pH in the presence of Mg^{2+} ions.

(*a*) G : G-C A : A-T T : A-T

(*b*)

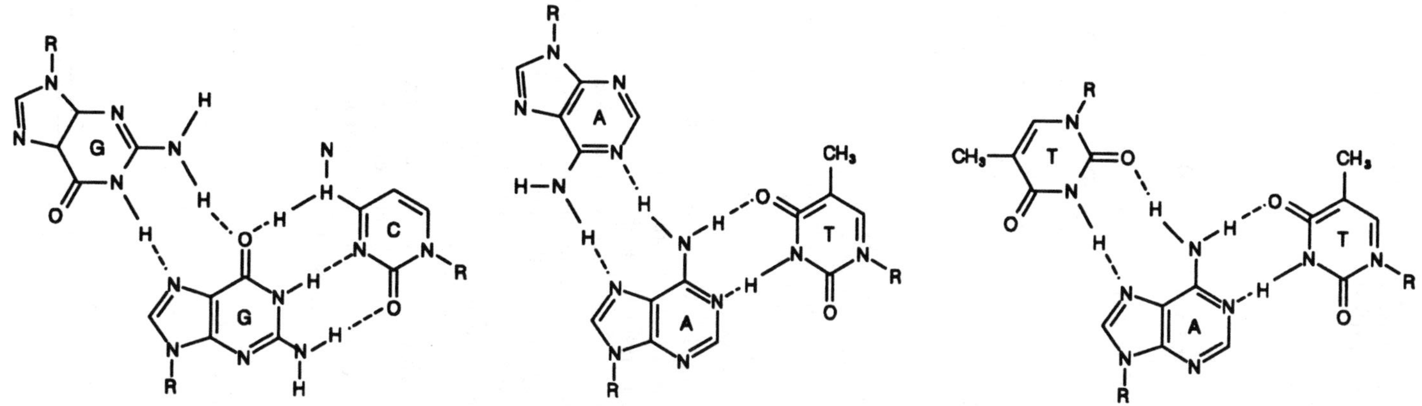

Figure 14.8 Triplex base pairing schemes favored by Mg^{2+} ions. *(a)* As written schematically. *(b)* Actual proposed structures.

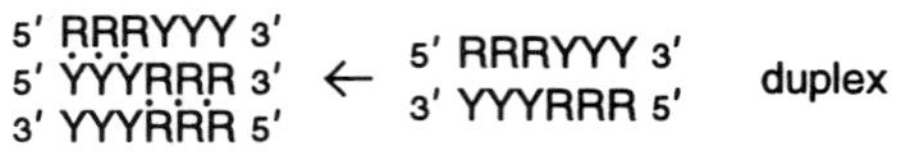

5′ RRRYYY 3′
5′ ẎẎẎRRR 3′ ← 5′ RRRYYY 3′   duplex
3′ YYYRRR 5′   3′ YYYRRR 5′

Figure 14.9 An example of a more complex triplex structure formed by alternating blocks of purines and pyrimidines. The third strand lies in the major groove of the Watson-Crick duplex. Its interactions with the duplex are indicated by dots.

TRIPLEX-MEDIATED DNA CLEAVAGE

The first application of triplexes to be discussed is their use in recognizing particular duplexes and rendering these susceptible to specific chemical or enzymatic cleavage. This potential was already described briefly in the previous section when chemical derivatives of the third strand were used to help analyze the structure of the triplex. The appeal of this approach is that it will be relatively easy to find or introduce a unique DNA sequence capable of forming triple strands into a target of interest. Subsequent cleavage at this sequence would represent the sort of cut that is extremely useful for executing any of the Smith-Birnstiel-like mapping strategies we described in Chapter 8.

Chemical cleavage agents that have been tried include Cu-phenanthroline complexes, iron-EDTA-ethidium bromide complexes, and others shown elsewhere in the chapter. The types of reactions one would like to be able to carry out with these modified oligonucleotides are shown schematically in Figure 14.10. Rather good yields and specificities have been observed when the chemical cleavage is used to cut the complementary strand of a duplex (Fig. 14.10*a*). Much less success has been had with direct triplex-mediated cleavage of a duplex (Fig. 14.10*b*). Generally, nicking of one strand of the duplex proceeds very well, but it is difficult to make the second cut needed to affect a true doublestrand cleavage. The reason for this is that many of the chemical agents used are stoichiometric rather than catalytic. They have to be reactivated or replaced by a fresh reagent in order to be able to perform a second strand cleavage. While elegant chemical methods have been proposed to circumvent this problem, to date, specific efficient duplex chemical cleavage has been an elusive goal. However, this has not proved to be a serious roadblock, because alternative methods for using triplexes to promote specific enzymatic cleavage of duplexes have been very successful.

Achilles's heel strategies are based on the general notion of using restriction methylases to protect all except a single or small set of protected recognition sites. This renders most of the potential sites in the sample resistant to the conjugate restriction nuclease. Then the protecting agent is removed, and the nuclease added. Cleavage only occurs at

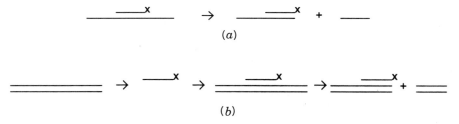

Figure 14.10 Triplex-mediated DNA cleavage using a DNA strand containing a chemically reactive group *(x)* that generates radicals. *(a)* Cleavage of a single strand. *(b)* Cleavage of a duplex.

the sites that escaped the initial protection. These strategies were named by one of their developers, Wacslaw Szbalski, at the University of Wisconsin, by analogy to the myth of Achilles. As an infant Achilles was dipped into the river Styx by his mother which rendered him immune to all physical harm except for his heel, which was masked from the effects of the Styx because his mother was holding him there. Two straightforward examples of Achilles's heel specific cleavage of DNA are shown in Figure 14.11. These approaches are applicable to very large DNA or even intact genomic DNA, since they can be carried out in situ in agarose. In one case it is necessary to find a tight binding site for a protein that masks a restriction site (Fig. 14.11a). Examples of suitable protein and binding sites are lac repressor, lambda repressor, *E. coli* lexA protein, or a host of eukaryotic transcription factors, particularly viral factors like the NFAT protein. To be useful, the site must contain an internal or nearby flanking restriction site. There is no guarantee that such a site will conveniently exist in a target of interest. However, given the large number of potentially useful restriction sites, there are many possibilities. If necessary, for some applications the desired site can always be designed and introduced.

Instead of proteins, triplexes can be used to mask restriction sites (Fig. 14.11b). It turns out that embedding a four-base restriction site in a homopurine-homopyrimidine

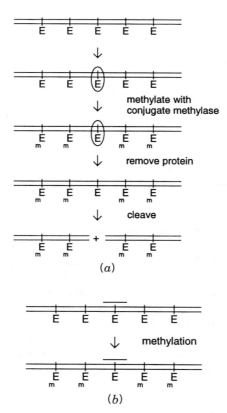

Figure 14.11 Achilles's heel strategies for specific DNA cleavage. *(a)* Blocking a restriction enzyme cleavage site *E* with a DNA binding protein. *(b)* Blocking a restriction enzyme cleavage site *E* with a triplex. *M* indicates methylation sites.

stretch destabilizes the resulting triplex only slightly. However, triplex formation renders the site totally unaccessible to the restriction methylase. After the remaining restriction sites have been methylated, conditions are altered to dissociate the triplex. Then the restriction enzyme is added, and cleavage is allowed to occur. This Achilles's heel approach works very well, even at the level of single sites in the human genome. However, it still suffers from the limitation that only a small subset of sequences within a target will be potential sites of triplex-mediated specific cleavage.

A generalization of the Achilles's heel approach is possible by the use of the *E. coli* recA protein. Developed by Camerini-Otero and elaborated by Szybalski (Koob et al., 1992), this method has been called recA-assisted restriction endonuclease (RARE) cleavage (Fig. 14.12). The method is applicable to genomic DNA because all of the steps can be carried out in agarose. The recA protein has a number of different activities. One of these is a cooperative binding to single-stranded DNA, leading to a completely coated complex containing about one recA monomer for every five bases (Fig. 14.12*a*). The coated complex will then interact with double-stranded DNA molecules in a search for sequences homologous to the single strand. In *E. coli* this process constitutes one of the early steps that eventually leads to strand invasion and recombination. In the test tube, without accessory nucleases, the reaction stops if a homologous duplex sequence is found, and the third strand remains complexed to this homolog, even if the recA protein is subsequently removed (Fig. 14.12*b*). The mechanism of the sequence search is unknown. Similarly the actual nature of the complexes formed with recA protein present or after recA protein removal are still not completely understood despite intense efforts to study these processes because of their importance in basic *E. coli* biology. Some sort of triple strand is believed to be involved, although this has never been proven. What is key, however, for Achilles's heel applications, is that the recA protein-mediated complex blocks the access of restriction methylases to duplex DNA sequences contained within it.

A schematic outline of RARE cleavage is given in Figure 14.12*c*. The technique has worked well to cut at two selected sites 200 to 500 kb apart in a target to generate a specific internal fragment, and generation of a 1.3-Mb telomeric DNA fragment that requires only a single RARE cleavage has been reported. In practice, it has been more efficient to use a six-base specific restriction system like rather than the four-base systems used with other Achilles's heel methods. The reason is that a common source of background in these approaches is incomplete methylation. This produces a diverse distribution of hemimethylated sites which are cut, albeit slowly, by the conjugate nuclease. The result is a significant background of nonspecific cleavage. This background can be markedly reduced by going to the six-base enzyme, since its sites are 16 times less frequent, on average. Since recA-mediated cleavage is applicable, in principle, to any selected DNA sequence, the rarity of six-base cleavage sites does not pose a particular obstacle.

The recA protein-coated single strands can be as short as 15 bases for RARE cleavage, although in practice targets two to four times this length are usually employed. One makes a trade-off between the increased efficiency and specificity obtained with longer complexes, and the lowered efficiency of their diffusion into agarose-embedded DNA samples. Yields of the desired duplex of 40% to 60% have been reported in early experiments. It remains to be seen how generally obtainable such high yields will be. The power of RARE cleavage in physical mapping is that given two DNA probes spaced within about 1 Mb, RARE cleavage should provide the DNA between these probes as a unique fragment free from major contamination by the remainder of the genome.

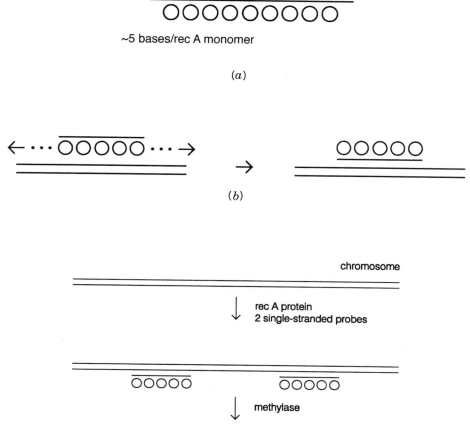

~5 bases/rec A monomer

(a)

(b)

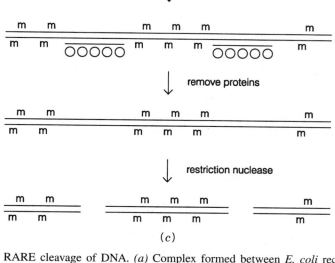

(c)

Figure 14.12 RARE cleavage of DNA. *(a)* Complex formed between *E. coli* recA protein and single-stranded DNA. *(b)* Complex between a recA-protein coated DNA strand and the homologous sequence in duplex DNA. *(c)* Outline of the procedure used to generate a large DNA fragment between two known sequences by RARE cleavage.

SEQUENCE-SPECIFIC DNA CAPTURE

In the majority of cases, the motivation behind attempts to develop ways for sequence-specific DNA cleavage is to provide access to particular DNA sequences, as just described above for RARE cleavage. A more direct approach would be to develop methods for the specific capture of DNA sequences of particular interest. There are a number of different approaches that have been used to purify DNAs based on sequence rather than size. Biological methods we have already described include PCR or differential cloning. These are extremely effective, but because the steps are usually carried out in solution and involve a number of enzymes, the targets of such purifications are usually limited in size. Large insert cloning systems have been described, but these are not yet applicable to differential cloning. For example, YAC cloning can be used to immortalize a set of large DNA fragments, but then this set has to be screened to identify the clone containing a particular sequence of interest. Such screening can be quite time-consuming. Instead of screening, we are concerned here with methods that select the target of interest. Purification techniques are a natural way to select molecules with desired properties such as specific DNA sequences.

In principle, a DNA sequence could be selected by cloning, by PCR, by binding to a specific protein, by binding to the complementary strand to form a duplex, or by binding to a third strand to form a triplex. In each of these approaches the desire is a simple sequence-specific purification method. In this section we will limit our attention to physical purifications based solely on DNA base interactions. Such methods have been called in the past, perhaps somewhat inelegantly, DNA fishing. Differential PCR and cloning methods will be described later in the chapter. The advantage of the pure, physical methods is that if successful, they are very easy to implement and easy to scale up to large numbers of samples or large quantities. A key obstacle in the use of such methods is that once the DNA target is captured by a probe, there either must be an efficient way to release it, or one has to have an efficient way of working directly with the immobilized DNA (as in solid state DNA sequencing, described earlier in Chapter 9).

TRIPLEX-MEDIATED DNA CAPTURE

The major advantage in using a third strand to capture a DNA duplex is that there is no need to melt the duplex. This avoids potential damage arising from any preexisting nicks in the DNA strands. It also helps avoid many of the complications caused by the extensive interspersed repeats in the DNA of higher organisms. There are a number of potential applications for triplex capture of DNA. For example, triplex formation could be used, in principle, to isolate from a library all those clones that contained specific homopurine-homopyrimidine sequences such as $(AG)_n$, $(AAG)_n$, or $(AAAG)_n$. These clones are all potentially useful genetic probes, since all are VNTRs. Such clones have been isolated in the past by screening libraries by hybridization with the repeated seqeunce. However, if one is interested in a large set of such clones, each one detected by hybridization has to be picked by hand for future use. It would be far simpler to isolate the entire set of VNTR-containing clones in a single physical purification step. A survey of simple repeated sequence motifs in the GenBank database several years ago is shown in Table 14.1. This indicates that a significant fraction of all simple sequence VNTRs known to date could be amenable to triplex-mediated capture because they involve homopurine repeats.

TABLE 14.1   Repeating Single Sequence Motifs in the Genbank Database (1991)

Dinucleotide	<u>AG</u>	24				
	AT	18				
	AC	86				
	TG	1				
Trinucleotide	AAC	14				
	CCG	12				
	<u>AGG</u>	9				
	AAT	7				
	AGC	10				
	<u>AAG</u>	3				
	ATC	2				
	ACC	2				
Tetranucleotide	<u>AAAG</u>	20	AGCC	1	AACT	1
	AAAT	22	AGAT	9	ACAT	1
	ACAG	3	ATAG	3	ACGC	1
	ACAT	2	AGCG	1	ACTG	2
	<u>AAGG</u>	8	ATCC	8	<u>AGGG</u>	1
	<u>AAGG</u>	13	AATC	4	AGCT	1
Lohger	<u>AAAAG</u>	1				
	<u>AAAAAAAAAG</u>	1				
	TTTTTG	1				

Source: Adapted from results summarized by Lincoln McBride.

Note: Underlined sequences are triplex selectable.

Two other potential applications for triplex capture concern the purification of a specific DNA sequence. If a genomic fragment contains a known specific homopurine sequence, one should be able to capture this sequence away from the entire genome by triplex formation on a solid support. Alternatively, such a sequence could be introduced into a desired section of a genome to allow subsequent capture of DNA from this region. Such approaches, once they have matured, could potentially expedite the analysis of selected megabase regions of DNA considerably. The second potential use of triplex capture would be to separate clones from host cells. Here the triplex forming sequence would be built into the vector used for cloning. After cell lysis, triplex capture would be used to retain the cloned DNA and discard the DNA of the host. This has many appealing features. Once host DNA is gone, direct analysis of cloned DNA is possible without the need for radiolabeled DNA probes or specific PCR primers. Host DNA leads to considerable background problems when low-copy number vectors like YACs are used. The sensitivity of hybridization or PCR analyses would be increased considerably if host DNA could be removed in advance.

A number of different ways to use triplexes to purify DNA have been explored. Although much more development work needs to be done, the preliminary findings are very promising. In our work, magnetic microbeads were used as the solid support. The power of this general approach was described earlier in applications for solid state DNA sequencing. Magnetic microbeads are available commercially that are coated with streptavidin. Streptavidin will bind specifically to virtually any biotinylated macromolecule. Any desired small DNA sequence can be prepared with a 5'-biotin by the use of biotin phosphoramidite in ordinary automated oligonucleotide synthesis. Longer 5'-biotinylated probes can be made by PCR using 5'-biotinylated primers. Alternatively, internally bi-otinylated DNA sequences can be made by PCR using biotinylated base analogs as dpppN substrates. Some of these compounds are incorporated quite efficiently by DNA polymerases. Once attached to the streptavidin microbeads, biotinylated DNA probes are still quite accessible for hybridization or for triplex formation. A schematic illustration of the way microbeads have been used to develop a simple triplex capture method of DNA purification is shown in Figure 14.13. The key point is that a permanent magnet can be used to hold the beads in place while supernatant is removed and exchanged. This allows very rapid and efficient exchanges of reagents. At pH 5 or 6 any duplex DNA captured by the complementary sequence on a bead will remain attached. At pH 8 the captured DNA will be readily released and removed with the supernatant.

Three different experiments have been used to test the ease and efficiency of triplex DNA capture with magnetic microbeads. In the first, an artificial mixture was made of two plasmids. One contained just a vector with no known forming capability, which in-cluded a lacZ gene. When this plasmid is transformed into *E. coli*, blue colonies are pro-duced on the appropriate indicator plates, which contain a substrate for the beta-galactosi-dase product of this gene that yields a blue-colored product. The second plasmid, initially

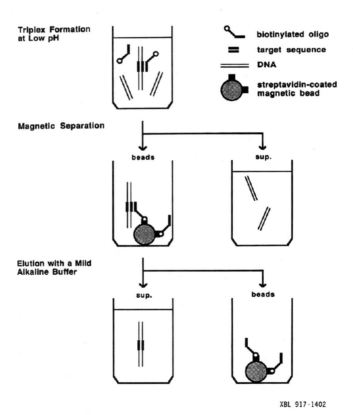

XBL 917-1402

Figure 14.13 Use of streptavidin-coated magnetic microbeads for affinity purification of triplex DNA. Adapted from Ito et al. (1992).

present at much lower concentrations, contained an insert of $(AG)_n$, cloned into the lacZ gene in such a way as to disrupt its translation. This results in white colonies after transformation of *E. coli*. Thus the ratio of plasmids in a mixture can be determined just by comparing the number of colonies of each color after transformation (making the reasonable assumption that the transformation efficiencies of the two almost identical plasmids are the same). When a mixture of the two types of DNA was subjected to a single cycle of purification via triplexes formed with magnetic beads containing $(TC)_n$, the result is a purification of the $(AG)_n$-containing plasmid by a factor of 140,000 fold with an overall yield of 80%, as shown in Table 14.2. Such a purification is sufficient for almost any need. However, if necessary the purity could surely be increased much further by a second cycle with the magnetic beads. The entire process took only a few hours. Recently Lloyd Smith and coworkers have demonstrated that a similar process can be carried out in only 15 minutes. G- and T-containing sequences can also be used in an analogous way to capture homopurine-homopyrimidine sequences in the presence of Mg^{2+} or other polyvalent cations (Takabatake et al., 1992).

A second test of triplex-mediated capture with magnetic microbeads is shown schematically in Figure 14.14. Here the challenge was to purify $(AG)_n$-containing DNA clones away from the remainder of a chromosome 21-specific library. The original bacteriophage library was first subcloned into a plasmid vector. Magnetic purification proceeded as described in Figure 14.13. Then the purified DNA was used to transform *E. coli*. Individual colonies were picked and screened for the presence of $(AG)_n$ by PCR between each of the vector arms and the internal $(AG)_n$ sequence. A positive signal should be seen for each $(AG)_n$-containing clone with one of the two vector arms depending on the orientation of the insert (Fig. 14.15). Some of the actual results obtained are shown in Figure 14.16. The effectiveness of the procedure is quite clear. Overall, 17 of the first 18 colonies tested showed vector-insert PCR signals, and each of these was different, implying that different genomic $(AG)_n$-containing clones had been selected.

TABLE 14.2   Triplex-Mediated Purification of Target Plasmids from a Reconstituted Library

Number of Colonies (%)	
White (pTC45)	Blue (pUC19)
Before Enrichment	
5.0×10^4	1.1×10^7
(0.5)	(99.5)
After Enrichment	
4.0×10^4	0.5×10^2
(99.9)	(0.1)

Source: Adapted from Ito et al. (1992).

Note: Plasmids prepared from a reconstituted library were used for transformation of *E. coli* with or without enrichment by triplex affinity capture. pTC45 (target) and pUC19 give white and blue colonies, respectively, on indicator plates. In this experiment the enrichment was 1.8×10^5-fold with a recovery of $\approx 80\%$.

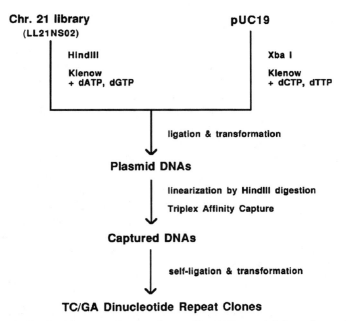

Figure 14.14 Outlined of the procedure used to purify $(AG)_n$-containing clones from a chromosome 21 library.

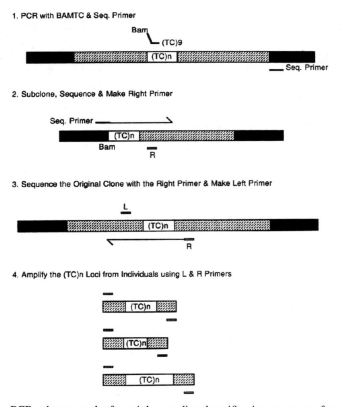

Figure 14.15 PCR scheme used, after triplex-mediated purification, to screen for clones that contain an $(AG)_n$ insert.

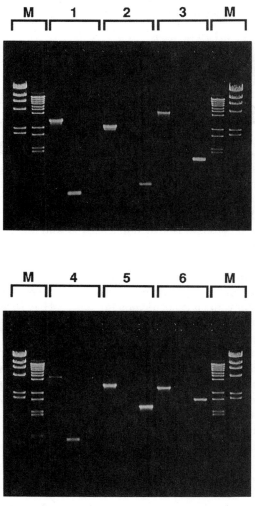

Figure 14.16 Results of the analysis of a number of triplex-purified clones by the PCR assay described in Figure 14.15. Adapted from Ito et al. (1992).

A third test of triplex capture took advantage of a strain of yeast that Peter Dervan and his co-workers had constructed, containing an insert of a 40-base homopurine-homopyrimidine sequence in yeast *(S. cerevisiae)* chromosome III. This strain was originally made to test the efficiency of triplex-mediated DNA cleavage. We were able to obtain and use the same strain to check the efficiency of triplex-mediated DNA capture. Instead of working with yeast genomic DNA, which would suffer considerable shear damage under the conditions used to manipulate the magnetic microbeads, we first subcloned this strain of yeast into a plasmid vector. Then plasmid DNAs were selected using magnetic microbeads containing the appropriate homopyrimidine sequence. In practice, 50% of the clones isolated contained the correct DNA insert as determined by DNA sequencing. The other contaminating clones had a similar, but not identical, sequence that was a natural component of the yeast genome. Presumably this contaminant could have been selected

against more efficiently by the use of slightly more stringent conditions for triplex formation and washing.

AFFINITY CAPTURE ELECTROPHORESIS

A major limitation with all of the magnetic microbead methods is that the shear damage
generated by liquid phase handling of DNA in this way restricts targets to DNAs less than
a few hundred kb in size. To work with larger DNAs, it is necessary that most or all manipulations take place in an anticonvective medium like agarose. This encourages using
electrophoresis in agarose to carry DNA past an immobilized triplex capture probe (Ito et
al., 1992). Such an approach has been termed affinity capture electrophoresis (ACE). To
test this method, we used streptavidin-containing microbeads (no need for magnets here)
embedded in agarose. A DNA sample containing a potential target for triplex formation
with the sequence on the beads was loaded in a sample well and electrophoresed at pH 5
past the potential capture zone, as shown in Figure 14.17. Then the pH was raised to 8,
and any material captured was released and analyzed in a subsequent electrophoresis step.

This simple procedure works, but at present, its efficiency is less than desired. The
poor efficiency arises from nonspecific binding between streptavidin and DNA at pH 5.
Streptavidin is a protein with an isolectric point of about pH 7. Therefore at pH 5 the protein is positively charged and, as such, binds nonspecifically to DNA. To destabliize these
electrostatic interactions, high ionic strength buffers can be used. Such buffers worked
very well with the magnetic separations described earlier. However, they lead to serious
complications in electrophoretic procedures because the high salt buffers have high conductivity, and as a result there is considerable heating and band broadening. These problems can probably be circumvented by changing the properties of the surface of streptavidin to introduce more negative charges. Since the three-dimensional structure of the
protein is known (Chapter 3), the molecular design and engineering needed to accomplish
this change in charge should be relatively straightforward.

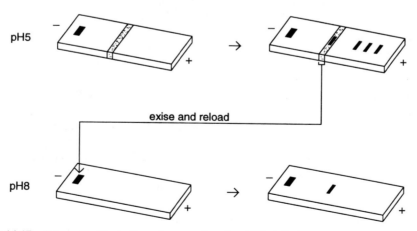

Figure 14.17 Schematic illustration of a procedure for ACE: Affinity capture electrophoresis. The
lightly shaded portion contains gel-embedded immobilized triplex-forming oligonucleotides corresponding to the target of interest.

To avoid the problems with the currently available form of triplex-mediated ACE, an alternative capture scheme was developed that could be implemented at neutral pH. This scheme lacks the attractive generality of triplex capture because it is applicable only to the ends of DNA fragments, and it requires that information be available about the DNA sequence at one end, or at least that a clone be available that overlaps the ends of the DNA to be captured. This is a major limitation, in principle, but it is less confining in practice. A major potential application for ACE is to try to purify specific large restriction fragments from complex mixtures. If a linking library that corresponds to the ends of these fragments is available, as described in Chapter 8, then the necessary clones or sequences will already be in hand.

The basic scheme of an alternative ACE procedure, termed end-capture electrophoresis, is shown in Figure 14.18. In this scheme, the ends of a long DNA duplex are treated with an exonuclease like *E. coli* Exonuclease III, or DNA polymerase in the absence of triphosphates, to remove a small portion of the 3'-ends of the duplex and expose the complementary 5' sequence as single strand. The affinity capture medium is made, as before, with the sequence complementary to the target. In this case the beads will contain the authentic 3'-end of the large DNA fragment, synthesized from known sequence or isolated as the appropriate half of a linking clone. Capture consists of ordinary duplex formation. The challenge is to find an efficient and nondisruptive way to release the target after it has been captured. An effective way to do this, illustrated in Figure 14.19, is to prepare the capture probe so that it contains the base dU instead of T. This still allows efficient, sequence-specific strand capture. However, subsequent treatment with Uracil DNA glycosidase, an enzyme that participates in the repair of DNA (Chapter 1), will release the captured target. The same general idea was described in Chapter 4 as a method for minimizing PCR contamination.

Figure 14.20 shows an actual example of end capture electrophoresis. Here, with DNA fragments of the order of 10 kb in size, the method works quite well. It still has not been successful in the capture of much larger targets. This may reflect the much slower rate at which these targets will find a probe during the electrophoresis. Because of excluded volume effects, the end of a large DNA will be accessible for hybridization only a small fraction of the time. If this is the problem, it should be resolvable in principle by using much slower electrophoresis rates.

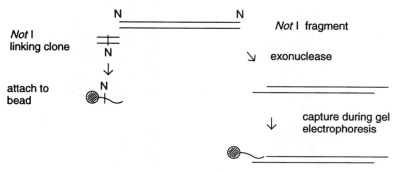

Figure 14.18 Schematic procedure for end capture electrophoresis. The appropriate single-stranded probes needed will be easily obtained from half-linking clones (see Chapter 8).

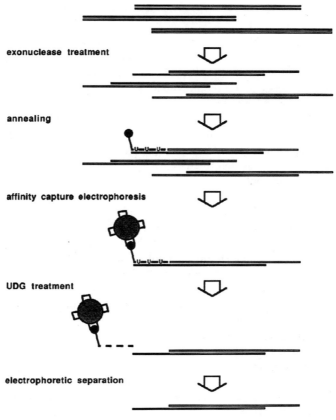

Figure 14.19 An end capture electrophoresis scheme that allows easy release of the captured DNA.

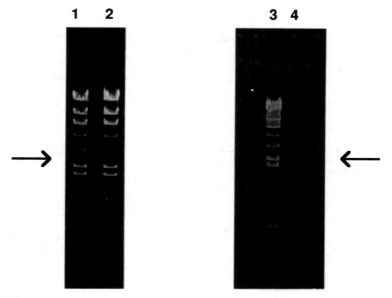

Figure 14.20 Example of the successful implementation of the scheme shown in Figure 14.19. Adapted from Ito et al. (1992). Lane 2 (arrow) shows a band removed by capture. Lane 4 (arrow) shows elution of the captured band.

USE OF BACKBONE ANALOGUES IN SEQUENCE-SPECIFIC DNA MANIPULATION

A number of backbone analogues of DNA have been described in recent years. The principle force motivating the development of such compounds is their presumed usefulness as DNA or RNA drugs. The expectation is that some of these analgoues will be resistant to intracellular enzymes that interfere with attempts to use natural RNA or DNA oligonucleotides to modulate ordinary cellular processes. We will discuss some of these efforts at the end of the chapter. One feature of a number of the backbone analogues that have been made is replacement of the highly charged phosphodiesters by neutral groups. The simplest approach is the use of phosphotriesters, in which the POO$^-$ group of the normal backbone is replaced by POO-R. The advantage conferred by such a substitution is that, in a duplex or triplex containing such an altered backbone, the natural electrostatic repulsion between the DNA strands will be minimized. Thus the resulting complexes ought to be much more stable, and one might expect that the backbone analog would preferentially form duplex or triplex at the expense of the normal strands present in a cell. This is why such compounds are attractive as potential drugs. Normal triplexes with two homopurine and one homopyrimidine strands, for example, are seriously destabilized at low-salt concentration because they have three intra-strand sets of electrostatic repulsion compared to only one for the duplex and a separated third strand (Fig. 14.21).

Phosphotriesters have a feature that considerably complicates their use. Since each phosphate now has four different substituents, as shown in Figure 14.22, it is optically active. This means that each alkyl group R can occupy one of two positions on each phosphate. The resulting number of different stereoisomers for a chain with n phosphates is 2^n. This is a very discouraging prospect for in vivo or in vitro studies, and phosphotriesters are likely to see rather limited use until the problem of synthesizing specific isomers can be solved. Until this has been accomplished, attention has focused on other backbone analogues which have the disadvantage of being less like natural nucleic acids but

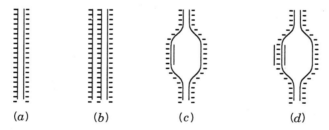

(a)   (b)   (c)   (d)

Figure 14.21 Electrostatic repulsion expected in various DNA structures. *(a)* Normal DNA duplex. *(b)* Normal DNA triplex. *(c)* DNA duplex with one strand displaced by an uncharged DNA analog. *(d)* DNA duplex with one strand displaced and a triplex formed at the other strand by binding of two uncharged DNA analogs.

Figure 14.22 Stereoisomeric pair of phosphotriesters.

which avoid the problem of multiple stereoisomers. Among the compounds that have been made and tested are polypeptide nucleic acids, in which a peptide backbone is used to replace the phosphodiester backbone, and polyamide nucleic acids (PNAs). The schematic structure of one example of this latter class of compounds is shown in Figure 14.23.

An amusing series of accidents clouded the first attempt to characterize the interactions of PNAs with duplex DNA. The actual compound used was R1-T10-R2. The notation T10 means that 10 thymine bases and backbone units were present. R1 was chosen to contain a positive charge for extra stability of interaction of the short PNA with DNA. R2 contained an intercalating acridine, also present to enhance binding stability, and a p-nitrobenzoylamide group. This latter promotes radical-induced cleavage of the DNA upon irradiation with near-UV (30 nm) light. This PNA was designed with the expectation that it would bind to a dA-dT stretch in a duplex and form a triple strand. What actually happened was more complex. A number of different chemical and enzymatic probes were used to examine the resulting PNA-DNA complex. All were consistent with the idea that the PNA had displaced the dT-containing strand of the natural duplex and formed a more stable duplex with the uncovered dA-stretch. A number of the results that led to this conclusion are shown in Figure 14.24a. The prospects raised by this outcome were extremely exciting because a duplex strand displacement mechanism would be applicable to any DNA sequence, not just to the more limited set capable of forming triplexes.

Further investigations of the properties of the PNA-DNA complex reveal an additional level of complication. It turns out that the complex contains not just one stoichiometric equivalent of PNA, but two instead. The result, as shown schematically in Figure 14.24b, is a complex in which displacement of one of the strands of the original DNA duplex has occurred, but the displaced strand is captured as a triplex with two PNAs. Thus this sort of reaction will be limited to sequences with triple-strand forming capabilities. Perhaps other backbone analogues will be found that work by displacing strands and capturing

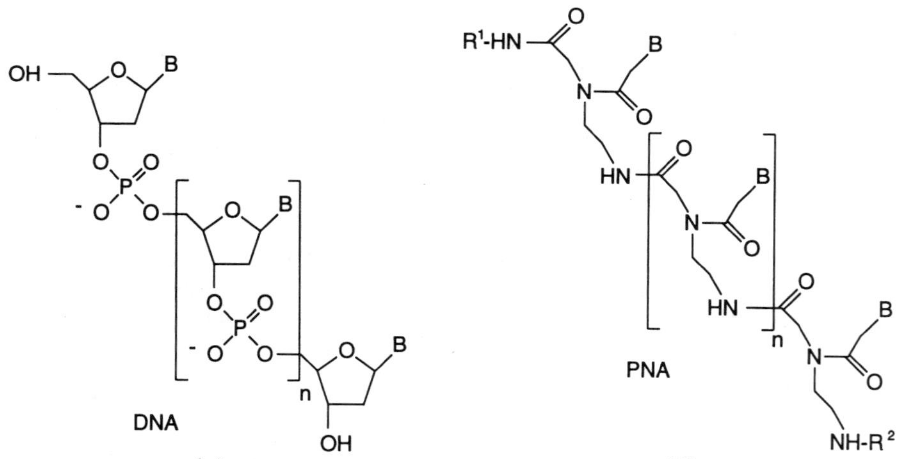

Figure 14.23 Chemical structures of the normal DNA backbone and polyamide nucleic acid (PNA) analogs. Adapted from Nielson et al. (1991).

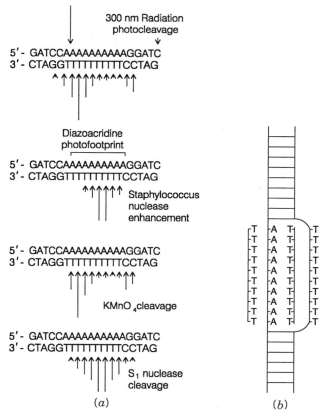

Figure 14.24 PNA binding to DNA. *(a)* Chemical evidence for the displacement of a (dT) 10 sequence by a corresponding PNA derivative. *(b)* The structure of the complex actually formed. Adapted from Nielson et al. (1991).

them as duplexes. An alternative scheme for in vitro work would be the use of recA protein-coated single strands. However, this is most unlikely to be useful in vivo.

In the past few years, a considerable amount of work has been done to explore the properties of PNAs and their potential usefulness in DNA analysis or clinical diagnostics. A few of these findings are briefly summarized here. The stability of PNA-DNA or DNA-PNA duplexes is essentially salt-independent (Wittung et al., 1994). Thus low salt can be used in hybridization procedures such as SBH to supress the interference caused by stable secondary structures in the target. PNAs are capable of forming sequence-specific duplexes that mimic the properties of double-stranded DNA except that the complexes are completely uncharged. Because there is no chirality in the PNA backbone, the duplexes are optically inactive; they have no preferred helical sense. However, attachment of a single chiral residue such as an amino acid at the end of the PNA strand leads to the formation of a helical duplex (Wittung et al., 1995). The ability of PNAs to bind tightly to specific homopurine, homopyrimidine duplexes leads to an effective form of Achilles's heel cleavage (Veselkov et al., 1996). Triplets that are located near restriction enzyme cleavage sites block these sites from recognition by the conjugate methylase. After removal of the triplex, the restriction nuclease will now cleave only at the sites that were previously protected, as in Figure 14.11*b*. The PNA-mediated protection appears to be quite efficient. A final novel use of PNAs is for hybridization prior to gel electrophoresis (Perry-O'Keefe et al., 1996). Since PNA is uncharged, it can be used to label ssDNA without interfering with subsequent high-resolution electrophoretic fractionations.

SEQUENCE-SPECIFIC CLONING PROCEDURES

Instead of physical isolation of particular DNA sequences, cloning or PCR procedures can be used to purify a desired component from a complex mixture. Direct PCR is very powerful if some aspect of the target DNA is known at the sequence level (Chapter 4). Where this is not the case, less direct methods must be used. Here several procedures will be described for specific cloning based indirect information about the desired DNA sequences to be purified. Several PCR procedures that take advantage of the possession of only a limited amount of DNA sequence information will be described later in the chapter.

Subtractive cloning is a powerful procedure that has played an important role in the search for genes. It can be carried out at the level of the full genome with much difficulty, or at the level of cDNAs with much greater ease. In subtractive cloning the goal is to isolate components of a complex DNA sample that are missing in a similar, but not identical, sample. One strategy for doing this, which illustrates the general principles, is shown in Figure 14.25. In this case, which is drawn from the search for the gene for Duchenne muscular dystrophy (DMD), two cell lines were available. One had a small deletion in the region of the X chromosome believed to contain the gene responsible for the disease. This deletion was actually found in a patient who displayed other inherited diseases in addition to DMD (the utility of such samples was discussed in Chapter 13). The objective of the

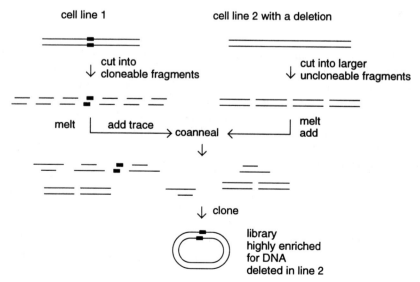

Figure 14.25 Differential cloning scheme originally used to obtain clones corresponding to the region of the genome deleted in Duchenne muscular dystrophy.

differential cloning was to find DNA probes that derived from the region that was deleted in this patient, since these would be candidate materials for the DMD gene itself.

A small amount of DNA from a normal individual was used as the target. This was cut with a restriction enzyme to give DNA fragments with cloneable ends. A large excess of DNA from the patient with the small deletion was prepared and cut into longer fragments than the target sample. This was done with an enzyme that would not give cloneable ends in the vector ultimately used. The two samples were melted and mixed together to co-anneal. Because the normal DNA was limiting, the DNA from the sample with the deletion acted as a driver. It rapidly formed duplexes with itself, and with corresponding fragments of the normal DNA. In contrast, DNA from the region of the deletion was present at very low concentrations in the mixture, and it renatured very slowly. Once renatured, however, the resulting duplexes had cloneable ends, unlike all of the rest of the DNA fragments in the sample. The mixture was then ligated into a vector and transformed into a suitable *E. coli* host. The resulting clones were, indeed, highly enriched for DNA from the desired deletion region.

The major difficulty inherent in the scheme shown in Figure 14.25, is that the desired DNA fragments are at low concentration and form duplex very slowly and inefficiently. In fact, to achieve an acceptable yield of clones, the renaturation had to be carried out in a phenol-water emulsion, which raises the effective DNA concentration markedly. This is not an easily managed or popular approach. More recent analogs of subtractive genomic cloning have been described that look powerful, and they should be more easy to adopt to a broad variety of problems (see Box 14.1). The potential power of such schemes is shown by the mathematical analysis of the kinetics of differential cloning in Box 14.2.

BOX 14.1
NEWER SCHEMES FOR DIFFERENTIAL GENOMIC DNA CLONING

Three schemes for cloning just the differences between two DNA samples will be described. The first two were designed to clone DNA corresponding to a region deleted in one available source but not in another. These schemes are similar to that described in Figure 14.25 except that they first use biotinylated driver DNA to facilitate the separation of target molecules from undesired contaminants, and then they use PCR to amplify the small amount of target molecules that remain uncaptured. In one scheme, developed by Straus and Ausubel (1990; Fig. 14.26), an excess of biotinylated driver DNA is used to capture and remove most of the target DNA by repeated cycles of hybridization and affinity purification with streptavidin-coated beads. Then the remaining desired target DNA is amplified and subsequently cloned by ligation of appropriate PCR adapters.

In a related scheme, developed by Eugene Sverdlov and co-workers, it is the target DNA that is biotinylated by filling in the ends of restriction fragments with dpppN derivatives (Fig. 14.27; Wieland et al., 1990). This target is then provided with PCR adapters by ligation. Excess driver DNA is used to deplete most of the target by cycles of hybridization and hydroxylapatite chromatography to remove any DNA duplexes formed. After several such cycles, streptavidin affinity chromatography is used to capture any biotinylated target remaining. The target molecules are then amplified by

(continued)

BOX 14.1 *(Continued)*

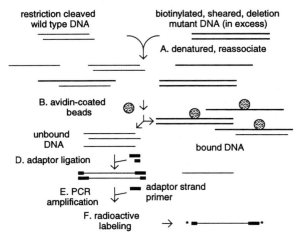

Figure 14.26   Differential cloning scheme based on repeated cycles of hybridization and biotin-affinity capture followed by PCR amplification. Adapted from Straus and Ausubel (1990).

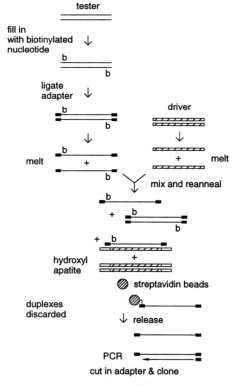

Figure 14.27   Differential cloning scheme based on repeated cycles of hybridization and hydroxyl apatite chromatography followed by biotin affinity capture and PCR amplification. Based on a method described by Wieland (1990).

(continued)

BOX 14.1 *(Continued)*

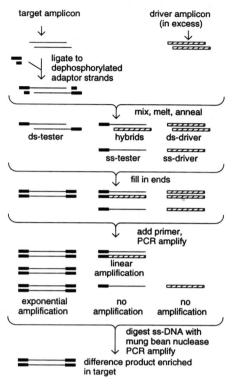

Figure 14.28 A method for cloning the differences between two genomes (RDA). Adapted from Lisityn et al. (1993).

primers complementary to the adapter sequences and cloned. Both of these methods appear to be quite satisfactory, and both can be enhanced, if necessary, by repeating the steps involved.

Recently a scheme has been described by the Lisityn et al. (1994) for cloning polymorphic restriction fragments. This scheme is illustrated in Figure 14.28. It has been called representation difference analysis (RDA). First, the complexity of both target and driver genomes is reduced by PCR to allow more effective subsequent hybridizations (Chapter 4). This is done by ligating on adapters and removing them after the PCR amplification. Then, as shown in Figure 14.28, the target is provided with new PCR adapters by ligation. Target is mixed with excess driver, melted, and reannealed. The ends of the duplexes formed are filled in with DNA polymerase. PCR is now used to amplify the entire reaction mixture. The key point is that target duplexes will show exponential amplification because they contain two adapters. Heteroduplexes will show only linear amplification, while driver DNA will not be amplified at all. Any single-stranded molecules remaining are destroyed by treatment with mung bean nuclease, a single-strand specific enzyme similar to S1. Then the cycles of hybridization and amplification are repeated.

BOX 14.2
SUBTRACTIVE HYBRIDIZATION

The purpose of subtractive hybridization is to purify a target DNA strand, symbolized by *T*, from other DNA, called tracer DNA, symbolized by *S*. This is accomplished by the use of driver DNA strands flanked by different primers, symbolized by *D*. The procedure is illustrated schematically below. In general, genomic or cDNA samples would be digested to completion with a restriction nuclease and ligated to splints to prepare sequences for subsequent PCR amplification.

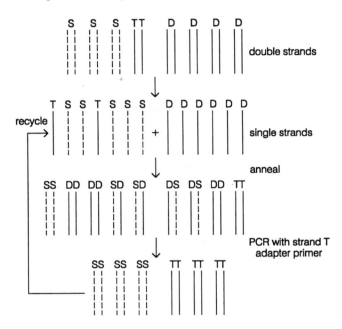

The mathematics behind this procedure is based on an equation developed in Chapter 3 to describe the kinetics of double-stranded DNA formation. If we call the initial concentration of single-stranded DNA segments is c_0, then the fraction of DNA that has formed double-stranded segments, f_{ds}, is given by the equation

$$f_{ds} = \frac{k_2 c_0 t}{1 + k_2 c_0 t}$$

where t is the time and k_2 is a constant for that particular sequence of DNA. Using this equation, we can determine the concentration of double-stranded segments by multiplying the initial concentration:

$$c_{ds} = f_{ds} c_0 = \frac{k_2 c_0^2 t}{1 + k_2 c_0 t}$$

(continued)

BOX 14.2 *(Continued)*

First Round of Subtraction

Consider two DNA samples. The first contains S- and T-type DNA; the second contains only S-type DNA. The sample containing only S-type DNA, however, is flanked by different primers, and will be designated by D. When these two samples are denatured and mixed, they will form double-stranded segments. The concentrations of various species can be determined by the equations above.

Since the T-type single strand will bind only with other T-type single strands, we can ignore the presence of S- and D-type strands in calculating the concentration of the double-stranded $T \cdot T$ duplexes formed. If c_T is the initial concentration of T single strands, the concentration of $T \cdot T$ double-stranded segments, is

$$(c_{T_{ds}})_{first\ round} = \frac{k_2 c_T^2 t}{1 + k_2 c_T t}$$

To calculate the concentration of the double-stranded tracer $S \cdot S$, we assume that the concentration of S is insignificant compared to the concentration of D. Double-strand formation of $S \cdot S$, $S \cdot D$, and $D \cdot D$ will occur indiscriminantly. Thus we can first compute the kinetics of $D \cdot D$ formation and then extract the amount of $S \cdot S$ by multiplying by the mole fraction of $S \cdot S$, denoted by $X_{S \cdot S}$:

$$X_{S \cdot S} = \frac{c_{S_1}^2}{c_{D_1}^2}$$

Where c_{S_1} is the initial concentration of S-type strands and c_{D_1} is the initial concentration of D-type strands. Therefore

$$(c_{S_{ds}})_{first\ round} = (c_{D_{ds}})_{first\ round} \times X_{S \cdot S} = \frac{k_2 c_{D_1}^2 t}{1 + k_2 c_{D_1} t} \times \frac{c_{S_1}^2}{c_{D_1}^2} = \frac{k_2 c_{S_1}^2 t}{1 + k_2 c_{D_1} t}$$

$(c_{D_{ds}})_{first\ round}$ and $(c_{S_{ds}})_{first\ round}$ are the concentrations of double-stranded $D \cdot D$ and $S \cdot S$ structures.

The next step in the subtraction protocol shown above is to amplify the strands by PCR. Since only the $T \cdot T$ and $S \cdot S$ strands have matching primers, only these strands will be amplified exponentially. This results in the effective removal of $S \cdot D$ and $D \cdot D$ strands, reducing the amount of S strand contamination, while increasing the concentration of T strands.

The ratio of concentration of $T \cdot T$ versus the concentration of $S \cdot S$ can now be calculated. This ratio is the enrichment resulting from the first subtraction step:

$$E_{first\ round} = \left(\frac{c_{T_{ds}}}{c_{S_{ds}}}\right)_{first\ round} = \frac{k_2 c_T^2 t}{1 + k_2 c_T t} \times \frac{1 + k_2 c_{D_1} t}{k_2 c_{S_1}^2 t} = \frac{c_T^2}{c_{S_1}^2} \times \frac{1 + k_2 c_{D_1} t}{1 + k_2 c_T t}$$

Since the initial round of subtraction is performed with samples directly from the genome, the concentration of T strands is the same as the concentration of S strands.

(continued)

BOX 14.2 *(Continued)*

This means that the ratio $c_T^2/c_{S_1}^2$ is equal to one for the first round. So, for large t, the enrichment ratio is

$$E_{first\ round} = \left(\frac{c_{T_{ds}}}{c_{S_{ds}}}\right)_{first\ round} = \frac{c_{D_1}}{c_T}$$

indicating that by the end of the first round, the ratio between T and S will be as large as the initial ratio of D and T DNAs used. Simply by using a much higher concentration of D strands than T strands, the presence of S-type strands can be significantly reduced.

Second Round of Subtraction

In following the procedure illustrated above for a second round of subtraction, it is important to note that the initial concentrations of S, T, and D for the second round are related to their concentrations at the end of the first round. The initial concentration of T strands for the second round can be assumed to be the same as the initial concentration of T strands for the first round. The concentration of S strands, however, will be less than that of the first round and will be denoted by c_{S_2}. Since the PCR amplification in the first round should not discriminate between S and T, we can assume that

$$\frac{c_{S_2}}{c_T} = \left(\frac{c_{S_{ds}}}{c_{T_{ds}}}\right)_{first\ round} = \frac{c_T}{c_{D_1}}$$

The concentration of D strands for the second round is much greater than either c_{S_2} or c_T. This value will be called c_{D_2}.

Calculations for the annealing kinetics in the second round proceed exactly the same way as in the first, except that we now use the values c_T, c_{S_2}, and c_{D_2}:

$$(c_{T_{ds}})_{second\ round} = \frac{k_2 c_T^2 t}{1 + k_2 c_{D_2} t}$$

$$(c_{S_{ds}})_{second\ round} = \frac{k_2 c_{S_2}^2 t}{1 + k_2 c_{D_2} t}$$

PCR amplification at this point replenishes the amount of T-type strands, while effectively removing some of the S-type strands. The ratio between the concentrations of T and S can be calculated, as in the first round, from the initial concentrations for the second round:

$$\left(\frac{c_{T_{ds}}}{c_{S_{ds}}}\right)_{second\ round} = \frac{k_2 c_T^2 t}{1 + k_2 c_T t} \times \frac{1 + k_2 c_{D_2} t}{k_2 c_{S_2}^2} = \frac{c_T^2}{c_{S_2}^2} \times \frac{1 + k_2 c_{D_2} t}{1 + k_2 c_T t}$$

(continued)

BOX 14.2 *(Continued)*

For large t, the final ratio of concentrations in the second round can be simplified to

$$E_{second\ round} = \left(\frac{c_{T_{ds}}}{c_{S_{ds}}}\right)_{second\ round} = \frac{c_T^2}{c_{S_2}^2} \times \frac{c_{D_2}}{c_T} = E_{first\ round}^2 \times \frac{c_{D_2}}{c_T}$$

Since c_{D_2} is chosen arbitrarily, if we use the value c_{D_1} again, the ratio of the final concentration of the second round simplifies even further:

$$E_{second\ round} = E_{first\ round}^2 \times \frac{c_{D_1}}{c_T} = E_{first\ round}^3$$

This is a remarkable result which shows that multiple subtraction protocols have a purification power that increases unexpectedly (adapted from notes provided by Eugene Sverdlov, as formulated by Ron Yaar).

If the starting DNA samples are genomic restriction fragments, the resulting amplified products eventually recovered will be those fragments in the subset of originally amplified material that had one restriction site that differed in the driver DNA. Such polymorphisms identified have been called polymorphic amplifiable restriction endonuclease fragments (PARFs). There are estimated to be around 1000 such *Bam*H I fragment differences between any two human genomes. Thus PARFs offer a potentially very powerful way to obtain useful genetic probes near preselected regions if DNA from appropriate individuals is available. For example, suppose that one has a population of individuals heterozygous for a dominant trait of interest. Subtraction of the DNAs of subsets of this population with DNAs from related individuals who lacked the trait should offer a reasonable chance of producing clones that contain polymorphisms linked to the trait or even responsible for the trait. A number of interesting variations on the original differential cloning scheme have been described (Yokata and Oishi, 1996; Rosenberg et al., 1995; Inoue et al., 1996). It remains to be seen how well such potentially very exciting new strategies actually perform in practice.

IDENTIFICATION OR CLONING OF SEQUENCES BASED ON DIFFERENCES IN EXPRESSION LEVEL

Once DNA sequences of potential interest have been identified, a frequent next step in understanding their function is to determine when and where in the organism they are expressed. For a simple sequence of interest, a suitable analytical method is the Northern blot. Here mRNAs from tissues or other samples of interest are fractionated by length using gel electrophoresis, transferred to a membrane and hybridized with a probe specific to the gene of interest. This is called a Northern blot, and it is a widely used procedure for accessing the expression level of individual genes. An alternative method is quantitative PCR (qRT-PCR). qRT-PCR requires much lower sample amounts but is difficult to standardize because of the intrinsically variable characteristics of PCR. Some protocols add a standard amount of target mimic to the PCR reaction. Since the mimic is usually shorter than the true target, the ratio of true target to mimic products can be determined. These approaches fall far short of the mark when the goal is to analyze many genes of interest. A number of interesting methods have been developed to sample mRNAs or corresponding cDNAs and look for expression differences or patterns in biological states. These are still evolving rapidly.

In differential display a degenerate short PCR primer is used to amplify the cellular population of cDNAs. This can be a poly dT complementary to the 3′ poly A sequence of messages, it can be an anchored dT_n (Liang et al., 1994), a short oligonucleotide, or a short sequence complementary to the end of DNA fragments generated by type II-S restriction enzymecleavage (Kato, 1995). Type II-S enzymes cut outside of their recognition sequences to yield a mixture of different single-stranded overhangs. A subset of this mixture can be captured by ligation to complementary overhangs, a technique that has been called molecular indexing (Unrau and Deugau, 1994). Regardless of the method used, the result is to reduce the complexity of the mRNA population and amplify the resulting cDNA to produce a discrete set of species to analyze quantitatively. The analysis can be performed by conventional or fluorescence-detected gel electrophoresis, or by hybridization (or two-color competitive hybridization) to an array of DNA probes analogous to the arrays used for SBH (Chapter 12).

These differential display methods are extremely powerful and broadly applicable. They suffer from a common limitation that whenever PCR amplification must be applied to a complex sample, the actual population of preexisting mRNAs will be distorted by differences in their ability to sustain multiple rounds of amplification. Two very different approaches for avoiding PCR-induced distortions are actively being explored. Picking cDNA clones at random and identifying them after sequencing is a powerful but expensive method. Alternatively, cDNAs can be sampled prior to amplification by cutting out a specific short fragment. The fragments are co-ligated into concatemers and amplified as a group. Then sequencing of cloned concatemers reveals relative abundances in a clever approach termed serial analyses of gene expression (SAGE).

An alternative to differential display is differential cloning. Differential cDNA cloning can proceed by schemes very similar to the one shown in Figure 14.25, or they can involve more elaborate techniques such as illustrated for the preparation of normalized cDNA libraries in Chapter 11. The objective is to recover cDNAs that represent messages present in one cell type but not another. The more carefully the cell types are selected, to differ just in the desired characteristics, the more efficient will be the search for the genes responsible for those characteristics. Differential cDNA cloning can be used to prepare genes specific for particular tissue types, developmental stages, chromosome origin, or even subchromosomal origin. These procedures are particularly effective when the target differences are very small and precisely defined. Examples are differences between unactivated and activated lymphocytes, to recover specific immune response genes, or differences between regenerating and nonregenerating tissue to isolate specific growth factors. As in genomic subtractive cDNA cloning, PCR can be used very effectively to solve most problems caused by small samples or rare messages. In principle, PCR should allow cDNAs to be made and subtracted from targets as small as single cells.

COINCIDENCE CLONING

Subtractive cloning allows the selective isolation of DNAs that differ in two samples. A related, but technically somewhat more demanding approach is coincidence amplification, involving either cloning or PCR, which is designed to allow the selective isolation of DNAs that are the same in two samples. In contrast to subtractive hybridization, which tries to purify unique homoduplexes (i.e., unique differences between samples), coincidence amplification targets unique similarities or homoduplexes between samples. Coincidence amplification will be most useful when two samples are available that contain only a small amount of DNA in common. A number of such situations exist. For example, suppose that one has isolated a chromosome or just a fragment of a chromosome in a hybrid cell. A successful coincidence cloning procedure would allow one to clone out just the human component by using DNA from normal human cells as the second sample. Two hybrids that contain only a small overlap region on a single human chromosome could allow the selective cloning of just that overlap region. Large DNA fragments cut out from a PFG fractionation could be used in coincidence cloning experiments with hybrid cell DNA to purify human components that lie, specifically, on a pre-chosen fragment size. Finally large DNA fragments could be used in coincidence with other large fragments to selectively clone just regions of overlap. Alternate PCR procedures exploiting human-specific interspersed repeating sequences may be used to isolate the human specific DNA. We will describe these methods in detail later in this chapter. However, they are not as general or as powerful, in principle, as coincidence cloning.

The basic task that has to be accomplished in any coincidence amplification is shown in Figure 14.29. DNAs from the two samples to be tested are melted, mixed, and allowed to reanneal. Most fragments in the samples will form homoduplexes because most of the DNA in two well-chosen samples will be different. Occasional heteroduplexes will be

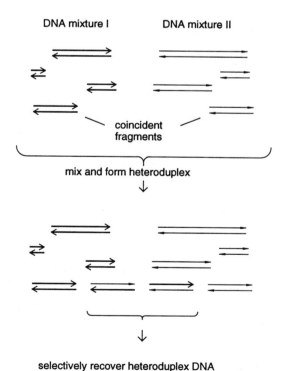

Figure 14.29   Basic requirements for coincidence cloning or coincidence PCR.

formed when sequences in the two samples match very accurately. What is needed, then, is a way of specifically cloning or amplifying the heteroduplexes. Note that this is a different problem than the heteroduplex detection we described in Chapter 13. In that case the desired heteroduplexes were those with one or more mismatches, and the mismatches were used as a specific handle to detect or capture the heteroduplexes. Here the desired heteroduplexes will in general be perfect matches. Quite a few different schemes have been tested for coincidence cloning (Box 14.3). None appear to work totally satisfactorily yet.

BOX 14.3
PROPOSED SCHEMES FOR COINCIDENCE AMPLIFICATION

Although these schemes for coincidence cloning are largely unproved, we will describe them in some detail because they illustrate some of the available arsenal of tricks for manipulating DNA sequences (Brooks and Porters, 1992). Similar schemes can be conceived of for coincidence PCR (Barley et al., 1993).

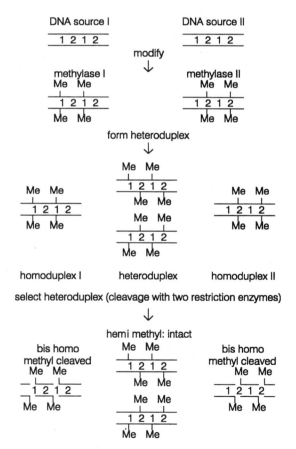

Figure 14.30   A scheme for coincidence cloning based on preferential cleavage of methylated homoduplexes. Adapted from Brooks and Porteus (1992).

(continued)

BOX 14.3 *(Continued)*

Figure 14.30 shows a scheme for differential cloning of heteroduplexes based on DNA methylation and the unusual properties of restriction enzymes like *Dpn* I. This enzyme has already been described in Chapter 8, where it was used to generate large DNA fragments by selective cutting. Here it will be used to destroy homoduplexes selectively. The scheme in Figure 14.30 requires two different restriction enzymes with the ability, like *Dpn* I, to cut only fully methylated DNA duplexes. DNA from one sample is methylated with one conjugate methylase; the second conjugate methylase is used to methylate DNA from the second sample. Then the two samples are mixed, denatured, and allowed to renature. The key feature of the resulting mixture is that all homoduplexes will be fully methylated at their respective sites, and they will be cut into small pieces when treated with a mixture of the two corresponding restriction nucleases. Only heteroduplexes will be hemi-methylated at all of the restriction sites in question, and so they should be much less sensitive targets for cleavage. Thus a size fractionation after the digestion should allow the preferential isolation and subsequent cloning or PCR of the heteroduplexes. The flaw in this scheme, as in the *Dpn* I-mediated specific DNA cleavage discussed in Chapter 8, is that it requires nucleases that do not cut hemi-methylated DNA. *Dpn* I at least does not have this necessary property.

A second scheme for coincidence cloning is illustrated in Figure 14.31. This scheme is based on the frequently used method of directional cloning. A vector is employed that requires two different ligatable ends for efficient cloning of a target. DNA from one source is amplified with a PCR primer with an extension that generates one of the necessary ends. DNA from the second source is amplified with the same PCR primer with an extension having different restriction enzyme cleavage sites. The best way to do this would be to start with separate libraries of the two source DNAs in the same vector. In this way tagged single primers corresponding to flanking vector sequence could be used for efficient PCR. Alternatively, tagged random primers or tagged short specific primers could be used (see Chapter 4). This is a more general approach that could be applied directly to genomic DNA, but it is likely to be much less efficient. After amplification the samples are treated with the restriction enzymes needed to cleave within the sites introduced by the primer extensions. Then the two samples are melted, mixed, and reannealed. In principle, only heteroduplexes will have the necessary ends required for efficient directional cloning.

In practice, this approach is likely to have problems of low yield and significant contamination with unwanted homoduplexes. Unless all of the cloned DNA is dephosphorylated, it will be quite common to co-clone homoduplex and heteroduplex fragments. Since the former are present in vast excess, they will contaminate most samples. Even if the restriction fragments are dephosphorylated prior to ligation to the vector arms, homoduplexes will be the major initial ligation product with the vast majority of the vector. This will consume most of the vector, and although the subsequent cloning of the resulting linear products will be relatively inefficient, it is likely to occur often enough to lead to a very serious background of unwanted homoduplex products.

A third strategy for coincidence cloning is presented in Figure 14.32. This appears to be potentially powerful, but it is also fairly complex. DNAs from two sources are separately treated with the same two restriction enzymes to form mixture of double

(continued)

BOX 14.3 *(Continued)*

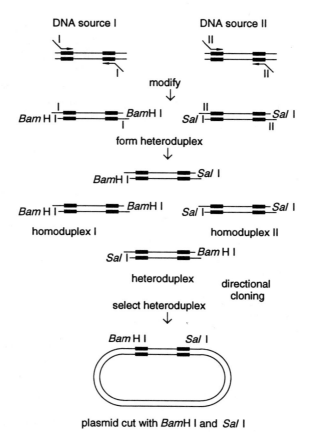

Figure 14.31   A scheme for coincidence cloning based on preparation of heteroduplexes to facilitate directed cloning. Adapted from Brooks and Porteus (1992).

digest products. One of the samples is left as restriction fragments. The other is directionally subcloned into the bacteriophage vector M13. Double-stranded flanking vector sequences, containing additional DNA segments, that will subsequently be used for PCR, are annealed to DNA from the M13 clones. Then both samples are melted, mixed, and allowed to reanneal. Heteroduplexes will be formed when restriction fragments from the first sample match clones from the second sample exactly. These are ligated to the tagged vector arms. Then PCR is used, with primers corresponding to the tagged sequences, to amplify the heteroduplexes specifically. The samples will be contaminated at this point by large amounts of M13 clones, but since these are all significantly larger than the PCR products, it should be possible to remove them by size selection.

(continued)

BOX 14.3 *(Continued)*

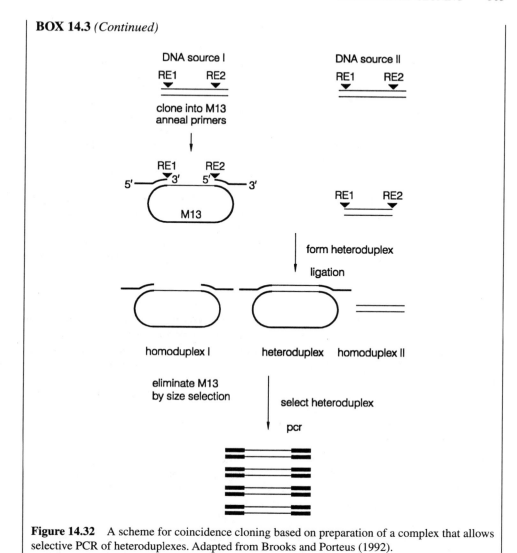

Figure 14.32   A scheme for coincidence cloning based on preparation of a complex that allows selective PCR of heteroduplexes. Adapted from Brooks and Porteus (1992).

There would be many interesting applications of robust coincidence cloning amplification procedures. For example, coincidence cloning of a cDNA library and a YAC should allow efficient capture of all of the genes on the YAC in a single step. Coincidence cloning of cDNAs from two very different tissue types should result in a highly enriched population of housekeeping genes that are not tissue specific. Finally coincidence cloning could be a very effective way to isolate very specific human DNAs from hybrid cell lines. An example of this is shown in Figure 14.33. Here probes of interest detect PFG-fractionated large human restriction fragments in a rodent background, and the goal is to obtain clones for the human DNAs contained on these fragments. The key point is that the human DNAs recognized by the same probes in the two different digests must have corresponding DNA sequences, while the rodent DNA that contaminates each human fraction

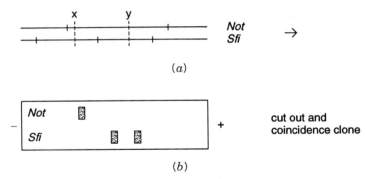

Figure 14.33 How coincidence cloning might be used to enrich the DNA from particular large re-striction fragments seen by hybridization in rodent-human hybrid cells. *(a)* Restriction map of the region; *X* and *Y* indicate two human-specific probes. *(b)* Southern blot of a PFG fractionation hybridized with the two available probes. Intact material from a duplicate PFG run would be used to cut out the regions detected by hybridization, and one digest would be used in coincidence cloning with the other.

is likely to be different. Thus coincidence cloning of DNAs from a pair of gel slices from the two different digests should preferentially yield the human material. Even more complex logical cloning schemes consisting of various separation and amplification steps can be conceived of. Their ultimate utility will depend on how effective the more simple straightforward procedures become.

HUMAN INTERSPERSED REPEATED DNA SEQUENCES

Most repeated sequence DNA in the human genome and other complex genomes appears to be interspersed with single-copy DNA. Although plenty of tandem repeats exist, such as VNTRs, except for centromeres these actually make up only a small fraction of the class of highly repeated sequences. The original demonstration that the bulk of the human repeats was interspersed was done in a series of classic experiments by Britten and Davidson. Subsequently the type of analysis they used has been refined and elaborated by Robert Moyzis and his colleagues. The basic scheme that underlies these approaches is shown in Figure 14.34*a*. A trace amount of labeled total human DNA is sheared randomly to various average lengths, *L*. This DNA is hybridized in solution with a vast excess of much shorter driver DNA. The driver DNA consists of unlabeled total human DNA isolated as all of the duplex that forms at a C_0t of less than or equal to 50. This sample will contain all significant high-copy number human repeats but will contain very little single-copy DNA or infrequent repeats such as those seen in gene families. The hybridization of the driver with the labeled DNA is carried out at a C_0t of 12.5. The goal of the experiment is to determine how much of the total labeled DNA can be captured by the repeated DNA driver during the hybridization.

The basic idea behind the experiment in Figure 14.34*a* is that all the DNA in clustered repeats will be captured very easily and efficiently, regardless of the length of DNA target used. In contrast, when *L* is small, most interspersed DNA will be captured without flanking single-copy sequences. As *L* increases, more and more single-copy DNA will be cap-

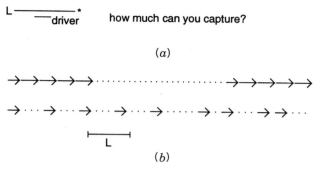

L ——————— *
        ——driver   how much can you capture?

(a)

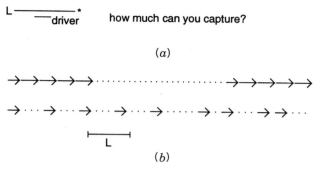

(b)

Figure 14.34 Analysis of the pattern of repeats in human DNA by renaturation kinetics. *(a)* Design of the basic experiment. Purified repeat is used in excess as the driver to capture indexed genomic DNA fragments of length *L*. *(b)* Two possible extreme patterns of repeating sequence.

tured by virtue of its neighboring repeated DNA (Figure 14.34*b*). A typical experimental result from such a procedure is shown in Figure 14.35. It is evident that only about a quarter of the genome is captured when *L* is small; this provides an estimate for the total amount of highly repeated DNA. Almost all of the genome is captured by the time the average fragment size reaches 8 kb. Thus many single-copy DNAs have a repeated sequence within 1 kb, and almost all single-copy DNAs have a repeat within 8 kb. The simplest fit to the data in Figure 14.35 suggests the occurrence of an interspersed repeat every 3 kb. A more elaborate and more accurate fitting procedure suggests that the distribution of repeats is bimodal. About 58% of the genome has a repeat on average every 1 kb; the remainder of the genome has a repeat on average every 8 kb. Of course this analysis does not reveal any information about the nature of the repeats or the number of different basic kinds of repeats and their particular distribution.

In fact there are just a few known major types of interspersed repeats in the human genome. Their properties are summarized in Table 14.3. By far the most common human repeat is a sequence called Alu because it contains two cutting sites for the restriction enzyme Alu I. Thus when human DNA is cut with Alu I and fractionated by size, the resulting material shows a bright, specific size band standing out from a background broad smear of other fragment sizes. There may be as many as 10^6 Alu sequences in the human genome. Very similar sequences are seen in other primates, but in more diverged species, like rodents, the Alu-like sequences are sufficiently different that Alu's can be used as

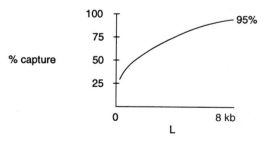

Figure 14.35 Fraction of total genomic DNA captured at low C_0t with excess repeated DNA driver, as a function of the length, *L*, of the genomic DNA used. Adapted from Moyzis et al. (1989).

TABLE 14.3   Major Known Human Repeats

Type	Abundance (Copies)	Hybridization Average Spacing	GenBank Average Spacing	Average Length	Total Mass (Est.)
3' line (L1)	5×10^4–10^5	30–60 kb	27 kb	1.1 kb	55–110 Mb
Intact L1	4×10^3–10^4	150–480 kb	—	6.4 kb	Trivial
$(GT)_n$	5×10^4–10^5	30–60 kb	54 kb	0.04 kb	Trivial
Alu	5×10^5–10^6	3–6 kb	4 kb	0.24 kb	120–240 Mb

general species-specific DNA probes. A typical Alu sequence is around 0.24 kb and actually consists of an approximate tandem repeat of a shorter sequence. This is illustrated in Figure 14.36, which shows the consensus among known Alu sequences. Alu is an example of a class of repeating DNA sequences called short interspersed repeat sequences (SINES). Alu sequences in the human are far from identical. Known sequences have been classified into at least five different families, and even within a family the average difference between two Alu's is roughly 10%.

```
HS CON   GGCCGGGCGC  GGTGGCTCAC  GCCTGTAATC  CCAGCACTTT  GGGAGGCCGA      50
pPD39    . . . . . . . . . .   . . . . . . . . . .   . . . . . . . . . .   . . . . . . . . . .   . . . . . . . . . .
BLUR 8   XXXXXXXXXX  XXXXXXXXXX  XXXX. . . . . .   . . . . . . . . . .   . . . . . . . .A.


HS CON   GGCGGGCGGA  TCACGAGGTC  AGGAGATCGA  GACCATCCCC  CCTAAAACGG     100
pPD39    . . . . . . . . . .   . . . . . . . . . .   . . . . . . . . . .   . . . . . . . . . .   . . . . . . . . . .
BLUR 8   . .A. . . .A. .   . . . .CT.AAGTC  . . . . .T.T. .   . . . . .G. .T.   . .C. .C.T. .


HS CON   TGAAACCCCG  TCTCTACTAA  AAATACAAAA  AATTAGCCGG  GCGTAGTGGC     150
pPD39    . . . . . . . . . .   . . . . . . . . . .   . . . . . . . . . .   . . . . . . . . . .   . . . . . . . . . .
BLUR 8   . . . . . .T. .A   . . . . . . . .G.   . . . . . . . . . .   .X. . . . . .A.   . .A.G. . .AT


HS CON   CGGCGCCTGT  AGTCCCAGCT  ACTTGGGAGG  CTGAGGCAGG  AGAATGGCGT     200
pPD39    . . . . . . . . . .   . . . . . . . . . .   . . . . . . . . . .   . . . . . . . . . .   . . . . . . . . . .
BLUR 8   .C.T. . . . .G   .A. . . . . . . .   . . . .A. . . . .   . . . . .A. . .A   . . . . .CC.T.


HS CON   GAACCCGGGA  GGCGGACCTT  GCAGTGAGCC  GAGATCCCGC  CACTGCACTC     250
pPD39    . . . . . . . . . .   . . . . . . . . . .   . . . . . . . . . .   . . . . . . . . . .   . . . . . . . . . .
BLUR 8   A. . . .AAX. .   . .T. . . .G. .   . . . . . . . . . .   . . . . . .G.A.   GG. . . . . . . .


HS CON   CAGCCTGGGC  GACAGAGCGA  GACTCCGTCT  CAAAAAAAAA                 290
pPD39    . . . . . . . . . .   . . . . . . . . . .   . . . . . . . . . .   . . . . . . . . . .  A12
BLUR 8   . . . . . . . .TX   . . . . . . . . . .   . . . . . .A. . .   . . . . . . . . .X
```

Figure 14.36   Sequence of a typical human Alu repeat. Shown are the consensus sequence for many known Alu's, and two particular clones often used as Alu probes. Taken from Batzer et al. (1994).

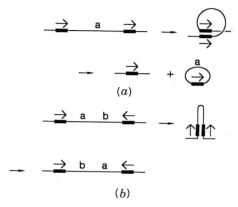

Figure 14.37 DNA rearrangements produced by recombination between interspersed repeats like the Alu sequence. *(a)* A direct repeat leads to a DNA deletion because the short fragment produced is unlikely to be retained or replicated. *(b)* An inverted repeat leads to an inversion in the DNA sequence.

The next most common human repeat is the L1 sequence. It sometimes called Kpn because it has characteristic restriction sites for the enzyme Kpn I. The Kpn family is quite complex. The total sequence is about 6.4 kb in length. However, only 10% to 20% of the repeats are full length. Most others retain only the 3'-terminal kb of the L1. These 3' L1's occur at about a tenth the frequency of the Alu sequence. L1's are more similar in primates and rodents than Alu's. Thus they must be used cautiously as species-specific probes. L1 sequences are an example of the class of repeats called long interspersed repeating sequences or LINES. Altogether the repeated sequences listed in Table 14.3 add up at most to 350 Mb of DNA. Centromeric satellite is estimated to account for another 150 to 300 Mb of DNA. This material is tandemly repeating (Chapter 2). Thus at least 100 Mb of the total estimated 750 Mb (25%) of repeated human DNA remains unaccounted for, and we may be missing as much half of all the repeated sequences.

Interspersed repeats appear to act as recombination hot spots. The sequence of most examples of Alu repeats is similar enough that recombination between pairs of such repeats is sometimes seen as a cause of disease alleles. As shown in Figure 14.37, depending on whether the repeats are head to head (tandem) or head to tail (inverted), the result of an intrachromosomal recombination event will be an inversion or a deletion. Recombinations between repeats on different chromosomes will produce translocations. The low-density lipoprotein receptor is one of the genes where a deletion caused by inter-Alu recombination has been seen. In this case the result is to produce a disease allele for familial hypercholesterolemia.

DISTRIBUTION OF REPEATS ALONG CHROMOSOMES

In Chapter 2 we illustrated the fact that light bands and dark bands of human chromosomes had some very different general properties. A commonly accepted mechanism for the spread of interspersed repeats is that at least some of the copies of these sequences are mobile elements and can spread themselves or copies to other sites. This has definitely been shown to be the case for the intact L1 repeat. Given this general picture, it is not surprising that certain types of repeats appear to cluster in certain genome regions. While the exact mechanism for this clustering is not understood, a formal mechanism to explain it would simply be to state that the transposition of the mobile elements favors genomic regions with certain properties.

Alu sequences are preferentially located in Giemsa light, G + C-rich bands. Note that the Alu's, themselves are G + C rich. Their average base composition is 56% G + C, and they have a CpG content that is only 64% of that expected statistically for such a G + C content. This is remarkable given the overall suppression of CpG throughout most of the genome. Thus Alu's behave a bit like HTF islands (Chapter 8). It is not surprising then, that like HTF islands. Alu's seem to preferentially associated with genes. The Giemsa light bands are very gene rich, and within many of these genes there are truly remarkable numbers of Alu repeats (in the introns, of course). These Alu's greatly complicate attempts to sequence genomic, gene-rich regions by shotgun strategies (Chapter 10) because of the difficulty of assembling the sequence.

L1 sequences occur preferentially in dark bands. These bands are A + T rich. The L1s themselves are A + T rich (58%). They are also extremely deficient in CpG sequences with only 13% of what would be expected statistically. Thus we can conclude from both the patterns of Alu's and L1's that like attracts like, but the mechanism behind this remains unclear. Earlier we indicated that the pattern of Alu distribution was really biphasic. Presumably the Alu-rich phase seen in renaturation experiments corresponds to DNA from light bands, but this has not been formally proved.

The final class of well-characterized interspersed repeats are the VNTRs. These are preferentially located in telomeric light Giemsa bands, although they are spread well enough through the genome to be generally useful as genetic markers. The reason why VNTRs cluster near the telomeres is unknown. However, it is worth noting that the frequency of meiotic recombination appears to be very high in the telomeric regions and that one mechanism of VNTR growth and shrinkage is recombination. There is no way to tell at present whether any causal relationships existing among these observations. However, they represent a tantalizing area for future study.

PCR BASED ON REPEATING SEQUENCES

Sequences in the Alu and L1 families are similar enough so that a single PCR primer can be used to initiate DNA synthesis within a large number of these elements. Some of the common Alu primers are summarized in Figure 14.38. These are chosen to try to focus on the most-conserved regions of the repeats within known human sequences without selecting sequences that are also conserved in rodents. Some Alu primers are tagged with extensions to allow more efficient amplification after the first few rounds of PCR where an inexact or very short match between primer and target template may be occurring (Chapter 4). The general situation in which these primers are applicable is shown in Figure 14.39. Neighboring copies of a repeat can have inverted configurations (head to head or tail to tail) or tandem configurations (head to tail). In the former case, a single PCR primer will serve to amplify the DNA between the repeats. In the latter case, two primers must be used. Inter-Alu PCR is a very powerful tool because so much of the human genome is dense in Alu sequences, and Alu sequences in humans are well diverged

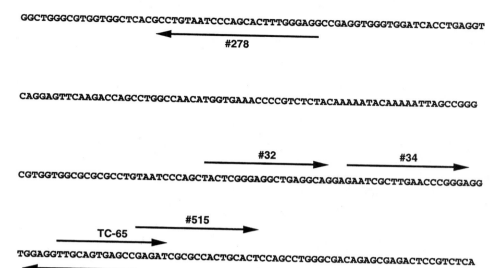

GGCTGGGCGTGGTGGCTCACGCCTGTAATCCCAGCACTTTGGGAGGCCGAGGTGGGTGGATCACCTGAGGT

#278

CAGGAGTTCAAGACCAGCCTGGCCAACATGGTGAAACCCCGTCTCTACAAAAATACAAAAATTAGCCGGG

#32   #34

CGTGGTGGCGCGCGCCTGTAATCCCAGCTACTCGGGAGGCTGAGGCAGGAGAATCGCTTGAACCCGGGAGG

#515

TC-65

TGGAGGTTGCAGTGAGCCGAGATCGCGCCACTGCACTCCAGCCTGGGCGACAGAGCGAGACTCCGTCTCA

#33

#517

Figure 14.38 DNA sequences of some of the primers commonly used to amplify DNA between repeated Alu sequences by PCR.

from those of rodents. Thus a significant fraction of the human genome is potentially amplifiable by inter-Alu PCR. The L1 repeat is less useful in this regard because it is rarer and because its sequence is more conserved in rodents and human. Despite this limitation inter-L1 PCR or PCR between L1 and Alu sequences can still be helpful tools.

It would take most of a chaper to describe the myriad applications of inter-Alu or inter-L1 PCR in detail. We will just list a number of the most prominent applications, and then illustrate a few in more detail. Inter-Alu PCR is helpful whenever one needs to selectively amplify the human component in a nonhuman background. Not all the human DNA will be amplified. In general, one can expect good amplification wherever two Alu sequences with close sequence homology to the primers used lie in the correct orientation within a few kb of each other. This will be quite frequent in Alu-rich regions of the genome, much rarer in other regions. For example, inter-Alu PCR will selectively amplify the human component in a rodent hybrid cell line. It will preferentially amplify the YAC DNA in a background of yeast DNA. Combining YAC vector primers and Alu primers

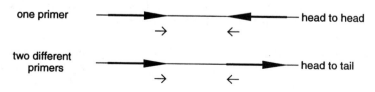

Figure 14.39 Some of the arrangements of interspersed repeats (arrows) that can be amplified by PCR using primers selected from the sequence of the repeat.

will preferentially amplify human DNA cloned near the ends of YAC inserts. These are exactly the samples most desirable as probes for YAC walking techniques. Single-sided Alu amplification procedures have also been described for regions where Alu's are too dilute to allow efficient inter-Alu PCR (Quereshi et al., 1994). Single-sided Alu PCR is also a very useful method to treat chromosome-specific hncDNA libraries, as described in Chapter 11.

The pattern of bands amplified by inter-Alu PCR from a whole human chromosome is often too complex to analyze (although inter-L1 PCR is helpful in such applications). Sometimes, with particular primers and conditions, the number of amplified bands can be reduced to of the order to 40. In such cases a significant fraction of these bands appears to be polymorphic in the population, and thus the Alu PCR provides a very convenient and easily used set of multiplexed genetic markers (Fig. 14.40).

In contrast to most attempts to amplify DNA from the whole genome, the pattern of bands from a few Mb of DNA is usually quite clear and diagnostic. Thus inter-Alu PCR is a powerful fingerprinting method. The added incentive is that the PCR-amplified bands seen in an electrophoretic analysis can be cut out and used as single-copy hybridization probes (competing any residual Alu sequences, as needed). Thus inter-Alu PCR finger-printing has been applied to YACs, chromosome fragments, radiation hybrids, and PFG gel slices, and it has been used to isolate single-copy DNA probes from all of these kinds of samples. The products of inter-Alu PCR reactions are very useful for FISH mapping. They are usually complex and concentrated enough to yield good results and allow the rough map position of the sample to be identified. Inter-Alu PCR products are also very useful for cross-connecting libraries.

An example of inter-Alu PCR applied to the analysis of PFG gel slices is shown in Figure 14.41. Here a *Not* I-digested hybrid cell line containing chromosome 21 as its only human component was fractionated by PFG under a number of different conditions to maximize the resolution of indivdiual size regions. Each gel lane was sliced into 40 frag-ments, and each fragment was subjected to inter-Alu PCR. In most cases lanes showed discrete patterns of several amplified bands, and adjacent slices often showed quite differ-ent patterns (Fig. 14.42). Thus the carryover of material from slice to slice during the PFG and subsequent steps is not so serious as to obscure the fractionation. This means that individual PCR products from analytical gels like the one shown in Figure 14.42 can be cut out, reamplified, and used as immortal specific single-copy DNA probes. Even if the PFG gel slice contained more than one genomic human *Not* I fragment, the individual PCR products from that slice are each likely to derive from only a single genomic frag-ment. Thus they constitute a very convenient source of new single-copy human DNA probes.

The method illustrated in Figures 14.41 and 14.42 is very helpful in the later stages of physical mapping where most fragments of a chromosome are located and the goal is to obtain new probes for unassigned bands as efficiently as possible. One problem with the kind of results shown in Figure 14.42 is that the number of new probes provided by a sin-gle experiment is very large, frequently a hundred or more. Before selecting probes for further study, usually one would like to know something about their regional location on the chromosome of interest. The standard way to do this is to take a probe of interest and hybridize it to a mapping panel of chromosome deletions as we described in Chapter 8. However, this is far too inefficient when a hundred or more probes must be mapped at once. An alternative approach, useful for YACs or slices of PFG fractionations, is shown in Figure 14.43. Ideally what one would like to do is take DNA from hybrid cell lines

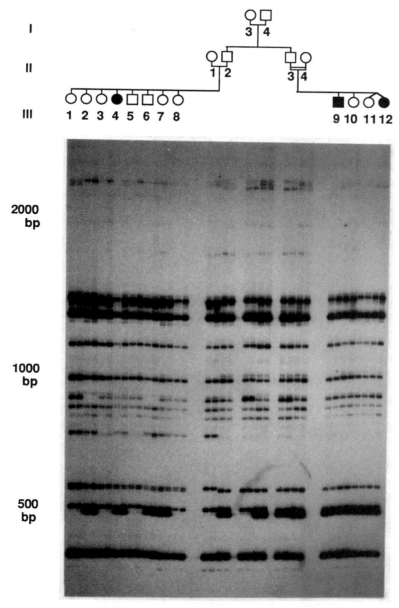

Figure 14.40 An example of Alumorphs: Polymorphic genomic DNA sequences amplified by inter-Alu PCR. Analysis of two pseudo–vitamin D-deficient rickets (PDDR) families (affected individuals are indicated by filled symbols). The ^{32}P-labeled products of PCR amplification using an Alu-specific primer were analyzed by electrophoresis in nondenaturing 6% polyacrylamide gel. Each individual from the pedigree shown on the top of the autoradiogram was analyzed in duplicate by two independent PCR amplifications, shown in two adjacent lanes on the gel. Molecular size markers are indicated at left. Taken from Zietkiewicz et al. (1992).

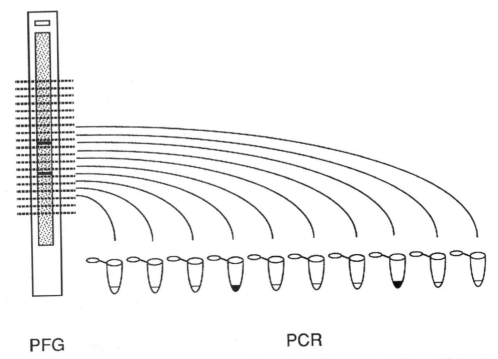

PFG   PCR

Figure 14.41 Schematic example of the use of inter-Alu PCR to preferentially amplify human DNA from PFG-fractionated large restriction fragments in a hybrid cell line.

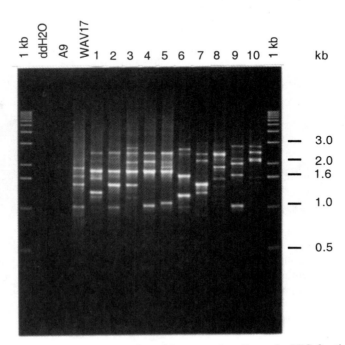

Figure 14.42 Inter-Alu PCR products from 10 consecutive slices of a PFG-fractionated *Not* I digest of a hybrid cell line containing chromosome 21 as its only human component. From Wang et al. (1995).

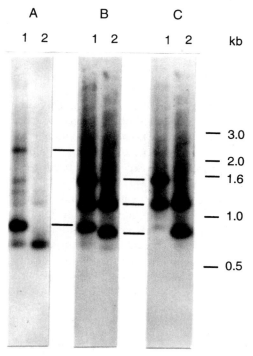

Figure 14.43 Example of how inter-Alu PCR products can be assigned, en masse, to chromosome regions. Inter-Alu probes generated from DNA from cell lines 8q$^-$ (A), R2-10W (B), or 21q$^+$ (C) were used to assign inter-Alu gel slice products regionally. Conventional gel lanes (lanes 1 and 2) containing inter-Alu products generated from template DNA contained in two different PFG slices are shown. Note that each cell line specific probe is hybridized to a different set of inter-Alu PCR products. From Wang et al. (1995).

containing human chromosome fragments and use this as a hybridization probe against a blot of a gel-fractionation of all the PCR products from a set of YACs or Not I fragments. The difficulty is that the complexity of the source DNA is too large, and one needs a way to reduce it and label the human component selectively in order to obtain efficient hybridization. However, inter-Alu PCR products from these cell lines provide the precise DNA subpopulation needed for efficient hybridization to inter-Alu PCR products from YACs or PFG fractions. By selectively amplifying the same segments of the DNA in both the hybridization probe and the hybridization target, one achieves enormously rapid and specific hybridizations. This same principle can be applied whenever inter-Alu PCR products are used for fingerprinting or for cross-connecting libraries.

The pattern of PCR products between repeating sequences can also be used to provide information about the distribution of repeats in the genome. With Alu, the patterns are too complex to analyze on a whole genome level. The situation is much more favorable with the 5' L1 sequence. To estimate the number of PCR products expected with a single 5' L1 primer, Yoshiyuki Sakaki assumed that there were about 3000 copies of the L1 repeat or one per Mb. This estimate is on the low side of the range reported by others (Table 14.3). Suppose that the PCR range is 2 kb. Only a quarter of the L1's within this range will be oriented head to head and thus amplified by the primer. So the expected frequency of PCR products per genome can be estimated as

Number $(3 \times 10^3) \times$ Spacing $(10^{-6}) \times$ Range $(2 \times 10^3) \times$ Orientation $(0.25) = 1.5$ per genome

In actuality about 20 genomic PCR products are seen. This presumably indicates that L1's are clustered, which is in accord with observations we have described previously. Note, however, that if we took a higher estimate for the number of 5' L1 sequences, say 1.5×10^4 copies, then the expected number of products would be 7.5 per genome, and the evidence for clustering, from this one experimental result alone, would be much less compelling.

PCR amplification schemes can also be based on tandemly repeating dinucleotide or trinucleotide sequences. These are too infrequent to allow amplification between repeats. Instead, single-sided amplification methods are used. Alternatively, repeating-sequence-containing fragments are captured by hybridization with an immobilized single-strand, and then the released repeats are amplified in a number of different ways. These procedures are quite efficient (Broude et al., 1997; Kandpal et al., 1994).

REPEAT EXPANSION DETECTION

A final example of the use of DNA amplification based on interspersed repeating sequences is shown in Figure 14.44. This illustrates a newly developed technique called repeat expansion detection (RED) which is designed as a way to specifically isolate very large tandemly repeating DNA sequences such as the expanded triplet repeats found in fragile X syndrome and other human disease alleles (Chapter 13). RED uses the ligase chain reaction (Chapter 4) instead of PCR. Oligonucleotide probes consisting of 11 to 17 tandem triplet repeats are annealed to target DNA in the presence of a thermostable DNA ligase. Repeated cycles of denaturation and renaturation are carried out. Long repeated triplet alleles in the target will promote more effective ligation of the probes than short alleles, and a more complex set of ligated products will result. This can be detected by electrophoresis after hybridization with the complementary triplet repeating sequence. This method is a very promising approach to the discovery of new genes, where unstable triplet repeats may be responsible for producing disease alleles. Other methods for repeat expansion detection may be based on the observation that PCR amplification of many long triplet repeats is inefficient at best and often fails completely (Broude et al., 1997).

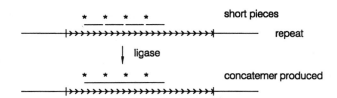

Figure 14.44 Schematic illustration of the repeat expansion detection (RED) procedure used to identify cells with large, potentially disease-causing repeated triplet alleles. Adapted from Schalling et al. (1993).

APTAMER SELECTION STRATEGIES

A relatively recently developed set of strategies combines physical purification and PCR to select DNAs (or RNAs or proteins) with desired sequences or binding properties. These methods appear to be powerful enough that in some cases one can start with all 4^n possible nucleic acids of length n and find the one or few with the optimal affinity for a given target. Among the potential applications are:

- Purification of DNA sequences with the highest affinity for a given protein, ligand, or drug (such molecules have been termed aptamers)
- Purification of RNA sequences with the highest affinity for a given DNA (via duplex or triplex), RNA, protein, ligand, or drug
- Purification of protein sequences with the highest affinity for a given receptor, ligand, or drug (also called aptamers)

It is relatively easy to make all 4^n possible DNAs of length n just by adding all four dpppN's at each step in automated DNA synthesis. More restricted mixtures can be made by an obvious extension of this approach.

The general principle behind select strategies is shown in Figure 14.45. A complex mixture is allowed to bind to an immobilized target of interest, usually at fairly low stringency. Those species that do bind are eluted, and PCR amplification is used to regenerate a population of molecules comparable to the initial total concentration. Now, however, this population should be enriched for molecules that have some affinity for the target. The cycles of affinity purification and amplification can be repeated as often as needed, until the complexity of the mixture becomes small enough to analyze. In the perfect case only a single species would remain, and if necessary, the stringency of the affinity step could be progressively increased during successive cycles. In actual cases a mixture of molecules will be seen, but this will eventually attain a small enough complexity so that individual components can be cloned and examined. Alternatively, a powerful approach is to sequence the mixture of molecules remaining after a large number of cycles of select purification. If certain positions within the DNA (or RNA or protein) are required for affinity, and others are not, the sequence of the mixture will show conserved residues at some positions, which will be unambiguously identified, and mixtures of residues (usually refractory to analysis) at other positions.

For the select approach to work, one needs a fairly good affinity purification with relatively little nonspecific background. Three potential implementations of the select strategy are shown in Figure 14.46. These allow for purification of DNAs, RNAs, or proteins with particular affinity properties. A number of highly successful examples of the application of selection strategies are summarized in Table 14.4. Such strategies are also effec-

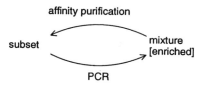

Figure 14.45 Basic principle behind a select strategy for purification of sequences with specific affinity properties.

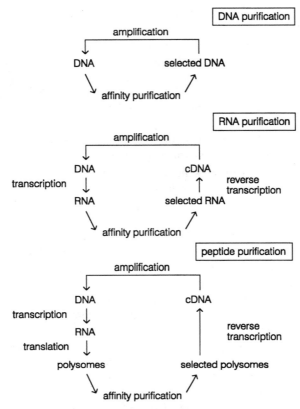

Figure 14.46   Select strategies that have been proposed for purifying DNAs, and DNAs coding for RNAs, and proteins with particular properties that can be converted to differential affinities. Adapted from Irvine et al. (1991).

TABLE 14.4   High-Affinity Aptamers

Target	Library	Nucleotides[a]	Rounds[b]	Motif	K_d (nM)
E. coli rho factor	RNA	30	8	Hairpin	1
E. coli metJ protein	RNA	40	15	Unknown	1
HIV-1 rev protein	RNA	32	10	Bulge	1
sPLA2	RNA	30	12	Complex	1
	Modified RNA	30	10	Pseudoknot	1
Basic fibroblast growth factor	RNA	30	13	Hairpin	0.20
Vascular endothelial growth factor	RNA	30	13	Hairpin/bulge	0.20
SLE monoclonal antibody	ssDNA	40	8	Unknown	1

Source: Adapted from Gold et al. (1995).

[a] Length of the random region.
[b] Rounds of selection used.

tive for finding consensus nucleic acid binding sequences to known proteins such as transcription factors (Pollack and Treisman, 1990; Nallur, et al., 1996). For proteins the select strategy shown may well be too cumbersome to use in practice. However, as an alternative to PCR amplification, one can use in vivo amplification for proteins instead. An elegant way to do this is bacteriophage display, shown schematically in Figure 14.47. Here the random coding sequence of interest is subcloned as a fusion with a surface coat protein of the bacteriophage M13. Either a minor coat protein is used with only five copies or the major coat protein with thousands of copies is used. Each bacteriophage plaque will be a clonal population representing one particular variant. Mixtures of plaques can be subjected to cycles of selection by physical affinity to the target of interest, and the successive populations of bacteriophage that remain will begin to be populated more and more with clones with the desired affinity properties.

Note that there is no reason why one must start with totally random sequences in select or bacteriophage display strategies. In many cases there will be existing structures with properties similar to the optimum behavior desired. In this case random mutagensis of just a small portion of an existing macromolecule can be used as a starting point try to select a more desirable variant. Despite the intrinsic attractiveness of bacteriophage display, this approach does have a number of limitations. Only monomeric target proteins can be examined by bacteriophage display. The protein targets must be able to fold properly within *E. coli,* and they must be oriented in the fusion so that the site that generates their affinity for the ligand or target is accessible.

A generalization of select strategies has been proposed by Sydney Brenner and Richard Lerner; it is called encoded combinatorial chemistry. The basic idea involves tagging linear oligomers of any type of residue, with a PCR-amplifiable specific DNA sequence that is a unique identifier of the particular oligomer. The general chemical structure needed is shown in Figure 14.48. Here the variable DNA identifier is placed between two constant PCR primers, and one of these is attached via a hub to the oligomer, which could be a nucleic acid, a peptide, an oligosaccharide, or really any kind of organic species that can be built up in a stepwise fashion. The number of different types of

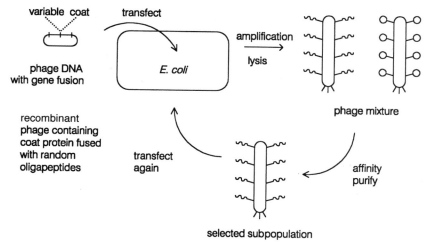

Figure 14.47 Bacteriophage display method for purifying DNA coding for proteins with selectable affinity properties.

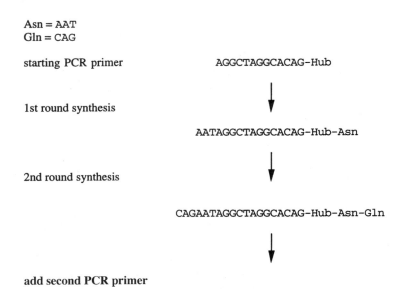

Asn = AAT
Gln = CAG

starting PCR primer   AGGCTAGGCACAG-Hub

1st round synthesis

                          AATAGGCTAGGCACAG-Hub-Asn

2nd round synthesis

                  CAGAATAGGCTAGGCACAG-Hub-Asn-Gln

add second PCR primer

Figure 14.48   General construct needed for encoded combinatorial chemistry. Adapted from Brenner and Lerner (1992).

monomeric units in the oligomer will determine the complexity of the coding scheme needed in the DNA. A particularly nice trick, if small numbers of monomer units are involved, is to use a comma-less code. For example to specify the 20 amino acids, only a subset of 20 of the 64 possible triplet codons is needed. One can choose these, for example, so that if AAT and CAG represent two different amino acids, which can occur in either order: AATCAG and CAGAAT, then ATC, TCA, AGA, and GAA are not assigned to any amino acids. This makes the code resistant to frame shift errors and other ambiguities.

The chimeric DNA-oligomer compounds (Fig. 14.48) are screened for whatever activity is desired in the oligomer; then PCR is used just as in the select strategy in order to identify those components that have the desired affinity for a target. It is relatively easy to synthesize the full set of sequence identifiers and oligomers in a systematic way. This is shown in Figure 14.49. The actual efficiency of such schemes needs to be tested experimentally. Undoubtedly new schemes and variations on existing schemes will proliferate. However, the important feature of all of these approaches is that they illustrate the immense power that DNA analysis can bring to conventional chemistry.

OLIGONUCLEOTIDES AS DRUGS

A large number of young biotechnology companies are betting their futures on the prospect that nucleic acids or nucleic acid analogs will function effectively as drugs. Most of this effort is not based on conventional ideas about gene therapy, where a underactive or inactive defective gene might be supplemented by an active one, or an overactive or inappropriately active gene might be substituted with a normal one. Such unconventional therapies are attractive, especially for many tissue-specific disorders, and such somatic

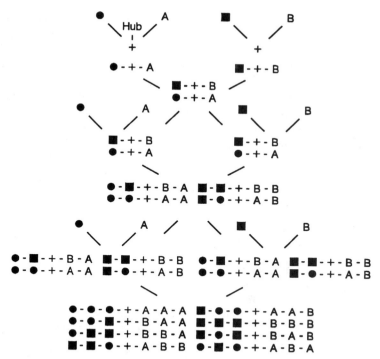

Figure 14.49 Split pool scheme for synthesizing a set of oligomeric compounds linked to their specific oligonucleotide identifiers. Adapted from Brenner and Lerner (1992).

gene therapy is already being tested in a few clinical trials. Here we are concerned with the much more limited and conventional approach of attempting to use short pieces of nucleic acids or their analogues as drugs.

The intrinsic attractiveness of nucleic acids as drugs is their sequence specificity. One can imagine that placed in the appropriate cell, an oligonucleotide could interfere with RNA or DNA function either by binding directly to these species or by competing with them for targets such as proteins. One approach is to simply use the sequence-specific binding affinity of the oligonucleotide to physically occlude a site or target. Here DNA is a potentially attractive target because it exists in very low copy numbers. An ideal antigene scenario would be to design an oligonucleotide that would form a very stable triplex under physiological conditions with an unwanted promoter and so turn off the transcription of the gene controlled by this promoter. Alternatively, the oligonucleotide could be used as an affinity reagent to carry a photochemical or other covalent modifier to a target of interest. Again this seems potentially most effective with a DNA target. An extreme version of oligonucleotide therapy would be to use catalytic RNAs to find and destroy multiple target molecules. If this can be realized, it will be an extremely effective way to deal with infections by viruses with RNA genomes, or to attack other RNA targets.

A number of obstacles must be overcome before successful oligonucleotide therapy will be achieved. First, the materials must be delivered effectively to the correct target cells and in sufficient quantities to be therapeutically active. Side reactions with other cells must be kept to a minimum. If cells lacked receptors for uptake of oligonucleotides, the problem would be to develop a targeting mechanism for the specific cells of interest.

Unfortunately, at least some cells in the body, T lymphocytes, have a natural pathway for oligonucleotide uptake. How general this phenomenon is remains to be seen. The implication is that unless cells with intrinsic uptake pathways are the desired target, this uptake may have to be suppressed, possibly by competition with a harmless oligonucleotide and possibly by shielding the therapeutic compound in some way.

Once cellular uptake is achieved, the oligonucleotide must then be targeted successfully to the desired intracellular location. This will be the nucleus for a reagent directed against DNA, and it might be the nucleus or endoplasmic reticulum for reagents directed against RNA. Such targeting is not a simple matter. Most extracellular macromolecules taken up by cells are automatically targeted to the lysosome where they are destroyed. This natural pathway must be interfered with to successfully deliver a nucleic acid elsewhere. There is no doubt that one should be able to do this by exploiting the same sorts of processes that various viruses use to enter cells and infect the nucleus or the cytoplasm. However, many of these processes are not yet well understood, and we may have to learn much more about them before successful oligonucleotide delivery mechanisms can be created.

Within the cell, or in intercellular fluids, a plethora of agents exist that can destroy or inactivate foreign nucleic acids. This is not surprising. Such agents must have evolved as antiviral defense mechanisms. To circumvent the action of these agents, oligonucleotide drugs would either have to be introduced in large quantities or rendered resistant to a variety of nucleases and other enzymes of nucleic acid metabolism. For example, antisense messenger RNAs have been proposed as therapeutic agents to interfere with the translation of an unwanted message, arising perhaps from a virus or a tumor cell. These would be expected to act by binding to the normal mRNA and inactivating it for translation by occlusion or by a more active destructive process. The difficulty in vivo is the presence of substantial amounts of RNA helicase activity. This is an enzyme that specifically recognizes double-stranded RNAs and unwinds the double helix. To be an effective drug under most circumstances, the backbone or bases of an antisense RNA will have to be altered so that this molecule is no longer recognized by RNA helicases.

Given the constrants mentioned above, the ideal oligonucleotide drug is probably likely to have an altered backbone to render it immune to normal nucleases and other enzymes and to increase its binding affinity to natural nucleic acids. Compounds with uncharged backbones like PNAs seem particularly attractive in this regard. It may also be desirable to equip potential oligonucleotide-analog drugs with additional chemical func-

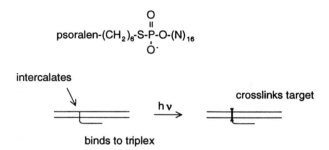

Figure 14.50 A potential oligonucleotide drug, designed by Claude Helene, that can bind to a DNA duplex and permanently inactivate it by photocrosslinking. For examples of results with this kind of approach, see Giovannangeli et al. (1992).

tionalities in order to further enhance their binding and effectiveness at inactivating the cellular target. An interesting example of such a potential drug is shown in Figure 14.50. Designed by Claude Helene, this compound consists of a triplex-forming oligonucleotide attached to a psoralen by a flexible chain. The psoralen enhances binding to DNA duplexes because it is an intercalator. More importantly, psoralen is a DNA photocrosslinking, so near-UV irradiation, after formation of the triplex, results in irreversible crosslinking of the target DNA duplex. This procedure has been demonstrated to work effectively in cells. It may be a prototype of the sorts of materials that we will eventually see in actual therapeutic use.

SOURCES AND ADDITIONAL READINGS

Bailey, D. M. D., Carter, N. P., de Vos, D., Leversha, M. A., Perryman, M. T., and Ferguson-Smith, M. A. 1993. Coincidence painting: A rapid method for cloning region specific DNA sequences. *Nucleic Acids Research* 21: 5117–5123.

Batzer, M. A., Alegria-Hartman, M., and Deininger, P. L. 1994. A consensus Alu repeat probe for physical mapping. *Genetic Analysis: Techniques and Applications* 11: 34–38.

Brenner, S., and Lerner, R. A. 1992. Encoded combinatorial chemistry. *Proceedings of the National Academy of Sciences USA* 89: 5381–5383.

Brookes, A. J., and Porteous, D. J. 1992. Coincident sequence cloning: A new approach to genome analysis. *Trends in Biotechnology* 10: 40–44.

DeRisi, J., Penland, P., Brown, P. O., Bittner, M. L., Meltzer, P. S., Ray, M., Chen, Y., Su, Y. A., and Trent, J. M. 1996. Use of a cDNA microarray to analyse gene expression patterns in human cancer. *Nature Genetics* 14: 457–460.

Eriksson, M., and Nielsen, P. E. 1996. PNA-nucleic acid complexes. Structure stability and dynamics. *Quarterly Review of Biophysics* 29: 369–394.

Feigon, J., Dieckmann, T., and Smith, F. W. 1996. Aptamer structures from A to Z. *Chemistry and Biology* 3: 611–617.

Ferguson, J. A., Bowles, T. C., Adams, C. P., and Walt, D. R. 1996. A fiber-optic DNA biosensor microarray for the analysis of gene expression. *Nature Biotechnology* 14: 1681–1684.

Ferrin, L. J., and Camerini-Otero, R. D. 1991. Selective cleavage of human DNA: RecA-assisted restriction endonuclease (RARE) cleavage. *Science* 254: 1494–1497.

Ferrin, L. J., and Camerini-Otero, R. D. 1994. Long-range mapping of gaps and telomeres with RecA-assisted restriction endonuclease (RARE) cleavage. *Nature Genetics* 6: 379–383.

Giovannangeli, C., Thuong, N. T., and Helene, C. 1992. Oligodeoxynucleotide-directed photo-induced cross-linking of HIV proviral DNA via triple-helix formation. *Nucleic Acids Research* 20: 4275–4281.

Gnirke, A., Iadonato, S. P., Kwok, P.-Y., and Olson, M. V. 1994. Physical calibration of yeast artificial chromosome contig maps by recA-assisted restriction endonuclease (RARE) cleavage. *Genomics* 24: 199–210.

Inque, S., Kiyama, R., and Oishi, M. 1996. Construction of highly extensive polymorphic DNA libraries by in-gel competitive reassociation procedure. *Genomics* 31: 271–276.

Irvine, D., Tuerk, C., and Gold, L. 1991. Selexion: Systematic evolution of ligands by exponential enrichment with integrated optimization by non-linear analysis. *Journal of Molecular Biology* 222: 739–761.

Ito, T., Kito, K., Adati, N., Mitsui, Y., Hagiwara, H., and Sakaki, Y. 1994. Fluorescent differential display: Arbitrarily primed RT-PCR fingerprinting on automated DNA sequencer. *FEBS Letters* 351: 231–236.

Ito, T., Smith, C. L., and Cantor, C. R. 1992. Sequence-specific DNA purification by triplex affinity capture. *Proceedings of the National Academy of Sciences USA* 89: 495–498.

Ito, T., Smith, C. L., and Cantor, C. R. 1992. Affinity capture electrophoresis for sequence-specific DNA purification. *Genetic Analysis: Techniques and Applications* 9: 96–99.

Ji, H., Smith, L. M., and Guilfoyle, R. A. 1994. Rapid isolation of cosmid insert DNA by triple-helix-mediated affinity capture. *Genetic Analysis: Techniques and Applications* 11: 43–47.

Kato, K. 1996. RNA fingerprinting by molecular indexing. *Nucleic Acids Research* 24: 394–395.

Kandpal, R. P., Kandpal, G., and Weissman, S. M. 1994. Construction of libraries enriched for sequence repeats and jumping clones, and hybridization selection for region-specific markers. *Proceedings of the National Academy of Sciences USA* 91: 88–92.

Kool, E. T. 1996. Circular oligonucleotides: New concepts in oligonucleotide design. *Annual Review of Biophysical and Biomolecular Structure* 25: 1–28.

Liang, P., and Pardee, A. B. 1992. Differential display of eukaryotic messenger RNA by means of the polymerase chain reaction. *Science* 257: 967–971.

Liang, P., Zhu, W., Zhang, X., Guo, Z., O'Connell, R. P. O. Averboukh, L., Wang, F., and Pardee, A. B. 1994. Differential display using one-base anchored oligo-dT primers. *Nucleic Acids Research* 22: 5763–5764.

Lisitsyn, N., Lisitsyn, N., and Wigler, M. 1993. Cloning the difference between two complex genomes. *Science* 259: 946–951.

Lockhart, D. J., Dong, H., Byrne, M. C., Follettie, M. T., Gallo, M. V., Chee, M. S., Mittmann, M., Wang, C., Kobayashi, M. Horton, H., and Brown, E. 1996. Expression monitoring by hybridization to high-density oligonucleotide arrays. *Nature Biotechnology* 14: 1675–1680.

Mathieu-Daude, F., Cheng, R., Welsh, J., and McClelland, M. 1996. Screening of differentially amplified cDNA products from RNA arbitrarily primed PCR fingerprints using single strand conformation polymorphism (SSCP) gels. *Nucleic Acids Research* 24: 1504–1507.

Moyzis, R. K., Torney, D. C., Meyne, J., Buckingham, J. M., Wu, J.-R., Burks, C., Sirotkin, K. M., and Goad, W. B. 1989. The distribution of interspersed repetitive DNA sequences in the human genome. *Genomics* 4: 273–289.

Nallur, G. N., Prakash, K., and Weissman, S. M. 1996. Multiplex selection techniques (MuST): An approach to clone transcription factor binding sites. *Proceedings of the National Academy of Sciences USA* 93: 1184–1189.

Nielson, P. E., Egholm, M., Berg, R. H., and Buchardt, O. 1991. Sequence-selective recognition of DNA by strand displacement with a thymine-substituted polyamide. *Science* 254: 1497–1500.

Perry-O'Keefe, H., Yai, X.-W., Coull, J. M., Fuchs, M., and Egholm, M. 1996. Peptide nucleic acid pre-gel hybridization: An alternative to Southern hybridization. *Proceedings of the National Academy of Sciences USA* 93: 14670–14675.

Pollock, R., and Treisman, R. 1990. A sensitive method for the determination of protein-DNA binding specificities. *Nucleic Acids Research* 18: 6197–6204.

Qureshi, S. J., Porteous, D. J., and Brookes, A. J. 1994. Alu-based vectorettes and splinkerettes more efficient and comprehensive polymerase chain reaction amplification of human DNA from complex sources. *Genetic Analysis: Techniques and Applications* 11: 95–101.

Rosenberg, M., Przybylska, M., and Straus, D. 1994. "RFLP subtraction." A method for making libraries of polymorphic markers. *Proceedings of the National Academy of Sciences USA* 91: 6113–6117.

Schena, M. Shalon, D., Davis, R. W., and Brown, P. O. 1995. Quantitative monitoring of gene expression patterns with a complementary DNA microarray. *Science* 270: 467–470.

Sosnowski, R. G., Tu, E., Butler, W. F., O'Connell, J. P., and Heller, M. J. 1997. Rapid determination of single base mismatch mutations in DNA hybrids by direct electric field control. *Proceedings of the National Academy of Sciences USA* 94: 1119–1123.

Straus, R., and Ausubel, F. M. 1990. Genomic subtraction for cloning DNA corresponding to deletion mutations. *Proceedings of the National Academy of Sciences USA* 87: 1889–1893.

Strobel, S. A., Doucette-Stamm, L. A., Riba, L., Housman, D. E., and Dervan, P. B. 1991. Site-specific cleavage of human chromosome 4 mediated by triple-helix formation. *Science* 254: 1639–1642.

Takabatake, T. et al. 1992. The use of purine-rich oligonucleotides in triplex-mediated DNA isolation and generation of unidirectional deletions. *Nucleic Acids Research* 20: 5853–5854.

Unrau, P., and Deugau, K. V. 1994. Non-cloning amplifications of specific DNA fragments from whole genomic DNA digests using DNA "indexers." *Gene* 145: 163–169.

Velculescu, V. E., Zhang, L., Vogelstein, B., and Kinzler, K. W. 1995. Serial analysis of gene expression. *Science* 270: 484–487.

Veselkov, A. G., Demidov, V. V., Nielsen, P. E., and Frank-Kamenetskii, M. D. 1996. A new class of genome rare cutters. *Nucleic Acids Research* 24: 2483–2487.

Wan, J. S., Sharp, S. J., Poirier, G. M.-C., Wagaman, P. C., Chambers, J., Pyati, J., Hom, Y.-L., Galindo, J. E., Huvar, A., Peterson, P. A., Jackson, M. R., and Erlander, M. G. 1996. Cloning differentially expressed mRNAs. *Nature Biotechnology* 14: 1685–1691.

Wittung, P., Kim, S. K., Buchart, O., Nielsen, P., and Norden, B. 1994. Interactions of DNA binding ligands with PNA-DNA hybrids. *Nucleic Acids Research* 22:5371–5377.

Wittung, P., Eriksson, M., Lyng, R., Nielsen, P., and Norden, B. 1995. Induced chirality in PNA-PNA duplexes. *Journal of the American Chemical Society* 117: 10167–10173.

Yokota, H., and Oishi, M. 1990. Differential cloning of genomic DNA: Cloning of DNA with an altered primary structure by in-gel competitive reassociation. *Proceedings of the National Academy of Sciences USA* 87: 6398–6402.

15 Results and Implications of Large-Scale DNA Sequencing

The accumulation of completed DNA sequences and the development and utilization of software tools to analyze these sequences are changing almost daily. It is extremely frustrating to attempt an accurate portrait of this area in the kind of static snapshot allowed by the written text. The authors know with certainty that much of this chapter will become obsolete in the time interval it takes to progress from completed manuscript to published textbook. This is truly an area where electronic rather than written communication must predominate. Hence, while a few examples of original projections of human genome project progress will be given, and a few examples of actual progress will be summarized, the emphasis will be on principles that underlie the analysis of DNA sequence. Here it is possible to be relatively brief, since a number of more complete and more advanced treatments of sequence analysis already exist (Ribskaw and Devereaux, 1991; Waterman, 1995). The interested reader is also encouraged to explore the databases and software tools available through the Internet (see the Appendix).

COSTING THE GENOME PROJECT

When the lectures that formed the basis of this book were given in the Fall 1992, there were more than 37 Mb of finished DNA sequence from the six organisms chosen as the major targets of the U.S. human genome project. The status in February 1997 of each of these efforts is contrasted with the status five years earlier in Table 15.1. The clear impression provided by Table 15.1 is that terrific progress has been made on *E. coli, S. cerevisiae,* and *C. elegans,* but we have a long way to go to complete the DNA sequence of any higher organism. In addition to the six organisms listed in Table 15.1, a few other organisms like the plant model system *Arabidopsis thaliana* and the yeast *S. pombe* will surely also be sequenced in the next decade along with a significant number of additional prokaryotic organisms. Other attractive targets for DNA sequencing would be higher plants with small genomes like rice, and higher animals with small genomes like the puffer fish. If methods are developed that allow efficient differential sequencing (see Chapter 12), methods that look just at sequence differences, some of the higher primates become of considerable interest. Although these genomes are as complex as the human genome, DNA sequences differences between them are only a few percent, and most of these will lie in introns. Thus a comparison between gorilla and human would make it very easy to locate small exons that might be missed by various computerized sequence search methods.

A different cast is provided by the figures in Table 15.2, which illustrate the average rate of DNA sequencing per investigator up until 1990 since the first 24 DNA bases were determined almost 30 years ago. The results in Table 15.2 show a steady rise in the rate of

TABLE 15.1   Progress Towards Completion of the Human Genome Project

| Organism | Finished DNA Sequence (Mb) | | | Comment |
	Complete Genome	June 1992	February 1997	
E. coli	4.6	3.4	4.6	Complete
S. cerevisiae	12.1	4.0	12.1	Complete
C. elegans	100	1.1	63.0	Cosmids[a]
D. melanogaster	165	3.0	4.3	Large contigs only
M. musculus	3000	8.2	24.0	Total assuming 2.5 × redundant
H. sapiens	3000	18.0	31.0	In contigs, > 10Kb
			116.0	Total assuming 2.5 × redundant

[a] Completion is expected at the end of 1998.

DNA sequencing. However, they understate this rise, since the data are derived from all DNA sequencing efforts. In practice, the majority of these sequencing efforts use simple, manual technology, and many are performed by individuals just learning DNA sequencing. The common availability and widespread use of automated DNA sequencing equipment probably had little impact on the results in Table 15.2 because these advances are too recent. Despite this fact, the general impression, based on DNA sequences deposited into databases, is that the total accumulation of DNA sequences is increasing nearly exponentially with time, and this trend certainly has continued to the present.

A stated goal of the human genome project in the United States is to complete the DNA sequence of one haploid-equivalent human genome by the year 2005. Two immediate cautionary notes must be struck. First, the sequence is unlikely to really be completed by then or perhaps by any time in the foreseeable future because it is unlikely anyone would want to, or could, sequence through millions of base pairs of centromeric tandem repeats, looking for occasional variations. Second, it may not be a single genome that is sequenced. A considerable amount of the material for DNA sequencing is likely to come from cell lines containing particular human chromosomes. In general, each line represents chromosomes from a different individual. Thus, to whatever degree of completion the first human DNA sequence is obtained, it will surely be a mosaic of many individual genomes. In view of the nature of the task, and its importance to humanity, this outcome actually seems quite appropriate. It also answers, once and for all, the often-asked question "who will be sequenced in the human genome project?"

Setting the above complications aside, the chance of achieving the DNA sequencing goals of the human genome project depends principally on three variables: the amount of money available for the project as a whole, the percent of it used to support large-scale genomic sequencing, and the efficiency of that sequencing, namely the cost per base pair and how it evolves over the course of the project. In most current genomic DNA sequencing efforts, labor still appears to be the predominant cost, and all other considerations can be scaled to the number of individuals working. This is changing with the incorporation of more and more highly automated methods; supplies and materials are becoming the dominant cost. However, since we cannot yet accurately estimate the impact of automation on sequencing costs, we will assume the continuation of dominant labor costs in order to make some projections.

TABLE 15.2 Summary of Progress Made in DNA Sequence Analysis Between 1967 and 1990

DNA Sequence Determined	Method(s)	Time Period	Nucleotides Determined	Number of Investigators Involved	Percent Time for Sequencing Steps	Nucleotides per Year, per Investigator	Relative Speed
Cohesive ends of λ DNA	Partial incorporation; partial digestion	1967–1970	24	2	80–90	4	1
Cohesive ends of 186p DNA	Same as above	1971–1972	38	2	80–90	12	3
φX174, f1	Mobility shift	1972–1973	140	5	40–50	28	7
φX174	Plus-and-minus	1973–1977	5000	9	20–30	138	35
Over 500 sequences	Dideoxy chain termination	1977–1982	66,000	350	20–30	380	94
Over 7000 sequences	Recombinant DNA M13 vectors, etc.	1982–1986	9,000,000	3,500	15–25	640	160
Over 40,000 sequences	Same as above, automatic DNA sequencer	1986–1989	27,000,000	9,000	10–20	1000	250
Over 23,000 sequences	Same as above	1989–1990	14,000,000	12,000	10–20	1160	290

Source: Adapted from Wu (1993).

528

In most U.S. academic or pure research settings, an average person working in the laboratory costs about $100,000 a year to support. This includes salary, fringe benefits, chemicals, supplies, and academic overhead such as administrative costs, light, heating, and amortization of laboratory space. Let's make the reasonably hard-nosed estimate that the average genome sequencer will work 250 days per year. Dividing this into the yearly cost results in a cost per working day of $400. A reasonable estimate of state-of-the-art DNA sequencing rates is about 10^4 raw base pairs per day per individual. Allowing for 20% waste, and dividing by a factor of 8 for the redundancy needed for shotgun strategies, we can estimate that the sequencing rate for finished base pairs per day is 10^3. When the daily cost is divided by the daily output, the result is a cost per finished base pair of $0.40. This is far lower than the current commercial charges for DNA sequencing services, which average several dollars per base, or the cost of more casual sequencing efforts which, based on the most recent rates shown in Table 15.2, weould be about $90 per base. The cost per finished base of the *H. influenza* project recently completed was $0.48 in direct supplies and labor. This is equivalent to at least $1.00/bp when overhead and instrument depreciation costs are added.

Although the relatively low cost of current automated DNA sequencing is impressive, compared to the cost of less systematic efforts, it falls far short of the economies that will need to be achieved to complete the human genome project within the allowable budget and time scale. The initial design of the U.S human genome project called for $3000 million to be spent over 15 years. This would translate into a cost of $1 available per human base pair if all one did was sequence. Such a plan would be ridiculous because it ignores mapping, which provides the samples to be sequenced, and it does not allow for any model organism work or technology development. The U.S. human genome budget in 1992 was $160 million per year. If we had ramped up to large-scale DNA sequencing immediately, starting in October 1993, we would have had 12 years to complete the project. A steady state sequencing rate model would have required sequencing at a rate of 250 million base pairs per year. At current best attainable costs this would require $100 million per year to be spent on human genomic DNA sequencing. So this way the project could be completed, but to proceed with such a plan that anticipates no enhancements in technology would be lunacy.

It is more sensible to scale up sequencing more gradually and to build in some assumptions about improved efficiency. The caveat to this approach is that the slower the scale up, the more efficient the final sequencing rates must become. Suppose that we arbitrarily limit the amount of funds committed to human genomic DNA sequencing to the $100 million annual costs required by the steady state model. Table 15.3 shows one set of cost projections, developed in 1992 by Robert Robbins. At first glance this may seem like an extremely optimistic scenario, since it starts with average costs that are $1.50 per base pair and requires only a factor of ten decrease in unit cost over a seven-year time period to reach $0.15 per finished base pair by 2001. However, these are average costs, and at these costs an average work force of 1000 individuals will be needed for the last five years of the project, just for human DNA sequencing. The bottom line is that the scenario in Table 15.3 seems reasonable and achievable, but one hopes that some of the potential improvements in DNA sequencing described in Chapters 11 and 12 will be realized and will result in considerably faster rates of sequence acquisition. From this point of view, the scenario in Table 15.3 is actually quite pessimistic.

One way to view the cost effectiveness of the genome program is to ask how much additional DNA sequencing will be accomplished beyond the stated goal of one human

TABLE 15.3   One Model for DNA Sequencing Costs in the Human Genome Project

Year	Finished, per-Base Direct Cost	Annual Sequencing Budget ($millions)	Genomic Sequence (Mb) Year	Cumulative	Percent of Genome Completed
1995	$1.50	16	11	11	0.33
1996	$1.20	25	21	32	0.96
1997	$0.90	35	39	71	2.15
1998	$0.60	50	84	155	4.71
1999	$0.45	75	168	324	9.81
2000	$0.30	100	337	660	20.01
2001	$0.15	100	673	1334	40.42
2002	$0.15	100	673	2007	60.82
2003	$0.15	100	673	2681	81.23
2004	$0.15	100	673	3354	101.63

genome. If the development of new sequencing methods proceeds very well, it may be possible to complete extensive portions of the mouse genome, and perhaps even other model organisms with large genomes, under the cost umbrella of funding for the human sequence. This is not inappropriate, since the more relevant model organism sequence data that we have available, the more powerful will be our ability to interpret the human DNA sequence.

Whether one adopts an optimistic or a pessimistic scenario, the inevitable conclusion is that by October 2005 or thereabouts, 3×10^9 base pairs of human genomic DNA encoding for something like 100,000 human genes will be thrust upon the scientific community. The challenge will be to find any genes in the sequence that are not already represented as sequenced cDNAs, translate the DNA sequence into protein sequence, make some preliminary guesses about the function of some of these proteins, and decide which ones to study first in more detail. The remainder of this chapter deals with these challenges. Because of the rapid advances in cDNA sequencing described earlier in Chapter 11, some of these challenges already confront us today.

FINDING GENES

There are two very different basic approaches to finding the genes encoded for by a sample of genomic DNA. The experimental approach is to use the DNA as a probe (or a source of probes or PCR primers) to screen available cDNA libraries. One can also use available DNA materials to look directly at the mRNA population in different cell types. This is most easily accomplished by Northern blots, as described in Chapter 13. None of these approaches require that the DNA sequence be completed, and except for the synthesis of PCR primers, none require any known DNA sequence. These pure experimental methods will succeed only if the mRNAs or cDNAs in question are present in available sources in sufficient quantities to be detected above the inevitable background caused by nonspecific hybridization or PCR artifacts.

The second basic approach is the one that we will concentrate on here. It requires that the sequence of the DNA be known. Genes have characteristic sequence properties that distinguish them, to a considerable extent, from nongenic DNA or from random sequences. The goal is to optimize the methods for making these discriminations. One trivial procedure should always be tried to an newly revealed piece of genomic DNA. This is to compare its sequence with all known sequences. A near or exact match could immediately reveal genes or other functional elements. For example, large amounts of partial cDNA sequence information are accumulating in publicly accessible databases. Frequently these cDNAs have already been characterized to some extent, such as their location on a human physical map or their pattern of expression in various tissues. Finding a genomic match to such a cDNA fragment clearly indicates the presence of a gene and provides a jump start to further studies. The problem that arises is what if the match between a portion of the genomic DNA sequence and other molecules with known sequence is not exact, or what if there is no significant detectable match at all?

A number of basic features of genomic DNA sequence can be examined to look for the location of genes. The most straightforward of these is to search for open reading frames (ORFs). This is illustrated in Figure 15.1. A given stretch of DNA sequence can potentially be translated into protein sequence in no less than six different ways. The triplet genetic code allows for three possible reading frames, and a priori either DNA strand (or both) could be coding. Computer analysis is used to translate the DNA sequence into protein in all six reading frames, and assemble these in order along the DNA. The issue is then to decide which represent actual segments of coding sequence and which are just noise.

Almost all known patterns of mRNA transcription occur in one direction along a template DNA strand. Segments of the mRNA precursor are then removed by splicing to make the mature coding mRNA. The exceptions to this pattern are fairly rare in general, and the few organisms in which they are more frequent are fairly restricted. Some parasitic protozoa and some worms have extensive trans-splicing where a mRNA is composed from units coded on different chromosomes. Editing of mRNAs, where DNA-templated bases are removed, individually, and bases not templated by DNA are added, individually, also appears to be a rare process concentrated in a few lower organisms. Thus it is almost always safe to assume that a true mRNA will be composed of one or more successive segments of DNA sequence organized, as shown schematically in Figure 15.2. The challenge is to predict where the boundaries of the possible exons are, and where individual genes actually begin and end. This requires the simultaneous examination of aspects of the RNA sequence as well as aspects of the coded protein sequence.

True exons must have arisen by splicing. Thus the sequences near each splice site must resemble consensus splicing signal sequences. While these are not that well defined, they have a few specific characteristics that we can look for. True exon coding segments can have no stop codons, except at the very end of the gene. Thus the presence of UAA, UAG, or UGA can usually eliminate most possible reading frames fairly quickly. A true gene must begin with a start codon. This is usually AUG, although it can also be GUG,

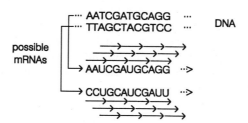

Figure 15.1 Six possible reading frames from a single continuous stretch of DNA sequence shown earlier in Figure 1.13.

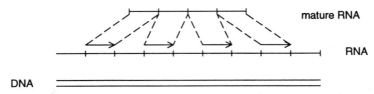

Figure 15.2 A message is read discontinuously from a DNA, but all of the examples known to date are read from the same strand, in a strict linear order (except for the very specialized case of transsplicing seen in some simple organisms).

UUG, or AUA. True start codons are context dependent. For example, a true AUG used for starting translation must have nearby sequences that are used by the ribosome or protein synthesis initiation factors (proteins) to bind the message and initiate translation. These residues usually lie just upstream from the AUG, but they can also extend a bit downstream.

The key factor in determining the correct points for the initiation of translation is that the context of starting AUGs is very species dependent. Thus, since the species of origin of a DNA will almost always be known in advance, a great deal of information can be brought to bear to recognize a true start. For example, in *E. coli,* a given mRNA can have multiple starting AUGs within a message because *E. coli* commonly makes and uses polycistronic messages. In eukaryotes, messages are usually monocistronic. Most frequently it is the first AUG in the message that signals the start of translation. This means that information about transcription starting signals or splicing signals must be used to help locate the true starting points for translation of eukaryotic DNA sequences.

A true mRNA must also have a transcription stop. This is relatively easy to find in prokaryotes. However, we still know very little about transcription termination in eukaryotes. Fortunately, most eukaryotic mRNAs are cleaved, and a polyA tail is added. The consensus sequence for this process is AATAAA. This is distinct, but it is very short, and such a sequence can easily occur by random fluctuations in A + T-rich mammalian DNA. This polyA addition signal is an example of a fairly general problem in sequence analysis. Many control sequences are quite short; they often work in concert with other control elements, and we do not yet know how to recognize the overall pattern.

MORE ROBUST METHODS FOR FINDING GENES BY DNA SEQUENCE ANALYSIS

If the methods just described represented all of the available information, the task of finding previously unknown genes by genomic DNA sequence analysis alone would be all but hopeless. Fortunately a great deal of additional information is available. Here we will describe some of the additional characteristics that distinguish between coding and noncoding DNA sequences. The real challenge is to figure out the optimal way to combine all of this additional knowledge into the most powerful prediction scheme. A simple place to start is to consider the expected frequency of each of the 20 amino acids. Some, like tryptophan, are usually very rare in proteins; others are much more common. A true coding sequence will, on average, contain relatively few rare amino acids. Similarly the overall average amino acid compositions of proteins vary, but they usually lie within certain bounds. A potential coding sequence that led to a very extreme amino acid composition would usually be rejected if an alternative model for the translation of this segment of the DNA led to a much more normal amino acid composition. Note that the argument we are using is one of plausibility, not certainty. Statistical weights have to be attached to such arguments to make them useful. The same sort of considerations will apply to all of the other measures of coding sequence that follow.

The genetic code is degenerate. All amino acids except methionine and tryptophan can be specified by multiple DNA triplet codons. Some amino acids have as many as six possible codons. The relative frequency at which these synonymous codons are used varies widely. In general, codon usage is very uneven and is very species specific. Even different classes of genes within a given species have characteristic usage patterns. For example in highly expressed *E. coli* genes the relative frequencies of arginine codon use per 1000 amino acids are

AGG	0.2	CGG	0.3
AGA	0.0	CGA	42.1
CGA	0.2	CGC	13.9

These nonrandom values have a significant effect on the selection of real open reading frames.

Some very interesting biological phenomena underlie the skewed statistics of codon usage. To some extent, the distribution of highly used codons must match the distribution of tRNAs capable of responding to these codons; otherwise, protein synthesis could not proceed efficiently. However, many other factors participate. Codon choice affects the DNA sequence in all reading frames; thus the choice of a particular codon may be mediated, indirectly, by the desire to avoid or promote a particular sequence in one of the other reading frames. For example, particular codon use near the starting point of protein synthesis can have the effect of strengthening or weakening the ribosome binding site of the mRNA. Rare codons usually correspond to rare tRNAs. This in turn will result in a pause in protein synthesis, while the ribosome-mRNA complex waits for such a tRNA to appear. This sort of pausing appears to be built into the sequence of many messages, since pausing at specific places, like domain boundaries, will assist the newly synthesized peptide chain to fold into its proper three-dimensional configuration.

Codon usage can also serve to encourage or avoid certain patterns of DNA sequence. For example, DNA sequences with successive A's spaced one helical repeat apart tend to be bent. This may be desirable in some regions and not in others. Inadvertent promoter-like sequences in the middle of actively transcribed genes are probably best avoided, as are accidental transcription terminators. The flexibility of the genetic code that results from its degeneracy allows organisms to synthesize whatever proteins they need, while avoiding these potential complications. For example, a continuous stretch of T's forms part of one of the transcription termination signals in *E. coli*. The resulting triplet, UUU, which codes for phenylalanine occurs only a third as often as the synonymous codon, UUC. Another example is seen with codons that signal termination of translation. UAA and UAG are two commonly used termination signals. The complements of these signals are UUA and CUA. These are both codons for leucine; however, CUA is the rarest leucine codon, and UUA is also rarely used.

There are many additional constraints on coding sequences beyond the statistics of codon usage. In real protein sequences there are some fairly strong patterns in the occurrence of adjacent amino acids. For example, in a beta sheet structure, adjacent amino acid side chains will point in opposite directions. Where the beta sheet forms part of the structural core of a protein domain, usually one side will be hydrophobic and face in toward the center of the protein, while the other side will be hydrophilic and face out toward the solvent. Thus codons that specify nonpolar residues will tend to alternate to some extent with codons that specify charged or polar residues. Alpha helices will have different patterns of alternation between polar and nonpolar residues because they have 3.4 residues per turn. In many protein structures one face of many of the alpha helices will be polar and one face will be nonpolar. These are not absolute rules; however, alpha helices and beta sheets are the predominant secondary structural motifs in proteins, and as such, they cast a strong statistical shadow over the sorts of codon patterns likely to be represented in a sequence that actually codes for a bona fide mRNA.

Quite a few other characteristics of known proteins seem to be general enough to affect the pattern of bases in coding DNA. Certain dipeptides like trp-trp and pro-pro are usually very rare; an exception occurs in collagenlike sequences, but then the pattern pro-pro-gly will be overwhelmingly dominant. Repeats and simple sequences tend to be rare inside of coding regions. Thus Alu sequences are unlikely to be seen within true coding regions; blocks like AAATTTCCCGGG . . . are also conspicuously rare. VNTRs are also usually absent in coding sequences, although there are some notable exceptions like the androgen receptor which contains three simple sequence repeats. Such exceptions may have important biological consequences, but we do not understand them yet. All of these statistically nonrandom aspects of protein sequence imply that we ought to be able to construct some rather elaborate and sophisticated algorithms for predicting ORFs and splice junctions. Seven of these that were used in the first successful algorithms for finding genes are described below.

Frame Bias

In a true open reading frame (ORF), the sequence is parsed so that every fourth base must be the beginning of a codon. If we represent a reading frame as $(\ )_n$, in a true reading frame the bases in each position should tend to be those consistent with the preferred codon usage in the particular species observed. The other possible reading frames should tend to be those with poor codon usage. In this way the possibility of accidentally reading a message in the wrong frame will be minimized.

Fickett Algorithm

This is an amalgam of several different tests. Some of these examine the 3-periodicity of each base versus the known properties of coding DNA. The 3-periodicity is the tendency for the same base to recur in the 1st, 4th, 7th, . . . positions. Other tests look at the overall base composition of the sequence.

Fractal Dimension

Some dinucleotides are rare, while others are common. The fractal dimension measures the extent to which common codons are clustered with other common ones, and rare codons are clustered with other rare ones. Clustering of similar codon classes is characterized by a low fractal dimension, while alternation will lead to a high fractal dimension. It turns out that exons have low fractal dimensions, while introns have high fractal dimensions. Thus this test combines some features of codon usage, common dipeptide sequences, and simple sequence rejection.

Coding Six-Tuple Word Preferences

A six-tuple is just a set of six continuous DNA bases. There are 4^6 possible six-tuples in DNA. Since we have tens of millions of base pairs of DNA to examine for some species, we can make reasonable projections of the likely occurrence of each of these six-tuples in coding sequences versus noncoding sequences. An appropriately weighted sum of these predictions will allow an estimate of the chances that a given segment is coding.

Coding Six-Tuple In-Frame Preferences

In this algorithm one computes the relative occurrence of preferred six-tuples in each possible reading frame. For true coding sequence, the real reading frame should show an excellent pattern of preferences, while in the other possible reading frames, when actual coding sequences have been examined, the six-tuple preferences appear to be fairly poor. This presumably aids in the selection and maintenance of the correct frame by the ribosome. This particular test turns out to be a very powerful one.

Word Commonality

This test is also based on the statistics of occurrences of six-tuples. Introns tend to use very common six-tuples; exons tend to use rare six-tuples. Note that here we are talking about the overall frequency of occurrence of six-tuples and not their relative frequency in coding or noncoding regions.

Repetitive Six-Tuple Word Preferences

This test looks specifically at the six-tuples that are common in the major classes of repeating DNA sequences. These six-tuples will also tend to be rare in true coding sequences.

The large list of tests just outlined raises an obvious dilemma: which one should be picked for best results? However, this is not an efficient way to approach such a problem. Instead, what one aims to do is find the optimal way to integrate the results of all of these tests to maximize the discrimination between coding and noncoding DNA. One must be prepared for the fact that the optimum measure will be species dependent; in addition it may well be dependent on the context of the particular sequence actually being examined. In other words, no simple generally applicable rule for combining the test results into a single score representing coding probability is likely to work. Instead, a much more sophisticated approach is needed. One such approach, which has been very successful, uses the methodology of artificial intelligence algorithms.

NEURAL NET ANALYSIS OF DNA SEQUENCES

A neural net is one form of artificial intelligence. It is so named because, with neural net algorithms, one attempts to mimic the behavioral characteristics of networks of neurons.

We know that such networks can be trained (i.e., adjusted) to respond to signals or stimuli and to integrate the input from many different sources or sensors. Here the basic properties of neural nets will be illustrated, and then examples of how they have been applied to the analysis of DNA sequences will be shown.

The basic element in a neural net is a node, as shown in Figure 15.3a. This node receives input from one or more sensors, and it delivers output to one or more other nodes or a detector. The behavior of nodes is quantized. The signal input from each sensor is continuously scanned. It is recorded as positive if is above some threshold; otherwise, it is scored as negative (Fig. 15.3b). An input can be stimulatory or inhibitory. A node receiving a stimulatory input will send out the same sign signal. A node receiving an inhibitory signal will send out the opposite sign signal. By analogy, a nerve cell receiving a stimulatory impulse fires, while one receiving an inhibitory impulse does not fire.

Neural nets are collections of nodes wired in particular ways. They are generalizations of simple logical circuits. The variables in a neural net are the signal thresholds and the nature of the response of the nodes. We will illustrate this with three cases of increasing complexity. Consider the simple two-input node shown in Figure 15.3. Suppose that it operates under the following rules: If both sensors are positive, the node sends a positive output. Otherwise, it sends a negative output. This node is operating as the logical and function. It is behaving like a neuron that needs two simultaneous positive inputs in order to fire.

As a second case, consider the same node in Figure 15.3, but now imagine that the node sends a positive output if either input or both inputs are positive. The only way the node sends a negative output is if both sensors are reading negative. This node is acting like the logical and/or function. It stimulates a nerve cell that needs only one positive stimulus to fire.

The third case we will consider is a node that sends a positive signal if either input sensor is positive but not if both sensor inputs are positive. It is difficult to represent this behavior by a single node with simple +/− binary logical properties. Instead, we can represent the behavior by a slightly more complex network with three nodes, as shown in Figure 15.4. Here the two sensors input their signal directly to two of the nodes. Each of these nodes views one input as stimulatory and the other input as inhibitory. Thus each node will fire if and only if it receives one positive and one negative signal. The two nodes feed stimulatory inputs into the third node. This node will be directed to fire if it receives a positive input from either one of the two nodes that precede it. One way to view the structure of the simple neural network shown in Figure 15.4 is that there is hidden

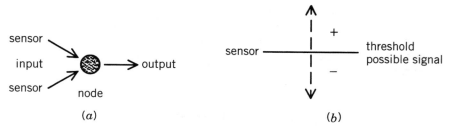

Figure 15.3   The simplest possible neural net. This net can perform the logical operations "and" and "and/or." *(a)* Coupling of two inputs to a single output. *(b)* Effect of sensor threshold on signal value.

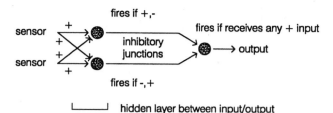

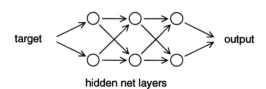

Figure 15.4 A more complex neural net which can perform the logical operation either but not both.

layer of nodes between the sensors and the final output node. In this particular case the hidden layer has a very simple structure; yet it is already capable of executing a complicated logical operation.

To use a neural net, one constructs a fairly general set of nodes and connections with one or more hidden layers, as shown in Figure 15.5. This is trained on sequences with known properties. The net is cycled through the training set of data, and weighting factors for each of the connections are adjusted to try to achieve the highest positive output scores for desired input characteristics and the lowest ones for undesired characteristics. A neural net could be used to examine DNA sequence directly, but this would take a very complex net, and the resulting training period would be computationally very intensive. Instead, what works quite satisfactorily is to use as sensor inputs, not individual bases, but instead the seven-sequence analysis algorithms described in the previous section. These sensors are each allowed to scan the DNA sequence over 10-base intervals. The net result of each scan is computed in a 99-base window. This is the length of sequence that is scanned and input into the net. Then the sequence is frameshifted by one base, and the analysis is repeated. The result is scaled, and then each sensor is fed into the neural net. The actual net structure used is shown in Figure 15.6. It consists of the 7 input sensors, 14 hidden nodes in a first layer, 5 hidden nodes in a second layer, and a single output node.

Edward Uberbacher and Robert Mural at Oak Ridge National Laboratory trained the neural net shown in Figure 15.6 on 240 kb of human DNA sequence data, adjusting thresholds, signs, and weighting until the performance of the net appeared to be optimum (1991). The result is a sequence analysis program called GRAIL. The detailed pattern of input into GRAIL from each of seven sensors for a particular DNA sequence is shown in Figure 15.7. Each plot shows the relative probability that the given 99-base window is an exon with coding potential. It is apparent that some sensors like coding six-tuple in frame preferences have much more powerful discrimination than others. However, when the input from all seven sensors is combined by the neural net, the result is a truly striking pattern of prediction of clear exons and introns. This is shown in Figure 15.8. GRAIL works

Figure 15.5 A still more complex neural net, with several hidden layers.

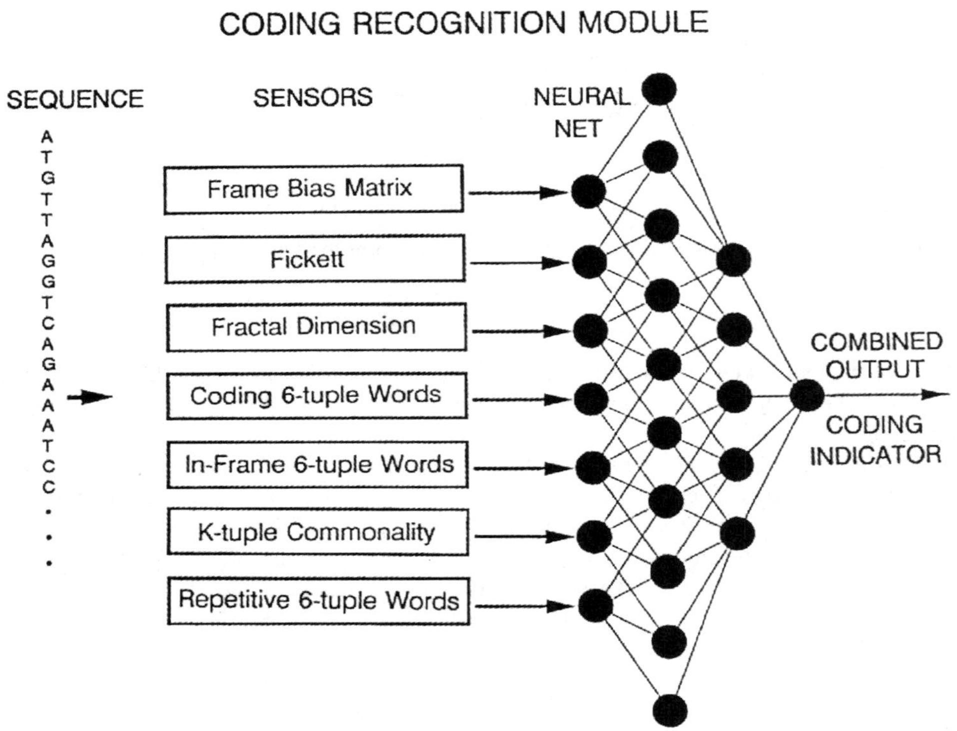

CODING RECOGNITION MODULE

Figure 15.6   The actual neural net used in GRAIL analysis of DNA sequences. Adapted from Uberbacher and Mural (1991).

on many different types of human proteins that were not included in the original training set. A number of examples are shown in Figure 15.9. Some caution is needed, however, because not all human genomic sequence is handled well by GRAIL. For example, the human T-cell receptor gene cluster is not readably amenable to GRAIL analysis. The program also has difficulty in finding very small exons, which is not surprising in view of the 99-base window used.

Neural net approaches similar to GRAIL appear to have great promise in other complex problems in biological and chemical analysis. These include prediction of protein secondary and tertiary structure, correction of DNA sequencing errors, and analysis of mass spectrometric chemical fragmentation data. Note, however, that neural nets are only one of a number of different types of algorithmic approaches applicable to such problems, and the vote is still out on which will eventually turn out to be the most effective for particular classes of analysis. However, for the past half-decade, GRAIL has proved to be an extremely useful tool for most applications to human DNA sequence analysis, and it is readily accessible via computer networks, to all interested users.

Since the introduction of GRAIL, improvements have been made on the original algorithms to produce GRAIL 2. Other approaches to gene finding have been proposed, including a linear discriminant method (Solovyev et al., 1994) and, most recently, a quadratic discriminant method (Zhang, 1997). These methods take into account additional factors like the compatibility of the reading frames of adjacent exons and consensus sequences to the intron segment that forms a branched structure as an intermediate step in

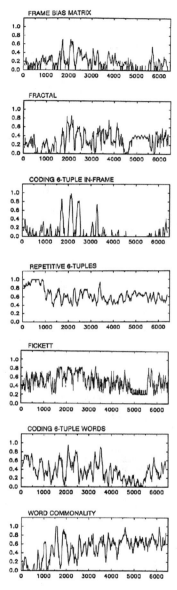

Figure 15.7   Performance of each of the seven sensors of the net shown in Figure 15.6 on one particular DNA sequence. The vertical axis indicates the probability that each sliding segment of DNA sequence is a coding exon. Taken from Uberbacher and Mural (1991).

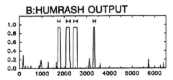

Figure 15.8   The output of the neural net, based on its optimal evaluation of the sensor results shown in Figure 15.7. Adapted from Uberbacher and Mural (1991).

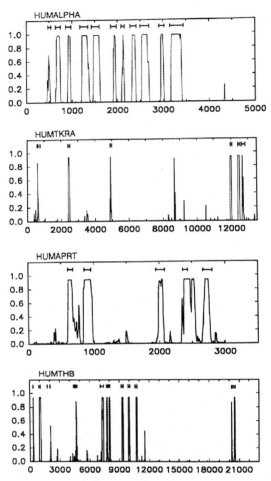

Figure 15.9   Examples of the performance of the neural net of Figure 15.6 on a set of different genomic DNA sequences. Adapted from Uberbacher and Mural (1991).

splicing. When tested in a large number of sequences, the three algorithms all perform well, but they are still far from perfect (Table 15.4).

TABLE 15.4   Success of Exon Prediction: Exons Found by Three Different Schemes

Scheme	Sensitivity TP/(TP + FN)	Specificity TP/(TP + FP)S
GRAIL 2	0.53	0.60
Linear discriminant analysis	0.73	0.75
Quadratic discriminant analysis	0.78	0.86

Source: Adapted from Hong (1997)

Note: True positives (TP) are true positives correctly predicted. False positives (FP) are true negatives predicted to be positive. False negatives (FN) are true positives predicted to be negative. Sensitivity is the fraction of true positives found. Specificity is the fraction of positives found that is true.

SURVEY OF PAST LARGE-SCALE DNA SEQUENCING PROJECTS

Most early large-scale DNA sequencing projects involved a pre-selected gene of particular interest. An example is the enzyme HPRT (57 kb). These projects are milestones in the history of DNA sequencing, but it is difficult to extrapolate the results of such projects to the situation that will apply in most genomic sequencing efforts. In such efforts, which will form the overwhelming bulk of the human genome project, one will be faced with large expanses of relatively uncharted DNA. While the regions selected may contain a few mapped genes, and many cDNA fragments, much of the rationale for looking at the particular region will have to come a posteriori, after the sequence has been completed. To try to get some impression of the difficulties in assembling the sequence, and making a first pass at its interpretation, it is useful to examine the first few efforts at sequencing segments of DNA without a strong functional pre-selection. Here we summarize results from seven projects: the complete sequence of *H. influenzae, M. genitalium,* partial sequences of *E. coli, S. cerevisiae, C. elegans,* and *D. melanogaster,* and several human cosmid DNAs. These sequence data and all other genomic sequence data currently reside in a set of publicly accessible databases. A description of these valuable resources, and how they can be accessed, is provided in the Appendix. A summary of all complete genome sequences publicly available in February 1997 is given in Table 15.5.

The complete DNA sequences of *Haemophilus influenzae* and *Mycoplasma genitalium* both correspond to relatively small bacterial genomes. As expected, they are very rich in genes, and they are especially rich in genes whose function can be surmised by comparison to other sequences in the available genome databases. *M. genetalium* has a 580,070 bp genome with 470 ORFs. These occur on average one per 1235 bp. The average ORF is 1040 bp. Overall the genome is 80% coding. Seventy-three percent of the ORF's correspond to previously known genes.

H. influenza has a genome size of 1,830,137 bp. This contains 1743 coding regions, an average of one every 1042 bp. The average gene is 900 bp long. Overall, 85% of the genome is coding. Currently 1007 (58%) of the coding regions can be assigned a functional role. Of the remainder, 385 are new genes that show no significant matches to the databases, while the others match known sequences of unknown function. At an average direct cost of $0.48 per base this project is probably representative of other large-scale efforts using similar technology.

Both the *H. influenzae* and *M. genetalium* sequencing projects were carried out at a single location totally by automated fluorescent DNA sequencing. In contrast, one of the

TABLE 15.5   Completed Genome Sequences

Species	DNA Molecules	kb DNA	Largest DNA (kb)	Open Reading Frames	Genes for RNA
M. genitalium	1	580	580	470	38
M. pneumonia	1	816	816	677	39
M. janneschii	3	1740	1665	1738	~45
H. influenza	1	1830	1830	1743	76
Synechoncystis sp.	1	3573	3573	3168	?
E. coli	1	4639	4639	4200	?
S. cerevisiae	16	12,068	1532	5885	455

efforts to sequence major sections of the *E. coli* genome, directed by Fred Blattner in Madison, Wisconsin, started as basically low-technology, manual DNA sequencing, employing a large number of relatively unskilled workers, and concentrated on relatively simple protocols. The initial result was a 91.4 kb contig. The region contained 82 predicted ORFs or roughly one per kb. The ORFs constituted about 84% of the total sequence. If we scale the properties of this region to the entire 4.7 Mb *E. coli* genome, we can predict that

$$\frac{4.7 \text{ Mb} \times 82 \text{ ORFs}}{0.0914 \text{ Mb}} = 4200 \text{ genes}$$

This is larger than estimates of the number of genes in *E. coli* based on the appearance of protein spots in two-dimensional electrophoretic separations. Past sampling of *E. coli* regions has revealed fairly uniform gene density except for areas around the terminus of replication. Hence the preliminary sequencing results on *E. coli* suggest that a significant number of new and interesting genes remain to be discovered. A more recent report of additional *E. coli* sequences is quite consistent with the earlier observations within a 338,500 base contig, 319 ORFs were found—one per 1060 bases. Of these, 46% are potentially new genes. The complete *E. coli* DNA sequence has just became available, and it contains 4300 genes, in 4.54 Mb, quite consistent with predictions based on partial sequencing results.

The early major accomplishments in *S. cerevisiae* sequencing derive from a very different organizational model than the work on *E. coli*. The approach was still mostly very low technology. It was mostly the result of a dispersed European effort among more than 30 different laboratories, coordinated through a common data collection center in France. The complete DNA sequence of one of the smallest *S. cerevisiae* chromosomes, number III, was the first one determined. At 315 kb it represented the longest continuous stretch of DNA sequence known at the time. The chromosome III sequence was originally reported to contain 182 ORFs. After this was corrected by a more rigorous examination, carried out by Christian Sander in Heidelberg, 176 ORFs remained. These occur at roughly one per 2 kb or half of the density seen in the three bacteria discussed above. The ORFs cover 70% of the DNA sequence; this is not too much lower than the total density of coding sequence in *E. coli*. We can make a rough estimate the number of genes in *S. cerevisiae* by scaling these results to the 12.1 Mb total size of the yeast genome. The result is

$$\frac{12.1 \text{ Mb} \times 176 \text{ ORFs}}{0.315 \text{ Mb}} = 6760 \text{ ORFs}$$

The total number of genes in *S. cerevisiae* will be slightly less than the number of ORFs because occasional genes in yeast consist of more than one exon. In addition, for both bacteria and yeast, we have to add in genes for rRNAs, tRNAs, and other nontranslated species (Table 15.5).

The complete DNA sequences of several other *S. cerevisiae* chromosomes reported were consistent with the results for chromosome III. For example, chromosome VIII has 562,698 bp. It contains 269 ORFs, or 1 per 2 kb. Of these, 124 (46%) corresponded to genes of known function. Chromosome VI has 270 kb. It contains 129 ORFs, again about 1 per 2kb. Of these, 76 (59%) correspond to genes with previously known function. The total sequence of *S. cerevisiae* is now completed. First estimates place the number of ORFs at 5885; doubtless this will change with further analysis.

In the case of *C. elegans* DNA sequencing, we are dealing not with continuous genomic sequence but with the sequence of selected cosmids. The effort, directed by John Sulston of Cambridge, England, and Robert Waterston of St. Louis, Missouri, is also state-of-the art fluorescent DNA sequencing technology with a great deal of automation. The strategy is mostly shotgun, with directed sequencing relegated mostly to closure of gaps between contigs. The first 21.14 Mb of *C. elegans* DNA sequence reported contained a total of 3980 genes of 1 per 4.8 kb on the autosomes and 1 per 6.6 kb on the X chromosome. Only 46% of these matched sequences already in the DNA databases. About 28% of the total DNA is coding; 50% of *C. elegans* is genes, including both exons and introns. This is a sharp drop from the density of coding sequences in simple organisms. The total number of genes in the nematode genome is estimated to be 13,000 ± 500. This is a number close to most contemporary expectations for the sizes of the genomes of typical multicellular, highly differentiated organisms like the nematode.

The remaining two DNA sequencing projects that we will discuss illustrate some of the frustrations in detailing with the genomes of higher organisms. The complete DNA sequence of a 338,234 bp region of *D. Melanogaster,* containing the bithorax complex, important in development, has been reported by groups at Caltech and Berkeley. This region is less than 2% coding. It contains only six genes. The final sequencing project we will discuss is a relatively early effort that involved several cosmids from the tip of the short arm of human chromosome 4, a region known to contain the gene responsible for Huntington's disease. The region is band 4p16.3. It is estimated to contain a total of 2.5 Mb of DNA. A 225-kb subset of this region was sequenced. This yielded 13 transcripts in 225 kb or one per 18 kb on average. Another estimate of gene density could be obtained by determining the number of HTF islands in the region. This will be a minimum estimate for the number of genes, since perhaps only half to two-thirds of all genes have HTF islands nearby. In fact, in the 225 kb region, one HTF island was found on average per 28 kb. By comparison, when HTF islands were mapped to a different section of chromosome 4, a 460 kb region near the marker D4S111, the frequency of occurrence of these gene-associated sequences was one per 30 kb. All of these estimates of gene density are remarkably consistent. If we scale these expected gene densities to the entire Huntington's disease region, we obtain an estimate of

$$\frac{2.5 \text{ Mb} \times 13 \text{ genes}}{0.225 \text{ Mb}} = 143 \text{ genes}$$

This makes it clear why finding the gene for Huntington's disease was not an easy task.

The first DNA sequencing effort in band 4p16.3 was carried out in Bethesda, Maryland, under the direction of Craig Ventor. It involved a total of 58 kb of DNA sequence in three cosmids. Three genes were found, each has an HTF island. The average gene density in this relatively small region is one per 19 kb, which is quite consistent with expectations. Less than 10% of the region is coding sequence. The number of Alu repeats in the region is 62, or roughly one per kb. This is comparable to what has been seen in the DNA sequence of two other gene rich, G + C-rich regions. In the human growth hormone region 0.7 Alu's were found per kb; in the HRPT region 0.9 Alu's were found per kb. In stark contrast, in the globin region which is G + C poor, there are only 0.1 Alu's per kb. These results illustrate the mosaic nature of the human genome rather dramatically.

Unlike simple genomes, with relatively uniform DNA compositions, mammalian genomes have mosaic compositions which is reflected in chromosome banding patterns. Scaling of a regional gene density to estimate the total number of genes, must take into account regional characteristics. Long before large-scale DNA sequencing or genome mapping was underway, Georgio Bernardi developed a method of fractionating genomes into regions with various G + C content. This was done by equilibrium ultracentrifugation in density gradients (Chapter 5). The resulting fractions were called isochores. Altogether, Bernardi obtained evidence for five distinct human DNA classes; these could be divided into three easily separated and manipulated fractions. Their properties are summarized below:

CLASS	GENE DENSITY	GENOME FRACTION	LOCATION
L1,L2	1	62%	Dark bands
H1,H2	2	31%	Light bands
H3	16	7%	Telomeric light bands

Several aspects of these results deserve comment. Gene density means the relative number of genes, based on cDNA library comparisons. The genome fraction is estimated from the total amount of material in the density-separated fractions. The telomeric light bands have very special properties, that we have alluded to before. Figure 15.10 illustrates the actual locations seen when DNAs from Bernardi's fraction H3 are mapped by FISH. The preferential location of these sequences on just a small subset of human chromosomal regions is really remarkable.

The Huntington's disease region is known to be a gene-rich light band, so we can pretty much exclude the L1 and L2 classes from consideration. In the Huntington's region, there is one gene on average per 18 kb. If this region is an H3 region, then we can estimate the number of genes in the human genome as

H3	11,700 genes
H1,H2	6500 genes
L1,L2	6500 genes

for a total of 24,700 genes. This estimate is less than twice the number of genes in *C. elegans,* which seems far too low. If we assume that the Huntington's disease region is an H1,H2 region, then the estimate of the number of genes in the human genome becomes

H3	92,000 genes
H1,H2	51,100 genes
L1,L2	51,100 genes for a total of 194,200 genes.

This is a depressingly large number, much larger than previous estimates. This example illustrates how difficult it is to know from very fragmentary data what the real target size of the human genome project is. Perhaps the Huntington's disease region is somewhere between the properties of the H3, and H1 plus H2 fractions, and the gene number somewhere mercifully between the two rather upsetting extremes we have computed. More recent estimates of the number of human genes range from 65,000 to 150,000, which is not too different from the average of our original estimates.

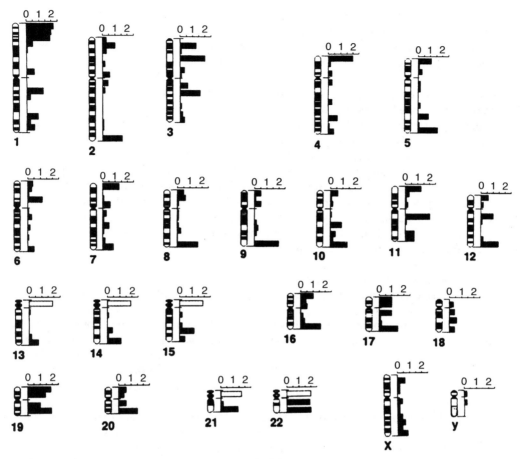

Figure 15.10 Distribution of extremely G + C-rich sequences in the human genome. Solid bars show relative hybridization of the H3 dark fraction. Open bars show rRNA-encoding DNA. Taken from Saccone et al. (1992).

FINDING ERRORS IN DNA SEQUENCES

Quite a few different kinds of errors contaminate data in existing DNA sequence banks. As the amount of data escalates, it will become increasingly important to audit these data continuously. Suspect data need to be flagged before they propagate and affect the results of many sequence comparisons or experimental scientific efforts. For example, an error in one of the earliest complete DNA sequences, the plasmid pBR322, produced a spurious stop codon in one of the proteins coded for by this plasmid. This confounded many researchers who were using this plasmid as a cloning and expression system, since a protein band with an unexplainable size was frequently seen.

Some common errors in DNA sequence data are quite easy to find and correct; others are almost impossible. A major class of error is incorporation of a totally inappropriate sequence. This can come about if, as is not uncommon, DNA samples are mixed up in the laboratory prior to sequencing. It can arise from cloning artifacts. A clone may have picked up bacterial DNA rather than the intended mammalian insert. A number of simple schemes exist that can help to find such errors. Putative genomic or cDNA sequences should be screened against all known common vector sequences. A very frequent error is to include pieces of vector inadvertently as part of the supposed insert. The presence of common repeats like Alu should be searched for in putative cDNAs or exons. Except in the rarest cases, these sequences should not be present there; finding them suggests that a cloning artifact may have occurred.

Rearrangements in cosmid clones and YACs are fairly common. The best way to find these errors at the DNA sequence level is to compare the sequences with other clones in available contigs. A major justification for the additional DNA sequencing required to examine a tiling set of cosmids is that there will be frequent overlaps which can help catch errors caused by rearrangements. Small sequencing errors are still about 1% in automated or manual sequencing. In many past efforts, considerable amounts of data were entered into sequence databases manually. It is vital that this be verified by a process of double entry and comparison. If not, except in the hands of the most compulsively careful individuals, typographical errors will abound.

When a single base is miscalled, either by misreading raw sequence data or by mistranscription in manipulating that data, the error is extremely difficult to detect. However, when a base is inserted or deleted, especially within an ORF, the error is sometimes easily caught. One way to do this is a procedure developed by Janos Posfai and Richard Roberts. In the course of searching a DNA database, to examine possible homology between a new sequence and all preexisting sequences, one can ask whether potential strong sequence homology (usually after the DNA has been translated into protein) is blocked by a frame shift. Where this occurs, a DNA sequencing error is almost always responsible. Several examples of the power of this approach in spotting sequencing errors are shown in Figure 15.11.

An unsolved problem is how to alert the community when errors are found. Given the size of the community and the complexity of the queries it makes against the sequence databases, this is an enormous problem. At some point the databases will have to be intel-

D: Transposase TNP A   E: Transposon Tn917

suspected DNA sequence error between positions 3998 and 4032

```
     473             570
D    LRNHIRSGDI.............PRIKLTDLLI
     LRNNIRSGDI........................
E    ..........................PRIKLSDLLI
     481             565
```

suspected DNA sequence error between nt positions 4091 and 4103

```
     587             616
D    LIEVASWTGF.............GTNIGLTKMA
     LIEVDSWTNF........................
E    ..........................GMNIGLEKMA
     582             611
```

N suspected DNA sequence error between nt positions 4374 and 4395

```
     681             732
D    GTTSSSDGMR...........LDGLLHHETD
     GKTTASDGMR........................
E    ..........................LDGLLYDETD
     676             728
```

suspected DNA sequence error between nt positions 4993 and 5056

```
     885             909
D    LNKGEAINAL.............ASALNIIINA
     LNKTEAINAL........................
E    ..........................ASALNVLINA
     881             904
```

Figure 15.11 Finding frameshift errors by comparing a new sequence with sequences preexisting in the databases. Adapted from Posfai and Roberts (1992).

ligent enough to be able to evaluate the effect of corrections on past queries and alert the initiators of those queries that might now be subject to altered outcomes. If this cannot be done, inevitably people will begin to repeat queries over and over again to guard against the effects of errors. A second potential unsolved problem is how to deal with fraudulent sequences. Research journals are increasingly reluctant to publish DNA sequence results, and it is almost impossible to publish the raw data supporting DNA sequencing results. Because of this, much sequence data are submitted directly to databases without editorial review of the actual experimental data. This entails the risk that databases might become contaminated willfully or accidentally by the deposit of sequences marred by artifacts or totally artificial. Just how these sequences could be detected and removed remains a serious dilemma. Ultimately it may be necessary to link the databases to archives of raw data so that validation of a suspected artifact is feasible.

SEARCHING FOR THE BIOLOGICAL FUNCTION OF DNA SEQUENCES

The major thrust of biological research is to understand function. From the viewpoint of the genome, this search for function can occur at two very different levels: individual genes or patterns of gene organization. We first discuss the genome from this latter vantage point. An overview of the arrangement of sequences in the genome may provide patterns of information that offer a clue to global aspects of function. These may be domains of gene activity or gene type that reflect biological processes we have not yet discovered. For example, most similar or related genes are not clustered. Some small clusters are seen, such as the globin genes (Fig. 2.10). The pattern of arrangement of the genes in these clusters presumably reflects an ancient gene duplication, which separated the alpha and beta families, and more recent duplications that evolved the more closely related members of these families. What is striking, and not yet explained, is that the order of the genes in each of these families accurately corresponds to the temporal order in which the genes are expressed during human development.

Another example of intriguing patterns of gene arrangement is the hox gene family in man and the mouse, shown in Figure 15.12. The genes in this family code for factors that determine the segmental pattern of organization of the developing embryo. The family is

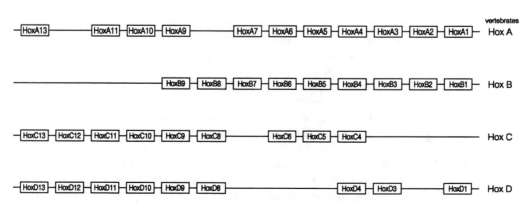

Figure 15.12 Organization of homeobox (hox) genes in the mammalian genomes. All genes in all four clusters are transcribed from left to right.

complex, and dispersed on a number of different chromosomes. Two aspects of the organization of the family are striking. First, it is so well conserved between the two species. Second, the spatial order of the genes within the family is the same as the order of the segments in the embryo that these genes affect. It is as if, for some totally unexplained reason, the map of the structure of the gene family is an image of the map of the function of that family.

A final example of functionally interesting gene arrangements is seen in a number of the members of the immunoglobulin superfamily including the light and heavy chains of antibodies, and several chains of the T-cell receptor. Here large numbers of related genes are grouped together, mostly in a single continuous segment of a chromosome. The reason for this is probably to assist the rearrangement of these genes, which takes place by DNA splicing to form mature expressed genes for antigen-specific proteins. If other regions of the genome are found with very large clusters of similar genes, one may well suspect that somatic DNA rearrangement or some other unusual biological mechanism will be at play with these genes.

A totally different view of global function afforded by complete physical maps and DNA sequences is the ability to compare these physical structures of DNA with the genetic map. An example is shown in Figure 15.13 for yeast chromosome III. There are clearly some regions where meiotic recombination is much more frequent than average and others where it is greatly suppressed. We do not yet understand the origin of these effects. One possibility is just the presence or absence of local DNA sequences that constitute recombination hot spots. However, there are other more global possibilities. Recombination may correlate with overall transcriptional activity, since highly transcribed chromatin is more open and accessible to all types of enzymes including those responsible for recombination. Thus there may be positional relationships between gene function and recombination, and thus gene evolution, that we still know nothing about today.

SEARCHING FOR THE BIOLOGICAL FUNCTION OF GENES

Most biologists, when they think of biological function in the context of the genome project, are referring to the function of individual genes. A common criticism of the genome project is that it is relatively useless to know the DNA sequences of genes without strong prior hints about their function. Most traditional molecular genetics begins with a function of interest and attempts to find the genes that determine or affect that function. This traditional view of biology is contrasted with the challenge posed by genome research in Figure 14.1, which can well serve as a paradigm for all of biology. In genome research we will discover DNA sequences with no a priori known function. Our current ability to translate these DNA sequences correctly into protein sequences is excellent, as we showed earlier, by using GRAIL or other powerful algorithms. Our current ability to take these protein sequences and draw immediate inferences about their possible function is well illustrated by the example in Figure 15.14a. Except for those rare readers of this book who are conversant in Dutch, this passage is largely unreadable. However, the frustrating aspect is that the passage is not totally unreadable. Because a number of scientific terms are cognates in Dutch and English, certain features stand out—one knows the passage has something to do with protein structure, but the full impact of the message is completely lost.

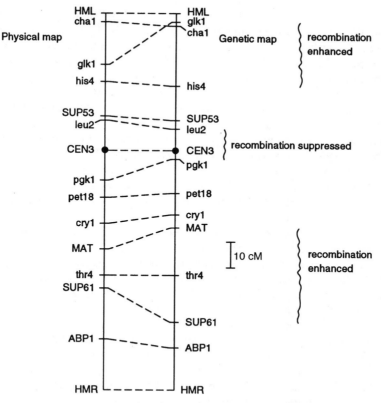

Figure 15.13   A comparison of the genetic and physical map of the yeast *S. cerevisiae.*

Ingewikkelde en grote biologische
macro-moleculen kunnen spontaan
in hun meest stabiele conformatie vouwen.
Helaas, ontbreekt ons de kennis om dit
proces te voorspellen want de gevouwen
strutuur kan belangrijke aanwijzingen
over die functie van het molecuul bevatten.

(*a*)

We know that large biological molecular can fold
into their most stable state spontaneously but
we really have little ability at present to predict
this folding. Our ignorance is most unfortunate
since the folded structure may contain
important clues on how the molecule functions.

(*b*)

Figure 15.14   An analogy for the current (*a*) and desired future (*b*) ability to interpret DNA sequence in terms of its likely biological function.

When the passage in Figure 15.14*a* is translated into English, it provides an important clue to one direction that can help find functional clues (Fig. 15.14*b*). Considerable experience to date shows that protein three-dimensional structures are better conserved during evolution than protein sequences. A great deal of current research effort is being devoted to improving our ability to infer possible protein structures from the sequences of sets of related proteins, provided that at least one of them has a known three-dimensional structure. As our ability to do this improves, and as the number of different classes of protein structures has one or more members successfully studied at high-resolution by X-ray crystallographic or nuclear magnetic resonance techniques, the prospects of stepping quickly from a sequence to a realistic, if not exact, model of the structure should improve markedly. However, just knowing a three-dimensional structure does not immediately provide definitive clues to function. It simply makes comparisons between a protein of unknown function and the set of proteins of known function more powerful and more likely to yield useful insights.

Today, when a new segment of DNA sequence is determined, the first thing that is almost always done with it is to compare it to all other known DNA sequences. The purpose is to see if it is related to anything already known. By related, we mean, that there is a statistically significant similarity to one or more preexisting DNA sequences. The definition of what statistically significant means in the context of sequence comparisons is not universally accepted despite decades of work in this area. Obviously, at one extreme, one may find that a new sequence is virtually identical to a preexisting one. Unless the two sequences derive from very similar but not identical organisms, the finding of near identity means true identity with the differences due to sequencing errors, or a new member of a gene family, or an example of proteins very strongly conserved in evolution, like the histones. At the other extreme, a new sequence may match nothing to within whatever local standards of minimal homology are considered operative.

Most often, however, when a new DNA sequence is compared with the current data base of more than 1000 Mb of DNA, some slight or significant sequence homology is found. For coding sequences, it is usually much more powerful to search after translation of DNA to protein. This translation loses very little functional information; it gains considerable statistical power because the noise caused by the degeneracy of the genetic code is blanked out. Thus consider, for example, two arginine codons like AGG and CGA in a corresponding place on two sequences; the only evidence for similarity is the G in position 2, which has roughly one chance in four of occurring randomly. In contrast, posing an arginine opposite an arginine at the same place in a protein sequence has, very crudely, only one chance in 20 of occurring randomly. (In reality the statistical differences are not this great because amino acids with six possible codons, like arginine, also tend to occur much more often than average.)

A statistically significant match between a new sequence and some preexisting sequence implies some or all of the following possibilities: similar function, similar structure, or evolutionary relatedness. It is not easy to sort out these different effects. However, one encouraging feature of such global sequence searches is that their effectiveness appears to be increasing markedly and rapidly as the database grows. Ten years ago Russell Doolittle noted that a new protein sequence had a 25% chance of matching something else in the databases. Currently the odds are considerably better than this. From the first bacterial sequencing projects described earlier in this chapter, between 54% and 78% of the ORFs found showed hints of homology in structure or function with something else in the data base. With the *S. cerevisiae* ORFs on chromosome III, 42% gave hints of homologous structure or function of which 14% were deemed really quite strong. In the case of *C. elegans,* where more extensive data are available, 45% of the ORFs were reported to be relatable to existing databases. It seems likely that in a few years it will be the odd new sequence that does not immediately match something known. While it is too early to be sure how rapidly this goal will be achieved, there is room for considerable optimism at present.

METHODS FOR COMPARING SEQUENCES

Entire books have been written about the relative merits of different approaches to aligning sequences and testing their relatedness (Waterman, 1995; Gribokow and Deveraux, 1991). The topic is actually quite complex because the nonrandom nature of natural DNA sequences greatly confounds attempts to construct simple statistical tests of relatedness. Here our goal will be to present the basic notions of how sequences are compared and what these comparisons mean. Sequences are strings of symbols. Any two strings can be compared by direct alignment and the use of scoring criteria for similarity. For two strings of length n and m there are $2(n + m - 1)$ possible continuous alignments, by which we mean that no gaps are allowed in either string. Of course many of these alignments are fairly trivial and uninteresting because the strings will barely overlap. The moment gaps are allowed on one or both strings, the number of alignments rises in a combinatorial manner to reach heights that can test the power of the fastest existing supercomputers if the problem is not handled intelligently.

An example of a very simple case in which two very similar DNA sequences are aligned is shown in Figure 15.15. In this case the alignment needed to maximize the apparent similarity between the two sequences is obvious. What is less obvious is the sort of score to give such an alignment. The simplest scoring scheme is black and white: Grade all identities the same and all differences the same. However, this makes little sense from either a biological or a statistical vantage point. As far as biology is concerned, if, for example, we are looking at the functional relatedness of proteins coded for by these sequences, or if we are looking at possible evolutionary relationships between them, transversions (interchange of a purine and a pyrimidine) should be weighted as more consequential differences than transitions (interchange between two pyrimidines or two purines). This is because the rate of transversion mutations is much less than the rate of transitions, and the genetic code appears to have evolved so that effects of transversions on the resulting amino acids are more functionally disruptive than the effect of transitions. For example, many synonymous codons are related by a transition in their third position. But the example goes much deeper; for example, codons for different hydrophobic amino acids are also related mostly by transitions.

AGCTTACGCAAACC
GCTCACGGTTGCCA

identities	I I I0I I I0I I0I I
mutations	0 00S000V00S00

Figure 15.15 A simple example of a comparison between two putatively related nucleic acid sequences and two ways in which their relatedness could be scored, S = transition and V = transversion.

To take statistical factors into account in estimating the significance of a mismatch or a match purely at the DNA level, we have to consider the relative frequency of each residue in the strings being compared. For example, sequences rich in A's will show large numbers of A's matched with A's, just by chance. In order to take this into account, and to add issues like transitions and transversions, one needs to employ a scoring matrix. This is illustrated in Figure 15.16a. The 4 × 4 scoring matrix for nucleic acid comparisons allows for any possible weight to be assigned to a particular set of bases at an alignment position. Generally, the same scoring matrix is used for every alignment position, although there is no reason why one should have to do this, nor is there any reason why it is desirable except for simplicity. Think ahead to the alignment of protein sequences where residues on exterior loops can be quite variable without perturbing the overall structure. Therefore, if one had some way of knowing a priori that a residue was in a loop as opposed to a helix or sheet, one could adjust the weighting factors accordingly. This example illustrates the complex interplay between sequence and structure information that really has to occur in very robust comparison algorithms.

The simplest possible DNA scoring matrix, corresponding to the rule used in Figure 15.15 is just a set of identities with no correction for overall base composition (Fig. 15.16b). The general case would consist of a set of elements a_{ij} that are all different, except that the matrix should be symmetrical; each $a_{ij} = a_{ji}$ since we have no way, in comparing just two proteins, to favor one sequence over another. The elements a_{ij} must incorporate all of our biological and statistical prejudices. When protein sequences are compared, the scoring matrices can become more complicated. First of all, the matrix must be 20 × 20 instead of 4 × 4. It can be as simple as an identity matrix, just as in the case for nucleic acids, but a much more accurate picture will incorporate statistical information about the relative frequency of amino acids. This immediately raises one serious problem: Does one use the amino acid composition of the two proteins in question to construct the scoring matrix, or does one use the amino acid compositions of all known proteins, or all known proteins from the particular species involved? One can elaborate the problem even further by asking whether the nonrandomness of dipeptide frequencies should be considered in making statistical evaluations for the scoring matrix. There are no simple answers to these questions.

Most commonly, with protein sequence comparisons, one incorporates information about amino acid physical properties into the values of the elements of the scoring matrix. Thus, for example, interchanges among ile, leu, and val, or ser and thr, among proteins known to be related in structure and function are very commonly seen and are presumably mostly innocuous. Examples of two real scoring matrices are shown in Figure 15.17.

Figure 15.16 Comparison matrices between two nucleic acid sequences. (*a*) A general matrix. (*b*) The simplest possible matrix.

	A	C	D	E	F	G	H	I	K	L	M	N	P	Q	R	S	T	V	W	Y
A	4	.	.	.	.	.	.	.	.	.	.	.	.	.	.	.	.	.	.	.
C	0	9	.	.	.	.	.	.	.	.	.	.	.	.	.	.	.	.	.	.
D	-2	-3	6	.	.	.	.	.	.	.	.	.	.	.	.	.	.	.	.	.
E	-1	-4	2	5	.	.	.	.	.	.	.	.	.	.	.	.	.	.	.	.
F	-2	-2	-3	-3	6	.	.	.	.	.	.	.	.	.	.	.	.	.	.	.
G	0	-3	-1	-2	-3	6	.	.	.	.	.	.	.	.	.	.	.	.	.	.
H	-2	-3	-1	0	-1	-2	8	.	.	.	.	.	.	.	.	.	.	.	.	.
I	-1	-1	-3	-3	0	-4	-3	4	.	.	.	.	.	.	.	.	.	.	.	.
K	-1	-3	-1	1	-3	-2	-1	-3	5	.	.	.	.	.	.	.	.	.	.	.
L	-1	-1	-4	-3	0	-4	-3	2	-2	4	.	.	.	.	.	.	.	.	.	.
M	-1	-1	-3	-2	0	-3	-2	1	-1	2	5	.	.	.	.	.	.	.	.	.
N	-2	-3	1	0	-3	0	1	-3	0	-3	-2	6	.	.	.	.	.	.	.	.
P	-1	-3	-1	-1	-4	-2	-2	-3	-1	-3	-2	-2	7	.	.	.	.	.	.	.
Q	-1	-3	0	2	-3	-2	0	-3	1	-2	0	0	-1	5	.	.	.	.	.	.
R	-1	-3	-2	0	-3	-2	0	-3	2	-2	-1	0	-2	1	5	.	.	.	.	.
S	1	-1	0	0	-2	0	-1	-2	0	-2	-1	1	-1	0	-1	4	.	.	.	.
T	0	-1	-1	-1	-2	-2	-2	-1	-1	-1	-1	0	-1	-1	-1	1	5	.	.	.
V	0	-1	-3	-2	-1	-3	-3	3	-2	1	1	-3	-2	-2	-3	-2	0	4	.	.
W	-3	-2	-4	-3	1	-2	-2	-3	-3	-2	-1	-4	-4	-2	-3	-3	-2	-3	11	.
Y	-2	-2	-3	-2	3	-3	2	-1	-2	-1	-1	-2	-3	-1	-2	-2	-2	-1	2	7

(a)

	A	C	D	E	F	G	H	I	K	L	M	N	P	Q	R	S	T	V	W	Y
A	4	.	.	.	.	.	.	.	.	.	.	.	.	.	.	.	.	.	.	.
C	-2	10	.	.	.	.	.	.	.	.	.	.	.	.	.	.	.	.	.	.
D	-1	-5	5	.	.	.	.	.	.	.	.	.	.	.	.	.	.	.	.	.
E	-1	-5	1	5	.	.	.	.	.	.	.	.	.	.	.	.	.	.	.	.
F	-2	-2	-5	-4	7	.	.	.	.	.	.	.	.	.	.	.	.	.	.	.
G	0	-4	-1	-2	-5	5	.	.	.	.	.	.	.	.	.	.	.	.	.	.
H	-2	-5	-1	-1	-1	-2	8	.	.	.	.	.	.	.	.	.	.	.	.	.
I	-2	-4	-4	-3	0	-5	-4	5	.	.	.	.	.	.	.	.	.	.	.	.
K	-1	-5	-1	1	-3	-2	0	-3	5	.	.	.	.	.	.	.	.	.	.	.
L	-2	-4	-5	-3	1	-4	-2	2	-2	5	.	.	.	.	.	.	.	.	.	.
M	-1	-2	-5	-2	1	-4	-2	1	-1	2	7	.	.	.	.	.	.	.	.	.
N	-1	-4	2	0	-3	-1	1	-4	0	-3	-2	5	.	.	.	.	.	.	.	.
P	-1	-5	-1	-1	-4	-2	-2	-3	-1	-4	-4	-2	7	.	.	.	.	.	.	.
Q	-1	-3	0	1	-3	-2	1	-3	1	-2	0	0	-2	6	.	.	.	.	.	.
R	-2	-3	-2	0	-4	-3	0	-3	2	-2	-2	-1	-2	1	7	.	.	.	.	.
S	0	-3	0	-1	-3	-1	-2	-4	-1	-3	-2	0	-1	0	-1	4	.	.	.	.
T	-1	-3	-1	-1	-3	-3	-2	-2	0	-2	-1	0	-1	0	-1	1	5	.	.	.
V	0	-2	-4	-2	-1	-4	-3	3	-2	1	0	-3	-3	-2	-3	-2	1	5	.	.
W	-3	-5	-5	-4	2	-4	-3	-2	-3	-1	0	-4	-5	-3	-1	-4	-5	-3	10	.
Y	-2	-4	-3	-2	3	-4	0	-1	-2	-1	0	-2	-3	-2	-2	-2	-2	-2	2	7
-	-7	-9	-7	-6	-8	-7	-8	-9	-7	-8	-8	-7	-6	-7	-7	-7	-8	-8	-9	-8

(b)

Figure 15.17 An example of actual scoring matrices for protein sequences that takes into account the similar properties of certain types of amino acids. *(a)* the Blosum G2 matrix used by BLAST (Henikoff and Henikoff, 1993). *(b)* The structural (STTR) matrix of Shpaer et al. (1996).

The values of these elements obviously vary over a wide range. However, despite their different origins, the two matrices are fairly similar.

There is still one additional complication that must be dealt with. This is especially serious when one wishes to estimate the evolutionary relatedness of two proteins or DNAs. Here a yardstick that is often used as a time scale for evolutionary divergence is the probable average number of mutations needed to convert one sequence into the other. Such comparisons among very similar proteins or nucleic acids are relatively simple. Differences seen are presumably real, and similarities are also presumed real. However, when more distant sequences are compared, an apparent similarity has an increasing chance of just being a statistical event, or a reversion. For example, as shown in Figure 15.18 two matching A's could be a true identity (no mutations) or a reversion (a minimum of two mutations). The more distantly related the two sequences, the more the latter possibility has to be weighted. Ways of doing this for simple identity comparison matrices were developed several decades ago by Jukes and Cantor, and later elaborated considerably to take into account statistical effects and similarities in residue properties. The kind of matrix needed in a very simple case is shown in Figure 15.19. It adjusts the relative weights of comparisons as a function of the average extent of differences between the two sequences. The problem of choosing an ideal comparison matrix, which deals with all of these interrelated issues, is still not a simple one.

Once a comparison matrix is chosen, it can be used to evaluate the relative similarity seen in all possible alignments between two strings. When gaps (caused by a putative insertion or deletion, or a pure statistical artifact) are allowed, the problem of actually enumerating and testing all possible comparisons becomes computationally extremely de-

Figure 15.18 Difficulties in sequence comparisons when the goal is to estimate the probable number of mutations that have occurred to derive one sequence from another (or both from a common ancestor).

	A	C	G	T
A	1-3a	a	a	a
C	a	1-3a	a	a
G	a	a	1-3a	a
T	a	a	a	1-3a

Figure 15.19 A simple scoring matrix that takes into account the average differences between two sequences and allows for the possibility of revertants. Where $a = 1/4(1 - e^{-4d/3})$. The parameter d is a measure of the true evolutionary distance between two sequences being compared. It is the average number of mutations per site that separate one sequence from the other. In the limit $d \to 0$ the matrix becomes equal to the right-hand panel of Figure 15-16. In the limit $d \to \infty$ all of the elements of the matrix become equal to 1/4. This means that the sequences have diverged so much that one is essentially comparing two random strings.

manding. Figure 15.20 shows a very simple example. The issue is how to test the likelihood that the postulated gap results in a statistically significant improvement in the alignment score of the two sequences. Obviously there must be a statistical penalty attached to the use of such a gap, since it greatly increases the number of possible comparisons, and thus the chance of finding, at random, a comparison with a score better than some arbitrary value.

From a practical point of view, it is impossible to test all possible gap numbers and locations. One way to deal with this problem is to compare two sequences through smaller windows, sets of successive residues, rather than globally (Fig. 15.21). With two strings of length n and m, and a window of length L, there are $(n - L + 1)(m - L + 1)$ possible comparisons to be done. This is not a major task for strings the sizes of typical genes. For each choice of window, two substrings of length L are compared, without gaps. The score for this comparison is calculated as the sum over the matrix elements a_{ij} for each of the L residues pairs. To provide a visual overview of the comparison, it is usually convenient to plot all scores above some threshold value as a dot in a rectangular field formed by writing one sequence along the horizontal axis and the other along the vertical axis. Any point in the field corresponds to an alignment of L residues positioned at particular residue positions in the two sequences. This kind of dot matrix plot is shown, schematically in Figure 15.22, and a real example of a sequence comparison at the DNA level for two closely related viruses, SV40 and polyoma is given in Figure 15.23. Any regions with

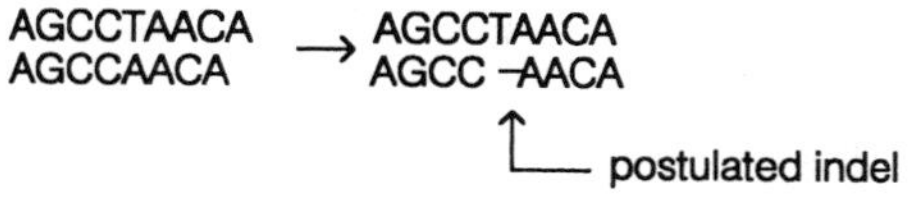

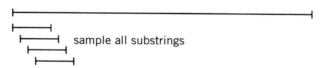

Figure 15.20   A simple case of two sequences potentially related by an insertion or a deletion.

Figure 15.21   Window selection on a single sequence assists in comparisons.

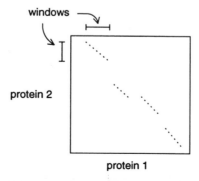

Figure 15.22   An example of the comparison of two proteins or DNAs using windows on each, evaluated with a scoring matrix. Shown as dots are all comparisons that score above a selected threshold.

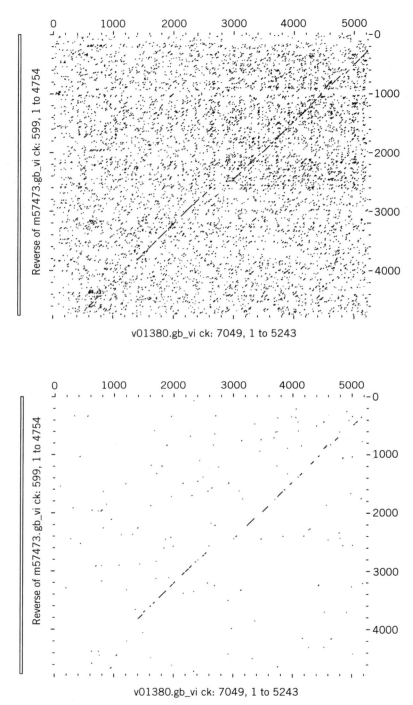

Figure 15.23   Results of two actual comparisons between polyoma and SV40 viral sequences, computed as described for the hypothetical example in Figure 15.22. Dots show DNA windows with more than 55% identity within a window (top) or more than 66% (bottom). Provided by Rhonda Harrison.

strong homology show up as a series of dots along a diagonal in these plots. One can piece together the set of diagonals found to construct a model for a global alignment involving all the residues. However, it is not easy to show if a particular global alignment reached in this manner is the best one possible.

DYNAMIC PROGRAMMING

Dynamic programming is a method that improves on the dot matrix approach just described because it has the power, in principle, to find an optimal sequence alignment. The method is computationally intensive, but not beyond the capabilities of current supercomputers or specialized computer architectures or hardware designed to perform such tasks rapidly and efficiently (Shpaer et al., 1996). The basic notion in dynamic programming is illustrated in Figure 15.24. Here an early stage in the comparison of two sequences is shown. Gaps are allowed, but a statistical penalty (a loss in total similarity score) is paid each time a gap is initiated. (In more complex schemes the penalty can depend on the size of the gap.) We already discussed the fiercely difficult combinatorial problem that results if one tries out all possible gaps. However, with dynamic programming, in the example shown in Figure 15.24, one argues that the best alignment achievable at point n (n is the total number of residues plus gaps that have been inserted into either of the sequences since start of the comparison) must contain the best alignment at point $n - 1$, plus the best scoring of the following three possibilities:

1. Align the two residues at position n.
2. Gap the top sequence.
3. Gap the bottom sequence.

The argument behind this approach is not totally rigorous. There is no reason why the optimal alignment of two sequences should be co-linear. DNA rearrangements can change the order of exons, invert short exons, or otherwise scramble the linear order of genetic information. Dynamic programming, as conventionally used, can only find the optimal co-linear alignment. However, one can circumvent this problem, in principle, by attempting the alignments starting at selected internal positions in both possible orientations, and determining if this alters the best match obtained.

Best alignment up to n-1	One of these three is the best alignment up to n
n-1	n
ACCG–AACGCCCA	T
TCCGTAATG–GGA	C
ACCG–AACGCCCA	–T
TCCGTAATG–GGA	C
ACCG–AACGCCCA	T
TCCGTAATG–GGA	–C

Figure 15.24 Basic methodology used in comparison of two sequences by dynamic programming.

Sequence alignment by dynamic programming is conveniently viewed by a plot as shown in Figure 15.25. The coordinates of the plot are the two sequences, just as in the dot matrix blot discussed earlier. However, what is plotted at each position is not a score from the comparison of two sequence windows; instead, it is a cumulative score for a path through the two sequences that represents a particular comparison. In the simple example in Figure 15.25, the scoring is just for identities. In a real case, the elements of a scoring matrix would be used. From the example in Figure 15.25a it can be seen that a particular alignment is just a path of arrows between adjacent residues. The three possible steps listed above at a given point in the comparison just correspond to:

1. Advancing along a diagonal
2. Advancing horizontally
3. Advancing vertically

In dynamic programming one must consider all possible paths through the matrix. A particular comparison is a continuous set of arrows. The best comparisons will give the highest scores. Usually there are multiple paths that give equal, optimal alignments. These can be found by starting with the point in the plot with the highest score and tracing back toward the beginning of the sequence comparison to find the various routes that can lead to this final point (Fig. 15.25b).

Rigorous dynamic programming is a very computationally intensive process because there are so many possible paths to be tested. Remember that for each new protein sequence, one wishes to test alignments with all previously known sequences. This is a very

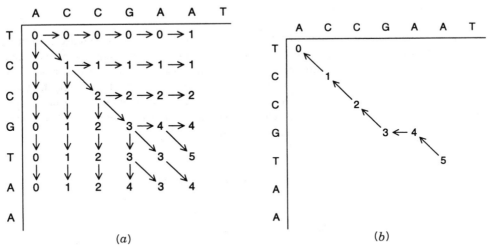

(a)   (b)

Figure 15.25 A simple example of dynamic programming. (a) Matrix used to describe the scores of various alignments between two sequences. Note that successive horizontal and vertical moves are forbidden, since this introduces a gap on both strands which is equivalent to a diagonal move— not a gap at all. In each cell only the route with the highest possible score is shown. In practice, each cell can be approached in three different ways—diagonally, horizontally, and vertically. (b) The trace back through the matrix from the highest scoring configuration to find all paths (alignments) consistant with that final configuration. Only the path with the highest score is kept.

large number of dynamic programming comparisons. A number of different procedures can be used to speed this up. Some pay the price of a loss of rigor, so no guarantee exists that the very best alignment will be found.

A very popular program for global database searching has been developed by David Lipman and his coworkers. This is called FASTA for nucleic acid comparisons, FASTP, for protein (or translated ORF) comparisons. The test protein is broken into k-tuple words. These are searched through the entire database, in looking for exact matches and scoring them. Some of these best scoring comparisons will form diagonal regions of dot plots with a high density of exact matches. The 10 best of these regions are rescored with a more accurate comparison matrix. They are trimmed to remove residues not contributing to the good score. Then nearby regions are joined, using gap penalties for the inevitable gaps required in this process. The resulting sets of comparisons are examined to find those with the best, provisional fits. Then each of these is examined by dynamic programming within a relatively narrow band of 32 residues around the sites of each of the provisional matches. Even faster algorithms exist, like Lipman's BLAST which looks only at short exact fits but makes a rigorous statistical assessment of their significance. This is used to winnow down the vast array of possible comparisons before more rigorous, and time-consuming, methods are employed.

A key point in favor of the dynamic programming method is that for each comparison cell, as shown in Figure 15.26, one only needs to consider three immediate precursors. This allows the computation to be broken into a set of steps that can be computed in parallel. The comparison can proceed down the diagonal of the plot in waves, with each cell computed independently of what is happening in the other cells along that diagonal (Fig. 15.26). Thus parallel processing computers are well adapted to the dynamic programming method. The result is an enormous increase in our power to do global sequence comparisons. Using a parallel architecture, it is possible, with the existing protein database, to do a complete inversion. That is, all sequences are compared with all sequences by dynamic programming. The resulting set of scores allows the database to be factored into sequence families, and new, unsuspected sets of families have been found in this way. As an alternative to parallel processing, specialized computer chips have been designed and built that perform dynamic programming sequences comparisons, but they do this very quickly because each sequence needs to be passed through the chip only once.

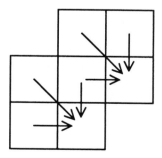

Figure 15.26   Parallel processing in dynamic programming. Only three preceding cells are needed to compute the contents of each new cell. Computation can be carried out on all cells of the next diagonal, simultaneously.

GAINING ADDITIONAL POWER IN SEQUENCE COMPARISONS

With increases in available computation power occurring almost continuously, it is undeniably tempting to carry sequence comparisons even further to make them more sensitive. The goal is to identify, as related, two or more sequences that lie at the edge of statistical certainty by ordinary methods. There are at least two different ways to attempt to do this. One can try to introduce information about possible protein three-dimensional structure for the reasons described earlier in this chapter. Alternatively, one can attempt to align more than two sequences at a time.

If two proteins are compared at the amino acid sequence level, but one of these proteins is of known three-dimensional structure, the sensitivity of the scoring used can be improved considerably, as we hinted earlier. One can postulate that if the two proteins are truly related, they should have a similar three-dimensional structure. In the interior of this structure, spatial constraints are very important. Insertions and deletions there will be very disruptive, and thus they should be assigned very costly comparison weights. The effects will be much less serious on the protein surface. Changes in any charged residues on the interior of the protein are also likely to be very costly. Interchanges among internal hydrophobic residues are likely to be tolerated, especially if there are compensating increases in size at some points and decreases in size elsewhere. Changes that convert internal hydrophobic residues to charged residues will be devastating. A small number of external hydrophobic residues may not be too serious. Changes in sequence that seriously destabilize prominent helices and beta sheets are obviously also costly.

A systematic approach to such structure-based sequence comparisons is called threading. Here the sequence to be tested is pulled through the structure of the comparison sequence. At each stage some structure variations are allowed to see if the new sequence is compatible in this alignment with the known structure. The effectiveness of threading procedures in improving homolog matching and structure prediction is still somewhat controversial. However, in principle, this appears to be a very powerful method as more and more proteins with known three-dimensional structures become available.

As originally implemented by Christian Sander, by attempting to fit proteins of unknown structure to the various known protein structures as just described above, one gained 5% to 10% in the fraction of the database that showed significant protein similarity matches. Clearly this approach will improve significantly as the body of known protein structures increase in size. Another aspect of the protein structure fitting problem deserves mention here. As more and more proteins with known three-dimensional structure become available, it is of interest to ask whether any pairs of proteins have similar three-dimensional structure, outside of any considerations about whether they have similar amino acid sequences. Sander has constructed algorithms that can look for similar structures, essentially by comparing plots of sets of residues that approach each other to within less than some critical distance. Such comparisons have revealed cases where two very similar three-dimensional structures had not been recognized before because they were being viewed from different angles that obscured visual recognition of their similarity. In the future we must anticipate that the number of structures will be so great that no visual comparison will suffice, unless there is first a detailed way to catalog or group the structures into potentially similar objects.

The second approach to enhanced protein sequence comparison is multiple alignments. This allows weak homology to be recognized if a family of related proteins is assembled. Suppose, for example, that one saw a sequence pattern like

TABLE 15.6 One-Letter Code for Amino Acids

One-Letter Code	Three-Letter Code	Amino Acid Name	Mnemonic
A	ala	alanine	*a*lanine
C	cys	cysteine	*c*ysteine
D	asp	aspartate	aspar*d*ate
E	glu	glutamate	glutamat*e*
F	phe	phenylalanine	*f*enylalanine
G	gly	glycine	*g*lycine
H	his	histidine	*h*istidine
I	ice	isoleucine	*i*soleucine
K	lys	lysine	*k* near l
L	leu	leucine	*l*eucine
M	met	methionine	*m*ethionine
N	asn	asparagine	asparagi*n*e
P	pro	proline	*p*roline
Q	gln	glutamine	"cute" amine
S	ser	serine	*s*erine
T	thr	threonine	*t*hreonine
V	val	valine	*v*aline
W	trp	tryptophan	t*w*yptophan
Y	tyr	tyrosine	t*y*rosine

$$—P—AHQ—L—$$

in two proteins. Here we use the convenient one letter code for amino acids summarized in Table 15.6. Such a pattern would be far too small to be scored as statistically significant. However, seeing this pattern in many proteins would make the case for all of them much stronger. Currently most multiple alignments are done by making separate pairwise comparisons among the proteins of interest. The extreme version of this is the database inversion described previously. The use of pairwise comparisons throws away a considerable amount of useful information. However, it saves massive amounts of computer time. The alternative, for rigorous algorithms, would be dynamic programming in multiple dimensions. This is not a task to be approached by the faint of heart (or short of budget) with current computer speeds. Future computer capabilities could totally change this.

DOMAINS AND MOTIFS

From an examination of the proteins of known three-dimensional structure, we know that much of these structures can be broken down into smaller elements. These are domains, independently folded units, motifs, and structural elements or patterns that describe the basic folding of a section of polypeptide chain, to form either a structural backbone or framework for a domain or a binding site for a ligand. Examples are various types of barrels formed by multiple-stranded beta sheets, zinc fingers and helix-turn-helix motifs, both of which are nucleic acid binding elements, Rossman folds and other ways of binding common small molecules like NAD and ATP, serine esterase active sites, transmembrane helices, kringles, and calcium binding sites. A few of these motifs are illustrated schematically in Figure 15.27. The motifs are not exact, but many can be recognized at the level of the amino acid sequence by conserved patterns of particular residues like cysteines or hydrophobic side chains.

Several aspects of the existence of protein motifs should aid our ability to analyze the function of newly discovered sequences once we understand these motifs better. First, there may well be just a finite number of different protein motifs, perhaps just several hundred, and eventually we will learn to recognize them all. Our current catalog of motifs derives from a rather biased set of protein structures. The current protein data base is composed mostly of materials that are stable, easy to crystallize, and obtainable in large quantities. Eventually the list should broaden considerably.

Each protein motif has a conserved three-dimensional structure, and so once we suspect a new protein of possessing a particular motif, we can use that structure for enhanced sequence comparisons as described above. If, as now seems likely, proteins are mostly composed of fixed motifs, any new sequence can be dissected into its motifs, and these can be analyzed one at a time. This will greatly simplify the process of categorizing new structures. However, the key aspect of motifs that makes them an attractive path from sequences to biological understanding is that motifs are not just structural elements, they

Proteins are assembled from motifs

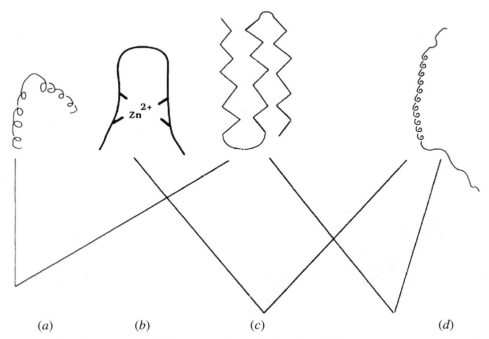

(a) (b) (c) (d)

Figure 15.27 Some of the striking structural and functional motifs found in protein structures. (a) Helix-turn-helix. (b) Zinc finger. (c) Beta barrel. (d) Transmembrane helical domain.

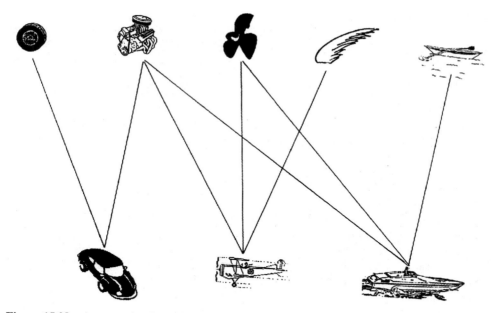

Figure 15.28   An example of how motifs (found in transportation vehicles) can be used to infer or relate the structures of composites of motifs.

are functional elements. There must be a relatively finite set of rules that determines what combinations of motifs can live together within a protein domain. Once we know these rules, our ability to make intelligent guesses about function from structural information alone should improve markedly.

One way to view motifs is to consider them as words in the language of protein function. What we do not yet know, in addition to many of the words in our dictionary, is the grammar that combines them. Another way to view protein motifs is the far-fetched, but perhaps still useful, analogy shown in Figure 15.28. This portrays various transportation vehicles and some of the structure-function motifs that are contained in them. If we knew the function of the individual motifs, and had some idea about how they could be combined, we might be able to infer the likely function of a new combination of motifs, even though we had never seen it before; for example, we should certainly be able to make a good de novo guess about the function of a seaplane.

INTERPRETING NONCODING SEQUENCE

Most of the genome is noncoding, and arguments about whether this is garbage, junk, or potential oil fields have been presented before. The parts of the noncoding region that we are most anxious to be able to interpret are the control sequences which determine when, where, and how much of a given gene product will be made. Our current ability to do this is minimal, and it is confounded by the fact that most of the functional control elements appear to be very short stretches of DNA sequence. These sequences are not translated, so we have only the four bases to use for comparisons or analyses. This leads to a situation with very poor signal to noise. In eukaryotes, control elements appeared to be used in complex combinations like the example shown schematically in Figure 15.29. To analyze of stretch of DNA sequence for all possible combinations of small elements that might be able to come together in such a complex seems intractable by current methods. Perhaps, after we have learned much more about the properties of some of these control regions, we will be able to do a better job. The true magnitude of this problem, currently, is well illustrated by the globin gene family where one key, short control element required for correct tissue-specific gene expression was found (experimentally, and after much effort) to be located more than 20-kb upstream from the start of transription. Thus finding control elements is definitely not like shooting fish in a barrel.

DIVERSITY OF DNA SEQUENCES

We cannot be interested in just the DNA sequence of a single individual. All of the intellectual interest in studying both inherited human diseases and normal human variations will come from comparisons among the sequences of different individuals. This is already done when searching for disease genes. Its ultimate manifestations may be the kind of large-scale diagnostic DNA sequencing postulated in Chapter 13.

How different are two humans? From recent DNA sequencing projects (hopefully on representative regions, but there is no way we can be sure of this) differences between any two homologous chromosomes occur at a frequency of about 1 in 500. Scaled to the entire human genome, this is 6×10^6 differences between any two chromosomes of a single individual, and potentially four times this number when two individuals are compared, since each homolog in one genome must be compared with each pair of homologs in the other. To simply make a catalog of all of the differences for any potential individual against an arbitrary reference haploid genome would require 12×10^6 entries.

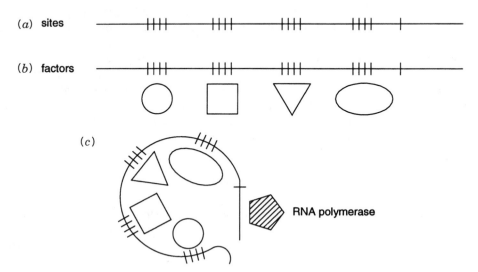

Figure 15.29   Structure of a typical upstream regulatory region for a eukaryotic gene. *(a)* Transcription factor-binding sequences. *(b)* Linear complex with transcription factors. *(c)* Three-dimensional structure of transcription factor-DNA complex that generates a binding site for RNA polymerase.

The current population of the earth is around 5×10^9 individuals. Thus, to record all of current human diversity would require a database with $12 \times 10^6 \times 5 \times 10^9 = 6 \times 10^{16}$ entries. By today's mass storage standards, this is a large database. The entire sequence of one human haploid genome could be stored on a single CD rom. On this scale, storage of the catalog of human diversity would require 2×10^7 CD roms, hardly something one could transport conveniently. However, computer mass storage has been improving at least by 10^4 per decade, and there is no sign that this rate of increase is slackening. As a result, by the time we have the tools to acquire the complete record of human diversity, some 15 to 20 years from now, we should be able to store it on whatever is the equivalent, then, of a few CD roms. Genome analysis is unlikely to be limited now, or in the future, by computer power.

Why would anyone want to view all of human diversity? There are two basic features that make this information compellingly interesting. First is the complexity of a number of interesting phenotypes that are undoubtedly controlled by many genes. Included are such items as personality, affect, and facial features. Given the limited ability for human experimentation, if we want to dissect some of the genetic contributions to these aspects of our human characteristics, we will have to be able to compare DNA sequences with a complex set of observations. The more sequence we have, and the broader the range of individuals represented, the more likely it seems we should be able to make progress on problems that today seem totally intractable.

The second motivation for studying the vast expanse of human diversity is a burgeoning field called molecular anthropology. A considerable amount of human prehistory has left its traces in our DNA. Within contemporary sequences should be a record of who mated with whom as far back as our species goes. Such a record, if we could untangle it, would provide information about our past that probably cannot be reconstructed so easily and accurately in any other way. Included will be information about mass migrations, like the colonization of the new world, population expansions in prehistory, catastrophes, like plagues, or major climatic fluctuations. A tool that is useful for such studies is called principle component analysis. It looks at DNA or protein sequence variations across geographic regions and attempts to factor out particular variants with localized geographic distribution, or with gradients of geographic distribution.

Many of the results obtained thus far in molecular anthropology are controversial and open to alternative explanations. As the body of sequence data that can be fed into such analyses grows, the robustness of the method and acceptance of its results will probably increase. Junk DNA may be especially useful for these kinds of analyses. It is much more variable between individuals than coding sequences, and thus the amount of information it contains about human prehistory should be all that greater. One final, intriguing source of information is available from occasional DNA samples found still intact in mummies, human remains trapped in ice, or other environments that preserve DNA molecules. Such information will provide benchmarks by which to test the extrapolations that otherwise have to be made from contemporary DNA samples.

SOURCES AND ADDITIONAL READINGS

Adams, M. D., Kelley, J. M., Gocayne, J. D., Dubnick, M., Polymeropoulos, M. H., et al. 1991. Complementary DNA sequencing: Expressed sequence tags and human genome project. *Science* 252: 1651–1656.

Berks, M. The C. elegans genome sequencing project. *C. elegans* Genome Mapping and Sequencing Consortium. 1995. *Genome Research* 5: 99–104.

Bult, C. J., White, O., Olsen, G. J., Zhou, L., Fleischmann, R. D., Sutton, G. G., Blake, J. A., FitzGerald, L. M., Clayton, R. A., Gocayne, J. D., et al. 1996. Complete genome sequence of the methanogenic archaeon, *Methanococcus jannaschii. Science* 273: 1058–1073

Burland, V., Plunkett III, G., Sofia, H. J., Daniels, D. L., and Blattner, F. R. 1995. Analysis of the *Escherichia coli* genome VI: DNA sequence of the region from 92.8 through 100 minutes. *Nucleic Acids Research* 23: 2105–2119.

Carlock, L., Wisniewski, D., Lorincz, M., Pandrangi, A., and Vo, T. 1992. An estimate of the number of genes in the Huntington disease gene region and the identification of 13 transcripts in the 4p16.3 segment. *Genomics* 13: 1108–1118.

Chothia, C. 1992. One thousand families for the molecular biologist. *Nature* 357: 543–544.

Daniels, D. L., Plunkett III, G., Burland, V., and Blattner, F. R. 1992. Analysis of the *Escherichia coli* genome: DNA sequence of the region from 84.5 to 86.5 minutes. *Science* 257: 771–778.

Fleischmann, R. D., Adams, M. D., White, O., Clayton, R. A., Kirkness, E. F., Kerlavage, A. R., Bult, C. J., Tomb, J. F., Dougherty, B. A., Merrick, J. M., et al. 1995. Whole-genome random sequencing and assembly of *Haemophilus influenzae. Science* 269: 496–512.

Fraser, C. M., Gocayne, J. D., White, O., Adams, M. D., Clayton, R. A., Fleischmann, R. D., Bult, C. J., Kerlavage, A. R., Sutton, G., Kelley, J. M., et al. 1995. The minimal gene complement of *Mycoplasma genitalium. Science* 270: 397–403.

Gardiner, K. 1996. Base composition and gene distribution: Critical patterns in mammalian genome organization. *Trends in Genetics* 12: 519–524.

Goffeau, A. 1995. Life with 482 genes. *Science* 270: 445–446.

Goffeau, A., Barrell, B. G., Bussey, H., Davis, R. W., Dujon, B., Feldman, H., Gailbert, F., Hoheisel, J. D., Jacq, C., Johnston, M., Louis, E. J., Mewes, H. W., Murakami, Y., Philippsen, P., Tettelin, H., and Oliver, S. G. 1996. Life with 6000 genes. *Science* 274: 546–567.

Gribskow, M., and Devereaux, J. 1991. *Sequence Analysis Primer.* Oxford: Oxford University Press.

Harper, R. 1994. Access to DNA and protein databases on the Internet. *Current Opinion in Biotechnology* 5: 4–18.

Himmelreich, R., Hilbert, H., Plagens, H., Pirkl, E., Li, B.-C., and Herrmann, R. 1996. Complete sequence analysis of the genome of the bacterium *Mycoplasma pneumoniae. Nucleic Acids Research* 24:4420–4449.

How, G.-F., Venkatesh, B., and Brenner, S. 1996. Conserved linkage between the puffer fish *(Fugu rubripes)* and human genes for platelet-derived growth factor receptor and macrophage colony-stimulating factor receptor. *Genome Research* 6: 1185–1191.

Johnston, M., Andrews, S., Brinkman, R., Cooper, J., Ding, H., Dover, J., Du, Z., Favello, A., Fulton, L., Gattung, S., et al. 1994. Complete nucleotide sequence of *Saccharomyces cerevisiae* chromosome VIII. *Science* 265: 2077–2082.

Kaneko, T., et al. 1996. Sequence analysis of the genome of the unicellular cyanobacterium *Synechocystis* sp. strain PCC 6803. II. Sequence determination of the entire genome and assignment of potential protein-coding regions. *DNA Research* 3: 109–136.

Karlin, S., and Mrazek, J. 1996. What drives codon choices in human genes? *Journal of Molecular Biology* 262: 459–472.

Lewis, E. B., Knafels, J. D., Mathog, D. R., and Celniker, S. E. 1995. Sequence analysis of the *cis*-regulatory regions of the bithorax complex of *Drosophila. Proceedings of the National Academy of Sciences USA* 92: 8403–8407.

Lopez, R., Larsen, F., and Prydz, H. 1994. Evaluation of the exon predictions of the GRAIL software. *Genomics* 24: 133–136.

Maier, W. G. 1997. Bonsai genomics: Sequencing the smallest eukaryotic genomes. *Trends in Genetics* 13: 46–49.

Martin, C. H., Mayeda, C. A., Davis, C. A., Ericsson, C. L., Knafels, J. D., Mathog, D. R., Celniker, S. E., Lewis, E. B., and Palazzolo, M. J. 1995. Complete sequence of the bithorax complex of *Drosophila*. *Proceedings of the National Academy of Sciences USA* 92: 8398–8402.

McCombie, W. R., Martin-Gallardo, A., Gocayne, J. D., FitzGerald, M., et al. 1992. Expressed genes, Alu repeats and polymorphisms in cosmids sequenced from chromosome 4p16.3. *Nature Genetics* 1: 348–353.

McFadden, G. I., Gilson, P. R., Douglas, S. E., Cavalier-Smith, T., Hofmann, C. J. B., and Murakami, Y., Naitou, M., Hagiwara, H., Shibata, T., Ozawa, M., Sasanuma, S.-I., Sasanuma, M., Tsuchiya, Y., Soeda, E., Yokoyama, K., Yamazaki, M., Tashiro, H., and Eki, T. 1995. Analysis of the nucleotide sequence of chromosome VI from *Saccharomyces cerevisiae*. *Nature Genetics* 10: 261–264.

Mushegian, A. R., and Koonin, E. V. 1996. A minimal gene set for cellular life derived by comparison of complete bacterial genomes. *Proceedings of the National Academy of Sciences USA* 93: 10268–10273.

Oliver, S. G., van der Aart, Q. J. M., Agostoni-Carbone, M. L., Aigle, M., et al. 1992. The complete DNA sequence of yeast chromosome III. *Nature* 357: 38–46.

Oliver, S. 1995. Size is important, but. . . . *Nature Genetics* 10: 253–256.

Oshima, T., et al. 1996. A 718-kb DNA sequence of the *Escherichia coli* K-12 genome corresponding to the 12.7-28.0 min region on the linkage map. *DNA Research* 3: 137–155.

Pearson, W. 1995. Comparison of methods for searching protein-sequence databases. *Protein Science* 4: 1145–1160.

Perlin, M. W., Lancia, G., and Ng, S.-K. 1995. Toward fully automated geneotyping: Genotyping microsatellite markers by deconvolution. *American Journal of Human Genetics.* 57: 1199–1210.

Peters, R., and Sikorski, R. S. 1996. Nucleic acid databases on the Web. *Nature Biotechnology* 14:1728–1729.

Posfai, J., and Robert, R. J. 1992. Finding errors in DNA sequences. *Proceedings of the National Academy of Sciences USA* 89: 4698–4702.

Rost, B., and Sander, C. 1996. Bridging the protein sequence-structure gap by structure predictions. *Annual Review of Biophysics and Biomolecular Structure* 25: 113–136.

Saccone, S., DeSario, A., Delle Valle, G., and Bernardi, G. 1992. The highest gene concentrations in the human genome are in telomeric bands of metaphase chromosomes. *Proceedings of the National Academy of Sciences USA* 89: 4913–4917.

Sharkey, M., Graba, Y., and Scott, M. P. 1997. Hox genes in evolution: Protein surfaces and paralog groups. *Trends in Genetics* 13: 145–151.

Shpaer, E. G., Robinson, M., Yee, D., Candlin, J. D., Mines, R., and Hunkapiller, T. 1996. Sensitivity and selectivity in protein similarity searchers: A comparison of Smith-Watermanin hardware to BLAST and FASTA. *Genomics* 38: 179–191.

Shevchenko, A., Jensen, O. N., Podtelejnikov, A. V., Sagliocco, F., Wilm, M., Vorm, O., Mortensen, P., Shevchenko, A., Boucherie, H., and Mann, M. 1996. Linking genome and proteome by mass spectrometry: Large-scale identification of yeast proteins from two dimensional gels. *Proceedings of the National Academy of Sciences USA* 93: 14440–14445.

Shoemaker, D. D., Lashkari, D. A., Morris, D., Mittman, M., and Davis, R. W. 1996. Quantitative phenotypic analysis of yeast deletion mutants using a highly parallel molecular bar-coding strategy. *Nature Genetics* 14: 450–456.

Smith, R. 1996. Perspective: Sequence data base searching in the era of large-scale genomic sequencing. *Genome Research* 6: 653–660.

Smith, V., Chou, K. N., Lashkari, D., Botstein, D., and Brown, P. O. 1996. Functional analysis of the genes of yeast chromosome V by genetic footprinting. *Science* 274: 2069–2074.

Sulston, J., Du, Z., Thomas, K., Wilson, R., et al. 1992. The *C. elegans* genome sequencing project: A beginning. *Nature* 356: 37–41.

Strauss, E. J., and Falkow, S. M. 1997. Microbial pathogenesis: Genomics and beyond. *Science* 276: 707–712.

Uberbacher, E. C., and Mural, R. J. 1991. Locating protein-coding regions in human DNA sequences by a multiple sensor-neural network approach. *Proceedings of the National Academy of Sciences USA* 88: 4698–4702.

Waterman, M. S. 1995. *Introduction to Computational Biology: Sequences, Maps and Genomes.* London: Chapman and Hall.

Wilson, R. et al. 1994. 2.2 Mb of contiguous nucleotide sequence from chromosome III of *C. elegans*. *Nature* 363: 32–36.

Wu, R. 1993. Development of enzyme-based methods for DNA sequence analysis and their applications in the genome proejcts. In A. Meister, ed., *Advances in Enzymology*. New York: Wiley.

Zhang, M. Q. 1997. Identification of protein coding regions in the human genome by quadratic discriminant analysis. *Proceedings of the National Academy of Sciences USA* 94: 565–568.

APPENDIX
Databases

GenBank: A public database built and distributed by the National Center for Biotechnology Information.
http://www.ncbi.nlm.nih.gov/
ftp:ncbi.nlm.nih.gov

EMBL Nucleotide Sequence Database: A central activity of the European Bioinformatics Institute; an EMBL outstation.
http://www.ebi.ac.uk
ftp:ftp.ebi.ac.uk

DNA Data Bank of Japan (DDBJ): Activities began in earnest in 1986 in collaboration with EMBL and GenBank.
http://www.ddbj.nig.ac.jp/
ftp:ftp2.ddbj.nig.ac.jp

Genome Sequence DataBase (GSDB): Designed to meet the community-wide challenges of managing, interpreting, and using DNA sequence data at an ever increasing rate.
http://www.ncgr.org/gsdb

Protein Information Resource (PIR) and the PIR-International Protein Sequence Database:
http://www.nbrf.georgetown.edu
ftp:nbrf.georgetown.edu

MIPS: A database for protein sequences, homology data, and yeast genome information:
www.mips.embnet.org

SWISS-PROT protein sequence data bank and its supplement TrEMBL:
www.nlm.nih.gov (or EBI server)

Metabolic pathway collection:
www.cme.msu.edu/WIT/

NRSub database:
ftp://biom3.univlyon1.fr/pub/nrsub
ftp://ftp.nig.ac.jp/pub/db/nrsub

GDB Human Genome Database:
http://gdbwww.gdb.org/
ftp://ftp.gdb.org/

Radiation Hybrid Database:
http://www.ebi.ac.uk/RHdb
ftp://ftp.ebi.ac.uk/pub/databases/RHdb

Mouse Genome Database (MGD): A comprehensive public resource of genetic, phenotypic, and genomic data.
ftp://www.informatics.jax.org/
ftp://ftp.informatics.jax.org/pub/

Molecular Probe Data Base (MPDB):
http://www.biotech.ist.unige.it/interlab/mpdb.html

Signal Recognition Particle Database (SPRDB):
http://pegasus.uthct.edu/SRPDB/SRPDB.html
ftp://diana.uthct.edu

Viroid and viroid-like sequence database: Addition of a hepatitis delta virus RNA section:
http://www.callisto.si.usherb.ca/~jpperra

Mutation spectra database for bacterial and mammalian genes:
http://info.med.yale.edu/mutbase/

PRINTS protein fingerprint database:
http://www.biochem.ucl.ac.uk/bsm/dbbrowser/
ftp://s-ind2.dl.ac.uk/pub/database/prints

PROSITE datases:
http://expasy.hcuge.ch/
ftp.ebi.ac.uk

Blocks Database servers:
http://blocks.fhcrc.org/
ftp://ncbi.nlm.nih.gov/repository/blocks

HSSP database of protein structure-sequence alignments:
http://www.sander.embl-heidelberg.de/
ftp://ftp.embl-ebi.ac.uk/pub/databases/

Dali/FSSP classification of the three-dimensional protein folds:
http://www.embl-heidelberg.de/dali/fssp/

DEF database of protein fold class predictions:
http://zeus.cs.uoi.gr/neural/biocomputing/def.html

SCOP: A structural classification of proteins database:
http://scop.mrc-lmb.cam.ac.uk/scop/

SBASE protein domain library: A collection of annotated protein sequence segments:
http://www.icgeb.triester.it
http://base.icgeb.triester.it/sbase/(using BLAST)

Codon usage tabulated from the international DNA sequence databases:
ftp://ftp.nig.ac.jp/pub/db/codon/current/

TransTerm: translational signal database:
http://biochem.otago.ac.nz:800/Transterm/homepage.html

REBASE: restriction enzymes and methylases:
http://www.neb.com/rebase

Ribonuclease P database:
http://www.mbio.ncsu.edu/RNaseP/home.html

TRANSFAC, TRRD, and COMPEL: Toward a federated database system on transcriptional regulation:
http://transfac.gbf-braunschweig.de
http://www.bionet.nsc.ru/TRRD

MHCPER: A database of MHC-binding peptides:
http://wehih.wehi.edu.au/mhcpep/
ftp.wehi.edu.au/pub/biology/mhcpep/

Histone and histone fold sequences and structures:
http://www.ncbi.nlm.nih.gov/Basevani/HISTONES/

Directory of P450-containing Systems:
http://www.icgeb.trieste.it/p450
http://p450.abc.hu

O-GLYCBASE: A revised database of O-glycosylated proteins:
http://www.cbs.dtu.dk/databases/OGLYCBASE/
ftp.cbs.dtu.dk/pub/Oglyc/Oglyc.base

Mitochondria

MITOMAP: Human mitochondrial genome database:
http://www.gen.emory.edu/mitomap.html

MmtDB: A Metazoa mitochondrial DNA variants database:
http://www.ba.cnr.it/~areamt08/MmtDBWWW.htm

IMGT: International ImMunoGeneTics database:
http://imgt.cnusc.fr:8104

SWISS-2DPAGE database of two-dimensional polyacrylamide gel electrophoresis:
http://expasy.hcuge.ch/

GRBase: A database linking information on proteins involved in gene regulation:
http://www.access.digex.net/~regulate
ftp://ftp.trevigen.com/pub/Tfactors

ENZYME data bank:
http://expasy.hcuge.ch/

E. coli

Compilation of DNA sequences of *E. coli* K12: Description of the interactive databases ECD and ECDC:
http://susi.bio.uni-giessen.de/usr/local/www/html/ecdc.html

EcoCyc: Encyclopedia of *E. coli* Genes and Metabolism:
http://www.ai.sri.com/ecocyc/ecocyc.html

GenProtEC: Genes and proteins of *E. coli* K12:
http://www.mbl.edu/html/ecoli.html

Yeast

Yeast Protein Database(YPD): A database for the complete proteome of *Saccharomyces cerevisiae:*
http://www.proteome.com/YPDhome.html

LISTA, LISTA-HOP, and LISTA-HON: A comprehensive compilation of protein encoding sequences and its associated homology databases from the yeast

Saccharomyces:

http://www.ch.embnet.org/
ftp://bioftp.unibas.ch

Drosophila

FlyBase: A *Drosophila* database:
http://flybase.bio.indiana.edu
ftp:flybase.bio.indiana.edu(in /flybase)

GIF-DB: A WWW database on gene interactions involved in *Drosophila melanogaster* development:
http://www-biol.univ-mrs.fr/~lgpd/GIFTS_home_page.html

RNA

Compilation of 5S rRNA and 5S rRNA gene sequences:
http://rose.man.poznan.pl/5SData/5SRNA.html
http://www.chemie.fu-berlin.de/fb_chemie/ibc/agerdmann/5S_rRNA.html
ftp.fu-berlin.de/science/biochem/db/5SrRNA

Small RNA database:
http://mbcr.bcm.tmc.edu/smallRNA/smallrna.html

uRNA database:
http://pegasus.uthct.edu/uRNADB/uRNADB.html

guide RNA database:
http://www.biochem.mpg.de/~goeringe/

Ribosomal Database Project (RDP):
http://rdpwww.life.uiuc.edu/
ftp://rdp.life.uiuc.edu

Database on the structure of small ribosomal subunit RNA:
http://rrna.uia.ac.be/ssu/

Database on the structure of large ribosomal subunit RNA:
http://rrna.uia.ac.be/lsu/

RNA modification database:
http://www-medlib.med.utah.edu/RNAmods/RNAmods.html
ftp://medlib.med.utah.edu/library/RNAmods

Expansion of the 16S and 23S ribosomal RNA mutation databases (16SMDB and 23SMDB):
http://www.fandm.edu/Departments/Biology/Databases/RNA.html

Compilation of tRNA sequences and sequences of tRNA genes:
ftp.ebi.ac.uk/pub/databases/tRNA

Specific gene Haemophilia:

Factor VIII Mutation Database on the WWW: The Haemophilia A Mutation, Search, Test, and Resource Site:
http://europium.mrc.rpms.ac.uk
ftp.ebi.ac.uk/pub/databases/hamsters

Haemophilia B: Database of point mutations and short additions and deletions:
http://www.umds.ac.uk/molgen/haemBdatabase

Database and software for the analysis of mutations in the human *p53* gene, the human *hpr*t gene, and both the *lacI,* and *lacZ* gene in transgenic rodents: http://sunsite.unc.edu/dnam/mainpage.html

p53 and APC gene mutations: Software and databases

Database of *p53* gene somatic mutations in human tumors and cell lines: http://www.ebi.ac.uk/srs/

Human

PAH mutation analysis consortium databases: http://www.cf.ac.uk/uwcm/mg/hgmd0.html

alpha/beta fold family of proteins database and the cholinesterase gene server ESTHER: http://www.ensam.inra.fr/cholinesterase

Marfan Database: Software and database for the analysis of mutations in the human FBN1 gene

BTKbase, mutation database for X-linked agammaglobulinemia (XLA): http://www.helsinki.fi/science/signal/btkbase.html

Human type I collagen mutation database: http://www.le.ac.uk/depts/ge/collagen/collagen.html

Receptors

Androgen receptor gene mutations database: http://www.mcgill.ca/andogendb/ ftp.ebi.ac.uk/pub/databases/androgen

Nuclear Receptor Resource Project: http://nrr.georgetown.edu/nrr.html

Glucocorticoid receptor resource: http://biochem1.basic-sci.georgetown.edu/grr/grr.html

INDEX

18 EIGHTEENTH EDITION

THE HODGES
ARBRACE
HANDBOOK

CHERYL GLENN
The Pennsylvania State University

LORETTA GRAY
Central Washington University

WADSWORTH
CENGAGE Learning

Australia · Brazil · Japan · Korea · Mexico · Singapore · Spain · United Kingdom · United States

WADSWORTH
CENGAGE Learning·

The Hodges Harbrace Handbook, Eighteenth Edition
Cheryl Glenn, Loretta Gray

Publisher: Lyn Uhl

Acquiring Sponsoring Editor: Kate Derrick

Development Editor: Michell Phifer

Editorial Assistant: Maggie Cross

Media Editor: Cara Douglass-Graff

Executive Marketing Manager: Stacey Purviance

Marketing Coordinator: Brittany Blais

Marketing Communications Manager: Linda Yip

Content Project Manager: Rosemary Winfield

Art Directors: Jill Ort, Marissa Falco

Manufacturing Planner: Betsy Donaghey

Rights Acquisition Specialist: Tom McDonough

Production Service: Lifland et al., Bookmakers

Cover and Text Designer: Anne Bell Carter

Compositor: PreMediaGlobal

For product information and technology assistance, contact us at Cengage Learning Customer & Sales Support, **1-800-354-9706**.

For permission to use material from this text or product, submit all requests online at **www.cengage.com/permissions**.
Further permissions questions can be emailed to **permissionrequest@cengage.com**.

Library of Congress Control Number: 2011942794

ISBN-13: 978-1-111-34670-6
ISBN-10: 1-111-34670-4

Wadsworth
20 Channel Center Street
Boston, MA 02210 USA

Cengage Learning is a leading provider of customized learning solutions with office locations around the globe, including Singapore, the United Kingdom, Australia, Mexico, Brazil, and Japan. Locate your local office at **international.cengage.com/region**.

Cengage Learning products are represented in Canada by Nelson Education, Ltd.

For your course and learning solutions, visit **www.cengage.com**.

Purchase any of our products at your local college store or at our preferred online store **www.cengagebrain.com**.

Instructors: Please visit **login.cengage.com** and log in to access instructor-specific resources.

Printed in the United States of America
1 2 3 4 5 6 7 15 14 13 12 11

Contents

M   PART 2   MECHANICS

P   PART 3   PUNCTUATION

R   PART 7   RESEARCH and DOCUMENTATION

Preface

Welcome to *The Hodges Harbrace Handbook,* Eighteenth Edition. The book in your hands has the longest history of any handbook on the market—it marked its seventieth anniversary in 2011. The original *Harbrace Handbook* included comprehensive, up-to-date, research-based coverage of essential topics for writers. This edition does the same. Reflecting current studies in composition and linguistics, its forty-five chapters help students at all stages of the writing process, whether they are choosing a topic, conducting research, organizing and revising their ideas, or proofreading drafts.

Like the original, this handbook has been class-tested. The students who used the materials for this edition told us that what they appreciate most in a handbook are solid and accessible explanations. Therefore, we provide guidance that shows, rather than just tells, students how to write. We also incorporate visuals, checklists, tip boxes, charts, examples, and annotated student papers that emphasize information discussed in the text. Skimming through the chapters will help you see what we mean.

The Eighteenth Edition both introduces the new and keeps the best of the old. As in previous editions, the first half of the handbook comprises chapters on writing clear and compelling sentences. The second half of the book includes chapters that support general writing assignments and that cover the integration of source material. Because of the nationwide emphasis on cross-disciplinary writing, we also provide chapters devoted to the conventions and documentation guides that will help students write for a range of classes.

How Have We Revised This Edition to Address "Framework for Success in Postsecondary Writing"?

Since the last edition of this handbook was published, the Council of Writing Program Administrators (CWPA), the National Council of Teachers of English (NCTE), and the National Writing Project (NWP) have jointly published "Framework for Success in Postsecondary Writing." This document stresses the need to help students develop their (1) rhetorical knowledge, (2) critical thinking, (3) understanding of the writing process, (4) knowledge of conventions, and (5) ability to craft prose in a range of contexts, including those that call for the use of technology. We have revised this edition of *The Hodges Harbrace Handbook* to ensure that each of these areas is fully covered.

Rhetorical knowledge

■ In **Part 1, "Grammar,"** Chapters 1 through 7 invite students to think rhetorically about grammar.

■ **Chapter 8, "Document Design,"** presents rhetorical principles for interpreting and incorporating visuals in both paper and electronic documents.

■ **Chapter 31, "Reading, Writing, and the Rhetorical Situation,"** introduces students to the elements of the rhetorical situation and to the steps for reading and writing rhetorically. The explanations of the rhetorical situation in this chapter (and throughout the book) have been simplified for easier understanding and application.

■ **Chapter 38, "Integrating Sources and Avoiding Plagiarism,"** reminds students to consider their rhetorical situation as they do research.

■ In **Part 7, "Research and Documentation,"** Chapters 39 through 42 cover the discipline-specific rhetorical knowledge that students need for informed research and successful writing.

Critical thinking

■ **Chapter 34, "Writing Arguments,"** demonstrates ways to analyze a text, reason logically, avoid rhetorical fallacies, incorporate evidence, and compose several types of arguments.

■ **Chapter 38, "Integrating Sources and Avoiding Plagiarism,"** provides students with the strategies and language they need to demonstrate proficiency at the levels of intellectual complexity noted in Bloom's Taxonomy of Learning Domains. Boxes provide students with sentence frames, or templates, that employ phrases commonly used to summarize, synthesize, and respond to sources.

Writing process

■ **Chapter 32, "Planning and Drafting Essays,"** helps students generate ideas for topics and organize the message of their essays.

■ **Chapter 33, "Revising and Editing Essays,"** introduces strategies for recursive writing, editing, and proofreading.

Knowledge of conventions

■ In **Part 2, "Mechanics,"** Chapters 9 through 11 include guidelines for other mechanical concerns: spelling, capitalization, and the use of italics, abbreviations, acronyms, and numbers.

■ In **Part 3, "Punctuation,"** Chapters 12 through 17 explain the conventions of punctuation.

■ In **Part 7, "Research and Documentation,"** Chapters 39 and 40 present the citation and documentation guidelines recommended by the Modern Language Association (MLA) and the American Psychology Association (APA), respectively.

Writing in multiple contexts
- **Chapter 8, "Document Design,"** guides students in the effective use and interpretation of visuals in their own essays and publications (and in works by others).

- **Chapter 35, "Online Writing,"** covers the skills needed to create types of online documents that may be called for in school or work environments.

- In **Part 7, "Research and Documentation,"** Chapters 39 through 42 describe discipline-specific expectations for written work.

What Is New to This Edition?
Based on our observations of the ever-changing student population, we have added new material to address current issues and challenges.

New Part 8, "Advice for Multilingual Writers"
- **Chapter 43, "Determiners, Nouns, and Adjectives,"** helps students decide when (and when not) to use determiners (such as articles) before nouns and adjectives.

- **Chapter 44, "Verbs and Verb Forms,"** provides detailed information on verb tenses, modal auxiliaries, phrasal and prepositional verbs, and participles used as adjectives.

- **Chapter 45, "Word Order,"** focuses on the ordering of adverbs and adjectives as well as on word order in embedded questions and in adjectival clauses.

New, fully annotated student papers
- **Chapter 33, "Revising and Editing Essays,"** includes a process-analysis essay that discusses the steps taken by student-athletes as they select a college. The writer conducted primary research in the form of interviews.

■ **Chapter 41, "Writing about Literature,"** presents a literary interpretation of the short story "The Yellow Wallpaper" from a feminist perspective.

Exercises
■ Over one-third of the handbook's exercises have been replaced to introduce fresh topics or to test newly introduced concepts.

The History

The Harbrace family of handbooks has the longest history of any set of handbooks in the United States. First published in 1941 by University of Tennessee English professor John C. Hodges, *The Harbrace Handbook of English* was a product of Hodges's classroom experience and his federally funded research, which comprised an analysis of twenty thousand student papers. Sixteen English professors from various regions of the United States marked those papers; they found a number of common mistakes, including (1) misplaced commas, (2) misspelling, (3) inexact language, (4) lack of subject-verb agreement, (5) superfluous commas, (6) shifts in tense, (7) misused apostrophes, (8) omission of words, (9) wordiness, and (10) lack of standard usage.

After collecting these data, Hodges worked with a cadre of graduate students to create a taxonomy of writing issues (from punctuation and grammar to style and usage) that would organize the first writing manual for American college students and teachers. This taxonomy still underpins the overall design and organization of nearly every handbook on the market today. Hodges's original handbook has evolved into *The Hodges Harbrace Handbook*, Eighteenth Edition, which continues to respond to the needs of students and writing instructors alike.

How to Use This Handbook

The Hodges Harbrace Handbook routinely receives praise for its comprehensive treatment of key topics for writers. Students have many questions in areas ranging from reading and writing rhetorically to punctuating sentences, and the answers are at their fingertips.

Brief Table of Contents

If you have a topic in mind, such as writing a thesis statement or using commas correctly, check the contents list inside the front cover. With each topic is a number-and-letter combination (2c or 20d, for example) that corresponds to the number and letter at the top of the relevant right-hand page(s) in the book.

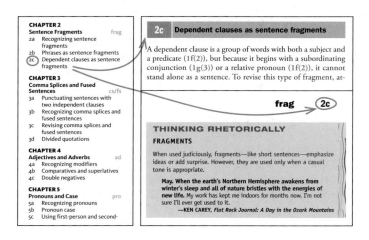

> **2c** **Dependent clauses as sentence fragments**
>
> A dependent clause is a group of words with both a subject and a predicate (1f(2)), but because it begins with a subordinating conjunction (1g(3)) or a relative pronoun (1f(2)), it cannot stand alone as a sentence. To revise this type of fragment, at-
>
> **frag** **2c**
>
> **THINKING RHETORICALLY**
> **FRAGMENTS**
>
> When used judiciously, fragments—like short sentences—emphasize ideas or add surprise. However, they are used only when a casual tone is appropriate.
>
> **May. When the earth's Northern Hemisphere awakens from winter's sleep and all of nature bristles with the energies of new life.** My work has kept me indoors for months now. I'm not sure I'll ever get used to it.
> —KEN CAREY, *Flat Rock Journal: A Day in the Ozark Mountains*

Tabs

Colored tabs, which correspond to the distinctive colors of Parts 1 through 8 and the two documentation chapters, are staggered down the outside edges of the book's pages. These tabs help orient you to the section of the handbook you are in as you look up information.

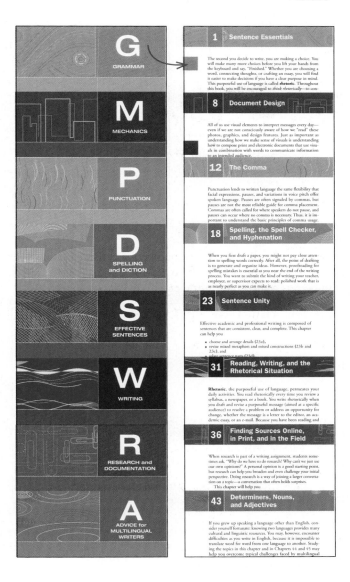

G
GRAMMAR

1   Sentence Essentials

The second you decide to write, you are making a choice. You will make many more choices before you lift your hands from the keyboard and say, "Finished." Whether you are choosing a word, connecting thoughts, or crafting an essay, you will find it easier to make decisions if you have a clear purpose in mind. This purposeful use of language is called **rhetoric**. Throughout this book, you will be encouraged to *think rhetorically*—to con-

M
MECHANICS

8   Document Design

All of us use visual elements to interpret messages every day—even if we are not consciously aware of how we "read" these photos, graphics, and design features. Just as important as understanding how we make sense of visuals is understanding how to compose print and electronic documents that use visuals in combination with words to communicate information to an intended audience.

P
PUNCTUATION

12   The Comma

Punctuation lends to written language the same flexibility that facial expressions, pauses, and variations in voice pitch offer spoken language. Pauses are often signaled by commas, but pauses are not the most reliable guide for comma placement. Commas are often called for where speakers do not pause, and pauses can occur where no comma is necessary. Thus, it is important to understand the basic principles of comma usage.

D
SPELLING and DICTION

18   Spelling, the Spell Checker, and Hyphenation

When you first draft a paper, you might not pay close attention to spelling words correctly. After all, the point of drafting is to generate and organize ideas. However, proofreading for spelling mistakes is essential as you near the end of the writing process. You want to submit the kind of writing your teacher, employer, or supervisor expects to read: polished work that is as nearly perfect as you can make it.

S
EFFECTIVE SENTENCES

23   Sentence Unity

Effective academic and professional writing is composed of sentences that are consistent, clear, and complete. This chapter can help you

- choose and arrange details (23a),
- revise mixed metaphors and mixed constructions (23b and 23c), and
- relate sentence parts (23d)

W
WRITING

31   Reading, Writing, and the Rhetorical Situation

Rhetoric, the purposeful use of language, permeates your daily activities. You read rhetorically every time you review a syllabus, a newspaper, or a book. You write rhetorically when you draft and revise a purposeful message (aimed at a specific audience) to resolve a problem or address an opportunity for change, whether the message is a letter to the editor, an academic essay, or an e-mail. Because you have been reading and

R
RESEARCH and DOCUMENTATION

36   Finding Sources Online, in Print, and in the Field

When research is part of a writing assignment, students sometimes ask, "Why do we have to do research? Why can't we just use our own opinions?" A personal opinion is a good starting point, but research can help you broaden and even challenge your initial perspective. Doing research is a way of joining a larger conversation on a topic—a conversation that often holds surprises. This chapter will help you

A
ADVICE for MULTILINGUAL WRITERS

43   Determiners, Nouns, and Adjectives

If you grew up speaking a language other than English, consider yourself fortunate: knowing two languages provides many cultural and linguistic resources. You may, however, encounter difficulties as you write in English. Because it is impossible to translate word for word from one language to another. Studying the topics in this chapter and in Chapters 44 and 45 may help you overcome typical challenges faced by multilingual

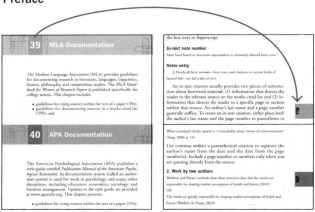

Index

You can also find information quickly by consulting the index at the back of the book, which provides chapter and section numbers as well as page numbers.

MLA and APA Directories

To find the format to use for citing a source or listing a source in a bibliography, refer to one of these style-specific directories. If you use one of these directories often, put a sticky note on its first page so that you can locate it in an instant.

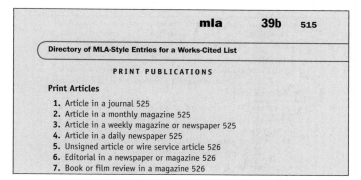

| | **mla** | **39b** | 515 |

Directory of MLA-Style Entries for a Works-Cited List

PRINT PUBLICATIONS

Print Articles

1. Article in a journal 525
2. Article in a monthly magazine 525
3. Article in a weekly magazine or newspaper 525
4. Article in a daily newspaper 525
5. Unsigned article or wire service article 526
6. Editorial in a newspaper or magazine 526
7. Book or film review in a magazine 526

Revision Symbols

Inside the back cover of the book is a list of revision symbols. The symbols can be used to provide feedback on papers, and the list identifies chapters or sections where pertinent rules, guidelines, or strategies are discussed in more detail.

coh		Coherence	" "	39	Quotation marks
	26d	modifiers	red	35a(1)	Redundant
	3c	paragraphs	ref	25d	Reference
:	37b	Colon	rep	31d, 35a	Repetition
,	36a–f	Comma	rev	3a–g	Revision
cs	23b–c	Comma splice	;	37a	Semicolon
con	35	Conciseness	sg		Singular
cst		Consistency	/	40h	Slash
	28a	verb tense	sp	41a–e	Spelling
	28b	point of view	sub	29a, 29c	Subordination
coor	29b–c	Coordination	[ ]	40f	Square brackets
—	40d	Dash	t	24b	Tense
ꞁ		Delete	trans	3d	Transition
dev		Development	∩		Transpose
	2g	essays	u		Unity
	2f	paragraphs		3c	paragraph
. . .	40g	Ellipsis points		27	sentence
emp	31	Emphasis		36g	Unnecessary comma

Glossary of Usage

This glossary includes definitions of words that are commonly confused or misused (such as *accept* and *except*). Organized like a dictionary, it provides not only common meanings for the words but also example sentences demonstrating usage.

---

a lot of *A lot of* is conversational for *many, much,* or *a great deal of:* They do not have ~~a lot of~~ **much** time. *A lot* is sometimes misspelled as *alot.*

a while, awhile *A while* means "a period of time." It is often used with the prepositions *after, for,* and *in:* We rested for **a while**. *Awhile* means "a short time." It is not preceded by a preposition: We rested **awhile**.

accept, except The verb *accept* means "to receive": I **accept** your apology. The verb *except* means "to exclude": The policy was to have everyone wait in line, but parents with mothers and small children were **excepted**. The preposition *except* means "other than": All **except** Joe will attend the conference.

advice, advise *Advice* is a noun: They asked their attorney for **advice**. *Advise* is a verb: The attorney **advised** us to save all relevant documents.

Teaching and Learning Resources
Enhanced InSite™ for *The Hodges Harbrace Handbook*
Printed Access Card (1 semester): 978-1-133-50773-4
Printed Access Card (2 semester): 978-1-133-50772-7
Instant Access Code (2 semester): 978-1-133-43573-0

This easy-to-navigate site allows users (you and your students) to manage the flow of papers online, check for originality, and conduct peer reviews. Students can access the interactive eBook for a text-specific workbook, private tutoring options, and writing resources that include tutorials for avoiding plagiarism and downloadable grammar podcasts. Enhanced InSite provides the tools and resources you and your students need *plus* the training and support you want. Learn more at **www.cengage.com/insite**. *(Access card/code is required.)*

Multimedia eBook
Printed Access Card: 978-0-495-91508-9
Instant Access Code: 978-1-133-50683-6

The Hodges Harbrace Handbook is available as a multimedia eBook. Now students can do all of their reading online or use the eBook as a handy reference while they are completing other coursework. The eBook includes the full text of the print version with interactive exercises and an integrated text-specific workbook; it also provides user-friendly navigation, search, and highlighting tools and links to videos that enhance the handbook content. *(Access card/code is required.)*

Online College Workbook
A great companion to the handbook, the Online College Workbook covers grammar, punctuation, usage, style, and writing. Also included are exercises with clear examples and explanations that supplement the exercises found in the handbook. The workbook is available as part of the multimedia eBook (described above) or as a password-protected, downloadable PDF for instructors at the handbook's companion website. The Answer Key is included in the Online Instructor's Resource Manual.

Online Instructor's Resource Manual
Available for easy download at the handbook's companion website, the password-protected Online Instructor's Resource Manual is designed to give instructors maximum flexibility in planning and customizing a course. The manual includes a variety of pedagogical questions (and possible solutions) relevant to those teaching a course with the handbook, sample syllabi with possible assignments for a semester-long course and for a quarter-long course, sample in-class collaborative learning activities, technology-oriented activities, and critical thinking and writing activities.

English Composition CourseMate

The Hodges Harbrace Handbook is complemented by English Composition CourseMate, which includes the following:

- an interactive eBook
- interactive teaching and learning tools, including quizzes, flashcards, videos, a workbook, and more
- Engagement Tracker, a first-of-its-kind tool that monitors student engagement in the course

Go to **login.cengage.com** and look for the icon ⌁CourseMate, which denotes a resource available within CourseMate.

WebTutor™

Blackboard® Printed Access Card: 978-1-133-43569-3
Instant Access Code: 978-1-133-43567-9

WebCT™ Printed Access Card: 978-1-133-43570-9
Instant Access Code: 978-1-133-43568-6

You can customize your course with text-specific content using this flexible course management system. It makes it easy to blend, add, edit, or delete content, media assets, quizzes, web links, discussion topics, interactive games and exercises, and more. For web-enabling a class or putting an entire course online, WebTutor delivers!

Personal Tutor

Printed Access Card: 978-1-133-43352-1
Instant Access Code: 978-1-133-43572-3

Access to Personal Tutor provides students with private tutoring as they write their papers. With this valuable resource, students gain access to multiple sessions in which they can either request tutoring services or submit a paper for review—whichever they need most.

InfoTrac® College Edition with InfoMarks™
Printed Access Card: 978-0-534-55853-6

This online research and learning center offers over twenty million full-text articles from nearly six thousand scholarly and popular periodicals. The articles cover a broad spectrum of disciplines and topics—providing an ideal resource for every type of research. Learn more at **www.cengage.com/infotrac**. *(Access code/card is required.)*

Turnitin®
Printed Access Card (1 semester): 978-1-4130-3018-1
Printed Access Card (2 semester): 978-1-4130-3019-8

This proven, online plagiarism-prevention tool promotes fairness in the classroom by helping students learn to cite sources correctly and allowing instructors to check for originality before reading and grading papers. *(Access code/card is required.)*

Additional Resources
***Merriam-Webster's Collegiate® Dictionary*, 11th Edition**
Casebound: 978-0-8777-9809-5   *Not available separately.*

***Merriam-Webster's Dictionary*, 2nd Edition**
Paperbound: 978-0-8777-9930-6   *Not available separately.*

Merriam-Webster's Dictionary/Thesaurus
Paperbound: 978-0-8777-9851-4   *Not available separately.*

Acknowledgments

The question was where to begin—so many people deserve thanks for their work on this book. We decided to start with our colleagues who reviewed previous editions and provided us with ideas for improving this edition. Their comments, questions, and occasional jokes made us see our work anew. We hope they will approve of the results of their feedback in these pages.

Elizabeth Aydelott, *Northwest Christian University*; Rachel Byrd, *Southern Adventist University*; James Crooks, *Shasta College*; Betty Davis, *Morgan State University*; Yasmin DeGout, *Howard University*; Cecile Anne de Rocher, *Dalton State College;* Jesse Airaudi, Rachel Webster, and Jared Wheeler, *Baylor University*; Jane Gamber, *Hutchinson Community College*; Bernadette Gambino, *University of North Florida*; Jan Geyer, *Hudson Valley Community College*; BJ Keeton and Eric Stalions, *Martin Methodist College*; Donna Kilgore, *Texas Southern University*; Larry Martin, *Hinds Community College*; James Mayo; *Jackson State Community College*; Patricia McLaughlin, *St. Ambrose University*; Richard Moye, *Lyndon State College*; Sherry Rosenthal, *College of Southern Nevada*; Blaine Wall, *Pensacola Junior College*; Ronald Wheeler, *Johnson Bible College*; and Stephen Whited, *Piedmont College*.

As we drafted (and revised), we depended on the assistance of our students. Penn State University PhD candidate Heather Brook Adams embodies grace, intelligence, and drive, making her the optimal traffic controller for large parts of this project; her tasks ranged from supervising the undergraduate contributions to composing the PowerPoint images. Another PhD candidate, Michael Faris, provided us with a screenshot from his course blog, for which we are grateful. In addition, we also received help from a number of undergraduates: Josephine Lee provided her comments from a course blog. Research interns Jordan Clapper, Kristin Ford, Ian Morgan, Jeremy Popkin, Salvia Sim, and Robert Turchick helped by researching and proofreading portions of the manuscript. Finally, two undergraduate students provided samples of their written work and enriched our understanding of the writing process. To Kristin Ford and

Mary LeNoir, we give our heartfelt thanks. Without such students, where would we instructors be?

The successful completion of this project would have been impossible without the members of the Cengage Learning/Wadsworth staff, whose patience, good humor, and innovative ideas kept us moving in the right direction. Tops on our thankyou list are PJ Boardman, Editor-in-Chief; Lyn Uhl, Senior Publisher; and Kate Derrick, Acquiring Sponsoring Editor. We are truly amazed by how they can make meetings so enjoyable and productive. For the scrupulous work that occurred after the project went into production, we are grateful to Rosemary Winfield, Senior Content Project Manager, and Jane Hoover, production and copy editor; for the striking interior design, we thank Anne Carter; and for helping us bring our work to you, we extend gratitude to Melissa Holt, Marketing Manager, and Stacey Purviance, Executive Marketing Manager.

Michell Phifer, Senior Development Editor and friend, deserves a paragraph of thanks all her own. She has been a member of our team for four editions, always there to arrange our schedules, respond to our drafts with insightful feedback, and give us encouragement as we desperately try to balance our teaching with our research and writing deadlines—and she does it all with generosity, brilliance, and her own touch of elegance. We cannot imagine producing this handbook without her editorial expertise and abiding friendship.

To all our friends and family members, we owe as much gratitude as a book could hold—at least a pageful for each day of our research and writing that they made possible. We know that they often changed their plans or sacrificed their own time so that we could complete this project. For their support, we are forever grateful.

Cheryl Glenn
Loretta Gray
December 2011

G

GRAMMAR

1 Sentence Essentials

The second you decide to write, you are making a choice. You will make many more choices before you lift your hands from the keyboard and say, "Finished." Whether you are choosing a word, connecting thoughts, or crafting an essay, you will find it easier to make decisions if you have a clear purpose in mind. This purposeful use of language is called **rhetoric**. Throughout this book, you will be encouraged to *think rhetorically*—to consider how to achieve your purpose with a given audience and within a specific context. This chapter covers grammar terms and concepts that will help you write clear, convincing sentences.

You will learn to

- identify the parts of speech (**1a**),
- recognize the essential parts of a sentence (**1b**),
- identify complements (**1c**),
- recognize basic sentence patterns (**1d**),
- recognize phrases (**1e**),
- recognize clauses (**1f**),
- connect clauses with conjunctions (**1g**), and
- identify sentence forms and functions (**1h**).

1a Parts of speech

When you look up a word in the dictionary, you will often find it followed by one or more of these labels: *adj., adv., conj., interj., n., prep., pron.,* and *v* (or *vb.*). These are the abbreviations for the traditional eight parts of speech: *adjective, adverb, conjunction, interjection, noun, preposition, pronoun,* and *verb*.

(1) Verbs

Verbs that indicate action (*walk, drive, study*) are called **action verbs**. Verbs that express being or experiencing are called **linking verbs**; they include *be, seem,* and *become* and the sensory verbs *look, taste, smell, feel,* and *sound.* Both action verbs and linking verbs are frequently accompanied by **auxiliary** or **helping verbs** that add shades of meaning, such as information about time (*will* study this afternoon), ability (*can* study), or obligation (*must* study). See Chapter 7 for more details about verbs.

The dictionary (base) form of most action verbs fits into this frame sentence:

> We should _____ (it). [With some verbs, *it* is not used.]

The dictionary (base) form of most linking verbs fits into this frame sentence:

> It can _____ good (terrible, fine).

THINKING RHETORICALLY

VERBS

Decide which of the following sentences evokes a clearer image.

> The team captain **was** absolutely ecstatic.
>
> Grinning broadly, the team captain **shot** both her arms into the air.

You probably chose the sentence with the action verb *shot* rather than the sentence with *was*. Most writers avoid using the verb *be* in any of its forms (*am, is, are, was, were,* or *been*) when their writing assignment calls for vibrant imagery. Instead, they use vivid action verbs.

(2) Nouns

Nouns usually name people, places, things, or ideas. **Proper nouns** are specific names. You can identify them easily because they are capitalized: *Bill Gates, Redmond, Microsoft Corporation*. **Common nouns** refer to any member of a class or category: *person, city, company*. There are three types of common nouns.

- **Count nouns** refer to people, places, things, and ideas that can be counted. They have singular and plural forms: *boy, boys; park, parks; car, cars; concept, concepts*.
- **Noncount nouns** refer to things or ideas that cannot be counted: *furniture, information*.
- **Collective nouns** are nouns that can be either singular or plural, depending on the context: *The **committee** published its report* [singular]. *The **committee** disagree about their duties* [plural]. (See **6a(7)**.)

Most nouns fit into this frame sentence:

(The) _____ is (are) important (unimportant, interesting, uninteresting).

THINKING RHETORICALLY

NOUNS

Nouns like *entertainment* and *nutrition* refer to concepts. They are called **abstract nouns**. In contrast, nouns like *guitar* and *apple* refer to things perceivable by the senses. They are called **concrete nouns**. When you use abstractions, balance them with tangible details conveyed through concrete nouns. For example, if you use the abstract nouns *impressionism* and *cubism* in an art history paper, also include concrete nouns that will enable readers to visualize the colors, shapes, and brushstrokes of the paintings you are discussing.

(3) Pronouns

Pronouns function as nouns, and most pronouns (*it, he, she, they,* and many others) refer to nouns that have already been mentioned. These nouns are called **antecedents** (6b).

My <u>parents</u> bought the cheap, decrepit <u>house</u> because **they** thought **it** had charm.

A pronoun and its antecedent may be found either in the same sentence or in separate, though usually adjacent, sentences.

The <u>students</u> collaborated on a research project last year. **They** even presented their findings at a national conference.

The pronouns in the preceding examples are called **personal pronouns**. For a detailed discussion of other types of pronouns, see Chapter 5.

(4) Adjectives

Adjectives most commonly modify nouns: *spicy* food, *cold* day, *special* price. Sometimes they modify pronouns: *blue* ones, anyone *thin*. Adjectives usually answer one of these questions: Which one? What kind of . . . ? How many? What color (or size or shape, and so on)? Although adjectives usually precede the nouns they modify, they occasionally follow them: *enough* time, time *enough*. Adjectives may also follow linking verbs such as *be, seem,* and *become*:

The <u>moon</u> is **full** tonight.   <u>He</u> seems **shy**.

When an adjective follows a linking verb, it modifies the subject of the sentence (1b).

Most adjectives fit into one of these frame sentences:

He told us about a/an _____ idea (person, place).

The idea (person, place) is very _____.

Articles are a subclass of adjectives because, like adjectives, they are used before nouns. There are three articles: *a, an,* and *the.* The article *a* is used before a consonant sound (**a** yard, **a** university, **a** VIP); *an* is used before a vowel sound (**an** apple, **an** hour, **an** NFL team).

For a detailed discussion of adjectives, see **4a**.

MULTILINGUAL WRITERS

ARTICLE USAGE

English has two types of articles: indefinite and definite. The **indefinite articles** *a* and *an* indicate that a singular noun is used in a general way, as when you introduce the noun for the first time or when you define a word:

Pluto is **a** dwarf <u>planet</u>.

There has been **a** <u>controversy</u> over the classification of Pluto.

A <u>planet</u> is a celestial body orbiting a star such as our sun.

The **definite article,** *the,* is used before a noun that has already been introduced or when a reference is obvious. *The* is also used before a noun that is related in form or meaning to a word mentioned previously.

Scientists distinguish between planets and <u>dwarf planets</u>. Three of **the** <u>dwarf planets</u> in our solar system are Ceres, Pluto, and Eris.

Scientists were not sure how to <u>classify</u> some celestial bodies. **The** <u>classification</u> of Pluto proved to be particularly controversial.

The definite article also appears before a noun considered unique, such as *moon, universe, solar system, sun, earth,* or *sky.*

The <u>moon</u> is full tonight.

For more information on articles, see Chapter **43**.

(5) Adverbs

Adverbs most frequently modify verbs. They provide information about time, manner, place, and frequency, thus answering one of these questions: When? How? Where? How often?

The conference <u>starts</u> **tomorrow.** [time]

I **rapidly** <u>calculated</u> the cost. [manner]

We <u>met</u> **here.** [place]

They **often** <u>work</u> late on Thursdays. [frequency]

Adverbs that modify verbs can often move from one position in a sentence to another.

He **carefully** removed the radio collar.

He removed the radio collar **carefully.**

Most adverbs that modify verbs fit into this frame sentence:

They _____ moved (danced, walked) across the room.

Adverbs also modify adjectives and other adverbs by intensifying or otherwise qualifying the meanings of those words.

I was **extremely** curious. [modifying an adjective]

The team played **surprisingly** well. [modifying an adverb]

For more information on adverbs, see 4a.

THINKING RHETORICALLY

ADVERBS

What do the adverbs add to the following sentences?

The scientist **delicately** places the slide under the microscope.

"You're late," he whispered **vehemently**.

She is **wistfully** hopeful.

Adverbs can help you portray an action, indicate how someone is speaking, and add detail to a description.

---

(6) Prepositions

A **preposition** is a word that combines with a noun and any of its modifiers to establish a relationship between the words or to provide additional detail—often by answering one of these questions: Where? When?

In the early afternoon, we walked **through** our old neighborhood. [answer the questions *When?* and *Where?*]

A preposition may also combine with a pronoun.

We walked **through** it.

SOME COMMON PREPOSITIONS

about	behind	except	of	through
above	beside	for	on	to
after	between	from	out	toward
around	by	in	over	under
as	despite	into	past	until
at	down	like	regarding	up
before	during	near	since	with

Phrasal prepositions consist of more than one word.

Except for the last day, it was a wonderful trip.

The postponement was **due to** inclement weather.

PHRASAL PREPOSITIONS

according to	except for	in spite of
as for	in addition to	instead of
because of	in case of	with regard to
due to	in front of	with respect to

MULTILINGUAL WRITERS

PREPOSITIONS IN IDIOMATIC COMBINATIONS

Some verbs, adjectives, and nouns combine with prepositions to form idiomatic combinations.

Verb + Preposition	Adjective + Preposition	Noun + Preposition
apply to	fond of	interest in
rely on	similar to	dependence on
trust in	different from	fondness for

(7) Conjunctions

Conjunctions are words that connect other words or groups of words. Each conjunction carries a specific meaning. The most common conjunction, *and,* signals addition.

> thunder **and** lightning [connecting words]

> The clap of thunder startled the adults **and** scared the children [connecting groups of words]

Another common conjunction is *but,* which signals contrast.

> startling **but** magnificent [connecting words]

> The lightning struck nearby, **but** no one was hurt [connecting groups of words]

These are only two examples of the many conjunctions used to show how ideas are linked. For a detailed discussion of coordinating conjunctions, correlative conjunctions, subordinating conjunctions, and conjunctive adverbs, see **1g**.

(8) Interjections

Interjections most commonly express a simple exclamation or an emotion such as surprise, dread, or resignation. Interjections that come before a sentence end in a period or an exclamation point.

> **Oh.** Now I understand.

> **Wow!** Your design is astounding.

Interjections that begin or interrupt a sentence are set off by commas.

> **Hey,** what are you doing?

> The solution, **alas,** was not as simple as I had hoped it would be.

EXERCISE 1

Identify the part of speech for each word in the sentences below.

1. Hey, are you a fan of both anime and manga?
2. You should join the Anime and Manga Club.
3. Every Tuesday at noon, we watch current anime.
4. Membership is free, but donations are always welcome.
5. You can easily find us in Smith Student Union.

1b Subjects and predicates

A sentence consists of two parts:

SUBJECT + PREDICATE

The **subject** is generally someone or something that either performs an action or is described. The **predicate** expresses the action initiated by the subject or gives information about the subject.

The <u>landlord</u> + <u>had</u> <u>renovated</u> the apartment.
[The subject performs an action; the predicate expresses the action.]

The <u>rent</u> + <u>seemed</u> reasonable.
[The subject is described; the predicate gives information about the subject.]

The central components of the subject and the predicate are often called the **simple subject** (the main noun or pronoun) and the **simple predicate** (the main verb and any auxiliary verbs). They are underlined in the examples above.

Compound subjects and **compound predicates** include a connecting word (conjunction) such as *and, or,* or *but.*

<u>The Republicans</u> **and** <u>the Democrats</u> are debating this issue. [compound subject]

The candidate <u>stated his views on abortion</u> **but** <u>did not discuss stem-cell research</u>. [compound predicate]

To identify the subject of a sentence, find the verb and then use it in a question beginning with *who* or *what,* as shown in the following examples.

Jennifer works at a clinic.   Meat contains cholesterol.
Verb: **works**   Verb: **contains**
Who works? **Jennifer**   *What* contains? **Meat**
(not the clinic) **works.**   (not cholesterol) **contains.**
Subject: **Jennifer**   Subject: **Meat**

Some sentences begin with an **expletive**—*there* or *it.* Such a word occurs in the subject position, forcing the true subject to follow the verb.

exp   v   s
There were **no exercise machines.**

A subject following the expletive *it* is often an entire clause. You will learn more about clauses in **1f.**

exp v   s
It is essential **that children learn about nutrition at an early age.**

SUBJECT AND PREDICATE

Generally, sentences have the pattern subject + predicate. However, writers often vary this pattern to provide cohesion, emphasis, or both.

He + elbowed his way into the lobby and paused.
[subject + predicate]

From a far corner came + shrieks of laughter.
[predicate + subject]

These two sentences are cohesive because the information in the predicate that begins the second sentence is linked to information in the first sentence. The reversed pattern in the second sentence (predicate + subject) also places emphasis on the subject: *shrieks of laughter.*

1c   Complements

Complements are parts of the predicate required by the verb to make a sentence complete. For example, the sentence *The chair of the committee presented* is incomplete without the complement *his plans.* A complement is generally a pronoun, a noun, or a noun with modifiers.

The chair of the committee introduced —
- **her.** [pronoun]
- **Sylvia Holbrook.** [noun]
- **the new <u>member</u>.** [noun with modifiers]

There are four different types of complements: direct objects, indirect objects, subject complements, and object complements.

(1) Direct object

A **direct object** follows an action verb and either receives the action of the verb or identifies the result of the action.

> Steve McQueen invented **the bucket seat** in 1960.

> I. M. Pei designed **the East Building of the National Gallery.**

Compound direct objects include a connecting word, usually *and.*

> Thomas Edison patented **the phonograph <u>and</u> the microphone.**

To identify a direct object, first find the subject and the verb; then use them in a question ending with *what* or *whom.*

Marie Curie discovered radium.	They hired a new engineer.
Subject and verb: **Marie Curie discovered**	Subject and verb: **They hired**
Marie Curie discovered *what*? **radium**	They hired *whom*? **a new engineer**
Direct object: **radium**	Direct object: **a new engineer**

A direct object may be a clause (**1f**).

> Researchers found **that patients benefited from the new drug.**

(2) Indirect object

Indirect objects typically name the person(s) receiving or benefiting from the action indicated by the verb. Verbs that often take indirect objects include *bring, buy, give, lend, offer, sell, send,* and *write.*

The supervisor gave **the new employees** <u>computers</u>.
[*To whom* were the computers given?]

She wrote **them** <u>recommendation letters</u>.
[*For whom* were the recommendation letters written?]

Like subjects and direct objects, indirect objects can be compound.

She offered **Elena and Octavio** <u>a generous benefits package</u>.

(3) Subject complement

A **subject complement** follows a linking verb (**1a(1)**) and re-names, classifies, or describes the subject. The most common linking verb is *be* (*am, is, are, was, were, been*). Other linking verbs are *become, seem,* and *appear* and the sensory verbs *feel, look, smell, sound,* and *taste.* A subject complement can be a pronoun, a noun, or a noun with modifiers. It can also be an adjective (**1a(4)**).

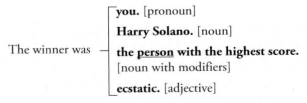

The winner was
- **you.** [pronoun]
- **Harry Solano.** [noun]
- **the <u>person</u> with the highest score.** [noun with modifiers]
- **ecstatic.** [adjective]

(4) Object complement

An **object complement** renames, classifies, or describes a direct object and helps complete the meaning of a verb such as *call, elect, make, name,* or *paint.* The object complement can be either a noun or an adjective, along with any modifiers.

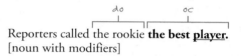

Reporters called the rookie **the best <u>player</u>.**
[noun with modifiers]

His recent performance left the fans **somewhat <u>disappointed</u>.**
[adjective with modifier]

EXERCISE 2

Identify the subject and the predicate in each sentence. Then, looking just at the predicate, identify the underlined complements.

1. A naturalist gave <u>us</u> <u>a short lecture on the Cascade Mountains</u>.
2. He showed <u>slides of mountain lakes and heather meadows</u>.
3. Douglas fir predominates in the Cascade forests.
4. Mountaineers and artists consider <u>the North Cascades the most dramatic mountains in the range</u>.
5. Timberlines are <u>low</u> because of the short growing season.
6. Many volcanoes are in the Cascades.
7. Mt. Rainier is <u>the highest volcano in the range</u>.
8. Many visitors to this area hike <u>the Pacific Crest Trail</u>.
9. My friend lent <u>me</u> <u>his map of the trail</u>.
10. The trail begins in southern California, passes through Oregon and Washington, and ends in British Columbia.

1d Basic sentence patterns

The six basic sentence patterns presented in the following box are based on three verb types: intransitive, transitive, and linking. Notice that *trans* in the words *transitive* and *intransitive* means "over or across." Thus, the action of a **transitive verb** carries across to an object, but the action of an **intransitive verb** does not. An intransitive verb has no complement, although it is often followed by an adverb (pattern 1). A transitive verb is followed by a direct object (pattern 2), by both a direct object and an indirect object (pattern 3), or by a direct object and an object complement (pattern 4). A linking verb (such as *be, seem, sound,* or *taste*) is followed by a subject complement (pattern 5) or by a phrase that includes a preposition (pattern 6).

BASIC SENTENCE PATTERNS

Pattern 1 SUBJECT + INTRANSITIVE VERB

Prices dropped.

Prices dropped precipitously.

Pattern 2 SUBJECT + TRANSITIVE VERB + DIRECT OBJECT

He writes detective stories.

Pattern 3 SUBJECT + TRANSITIVE VERB + INDIRECT OBJECT + DIRECT OBJECT

My father sent me a care package.

(*continued on page 18*)

(continued from page 17)

Pattern 4 SUBJECT + TRANSITIVE VERB + DIRECT OBJECT
+ OBJECT COMPLEMENT

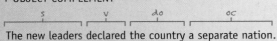

$$\underset{\text{s}}{\text{The new leaders}} \quad \underset{\text{v}}{\text{declared}} \quad \underset{\text{do}}{\text{the country}} \quad \underset{\text{oc}}{\text{a separate nation.}}$$

The new leaders declared the country a separate nation.

Pattern 5 SUBJECT + LINKING VERB + SUBJECT COMPLEMENT

Dr. Vargas is the discussion leader.

Pattern 6 SUBJECT + LINKING VERB + PREPOSITIONAL PHRASE

They are in the library.

When declarative sentences, or statements, are turned into questions, the subject and the auxiliary verb are usually inverted; that is, the auxiliary verb is moved to the beginning of the sentence, before the subject.

Statement: A Chinese skater (has) won a gold medal.

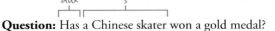

Question: Has a Chinese skater won a gold medal?

Often, a question word such as *what* or *why* opens a question. As long as the question word is *not* the subject of the sentence, the auxiliary verb precedes the subject.

Question: What has a Chinese skater won? [*What* is the object of *has won.*]

COMPARE: Who has won a gold medal? [*Who* is the subject of the sentence.]

If a statement does not include an auxiliary verb or a form of the linking verb *be,* a form of *do* is added to produce the corresponding question. Again, the auxiliary verb is placed in front of the subject.

Statement: A Chinese skater won a gold medal.

Question: Did a Chinese skater win a gold medal?

THINKING RHETORICALLY

SENTENCE PATTERNS

If you want to emphasize a contrast or intensify a feeling, alter the sentence pattern by placing the direct object at the beginning of the sentence.

I acquired English at home. I learned **French** on the street.

I acquired English at home. **French** I learned on the street.

A comma is sometimes used after the direct object in such sentences.

They loved the queen. They despised **the king.**

They loved the queen. **The king,** they despised.

As you study sentences more closely, you will find patterns other than the six presented in this section. For example, another pattern requires mention of a destination or location. The sentence *I put the documents* is incomplete without a phrase such as *on your desk.* Other sentences have phrases that are not essential but do add pertinent information. These phrases can sometimes be moved. For example, the phrase *on Friday* can be placed either at the beginning or at the end of the following sentence.

I finished my assignment **on Friday.**
On Friday, I finished my assignment.

To learn how to write effective sentences by varying their structure, see Chapter **30**.

EXERCISE 3

1. Identify the basic pattern of each sentence in Exercise 2.
2. Write a question corresponding to each of the sentences. Put a check mark next to those questions in which the subject and the verb are inverted.

1e   Phrases

A **phrase** is a sequence of grammatically related words without a subject, a predicate, or both. A phrase is categorized according to its most important word. This section introduces noun phrases, verb phrases, verbal phrases, prepositional phrases, appositives, and absolute phrases.

(1) Noun phrases

A noun phrase consists of a main noun and its modifiers. It can serve as a subject (**1b**) or as a complement (**1c**). It can also be the object of a preposition such as *in, of, on, at,* or *to.* (See **1a(6)** for a longer list of prepositions.)

The heavy frost killed **many fruit trees.** [subject and direct object]

> **My cousin** is **an organic farmer.** [subject and subject complement]

> **His farm** is in **eastern Oregon.** [subject and object of the preposition *in*]

THINKING RHETORICALLY

NOUN PHRASES

In the preceding example sentences, the adjectives *heavy, organic,* and *eastern* add specificity. For example, the noun phrase *an organic farmer* tells the reader more than *farmer* alone would. By composing noun phrases carefully, you will make your sentences more precise.

Much of Greenland lies within the Arctic Circle. ~~The area~~ is [This large island] owned by Denmark. Its [native] name is Kaballit Nunaat.

[*The area* could refer to either Greenland or the area within the Arctic Circle. *This large island* clearly refers to Greenland. *Its native name* is more precise than just *Its name.*]

MULTILINGUAL WRITERS

NUMBER AGREEMENT IN NOUN PHRASES

Some words must agree in number with the nouns they precede. The words *a, an, this,* and *that* are used before singular nouns; *some, few, these, those,* and *many* are used before plural nouns:

> **an/that** opportunity [singular noun]

> **some/few/those** opportunities [plural noun]

The words *less* and *much* precede nouns representing abstract concepts or masses that cannot be counted (noncount nouns; **1a(2)**):

> **less** freedom, **much** water

For more information, see Chapter **43**.

(2) Verb phrases
A verb is essential to the predicate of a sentence (**1b**). It generally expresses action or a state of being. Besides a main verb, a
verb phrase includes one or more **auxiliary verbs**, sometimes
called *helping verbs,* such as *be, have, do, will,* and *should.*

> The passengers **have deplaned.** [auxiliary verb + main verb]
>
> The flight **will be departing** at 7:00 p.m.
> [two auxiliary verbs + main verb]

For a comprehensive discussion of verbs, see Chapter 7.

(3) Verbal phrases (gerund, participial, and infinitive phrases)
A **verbal phrase** differs from a verb phrase (**1e(2)**) in that the
verb form in a verbal phrase serves as a noun or a modifier
rather than as a verb.

> He <u>was</u> **reading** the story aloud. [*Reading* is part of the verb
> phrase *was reading.*]
>
> **Reading** is fundamental to academic success. [*Reading* serves
> as a noun. COMPARE: **It** is fundamental to academic
> success.]
>
> The student **reading** aloud is an education major. [*Reading
> aloud* modifies *the student.*]

Because of their origin as verbs, verbals in phrases often have
their own complements (**1c**) and modifiers (Chapter 4).

> He decided **to read the story aloud.** [The object of the verbal
> *to read* is *the story. Aloud* is a modifier.]

Verbal phrases are divided into three types: gerund phrases,
participial phrases, and infinitive phrases.

Gerund phrases include a verb form ending in *-ing* (see 7a(1)). A gerund phrase serves as a noun, usually functioning as the subject (1b) or the object (1c) in a sentence.

<u>**Writing** a bestseller</u> was her only goal. [subject]

My neighbor enjoys <u>**writing** about distant places.</u> [object]

Because gerund phrases act as nouns, pronouns can replace them.

That was her only goal.

My neighbor enjoys **it.**

THINKING RHETORICALLY
GERUNDS

What is the difference between the following sentences?

They bundle products together, which often results in higher consumer costs.

Bundling products together often results in higher consumer costs.

In the first sentence, the actor, *they,* is the focus. In the second sentence, the action of the gerund phrase, *bundling products together,* is the focus. As you draft or revise, ask yourself whether you want to emphasize actors or actions.

Participial phrases include either a present participle (a verb form ending in *-ing*) or a past participle (a verb form ending in *-ed* for regular verbs or another form for irregular verbs). (See 7a for more information on verb forms.)

<u>**Planning** her questions carefully,</u> she was able to hold fast-paced and engaging interviews. [present participle]

<u>**Known** for her interviewing skills,</u> she was asked to host her own radio program. [past participle]

Participial phrases function as modifiers (**4a(2)**). They may appear at the beginning, in the middle, or at the end of a sentence. In the following sentences, the participial phrases modify *farmers.*

> **Fearing a drought,** the farmers used less irrigation water.

> The farmers, **fearing a drought,** used less irrigation water.

> The farmers conserved water, **fearing a drought.**

Remember that gerund phrases and participial phrases have different functions. A gerund phrase functions as a noun; a participial phrase functions as a modifier.

> **Working together** can spur creativity. [gerund phrase]

> **Working together,** the students designed their own software. [participial phrase]

For advice on using punctuation with participial phrases, see **12d**.

Infinitive phrases serve as nouns (**1a(2)**) or as modifiers (Chapter **4**). The form of an infinitive is distinct—the infinitive marker *to* is followed by the base form of the verb.

> The company intends **to hire twenty new employees.** [noun]

> We discussed his plan **to use a new packing process.** [modifier of the noun *plan*]

> **To attract customers,** the company changed its advertising strategy. [modifier of the verb *changed*]

Some instructors advise against putting words between the infinitive marker *to* and the base form of the verb.

> Under the circumstances, the
> ∧ ~~The~~ jury was unable to, ~~under the circumstances,~~ convict the defendant.

This is good advice to remember if the intervening words create a cumbersome sentence. However, most writers today recognize that a single word splitting an infinitive can provide emphasis.

He did not expect **to** actually **publish** his work.

MULTILINGUAL WRITERS

VERBS FOLLOWED BY GERUNDS AND/OR INFINITIVES

Some verbs in English can be followed by a gerund, some can be followed by an infinitive, and some can be followed by either.

Verbs Followed by a Gerund

admit   avoid   consider   deny   dislike   enjoy   finish   suggest

Example: She **enjoys playing** the piano.

Verbs Followed by an Infinitive

agree   decide   deserve   hope   need   plan   promise   seem

Example: She **promised to play** the piano for us.

Verbs Followed by Either a Gerund or an Infinitive

begin   continue   like   prefer   remember   stop   try

Examples: She **likes to play** the piano. She **likes playing** the piano.

For more information on verbs and verb forms, see Chapters 4 and 44.

(4) Prepositional phrases

Prepositional phrases are modifiers. They provide information about time, place, cause, manner, and so on. They can also answer one of these questions: Which one? What kind of . . . ?

With great feeling, Martin Luther King expressed his dream **of freedom.**
[*With great feeling* describes the way the speech was delivered, and *of freedom* specifies the kind of dream.]

King delivered his most famous speech **at a demonstration in Washington, DC.**
[Both *at a demonstration* and *in Washington, DC* provide information about place.]

A **prepositional phrase** consists of a **preposition** (a word such as *at, of,* or *in*) and a pronoun, noun, or noun phrase (called the **object of the preposition**). A prepositional phrase modifies another element in the sentence.

Everyone **in class** went to the play. [modifier of the pronoun *everyone*]

Some students met the professor **after the play.** [modifier of the verb *met*]

A prepositional phrase sometimes consists of a preposition and an entire clause (**1f**).

They will give the award **to whoever produces the best set design.**

A grammar rule that has been controversial advises against ending a sentence with a preposition. Most professional writers now follow this rule only when they adopt a formal tone. If their assignment calls for an informal tone, they will not hesitate to place a preposition at the end of a sentence.

He found friends **on** whom he could depend. [formal]

He found friends he could depend **on.** [informal]

(5) Appositives
An **appositive** is most often a noun or a noun phrase that refers to the same person, place, thing, or idea as a preceding

noun or noun phrase but in different words. The alternative wording either clarifies the reference or provides extra details. When the appositive simply specifies the referent, no commas are used.

> Cormac McCarthy's novel *The Road* won a Pulitzer Prize.
> [The appositive specifies which of McCarthy's novels won the award.]

When the appositive provides extra details, commas set it off.

> *The Road*, **a novel by Cormac McCarthy,** won a Pulitzer Prize.
> [The appositive provides extra details about the book.]

For more information on punctuating nonessential appositives, see **12d(2)**.

(6) Absolute phrases

An **absolute phrase** is usually a noun phrase modified by a prepositional phrase or a participial phrase (**1e(3)**). It provides descriptive details or expresses a cause or condition.

> <u>**Her guitar** in the front seat,</u> she pulled away from the curb.

> She left town at dawn, <u>**all her belongings** packed into a **Volkswagen Beetle.**</u>

The preceding absolute phrases provide details; the following absolute phrase expresses cause.

> **Her friend's directions lacking clarity,** she frequently checked her map.

Be sure to use commas to set off absolute phrases.

EXERCISE 4

Label the underlined phrases in the following sentences as noun phrases, verb phrases, prepositional phrases, or verbal phrases. For verbal phrases, specify the type: gerund, participial, or infinitive. When a long phrase includes a short phrase, identify just the long phrase. Finally, identify any appositive phrases or absolute phrases in the sentences.

1. <u>After the Second World War,</u> <u>fifty-one countries</u> formed <u>the United Nations,</u> <u>an international organization dedicated to peace, tolerance, and cooperation.</u>
2. <u>The Charter of the United Nations</u> <u>was written</u> <u>in 1945.</u>
3. <u>According to this charter,</u> the United Nations <u>may address</u> a wide range <u>of issues.</u>
4. <u>The United Nations</u> devotes most of its energies to <u>protecting human rights,</u> <u>maintaining peace,</u> and <u>encouraging social development.</u>
5. <u>To reach its goals,</u> the United Nations depends on funding <u>from its member states.</u>
6. <u>Its blue flag easily recognized everywhere,</u> the United Nations now includes <u>192 countries.</u>
7. <u>Symbolizing peace,</u> the emblem <u>on the flag</u> is a map <u>enclosed by olive branches.</u>

1f	**Clauses**

(1) Independent clauses

A **clause** is a group of related words that contains a subject and a predicate. An **independent clause,** sometimes called a *main clause,* has the same grammatical structure as a simple sentence: both contain a subject and a predicate (see **1b**).

The students earned high grades.

An independent clause can stand alone as a complete sentence. Other clauses can be added to an independent clause to form a longer, more detailed sentence.

(2) Dependent clauses

A **dependent clause** also has a subject and a predicate (**1b**). However, it cannot stand alone as a complete sentence because of the word introducing it—usually a relative pronoun or a subordinating conjunction.

The athlete **who placed first** grew up in Argentina. [relative pronoun]

She received the gold medal **because she performed flawlessly.** [subordinating conjunction]

If it is not connected to an independent clause, a dependent clause is considered a sentence fragment (**2c**).

(a) Noun clauses

Dependent clauses that serve as subjects (**1b**) or objects (**1c**) are called **noun clauses** (or **nominal clauses**). They are introduced by *if, that,* or a *wh-* word such as *what* or *why.* Notice the similarity in usage between noun phrases and noun clauses.

Noun phrases	Noun clauses
The testimony may not be true. [subject]	**What the witness said** may not be true. [subject]
We do not understand **their motives.** [direct object]	We do not understand **why they did it.** [direct object]

When no misunderstanding would result, the word *that* can be omitted from the beginning of a noun clause.

> The scientist said **she was moving to Australia.**
> [*That* has been omitted.]

However, *that* should always be retained when there are two noun clauses.

> The scientist said **that she was moving to Australia** and **that her research team was planning to accompany her.**
> [*That* is retained in both noun clauses.]

(b) Adjectival (relative) clauses

An **adjectival clause**, or **relative clause**, follows a pronoun, noun, or noun phrase and answers one of these questions: Which one? What kind of . . . ? Such a clause usually begins with a **relative pronoun** (*who, whom, that, which,* or *whose*). Notice the similarity in usage between adjectives and adjectival clauses.

Adjectives	Adjectival clauses
Effective supervisors give clear directions. [answers the question *Which supervisors?*]	Supervisors **who give clear directions** earn the respect of their employees. [answers the question *Which supervisors?*]
Long, complicated directions confuse employees. [answers the question *What kind of directions?*]	Directions **that are long and complicated** confuse employees. [answers the question *What kind of directions?*]

An **essential (restrictive) adjectival clause** contains information that is necessary to identify the main noun that precedes the clause. Such a clause is *not* set off by commas. The essential adjectival clause in the following sentence is needed for the reader to know which state carries a great deal of influence in a presidential election.

> The state **that casts the most electoral votes** greatly influences the outcome of a presidential election.

A **nonessential (nonrestrictive) adjectival clause** provides extra details that, even though they may be interesting, are not needed for identifying the preceding noun. An adjectival clause following a proper noun (**1a(2)**) is almost always nonessential. A nonessential adjectival clause should be set off by commas.

> California**, which has fifty-five electoral votes,** greatly influences the outcome of any presidential election.

Many writers use *that* to begin essential clauses and *which* to begin nonessential clauses. Follow this convention if you are required to use the style guidelines of the American Psychological Association (APA) or the Modern Language Association (MLA) (although the MLA accepts *which* instead of *that* in essential clauses). For more information on the use of *which* and *that*, see **5a(3)**.

A relative pronoun can be omitted from an adjectival clause as long as the meaning of the sentence is still clear.

> Mother Teresa was someone **the whole world admired.**
> [*Whom,* the direct object of the clause, has been omitted: the whole world admired *whom.*]

> She was someone **who cared more about serving than being served.**
> [*Who* cannot be omitted because it is the subject of the clause.]

The relative pronoun is not omitted when the adjectival clause is set off by commas (that is, when it is a nonessential clause).

Mother Teresa, **whom the whole world admired,** cared more about serving than being served.

(c) Adverbial clauses

An **adverbial clause** usually answers one of the following questions: Where? When? How? Why? How often? In what manner? Adverbial clauses are introduced by subordinating conjunctions such as *because, although,* and *when.* (For a list of subordinating conjunctions, see **1g(3)**.) Notice the similarity in usage between adverbs and adverbial clauses.

Adverbs	**Adverbial clauses**
Occasionally, the company hires new writers. [answers the question *How often does the company hire new writers?*]	**When the need arises,** the company hires new writers. [answers the question *How often does the company hire new writers?*]
She acted **selfishly.** [answers the question *How did she act?*]	She acted **as though she cared only about herself.** [answers the question *How did she act?*]

Adverbial clauses can appear at various points in a sentence. Use a comma or commas to set off an adverbial clause placed at the beginning or in the middle of a sentence.

Because they disagreed, the researchers made little progress.

The researchers, **because they disagreed,** made little progress.

An adverbial clause at the end of a sentence is rarely preceded by a comma.

The researchers made little progress **because they disagreed.**

However, if a final adverbial clause in a sentence contains an extra detail—information you want the reader to pause before—use a comma to set it off.

I slept soundly that night, **even though a storm raged outside.**

THINKING RHETORICALLY

ADVERBIAL CLAUSES

In an adverbial clause that refers to time or establishes a fact, both the subject and any form of the verb *be* can be omitted. Using such **elliptical clauses** will make your writing more concise.

While fishing, he saw a rare owl.
[COMPARE: **While he was fishing,** he saw a rare owl.]

Though tired, they continued to study for the exam.
[COMPARE: **Though they were tired,** they continued to study for the exam.]

Be sure that the omitted subject of an elliptical clause is the same as the subject of the independent clause. Otherwise, revise either the adverbial clause or the main clause.

While ‸ reviewing your report, a few questions occurred to me.
I was

OR

While reviewing your report, ‸ a few questions ~~occurred to~~ me.
I thought of

For more information on the use of elliptical constructions, see **21b**.

EXERCISE 5

Identify the dependent clauses in the following paragraph.

¹If you live by the sword, you might die by the sword.
²However, if you make your living by swallowing swords, you
will not necessarily die by swallowing swords. ³At least, this is
the conclusion Brian Witcombe and Dan Meyer reached after
they surveyed forty-six professional sword swallowers. ⁴(Brian
Witcombe is a radiologist, and Dan Meyer is a famous sword
swallower.) ⁵Some of those surveyed mentioned that they had
experienced either "sword throats" or chest pains, and others who
let their swords drop to their stomachs described perforation of
their innards, but the researchers could find no listing of a sword-
swallowing mortality in the medical studies they reviewed. ⁶The
researchers did not inquire into the reasons for swallowing swords
in the first place.

1g   Conjunctions and conjunctive adverbs

Conjunctions are connectors; they fall into four categories:
coordinating, correlative, subordinating, and adverbial.

(1) Coordinating conjunctions

A **coordinating conjunction** connects similar words or groups
of words; that is, it generally links a word to a word, a phrase
to a phrase (**1e**), or a clause to a clause (**1f**).

English **and** Spanish [*And* joins two words and signals addition.]

in school **or** at home [*Or* joins two phrases and marks them as
alternatives.]

We did not share a language, **but** somehow we communicated.
[*But* joins two independent clauses and signals contrast.]

There are seven coordinating conjunctions. Use the made-up word *fanboys* to help you remember them.

F	A	N	B	O	Y	S
for	and	nor	but	or	yet	so

A coordinating conjunction such as *but* may also link independent clauses (**1f(1)**) that stand alone as sentences.

> The momentum in the direction of globalization seems too powerful to buck, the economic logic unmatchable. **But** in a region where jobs are draining away, and where an ethic of self-reliance remains a dim, vestigial, but honored memory, it seems at least an outside possibility.
>
> —BILL McKIBBEN, "Small World"

(2) Correlative conjunctions

A **correlative conjunction** (or **correlative**) consists of two parts. The most common correlatives are *both . . . and, either . . . or, neither . . . nor,* and *not only . . . but also.*

> **either** Pedro **or** Sue [*Either . . . or* joins two words and marks them as alternatives.]

> **neither** on the running track **nor** in the pool [*Neither . . . nor* joins two phrases and marks them both as false or impossible.]

> **Not only** did they run ten miles, **but** they **also** swam twenty laps. [*Not only . . . but also* joins two independent clauses and signals addition.]

As the preceding examples show, correlative conjunctions join words, phrases, or clauses, but they do not join sentences. Generally, a correlative conjunction links similar structures.

The following sentence needed to be revised because the correlative conjunction was linking a phrase to a clause:

Not only ~~saving~~ _{did he save} the lives of the accident victims, **but** he **also** prevented many spinal injuries.

(3) Subordinating conjunctions

A **subordinating conjunction** introduces a dependent clause (**1f(2)**). It also carries a specific meaning; for example, it may indicate cause, concession, condition, purpose, or time. A dependent clause that begins a sentence is followed by a comma.

Unless the project receives more funding, the research will stop.
[*Unless* signals a condition.]

The project continued **because** it received additional funding.
[*Because* signals a cause.]

SUBORDINATING CONJUNCTIONS

after	how	than
although	if	though
as if	in case	unless
as though	in that	until
because	insofar as	when, whenever
before	once	where, wherever
even if	since	whether
even though	so that	while

The word *that* can be omitted from the subordinating conjunction *so that* if the meaning remains clear.

I left ten minutes early **so** I would not be late. [*That* has been omitted.]

However, when *that* is omitted, the remaining *so* can be easily confused with the coordinating conjunction *so*.

I had some extra time, **so** I went to the music store.

Because sentences with subordinating conjunctions are punctuated differently from sentences with coordinating conjunctions, be careful to distinguish between them. If *so* stands for "so that," it is a subordinating conjunction. If *so* means "thus," it is a coordinating conjunction.

(4) Conjunctive adverbs

Conjunctive adverbs—such as *however, nevertheless, then,* and *therefore*—link independent clauses (**1f(1)**). These conjunctions, also called **adverbial conjunctions**, signal relationships such as cause, condition, and contrast. Conjunctive adverbs are set off by commas. An independent clause preceding a conjunctive adverb may end in a semicolon instead of a period.

The senator supported the plan**; however,** the voters did not.
. **However,** the voters did not.
. The voters, **however,** did not.
. The voters did not, **however.**

CONJUNCTIVE ADVERBS

also	however	moreover	still
consequently	instead	nevertheless	then
finally	likewise	nonetheless	therefore
furthermore	meanwhile	otherwise	thus

1h   Sentence forms

You can identify the form of a sentence by noting the number of clauses it contains and the type of each clause. There are four sentence forms: simple, compound, complex, and compound-complex.

(1) Simple sentences

> ONE INDEPENDENT CLAUSE

A **simple sentence** is equivalent to one independent clause; thus, it must have a subject and a predicate.

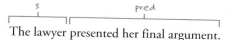

The lawyer presented her final argument.

However, you can expand a simple sentence by adding one or more verbal phrases (1e(3)) or prepositional phrases (1e(4)).

Encouraged by the apparent sympathy of the jury, the lawyer presented her final argument. [The verbal phrase adds detail.]

The lawyer presented her final argument **in less than an hour.** [The prepositional phrase adds information about time.]

(2) Compound sentences

> INDEPENDENT CLAUSE + INDEPENDENT CLAUSE

A compound sentence consists of at least two independent clauses but no dependent clauses. The independent clauses of a compound sentence are most commonly linked by a

coordinating conjunction. However, punctuation may sometimes serve the same purpose (**14a**).

> The Democrats proposed a new budget, **but** the Republicans
> opposed it.
> [The coordinating conjunction *but* links two independent
> clauses and signals contrast.]

> The Democrats proposed a new budget; the Republicans
> opposed it.
> [The semicolon serves the same purpose as the coordinating
> conjunction.]

(3) Complex sentences

> INDEPENDENT CLAUSE + DEPENDENT CLAUSE

A complex sentence consists of one independent clause and at
least one dependent clause. A dependent clause in a complex
sentence can be a noun clause, an adjectival clause, or an adverbial clause (**1f(2)**).

> **Because he was known for architectural ornamentation,** no
> one predicted **that the house <u>he designed for himself</u> would
> be so plain.** [This sentence has three dependent clauses. *Because
> he was known for architectural ornamentation* is an adverbial
> clause. *That the house he designed for himself would be so plain* is
> a noun clause, and *he designed for himself* is an adjectival clause
> within the noun clause. The relative pronoun *that* has been
> omitted from the beginning of the embedded adjectival clause.]

(4) Compound-complex sentences

> INDEPENDENT CLAUSE + INDEPENDENT CLAUSE
> + DEPENDENT CLAUSE

The combination of a compound sentence and a complex sentence is called a **compound-complex sentence**. A compound-complex sentence consists of at least two independent clauses and at least one dependent clause.

Conflict is essential to good storytelling, **so** fiction writers often create a character **who faces a major challenge.** [The coordinating conjunction *so* joins the two independent clauses; the relative pronoun *who* introduces the dependent clause.]

EXERCISE 6

Identify each sentence in the paragraph in Exercise 5 as simple, compound, complex, or compound-complex.

EXERCISE 7

Vary the sentence forms in the following paragraph. Add details as needed.

Most people write on a computer. Many still keep a pencil nearby. They most likely use it to jot notes. They rarely think about its role in history. This common writing instrument was invented during the sixteenth century. It was used by George Washington while surveying the Ohio Territory. It was used by Meriwether Lewis and William Clark during their expedition to the Northwest. And Ulysses S. Grant used a pencil to make battle plans. The authors Henry David Thoreau, John Steinbeck, and Ernest Hemingway were other well-known pencil users. The graphite pencil began as an alternative to a stylus made of lead. In fact, we still speak of a pencil lead because the stylus contained this compound. However, the marks a pencil makes are nothing more than flecks of graphite.

1i   Sentence functions

Sentences serve a number of functions. Writers commonly state facts or report information with **declarative sentences**. They give instructions with **imperative sentences (commands)**. They use questions, or **interrogative sentences**, to elicit information or to introduce topics. And they express emotion with **exclamatory sentences (exclamations)**.

Declarative	The runners from Kenya won the race.
Imperative	Compare their times with the record.
Interrogative	What were their times?
Exclamatory	The runners from Kenya won the race! Check their times! What an incredible race that was!

Courtesy of United Airlines

Expect great things.

Now more daily nonstops to China from the U.S. than any other airline.

■ UNITED
It's time to fly.

Advertisers often use imperatives to attract the reader's attention.

Although most of the sentences you are likely to write will be declarative, an occasional command, question, or exclamation will add variety (**30c**).

Taking note of end punctuation can help you identify the function of a sentence. Generally, a period indicates the end of a declarative sentence or an imperative sentence, and a question mark ends an interrogative sentence. An exclamation point indicates that a sentence is exclamatory. To distinguish between an imperative sentence and a declarative sentence, look for a subject (**1b**). If you cannot find one, the sentence is imperative. Because an imperative is directed to another person or persons, the subject *you* is implied:

Look over there.
[COMPARE: You look over there.]

THINKING RHETORICALLY

QUESTIONS

One type of interrogative sentence, the **rhetorical question**, is not a true question, because an answer is not expected. Instead, like a declarative sentence, it is used to state an opinion. However, a positive rhetorical question can correspond to a negative assertion, and vice versa.

Rhetorical questions	Equivalent statements
Should we allow our rights to be taken away?	We should not allow our rights to be taken away.
Isn't it time to make a difference?	It's time to make a difference.

Because they are more emphatic than declarative sentences, rhetorical questions focus the reader's attention on major points.

EXERCISE 8

Identify each sentence type in the passage below. What type is used most often? Why?

¹Think of the thousand cartoons you have seen—in the *New Yorker* or a multitude of other places—of the marooned human or pair of humans (in whatever combination of sexes) on some microscopic tropical atoll, a little sand, one palm tree, one rock, the vastness of the sea. ²Humor and pathos live together in these scenes. ³Here at last, we think, life is cut down to the bone so that we can see what stuff it is made of. ⁴If two men, they are The Odd Couple; if man and woman, they will find the roots of the old sex wars and quarrel, as they might on a street in New York; if only one [person], there is a message in a bottle, generally with cheerless news. ⁵Do we love this cartoon scene because we imagine we can discover the bedrock of human nature inside it?

—BILL HOLM, *Eccentric Islands: Travels Real and Imaginary*

TECH SAVVY

Using a Grammar Checker

Most word-processing programs have a tool that helps writers identify grammar errors as well as problems with usage and style, but any grammar checker has significant limitations. A grammar checker will usually identify

- fused sentences, sometimes called run-on sentences (Chapter 3),
- wordy or overly long sentences (30a and 21a), and
- missing apostrophes in contractions (15b).

However, a grammar checker can easily miss

- sentence fragments (Chapter 2),
- misplaced or dangling modifiers (24d and 24e),
- problems with pronoun-antecedent agreement (5c),
- errors in subject-verb agreement (7e), and
- misused or missing commas (Chapter 12).

Because these errors can weaken your credibility as a writer, you should never rely solely on a grammar checker to find them. Furthermore, grammar checkers can mark as wrong words or phrases you have chosen deliberately (Chapter 31).

Used carefully, a grammar checker can be a helpful tool, but keep the following advice in mind:

- Use a grammar checker only in addition to your own editing and proofreading.
- Always evaluate any sentences flagged by a grammar checker to determine whether there is, in fact, a problem.
- Adjust the settings on your grammar checker to look for specific types of errors.
- Carefully review the revisions proposed by a grammar checker before accepting them. Proposed revisions may create new errors.

2 Sentence Fragments

As its name suggests, a **sentence fragment** is only a piece of a sentence; it is not complete. This chapter can help you

- recognize sentence fragments (**2a**) and
- revise fragments resulting from incorrectly punctuated phrases and dependent clauses (**2b** and **2c**).

2a Recognizing sentence fragments

A sentence is considered a fragment when it is incomplete in any of the following ways:

- It is missing a subject or a verb.

  Derived from a word meaning "nervous sleep." Hypnotism actually refers to a type of focused attention. [no subject]

  Alternative medical treatment may include hypnosis. **The placement of a patient into a relaxed state.** [no verb]

- It is missing both a subject and a verb.

  The hypnotic state differs from sleep. **Contrary to popular belief.**

- It is a dependent clause.

  Most people can be hypnotized easily. **Although the depth of the trance for each person varies.**

Note that imperative sentences (**1i**) are not considered fragments. In these sentences, the subject, *you,* is not stated explicitly. Rather, it is implied.

Find out as much as you can about alternative treatments. [COMPARE: You find out as much as you can about alternative treatments.]

FOUR METHODS FOR IDENTIFYING FRAGMENTS

If you have trouble recognizing fragments in your own writing, try one or more of these methods:

1. Read each paragraph backwards, sentence by sentence. When you read your sentences out of order, you may more readily note the incompleteness of a fragment.

2. Locate the essential parts of each sentence. First, find the main verb and any accompanying auxiliary verbs. Remember that gerunds and participles cannot function as main verbs (**1e(3)**). After you find the main verb, identify the subject (**1b**). Finally, make sure that the sentence does not begin with a relative pronoun (**1f(2)**) or a subordinating conjunction (**1a(7)**).

   Test sentence 1: The inventor of the Frisbee.

   Test: Main verb? *None.*
   [Because there is no verb, this test sentence is a fragment.]

   Test sentence 2: Walter Frederick Morrison invented the Frisbee.

   Test: Main verb? *Invented.*
   Subject? *Walter Frederick Morrison.*
   Relative pronoun or subordinating conjunction? *None.*
   [The test sentence is complete: it contains a subject and a verb and does not begin with a relative pronoun or a subordinating conjunction.]

3. Put any sentence you think might be a fragment into this frame sentence:

   They do not understand the idea that _____.

   Only a full sentence will make sense in this frame sentence. If a test sentence, other than an imperative, does not fit into the frame sentence, it is a fragment.

Test sentence 3: Because it can be played almost anywhere.

Test: They do not understand the idea that *because it can be played almost anywhere.*
[The frame sentence does not make sense, so the test sentence is a fragment.]

Test sentence 4: Ultimate Frisbee is a popular sport because it can be played almost anywhere.

Test: They do not understand the idea that *Ultimate Frisbee is a popular sport because it can be played almost anywhere.*
[The frame sentence makes sense, so the test sentence is complete.]

4. Rewrite any sentence you think might be a fragment as a question that can be answered with *yes* or *no*. Only complete sentences can be rewritten this way.

Test sentence 5: That combines aspects of soccer, football, and basketball.

Test: *Is that combines aspects of soccer, football, and basketball?*
[The question does not make sense, so the test sentence is a fragment.]

Test sentence 6: Ultimate Frisbee is a game that combines aspects of soccer, football, and basketball.

Test: *Is Ultimate Frisbee a game that combines aspects of soccer, football, and basketball?*
[The question makes sense, so the test sentence is complete.]

2b Phrases as sentence fragments

A phrase is a group of words without a subject and/or a predicate (1e). When punctuated as a sentence (that is, with a period, a question mark, or an exclamation point at the end), a phrase becomes a fragment. You can revise such a fragment by attaching it to a related sentence, usually the one preceding it.

Verbal phrase as a fragment

Early humans valued color. ~~Creating~~ , creating ∧ **permanent colors with natural pigments.**

Prepositional phrase as a fragment

For years, the Scottish have dyed sweaters with soot. ~~Originally~~ , originally ∧ **from the chimneys of peat-burning stoves.**

Compound predicate as a fragment

Arctic foxes turn white when it snows. ~~And~~ and ∧ **thus conceal themselves from prey.**

Appositive phrase as a fragment

During the Renaissance, one of the most highly valued pigments was ultramarine. ~~An~~ —an ∧ **extract from lapis lazuli.**

Appositive list as a fragment

To derive dyes, we have always experimented with what we find in nature. ~~Shells,~~ : shells, ∧ **roots, insects, flowers.**

Absolute phrase as a fragment

The deciduous trees of New England are known for their brilliant autumn color. ~~Sugar~~ , sugar ∧ **maples dazzling tourists with their orange and red leaves.**

Instead of attaching a fragment to the preceding sentence, you can recast the fragment as a complete sentence. This method of revision elevates the importance of the information conveyed in the fragment.

Fragment	Humans painted themselves for a variety of purposes. **To attract a mate, to hide themselves from game or predators, or to signal aggression.**
Revision	Humans used color for a variety of purposes. For example, they painted themselves to attract a mate, to hide themselves from game or predators, or to signal aggression.

EXERCISE 1

Revise each fragment by attaching it to a related sentence or by recasting it as a complete sentence.

1. A brilliant twenty-three-year-old Englishman. Isaac Newton was the first person to study color.

2. By passing a beam of sunlight through a prism. Newton showed that white light comprised all the visible colors of the spectrum.

3. White light passed through the prism. And separated into the colors of the rainbow.

4. Rainbows are arcs of color. Caused by water droplets in the air.

5. Sometimes rainbows contain all the spectrum colors. Red, orange, yellow, green, blue, indigo, and violet.

6. Particles of spray in waterfalls can act as prisms. Producing a variety of colors.

7. Our brains easily fooled. We sometimes see more colors than are actually present.

| **2c** | **Dependent clauses as sentence fragments** |

A dependent clause is a group of words with both a subject and a predicate (**1f(2)**), but because it begins with a subordinating conjunction (**1g(3)**) or a relative pronoun (**1f(2)**), it cannot stand alone as a sentence. To revise this type of fragment, attach it to a related sentence, usually the sentence preceding it.

The iceberg was no surprise. ~~**Because**~~ ^because^ the *Titanic*'s wireless **operators had received reports of ice in the area.**

More than two thousand people were aboard the *Titanic*. ~~**Which**~~ ^, which^ **was the largest ocean liner in 1912.**

You can also recast the fragment as a complete sentence by removing the subordinating conjunction or relative pronoun and supplying any missing elements.

The iceberg was no surprise. The *Titanic*'s wireless operators had received reports of ice in the area.

More than two thousand people were aboard the *Titanic*. In 1912, this ocean liner was the world's largest.

You can also reduce a clause that is a fragment to a phrase (**1f**) and then attach it to a related sentence.

More than two thousand people were aboard the *Titanic*, the largest ocean liner in 1912. [fragment reduced to an appositive phrase]

If you are unsure of the punctuation to use with phrases or dependent clauses, see Chapter **12**.

THINKING RHETORICALLY

FRAGMENTS

When used judiciously, fragments—like short sentences—emphasize ideas or add surprise. However, they are used only when a casual tone is appropriate.

> **May. When the earth's Northern Hemisphere awakens from winter's sleep and all of nature bristles with the energies of new life.** My work has kept me indoors for months now. I'm not sure I'll ever get used to it.
> —KEN CAREY, *Flat Rock Journal: A Day in the Ozark Mountains*

EXERCISE 2

Follow the guidelines in this chapter to locate and revise the fragments in the following paragraph. If you find it necessary, make other improvements as well. Be prepared to explain your revisions.

¹One of the most popular rides at any county fair or amusement park is the Ferris wheel. ²The original Ferris wheel, designed by George Washington Gale Ferris, Jr., for a national exposition in 1893. ³Rose to a height of 264 feet. ⁴And accommodated 2,140 passengers. ⁵Ferris's goal was to build something that would surpass in effect the Eiffel Tower. ⁶Which was constructed just a few years earlier. ⁷Though Ferris's plans were not immediately accepted. ⁸Once they were, and the wheel opened to the public, it became an immediate success. ⁹At times carrying 38,000 passengers a day. ¹⁰Since the nineteenth century. ¹¹Engineers have designed taller and taller Ferris wheels. ¹²The 541-foot Singapore Flyer holds the record, but the Beijing Great Wheel, currently under construction. ¹³Will be over 100 feet taller.

3 Comma Splices and Fused Sentences

Comma splices and fused sentences are sentence-level mistakes resulting from incorrect or missing punctuation. Both are punctuated as one sentence when they should be punctuated as two sentences (or two independent clauses). By revising comma splices and fused sentences, you indicate sentence boundaries and thus make your writing easier to read. This chapter will help you

- review the rules for punctuating independent clauses (**3a**),
- recognize comma splices and fused sentences (**3b**), and
- learn ways to revise them (**3c** and **3d**).

3a Punctuating sentences with two independent clauses

In case you are unfamiliar with or unsure about the conventions for punctuating sentences with two independent clauses, here is a short review.

A comma and a coordinating conjunction can join two independent clauses (**12a**). The coordinating conjunction indicates the relationship between the two clauses. For example, *and* signals addition, whereas *but* and *yet* signal contrast. The comma precedes the conjunction.

INDEPENDENT CLAUSE, **and** INDEPENDENT CLAUSE.

The new store opened this morning, **and** the owners greeted everyone at the door.

A semicolon can join two independent clauses that are closely related. A semicolon generally signals addition or contrast.

INDEPENDENT CLAUSE**;** INDEPENDENT CLAUSE.

One of the owners comes from this area**;** the other grew up in Cuba.

A semicolon may also precede an independent clause that begins with a conjunctive adverb (adverbial conjunction) such as *however* or *nevertheless*. Notice that a comma follows this type of connecting word.

The store will be open late on Fridays and Saturdays**;** **however,** it will be closed all day on Sundays.

A colon can join two independent clauses. The second clause usually explains or elaborates on the first.

INDEPENDENT CLAUSE**:** INDEPENDENT CLAUSE.

The owners have announced a special offer**:** anyone who makes a purchase during the opening will receive a 10 percent discount.

If you are following MLA guidelines, capitalize the first word of a clause following a colon when the clause expresses a rule or principle (**17d(1)**).

A period separates clauses into distinct sentences.

INDEPENDENT CLAUSE**.** INDEPENDENT CLAUSE.

You can also find comma splices and fused sentences by remembering that they commonly occur in certain contexts.

- With transitional words and phrases such as *however, therefore,* and *for example* (see also **3c(5)**)

  Comma splice: The director is unable to meet with you this

  week, however , next week she will have time on Tuesday.

  (semicolon ; marked above the comma after "week"; carets below "week," and "however")

  [Notice that a semicolon replaces the comma.]

- When an explanation or an example is given in the second sentence

  Fused sentence: The cultural center has a new collection of

  spear points . Many ~~many~~ of them were donated by a retired

  anthropologist.

- When a clause that includes *not* is followed by one without this word, or vice versa

  Comma splice: A World Cup victory is not just an everyday

  sporting event~~, it~~ . It is a national celebration.

- When the subject of the second clause is a pronoun whose antecedent is in the preceding clause

  Fused sentence: Lake Baikal is located in southern Russia~~it~~ . It is 394 miles long.

3c Revising comma splices and fused sentences

If you find comma splices or fused sentences in your writing, try one of the following methods to revise them.

(1) Linking independent clauses with a comma and a coordinating conjunction

By linking clauses with a comma and a coordinating conjunction (such as *and* or *but*), you signal the relationship between the clauses (addition or contrast, for example).

Fused sentence: The diplomats will end their discussions on

Friday∧they will submit their final decisions on Monday.
      , and

Comma splice: Some diplomats applauded the treaty,∧others opposed it vehemently.
                                                    but

(2) Linking independent clauses with a semicolon or a colon or separating them with a period

When you link independent clauses with a semicolon, you signal their connection indirectly. There are no explicit conjunctions to use as cues. The semicolon usually indicates addition or contrast. When you link clauses with a colon, the second clause serves as an explanation or an elaboration of the first. A period indicates that each clause is a complete and separate sentence.

Comma splice: Our division's reports are posted on our web page, hard copies are available by request.

Revision 1: Our division's reports are posted on our web page; hard copies are available by request.

Revision 2: Our division's reports are posted on our web page. Hard copies are available by request.

Fused sentence: Our mission statement is simple∧we aim to provide athletic gear at affordable prices.
                                              :

(3) Recasting an independent clause as a dependent clause or as a phrase

A dependent clause (**1f(2)**) includes a subordinating conjunction such as *although* or *because,* which indicates how the dependent and independent clauses are related (in a cause-and-consequence relationship, for example). A prepositional phrase (**1e(4)**) includes a preposition such as *in, on,* or *because of* that may also signal a relationship directly. Verbal, appositive, and absolute phrases (**1e(3)**, **1e(5)**, and **1e(6)**) suggest relationships less directly because they do not include connecting words.

> **Comma splice:** The wind had blown down trees and power lines, the whole city was without electricity for several hours.
>
> **Revision 1: Because the wind had blown down power lines,** the whole city was without electricity for several hours. [dependent clause]
>
> **Revision 2: Because of the downed power lines,** the whole city was without electricity for several hours. [prepositional phrase]
>
> **Revision 3: The wind having blown down power lines,** the whole city was without electricity for several hours. [absolute phrase]

(4) Integrating one clause into the other

When you integrate clauses, you will generally retain the important details but omit or change some words.

> **Fused sentence:** The proposal covers all but one point it does not describe how the project will be assessed.
>
> **Revision:** The proposal covers all the points except assessment procedures.

(5) Using transitional words or phrases to link independent clauses

Another way to revise fused sentences and comma splices is to use transitional words and phrases such as *however, on the contrary,* and *in the meantime.* (For other examples, see the list on page 365.)

Fused sentence: Sexual harassment is not just an issue for

women *. After all,* men can be sexually harassed too.

Comma splice: The word *status* refers to relative position within

a group *; however,* it is often used to indicate only positions of prestige.

If you have questions about punctuating sentences that contain transitional words and phrases, see **14a**.

As you edit fused sentences and comma splices, you will refine the connections between your sentences and thereby help your readers follow your train of thought. The following checklist will help you find and fix comma splices and fused sentences.

CHECKLIST for Comma Splices and Fused Sentences

1 Common Sites for Comma Splices or Fused Sentences

- With transitional words such as *however* and *therefore*
- When an explanation or an example occurs in the second clause
- When a clause that includes *not* is followed by one without this word, or vice versa
- When the subject of the second clause is a pronoun whose antecedent is in the first clause

2 Ways to Fix Comma Splices and Fused Sentences

- Link the clauses with a comma and a coordinating conjunction.
- Link the clauses, using a semicolon or a colon.
- Separate the clauses by punctuating each as a sentence.
- Make one clause dependent.
- Reduce one clause to a phrase.
- Rewrite the sentence, integrating one clause into the other.
- Use a transitional word or phrase.

| **3d** | **Divided quotations** |

When dividing quotations with signal phrases such as *he said* or *she asked,* use a period between independent clauses.

Comma splice: "Beauty brings copies of itself into being," states Elaine Scarry, "it makes us draw it, take photographs of it, or

describe it to other people."

[Both parts of the quotation are complete sentences, so the signal phrase is attached to the first, and the sentence is punctuated with a period.]

A comma separates two parts of a single quoted sentence.

"Musing takes place in a kind of meadowlands of the imagination," writes Rebecca Solnit, "a part of the imagination that has not yet been plowed, developed, or put to any immediately practical use."

[Because the quotation is a single sentence, a comma is used.]

EXERCISE 1

Revise each comma splice or fused sentence in the following paragraph. Some sentences may not need revision.

[1]In *The Politics of Happiness,* Derek Bok, former president of Harvard University, discusses recent findings that researchers studying well-being have reported. [2]He mentions, for example, research showing that measurements of happiness in the United States have not risen much in the last fifty years, people are responding to survey questions about their levels of happiness in much the same way as they did in 1960. [3]Even though average incomes have grown, levels of happiness have not. [4]Bok believes that people become accustomed to higher standards of living they do not realize how quickly they adapt and so do not become happier. [5]Bok recognizes that not everyone's income has increased but notes that, strangely enough, the disparity between rich and poor has not caused increased dissatisfaction among the poor, he cites further studies showing that citizens in countries with costly welfare programs are not necessarily happier than citizens in countries with welfare programs that are not as generous. [6]Because of these studies, Bok suggests that our government not focus on economic growth alone as an indicator of well-being and instead take into account current research on what makes people happy. [7]This discussion "is bound to contribute to the evolution of society and the refinement of its values," he explains, "that alone will be an accomplishment of enduring importance to humankind" (212).

4 · Adjectives and Adverbs

Adjectives and **adverbs** are **modifiers,** words that qualify or limit the meaning of other words. Phrases (**1e**) and clauses (**1f**) can also be modifiers. For example, if you were to describe a sandwich as "humdrum," as "lacking sufficient mustard," or as something "that might have tasted good two days ago," you would be using a word, a phrase, or a clause to modify *sandwich.* When used effectively, modifiers enliven writing with details and enhance its coherence. This chapter will help you

- recognize modifiers (**4a**),
- use conventional comparative and superlative forms (**4b**), and
- revise double negatives (**4c**).

4a   Recognizing modifiers

The most common modifiers are adjectives and adverbs. You can distinguish an adjective from an adverb by determining what type of word is modified. **Adjectives** modify nouns and pronouns (**1a(4)**); **adverbs** modify verbs, adjectives, and other adverbs (**1a(5)**).

Adjective	**Adverb**
She looked **curious.**	She looked at me **curiously.**
[modifies pronoun]	[modifies verb]

Adjectives	**Adverbs**
productive meeting [modifies noun]	**highly** productive meeting [modifies adjective]
a **quick** lunch [modifies noun]	**very** quickly [modifies adverb]

You can also identify a modifier by considering its form. Many adjectives end with one of these suffixes: *-able, -al, -ful, -ic, -ish, -less,* or *-y.*

accept**able**   ren**tal**   event**ful**   ange**lic**
sheep**ish**   effort**less**   sleep**y**

Present and past participles (7a(5)) can be used as adjectives.

a **determining** factor   a **determined** effort
[present participle]   [past participle]

Be sure to include the complete *-ed* ending of a past participle.

Please see the ˄enclose documents for more details.
                enclosed

THINKING RHETORICALLY

ADJECTIVES

When your writing assignment calls for vivid images or emotional intensity, choose appropriate adjectives to convey these qualities. That is, instead of describing a movie you did not like with the overused adjective *boring*, you could say that it was *tedious* or *mind-numbing*. When you sense that you might be using a lackluster adjective, search for an alternative in a thesaurus. If any of the words listed there are unfamiliar, be sure to look them up in a dictionary so that you use them correctly.

Movie posters often include descriptive adjectives.

The easiest type of adverb to identify is the adverb of manner (**1a(5)**). It is formed by adding *-ly* to an adjective.

careful**ly**   unpleasant**ly**   silent**ly**

However, not all words ending in *-ly* are adverbs. Certain adjectives related to nouns also end in *-ly* (*friend, friendly; hour, hourly*). In addition, not all adverbs end in *-ly*. Adverbs that indicate time or place (*today, tomorrow, here, there*) do not have the *-ly* ending; neither does the negator *not*. A few words—for example, *fast* and *well*—can function as either adjectives or adverbs.

They like **fast** cars. [adjective]

They ran **fast** enough to catch the bus. [adverb]

(1) Modifiers of linking verbs and action verbs

An adjective used after a sensory linking verb (*look, smell, taste, sound,* or *feel*) modifies the subject of the sentence (**1b**). A common error is to use an adverb after this type of linking verb.

I felt _∧~~badly~~ about missing the rally. [The adjective *bad*
modifies *I*.]

(*bad* written above *badly*)

However, when *look, smell, taste, sound,* or *feel* is used as an action verb (**1a(1)**), it can be modified by an adverb.

She looked **angrily** at the referee. [The adverb *angrily*
modifies *looked*.]

BUT She looked **angry**. [The adjective *angry* modifies *she*.]

The words *good* and *well* are easy to confuse. In academic writing, *good* is considered an adjective and so is not used with action verbs.

The whole team played _∧~~good~~.

(*well* written above *good*)

Another frequent error is the dropping of *-ly* endings from adverbs. Although you may not hear the ending when you speak, be sure to include it when you write.

They bought only _∧~~local~~ grown vegetables.

(*locally* written above *local*)

EXERCISE 1

Revise the following sentences so that all adjectives and adverbs are used in ways considered conventional in academic writing.

1. Relaxation techniques have been developed for people who feel uncomfortably in some way.

2. Meditation is one technique that is real helpful in relieving stress.

3. People searching for relief from tension have found that a breathing meditation works good.

4. They sit quiet and concentrate on both inhaling and exhaling.

5. They concentrate on breathing deep.

(2) Phrases and clauses as modifiers

Participial phrases, prepositional phrases, and some infinitive phrases are modifiers (**1e(3)** and **1e(4)**).

> **Growing in popularity every year,** mountain bikes now dominate the market. [participial phrase modifying the noun *bikes*]

> Mountain bikes first became popular **in the 1980s.** [prepositional phrase modifying the verb *became*]

> Some people use mountain bikes **to commute to work.** [infinitive phrase modifying the verb *use*]

Adjectival (relative) clauses and adverbial clauses are both modifiers (see **1f(2)**).

> BMX bicycles have frames **that are relatively small.** [adjectival clause modifying the noun *frames*]

> **Although mountain bikes are designed for off-road use,** many people use them on city streets. [adverbial clause modifying the verb *use*]

4b   Comparatives and superlatives

Many adjectives and adverbs change form to show degrees of quality, quantity, time, distance, manner, and so on. The **positive form** of an adjective or adverb is the word you would look for in a dictionary: *hard, urgent, deserving.* The **comparative form**, which either ends in *-er* or is preceded by *more* or *less*, compares two elements: *I worked **harder** than I ever had before.* The **superlative form**, which either ends in *-est* or is preceded by *most* or *least*, compares three or more elements: *Jeff is the **hardest** worker I have ever met.*

Positive	Comparative	Superlative
hard	harder	hardest
urgent	more/less urgent	most/least urgent
deserving	more/less deserving	most/least deserving

(1) Complete and logical comparisons

When you use the comparative form of an adjective or an adverb, be sure to indicate what two elements you are comparing. The revision of the following sentence makes it clear that a diesel engine and a gas engine are being compared:

A diesel engine is **heavier** ^*than a gas engine*.

Occasionally, the second element in a comparison is implied. The word *paper* does not have to be included after *second* in the sentence below. The reader can infer that the grade on the second paper was better than the grade on the first paper.

She wrote **two** papers; the instructor gave her a **better** grade on the second.

A comparison should also be logical. The following example illogically compares *population* and *Wabasha*:

The **population** of Winona is larger than **Wabasha.**

Here are two common ways to revise this type of faulty comparison:

- Repeat the word that refers to what is being compared.

  The **population** of Winona is larger than the **population** of Wabasha.

- Use a pronoun that corresponds to the first element in the comparison.

The **population** of Winona is larger than **that** of Wabasha.

(2) Double comparatives or superlatives

Use either an ending (*-er* or *-est*) or a preceding qualifier (*more* or *most*), not both, to form a comparative or superlative.

The first bridge is **more narrower** than the second.

The **most narrowest** bridge is in the northern part of the state.

Some modifiers have *absolute meanings*. These modifiers name qualities that are either present in full or not at all. Expressing degrees of such modifiers is illogical, so their comparative and superlative forms are rarely used in academic writing.

a **more** perfect society   the **most** unique campus

EXERCISE 2

Provide the correct comparative or superlative form of each modifier within parentheses.

1. Amphibians can be divided into three groups. Frogs and toads are in the (common) group.

2. Because they do not have to maintain a specific body temperature, amphibians eat (frequently) than mammals do.

3. Reptiles may look like amphibians, but their skin is (dry).

4. During the Devonian period, the (close) ancestors of amphibians were fish with fins that looked like legs.

5. In general, amphibians have (few) bones in their skeletons than other animals with backbones have.

6. Color markings on amphibians vary, though the back of an amphibian is usually (dark) than its belly.

4c   Double negatives

The term **double negative** refers to the use of two negative words to express a single negation. Unless you are portraying dialogue, revise any double negatives you find in your writing.

He did**n't** keep_∧ ^{any} ~~no~~ records.

OR

He_∧ ^{kept} ~~did**n't** keep~~ **no** records.

Using *not* or *nothing* with *hardly, barely,* or *scarcely* creates a double negative. The following examples show how sentences containing such double negatives can be revised:

I could~~**n't**~~ **hardly** quit in the middle of the job.

OR

I could**n't** ~~**hardly**~~ quit in the middle of the job.

MULTILINGUAL WRITERS

NEGATION IN OTHER LANGUAGES

The use of two negative words in one sentence is common in languages such as Spanish:

> *Yo **no** compré **nada**.* ["I didn't buy anything."]

If your primary language allows this type of negation, be especially careful to check for and revise any double negatives you find in your English essays.

5 Pronouns and Case

When you use pronouns effectively, you add clarity and coherence to your writing. However, if you do not provide the words, phrases, or clauses that make your pronoun references clear, you might unintentionally cause confusion. This chapter will help you

- recognize various types of pronouns (**5a**) and
- use them appropriately (**5b** and **5c**).

5a   Recognizing pronouns

A **pronoun** is commonly defined as a word used in place of a noun that has already been mentioned—its **antecedent**.

John said **he** would guide the trip.

A pronoun may also substitute for a group of words acting as a noun (see **1f(2)**).

The participant with the most experience said **he** would guide the trip.

Most pronouns refer to nouns, but some modify nouns.

This man is our guide.

Pronouns are categorized as personal, reflexive/intensive, relative, interrogative, demonstrative, and indefinite.

(1) Personal pronouns

To understand the uses of personal pronouns, you must first be able to recognize person, number, and case. **Person** indicates whether a pronoun refers to the writer (**first person**), to the reader (**second person**), or to another person, place, thing, or idea (**third person**). **Number** reveals whether a pronoun is singular or plural. **Case** refers to the form a pronoun takes to indicate its function in a sentence. Pronouns can be subjects, objects, or possessives. When they function as subjects (**1b(1)**), they are in the subjective case; when they function as objects (**1b(2)**), they are in the objective case; and when they indicate possession or a related meaning (**15a**), they are in the possessive case. (See **5b** for more information on case.) Possessives can be divided into two groups based on whether they are followed by nouns: *my, your, his, her, its, our,* and *their* are all followed by nouns; *mine, yours, his, hers, ours,* and *theirs* are not. (Notice that *his* is in both groups.)

Their budget is higher than **ours.**
[*Their* is followed by a noun; *ours* is not.]

CASE:	Subjective		Objective		Possessive	
NUMBER:	Singular	Plural	Singular	Plural	Singular	Plural
First person	I	we	me	us	my mine	our ours
Second person	you	you	you	you	your yours	your yours
Third person	he, she, it	they	him, her, it	them	his, her, hers, its	their theirs

(2) Reflexive/intensive pronouns

Reflexive pronouns direct the action back to the subject (*I saw myself*); **intensive pronouns** are used for emphasis (*I myself questioned the judge*). *Myself, yourself, himself, herself, itself, ourselves, yourselves,* and *themselves* are used as either reflexive pronouns or intensive pronouns. Both types of pronouns are objects and must be accompanied by subjects.

Reflexive pronoun	**He** was always talking to **himself.**
Intensive pronoun	**She herself** delivered the letter.

Avoid using a reflexive pronoun as a subject. A common error is using *myself* in a compound subject.

Ms. Palmquist and ^I~~myself~~ discussed our concern with the senator.

Hisself, themself, and *theirselves* are inappropriate in college or professional writing. Instead, use *himself* and *themselves.*

(3) Relative pronouns

An adjectival clause (or relative clause) ordinarily begins with a relative pronoun: *who, whom, which, that,* or *whose.* To provide a link between this type of dependent clause and the main clause, the relative pronoun corresponds to its **antecedent**—a word or phrase in the main clause.

The students talked to **a reporter who** had just returned from overseas.

Notice that if you rewrite the dependent clause as a separate independent clause, you use the antecedent in place of the relative pronoun.

A reporter had just returned from overseas.

Who, whose, and *whom* ordinarily refer to people; *which* refers to things; *that* refers to things and, in some contexts, to people. The possessive *whose* (used in place of the awkward *of which*) usually refers to people but sometimes refers to things.

The poem, **whose** author is unknown, has recently been set to music.

	Refers to people	Refers to things	Refers to either
Subjective	who	which	that
Objective	whom	which	that
Possessive			whose

Knowing the difference between an essential clause and a nonessential clause will help you decide whether to use *which* or *that* (see **1f(2)**). A clause that a reader needs in order to identify the antecedent correctly is an **essential clause**.

```
      ant              ess cl
┌──────────┐┌──────────────────────────┐
```
The person who presented the award was last year's winner.

If the essential clause were omitted from this sentence, the reader would not know which person was last year's winner.

A **nonessential clause** is *not* needed for correct identification of the antecedent and is thus set off by commas. A nonessential clause often follows a proper noun (a specific name).

```
      ant              noness cl
┌──────────┐┌──────────────────────────┐
```
Andrea Bowen, who presented the award, was last year's winner.

Notice that if the nonessential clause were removed from this sentence, the reader would still know the identity of last year's winner.

According to a traditional grammar rule, *that* is used in essential adjectival clauses, and *which* is used in nonessential adjectival clauses.

I need a job **that** pays well.

For years, I have had the same job, **which** pays well enough.

However, some professional writers do not follow both parts of this rule. Although they will not use *that* in nonessential clauses, they will use *which* in essential clauses. See **1f(2)** for more information about the use of *which* and *that*.

(4) Interrogative pronouns

The **interrogative pronouns** *what, which, who, whom,* and *whose* are question words. Be careful not to confuse *who* and *whom* (see **5b(5)**). *Who* functions as a subject; *whom* functions as an object.

Who won the award? [COMPARE: **He** won the award.]

Whom did you see at the ceremony? [COMPARE: I saw **him**.]

(5) Demonstrative pronouns

The **demonstrative pronouns**, *this* and *these*, indicate that someone or something is close by in time, space, or thought. *That* and *those* signal remoteness.

These are important documents; **those** can be thrown away.

Demonstrative pronouns sometimes modify nouns.

These documents should be filed.

THINKING RHETORICALLY

PRONOUNS

Why is the following passage somewhat unclear?

> The study found that students succeed when they have clear directions, consistent and focused feedback, and access to help. This led administrators to create a tutoring center at our university.

The problem is that the pronoun *this* at the beginning of the second sentence could refer to all of the information provided by the study or just to the single finding that students need access to help. If you discover that one of your pronouns lacks a clear antecedent, replace the pronoun with more specific words.

The results of this study led administrators to create a tutoring center at our university.

(6) Indefinite pronouns

Indefinite pronouns usually do not refer to specific persons, objects, ideas, or events.

anyone	anybody	anything
everyone	everybody	everything
someone	somebody	something
no one	nobody	nothing
each	either	neither

Indefinite pronouns do not refer to an antecedent. In fact, some indefinite pronouns *serve* as antecedents.

Someone forgot **her** purse.

5b	Pronoun case

The term *case* refers to the form a pronoun takes to indicate its function in a sentence. There are three cases: subjective, objective, and possessive. The following sentence includes all three.

He [subjective] wants **his** [possessive] legislators to help **him** [objective].

(1) Pronouns in the subjective case

A pronoun that is the subject of a sentence is in the subjective case. To determine which pronoun form is correct in a compound subject (a noun and a pronoun joined by *and*), say the sentence using the pronoun alone, omitting the noun. For the following sentence, notice that "*Me* solved the problem" is not Standard English, but "*I* solved the problem" is.

~~Me and~~ Marisa ∧solved the problem.
　　　　　　and I

Place the pronoun last in the sequence. If the compound subject contains two pronouns, test each one by itself.

　　　He
∧~~Him~~ and I confirmed the results.

Pronouns following a *be* verb (*am, is, are, was, were, been*) should also be in the subjective case.

The first presenters were Kevin and ∧~~me~~.
　　　　　　　　　　　　　　　　　I

(2) Pronouns in the objective case

Whenever a pronoun follows an action verb or a preposition, it takes the **objective case**.

Direct object　　　　　　The whole staff admired **him.**

Object of a preposition　The staff depended on **him.**

Pronouns in compound objects are also in the objective case.

> They will appoint Tom or ~~I~~. [direct object]
>
> [above "I": me]

> The manager sat between Tom and ~~I~~ at the meeting. [object of the preposition]
>
> [above "I": me]

To determine whether to use the subjective or objective case, remember to say the sentence with just the pronoun. Notice that "They will appoint *I*" does not sound right. Another test is to substitute *we* and *us*. If *we* sounds fine, use the subjective case. If *us* sounds better, use the objective case, as in "The manager sat between *us*."

(3) Possessive forms

Its, their, and *whose* are possessive forms. Be sure not to confuse them with common contractions: *it's* (*it is* or *it has*), *they're* (*they are*), and *who's* (*who is* or *who has*).

(4) Appositive pronouns

Appositive pronouns are in the same case as the nouns they rename. In the following sentence, *the red team* is the subject, so the appositive pronoun should be in the subjective case.

> The red team—Rebecca, Leroy, and ~~me~~—won by only one point.
>
> [above "me": I]

In the next sentence, *the red team* is the object of the preposition *to,* so the appositive pronoun should be in the objective case.

> A trophy was presented to the red team—Rebecca, Leroy, and ~~I~~.
>
> [above "I": me]

(5) *Who/whoever* and *whom/whomever*

To choose between *who* and *whom* or between *whoever* and *whomever,* you must first determine whether the word is functioning as a subject (**1b**) or an object (**1c**). A pronoun functioning as the subject takes the subjective case.

> **Who** won the award? [COMPARE: **She** won the award.]
>
> The teachers know **who** won the award.
>
> **Whoever** won the award deserves it.

When the pronoun is an object, use *whom* or *whomever.*

> **Whom** did they hire? [COMPARE: They hired **him.**]
>
> The student **whom** they hired graduated in May.
>
> **Whomever** they hired will have to work hard this year.

Whom may be omitted in sentences when no misunderstanding would result.

> The friend he relied on moved away.
>
> [*Whom* has been omitted after *friend.*]

(6) Pronouns with infinitives and gerunds

A pronoun grouped with an infinitive (*to* + the base form of a verb) takes the objective case, whether it comes before or after the infinitive.

> The director wanted **me** to help **him.**

A gerund (*-ing* verb form functioning as a noun) is preceded by a possessive pronoun.

> I appreciated **his** helping Denise. [COMPARE: I appreciated **Tom's** helping Denise.]

Notice that a possessive pronoun is used before a gerund but not before a present participle (*-ing* verb form functioning as an adjective).

> I saw **him** helping Luke.

(7) Pronouns in elliptical constructions

The words *as* and *than* frequently introduce **elliptical constructions**—clauses in which the writer has intentionally omitted words. To check whether you have used the correct case in an elliptical construction, read the written sentence aloud, inserting any words that have been omitted from it.

She admires Clarice as much as **I.** [subjective case]

Read aloud: She admires Clarice as much as *I do.*

She admires Clarice more than **me.** [objective case]

Read aloud: She admires Clarice more than *she admires me.*

EXERCISE 1

Revise the following paragraph, using appropriate pronouns. Some sentences may not require editing.

¹When me and my brother were in middle school, we formed a band with our friends Jason and Andrew. ²My grandmother had given Jake a guitar and I a drum kit for Christmas. ³We practiced either alone or together for the rest of the winter. ⁴Then, in the spring, we met up with Jason, who we had known for years. ⁵Him and his cousin Andrew, whom we later called Android, were excited to join me and Jake. ⁶Jason already had a guitar, and Andrew could sing. ⁷After we played together one afternoon, we decided to call ourself The Crash. ⁸Jason and Andrew came over to our house to jam whenever they're parents let them—which was most of the time. ⁹Our parents did not mind our noise at all. ¹⁰My dad said us playing reminded him of his own teenage garage band.

5c   Use of first-person and second-person pronouns

Using *I* is appropriate when you are writing about personal experience. In academic and professional writing, the use of the first-person singular pronoun is also a way to distinguish your own views from those of others. However, if you frequently repeat *I feel* or *I think,* your readers may suspect that you do not understand much beyond your own experience.

We, the first-person plural pronoun, is trickier to use correctly. When you use it, make sure that your audience can tell which individuals are included in this plural reference. For example, if you are writing a paper for a college course, does *we* mean you and the instructor, you and your fellow students, or some other group (such as all Americans)? Because you may inadvertently use *we* in an early draft to refer to more than one group of people, as you edit, check to see that you have used this first-person plural pronoun consistently.

If you decide to address readers directly, you will undoubtedly use the second-person pronoun *you* (as we, the authors of this handbook, have done). There is some disagreement, though, over whether to permit the use of the indefinite *you* to mean "a person" or "people in general." Check with your instructor about this usage. If you are told to avoid using the indefinite *you*, recast your sentences. For example, use *one* instead of *you.*

Even in huge, anonymous cities, ~~you find~~ ^{one finds} community spirit.

If the use of *one* is too formal, try changing the word order and/or using different words.

Community spirit arises even in huge, anonymous cities.

EXERCISE 2

Revise the following paragraph to eliminate the use of the first- and second-person pronouns.

¹In my opinion, some animals should be as free as we are. ²For example, I think orangutans, African elephants, and Atlantic bottlenose dolphins should roam freely rather than be held in captivity. ³We should neither exhibit them in zoos nor use them for medical research. ⁴If you study animals such as these you will see that, like us, they show emotions, self-awareness, and intention. ⁵You might even find that some use language to communicate. ⁶It is clear to me that they have the right to freedom.

6   Agreement

Hearing the word *agree,* you might think of accord between two or more people. Perhaps they like the same kind of movies or support the same political candidate. Grammatical **agreement** is also about sameness—regarding number (singular or plural), person (first, second, or third), or gender (masculine, feminine, or neuter). This chapter will help you ensure that your sentences show agreement

- between subjects and verbs (**6a**) and
- between pronouns and antecedents (**6b**).

6a   Subject-verb agreement

To say that a verb *agrees* with a subject means that the form of the verb (*-s* form or base form) is appropriate for the subject. For example, if the subject refers to one person or thing (*an athlete, a computer*), the *-s* form of the verb is appropriate (*runs*). If the subject refers to more than one person or thing (*athletes, computers*), the base form of the verb is appropriate (*run*). Notice in the following examples that the singular third-person subjects take a singular verb (*-s* form) and all the other subjects take the base form

He, she, it, Joe, a student	has, looks, writes
I, you, we, they, the Lees, the students	have, look, write

The verb *be* has three different present-tense forms and two different past-tense forms:

I	am/was
He, she, it, Joe, a student	is/was
You, we, they, the Lees, the students	are/were

The following subsections offer guidance on subject-verb agreement in particular situations.

(1) Subject and verb separated by one or more words
When phrases such as the following occur between the subject and the verb, they do not affect the number of the subject or the form of the verb:

accompanied by	in addition to	not to mention
along with	including	together with
as well as	no less than	

Her **salary,** together with tips, **is** just enough to live on.

Tips, together with her salary, **are** just enough to live on.

(2) Subjects joined by *and*
A compound subject (two nouns joined by *and*) that refers to a single person or thing takes a singular verb.

The **founder <u>and</u> president** of the art association **was** elected to the board of the museum.

Red beans <u>and</u> rice is the specialty of the house.

(3) Subjects joined by *or, either . . . or*, or *neither . . . nor*
When singular subjects are linked by *or, either . . . or,* or *neither . . . nor,* the verb is singular as well.

The **provost <u>or</u>** the **dean** usually **presides** at the meeting.

<u>Either</u> his **accountant <u>or</u>** his **lawyer has** the will.

If the subjects linked by one of these conjunctions differ in person or number, the verb agrees in number with the subject closer to the verb.

> Neither the basket nor the **apples were** expensive. [plural]
>
> Neither the apples nor the **basket was** expensive. [singular]

(4) Inverted word order

In most sentences, the subject precedes the verb.

> The large **cities** of the Northeast **were** the hardest hit by the winter storm.

The subject and verb can sometimes be inverted for emphasis; however, they must still agree.

> The hardest hit by the winter storm **were** the large **cities** of the Northeast.

When *there* begins a sentence, the subject and verb are always inverted (**1b**); the verb still agrees with the subject, which follows it.

> There **are** several **cities** in need of federal aid.

(5) Clauses with relative pronouns

In an adjectival (relative) clause (**1f(2)**), the subject is generally a relative pronoun (*that, who,* or *which*). To determine whether the relative pronoun is singular or plural, you must find its **antecedent** (the word or words it refers to). When the antecedent is singular, the relative pronoun is singular; when the antecedent is plural, the relative pronoun is plural.

In essence, the verb in the adjectival clause agrees with the antecedent.

sing ant   *sing v*

The person <u>who</u> reviews applications is out of town this week.

pl ant   *pl v*

The director met with the **students <u>who</u> are** studying abroad next quarter.

According to rules of traditional grammar, in sentences that contain the pattern *one of* + plural noun + relative pronoun + verb, the antecedent for the relative pronoun is the plural noun. The verb is then plural as well.

pl ant   *pl v*

Julie is one of the **students <u>who</u> plan** to study abroad.

However, professional writers often consider *one*, instead of the plural noun, to be the antecedent of the relative pronoun and thus use the singular verb:

sing ant   *sing v*

Julie is **one** of the students **<u>who</u> plans** to study abroad.

(6) Indefinite pronouns

The indefinite pronouns *each, either, everybody, everyone,* and *anyone* are considered singular and so require singular verb forms.

<u>**Each**</u> of them **is willing** to lead the discussion.

<u>**Everybody**</u> in our class **takes** a turn giving a presentation.

Other indefinite pronouns, such as *all, any, some, none, half,* and *most,* can be either singular or plural, depending on whether they refer to a unit or quantity (singular) or to individuals (plural).

My sister collects comic **books; <u>some</u> are** quite valuable.

My sister collects antique **jewelry; <u>some</u>** of it **is** quite valuable.

When an indefinite pronoun is followed by a prepositional phrase beginning with the preposition *of,* the verb agrees in number with the object of the preposition.

More than **<u>half</u>** of the **population** in West Texas **is** Hispanic.

More than **<u>half</u>** of the **people** in West Texas **are** Hispanic.

(7) Collective nouns and measurement words
Collective nouns (**1a(2)**) and measurement words require singular verbs when they refer to groups or units. They require plural verbs when they refer to individuals or parts.

Singular (regarded as a group or unit)	Plural (regarded as individuals or parts)
The **majority rules.**	The **majority** of us **are** in favor.
Ten million gallons of oil **is** more than enough.	**Ten million gallons** of oil **were spilled.**
The **number** of errors **is** insignificant.	A **number** of workers **were** absent.

(8) Words ending in -s
Titles of works that are plural in form (for example, *Star Wars* and *Dombey and Son*) are treated as singular because they refer to a single book, movie, recording, or other work.

> ***The Three Musketeers* is** one of the films she discussed in her paper.

A reference to a word is also considered singular.

> ***Beans* is** slang for "the least amount": I don't know beans about football.

Some nouns ending in *-s* are actually singular: *linguistics, news,* and *Niagara Falls.*

> The **news is** encouraging.

Nouns such as *athletics, politics,* and *electronics* can be either singular or plural, depending on their meanings.

> **Statistics is** an interesting subject. [singular]
>
> **Statistics are** often misleading. [plural]

(9) Subjects and subject complements
Some sentences may have a singular subject (**1b**) and a plural subject complement (**1c**), or vice versa. In either case, the verb agrees with the subject.

> Her primary **concern is** rising health-care **costs.**
>
> **Rising health-care costs are** her primary **concern.**

THINKING RHETORICALLY

AGREEMENT OF RELATED SINGULAR AND PLURAL NOUNS

When a sentence has two or more nouns that are related, use either the singular form or the plural form consistently.

> The **student** raised her **hand.**
>
> The **students** raised their **hands.**

Occasionally, you may have to use a singular noun to retain an idiomatic expression or to avoid ambiguity.

> **They** kept their **word.**
>
> The **participants** were asked to name their favorite **movie.**

(10) Subjects beginning with *what*

In noun clauses (**1f(2)**), *what* may be understood as either "the thing that" or "the things that." If it is understood as "the thing that," the verb in the main clause is singular.

> What we need **is** a new policy.
> [*The thing that* we need is a new policy.]

If *what* is understood as plural (the things that), the verb in the main clause is plural.

> What we need **are** new guidelines.
> [*The things that* we need are new guidelines.]

According to a traditional grammar rule, a singular verb should be used in both the noun clause beginning with *what* and the main clause.

> What **is** needed **is** new guidelines.

However, many writers and editors today consider this rule outmoded.

EXERCISE 1

In each sentence, choose the correct form of the verb in parentheses. Make sure that the verb agrees with its subject according to the conventions for academic and professional writing.

1. There (is/are) at least two good reasons for changing motor oil: risk of contamination and danger of additive depletion.

2. Reasons for not changing the oil (include/includes) the cost to the driver and the inconvenience of the chore.

3. What I want to know (is/are) the number of miles I can drive before changing my oil.

4. Each of the car manuals I consulted (recommends/recommend) five-thousand-mile intervals.

5. Neither the automakers nor the oil station attendants (know/knows) how I drive, however.

6. My best friend and mechanic (says/say) four thousand miles.

7. But my brother says three thousand miles (is/are) not long enough.

6b Pronoun-antecedent agreement

A pronoun and its antecedent (the word or word group to which it refers) agree in number (both are singular or both are plural).

> The **supervisor** said **he** would help.
> [Both antecedent and pronoun are singular.]

> My **colleagues** said **they** would help.
> [Both antecedent and pronoun are plural.]

A pronoun also agrees with its antecedent in gender (masculine, feminine, or neuter).

> **Joseph** claims **he** can meet the deadline.
> [masculine antecedent]

> **Anna** claims **she** can meet the deadline.
> [feminine antecedent]

> The **committee** claims **it** can meet the deadline.
> [neuter antecedent]

MULTILINGUAL WRITERS

POSSESSIVE PRONOUNS

A possessive pronoun (*my, your, our, his, her, its,* or *their*), also called a **possessive determiner**, agrees with its antecedent, not with the noun it precedes.

> Ken Carlson brought ^*his* her young daughter to the office today.

> [The possessive pronoun *his* agrees with the antecedent, *Ken Carlson*, not with the following noun, *daughter*.]

(1) Indefinite pronouns

Although most antecedents for pronouns are nouns, antecedents can be indefinite pronouns (5a(6)). Notice that an indefinite pronoun such as *everyone, someone,* or *anybody* takes a singular verb form.

Everyone **has** [not *have*] the right to an opinion.

Difficulties arise, however, because words like *everyone* and *everybody* seem to refer to more than one person even though they take a singular verb. Thus, the definition of grammatical number and our everyday notion of number conflict. In conversation and informal writing, a plural pronoun (*they, them,* or *their*) is often used with the singular *everyone.* Nonetheless, when you write for an audience that expects you to follow traditional grammar rules, make sure to use a third-person singular pronoun.

Everyone has the combination to ∧~~their~~ private locker. [*his or her*]

You can avoid the awkwardness of using *his or her* by using an article instead, by making both the antecedent and the possessive pronoun plural, or by rewriting the sentence using the passive voice (7c).

Everyone has the combination to **a** private locker. [article]

Students have combinations to **their** private lockers. [plural antecedent and plural possessive pronoun]

The combination to a private locker **is issued** to everyone. [passive voice]

(2) Referring to both genders

When an antecedent can refer to people of either gender, rewrite the sentence to make the antecedent plural or use *he or she* or *his or her* if doing so is not too cumbersome.

Lawyers *their*
∧A lawyer represents ∧his clients. [plural pronoun and plural antecedent]

A lawyer represents the clients **he or she** has accepted.

A lawyer represents **his or her** clients.

(See **19c** for more information on using inclusive language.)

(3) Two antecedents joined by *or* or *nor*

If a singular and a plural antecedent are joined by *or* or *nor,* place the plural antecedent second and use a plural pronoun.

Either the senator **or** her <u>assistants</u> will explain how <u>they</u> devised the plan for tax reform.

Neither the president **nor** the <u>senators</u> stated that <u>they</u> would support the proposal.

(4) Collective nouns

When an antecedent is a collective noun (**1a(2)**) such as *team, faculty,* or *committee,* determine whether you intend the noun to be understood as singular or plural and then make sure that the pronoun agrees in number with the noun.

it
The choir decided that∧they would tour during the winter. [Because the choir decided as a group, *choir* should be considered singular. The singular form, *it,* replaces the plural, *they.*]

they
The committee may disagree on methods, but∧it must agree on basic aims. [Because the committee members are behaving as individuals, *committee* is regarded as plural. The plural form, *they,* replaces the singular, *it.*]

EXERCISE 2

Revise the following sentences so that pronouns and antecedents agree.

1. A researcher relies on a number of principles to help him make ethical decisions.

2. Everyone should have the right to participate in a study only if they feel comfortable doing so.

3. A team of researchers should provide its volunteers with consent forms, in which they describe to the volunteers the procedures and risks involved in participation.

4. Every participant should be guaranteed that the information they provide will remain confidential.

5. Institutions of higher education require that a researcher address ethical issues in their proposal.

7 Verbs

Choosing verbs to convey your message precisely is the first step toward writing clear sentences. The next step is to ensure that the verbs you choose conform to the conventions your audience expects you to follow. This chapter will help you

- identify conventional verb forms (**7a**),
- use verb tenses to provide information about time (**7b**),
- distinguish between the active voice and the passive voice (**7c**), and
- use verbs to signal the factuality or likelihood of an action or event (**7d**).

7a Verb forms

Most English verbs have four forms, following the model for *walk:*

walk, walks, walking, walked

However, English also includes irregular verbs, which may have as few as three forms or as many as eight:

let, lets, letting *be, am, is, are, was, were, being, been*

(1) Regular verbs
Regular verbs have four forms. The **base form** is the form you can find in a dictionary. *Talk, act, change,* and *serve* are all base forms.

The second form of a regular verb is the **-s form**. To derive this form, add to the base form either -s (*talks, acts, changes, serves*) or, in some cases, -es (*marries, carries, tries*). See **18d** for information on changing *y* to *i* before adding -es.

The third form of a regular verb is the **-ing form**, also called the **present participle**. It consists of the base form and the ending -ing (*talking, acting*). Depending on the verb, a spelling change may be required when the suffix is added (*changing, chatting*) (**18d**).

The fourth form of a regular verb consists of the base form and the ending -ed (*talked, acted*). Again, spelling may vary when the suffix is added (*changed, chatted*) (**18d**). The -ed form has two names. When this form is used without a form of the auxiliary verb *have* or *be*, it is called the **past form**: We *talked* about the new plan. When the -ed form is used with one of these auxiliary verbs, it is called the **past participle**: We *have talked* about it several times. A committee *was formed* to investigate the matter.

	Verb Forms of Regular Verbs		
Base Form	**-s Form (Present Tense, Third Person, Singular)**	**-ing Form (Present Participle)**	**-ed Form (Past Form or Past Participle)**
work	works	working	worked
watch	watches	watching	watched
apply	applies	applying	applied
stop	stops	stopping	stopped

CAUTION

When verbs are followed by words with similar sounds, you may find their endings (-*s* or -*ed*) difficult to hear. In addition, these verb endings may seem unfamiliar because your dialect does not have them. Nonetheless, you should use -*s* and -*ed* when you write for an audience that expects you to include these endings.

She ~~seem~~ satisfied with the report. [*seems* inserted above]

We were ~~suppose~~ to receive the results yesterday. [*supposed* inserted above]

(2) Irregular verbs

Most irregular verbs, such as *write,* have forms similar to some of those for regular verbs: base form (*write*), -*s* form (*writes*), and -*ing* form (*writing*). However, the past form (*wrote*) and the past participle (*written*) vary from those of the regular verbs. In fact, some irregular verbs have two acceptable past forms and/ or past participles (see *awake, dive, dream,* and *get* in the following chart). Other irregular verbs have only three forms because the same form serves as the base form, the past form, and the past participle (see *set* in the chart). If you are unsure about verb forms not included in the chart, consult a dictionary.

	Verb Forms of Irregular Verbs			
Base Form	**-*s* Form (Present Tense, Third Person, Singular)**	**-*ing* Form (Present Participle)**	**Past Form**	**Past Participle**
awake	awakes	awaking	awaked, awoke	awaked, awoken
begin	begins	beginning	began	begun
break	breaks	breaking	broke	broken

(continued on page 96)

(continued from page 95)

Base Form	-s Form (Present Tense, Third Person, Singular)	-ing Form (Present Participle)	Past Form	Past Participle
bring	brings	bringing	brought	brought
choose	chooses	choosing	chose	chosen
come	comes	coming	came	come
dive	dives	diving	dived, dove	dived
do	does	doing	did	done
dream	dreams	dreaming	dreamed, dreamt	dreamed, dreamt
drink	drinks	drinking	drank	drunk
eat	eats	eating	ate	eaten
forget	forgets	forgetting	forgot	forgotten
get	gets	getting	got	gotten, got
give	gives	giving	gave	given
go	goes	going	went	gone
hang (suspend)	hangs	hanging	hung	hung
hang (execute)	hangs	hanging	hanged	hanged
know	knows	knowing	knew	known
lay (see the Glossary of Usage)	lays	laying	laid	laid

Base Form	-s Form (Present Tense, Third Person, Singular)	-ing Form (Present Participle)	Past Form	Past Participle
lead	leads	leading	led	led
lie (see the Glossary of Usage)	lies	lying	lay	lain
lose	loses	losing	lost	lost
pay	pays	paying	paid	paid
rise (see the Glossary of Usage)	rises	rising	rose	risen
set (see the Glossary of Usage)	sets	setting	set	set
sink	sinks	sinking	sank	sunk
sit (see the Glossary of Usage)	sits	sitting	sat	sat
swim	swims	swimming	swam	swum
take	takes	taking	took	taken
wear	wears	wearing	wore	worn
write	writes	writing	wrote	written

The verb *be* has eight forms:

be	**Be** on time!
am	I **am** going to arrive early tomorrow.
is	Time **is** of the essence.
are	They **are** always punctual.
was	The meeting **was** scheduled for 10 a.m.
were	We **were** only five minutes late.
being	He is **being** delayed by traffic.
been	How long have we **been** here?

(3) Prepositional verbs and phrasal verbs

A **prepositional verb** is a frequently occurring combination of a verb and a preposition. *Rely on, think about, look like,* and *ask for* are all prepositional verbs. A **phrasal verb** is a combination of a verb and a particle such as *up, out,* or *on.* A **particle** resembles an adverb or a preposition, but it is so closely associated with a verb that together they form a unit of meaning. *Carry out, go on, make up, take on,* and *turn out* are phrasal verbs commonly found in college-level writing. Notice that these five phrasal verbs have meanings that can be expressed in one word: *do, continue, form, accept,* and *attend.*

MULTILINGUAL WRITERS

PHRASAL VERBS

Definitions of phrasal verbs are often not obvious. For example, *find out* means "to discover." To learn more about phrasal verbs, see Chapter 44.

(4) Auxiliary verbs

The auxiliary verbs *be, do,* and *have* combine with main verbs, both regular and irregular.

be	am, is, are, was, were surprised
	am, is, are, was, were writing
do	does, do, did call
	doesn't, don't, didn't spend
have	has, have, had prepared
	has, have, had read

When you combine auxiliary verbs with main verbs, you alter the meanings of the main verbs in subtle ways. The resulting verb combinations may provide information about time, emphasis, or action in progress.

Be, do, and *have* are not just auxiliary verbs, though. They may be used as main verbs as well.

be	I **am** from Texas.
do	He **does** his homework early in the morning.
have	They **have** an apartment near a park.

A sentence may even include one of these verbs as both an auxiliary and a main verb.

They **are being** careful.

Did you **do** your taxes by yourself?

She **has** not **had** any free time this week.

Another type of auxiliary verb is called a **modal auxiliary.** There are nine modal auxiliaries: *can, could, may, might, must, shall, should, will,* and *would.* More information on the use of modal auxiliaries is presented in **44b(2)**.

CAUTION

When a modal auxiliary occurs with the auxiliary *have* (*must have forgotten, should have known*), *have* frequently sounds like the word *of*. When you proofread, be sure that modal auxiliaries are not followed by *of*.

They **could~~of~~ taken** another route.

[have inserted above "of"]

Writers generally do not combine modal auxiliaries unless they want to portray a regional dialect.

We **might ~~could~~** plan the meeting for after the holidays.

[be able to inserted above "could"]

(5) Participles

Present participles (*-ing* verb forms) are used with a form of the auxiliary verb *be*: We *are waiting* for the next flight. It *is arriving* sometime this afternoon. Depending on the intended meaning, past participles can be used with either *be* or *have*: The first flight *was canceled*. We *have waited* for an hour. If a sentence contains only a participle, it is probably a fragment (**2b**).

I sit on the same bench every day~~.~~ ~~Dreaming~~ of far-off places.

[, dreaming inserted]

When a participle is part of a verbal phrase, it often appears without an auxiliary verb (**1a(4)**).

Swatting at mosquitoes and **cursing** softly, we packed our gear. [COMPARE: We **were swatting** at mosquitoes and **cursing** softly as we packed our gear.]

EXERCISE 1

Revise the following sentences. Explain any changes you make.

1. Any expedition into the wilderness suffer its share of mishaps.
2. The Lewis and Clark Expedition began in May 1804 and end in September 1806.

3. Fate must of smiled on Meriwether Lewis and William Clark, for there were no fatalities under their leadership.

4. By 1805, the Corps of Discovery, as the expedition was call, included thirty-three members.

5. The Corps might of lost all maps and specimens had Sacajawea, a Native American woman, not fish them from the Missouri River.

6. The success of the expedition depend on its members' willingness to help one another.

7b Verb tenses

Verb tenses provide information about time. For example, the tense of a verb may indicate that an action took place in the past or that an action is ongoing. Verb tenses are labeled as present, past, or future; they are also labeled as simple, progressive, perfect, or perfect progressive. The following chart shows how these labels apply to the tenses of *walk*.

Verb Tenses			
	Present	**Past**	**Future**
Simple	I/you/we/they **walk.** He/she/it **walks.**	walked	will walk
Progressive	I **am walking.** He/she/it **is walking.** You/we/they **are walking.**	I/he/she/it **was walking.** You/we/they **were walking.**	will be walking
Perfect	I/you/we/they **have walked.** He/she/it **has walked.**	had walked	will have walked
Perfect progressive	I/you/we/they **have been walking.** He/she/it **has been walking.**	had been walking	will have been walking

Some of the tenses have more than one form because they depend on the person and the number of the subject (generally, the main noun or the pronoun that precedes the verb). **Person** refers to the role of the subject. First person (expressed by the pronoun *I* or *we*) indicates that the subject of the verb is the writer or writers. Second person (*you*) indicates that the subject is the audience. Third person (*he, she, it,* or *they*) indicates that the subject is someone or something other than the writer or audience. First- and second-person references are pronouns, but third-person references can be either pronouns or nouns (such as *book* or *books*). **Number** signals whether the subject is singular (referring to just one person or thing) or plural (referring to more than one person or thing).

7c Voice

Voice indicates the relationship between a verb and its subject. When a verb is in the **active voice**, the subject is generally a person or thing performing the action indicated by the verb. When a verb is in the **passive voice**, the subject is usually the *receiver* of the action.

Jen Wilson **wrote** the essay. [active voice]

The essay **was written** by Jen Wilson. [passive voice]

Notice that the actor, Jen Wilson, appears in a prepositional phrase beginning with *by* in the passive sentence. Some sentences, however, do not include a *by* phrase because the actor is unknown or unimportant.

Jen Wilson's essay **was published** in the student newspaper.

In the sentence above it is not important to know who accepted Jen's essay for publication, only that it was published. The best way to decide whether a sentence is in the passive voice is to examine its verb phrase.

(1) Verbs in the passive voice

The verb phrase in a sentence written in the passive voice consists of a form of the auxiliary verb *be* (*am, is, are, was, were, been*) and a past participle (7a(1)). Depending on the verb tense, other auxiliaries such as *have* and *will* may appear as well. The following sentences in the passive voice show which auxiliaries are used with *called*:

Simple present	The meeting *is called* to order.
Simple past	The recruits *were called* to duty.
Present progressive	The council *is being called* to act on the proposal.
Past perfect	Ms. Jones *had been called* for jury duty twice last year, but she was glad to serve again.

If a verb phrase does not include both a form of the auxiliary verb *be* and a past participle, it is in the active voice.

(2) Choosing between the active and the passive voice

Sentences in the active voice are generally clearer and more vigorous than their passive counterparts. To use the active voice for emphasizing an actor and an action, first make the actor the subject of the sentence and then choose verbs that will help your readers see what the actor is doing. Notice how the following sentences in the active voice emphasize the role of the students:

Active voice	A group of students planned the graduation ceremony. They invited a well-known columnist to give the graduation address.
Passive voice	The graduation ceremony was planned by a group of students. A well-known columnist was invited to give the graduation address.

Use the passive voice when you want to stress the recipient of the action, rather than the actor, or when the actor's identity is unimportant or unknown. For example, you may want to emphasize the topic of a discussion.

Tuition increases **were discussed** at the board meeting.

Or you may be unable to identify the actor who performed some action.

The lights **were left** on in the building last night.

EXERCISE 2

Identify the voice in each sentence as active or passive.

1. In a *National Geographic* article, Zahi Hawass discusses recent information regarding the life and death of King Tut.

2. King Tut was enthroned at the age of nine.

3. Originally, he was called Tutankhaten.

4. Later he changed his name to Tutankhamun.

5. At nineteen years of age, King Tut died.

6. His mummy was discovered in 1922.

7. Recently, King Tut's DNA was obtained.

8. The findings of the genetic testing reveal that King Tut may have died of malaria.

EXERCISE 3

Rewrite the sentences in Exercise 2, making sentences in the active voice passive, and vice versa. Add or delete actors when necessary. If one version of a sentence is better than the other, explain why.

The **mood** of a verb indicates the writer's attitude concerning the factuality of what is being expressed. The **indicative mood** is used for statements and questions regarding fact or opinion. The **imperative mood** is used to give commands or directions. The **subjunctive mood** is used to state requirements, make requests, and express wishes.

Indicative	I am on the board of directors.
	Were you on the board last year?
	The board will meet in two weeks.
Imperative	Plan on attending the meeting.
	Be on time!
Subjunctive	She suggests that you come early.
	If you came to more meetings, you would understand the issues.
	If I had attended regularly, I would have voted for the plan.

The subjunctive mood is also used to signal hypothetical situations—situations that are not real or not currently true (for example, *If I were president, . . .*).

Verb forms in the subjunctive mood serve a variety of functions. The **present subjunctive** is the base form of the verb. It is used to express necessity.

The manager suggested that he **pay** for his own travel.

The **past subjunctive** has the same form as the simple past (for example, *had, offered,* or *wrote*). However, the past subjunctive form of *be* is *were*, regardless of person or number. This form is used to present hypothetical situations.

If they **offered** me the job, I would take it.

Even if I **were promoted,** I would not change my mind.

The **perfect subjunctive** has the same form as the past perfect tense: *had* + past participle. The perfect subjunctive signals that the action did not occur.

She wishes she **had participated** in the scholarship competition.

The following guidelines should help you avoid pitfalls when using the subjunctive.

TIPS FOR USING THE SUBJUNCTIVE

■ In clauses beginning with *as if* and *as though*, use the past subjunctive or the perfect subjunctive:

He acts as if he ^were ~~was~~ the owner.

She looked at me as though she ^had **heard** this story before.

■ In a dependent clause that begins with *if* and refers to a condition or action that did not occur, use the past subjunctive or the perfect subjunctive. Avoid using *would have* in such an *if* clause.

If I ^were ~~was~~ rich, I would buy a yacht.

If the driver ^had **~~would have~~ checked** his rearview mirror, the accident would not have happened.

Notice that an indicative clause beginning with *if* may describe a condition or action that can occur.

If it is sunny tomorrow, I'm going fishing. [indicative mood]

■ In dependent clauses following verbs that express wishes, requirements, or requests, use the past subjunctive or the perfect subjunctive.

I wish I ^were ~~was~~ taller.

My brother wishes he ^had **studied** harder years ago.

EXERCISE 4

Use subjunctive verb forms to revise the following sentences.

1. The planners of Apollo 13 acted as if the number 13 was a lucky number.

2. Superstitious people think that if NASA changed the number of the mission, the astronauts would have had a safer journey.

3. They also believe that if the lunar landing would have been scheduled for a day other than Friday the Thirteenth, the crew would not have encountered any problems.

4. The crew used the lunar module as though it was a lifeboat.

5. If NASA ever plans a space mission on Friday the Thirteenth again, the public would object.

8 Document Design

All of us use visual elements to interpret messages every day—even if we are not consciously aware of how we "read" these photos, graphics, and design features. Just as important as understanding how we make sense of visuals is understanding how to compose print and electronic documents that use visuals in combination with words to communicate information to an intended audience.

In this chapter, you will learn the rhetorical principles of combining visual elements with text, the genres of visual documents, and the conventions of layout—all of which will help you achieve your rhetorical purpose. More specifically, this chapter will help you

- understand visual documents in terms of the rhetorical situation (**8a**),
- employ the design principles of visual rhetoric (**8b**), and
- combine visual and verbal elements effectively (**8c**).

8a Visual documents and the rhetorical situation

Opportunity, audience, purpose, message, and context—the rhetorical elements underlying the interpretation and composition of verbal texts—apply to visual documents as well. **Visual documents** combine visual elements (images or graphics) with verbal text, in order to respond to a rhetorical opportunity, express meaning, and deliver a message to an intended audience. In addition to images and **graphics** (such as diagrams and tables), visual elements include the design features and the layout of a document. Whether their purpose is expressive, expository, or argumentative (**31c**), visual documents—ranging from magazine advertisements to posters, billboards, brochures, newsletters, and websites—must always take into account the relationship between purpose and audience.

Consider the brochure in Figure 8.1, which features "winning solutions" to the problem of global warming. This brochure serves a distinct purpose and employs rhetorical

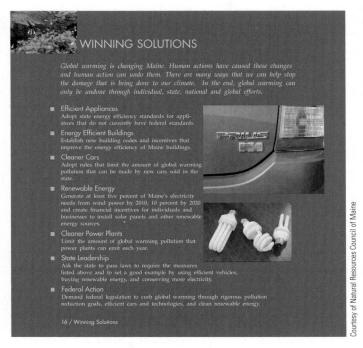

WINNING SOLUTIONS

Global warming is changing Maine. Human actions have caused these changes and human action can undo them. There are many ways that we can help stop the damage that is being done to our climate. In the end, global warming can only be undone through individual, state, national and global efforts.

■ Efficient Appliances
Adopt state energy efficiency standards for appliances that do not currently have federal standards.

■ Energy Efficient Buildings
Establish new building codes and incentives that improve the energy efficiency of Maine buildings.

■ Cleaner Cars
Adopt rules that limit the amount of global warming pollution that can be made by new cars sold in the state.

■ Renewable Energy
Generate at least five percent of Maine's electricity needs from wind power by 2010, 10 percent by 2020 and create financial incentives for individuals and businesses to install solar panels and other renewable energy sources.

■ Cleaner Power Plants
Limit the amount of global warming pollution that power plants can emit each year.

■ State Leadership
Ask the state to pass laws to require the measures listed above and to set a good example by using efficient vehicles, buying renewable energy, and conserving more electricity.

■ Federal Action
Demand federal legislation to curb global warming through rigorous pollution reduction goals, efficient cars and technologies, and clean renewable energy.

16 / Winning Solutions

Courtesy of Natural Resources Council of Maine

Figure 8.1. An effective brochure that is aimed at a specific audience.

strategies that appeal to a specific audience: readers with a vested interest in the topic who are already predisposed to the message. This audience likely sought out the brochure and shares at least some of its creator's views and opinions about the importance of reversing global warming. Therefore, the brochure need only respond to (rather than elicit) the audience's interest. With no need to argue the importance of paying attention to global warming, the brochure's creator could concentrate instead on outlining various strategies for slowing the progress of the problem. In other words, the purpose of the brochure is not to argue a point but to deliver new information to a specific audience. Thus, the intended audience and the rhetorical purpose of the brochure are linked by the seven winning solutions.

The rhetorical situation also influences the genre of visual document that is chosen to deliver a message. Most posters, billboards, and advertisements contain a small amount of text, allowing the audience to absorb the message visually, with only a brief glance. The ease of access helps these predominantly visual documents reach a large, diverse audience. By contrast, the creator of the global warming brochure has assumed that those in the intended audience will take the time to read the more extensive text. The volume of information in a brochure, as well as the specialized focus, makes this genre particularly appropriate for an educated, already interested audience.

EXERCISE 1

Select a visual document that has caught your attention. Write for five to ten minutes in response to that document. Then, working with one or two classmates, analyze the document in terms of the rhetorical opportunity it addresses, the components of the rhetorical situation, and the relationship between words and images. Be prepared to share your document and analysis with the rest of the class.

8b   The design principles of visual rhetoric

After considering the context of a visual document as a whole, you can analyze how the various elements work together to create a coherent message. Just as writers organize words into sentences and paragraphs, designers structure the visual elements of their documents in order to achieve coherence, develop ideas, and make a point. Experienced designers know to stand back several feet from a document in order to see which elements draw their attention. If the visual elements compete, no part of the document gets sufficient attention. Thus, presenting complicated material in visual form requires a set of strategies different from those used in writing academic papers. Rather than relying on paragraph breaks and topic sentences, designers of visual documents call on four important principles to organize, condense, and develop ideas: alignment, proximity, contrast, and repetition. These four design principles will help you organize complex information, making it visually appealing and easily accessible to your audience.

(1) The principle of alignment

The principle of **alignment** involves the use of an invisible grid system, running vertically and horizontally, to place and connect elements on a page. The fewer the invisible lines, the stronger the document design. For instance, the poster in Figure 8.2 has two obvious sets of primary lines: one set that moves from left to right over the top half of the poster and a strong line down the center of the bottom half. These lines organize and unify the poster and give it a sharp, clean look, directing the viewer's eye to the smiling young people leaning on the fence. The words along the bottom of the poster, "RURAL ELECTRIFICATION ADMINISTRATION," reveal that the

Figure 8.2. This Rural Electrification Administration poster by Lester Beall (1934) was purposefully designed to herald progress.

happy expressions are related to the expectation of electrifi-cation. The broad red-and-white and blue-and-white stripes affirm the patriotism of the federal program for rural electrifi-cation. Overall, the poster communicates that rural Americans can look forward to a better future because of electricity.

(2) The principle of proximity

The principle of **proximity** requires the grouping of related textual or visual elements, such as the horizontal stripes and the fence rails in the poster in Figure 8.2. Dissimilar elements are separated by **white space** (blank areas around blocks of text or around graphics or images). The audience perceives each grouping (or chunk) of elements in a well-designed visual document as a single unit and interprets it as a whole before moving on to the next group. In other words, the chunks serve a function similar to sections in a written document, organizing the page and reducing clutter. In the poster in Figure 8.2, the proximity of the text to the image of the young people links the textual and visual elements and allows them to be interpreted together.

(3) The principle of contrast

The principle of **contrast** establishes a visual hierarchy, providing clear clues as to which elements are most important and which are less so. The most salient textual or visual elements (such as the red, white, and blue stripes in the poster in Figure 8.2) stand out from the rest of the document, while other elements (the line of text, for instance) are not as noticeable. The most significant elements of a document are generally contrasted with other elements by differences in size, color, or typeface. Academic and professional documents, for example, usually have their headings in bold or italic type or capital letters to distinguish them from the rest of the text. The brochure in Figure 8.1 (on page 111) features a large title in capital letters, "WINNING SOLUTIONS," which dominates the page; the identical size and typeface of all the headings indicate that they are of equal importance but subordinate to the title. Just a brief glance at this brochure allows the viewer to determine the hierarchy of information and the basic structure.

(4) The principle of repetition

The principle of **repetition** has to do with the replication throughout a document of specific textual or visual elements, such as the headings in the brochure in Figure 8.1 and the stripes in the poster in Figure 8.2. For example, nearly all academic and professional papers use a consistent typeface for large blocks of text, which creates a unified look throughout these documents. Visual documents follow a similar strategy, purposefully limiting the number of typefaces, colors, and graphics in order to enhance coherence with repetition. The repeated bullets and the repeated typefaces in the headings and text of the brochure on global warming as well as the stripes in the poster for rural electrification structure these visual documents and reinforce their unity.

8c Combining visual and verbal elements

Although words or images alone can have a tremendous impact on an audience, the combination of the two is often necessary, especially when neither verbal nor visual elements alone can successfully respond to a rhetorical opportunity, reach an intended audience, or fulfill the rhetorical purpose. As you know, newspaper and magazine articles often include powerful images to heighten the emotional impact of the text. Diagrams that accompany a set of product-assembly instructions reinforce the process analysis **32g(3)** and make assembly easier for the reader. Thus, when used together, words and images can reinforce each other to deliver a message—or even deliberately contradict each other to establish an ironic tone.

(1) Graphics

Many academic and professional documents that are primarily composed of text also include visual displays, or **graphics**, to clarify written material. Graphics can be used to illustrate a concept, present data, provide visual relief, or simply attract readers' attention. Different types of graphics—tables, charts or graphs, and pictures—serve different purposes, and some may serve multiple purposes in a given document. Any of these types of graphics can enable readers to absorb a message more quickly than they would by reading long sections of text. However, if there is any chance that readers might not receive the intended message, it is a good idea to supplement graphics with textual discussion.

(a) Tables

Tables use a row-and-column arrangement to organize data (numbers or words) spatially; they are especially useful for presenting great amounts of numerical information in a small space, enabling the reader to draw direct comparisons among pieces of data or even to locate specific items. When you design a table, be sure to label all of the columns and rows accurately and to provide both a title and a number for the table. In Table 8.1, you can see that the columns, labeled "Season," "Season Total," and "Biggest One-Day Snowfall for the Season and the Amount," contain important information for people interested in Great Lakes snow-belt weather conditions. The table number and title traditionally appear above the table body, as Table 8.1 demonstrates, and any notes or source information are placed below it.

Most word-processing programs have settings that let you insert a table wherever you need one. You can determine how many rows and columns the table will have, and you can also size each row and each column appropriately for the information it will hold.

(b) Charts and graphs

Like tables, charts and graphs also display relationships among statistical data in visual form; unlike tables, they do so using

Table 8.1. Snowfall in Cleveland, Ohio, by Season (in inches)

Season	Season Total	Biggest One-Day Snowfall for the Season and the Amount	
2010–2011	69.5	February, 25, 2011	8.9
2009–2010	59.8	January 4, 2010	4.5
2008–2009	80.0	February 4, 2009	10.9
2007–2008	77.2	March 8, 2008	10.8
2006–2007	76.5	February 13, 2007	10.4
2005–2006	50.6	February 8, 2006	6.9
2004–2005	117.9	December 22, 2004	9.4
2003–2004	91.2	March 16, 2004	7.1
2002–2003	94.9	December 25, 2002	10.2
2001–2002	45.8	March 25, 2002	6.9
2000–2001	74.3	March 25, 2001	7.7
1999–2000	59.2	January 26, 2000	6.5
1998–1999	61.6	February 13, 1999	6.5
1997–1998	33.7	December 6, 1997	5.6
1996–1997	55.4	November 11, 1996	7.2
1995–1996	101.1	December 19, 1995	12.0

Source: http//:blog.cleveland.com/datacentral/index.ssf/2010/07/annual_cleveland_snowfall_total.html

lines, bars, or other visual elements rather than just letters and numbers. Data can be displayed in several different graphic forms: pie charts, line graphs, and bar charts are the most common examples.

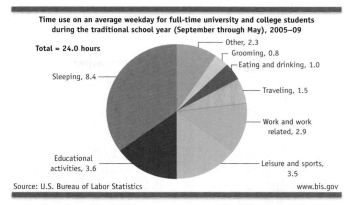

Time use on an average weekday for full-time university and college students during the traditional school year (September through May), 2005–09

Total = 24.0 hours

Other, 2.3
Grooming, 0.8
Eating and drinking, 1.0
Sleeping, 8.4
Traveling, 1.5
Work and work related, 2.9
Educational activities, 3.6
Leisure and sports, 3.5

Source: U.S. Bureau of Labor Statistics   www.bis.gov

Figure 8.3. This easy-to-read pie chart shows how the average full-time university or college student spends time on an average weekday.

Pie charts are especially useful for showing the relationship of parts to a whole (see Figure 8.3), but these graphics can only be used to display sets of data that add up to 100 percent. (In the chart in Figure 8.3, twenty-four hours represents 100 percent of a day.)

Line graphs show the change in the relationship between one variable (indicated as a value on the vertical axis, or *y* axis) and another variable (indicated as a value on the horizontal axis, or *x* axis). The most common *x*-axis variable is time. Line graphs are very good at showing how a variable changes over time. A line graph might be used, for example, to illustrate the progression of sleep stages during one night, increases or decreases in student achievement from semester to semester, or trends in financial markets over a number of years.

Bar charts show correlations between two variables that do not involve smooth changes over time. For instance, a bar chart might illustrate gross national product for several

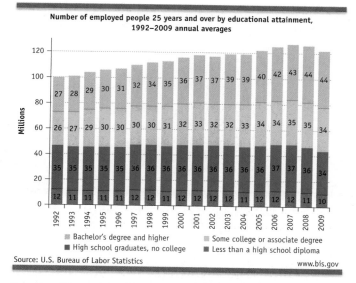

Number of employed people 25 years and over by educational attainment, 1992–2009 annual averages

Bachelor's degree and higher
Some college or associate degree
High school graduates, no college
Less than a high school diploma

Source: U.S. Bureau of Labor Statistics

www.bis.gov

Figure 8.4. This bar chart illustrates the composition of the US workforce with respect to level of education.

nations, the relative speeds of various computer processors, or statistics about the composition of the US workforce (see Figure 8.4).

(c) Pictures

Pictures include photos, sketches, technical illustrations, paintings, icons, and other visual representations. Photographs are often used to reinforce textual descriptions or to show a reader exactly what something looks like. Readers of a used-car ad, for instance, will want to see exactly what the car looks like, not an artistic interpretation of its appearance. Likewise, a travel brochure about Costa Rica needs to contain lots of full-color photos of dazzling beaches, verdant forests, and azure water.

 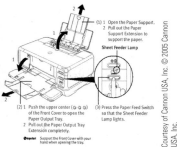

Figure 8.5. A photo and a drawing of the same printer.

But photographs are not always the most informative type of picture. Compare the two images in Figure 8.5. Although the photograph is a more realistic image of the actual printer, the illustration more clearly shows the printer's important features: buttons, panels, and so forth. With its simple lines and clear labels, the illustration suits its purpose: to help the viewer set up and use the printer. Line drawings enable the designer of a document such as a user manual to highlight specific elements of an object while deemphasizing or eliminating unnecessary information. The addition of arrows, pointers, and labels adds useful detail to such an illustration.

(2) Effective integration of visual and verbal elements

To integrate visual elements into written text, you want to position them purposefully, whether you place images close together in a document or you put an image at the beginning or end of a document so as not to disrupt the text.

(a) Considering proximity

Proximity—placing an image as close as possible to the text that refers to it—is one way of establishing a connection between the verbal and visual elements. Think of how helpful it can be to have images accompanying printed instructions for assembling a piece of furniture, setting up a new computer, or following a complicated recipe. When an image aligns with verbal instructions, the document is effective and instructive. Wrapping the text around an image also serves to integrate visual and verbal elements; wrapping places an image and its corresponding text in very close proximity. In addition, cropping unnecessary elements from an image strengthens connections between it and the textual components of a document, highlighting what is most important while preserving what is authentic. When a visual element (a graph or table) is too large to fit on the same page as the related text, it is placed on a separate page or moved to an appendix. Thus, proximity must occasionally be forfeited for the sake of in-depth explanation or detailed support.

(b) Including captions and labels

Captions and labels are also crucial to the integration of visual and verbal components of a document. In academic texts and professional documents, figures and tables are labeled by being numbered consecutively and separately. Each label is followed by a captions. Moreover, the body of an academic text or professional documents includes **anchors**, specific references to each image or graphic used, such as "see Figure 5" and "as shown in Table 2." In popular magazines and newspapers, on the other hand, text-enhancing visuals are usually integrated into the text through the use of captions and layout.

9 Capitals

When you look at an advertisement, an e-mail message, or a paragraph in this book, you can easily pick out capital letters. These beacons draw your attention to significant details—for example, the beginnings of sentences or the names of particular people, places, and products. Although most capitalization conventions apply to any rhetorical situation, others are specific to a discipline or a profession. In this chapter, you will learn the conventions that are followed in most academic and professional settings. This chapter will help you

- use capitals for proper names (**9a**);
- capitalize words in titles and subtitles of works (**9b**);
- capitalize the first letter of a sentence (**9c**);
- use capitals for computer keys, menu items, and icon names (**9d**); and
- avoid unnecessary capitalization (**9e**).

9a Proper names

When you capitalize a word, you emphasize it. That is why names of people and places are capitalized, even when they are used as modifiers (*Mexico, Mexican government*). Some words, such as *college, company, park,* and *street,* are capitalized only if they are part of a name (*a university* but *University of Pennsylvania*). The following names and titles should be capitalized.

Advertisers often highlight important words by capitalizing them.

(1) Names of specific persons or things

Zora Neale Hurston	Flight 224	Honda Accord
John Paul II	Academy Award	USS *Cole*
Skylab	Nike	Microsoft Windows

For a brand name such as eBay or iPod that begins with a lowercase letter, do not change that letter to a capital when the name begins a sentence.

Many people like to shop on eBay.

eBay attracts many shoppers.

A word denoting a family relationship is capitalized only when it substitutes for the person's proper name.

I told **Mom** about the event. [I told Virginia about the event.]

I told my **mom** about the event. [NOT I told my Virginia about the event.]

(2) Titles accompanying proper names

A title is capitalized when it precedes the name of a person but not when it follows the name or stands alone.

Governor Bill Haslam	Bill Haslam, the governor
Captain Ray Machado	Ray Machado, our captain
Aunt Helen	Helen, my aunt
President Lincoln	Abraham Lincoln, the president of the United States

(3) Names of ethnic or cultural groups and languages

Asians	African Americans	Latinos/Latinas	Poles
Arabic	English	Korean	Spanish

(4) Names of bridges, buildings, monuments, and geographical features

Golden Gate Bridge	Empire State Building	Lincoln Memorial
Arctic Circle	Mississippi River	Grand Canyon

When referring to two or more geographical features, do not capitalize the generic term: *Lincoln and Jefferson memorials, Yellowstone and Olympic national parks.*

(5) Names of organizations, government agencies, institutions, and companies

B'nai B'rith	National Institutes of Health
Phi Beta Kappa	Internal Revenue Service
Howard University	Ford Motor Company

When used as common nouns, *service, company,* and *university* are not capitalized. However, universities and other organizations often capitalize these words when they are used as shortened forms of the institutions' full names.

> The policies of Hanson **U**niversity promote the rights of all individuals to equal opportunity in education. The **U**niversity complies with all applicable federal, state, and local laws.

(6) Names of days of the week, months, and holidays

Wednesday	August	Fourth of July

The names of the seasons—spring, summer, fall, winter—are not capitalized.

MULTILINGUAL WRITERS

CAPITALIZING DAYS OF THE WEEK

Capitalization rules vary according to language. For example, in English, the names of days and months are capitalized, but in some other languages, such as Spanish and Italian, they are not.

(7) Designations for historical documents, periods, events, movements, and styles

Declaration of Independence   Renaissance

Industrial Revolution   Prohibition

A historical period that includes a number is not capitalized unless it is considered a proper name.

twentieth century   the Roaring Twenties

the seventies   the Gay Nineties

The name of a cultural movement or style is capitalized if it is derived from a person's name or if capitalization distinguishes the name of the movement or style from the ordinary use of the word or phrase.

Platonism   Reaganomics   New Criticism

Most names of cultural movements and styles are not capitalized.

art deco   impressionism   realism   deconstruction

(8) Names of religions, their adherents, holy days, titles of holy books, and words denoting the Supreme Being

Buddhism, Christianity, Islam, Judaism

Buddhist, Christian, Muslim, Jew

Bodhi Day, Easter, Ramadan, Yom Kippur

Sutras, Bible, Koran, Talmud BUT biblical, talmudic

Buddha, God, Allah, Yahweh

Some writers always capitalize personal pronouns (5a(1)) that refer to the Supreme Being; others capitalize such words only when capitalization is needed to prevent ambiguity:

The Lord commanded the prophet to warn His people.

(9) Words derived from proper names

Americanize [verb]   Orwellian [adjective]   Marxism [noun]

When a proper name becomes the name of a general class of objects or ideas, it is no longer capitalized. For example, the word *zipper* was originally the trademarked name of the fastening device and was capitalized; it now refers to the class of such devices and is written with a lowercase letter. A word derived from a brand name, such as *Xerox* or *Kleenex*, should be capitalized. If possible, avoid using brand names and choose generic terms such as *photocopy* and *tissue* instead. If you are not sure whether a proper name or derivative has come to stand for a general class, look up the word in a dictionary.

(10) Abbreviations and acronyms

These forms are derived from the initial letters of capitalized word groups:

AMEX   AT&T   CBS   NFL   UNICEF   YMCA

(See also Chapter **11** and **17a(2)**.)

(11) Military terms

Names of forces and special units are capitalized, as are names of wars, battles, revolutions, and military awards.

United States Army	Marine Corps	Secret Service
Green Berets	Russian Revolution	Purple Heart

Military words such as *army*, *navy*, and *war* are not capitalized when they stand alone.

My sister joined the navy in 2008.

BUT

My sister joined the United States Navy in 2008.

STYLE SHEET FOR CAPITALIZATION

Capitals	No capitals
the West [geographical region]	driving west [compass point]
a Chihuahua [a breed of dog named after a state in Mexico]	a poodle [a breed of dog]
Washington State University [a specific institution]	a state university
Revolutionary War [a specific war]	an eighteenth-century war
US Army [a specific army]	a peacetime army
Declaration of Independence [title of a document]	a declaration of independence
May [specific month]	spring [general season]
Memorial Day [specific day]	a holiday
two Democratic candidates [refers to a political party]	democratic procedures [refers to a form of government]
a Ford tractor [brand name]	a farm tractor
Parkinson's disease [a disease named for a person]	flu, asthma, leukemia
Governor Clay [a person's title]	the governor of this state

9b   Titles and subtitles

The first and last words in titles and subtitles are capitalized, as are major words—that is, all words other than articles (*a, an,* and *the*), coordinating conjunctions (*and, but, for, nor, or, so,* and *yet*), prepositions (see the list on page 9), and the infinitive marker *to.* (For more information on titles, see **10a** and **16b**.)

From Here to Eternity

"To Be a Student or Not to Be a Student"

APA guidelines differ slightly from other style guidelines: APA recommends capitalizing any word in a title, including a preposition, that has four or more letters.

Southwestern Pottery from Anasazi to Zuni [MLA]

Southwestern Pottery From Anasazi to Zuni [APA]

MLA and APA advise capitalizing all words in a hyphenated compound, except for articles, coordinating conjunctions, and prepositions.

"The Arab-Israeli Dilemma" [compound proper adjective]

"Stop-and-Go Signals" [lowercase for the coordinating conjunction]

When a hyphenated word containing a prefix appears in a title or subtitle, capitalize both elements when the second element is a proper noun (**1a(2)**) or adjective (*Pre-Columbian*). However, if the word following the prefix is a common noun (as in *anti-independence*), capitalize it only if you are following APA guidelines.

"Pre-Columbian Artifacts in Peruvian Museums" [MLA and APA]

"Anti-Independence Behavior in Adolescents" [APA]

"Anti-independence Behavior in Adolescents" [MLA]

9c   Beginning a sentence

It is not difficult to remember that a sentence begins with a capital letter, but there are certain types of sentences that deserve special note.

(1) A quoted sentence

If a direct quotation is a full sentence, the first word should be capitalized.

> When asked to name the books she found most influential, Nadine Gordimer responded, "**I**n general, the works that mean most to one—change one's thinking and therefore maybe one's life—are those read in youth."

Even if you interrupt a quoted sentence with commentary, only the first letter should be capitalized.

> "**O**ddly," states Ved Mehta, "**l**ike my earliest memories, the books that made the greatest impression on me were the ones I encountered as a small child."

However, if you integrate someone else's sentence into a sentence of your own, the first letter should be lowercase—and placed in brackets if you are following MLA guidelines.

> Nadine Gordimer believes that "[**i**]n general, the works that mean most to one—change one's thinking and therefore maybe one's life—are those read in youth" (102).

(2) A freestanding parenthetical sentence

If you place a full sentence inside parentheses, and it is not embedded in a sentence of your own, be sure to capitalize the first word.

> Lance Armstrong won the Tour de France a record-breaking seven times. (**P**revious record holders include Jacques Anquetil, Bernard Hinault, Eddy Merckx, and Miguel Indurain.)

If the sentence inside the parentheses occurs within a sentence of your own, the first word should not be capitalized.

> Lance Armstrong won the Tour de France a record-breaking seven times (**p**reviously, he shared the record with four other cyclists).

(3) An independent clause following a colon

According to one style convention, if there is only one independent clause (**1f(1)**) following a colon, the first word should be lowercased. However, if two or more independent clauses follow the colon, the first word of each clause is capitalized.

> The ear thermometer is used quite frequently now: this type of thermometer records a temperature more accurately than a glass thermometer.

> Two new thermometers are replacing the old thermometers filled with mercury: The digital thermometer uses a heat sensor to determine body temperature. The ear thermometer is actually an infrared thermometer that detects the temperature of the eardrum.

The APA manual recommends capitalizing the first word of any independent clause following a colon. The MLA manual advises capitalizing the first word only if the independent clause is a rule or principle.

> Think of fever as a symptom, not as an illness: It is the body's response to infection. [APA]

> He has two basic rules for healthy living: Eat sensibly and exercise strenuously at least three times a week. [APA and MLA]

A grammar checker will flag a word at the beginning of a sentence that should be capitalized, but it will not be able to determine whether a word following a colon should be capitalized.

(4) Abbreviated questions

In a series of abbreviated questions, the first words of all the questions are capitalized when the intent is to draw attention to the individual questions. Otherwise, questions in a series begin with lowercase letters.

How do we distinguish the legal codes for families? For individuals? For genetic research?

Did you remember to include your application? résumé? letters of recommendation?

9d Computer keys, menu items, and icon names

When referring to a specific computer key, menu item, or icon name, capitalize the first letter.

To find the thesaurus, press Shift and the function key F7.

Instead of choosing Copy from the Edit menu, you can press Ctrl+C.

For additional information, click on Resources.

9e Unnecessary capitals

(1) Capitalizing common nouns

Many nouns can be either common or proper, depending on the context. A **proper noun** (1a(2)), also called a *proper name*, identifies a specific entity. A **common noun** (1a(2)), which is usually preceded by a word such as *the, a, an, this,* or *that,* is not capitalized.

a speech course in theater and television [COMPARE: Speech 324: Theater and Television]

a university, this high school [COMPARE: University of Michigan, Elgin High School]

(2) Overusing capitalization to signal emphasis

Occasionally, a common noun is capitalized for emphasis.

> Some politicians will do anything they can for Power.

If you use capitals for emphasis, do so sparingly; overuse will weaken the effect. For other ways to achieve emphasis, see Chapter **29**.

(3) Signaling emphasis online

For online writing in academic and professional contexts, capitalize as you normally do. Be careful not to capitalize whole words for emphasis because your reader may feel as though you are SHOUTING (which is the term used for this rude and undesirable practice).

EXERCISE 1

Edit the capitalization errors in the following paragraph. Be prepared to explain any changes that you make.

[1]Diana taurasi (Her teammates call her dee) plays basketball for the Phoenix mercury. [2]She has all the skills she needs to be a Star Player: She can pass and shoot, as well as rebound, block, and steal. [3]While playing for the university of connecticut huskies, she won the Naismith award twice and ranked in the majority of the big east's statistical categories. [4]Shortly after the huskies won their third straight ncaa title, taurasi was drafted first overall by the Phoenix mercury. [5]In april of 2004, taurasi played on the american national team against japan, and, in the Summer of 2004, she made her olympic debut in Athens.

10 | Italics

Italics indicate that a word or a group of words is being used in a special way. For example, the use of italics can clear up the ambiguity in the following sentence:

The linguistics students discussed the word stress.

Does this sentence mean that the students discussed a particular word or that they discussed the correct pronunciation of words? By italicizing *stress,* the writer indicates that it was the word itself, not an accent pattern, that the students discussed.

The linguistics students discussed the word *stress.*

This chapter will help you use italics for

- the titles of separate works (**10a**);
- foreign words (**10b**);
- the names of legal cases (**10c**);
- the names of ships, submarines, aircraft, spacecraft, and satellites (**10d**);
- words, letters, or numerals used as such or letters used in mathematical expressions (**10e**); and
- words receiving emphasis (**10f**).

Word-processing programs make it easy to use italics. In handwritten documents, you can indicate italics by underlining.

Paul Harding's novel <u>Tinkers</u> won the 2010 Pulitzer Prize for literature.

The use of italics instead of underlining is now widely accepted in business writing and academic writing. Both MLA and APA call for italics.

TECH SAVVY

Remember that in e-mail messages and on web pages, an underlined word or phrase often indicates a hyperlink. If you are not able to format your e-mails or other electronic text with italics, use an underscore before and after words you would normally italicize.

Paul Harding's novel _Tinkers_ won the 2010 Pulitzer Prize for literature.

10a Titles of works published or produced separately

Italics indicate the title of a longer work, while quotation marks indicate the title of a shorter work. For instance, the title of a collection of poetry published as a book (a longer work) is italicized (or underlined), and the title of any poem (shorter work) included in the book is enclosed in quotation marks (**16b**). These conventions help readers recognize the nature of a work and sometimes its relationship to another work.

Walt Whitman's "I Sing the Body Electric" first appeared in 1855 in the collection *Leaves of Grass*.

The titles of the following kinds of works are italicized:

Books	*The Little Prince*	*Huck Finn*
Magazines	*Wired*	*Rolling Stone*
Newspapers	*USA Today*	*Wall Street Journal*

Plays, films, DVDs	*The Lion King*	*Akeelah and the Bee*
Television and radio shows	*Mad Men*	*A Prairie Home Companion*
Recordings	*Can't Be Tamed*	*Great Verdi Overtures*
Works of art	*American Gothic*	*David*
Long poems	*Paradise Lost*	*The Divine Comedy*
Pamphlets	*Saving Energy*	*Tips for Gardeners*
Comic strips	*Peanuts*	*Doonesbury*

When an italicized title includes the title of a longer work within it, the embedded title is not italicized.

Modern Interpretations of Paradise Lost

If the italicized title includes the title of a short work within it, both titles are italicized, and the short work is also enclosed in quotation marks.

Willa Cather's "Paul's Case"

Titles are not placed in italics or between quotation marks when they stand alone on a title page, a book cover, or a newspaper page. Furthermore, neither italics nor quotation marks are used for titles of major historical documents, religious texts, or websites.

The Bill of Rights contains the first ten amendments to the US Constitution.

The Bible, a sacred text just as the Koran or the Torah is, begins with the Book of Genesis.

Instructions for making a cane-and-reed basket can be found on Catherine Erdly's website, Basket Weaving.

According to MLA guidelines, an initial *the* in a newspaper or periodical title is not italicized. Nor is it capitalized, unless it begins a sentence.

The story was published in the *New York Times.*

Also recommended is the omission of an article (*a, an,* or *the*) at the beginning of such a title when it would make a sentence awkward.

The report will appear in Thursday's ~~the~~ *Wall Street Journal.*

10b   Foreign words

Use italics to indicate foreign words.

Japan has a rich store of traditional folktales, *mukashibanashi,* "tales of long ago."   —**GARY SNYDER,** *Back on the Fire*

A foreign word used frequently in a text should be italicized only once—at its first occurrence.

The Latin words used to classify plants and animals according to genus and species are italicized.

Homo sapiens   *Rosa setigera*   *Ixodes scapularis*

Countless words borrowed from other languages have become part of English and are therefore not italicized.

bayou (Choctaw)   karate (Japanese)   arroyo (Spanish)

If you are not sure whether a word has been accepted into English, look for it in a standard dictionary (**19d**).

10c Legal cases

Italics identify the names of legal cases.

Miranda v. Arizona   *Roe v. Wade*

The abbreviation *v.* (for "versus") may appear in either italic or nonitalic type, as long as the style is used consistently. Italics are also used for the shortened name of a well-known legal case.

According to the *Miranda* decision, suspects must be informed of their right to remain silent and their right to legal advice.

Italics are not used to refer to a case by other than its official name.

All the major networks covered the O. J. Simpson trial.

10d Names of ships, submarines, aircraft, spacecraft, and satellites

Italicize the names of specific ships, submarines, aircraft, spacecraft, and satellites.

USS *Enterprise*   USS *Hawkbill*   *Enola Gay*   *Atlantis*   *Aqua*

The names of trains, the models of vehicles, and the trade names of aircraft are not italicized.

Orient Express   Ford Mustang   Boeing 747

| 10e | Words, letters, or numerals referred to as such and letters used in mathematical expressions |

When you refer to a specific word, letter, or numeral as itself, you should italicize it.

The word *love* is hard to define. [COMPARE: They were in love.]

The *b* in *bat* is not aspirated. [COMPARE: He earned a B+.]

The *2* on the sign has faded, and the *5* has disappeared. [COMPARE: She sent 250 cards.]

Statistical symbols and variables in algebraic expressions are also italicized.

The Pythagorean theorem is expressed as $a^2 + b^2 = c^2$.

| 10f | Words receiving emphasis |

Used sparingly, italics can signal readers to stress certain words.

These *are* the right files. [The verb *are* receives more emphasis than it normally would.]

Italics can also emphasize emotional content.

We have to go *now*. [The italicized word signals urgency.]

If overused, italics will lose their impact. Instead of italicizing words, substitute more specific words (Chapter **20**) or vary sentence structures (Chapter **30**).

EXERCISE 1

Identify all words that should be italicized in the following sentences. Explain why italics are necessary in each case.

1. Information about museum collections and exhibits can be found in art books, museum websites, and special sections of magazines and newspapers such as Smithsonian Magazine and the New York Times.

2. The website for the Metropolitan Museum of Art has pictures of Anthony Caro's sculpture Odalisque and Charles Demuth's painting The Figure 5 in Gold.

3. This book includes a photograph of a beautiful script used in the Koran; the script is known as the maghribi, or Western, style.

4. The large Tyrannosaurus rex discovered by Sue Hendrickson in South Dakota is on display at the Field Museum.

5. The Great Train Robbery, It Happened One Night, and Grand Illusion are in the collection at the Celeste Bartos Film Preservation Center.

11 | Abbreviations, Acronyms, and Numbers

Abbreviations, acronyms, and numbers facilitate easy recognition and effective communication in both academic papers and business documents. An **abbreviation** is a shortened version of a word or phrase: *assn.* (association), *dept.* (department), *et al.* (*et alii,* or "and others"). An **acronym** is a word formed by combining the initial letters and/or syllables of a series of words: *AIDS* (**a**cquired **i**mmune **d**eficiency **s**yndrome), *sonar* (**so**und **na**vigation **r**anging). This chapter will help you learn

- how and when to abbreviate (**11a–d**),
- when to explain an acronym (**11e**), and
- whether to spell out a number or use numerals (**11f** and **11g**).

Courtesy of UPS Worldwide, Inc.

Abbreviated brand names create instant recognition for products or services.

11a Abbreviations with names

The abbreviations *Ms., Mr., Mrs.,* and *Dr.* appear before names, whether given as full names or only surnames.

Ms. Sandy Scharnhorst   **Mrs.** Campbell

Mr. Alfredo Luján   **Dr.** Bollinger

Civil or military titles should not be abbreviated in academic writing.

Senator Bob Corker   Captain Derr   Professor Sue Li

Abbreviations such as *Jr., Sr.,* and *MD* appear after names.

Samuel Levy **Jr.**   Imogen Hickey, **MD**

Mark Ngo **Sr.**   Joan Richtsmeier, **PhD**

In the past, periods were customarily used in abbreviations for academic degrees, but APA and MLA now recommend omitting periods from abbreviations such as *MA, PhD,* and *MD.* Although MLA still follows the convention calling for commas to set off *Jr.* or *Sr.,* these abbreviations are increasingly considered part of the names they follow and thus need not be set off by commas unless you are following MLA style.

Note that when two designations are possible, only one should be used.

Dr. Kristin Grine OR Kristin Grine, **MD**
[NOT Dr. Kristin Grine, MD]

Abbreviations of plural proper nouns are often formed by simply adding *s* before the period: *Drs.* Grine and Hickey. But there are exceptions: the plural of *Mr.* is *Messrs.,* and the plural of *Mrs.* is *Mesdames,* for which there is no abbreviated form.

11b Addresses in correspondence

The names of states and words such as *Street, Road, Company,* and *Corporation* are usually written out when they appear in a sentence or in the letterhead on stationery. However, such words may be abbreviated when used in an address on an envelope.

Sentence   Derson Manufacturing Company is located on Madison Street in Watertown, Minnesota.

Address   Derson Manufacturing Co.
                  200 Madison St.
                  Watertown, MN 55388

When addressing correspondence within the United States, use the two-letter state abbreviations established by the US Postal Service. (No period follows these abbreviations.) If you do not know an appropriate state abbreviation or zip code, you can find it on the Postal Service's website.

11c Abbreviations in source documentation

Abbreviations are commonly used when citing research sources in bibliographies, footnotes, and endnotes. Common abbreviations include the following (not all citation styles accept all of these abbreviations).

Bibliographies and Notes

anon., Anon.	anonymous, Anonymous
biog.	biography, biographer, biographical
bull.	bulletin
c. or ca.	circa, about (for example, *c. 1920*)
col., cols.	column, columns

cont.	contents OR continues, continued
ed., eds.	editor, editors
et al.	*et alii* ("and others")
fig.	figure
fwd.	foreword, foreword by
illus.	illustrated by, illustrator, illustration
inc., Inc.	including, Incorporated
intl.	international
introd.	introduction, introduction by
ms., mss.	manuscript, manuscripts
natl.	national
n.d.	no date, no date of publication
n.p.	no place of publication, no publisher
n. pag.	no pagination
no., nos.	number, numbers
p., pp.	page, pages
P, Pr.	Press
pref.	preface
pt., pts.	part, parts
trans. or tr.	translation, translated by
U, Univ.	University

Computer Terms

FTP	file transfer protocol
HTML	hypertext markup language
http	hypertext transfer protocol
KB	kilobyte
MB	megabyte
MOO	multiuser domain, object-oriented

(*continued on page 146*)

(continued from page 145)

PDF	portable document format
URL	uniform resource locator

Divisions of Government

Cong.	Congress
dept.	department
div.	division
govt.	government
GPO	Government Printing Office
HR	House of Representatives

11d   Acceptable abbreviations in academic and professional writing

Abbreviations are usually too informal for use in sentences, although some have become so familiar that they are considered acceptable substitutes for full words.

(1) Abbreviations for special purposes

The names of months, days of the week, and units of measurement are usually written out (not abbreviated) when they are included in sentences, as are words such as *Street* and *Corporation*.

On a Tuesday in September, we drove ninety-nine miles to San Francisco, California, where we stayed in a hotel on Market Street.

Words such as *volume, chapter,* and *page* are abbreviated (*vol., ch.,* and *p.*) in bibliographies and in citations of research sources, but they are written out within sentences.

I read the introductory chapter and the three final pages in the first volume of the committee's report.

(2) Clipped forms

A word shortened by common usage, a **clipped form**, does not end with a period. Some clipped forms—such as *rep* (for *representative*), *exec* (for *executive*), and *info* (for *information*)—are too informal for use in college writing. Others—such as *exam, lab*, and *math*—have become acceptable because they have been used so frequently that they no longer seem like shortened forms.

(3) Abbreviations for time periods and zones

82 BC for *before Christ* [OR 82 BCE for *before the Common Era*]

AD 95 for *anno Domini,* "in the year of our Lord" [OR 95 CE for *of the Common Era*]

7:40 a.m. for *ante meridiem,* "before noon"

4:52 EST for *Eastern Standard Time*

Words designating units of time, such as *minute* and *month,* are written out when they appear in sentences. They can be abbreviated in tables or charts.

sec.   min.   hr.   wk.   mo.   yr.

(4) The abbreviation U.S. or US as an adjective

The abbreviation *U.S.* or *US* should be used only as an adjective in academic and professional writing. When using *United States* as a noun, spell it out. The choice of *U.S.* or *US* will depend on the discipline in which you are writing: MLA lists US as the preferred form, but APA uses U.S.

the U.S. Navy, the US economy
[COMPARE: They moved to the United States in 1990.]

(5) Individuals known by their initials

JFK   LBJ   E. B. White   B. B. King

In most cases, however, first and last names should be written out in full.

Oprah Winfrey   Peyton Manning   Donald Trump

(6) Some abbreviations for Latin expressions
Certain abbreviations for Latin expressions are common in academic writing.

cf. [compare]   et al. [and others]   i.e. [that is]

e.g. [for example]   etc. [and so forth]   vs. OR v. [versus]

11e   Acronyms

The ability to identify a particular acronym will vary from one audience to another. Some readers will know that NAFTA stands for the North American Free Trade Agreement; others may not. By spelling out acronyms the first time you use them, you are being courteous and clear. Introduce the acronym by placing it in parentheses after the group of words it stands for.

The Federal Emergency Management Administration (FEMA) was criticized by many after Hurricane Katrina.

MULTILINGUAL WRITERS

USING ARTICLES WITH ABBREVIATIONS, ACRONYMS, OR NUMBERS

When you use an abbreviation, an acronym, or a number, you sometimes need an indefinite article. Choose *a* or *an* based on the pronunciation of the initial sound of the abbreviation, acronym, or number: use *a* before a consonant sound and *an* before a vowel sound.

A picture of **a UN** delegation is on the front page of today's newspaper. [*UN* begins with a consonant sound (it is pronounced "you en").]

I have **an IBM** computer. [*IBM* begins with a vowel sound.]

The reporter interviewed **a NASA** engineer. [*NASA* begins with a consonant sound.]

My friend drives **a 1964** Mustang. [*1964* begins with a consonant sound.]

EXERCISE 1

Decide whether the following sentences use forms appropriate for academic writing. Correct any usage that is not appropriate.

1. I always wake up before 6 a.m.
2. The pope was buried in 670 anno Domini.
3. The Walkers live on Sandy Ridge Rd.
4. We can meet at the UPS shipping store.
5. She prefers to be addressed as Ms. Terry Campbell.

11f   General uses of numbers

Depending on their uses, numbers are treated in different ways. MLA recommends spelling out numbers that are expressed in one or two words (*nine, ninety-one, nine hundred, nine million*). A numeral is used for any other number (*9½, 9.9, 999*), unless it begins a sentence.

> I scored eighty-nine on the chemistry quiz.
>
> The population of Pennsylvania is almost thirteen million.
>
> The register recorded 164 names.

APA advises spelling out numbers below ten, common fractions, and numbers that are spelled out in universally accepted usage (for example, the Twelve Apostles). Both MLA and APA recommend using words rather than numerals at the beginning of a sentence.

> One hundred sixty-four names were recorded in the register. [Notice that *and* is not used in numbers greater than one hundred. NOT One hundred and sixty-four names . . . .]

When numbers or amounts refer to the same entities throughout a passage, use numerals when any of the numbers would be more than two words long if spelled out.

> Only 5 of the 134 delegates attended the final meeting. The remaining 129 delegates will be informed by e-mail.

In scientific or technical writing, numerals are used before abbreviations of units of measurements (*2 L, 30 cc*).

11g Special uses of numbers

(1) Expressing specific times of day in either numerals or words

Numerals or words can be used to express times of day. They should be used consistently.

> 4 p.m. OR four o'clock in the afternoon
>
> 9:30 a.m. OR half-past nine in the morning OR nine-thirty in the morning [Notice the use of hyphens.]

(2) Using numerals and words for dates

Months are written as words, years as numerals, and days and decades as either words or numerals. However, 9/11 is an acceptable alternative to September 11, 2001.

> May 20, 1976 OR 20 May 1976 [NOT May 20th, 1976]
>
> the fourth of December OR December 4
>
> the fifties OR the 1950s
>
> from 1999 to 2003 OR 1999–2003 [Use an en dash, not a hyphen, in number ranges.]

(3) Using numerals in addresses

Numerals are commonly used in street addresses and for zip codes.

> 25 Arrow Drive, Apartment 1, Columbia, MO 78209
>
> OR, for a mailing envelope, 25 Arrow Dr., Apt. 1
> Columbia, MO 78209

(4) Using numerals for identification

A numeral may be used as part of a proper noun (**1a(2)**).

> Channel 10   Edward III   Interstate 40   Room 311

(5) Referring to pages and divisions of books and plays

Numerals are used to designate pages and other divisions of books and plays.

　page 15　　Chapter 8　　Part 2

　in act 2, scene 1 OR in Act II, Scene 1

(6) Expressing decimals and percentages numerically

Numerals are used to express decimals and percentages.

　a 2.5 average　　12 percent　　0.853 metric ton

(7) Using numerals for large fractional numbers

Numerals with decimal points can be used to express large fractional numbers.

　5.2 million inhabitants　　1.6 billion years

MULTILINGUAL WRITERS

COMMAS AND PERIODS WITH NUMERALS

Cultures differ in their use of the period and the comma with numerals. In American usage, a decimal point (period) indicates a number or part of a number that is smaller than one, and a comma divides large numbers into units of three digits.

　　7.65 (seven and sixty-five　　10,000
　　one-hundredths)　　　　　　　(ten thousand)

In some other cultures, these usages of the decimal point and the comma are reversed.

　　7,65 (seven and sixty-five　　10.000
　　one-hundredths)　　　　　　　(ten thousand)

(8) Different ways of writing monetary amounts

Monetary amounts should be spelled out if they occur infrequently in a piece of writing. Otherwise, numerals and symbols can be used.

two million dollars	$2,000,000
ninety-nine cents	99¢ OR $0.99

EXERCISE 2

Edit the following sentences to correct the usage of abbreviations and numbers.

1. A Natl. Historic Landmark, Hoover Dam is located about 30 miles s.e. of Las Vegas, Nev.

2. Built by the fed. gov. between nineteen thirty-three and 1935, this dam is still considered one of the greatest achievements in the history of civ. engineering.

3. Construction of the dam became possible after several states in the Southwest (namely, AZ, CA, CO, NV, NM, UT, and WY) agreed on a plan to share water from the river.

4. 3,500 men worked on the dam during an average month of construction; this work translated into a monthly payroll of $500,000.

5. A popular tourist attraction, Hoover Dam was closed to the public after terrorists attacked the U.S. on 9/11.

P

PUNCTUATION

12 The Comma

Punctuation lends to written language the same flexibility that facial expressions, pauses, and variations in voice pitch offer spoken language. Pauses are often signaled by commas, but pauses are not the most reliable guide for comma placement. Commas are often called for where speakers do not pause, and pauses can occur where no comma is necessary. Thus, it is important to understand the basic principles of comma usage. This chapter will help you use commas to

- separate independent clauses joined by coordinating conjunctions (**12a**),
- set off introductory clauses and phrases (**12b**),
- separate items in a series (**12c**),
- set off nonessential (nonrestrictive) elements (**12d**),
- set off geographical names and items in dates and addresses (**12e**), and
- set off direct quotations (**12f**).

12a Before a coordinating conjunction linking independent clauses

Use a comma before a coordinating conjunction (*and, but, for, nor, or, so,* or *yet*) that links two independent clauses. (Some people use *fanboys*, a made-up word formed of the first letters of the coordinating conjunctions, as an aid to remembering them; see **1g(1)**.) An **independent clause** is a group of words that can stand as a sentence; that is, it has a subject and a predicate (**1f(1)**).

INDEPENDENT CLAUSE,	COORDINATING CONJUNCTION	INDEPENDENT CLAUSE.
	and	
	but	
	for	
Subject + predicate,	**nor**	subject + predicate.
	or	
	so	
	yet	

The Iditarod Trail Sled Dog Race begins in March, **but** training starts much sooner.

In the 1960s, Dorothy Page wanted to spark interest in the role of dog sledding in Alaskan history, **so** she proposed staging a long race.

No matter how many clauses are in a sentence, a comma comes before each coordinating conjunction.

The race takes several days to complete, **and** training is a year-round activity, **but** the mushers do not complain.

When the independent clauses are short, the comma is often omitted before *and, but,* or *or.*

My friend races **but** I don't.

If a coordinating conjunction joins two parts of a compound predicate (which means there is only one subject), a comma is not normally used before the conjunction. (See **1b** and **13c.**)

The race starts in Anchorage and ends in Nome.

A semicolon, instead of a comma, precedes a conjunction joining two independent clauses when at least one of the clauses already contains a comma. (See also **14a**.)

When running long distances, sled dogs burn more than ten thousand calories a day**;** **so** they must be fed well.

EXERCISE 1

Combine each of the following pairs of sentences by using coordinating conjunctions and inserting commas where appropriate. (Remember that not all coordinating conjunctions link independent clauses and that *but, for, so,* and *yet* do not always function as coordinating conjunctions.) Explain why you used each of the conjunctions you inserted.

1. Dinosaurs lived for 165 million years. Then they became extinct.

2. No one knows why dinosaurs became extinct. Several theories have been proposed.

3. Some theorists believe that a huge meteor hit the earth. The climate may have changed dramatically.

4. Another theory suggests that dinosaurs did not actually become extinct. They simply evolved into lizards and birds.

5. Yet another theory suggests that they just grew too big. Not all of the dinosaurs were huge.

12b   After introductory clauses, phrases, or words

(1) Following an introductory dependent clause

If you begin a sentence with a dependent (subordinate) clause (**1f(2)**), place a comma after it to set it off from the independent (main) clause (**1f(1)**).

> INTRODUCTORY CLAUSE, INDEPENDENT CLAUSE.

Although the safest automobile on the road is expensive, the protection it offers justifies the cost.

(2) Following an introductory phrase

Place a comma after an introductory phrase to set it off from the independent clause.

> INTRODUCTORY PHRASE, INDEPENDENT CLAUSE.

(a) Introductory prepositional phrases

Despite a downturn in the national economy, the number of students enrolled in this university has increased.

If you begin a sentence with a short introductory prepositional phrase (1e(4)), you may omit the comma as long as the resulting sentence is not difficult to read.

In 2009 the enrollment at the university increased.

BUT

In 2009, 625 new students enrolled in courses.
[A comma separates two numbers.]

A comma is not used after a prepositional phrase that begins a sentence in which the subject and predicate (1b) are inverted.

With children came responsibilities.
[The subject of the sentence is *responsibilities*: Responsibilities came with children.]

(b) Other types of introductory phrases

If you begin a sentence with a participial phrase (**1e(3)**) or an absolute phrase (**1e(6)**), place a comma after the phrase.

> **Having never left home,** she imagined the outside world to be fantastic, almost magical. [participial phrase]

> **The language difference aside,** life in Germany did not seem much different from life in the United States. [absolute phrase]

(3) Following an introductory word

> **INTRODUCTORY WORD,** INDEPENDENT CLAUSE.

Use a comma to set off an interjection, a **vocative** (a word used to address someone directly), or a transitional word that begins a sentence.

> **Yikes,** I forgot to pick him up from the airport. [interjection]

> **Bob,** I want you to know how very sorry I am. [vocative]

> **Moreover,** I insist on paying for your taxi. [transitional word]

When there is no risk of misunderstanding, some introductory adverbs and transitional words do not need to be set off by a comma (see also **14a**).

> **Sometimes** even a good design is rejected by the board.

EXERCISE 2

Insert commas where necessary in the following paragraph. Explain why each comma is needed. Some sentences may not require editing.

[1]If you had to describe sound would you call it a wave? [2]Although sound cannot be seen people have described it this way for a long time. [3]In fact the Greek philosopher Aristotle

believed that sound traveling through air was like waves in the sea. [4]Envisioning waves in the air he hypothesized that sound would not be able to pass through a vacuum because there would be no air to transmit it. [5]Aristotle's hypothesis was not tested until nearly two thousand years later. [6]In 1654 Otto von Guericke found that he could not hear a bell ringing inside the vacuum he had created. [7]Thus Guericke established the necessity of air for sound transmission. [8]However although most sound reaches us through the air it travels faster through liquids and solids.

12c Separating elements in a series or coordinate adjectives

A **series** contains three or more parallel elements. To be parallel, elements must be grammatically equal; all of them must be words, phrases, or clauses. (See Chapter 26.)

(1) Words, phrases, or clauses in a series
A comma appears after each item in a series except the last one.

Ethics are based on **moral, social,** or **cultural values.** [words in a series]

The company's code of ethics encourages **seeking criticism of work, correcting mistakes,** and **acknowledging the contributions of everyone.** [phrases in a series]

Several circumstances can lead to unethical behavior: **people are tempted by a desire to succeed, they are pressured by others into acting inappropriately,** or **they are simply trying to survive.** [clauses in a series]

If elements in a series contain internal commas, you can prevent misreading by separating the items with semicolons.

> According to their code of ethics, researchers must disclose all results, without omitting any data**;** indicate various interpretations of the data**;** and make the data and methodology available to other researchers, some of whom may choose to replicate the study.

THINKING RHETORICALLY
SERIES COMMAS VERSUS CONJUNCTIONS

How do the following sentences differ?

> We discussed them all: life**,** liberty**,** **and** the pursuit of happiness.
> We discussed them all: life **and** liberty **and** the pursuit of happiness.
> We discussed them all: life**,** liberty**,** the pursuit of happiness.

The first sentence follows conventional guidelines; that is, a comma and a conjunction precede the last element in the series. The less conventional second and third sentences do more than convey information. With two conjunctions and no commas, the second sentence slows down the pace of the reading, causing stress to be placed on each of the three elements in the series. In contrast, the third sentence, with commas but no conjunctions, speeds up the reading, as if to suggest that the rights listed do not need to be stressed because they are so familiar. To get a sense of how your sentences will be read and understood, try reading them aloud.

(2) Coordinate adjectives

Two or more adjectives that precede the same noun are called **coordinate adjectives**. To test whether adjectives are coordinate, either interchange them or put *and* between them. If the altered version of the adjectives-and-noun combination is acceptable, the adjectives are coordinate and should be separated by a comma or commas.

> Crossing the **rushing, shallow** creek, I slipped off a rock and fell into the water.
> [COMPARE: a rushing and shallow creek OR a shallow, rushing creek]

The adjectives in the following sentence are not separated by a comma. Notice that they cannot be interchanged or joined by *and*.

> Sitting in the water, I saw an **old wooden** bridge.
> [NOT a wooden old bridge OR an old and wooden bridge]

12d With nonessential elements

Nonessential (nonrestrictive) elements provide supplemental information, that is, information a reader does not need in order to identify who or what is being discussed (see also 1f(2)). Use commas to set off a nonessential word or word group: one comma precedes a nonessential element at the end of a sentence; two commas set off a nonessential element in the middle of a sentence.

> The Annual Hilltop Folk Festival**, planned for late July,** should attract many tourists.

In the preceding sentence, the phrase placed between commas, *planned for late July*, conveys nonessential information: the reader knows which festival will attract tourists without being told when it will be held. When a phrase follows a proper noun (**1a(2)**), such as *The Annual Hilltop Folk Festival,* it is usually nonessential. Note, however, that in the following sentence, the phrase *planned for late July* is necessary for the reader to identify the festival as the one scheduled to occur in late July, not another time:

> The festival **planned for late July** should attract many tourists.

In the preceding sentence, the phrase is an **essential (restrictive) element** because, without it, the reader will not know which festival the writer has in mind. Essential elements are not set off by commas; they are integrated into sentences (**1f(2)**).

(1) Setting off nonessential elements used as modifiers

(a) Adjectival clauses

Nonessential modifiers are often **adjectival (relative) clauses**, which are usually introduced by a relative pronoun, *who, which,* or *that* (**1f(2)**). In the following sentence, a comma sets off the adjectival clause because the reader does not need the content of that clause in order to identify the mountain:

> We climbed Mt. McKinley, **which is over 15,000 feet high.**

(b) Participial phrases

Nonessential modifiers also include **participial phrases** (phrases that begin with a present or past participle of a verb) (**1e(3)**).

> Mt. McKinley, **towering above us,** brought to mind our abandoned plan for climbing it. [participial phrase beginning with a present participle]

My sister, **slowed by a knee injury,** rarely hikes anymore. [participial phrase beginning with a past participle]

(c) Adverbial clauses

An **adverbial clause** (1f(2)) begins with a subordinating conjunction that signals cause (*because*), purpose (*so that*), or time (*when, after, before*). This type of clause is usually considered essential and thus is not set off by a comma when it appears at the end of a sentence.

Dinosaurs may have become extinct **because their habitat was destroyed.**

In contrast, an adverbial clause that provides nonessential information, such as an aside or a comment, should be set off from the main clause.

Dinosaurs are extinct, **though they are alive in many people's imaginations.**

(2) Setting off nonessential appositives

Appositives refer to the same person, place, object, idea, or event as a nearby noun or noun phrase but in different words (1e(5)). Nonessential appositives provide extra details about nouns or noun phrases (1e(1)) and are set off by commas; essential appositives are not. In the following sentence, the title of the article is mentioned, so the reader does not need the information provided by the appositive in order to identify the article. The appositive is thus set off by commas.

"Living on the Line," **Joanne Hart's most recent article,** describes the lives of factory workers in China.

In the next sentence, *Joanne Hart's article* is nonspecific, so an essential appositive containing the specific title of the article is integrated into the sentence. It is not set off by commas. Without the appositive, the reader would not know which of Hart's articles describes the lives of factory workers in China.

> Joanne Hart's article "Living on the Line" describes the lives of factory workers in China.

If Hart had written only this one article, the title would be set off by commas. The reader would not need the information in the appositive to identify the article.

Abbreviations of titles or degrees appearing after names are treated as nonessential appositives.

> Was the letter from Frances Evans **,** PhD **,** or from Francis Evans **,** MD?

Increasingly, however, *Jr., Sr., II,* and *III* are considered part of a name, and the comma is thus often omitted (see **11a**).

> William Homer Barton **,** Jr.   OR   William Homer Barton Jr.

EXERCISE 3

Set off nonessential clauses, phrases, and appositives with commas.

1. Maine Coons long-haired cats with bushy tails have adapted to a harsh climate.

2. These animals which are extremely gentle despite their large size often weigh twenty pounds.

3. Most Maine Coons have exceptionally high intelligence for cats which enables them to recognize language and even to open doors.

4. Unlike most cats Maine Coons will play fetch with their owners.

5. According to a legend later proven to be false Maine Coons developed from interbreeding between wildcats and domestic cats.

(3) Setting off absolute phrases

An **absolute phrase** (the combination of a noun and a modifying word or phrase; see 1e(6)) provides nonessential details and so should always be set off by a comma or commas.

The actor, **his hair wet and slicked back,** began his audition.

The director stared at him, **her mind flipping through the photographs she had viewed earlier.**

(4) Setting off transitional expressions and other parenthetical elements

Commas customarily set off transitional words and phrases such as *for example, that is,* and *namely.*

An airline ticket, **for example,** can be delivered electronically.

Some transitional words and short phrases such as *also, too, at least,* and *thus* need not be set off by commas.

Traveling has **thus** become easier in recent years.

Use commas to set off other parenthetical elements, such as words or phrases that provide commentary you wish to stress.

Over the past year, my flights have, **miraculously,** been on time.

(5) Setting off contrasted elements

Commas set off sentence elements in which words such as *never* and *unlike* express contrast.

A planet, **unlike** a star, reflects rather than generates light.

In sentences in which contrasted elements are introduced by *not only . . . but also,* place a comma before *but* if you want to emphasize what follows it. Otherwise, leave the comma out.

Planets **not only** vary in size, **but also** travel at different speeds. [Comma added for emphasis.]

12e With geographical names and items in dates and addresses

Use commas to make geographical names, dates, and addresses easy to read.

(1) City and state

Nashville, Tennessee, is the capital of country-and-western music in the United States.

(2) Day and date

Martha left for Peru on **Wednesday, February 12, 2009,** and returned on March 12.

OR

Martha left for Peru on **Wednesday, 12 February 2009,** and returned on 12 March.

In the style used in the second sentence (which is not as common in the United States as the style in the first example), one comma is omitted because *12* precedes *February* and is thus clearly separate from *2009.*

(3) Addresses

In a sentence containing an address, the name of the person or organization, the street address, and the name of the town or city are all followed by commas, but the abbreviation for the state is not.

> I had to write to **Ms. Melanie Hobson, Senior Analyst, Hobson Computing, 2873 Central Avenue, Orange Park, FL 32065.**

12f With direct quotations

Many sentences containing direct quotations also contain signal phrases (or attributive tags) such as *The author claims* or *According to the author* (38d(2)). Use commas to set off these phrases whether they occur at the beginning, in the middle, or at the end of a sentence.

(1) Signal phrase at the beginning of a sentence

Place the comma directly after the signal phrase, before the quotation marks.

> As Jacques Barzun claims, "It is a false analogy with science that makes one think latest is best."

(2) Signal phrase in the middle of a sentence
Place the first comma inside the quotation marks that precede the signal phrase; place the second comma directly after the phrase, before the next set of quotation marks.

"It is a false analogy with science," claims Jacques Barzun, "that makes one think latest is best."

(3) Signal phrase at the end of a sentence
Place the comma inside the quotation marks before the signal phrase.

"It is a false analogy with science that makes one think latest is best," claims Jacques Barzun.

13 Unnecessary or Misplaced Commas

Although a comma may signal a pause, not every pause calls for a comma. As you read the following sentence aloud, you may pause naturally at several places, but no commas are necessary.

> Heroic deeds done by ordinary people inspire others to act in ways that are not only moral but courageous.

This chapter will help you recognize unnecessary or misplaced commas that

- separate a subject and its verb or a verb and its object (**13a**);
- follow a coordinating conjunction (**13b**);
- separate elements in a compound predicate (**13c**);
- set off essential words, phrases, or clauses (**13d**); and
- precede the first item of a series or follow the last (**13e**).

13a No comma between a subject and its verb or a verb and its object

Although speakers often pause after the subject (**1b**) or before the object (**1c**) of a sentence, such a pause should not be indicated by a comma.

> In this climate, rain at frequent intervals produces mosquitoes. [no separation between the subject (*rain*) and the verb (*produces*)]

> The forecaster said that rain was likely. [no separation between the verb (*said*) and the direct object (the noun clause *that rain was likely*)]

13b No comma following a coordinating conjunction

Avoid using a comma after a coordinating conjunction (*and, but, for, nor, or, so,* or *yet*).

We worked very hard on her campaign for state representative, but⌇the incumbent was too strong to defeat in the northern^ districts.

13c No comma separating elements in a compound predicate

In general, avoid using a comma between two elements of a compound predicate (**1b**).

I read the comments carefully⌇and then started my revision.

However, if you want to place stress on the second element in a compound predicate, you may place a comma after the first element. Use this option sparingly, or it will lose its effect.

I read the comments word by word, and despaired.

13d No comma setting off essential words, phrases, and clauses

In the following sentences, the elements in boldface are essential and so should not be set off by commas.

Zoe was born⌇**in Chicago during the Great Depression.**

Perhaps⌇the thermostat is broken.

Everyone⌇**who has a mortgage**⌇is required to have fire insurance.

Someone⌇**wearing an orange wig**⌇greeted us at the door.

13e No comma preceding the first item of a series or following the last

Make sure that you place commas only between elements in a series, not before or after the series.

She was known for⟲her photographs, sketches, and engravings.

The exhibit included her most exuberant, exciting, and expensive⟲photographs.

EXERCISE 1

Explain the use of each comma in the following paragraph.

¹There is some evidence that musical training may enhance performance on some tests of mental abilities, but the effects are not great. ²To some extent, this is another chicken-and-egg problem. ³Does musical training enhance performance on the tests, or do children who take musical training exhibit enhanced performance on tests because of their particular interests and abilities? ⁴But early musical training perhaps increases the possibility the child will have perfect pitch, as we noted, presumably enhancing later musical capabilities. ⁵Actually, one recent and carefully done study found a greater increase in the IQ scores of children after taking music lessons than after taking drama or no lessons.

—RICHARD F. THOMPSON AND STEPHEN A. MADIGAN,
Memory: The Key to Consciousness

14 | The Semicolon

The semicolon indicates that the phrases or clauses on either side of it are closely related. It most frequently connects two independent clauses when the second clause supports or contrasts with the first, but it can be used for other purposes as well. This chapter will help you understand that semicolons

- link closely related independent clauses (**14a**) and
- separate parts of a sentence containing internal commas (**14b**) but
- do not connect independent clauses to phrases or dependent clauses (**14c**).

14a | Connecting independent clauses

A semicolon placed between two independent clauses indicates that they are closely related. The second of the two clauses generally supports or contrasts with the first.

For many cooks, basil is a key ingredient; it appears in recipes worldwide. [support]

Sweet basil is used in many Mediterranean dishes; Thai basil is used in Asian and East Indian recipes. [contrast]

Although *and, but*, and similar words can signal these kinds of relationships, consider using an occasional semicolon for variety.

Sometimes, a transitional expression such as *for example* or *however* (**3c(5)**) accompanies a semicolon and further establishes the exact relationship between the ideas in the linked clauses.

> Basil is omnipresent in the cuisine of some countries**;** **for example,** Italians use basil in salads, soups, and many vegetable dishes.

> The culinary uses of basil are well known**;** **however,** this herb also has medicinal uses.

A comma is usually inserted after a transitional word, but it can be omitted if doing so will not lead to a misreading.

> Because *basil* comes from a Greek word meaning "king," it suggests royalty**;** **indeed** some cooks accord basil royal status among herbs and spices.

14b | Separating elements that contain commas

In a series of phrases or clauses (**1e** and **1f**) that contain commas, semicolons indicate where each phrase or clause ends and the next begins. In the following sentence, the semicolons help the reader distinguish three separate phrases.

> To survive, mountain lions need a large area in which to range**;** a steady supply of deer, skunks, raccoons, foxes, and opossums**;** and the opportunity to find a mate, establish a den, and raise a litter.

EXERCISE 1

Revise the following sentences, using semicolons to separate independent clauses or elements that contain internal commas.

1. Soccer is a game played by two opposing teams on a rectangular field, each team tries to knock a ball, roughly twenty-eight inches in circumference, through the opponent's goal.

(continued on page 176)

(continued from page 175)

2. The game is called *soccer* only in Canada and the United States, elsewhere it is known as *football*.

3. Generally, a team consists of eleven players: defenders (or fullbacks), who defend the goal by trying to win control of the ball, midfielders (or halfbacks), who play both defense and offense, attackers (or forwards), whose primary responsibility is scoring goals, and a goalkeeper (or goalie), who guards the goal.

4. Soccer players depend on five skills: kicking, which entails striking the ball powerfully with the top of the foot, dribbling, which requires tapping or rolling the ball while running, passing, which is similar to kicking but with less power and more control, heading, which involves striking the ball with the forehead, and trapping, which is the momentary stopping of the ball.

14c   Revising common semicolon errors

Semicolons do not set off phrases (**1e**) or dependent clauses (**1f(2)**) unless those elements contain commas. Use commas to set off a phrase or a dependent clause.

We consulted Alinka Kibukian;, the local horticulturalist.

Needing summer shade;, we planted two of the largest trees we could afford.

We learned that young trees need care;, which meant we had to do some extra chores after dinner each night.

Our trees have survived;, even though we live in a harsh climate.

EXERCISE 2

Use a comma to replace any semicolon that sets off a phrase or a dependent clause in the following sentences. Do not change properly used semicolons.

1. The Vice President of the United States walks along the fenced-in crowd; regularly stopping to shake a hand, pose for the camera, or hold a baby; a man who has held elected office for over forty years, he relishes being in the public eye.

2. Joe Biden has gone into the Grange Hall, where he sits with a small group of family farmers; peppering them with questions about the future of the farm, whether the corporate farm, the family farm, or some combination of the two bodes best for our nation's future; he leaves the hall, thanks his hosts, and makes his way to Air Force Two and his next commitment.

15 The Apostrophe

Apostrophes serve a number of purposes. You can use them to show that someone owns something (*my neighbor's television*), that someone has a specific relationship with someone else (*my neighbor's children*), or that someone has produced or created something (*my neighbor's recipe*). Apostrophes are also used in contractions (*can't, don't*) and in a few plural forms (*x's* and *y's*). This chapter will help you use apostrophes to

- indicate ownership and other relationships (**15a**),
- mark omissions of letters or numbers (**15b**), and
- form certain plurals (**15c**).

15a Indicating ownership and other relationships

An apostrophe, often followed by an *s*, signals the possessive case of nouns. (For information on case, see **5a(1)** and **5b**.) Possessive nouns are used to express a variety of meanings.

Ownership	**Dyson's** sermon, the **minister's** robe
Origin	**Leakey's** research findings, the **guide's** decision
Human relationships	**Helen's** sister, the **teacher's** students
Possession of physical or psychological traits	**Mona Lisa's** smile, the **team's** spirit
Specification of amounts	a **day's** wages, an **hour's** delay

Association between abstractions and attributes	**democracy's** struggles, **tyranny's** influence
Identification of documents and credentials	**driver's** license, **bachelor's** degree
Identification of things named after people	**St. John's** Cathedral, **Valentine's** Day

MULTILINGUAL WRITERS

WORD WITH APOSTROPHE AND S VERSUS PHRASE BEGINNING WITH *OF*

In many cases, to indicate ownership, origin, and other meanings discussed in this chapter, you can use either a word with an apostrophe and an *s* or a prepositional phrase beginning with *of*.

Henning Mankell**'s** novels
OR
the novels **of** Henning Mankell

the plane**'s** arrival   OR   the arrival **of** the plane

However, the ending -*'s* is more commonly used with nouns referring to people, and a phrase beginning with *of* is used with most nouns referring to location.

my **uncle's** workshop, **Edward's** truck, the **student's** paper [nouns referring to people]

the **end of** the movie, the **middle of** the day, the **front of** the building [nouns referring to location]

> *(continued from page 181)*
>
> Likewise, to form the plural of a family name, use *-s* or *-es*, not an apostrophe.
>
> Johnsons
> The⸌ ~~Johnson's~~ participated in the study.
>
> [COMPARE: The Johnsons' participation in the study was crucial.]
>
> Jameses
> The⸌ ~~James's~~ live in the yellow house on the corner.
>
> [COMPARE: The Jameses' house is the yellow one on the corner.]

(3) Showing collaboration or joint ownership

In the first example below, the ending *-'s* has been added to the second singular noun (*plumber*). In the second example, just an apostrophe has been added to the second plural noun (*Lopezes*), which already ends in *s*.

the carpenter and the **plumber's** decision [They made the decision collaboratively.]

the Becks and the **Lopezes'** cabin [They own one cabin jointly.]

(4) Showing separate ownership or individual contributions

In the examples below, the possessive form of each plural noun ends with an apostrophe, and that of each singular noun has the ending *-'s*.

the **Becks'** and the **Lopezes'** cars [Each family owns a car.]

the **carpenter's** and the **plumber's** proposals [They each made a proposal.]

(5) Forming the possessive of a compound noun

Add *-'s* to the last word of a compound noun.

my brother-in-**law's** friends, the attorney **general's** statements [singular] my brothers-in-**law's** friends, the attorneys **general's** statements [plural]

To avoid awkward constructions such as the last two, consider using a prepositional phrase beginning with *of* instead: *a friend of my brothers-in-law* and *the statements of the attorneys general.*

(6) Forming the possessive of a noun preceding a gerund
Depending on its number, a noun that precedes a gerund takes either -'s or just an apostrophe.

> **Lucy's having** to be there seemed unnecessary. [singular noun preceding a gerund]

> The family appreciated the **lawyers' handling** of the matter. [plural noun preceding a gerund]

MULTILINGUAL WRITERS

GERUND PHRASES

When a gerund appears after a possessive noun, the noun is the subject of the gerund phrase.

> **Lucy's having to be there** [COMPARE: **Lucy** has to be there.]

> **The lawyers' handling of the matter** [COMPARE: **The lawyers** handled the matter.]

A gerund phrase may serve as the subject or the object in a sentence (**1e(3)**).

s
Lucy's having to be there seemed unnecessary.

obj
The family appreciated **the lawyers' handling of the matter.**

Sometimes you may find it difficult to distinguish between a gerund and a present participle (**1e(3)**). A good way to tell the difference is to note whether the emphasis is on an action or on a person. In a sentence containing a gerund, the emphasis is on the action; in a sentence containing a participle, the emphasis is on the person.

Our successful completion of the project depends on **Terry's providing** the illustrations. [gerund; the emphasis is on the action, *providing*]

I remember **my brother telling** me the same joke last year. [participle; the emphasis is on the person, *my brother*]

(7) Naming products and geographical locations

Follow an organization's preference for the use of an apostrophe in its name or the name of a product. Follow local conventions for an apostrophe in the name of a geographical location.

Consumers Union	Actors' Equity	Taster's Choice
Devil's Island	Devils Tower	Devils Mountain

© N. Reed of QEDimages/Alamy

REUTERS/Joe Skipper/Landov

Whether an apostrophe is used in a brand name is determined by the organization that owns that name.

EXERCISE 1

Following the pattern of the examples, change the modifier after each noun to a possessive form that precedes the noun.

EXAMPLES

proposals made by the committee   *the committee's proposals*

poems written by Keats   *Keats's poems*

1. the day named after St. Patrick
2. a leave of absence lasting six months
3. the report given by the eyewitness
4. a new book coauthored by Pat and Alan
5. the weights of the children

15b   Marking omissions of letters or numbers

Apostrophes signal contractions and other omissions in numbers and in words representing speech.

they're [they are]   y'all [you all]

class of '11 [class of 2011]   singin' [singing]

Contractions are not always appropriate for formal contexts. Your audience may expect you to use full words instead (for example, *cannot* instead of *can't* and *will not* instead of *won't*).

15c   Forming certain plurals

In the past, an apostrophe and an *s* were used to form the plural of a number, an abbreviation, or a word referred to as a term. Today, the apostrophe is rarely used in plural forms except for those of abbreviations that take periods or that contain lowercase letters or symbols.

The following plurals are generally formed by simply adding an *s:*

1990s	fours and fives	YWCAs
two *and*s	the three Rs	

A few plural forms still include an apostrophe:

x's and y's   +'s and −'s

The MLA also recommends that an apostrophe be used to form the plural of an uppercase letter (A's and B's).

EXERCISE 2

Insert apostrophes where needed in the following sentences. Be prepared to explain why they are necessary.

1. Whos in charge here?

2. Hansons book was published in the early 1920s.

3. They hired a rock n roll band for their engagement party.

4. NPRs fund drive begins this weekend.

5. Youll have to include the ISBNs of the books youre going to purchase.

6. Only three of the proposals are still being considered: yours, ours, and the Wilbers.

7. Its always a big deal when your children leave for college.

8. Not enough students enrolled in the summer 11 course.

9. The students formed groups of twos and threes.

10. Laquisha earned two As and two Bs this semester.

16 Quotation Marks

Quotation marks enclose sentences or parts of sentences that play a special role. They can indicate that the words between them were first written or spoken by someone else or that they are being used in an unconventional way. This chapter will help you use quotation marks

- with direct quotations (**16a**),
- with titles of short works (**16b**),
- to indicate that words or phrases are used ironically or unconventionally (**16c**), and
- in combination with other punctuation marks (**16d**).

16a Direct quotations

Double quotation marks set off direct quotations, including those in dialogue. Single quotation marks set off a quotation within a quotation.

(1) Double quotation marks with direct quotations

Quotation marks enclose only a direct quotation, not any accompanying signal phrase such as *she said* or *he replied*. When a sentence ends with quoted material, place the period inside the quotation marks. For guidelines on comma placement, see **16d(1)**.

> "I believe that we learn by practice," writes Martha Graham in "An Athlete of God." "Whether it means to learn to dance by practicing dancing or to learn to live by practicing living, the principles are the same."

When using direct quotations, reproduce all quoted material exactly as it appears in the original, including capitalization and punctuation. To learn how to set off long quotations as indented blocks, see 39a(2).

(2) No quotation marks for indirect quotations or paraphrases
Indirect quotations and paraphrases (38d(3)) are restatements of what someone else has said or written.

> Martha Graham believes that practice is necessary for learning, regardless of what we are trying to learn.

(3) Single quotation marks for quotations within quotations
If the quotation you are using includes another direct quotation, use single quotation marks with the embedded quotation.

> According to Anita Erickson, "when the narrator says, 'I have the right to my own opinion,' he means that he has the right to his own delusion."

However, if an embedded quotation appears within a block quotation, it should be enclosed in double quotation marks. (Keep in mind that double quotation marks are not used at the beginning or end of a block quotation.)

> Anita Erickson claims that the narrator uses the word *opinion* deceptively.
>> Later in the chapter, when the narrator says, "I have the right to my own opinion," he means that he has the right to his own delusion. Although it is tempting to believe that the narrator is making decisions based on a rational belief system, his behavior suggests that he is more interested in deception. With poisonous lies, he has already deceived his business partner, his wife, and his children.

(4) Dialogue in quotation marks
When creating or reporting a dialogue, enclose in quotation marks what each person says, no matter how short. Use

a separate paragraph for each speaker, beginning a new paragraph whenever the speaker changes. Narrative details can be included in the same paragraph as a direct quotation.

> Farmer looked up, smiling, and in a chirpy-sounding voice he said, "But that feeling has the disadvantage of being . . . " He paused a beat. "Wrong."
> "Well," I retorted, "it depends on how you look at it."
> —TRACY KIDDER, *Mountains Beyond Mountains*

When quoting more than one paragraph by a single speaker, put quotation marks at the beginning of each paragraph. However, do not place closing quotation marks at the end of each paragraph—only at the end of the last paragraph.

(5) Thoughts in quotation marks

Quotation marks set off thoughts that resemble speech.

> "He's already sulking about the outcome of the vote," I thought, as I watched the committee chair work through the agenda.

Thoughts are usually marked by such phrases as *I thought, he felt,* and *she believed.* Remember, though, that quotation marks are not used with thoughts that are reported indirectly (16a(2)).

> I wondered why he had not responded to my memo.

(6) Short excerpts of poetry within a sentence in quotation marks

When quoting fewer than four lines of poetry, enclose them in quotation marks and use a slash (17i) to indicate the line division.

> Together, mother and daughter recited their favorite lines: "Shall I compare thee to a summer's day? / Thou art more lovely and more temperate."

To learn how to format longer quotations of poetry, see 41e(4).

16b Titles of short works

Quotation marks enclose the title of a short work, such as a story, an essay, a poem, or a song. The title of a longer work, such as a book, a magazine, a newspaper, or a play, should be italicized (see **10a**).

"The Girls of Summer" first appeared in the *New Yorker*.

Short story	"The Lottery"	"A Good Man Is Hard to Find"
Essay	"Walden"	"A Modest Proposal"
Article	"Small World"	"Arabia's Empty Quarter"
Book chapter	"Rain"	"Cutting a Dash"
Short poem	"Orion"	"Where the Sidewalk Ends"
Song	"Lazy River"	"Like a Rolling Stone"
TV episode	"Show Down!"	"The Last Time"

Use double quotation marks around the title of a short work embedded in a longer italicized title.

Interpretations of "*Young Goodman Brown*" [book about a short story]

Use single quotation marks for a title within a longer title that is enclosed in double quotation marks.

"Irony in 'The Road Not Taken'" [article about a poem]

MULTILINGUAL WRITERS

DIFFERING USES OF QUOTATION MARKS

In works published in Great Britain, the use of quotation marks differs in some ways from the U.S. style presented here. For example, single quotation marks are always used to set off direct quotations and the titles of short works. A period is placed outside a quotation mark ending a sentence. Double quotation marks indicate a quotation within a quotation. When writing in the United States, be sure to follow the rules for American English.

British usage	In class, we compared Wordsworth's 'Upon Westminster Bridge' with Blake's 'London'.
American usage	In class, we compared Wordsworth's "Upon Westminster Bridge" with Blake's "London."

16c For ironic tone or unusual usage

Writers sometimes use quotation marks to indicate that they are using a word or phrase ironically. The word *gourmet* is used ironically in the following sentence.

His "gourmet" dinner turned out to be processed turkey and instant mashed potatoes.

CAUTION

Avoid using quotation marks around words that may not be appropriate for your rhetorical situation. Instead, take the time to choose suitable words. The revised sentence in the following pair is more effective than the first.

(*continued on page 192*)

(continued from page 191)

| Ineffective | He is too much of a "wimp" to be a good leader. |
| Revised | He is too indecisive to be a good leader. |

Similarly, putting a cliché (20b) in quotation marks may make readers conclude that you do not care enough about conveying your meaning to think of a fresh expression.

16d   With other punctuation marks

To decide whether to place some other punctuation mark inside or outside quotation marks, determine whether the particular mark functions as part of the quotation or part of the surrounding context.

(1) With commas and periods

Quoted material is usually accompanied by a signal phrase such as *she said, he replied*, or *the author argued*. When a sentence starts with such an expression, place a comma after it to separate the signal phrase from the quotation.

> She replied, "There's more than one way to slice a pie."

If the sentence starts with the quotation, place the comma inside the closing quotation marks.

> "There's more than one way to slice a pie," she replied.

Place a period inside closing quotation marks, whether single or double, if a quotation ends a sentence.

> Jeff responded, "I didn't understand 'An Algorithm for Life.'"

When quoting material from a source, provide the relevant page number(s). If you are following MLA guidelines, note the page number(s) in parentheses after the final quotation marks.

Place the period that ends the sentence after the final parenthesis, unless the quotation is a block quotation (39a(2)).

> According to Diane Ackerman, "Love is a demanding sport involving all the muscle groups, including the brain" (86).

CAUTION

Do not put a comma after *that* when it precedes a quotation.

> Diane Ackerman claims that "[l]ove is a demanding sport involving all the muscle groups, including the brain" (86).

(2) With semicolons and colons
Place semicolons and colons outside quotation marks.

> His favorite song was "Cyprus Avenue"; mine was "Astral Weeks."

> Because it is repeated, one line stands out in "The Conductor": "We are never as beautiful as now."

(3) With question marks, exclamation points, and dashes
If the direct quotation includes a question mark, an exclamation point, or a dash, place that punctuation *inside* the closing quotation marks.

> Jeremy asked, "What is truth?"

> Gordon shouted "Congratulations!"

> Laura said, "Let me tell—" Before she could finish her sentence, Dan walked into the room.

Use just one question mark inside the quotation marks when a question you write ends with a quoted question.

> Why does the protagonist ask, "Where are we headed?"

If the punctuation is not part of the quoted material, place it *outside* the closing quotation marks.

Who wrote "The Figure a Sentence Makes"?

You have to read "Awareness and Freedom"!

She called me a "toaster head"—perhaps justifiably under the circumstances.

EXERCISE 1

Revise sentences in which quotation marks are used incorrectly and insert quotation marks where they are needed. Do not alter sentences that are written correctly. (The numbers in parentheses are page numbers, placed according to MLA guidelines.)

1. Have you read Susan B. Anthony's essay On Women's Right to Vote?

2. Anthony states, [I] exercised my citizen's rights (2).

3. However, she realizes that she has been indicted for "the alleged crime of having voted in the last presidential election" (2).

4. Anthony suggests that not allowing women to vote is a violation of 'the supreme law of the land' (3).

5. According to the author, "We, the whole people, . . . formed the Union."

6. She goes on to argue that Webster, Worcester, and Bouvier all define a citizen to be a person in the United States, entitled to vote and hold office.

7. Anthony maintains, Being persons, then, women are citizens; and no state has a right to make any law . . . that shall abridge their privileges or immunities (3).

17 The Period and Other Punctuation Marks

To indicate the end of a sentence, you can use one of three punctuation marks: the period, the question mark, or the exclamation point. Which one you use depends on your meaning: do you want to make a statement, ask a question, or express an exclamation?

Everyone passed the exam.

Everyone passed the exam? [informal usage]

Everyone passed the exam!

Within sentences, you can use colons, dashes, parentheses, square brackets, ellipsis points, and slashes to emphasize, downplay, or clarify the information you want to convey. (For use of the hyphen, see **18f.**) This chapter will help you use

- end punctuation marks (the period (**17a**), the question mark (**17b**), and the exclamation point (**17c**)),
- the colon (**17d**),
- the dash (**17e**),
- parentheses (**17f**),
- square brackets (**17g**),
- ellipsis points (**17h**), and
- the slash (**17i**).

To accommodate computerized typesetting, APA guidelines call for only one space after a period, a question mark, an exclamation point, and a colon. According to this manual, there should be no space preceding or following a hyphen or a dash.

The MLA style manual recommends using only one space after end punctuation marks but allows two spaces if they are used consistently.

17a   The period

(1) Marking the end of a sentence

Use a period at the end of a declarative sentence.

> Many adults in the United States are overfed yet undernourished.

> Soft drinks account for 7 percent of their average daily caloric intake.

In addition, place a period at the end of an instruction or recommendation written as an imperative sentence (1i).

> Eat plenty of fruits and vegetables. Drink six to eight glasses of water a day.

Indirect questions are phrased as statements, so be sure to use a period, rather than a question mark, at the end of such a sentence.

> The researcher explained why people eat so much junk food. [COMPARE: Why do people eat so much junk food?]

(2) Following some abbreviations

> Dr.   Jr.   a.m.   p.m.   vs.   etc.   et al.

Only one period follows an abbreviation that ends a sentence.

> The tour begins at 1:00 p.m.

Periods are not used with many common abbreviations, such as *MVP, mph,* and *FM* (see Chapter 11). A dictionary lists the conventional form of an abbreviation as well as any alternatives.

17b The question mark

Place a question mark after a direct question.

How does the new atomic clock work? Who invented
this clock?

Use a period, instead of a question mark, after an indirect
question—that is, a question embedded in a statement.

I asked whether the new atomic clock could be used in
cell phones.
[COMPARE: Can the new atomic clock be used in cell phones?]

MULTILINGUAL WRITERS

INDIRECT QUESTIONS

In English, indirect questions are written as declarative sentences.
The subject and verb are not inverted as they would be in the
related direct question.

We do not know when ~~will~~ the meeting _∧end. (will)

[COMPARE: When will the meeting end?]

For more on word order and questions, see Chapter 45.

Place a question mark after each question in a series of
related questions, even when they are not full sentences.

Will the new atomic clock be used in cell phones? word
processors? car navigation systems?

If a direct quotation is a question, place the question mark
inside the final quotation marks.

Tony asked, "How small is this new clock?"

In contrast, if you include quoted material in a question of your own, place the question mark outside the final quotation marks.

Is the clock really "no larger than a sugar cube"?

If you embed in the middle of a sentence a question not attributable to anyone in particular, place a comma before it and a question mark after it.

When the question, how does the clock work? arose, the researchers described a technique used by manufacturers of computer chips.

The first letter of such a question should not be capitalized unless the question is extremely long or contains internal punctuation.

To indicate uncertainty about a fact such as a date of birth, place a question mark inside parentheses directly after the fact in question.

Chaucer was born in 1340 (?) and died in 1400.

17c   The exclamation point

An exclamation point often marks the end of a sentence, but its primary purpose is rhetorical—to create emphasis.

Wow! What a game!

When a direct quotation ends with an exclamation point, no comma or period is placed immediately after it.

"Get a new pitcher!" he yelled.

He yelled, "Get a new pitcher!"

Use the exclamation point sparingly so that you do not diminish its impact. If you do not intend to signal strong emotion, place a comma after an interjection and a period at the end of the sentence.

Well, no one seriously expected this victory.

EXERCISE 1

Compose and punctuate brief sentences of the following types.

1. a declarative sentence containing a quoted exclamation
2. a sentence beginning with an interjection
3. a direct question
4. a declarative sentence containing an indirect question
5. a declarative sentence containing a direct question

17d   The colon

A colon calls attention to what follows, whether the grammatical unit is a clause, a phrase, or words in a series. It also separates numbers in parts of scriptural references and titles from subtitles. Leave only one space after a colon.

(1) Directing attention to an explanation, a summary, or a quotation

When a colon appears between two independent clauses, it signals that the second clause will explain or expand on the first.

No one expected the game to end as it did: after seven extra innings, the favored team collapsed.

A colon is also used after an independent clause to introduce a direct quotation.

> The Fourteenth Dalai Lama explained the importance of forgiveness: "When other beings, especially those who hold a grudge against you, abuse and harm you out of envy, you should not abandon them, but hold them as objects of your greatest compassion and take care of them."

CAUTION

The rules for using an uppercase or a lowercase letter to begin the first word of an independent clause that follows a colon vary across style manuals.

MLA   The first letter should be lowercase unless (1) it begins a word that is normally capitalized, (2) the independent clause is a quotation, or (3) the clause expresses a rule or principle.

APA   The first letter should be uppercase.

A colon at the end of an independent clause is sometimes followed by a phrase rather than another clause. This use of the colon puts emphasis on the phrase.

> I was finally confronted with what I had dreaded for months: the due date for the final balloon payment on my car loan.

All the style manuals advise using a lowercase letter to begin a phrase following a colon.

(2) Signaling that a list follows

Writers frequently use colons to introduce lists (which add to or clarify the information preceding the colon).

> Three students received internships: Asa, Vanna, and Jack.

Avoid placing a colon between a verb and its complement (**1c**) or after the words *including* and *such as*.

The winners were⊙Asa, Vanna, and Jack.

Many vegetarians do not eat dairy products such as⊙butter and cheese.

(3) Separating a title and a subtitle

Use a colon between a work's title and its subtitle.

Collapse: *How Societies Choose to Fail or Succeed*

(4) In reference numbers

Colons are often used between numbers in scriptural references.

Ps. 3:5   Gen. 1:1

However, MLA prefers periods instead of colons in such references.

Ps. 3.5   Gen. 1.1

(5) Specialized uses in business correspondence

A colon follows the salutation of a business letter and any notations.

Dear Dr. Hodges: Dear Imogen: Encl:

A colon introduces the headings in a memo.

To: From: Subject: Date:

EXERCISE 2

Insert colons where they are needed in the following sentences.

1. Before we discuss marketing, we need to outline the behavior of consumers consumer behavior is the process individuals go through as they select, buy, or use products or services to satisfy their needs and desires.

(continued on page 202)

(continued from page 201)

2. The process consists of six stages recognizing a need or desire, finding information, evaluating options, deciding to purchase, purchasing, and assessing purchases.

3. Many consumers rely on one popular publication for product information *Consumer Reports.*

4. When evaluating alternatives, a consumer uses criteria; for example, a house hunter might use some of the following price, location, size, age, style, and landscaping design.

5. The post-purchase assessment has one of two results satisfaction or dissatisfaction with the product or service.

17e The dash

A dash (or em dash) marks a break in thought, sets off a nonessential element for emphasis or clarity, or follows an introductory list or series. The short dash (or en dash) is used mainly in number ranges (**11g(2)**).

TECH SAVVY

To use your keyboard to create a dash, type two hyphens with no spaces between, before, or after them. Most word-processing programs can be set to convert these hyphens automatically to an em dash. In Microsoft Word, you can also hold down the Option and Shift keys while typing a hyphen.

(1) To mark a break in the normal flow of a sentence
Use a dash to indicate a shift in thought or tone.

I was awed by the almost superhuman effort Stonehenge represents—but who wouldn't be?

(2) To set off a nonessential element

A dash or a pair of dashes sets off a nonessential element for emphasis or clarity.

> Dr. Kruger's specialty is mycology—the study of fungi.

> The trail we took into the Grand Canyon—steep, narrow, winding, and lacking guardrails—made me wonder whether we could call a helicopter to fly us out.

(3) To set off an introductory list or series

If you decide to place a list or series at the beginning of a sentence in order to emphasize it, the main part of the sentence (after the dash) should sum up the meaning of the list or series.

> Eager, determined to succeed, and scared to death—all of these describe how I felt on the first day at work.

THINKING RHETORICALLY

COMMAS, DASHES, AND COLONS

Although a comma, a dash, or a colon may be followed by an explanation, an example, or an illustration, the rhetorical impact of each of these punctuation marks differs.

> He never failed to mention what was most important to him, the bottom line.
> He never failed to mention what was most important to him—the bottom line.
> He never failed to mention what was most important to him: the bottom line.

The comma, one of the most common punctuation marks, barely draws attention to what follows it. The dash, in contrast, signals a longer pause and so causes more emphasis to be placed on the information that follows. The colon is more direct and formal than either of the other two punctuation marks. (See **17d** for more about the colon.)

17f   Parentheses

Use parentheses to set off information that is not closely related to the main point of a sentence or paragraph but that provides an interesting detail, an explanation, or an illustration.

> One of the most striking peculiarities of the human brain is the great development of the frontal lobes—they are much less developed in other primates and hardly evident at all in other mammals. They are the part of the brain that grows and develops most after birth (and their development is not complete until about the age of seven).
>
> —OLIVER SACKS, *An Anthropologist on Mars*

Place parentheses around an acronym or an abbreviation when introducing it after its full form.

> The Search for Extraterrestrial Intelligence (SETI) uses the Very Large Array (VLA) outside Sicorro, New Mexico, to scan the sky.

If you use numbers or letters in a list within a sentence, set them off by placing them within parentheses.

> Your application should include (1) a current résumé, (2) a statement of purpose, and (3) two letters of recommendation.

For information on the use of parentheses in bibliographies and in-text citations, see Chapters **39** and **40**.

THINKING RHETORICALLY

DASHES AND PARENTHESES

Dashes and parentheses are both used to set off part of a sentence, but they differ in the amount of emphasis they signal. Whereas dashes call attention to the material that is set off, parentheses usually deemphasize such material.

Her grandfather—born during the Great Depression—was appointed by the president to the Securities and Exchange Commission.

Her grandfather (born in 1930) was appointed by the president to the Securities and Exchange Commission.

17g Square brackets

Square brackets set off additions or alterations that clarify direct quotations. In the following example, the bracketed noun specifies what is meant by the pronoun *They*:

"They [hyperlinks] are what turn the Web from a library of pages into a web" (Weinberger 170).

Square brackets also indicate that a letter at the beginning of a quotation has been changed from uppercase to lowercase, or vice versa.

David Weinberger claims that "[e]ven our notion of self as a continuous body moving through a continuous map of space and time is beginning to seem wrong on the Web" (10).

To avoid the awkwardness of using brackets in this way, you may be able to quote only part of a sentence and thus not need to change the capitalization.

David Weinberger claims that "our notion of self as a continuous body moving through a continuous map of space and time is beginning to seem wrong on the Web" (10).

Within parentheses, square brackets are used because having two sets of parentheses could be confusing.

People frequently provide personal information online. (See, for example, David Weinberger's *Small Pieces Loosely Joined* [Cambridge: Perseus, 2002].)

Angle brackets (< >) are used to enclose any web address included in an MLA works-cited list (**39b**) so that the period at the end of an entry is not confused with the dot(s) in the URL: <http://www.mla.org>.

17h   Ellipsis points

Ellipsis points indicate an omission from a quoted passage or a reflective pause or hesitation.

(1) To mark an omission within a quoted passage
Whenever you omit anything from material you quote, replace the omitted material with ellipsis points—three equally spaced periods. Be sure to compare your quoted sentence to the original, checking to see that your omission does not change the meaning of the original.

To avoid excessive use of ellipses, replace some direct quotations with paraphrases (**38d(3)**).

The following examples illustrate how to use ellipsis points in quotations from a passage by Patricia Gadsby.

Original
Cacao doesn't flower, as most plants do, at the tips of its outer and uppermost branches. Instead, its sweet white buds hang from the trunk and along a few fat branches, popping out of patches of bark called cushions, which form where leaves drop off. They're tiny, these flowers. Yet once pollinated by midges, no-see-ums that flit in the leafy detritus below, they'll make pulp-filled pods almost the size of rugby balls.

—PATRICIA GADSBY, "Endangered Chocolate"

(a) Omission within a quoted sentence

Patricia Gadsby notes that cacao flowers "once pollinated by midges . . . make pulp-filled pods almost the size of rugby balls."

Retain a comma, colon, or semicolon that appears in the original text if it makes a quoted sentence easier to read. If no misreading will occur, the punctuation mark can be omitted.

Patricia Gadsby describes the outcome of pollinating the cacao flowers: "Yet once pollinated by midges, . . . they'll make pulp-filled pods almost the size of rugby balls." [The comma after "midges" is retained.]

According to Gadsby, "Cacao doesn't flower . . . at the tips of its outer and uppermost branches." [The comma after "flower" is omitted.]

(b) Omission at the beginning of a quoted sentence

Do not use ellipsis points to indicate that you have deleted words from the beginning of a quotation, whether it is run into the text or set off in a block. The opening part of the original sentence has been omitted in the following quotation.

According to Patricia Gadsby, cacao flowers will become "pulp-filled pods almost the size of rugby balls."

Note that the first letter of the integrated quotation is not capitalized.

(c) Omission at the end of a quoted sentence

To indicate that you have omitted words from the end of a sentence, put a single space between the last word and the set of three spaced ellipsis points. Then add the end punctuation mark (a period, a question mark, or an exclamation point). If the quoted material is followed by a parenthetical source or page reference, the end punctuation comes after the second parenthesis.

> Claiming that cacao flowers differ from those of most plants, Patricia Gadsby describes how "the sweet white buds hang from the trunk and along a few fat branches • • •."
> OR ". . . a few fat branches • •." (2).

(d) Omission of a sentence or more

To signal the omission of a sentence or more (even a paragraph or more), place an end punctuation mark (usually a period) before the ellipsis points.

> Patricia Gadsby describes the flowering of the cacao plant: "Its sweet white buds hang from the trunk and along a few fat branches, popping out of patches of bark called cushions, which form where leaves drop off. • • • Yet once pollinated by midges, no-see-ums that flit in the leafy detritus below, they'll make pulp-filled pods almost the size of rugby balls."

(e) Omission of a line or more of a poem

To signal the omission of a full line or more in quoted poetry, use spaced periods extending the length of either the line above it or the omitted line.

> The yellow fog that rubs its back upon the window-panes,
> •
> Curled once about the house, and fell asleep.
>
> —T. S. ELIOT, "The Love Song of J. Alfred Prufrock"

(2) To mark an incomplete sentence

Ellipsis points show that a sentence has been intentionally left incomplete.

> Read aloud the passage that begins "The yellow fog • • •" and explain the imagery.

(3) To mark hesitation in a sentence

Ellipsis points can mark a reflective pause or a hesitation.

> Keith saw four menacing youths coming toward him • • • and ran.

A dash can also be used to indicate this type of a pause.

17i The slash

A slash between words, as in *and/or*, *young/old*, and *heaven/hell*, indicates that either word is applicable in the given context. There are no spaces before and after a slash used in this way. Because extensive use of the slash can make writing choppy, use it judiciously and sparingly. (If you are following APA or MLA guidelines, avoid using *he/she*, *him/her*, and so on.)

A slash is also used to mark line divisions in quoted poetry. A slash used in this way is preceded and followed by a space.

> Wallace Stevens refers to the listener who, "nothing himself, beholds **/** Nothing that is not there and the nothing that is."

EXERCISE 3

Add appropriate dashes, parentheses, square brackets, and slashes to the following sentences. Be ready to explain the reason for each mark you add.

1. The three recognized autism spectrum disorders ASDs are autism, Asperger syndrome, and pervasive developmental disorder–not otherwise specified PDD-NOS.

2. Disagreement concerning the causes of autism environmental, medical, and or genetic continues to flourish.

3. The rise in diagnoses of autism might be due to better diagnostic practices or an increase in the disorder itself.

EXERCISE 4

Punctuate the following sentences with appropriate end marks, commas, colons, dashes, and parentheses. Do not use unnecessary punctuation. Give a justification for each mark you add, especially where more than one type of mark (for example, commas, dashes, or parentheses) is acceptable.

1. Many small country towns are very similar a truck stop a gas station a crowded diner and three bars

2. The simple life a nonexistent crime rate and down-home values these are some of the advantages these little towns offer

3. Why do we never see these quaint examples of pure Americana when we travel around the country on the interstates

4. Rolling across America on one of the big interstates I-20 I-40 I-70 I-80 or I-90 you are likely to pass within a few miles of a number of these towns

5. Such towns almost certainly will have a regional or perhaps an ethnic flavor Hispanic in the southwest Scandinavian in the north

6. When I visit one of these out-of-the-way places I always have a sense of well really a feeling of safety

7. There's one thing I can tell you small-town life is not boring

8. My one big question however is what do you do to earn a living in these towns

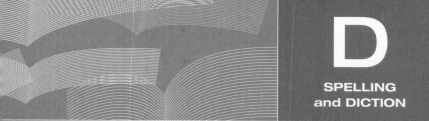

D

SPELLING and DICTION

18 | Spelling, the Spell Checker, and Hyphenation

When you first draft a paper, you might not pay close attention to spelling words correctly. After all, the point of drafting is to generate and organize ideas. However, proofreading for spelling mistakes is essential as you near the end of the writing process. You want to submit the kind of writing your teacher, employer, or supervisor expects to read: polished work that is as nearly perfect as you can make it.

You can train yourself to be a good proofreader by checking a dictionary every time you question the spelling of a word. If two spellings are listed, such as *fulfill* and *fulfil*, either form is correct, although the first option provided is generally considered more common. Once you choose between such options, be sure to use the spelling you pick consistently. You can also learn to be a better speller by studying a few basic strategies. This chapter will help you

- use a spell checker (**18a**),
- spell words according to pronunciation (**18b**),
- spell words that sound alike (**18c**),
- understand how prefixes and suffixes affect spelling (**18d**),
- use *ei* and *ie* correctly (**18e**), and
- use hyphens to link and divide words (**18f**).

18a | Spell checker

The spell checker is a wonderful invention, but only when you use it with care. A spell checker will usually flag

- misspellings of common words,
- some commonly confused words (such as *affect* and *effect*), and
- obvious typographical errors (such as *tge* for *the*).

However, a spell checker generally will *not* detect

- specialized vocabulary or foreign words not in its dictionary,
- typographical errors that are still correctly spelled words (such as *was* for *saw*), and
- misuses of words that sound alike but are not on the spell checker's list of commonly confused words.

The following strategies can help you use a spell checker effectively.

TIPS FOR USING A SPELL CHECKER

- If a spell checker regularly flags a word that is not in its dictionary but is spelled correctly, add that word to its dictionary by clicking on the Add button. From that point on, the spell checker will accept the word.
- Reject any offers the spell checker makes to correct all instances of a particular error.
- Use a dictionary to evaluate the alternative spellings the spell checker provides; some of them may be erroneous.

18b Spelling and pronunciation

Many words in English are not spelled the way they are pronounced, so pronunciation is not a reliable guide to correct spelling. Sometimes, people skip over an unstressed syllable, as when *February* is pronounced "Febuary," or they slide over a sound that is hard to articulate, as when *library* is pronounced "libary." Other times, people add a sound—for instance, when they pronounce *athlete* as "athalete." And people may switch sounds around, as in "irrevelant" for *irrelevant.* Such mispronunciations can lead to misspellings.

You can help yourself remember the spellings of some words by considering the spellings of their root words—for example, the root word for *irrelevant* is *relevant.* You can also teach yourself the correct spellings of words by pronouncing them the way they are spelled, that is, by pronouncing each letter mentally so that you "hear" even silent letters. You are more likely to remember the *b* in *subtle* if you pronounce it when spelling that word. Here are a few words typically misspelled because they include unpronounced letters:

condem*n*   for*ei*gn   lab*o*ratory   mus*c*le   solem*n*

Here are a few more that include letters that are often not heard in rapid speech, though they can be heard when carefully pronounced:

can*d*idate   diff*e*rent   enviro*n*ment   gover*n*ment   sep*a*rate

CAUTION

The words *and, have,* and *than* are often not stressed in speech and are thus likely misspelled.

> They would rather ~~of~~ *have* written two papers ~~then~~ *than* taken midterm ~~an~~ *and* final exams.

Watch for these misspellings when you proofread your papers.

18c Words that sound alike

Pairs of words such as *forth* and *fourth* or *sole* and *soul* are **homophones**: they sound alike but have different meanings and spellings. Some words that have different meanings sound exactly alike (*break/brake*); others sound alike in certain dialects (*marry/merry*). If you are unsure about the difference in meaning between any two words that sound alike, consult a dictionary. A number of frequently confused words are listed with explanations in this handbook's **Glossary of Usage**.

Also troublesome are two-word sequences that can be written as compound words or as separate words. The following are examples:

Everyday life was grueling.	She attended class **every day.**
They do not fight **anymore.**	They could not find **any more** evidence.

Other examples are *awhile/a while*, *everyone/every one*, *maybe/may be*, and *sometime/some time*.

A lot and *all right* are still spelled as two words. *Alot* is always considered incorrect; *alright* is also considered incorrect except in some newspapers and magazines. (See the **Glossary of Usage**.)

Singular nouns ending in *-nce* and plural nouns ending in *-nts* are easily confused.

Assistance is available.	I have two **assistants.**
His **patience** wore thin.	Some **patients** waited for hours.

Contractions and possessive pronouns are also often confused. In contractions, an apostrophe indicates an omitted letter (or letters). In possessive pronouns, there is no apostrophe. (See also **5b** and **15a(1)**.)

Contraction	Possessive
It's my turn next.	Each group waited **its** turn.
You're next.	**Your** turn is next.
There's no difference.	**Theirs** is no different.

TIPS FOR SPELLING WORDS THAT SOUND ALIKE

- Be alert for words that are commonly confused (*accept/except*).
- Distinguish between two-word sequences and single words that sound similar (*may be/maybe*).
- Use *-nts,* not *-nce,* for plural words (*instants/instance*).
- Mark contractions, but not possessive pronouns, with apostrophes (*who's/whose*).

18d   **Prefixes and suffixes**

When a prefix is added to a base word (often called the **root**), the spelling of the base word is unaffected.

necessary, **un**necessary   moral, **im**moral

However, adding a suffix to the end of a base word often changes the spelling.

beauty, beauti**ful**   describe, descri**ption**
BUT   resist, resist**ance**

Although spellings of words with suffixes are irregular, they follow certain conventions.

(1) Dropping or retaining a final e

- If a suffix begins with a vowel, the final *e* of the base word is dropped: bride, brid**al**; come, com**ing**; combine, combin**ation**; prime, prim**ary**. However, to keep the /s/ sound of *ce* or the /j/ sound of *ge*, retain the final *e* before -*able* or -*ous:* courag**eous**, manage**able**, notic**eable**.
- If a suffix begins with a consonant, the final *e* of the base word is usually retained: entire, entire**ly**; rude, rude**ness**; place, place**ment**; sure, sure**ly**. Exceptions include *argument, awful, ninth, truly,* and *wholly.*

(2) Doubling a final consonant when a suffix begins with a vowel

- If a one-syllable word with a single vowel or a stressed syllable with a single vowel ends with a consonant, double the final consonant: stop, sto**pped**, sto**pping**; omit, omi**tted**, omi**tting**.
- If there are two vowels before the consonant, the consonant is not doubled: seat, seat**ed**, seat**ing**; remain, remain**ed**, remain**ing**.
- If the final syllable is not stressed, the consonant is not doubled: edit, edit**ed**, edit**ing**; picket, picket**ed**, picket**ing**.

(3) Changing or retaining a final y

- Change a final *y* following a consonant to *i* when adding a suffix (except -*ing*): lazy, laz**ily**; defy, def**ies**, def**ied**, def**iance** BUT defy**ing**; modify, modif**ies**, modif**ied**, modif**ier** BUT modify**ing**.
- Retain a final *y* when it follows a vowel: gray, gray**ish**; stay, stay**s**, stay**ed**; obey, obey**s**, obey**ed**.
- Some verb forms are irregular and thus can cause difficulties: *lays, laid; pays, paid.* For a list of irregular verbs, see pages 95–97.

(4) Retaining a final l when -ly is added

cool, coo**lly**   formal, forma**lly**   real, rea**lly**   usual, usua**lly**

EXERCISE 1

Add the specified suffixes to the words that follow. Be prepared to explain the reason for the spelling of each resulting word.

EXAMPLE

-ly: late, casual, psychological   *lately   casually   psychologically*

1. -ing: put, admit, write, use, try, play

2. -ment: manage, commit, require, argue

3. -ous: continue, joy, acrimony, libel

4. -ed: race, tip, permit, carry, pray

5. -able: desire, read, trace, knowledge

6. -ly: true, sincere, normal, general

(5) Making a noun plural by adding -s or -es to the singular form

- If the sound in the plural form of a noun ending in *f* or *fe* changes from /f/ to /v/, change the ending to *-ve* before adding *-s:* thie**f**, thie**ves**; li**fe**, li**ves** BUT roo**f**, roo**fs**.
- Add *-es* to most nouns ending in *s, z, ch, sh,* or *x:* box, box**es**; peach, peach**es**.
- If a noun ends in a consonant and *y*, change the *y* to *i* and add *-es:* company, compan**ies**; ninety, ninet**ies**; territory, territor**ies**. (See also **18d(3)**.)
- If a noun ends in a consonant and *o*, add *-es:* hero, hero**es**; potato, potato**es**. However, note that sometimes just *-s* is added (photo, photo**s**; memo, memo**s**) and other times either *-s* or *-es* can be added (motto**s**, motto**es**; zero**s**, zero**es**).
- Certain nouns have irregular plural forms: woman, wom**en**; child, child**ren**; foot, f**eet**.
- Add *-s* to most proper nouns: the Lee**s**; the Kennedy**s**. Add *-es* to most proper nouns ending in *s, z, ch, sh,* or *x:* the Rodriguez**es**, the Jones**es** BUT the Bach**s** (in which *ch* is pronounced /k/).

CAUTION

Words borrowed from Latin or Greek generally form their plurals as they did in the original language.

Singular	criterion	alumnus, alumna	analysis	datum	species
Plural	criteria	alumni, alumnae	analyses	data	species

Many words with such origins gradually come to be considered part of the English language, and during this process, two plural forms will be listed in dictionaries as acceptable: *syllabus/syllabuses, syllabi.* Be sure to use only one of the acceptable plural forms in a paper you write.

EXERCISE 2

Provide the plural forms for the following words. If you need extra help, check a dictionary.

1. virus
2. committee
3. phenomenon
4. copy
5. hero
6. embargo
7. shelf
8. belief
9. foot
10. portfolio
11. cactus
12. census

18e   Confusion of *ei* and *ie*

An old rhyme will help you remember the order of letters in most words containing *e* and *i:*

Put *i* before *e*
Except after *c*
Or when sounded like *a*
As in *neighbor* and *weigh.*

Words with *i* before *e:* bel**ie**ve, ch**ie**f, pr**ie**st, y**ie**ld

Words with *e* before *i,* after *c:* conc**ei**t, perc**ei**ve, rec**ei**ve

Words with *ei* sounding like *a* in *cake:* **ei**ght, r**ei**n, th**ei**r, h**ei**r

Words that are exceptions to the rules in the rhyme include *either, neither, species, foreign*, and *weird*.

MULTILINGUAL WRITERS

AMERICAN AND BRITISH SPELLING DIFFERENCES

Although most words are spelled the same in both the United States and Great Britain, some are spelled differently, including the following:

American	check	realize	color	connection
British	cheque	realise	colour	connexion

Use the American spellings when writing for an audience in the United States.

18f   Hyphens

Hyphens link two or more words functioning as a single word and separate word parts to clarify meaning. They also have many conventional uses in numbers, fractions, and measurements. (Do not confuse the hyphen with a dash; see **11g(2)** and **17e**.)

(1) Between two or more words that form a compound

Some compounds are listed in the dictionary with hyphens (*eye-opener, cross-examination*), others are written as two words (*eye chart, cross fire*), and still others appear as one word (*eyewitness, crossbreed*). If you have questions about the spelling of a compound word, a dictionary is a good resource. However, it is also helpful to learn a few basic patterns.

- If two or more words serve as a single adjective before a noun, they should be hyphenated. If the words follow the noun, they are not hyphenated.

You submitted an **up‑to‑date** report. The report was
up to date.

A **well‑known** musician is performing tonight. The
musician is **well known.**

- When the second word in a hyphenated expression is omit‑
ted, the first word is still followed by a hyphen.

They discussed both **private‑** and **public‑sector** partnerships.

- A hyphen is not used after adverbs ending in *-ly* (*poorly
planned event*), in names of chemical compounds (*sodium
chloride solution*), or in modifiers with a letter or numeral as
the second element (*group C homes, type IV virus*).

(2) Between a prefix and a word to clarify meaning

- To avoid ambiguity or an awkward combination of letters
or syllables, place a hyphen between the base word and its
prefix: *anti‑intellectual, de‑emphasize, re‑sign the petition*
[COMPARE: *resign the position*].
- Place a hyphen between a prefix and a word beginning with
a capital letter and between a prefix and a word already con‑
taining a hyphen: *anti‑American, non‑self‑promoting.*
- Place a hyphen after the prefix *all-, e-, ex-,* or *self-*: *all‑inclu‑
sive, e‑commerce, ex‑husband, self‑esteem.* Otherwise, most
words with prefixes are not hyphenated. (The use of the
unhyphenated *email* has become very common, but *e‑mail*
is the spelling preferred by APA and MLA. The prefix *e-* is
sometimes used without a hyphen in trade names, such as
eBay.)

(3) In numbers, fractions, and units of measure

- Place a hyphen between two numbers when they are spelled
out: *thirty‑two, ninety‑nine.* However, no hyphen is used
before or after the words *hundred, thousand,* and *million: five
hundred sixty‑three, forty‑one million.*

- Hyphenate fractions that are spelled out: *three-fourths, one-half.*
- When you form a compound modifier that includes a number and a unit of measurement, place a hyphen between them: *twenty-first-century literature, twelve-year-old boy, ten-year project.*

EXERCISE 3

Convert the following groups of words into hyphenated compounds.

EXAMPLE
a movie lasting two hours   *a two-hour movie*

1. a boss who is well liked
2. a television screen that is forty-eight inches across
3. a highway with eight lanes
4. a painting from the seventeenth century
5. a chemist who won the Nobel Prize
6. a virus that is food borne

19 | Good Usage

Using the right words at the right time can make the difference between having your ideas taken seriously and seeing them brushed aside. Keeping your readers in mind will help you choose words they understand and consider appropriate. This chapter will help you

- write in a clear, straightforward style (**19a**);
- choose words that are appropriate for your audience, purpose, and context (**19b**);
- use inclusive language (**19c**); and
- find information in dictionaries (**19d**) and thesauruses (**19e**).

19a | Clear style

Although different styles are appropriate for different situations, you should strive to make your writing easy to read. To achieve a clear style, first choose words that your audience understands and that are appropriate for the occasion.

Ornate The majority believes that achievement derives primarily from the diligent pursuit of allocated tasks.

Clear Most people believe that success results from hard work.

Using words that are precise (**20a**) and sentences that are concise (Chapter **21**) can also help you achieve a clear style.

EXERCISE 1

Revise the following sentences for an audience that prefers a clear, straightforward style.

1. Expert delineation of character in a job interview is a goal that is not always possible to achieve.

2. In an employment situation, social pleasantries may contribute to the successful functioning of job tasks, but such interactions should not distract attention from the need to complete all assignments in a timely manner.

3. Commitment to an ongoing and carefully programmed schedule of physical self-management can be a significant resource for stress reduction in the workplace.

19b Appropriate word choice

Unless you are writing for a specialized audience and have good reason to believe that this audience will welcome slang, colloquial expressions, or jargon, the following advice can help you determine which words to use and which to avoid.

(1) Slang

The term **slang** covers a wide range of words or expressions that are used in informal situations or are considered fashionable by people in a particular age group, locality, or profession. Although such words are often used in private conversation or in writing intended to mimic conversation, they are usually out of place in academic or professional writing.

(2) Conversational (or colloquial) words

Words labeled *colloquial* in a dictionary are fine for casual conversation and for written dialogues or personal essays on a light topic. Such words are sometimes used for special effect in academic writing, but you should usually replace them with more appropriate words. For example, the conversational words *dumb* and *kid around* could be replaced by *illogical* and *tease*.

(3) Regionalisms

Regionalisms—such as *tank* for "pond" and *sweeper* for "vacuum cleaner"—can make writing lively and distinctive, but they are often considered too informal for academic and professional writing.

(4) Technical words or jargon

When writing for a diverse audience, an effective writer will not refer to the need for bifocals as *presbyopia*. However, technical language is appropriate when the audience can understand it (as when one physician writes to another) or when the audience would benefit by learning the terms in question.

19c | Inclusive language

By choosing words that are inclusive rather than exclusive, you invite readers into your writing. Prejudiced or derogatory language has no place in academic or professional writing; using it undermines your authority and credibility. It is best to use language that will engage, not alienate, your readers.

(1) Nonsexist language

Effective writers show equal respect for men and women. For example, they avoid using *man* to refer to people in general because they understand that the word excludes women.

Achievements [OR Human achievements]
∧ ~~Man's achievements~~ in science are impressive.

Photographs and statements on the websites of many companies indicate a commitment to an inclusive work environment.

Women, like men, can be *firefighters* or *police officers*—words that have become gender-neutral alternatives to *firemen* and *policemen*. Use the following tips to ensure that your writing is respectful.

TIPS FOR AVOIDING SEXIST LANGUAGE

When reviewing drafts, check for and revise the following types of sexist language.

- **Generic *he*:** A doctor should listen to *his* patients.

  A doctor should listen to **his or her** patients. [use of the appropriate form of *he or she*]

  Doctors should listen to **their** patients. [use of plural forms]

By listening to patients, **doctors obtain important diagnostic information.** [elimination of *his* by revising the sentence]

- **Occupational stereotype:** Glenda James, a *female* engineer at Howard Aviation, won the best-employee award.

  Glenda James, an engineer at Howard Aviation, won the best-employee award. [removal of the unnecessary gender reference]

- **Terms such as** *man* **and** *mankind* **or those with** *-ess* **or** *-man* **endings:** Labor laws benefit the common *man*. *Mankind* benefits from philanthropy. The *stewardess* brought me some orange juice.

  Labor laws benefit **working people.** [replacement of the stereotypical term with a gender-neutral term]

  Everyone benefits from philanthropy. [use of an indefinite pronoun]

  The **flight attendant** brought me some orange juice. [use of a gender-neutral term]

- **Stereotypical gender roles:** I was told that the university offers free tuition to faculty *wives*. The minister pronounced them *man* and *wife*.

  I was told that the university offers free tuition to faculty **spouses.** [replacement of the stereotypical term with a gender-neutral term]

  The minister pronounced them **husband** and wife. [use of a term equivalent to *wife*]

- **Unstated gender assumption:** Have your *mother make your costume* for the school pageant.

  Have your **parents provide you with a costume** for the school pageant. [replacement of the stereotypical words with gender-neutral ones]

EXERCISE 2

Make the following sentences inclusive by eliminating sexist language.

1. A special code of ethics guides a nurse in fulfilling her responsibilities.

2. According to the weatherman, this summer will be unseasonably cold.

3. Professor Garcia mapped the journey of modern man.

4. While in college, she worked as a waitress in a diner.

(2) Nonracist language

Rarely is it necessary to identify anyone's race or ethnicity in academic or professional writing. However, you may need to use appropriate racial or ethnic terms if you are writing a demographic report, an argument against existing racial inequities, or a historical account of a particular event involving ethnic groups. Determining which terms a particular group prefers can be difficult because preferences sometimes vary within a group and change over time. One conventional way to refer to Americans of a specific descent is to include an adjective before the word *American*: *African American, Asian American, European American, Latin American, Mexican American, Native American.* These words are widely used; however, members of a particular group may identify themselves in more than one way. In addition to *African American* and *European American, Black* (or *black*) and *White* (or *white*) have long been used. People of Spanish-speaking descent may prefer *Chicano/Chicana, Hispanic, Latino/Latina, Puerto Rican,* or other terms. Members of cultures that are indigenous to North America may prefer a specific name such as *Cherokee* or *Haida,* though some also accept *American Indians* or *Native People.*

An up-to-date dictionary that includes notes on usage can help you choose appropriate terms.

(3) Respectful language about differences

If a writing assignment requires you to distinguish people based on age, ability, geographical area, religion, or sexual orientation, show respect to the groups or individuals you discuss by using the terms they prefer.

(a) Referring to age

Although some people object to the term *senior citizen,* a better alternative has not emerged. When used respectfully, the term refers to a person who has reached the age of retirement (but may not have decided to retire) and is eligible for certain privileges granted by society. However, if you know your audience would object to this term, find out which alternative is preferred.

(b) Referring to disability or illness

A current recommendation for referring to disabilities and illnesses is "to put the person first." In this way, the focus is placed on the individual rather than on the limitation. Thus, *persons with disabilities* is preferred over *disabled persons.* For your own writing, you can find out whether such person-first expressions are preferred by noting whether they are used in the articles and books (or by the people) you consult.

(c) Referring to geographical areas

Certain geographical terms need to be used with special care. Though most frequently used to refer to people from the United States, the term *American* may also refer to people from Canada, Mexico, Central America, or South America. If your audience may be confused by this term, use *people from the United States* or *US citizens* instead.

The term *Arab* refers to people who speak Arabic. If you cannot use specific terms such as *Iraqi* or *Saudi Arabian,* be sure you know that a country's people speak Arabic and not another language. Iranians, for example, are not Arabs because they speak Farsi.

British, rather than *English,* is the preferred term for referring to people from the island of Great Britain or from the United Kingdom.

(d) Referring to religion

Reference to a person's religion should be made only if it is relevant. If you must mention religious affiliation, use only those terms considered respectful. Because religions have both conservative and liberal followers, be careful not to make generalizations (**34i(12)**) about political stances.

(e) Referring to sexual orientation

If your writing assignment calls for identifying sexual orientation, choose terms used by the people you are discussing.

19d   Dictionaries

A good dictionary is an indispensable tool for a writer. It does much more than provide the correct spellings of words; it also gives meanings, parts of speech, plural forms, and verb tenses, as well as information about pronunciation and origin. In addition, a reliable dictionary includes labels that can help you decide whether words are appropriate for your purpose, audience, and context. Words labeled *dialect, slang, colloquial, non-standard,* or *unconventional,* as well as those labeled *archaic* or *obsolete* (meaning that they are no longer in common use), are generally inappropriate for college and professional writing. If a word has no label, you can safely assume that it can be used in writing for school or work. Whether a word is appropriate, however, depends on the precise meaning a writer wants to convey (20a). Because meanings of words change and because new words are constantly introduced into English, it is important to choose a dictionary, whether print or electronic, that has a recent copyright date.

(1) Unabridged or specialized dictionaries
An **unabridged dictionary** provides a comprehensive survey of English words, including detailed information about their origins. A **specialized dictionary** presents words related to a specific discipline or to some aspect of usage.

Unabridged Dictionaries

The Oxford English Dictionary. 2nd ed. 20 vols. 1989–. CD-ROM. 2005.

Webster's Third New International Dictionary of the English Language. CD-ROM. 2002.

These dictionaries also have regularly updated online versions.

Specialized Dictionaries

The American Heritage Guide to Contemporary Usage and Style. 2005.

The Cambridge Guide to English Usage. 2004.

MULTILINGUAL WRITERS

DICTIONARIES AS RESOURCES

The following dictionaries are recommended for nonnative speakers of English.

Longman Advanced American English. 2007.

Merriam-Webster's Advanced Learner's English Dictionary. 2008.

(2) Dictionary entries
Dictionary entries provide a range of information. Figure 19.1 shows sample entries from the eleventh edition of *Merriam-Webster's Collegiate Dictionary.* Notice that *cool* is listed four times: as an adjective, a verb, a noun, and an adverb. The types of information these entries provide can be found in almost all desk dictionaries, though sometimes in a different order.

TYPES OF INFORMATION PROVIDED BY DICTIONARY ENTRIES

- **Spelling, syllabication (word division), and pronunciation.**

- **Parts of speech and word forms.** Dictionaries identify parts of speech—for instance, with *n* for "noun" or *vi* for "intransitive verb." Meanings will vary depending on the part of speech identified. Dictionaries also indicate irregular forms of verbs, nouns, and adjectives: *fly, flew, flown, flying, flies; child, children; good, better, best.*

- **Word origin.** A dictionary entry indicates whether a word has roots in an older version of English, has even deeper roots in another language, such as Greek or Latin, or has been added to English more recently from another language.

- **Date of first occurrence.** Most entries include the date when the use of the word was initially recorded.

- **Definition(s).** Generally, the oldest meaning is given first. However, meanings can also be ordered according to frequency of usage, with the most common usage listed first.

- **Usage.** Quotations show how the word can be used in various contexts. Sometimes a comment on usage problems is placed at the end of an entry.

- **Idioms.** When the word is part of a common idiom (20c), the idiom is listed and defined, usually at the end of the entry.

- **Synonyms.** Some dictionaries provide explanations of subtle differences in meaning among a word's synonyms.

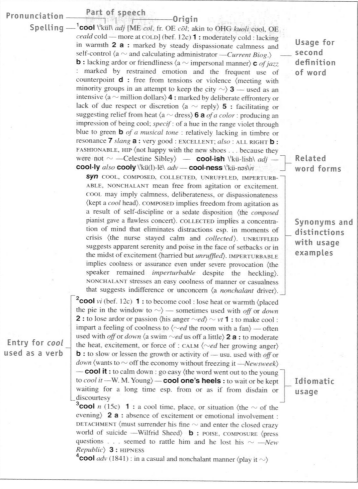

Pronunciation
Spelling — Part of speech
Origin

Usage for second definition of word

Related word forms

Synonyms and distinctions with usage examples

Entry for *cool* used as a verb

Idiomatic usage

1cool \'kül\ *adj* [ME *col*, fr. OE *cōl*; akin to OHG *kuoli* cool, OE *ceald* cold — more at COLD] (bef. 12c) **1 :** moderately cold : lacking in warmth **2 a :** marked by steady dispassionate calmness and self-control ⟨a ∼ and calculating administrator —*Current Biog.*⟩ **b :** lacking ardor or friendliness ⟨a ∼ impersonal manner⟩ **c** *of jazz* : marked by restrained emotion and the frequent use of counterpoint **d :** free from tensions or violence ⟨meeting with minority groups in an attempt to keep the city ∼⟩ **3** — used as an intensive ⟨a ∼ million dollars⟩ **4 :** marked by deliberate effrontery or lack of due respect or discretion ⟨a ∼ reply⟩ **5 :** facilitating or suggesting relief from heat ⟨a ∼ dress⟩ **6 a** *of a color* : producing an impression of being cool; *specif* : of a hue in the range violet through blue to green **b** *of a musical tone* : relatively lacking in timbre or resonance **7** *slang* **a :** very good : EXCELLENT; *also* : ALL RIGHT **b :** FASHIONABLE, HIP ⟨not happy with the new shoes . . . because they were not ∼ —*Celestine Sibley*⟩ — **cool·ish** \'kü-lish\ *adj* — **cool·ly** *also* **cooly** \'kü(l)-lē\ *adv* — **cool·ness** \'kül-nəs\ *n*

syn COOL, COMPOSED, COLLECTED, UNRUFFLED, IMPERTURBABLE, NONCHALANT mean free from agitation or excitement. COOL may imply calmness, deliberateness, or dispassionateness ⟨kept a *cool* head⟩. COMPOSED implies freedom from agitation as a result of self-discipline or a sedate disposition ⟨the *composed* pianist gave a flawless concert⟩. COLLECTED implies a concentration of mind that eliminates distractions esp. in moments of crisis ⟨the nurse stayed calm and *collected*⟩. UNRUFFLED suggests apparent serenity and poise in the face of setbacks or in the midst of excitement ⟨harried but *unruffled*⟩. IMPERTURBABLE implies coolness or assurance even under severe provocation ⟨the speaker remained *imperturbable* despite the heckling⟩. NONCHALANT stresses an easy coolness of manner or casualness that suggests indifference or unconcern ⟨a *nonchalant* driver⟩.

2cool *vi* (bef. 12c) **1 :** to become cool : lose heat or warmth ⟨placed the pie in the window to ∼⟩ — sometimes used with *off* or *down* **2 :** to lose ardor or passion ⟨his anger ∼*ed*⟩ — *vt* **1 :** to make cool : impart a feeling of coolness to ⟨∼*ed* the room with a fan⟩ — often used with *off* or *down* ⟨a swim ∼*ed* us off a little⟩ **2 a :** to moderate the heat, excitement, or force of : CALM ⟨∼*ed* her growing anger⟩ **b :** to slow or lessen the growth or activity of — usu. used with *off* or *down* ⟨wants to ∼ off the economy without freezing it —*Newsweek*⟩ — **cool it :** to calm down : go easy ⟨the word went out to the young to *cool it* —*W. M. Young*⟩ — **cool one's heels :** to wait or be kept waiting for a long time esp. from or as if from disdain or discourtesy

3cool *n* (15c) **1 :** a cool time, place, or situation ⟨the ∼ of the evening⟩ **2 a :** absence of excitement or emotional involvement : DETACHMENT ⟨must surrender his fine ∼ and enter the closed crazy world of suicide —*Wilfrid Sheed*⟩ **b :** POISE, COMPOSURE ⟨press questions . . . seemed to rattle him and he lost his ∼ —*New Republic*⟩ **3 :** HIPNESS
4cool *adv* (1841) : in a casual and nonchalant manner ⟨play it ∼⟩

Figure 19.1. Examples of dictionary entries.

19e Thesauruses

A **thesaurus** provides alternatives for frequently used words. Unlike a dictionary, which explains what a word means and how it evolved, a thesaurus provides only a list of words that serve as possible synonyms for each term it includes. A thesaurus can be useful, especially when you want to jog your memory about a word you know but cannot recall. You may, however, use a word incorrectly if you simply pick it from a list in a thesaurus. If you find an unfamiliar yet intriguing word, make sure that you are using it correctly by looking it up in a dictionary.

EXERCISE 3

Find definitions for the pairs of words in parentheses. Then choose the word you think best completes each sentence. Be prepared to explain your answers.

1. Sixteen prisoners on death row were granted (mercy/clemency).

2. The outcome of the election (excited/provoked) a riot.

3. The young couple was (covetous/greedy) of their neighbors' estate.

4. While she was traveling in Muslim countries, she wore (modest/chaste) clothing.

5. The president of the university (authorized/confirmed) the rumor that tuition would be increasing next year.

20 Exactness

Make words work for you. By choosing the right word and putting it in the right place, you can communicate exactly what you mean and make your writing distinctive. This chapter will help you

- choose words appropriate for your purpose, audience, and context (**20a**);
- create fresh, clear expressions (**20b**);
- use idioms and collocations (**20c**); and
- compose clear definitions (**20d**).

20a Accurate and precise word choice

(1) Denotations and connotations

Denotations are definitions of words, such as those that appear in dictionaries. For example, the noun *beach* denotes a sandy or pebbly shore. However, some words have more than one definition or one definition that can be interpreted in a number of ways. Select words whose denotations convey your point exactly.

Padre Island National Seashore~~is really great.~~ astounds even an indifferent tourist like me.

[Because *great* can mean "extremely large" as well as "outstanding" or "powerful," its use in this sentence is imprecise.]

The **Glossary of Usage** at the back of this book includes the definitions of many words that are commonly confused.

Connotations are the associations evoked by a word. *Beach,* for instance, may connote natural beauty, surf, shells, swimming, tanning, sunburn, and/or crowds. The context in which a word appears affects the associations it evokes. In a treatise on shoreline management, *beach* has scientific and geographic connotations; in a fashion magazine, this word is associated with bathing suits, sunglasses, and sunscreen. The challenge for writers is to choose the words that are most likely to spark the appropriate connotations in their readers' minds.

> resilience
> The ~~obstinacy~~ of the Kemp's ridley sea turtle has delighted park rangers.
>
> [*Obstinacy* has negative connotations, which make it an unlikely quality to cause delight.]

(2) Specific, concrete words

A **general word** is all-inclusive, indefinite, and sweeping in scope. A **specific word** is precise, definite, and limited in scope.

General	Specific	More Specific/Concrete
food	fast food	cheeseburger
media	newspapers	*The Miami Herald*
place	city	Atlanta

An **abstract word** refers to a concept or idea, a quality or trait, or anything else that cannot be touched, heard, or seen. A **concrete word** signifies a particular object, a specific action, or anything that can be touched, heard, or seen.

Abstract   democracy, evil, strength, charity

Concrete   mosquito, hammer, plastic, fog

As you select words to fit your context, be as specific and concrete as you can. For example, instead of the word *bad,* consider using a more precise adjective.

bad neighbors: rowdy, snobby, nosy, fussy, sloppy, threatening

bad meat: tough, tainted, overcooked, undercooked, contaminated

bad wood: rotten, warped, scorched, knotty, termite-ridden

(3) Figurative language

Figurative language is the use of words in an imaginative rather than a literal sense. Similes and metaphors are the chief **figures of speech**. A **simile** is a comparison of dissimilar things using *like* or *as*. A **metaphor** is an implied comparison of dissimilar things, without *like* or *as*.

Similes

Norms live in the culture **like genes, manifesting themselves unexpectedly, the way a child's big ears appear from an ancestor of whom no picture or name remains.**

—CHARLES WOHLFORTH, "Conservation and Eugenics: The Environmental Movement's Dirty Secret"

When **her body was hairless as a baby's,** she adjusted the showerhead so that the water burst forth in pelting streams.

—LOIDA MARITZA PÉREZ, *Geographies of Home*

Metaphors

His **money was a sharp pair of scissors** that snipped rapidly through tangles of red tape.   —HISAYE YAMAMOTO, "Las Uegas Charley"

Making tacos is a graceful dance.

—DENISE CHÁVEZ, *A Taco Testimony*

Single words can be used metaphorically.

> These roses must be **planted** in good soil. [literal]
>
> Keep your life **planted** wherever you can put down the most roots. [metaphorical]

Similes and metaphors are especially valuable when they are concrete and describe or evoke essential relationships that cannot otherwise be communicated. Similes or metaphors can be extended throughout a paragraph of comparison, but be careful not to mix them (**23d**).

EXERCISE 1

Study the passage below, and prepare to discuss the author's use of exact and figurative language to communicate her ideas.

> ¹The kitchen where I'm making dinner is a New York kitchen. ²Nice light, way too small, nowhere to put anything unless the stove goes. ³My stove is huge, but it will never go. ⁴My stove is where my head clears, my impressions settle, my reporter's life gets folded into my life, and whatever I've just learned, or think I've learned—whatever it was, out there in the world, that had seemed so different and surprising—bubbles away in the very small pot of what I think I know and, if I'm lucky, produces something like perspective.   —**JANE KRAMER,** "The Reporter's Kitchen"

20b   Clichés and euphemisms

When forced or overused, certain expressions lose their impact. For example, the expressions *bite the dust, breath of fresh air,* and *smooth as silk* were once striking and thus effective. Excessive use, though, has drained them of their original force and made them **clichés.** Newer expressions such as *put a spin* *on something* and *think outside the box* have also lost their vitality because of overuse. Nonetheless, clichés are so much a part of the language, especially the spoken language, that nearly every writer uses them from time to time. But effective writers often give a fresh twist to an old saying.

> I seek a narrative, a fiction, to order days like the one I spent several years ago, on a gray June day in Chicago, when I took a roller-coaster ride on the bell curve of my experience.
>
> —**GAYLE PEMBERTON**, "The Zen of Bigger Thomas"

[Notice how much more effective this expression is than a reference to "being on an emotional roller coaster."]

Sometimes writers coin new expressions to substitute for words that have coarse or unpleasant connotations. These expressions, called **euphemisms**, occasionally become standardized. To avoid the word *dying*, for example, a writer might say that someone was *terminally ill*. Although euphemisms sound more pleasant than the words they replace, they can sometimes obscure facts. Euphemisms such as *revenue enhancement* for *tax hike* and *pre-owned* for *used* are considered insincere or deceitful.

EXERCISE 2

Replace the following overused expressions with carefully chosen words. Then use the replacements in sentences.

EXAMPLE

beyond the shadow of a doubt
undoubtedly   OR   with total certainty

1. reality check
2. global village
3. bottom line

4. over the top
5. call the shots

20c   Idioms and collocations

Idioms are fixed expressions whose meanings cannot be entirely determined by knowing the meanings of their parts—examples are *bear in mind, fall in love,* and *stand a chance.* **Collocations** are combinations of words that frequently occur together. Unlike idioms, they have meanings that *can* be determined by knowing the meanings of their parts—think of *depend on, fond of, little while,* or *right now.* Regardless of whether you are using an idiom or a collocation, if you make even a small change to the expected wording, you may distract or confuse your readers.

She tried to keep a ~~small~~ low profile.

They had ~~an invested~~ a vested interest in the project.

Because prepositions are often small, unstressed words, writers sometimes confuse them. The following is a list of common collocations containing prepositions.

CHOOSING THE RIGHT PREPOSITION

according **to** the source

accused **of** the crime

based **on** the novel

bored **by** it

conform **to/with** standards

connected **to** each other

consists **of** cards and letters

different **from** the first draft

happened **by** accident

intend **to** finish his degree

opposition **to** the idea

plan **to** attend

sure **to** see the movie

try **to** be on time

EXERCISE 3

Write a sentence using each idiom or collocation correctly.

1. do one's best, do one's part, do one's duty

2. cut down, cut back, cut corners

3. make time, make sure, make sense

20d Clear definitions

When words have more than one meaning, establish which meaning you have in mind. A definition can set the terms of the discussion.

> In this paper, I use the word *communism* **in the Marxist sense of social organization based on the holding of all property in common.**

A **formal definition** first states the term to be defined, then puts it into a class, and finally differentiates it from other members of that class.

> A *phosphene* [term] is **a luminous visual image** [class] that **results from applying pressure to the eyeball** [differentiation].

A short dictionary definition may be adequate when you need to convey the meaning of a word unfamiliar to readers.

> Here, *galvanic* means **"produced as if by electric shock."**

Giving a synonym may also clarify the meaning of a term. Such synonyms are often used as appositives (**1e(5)**).

> *Machismo,* **confidence with an attitude,** can be a pose rather than a reality.

Writers frequently show—rather than tell—what a word means by giving examples.

Many homophones (**such as** *be* **and** *bee* **or** *see* **and** *sea*) are not spelling problems.

Sometimes, your own definition can clarify a concept.

Clichés could be defined as **thoughts that have hardened.**

EXERCISE 4

Using your own words, define each of the following terms in full sentences.

1. audacity
2. professionalism
3. dilemma
4. indifference
5. ambiguity
6. equal opportunity

21 Conciseness

To facilitate readers' understanding, effective writers convey their thoughts clearly and efficiently. This does not mean that they always write short sentences; rather, they use each word wisely. This chapter will help you

- make each word count (**21a**) and
- use elliptical constructions (**21b**).

21a Eliminating redundancy and wordiness

After writing a first draft, review your sentences to make sure that they contain only the words necessary to make your point.

(1) Redundancy

Restating a key point in different words can help readers understand it. But if you rephrase readily understood terms, your work will suffer from **redundancy**—repetition for no good reason.

Each student had a unique talent ~~and ability that he or she used in his or her~~ acting. *(for)*

You should also avoid grammatical redundancy, as in double subjects (*my sister [she] is*), double comparisons (*[more] easier than*), and double negatives (*could[n't] hardly*).

(2) Wordiness

As you edit a draft, look for ways to rewrite sentences in fewer words, without risking the loss of important details. One exact word often says as much as several inexact ones.

Some unscrupulous brokers are ~cheating~ ~~taking money and savings from~~ elderly people ~out of their pensions.~ ~~who need that money because they planned to~~ ~~use it as a retirement pension.~~

In addition, watch for vague words such as *area, aspect, factor, feature, kind, situation, thing,* and *type.* They may signal wordiness.

~Effective~ ~~In an employment situation, effective~~ communication is essential at work.

REPLACEMENTS FOR WORDY EXPRESSIONS

Instead of	Use
at this moment (point) in time	now, today
due to the fact that	because
for the purpose of	for
it is clear (obvious) that	clearly (obviously)
there is no question that	unquestionably, certainly
without a doubt	undoubtedly
in this day and age	today
in the final analysis	finally

(3) *There are* and *it is*

There or *it* may function as an **expletive**—a word that signals that the subject of the sentence will follow the verb, usually a form of *be* (**1b**). Writers use expletives to emphasize words that would not be emphasized in the typical subject-verb order. Notice the difference in rhythm between the following sentences:

Two children were playing in the yard. [typical order]

There were two children playing in the yard. [use of expletive]

However, expletives are easily overused. If you find that you have drafted several sentences that begin with expletives, revise a few of them.

Hundreds   were
~~There were hundreds~~ of fans crowding onto the field.

Joining the crowd
~~It~~ was frightening ~~to join the crowd~~.

(4) Relative pronouns

The relative pronoun *who*, *which*, or *that* can frequently be deleted without affecting the meaning of a sentence. If one of these pronouns is followed by a form of the verb *be* (*am, is, are, was,* or *were*), you can often omit the pronoun and sometimes the verb as well.

The change ~~that~~ the young senator proposed yesterday angered most legislators.

The Endangered Species Act, ~~which was~~ passed in 1973, protects the habitat of endangered plants and animals.

When deleting a relative pronoun, you might have to make other changes to a sentence as well.

Nations ~~that provide~~ ^{providing} protection for endangered species often create preserves and forbid hunting of these species.

MULTILINGUAL WRITERS

USING RELATIVE PRONOUNS

Review your sentences to make sure that no clause includes both a personal pronoun (**5a(1)**) and a relative pronoun (**5a(3)**) referring to the same antecedent (**6b**).

The drug **that** we were testing ~~it~~ has not been approved by the Food and Drug Administration.

For more information on relative (adjectival) clauses, see Chapter **45**.

21b   Using elliptical constructions

An **elliptical construction** is one that deliberately omits words that can be understood from the context.

Speed is the goal for some swimmers, endurance ~~is the goal~~ for others, and relaxation ~~is the goal~~ for still others.

Sometimes, as an aid to clarity, commas mark omissions in elliptical constructions.

My family functioned like a baseball team: my mom was the coach; my brother, the pitcher; and my sister, the shortstop. [Use semicolons to separate items with internal commas (**14a**).]

EXERCISE 1

Revise this paragraph to eliminate wordiness and needless repetition.

[1]Founded in the year 1967, *Rolling Stone* has become well known for covering culture and politics considered popular. [2]Back in the day, the *Rolling Stone's* magazine's original focus was rock and roll. [3]In its first year, I think it helped popularize musicians such as John Lennon, Mick Jagger, and also the guitarist Pete Townshend. [4]Cover photographs featured the Beatles, Jimi Hendrix, Tina Turner, and the vocalist Jim Morrison, among others. [5]The magazine also carried news reports related to the music of the 1960s period. [6]The cost of the Monterey Pop Festival and additionally a raid at the Grateful Dead's house were two of the breaking news stories in 1967. [7]Since that day and age, the magazine has become a mainstay on newsstands everywhere all over the world. [8]Writers, as well as musicians, have gained nationwide recognition: in the early days, Hunter S. Thompson gained recognition for "Fear and Loathing in Las Vegas," and Tom Wolfe gained recognition for presenting the inner world of the Mercury astronauts; more recently, Michael Hastings gained recognition for reporting on military leadership. [9]Though articles and interviews on topics related to music and politics continue to attract readers, reviews of films, books, and other types of entertainment have now become standard fare also. [10]Always eye-catching and really impressive, the cover of the magazine now spotlights all kinds of people in the news, not just musicians: politicians, actors, actresses, models, comedians, movie directors, and even fictional characters (Darth Vader and Waldo) have appeared under the distinctive and colorful letters of the name *Rolling Stone*.

22 | Clarity and Completeness

Clarity in writing depends on more than grammar. Clarity results as much from critical thinking, logical development, and exact diction as it does from correct grammar. However, grammatical slips can mar what would otherwise be clear writing. This chapter will help you

- include all necessary words in a sentence (22a) and
- complete comparisons (22b) and intensifiers (22c).

22a | Including necessary words

When we speak or write quickly, we often omit small words. As you revise, be sure to include all necessary articles, prepositions, verbs, and conjunctions.

The meeting took place in ˄an auditorium. [missing an article]

We discussed a couple ˄of issues. [missing a preposition]

When a sentence has a **compound verb** (two verbs linked by a conjunction), you may need to supply a different preposition for each verb to make your meaning clear.

He neither **believes** ˄in nor **approves of** the plan.

All verbs, both auxiliary and main (7a(4)), should be included to make sentences complete.

She *has* spoken with all the candidates.

Voter turnout has never *been* and will never be 100 percent.

Include the word *that* before a clause when it makes the sentence easier to read. Without the added *that* in the following sentence, a reader may stumble over *discovered the fossil* before understanding that *the fossil* is linked to *provided*.

The paleontologists discovered *that* the fossil provided a link between the dinosaur and the modern bird.

When a sentence has two *that* clauses, *that* should begin each one.

The graph indicated **that the population had increased** but **that the number of homeowners had not.**

22b   Completing comparisons

A comparison has two parts: someone or something is compared to someone or something else. As you revise your writing, make sure that your audience knows who or what is being compared. To revise incomplete comparisons, add necessary words, phrases, or clauses.

Printers today are quite different *from those sold in the early 1990s*.

His first novel was better *than the one just published*.

After you are sure that your comparisons are complete, check to see that they are also logical.

THINKING RHETORICALLY

SENTENCE STYLE

Most professional writers and readers use the following words to describe effective sentences.

- *Exact*. Precise words and word combinations ensure exactness and enable readers to come as close as they can to a full understanding of the writer's message.
- *Conventional*. Sentences are conventional when they conform to the usage expectations of a particular community. For most academic assignments, you will be expected to use Standardized English.
- *Consistent*. A consistent writing style is characterized by the use of the same types of words and grammatical structures throughout a piece of writing. A style that is inconsistent jars the reader's expectations.
- *Parallel*. Related to consistency, parallelism refers to the placement of similar ideas into similar grammatical structures.
- *Concise*. Concise prose is free of redundancies.
- *Coherent*. Coherence refers to clear connections between adjacent sentences and paragraphs.
- *Varied*. To write appealing paragraphs, a writer uses both short and long sentences. When sentences vary in length, they usually also vary in structure, rhythm, and emphasis.

In the following chapters, you will learn to identify the options considered effective by most academic and professional writers. Remember, though, that appropriateness depends on your audience and purpose. You may find that it does not make sense to apply a general rule such as "Use the active voice" in all circumstances. For example, you may be expected to write a vigorous description of an event, detailing exactly what happened, but find that you need to use the passive voice when you do not know who was responsible for the event: Several of the campaign signs *were defaced*. Determining your audience and purpose will help you write sentences that engage your readers.

23 Sentence Unity

Effective academic and professional writing is composed of sentences that are consistent, clear, and complete. This chapter can help you

- choose and arrange details (**23a**),
- revise mixed metaphors and mixed constructions (**23b** and **23c**), and
- relate sentence parts (**23d**).

23a Choosing and arranging details

Well-chosen details add interest and credibility to your writing. As you revise, you may occasionally notice a sentence that would be clearer and more believable with the addition of a phrase or two about time, location, or cause.

| **Missing important detail** | An astrophysicist from the Harvard-Smithsonian Center has predicted a galactic storm. |
| **With detail added** | An astrophysicist from the Harvard-Smithsonian Center has predicted **that** a galactic storm **will occur within the next 10 million years.** |

Without the additional information about time, most readers would wonder when the storm was supposed to occur. The added detail makes the sentence clearer.

The details you choose will help your readers understand your message. If you provide too many details within a single sentence, though, your readers may lose sight of your main point. The writer of the following sentence deleted the mention of her uncle as she revised because this information was irrelevant to the main idea of her essay.

> When I was only sixteen, I left home to attend a college in California ~~that my uncle had graduated from twenty years earlier~~.

Besides choosing details purposefully, you also need to indicate a clear connection between the details and the main idea of your sentence.

Unrelated	Many tigers facing possible extinction live in India, **where there are many people.**
Related	Many tigers facing possible extinction live in India, **where their natural habitat is shrinking because of population pressure.**

23b   Revising mixed metaphors

When you use language that evokes images, make sure that the images are meaningfully related. Unrelated images that appear in the same sentence are called **mixed metaphors**. The following sentence includes incompatible images.

> As he climbed the corporate ladder, he ~~sank into a sea of~~ ^{incurred a large} debt.

The combination of two images—climbing a ladder and sinking into a sea—could create a picture in the reader's mind of a man hanging onto a ladder as it disappears into the water. To revise such a sentence, replace the words evoking one of the conflicting images.

23c Revising mixed constructions

A sentence that begins with one kind of grammatical structure and shifts to another is a **mixed construction**. To untangle a mixed construction, make sure that the sentence includes a conventional subject—a noun, a noun phrase, a gerund phrase, an infinitive phrase, or a noun clause. Prepositional phrases and adverbial clauses are not typical subjects.

Practicing
~~By practicing~~ a new language daily will help you become proficient. [A gerund phrase replaces a prepositional phrase.]

Her scholarship award
~~Although she won a scholarship~~ does not give her the right to skip classes. [A noun phrase replaces an adverbial clause.]

If you find a sentence that has a mixed construction, you can either revise the subject, as in the previous examples, or leave the beginning of the sentence as a modifier and add a new subject after it.

By practicing a new language daily, **you** will become more proficient.

Although she won a scholarship, **it** does not give her the right to skip classes.

EXERCISE 1

Revise the following sentences so that details clearly support the main idea. Correct any mixed metaphors or mixed constructions.

1. In the United States, each person has one vote, but there may be problems at the polling booths.

2. Everyone's voting rights should be protected. The federal government has funded the replacement of the punch-card ballot.

3. Many states use optical scanners, which were also used on the standardized tests we took in high school. These scanners sort readable from unreadable ballots.

4. Some voters question the use of touch-screen voting systems. These systems leave no paper trail of all the ballots election officers need to swim through during a recount.

5. By providing educational materials helps citizens learn where and how to vote.

23d   Relating sentence parts

When drafting, writers sometimes compose sentences in which the subject is said to be something or to do something that is not logically possible. This breakdown in meaning is called **faulty predication**. Similarly, mismatches between a verb and its complement can obscure meaning.

(1) Mismatch between subject and verb

The joining of a subject and a verb must create a meaningful idea.

Mismatch	The absence of detail screams out at the reader. [An *absence* cannot scream.]
Revision	The reader immediately notices the absence of detail.

(2) Illogical equation with *be*

When a form of the verb *be* joins two parts of a sentence (the subject and the subject complement), these two parts need to be logically related.

> Free speech
> ∧~~The importance of free speech~~ is essential to a democracy. [*Importance* cannot be essential.]

(3) Mismatches in definitions

When you write a sentence that states a formal definition, the term you are defining should be followed by a noun or a noun phrase, not an adverbial clause (**1f(2)**). Avoid using *is when* or *is where*.

> *Ecology* is∧ the ~~when you~~ study ∧ of the relationships among living organisms and between living organisms and their environment.

> *Exploitative competition* is∧ the contest between ~~where~~ two or more organisms ∧ vying ~~vie~~ for a limited resource such as food.

(4) Mismatch of *reason* with *is because*

You can see why *reason* and *is because* are a mismatch by looking at the meaning of *because*: "for the reason that." Saying "the reason is for the reason that" is redundant. Be sure to revise any sentence containing the construction *the reason is … because*.

> The ~~reason the~~ old train station was closed ~~is~~ because it had fallen into disrepair.

(5) Mismatch between verb and complement

A verb and its complement should fit together meaningfully.

| **Mismatch** | Only a few students used the incorrect use of *there*. [To "use an incorrect use" is not logical.] |
| **Revision** | Only a few students used *there* incorrectly. |

To make sure that a relative pronoun in the object position is connected logically to a verb, replace the pronoun with its antecedent. Then check that the subject and verb have a logical connection. In the following sentence, *the inspiration* is the antecedent for *that*.

Mismatch The inspiration that the author created touched young writers. ["The author created the inspiration" does not make sense.]

Revision The author inspired young writers.

Verbs used to integrate information appear in *signal phrases* and are often followed by specific types of complements. Some of the verbs used in this way are listed with their typical complements. (Some verbs fall into more than one category.)

VERBS FOR SIGNAL PHRASES AND THEIR COMPLEMENTS

Verb + *that* **noun clause**

agree	claim	explain	report	suggest

Example: The researcher **reported** that the weather patterns had changed.

Verb + noun phrase + *that* **noun clause**

assure	convince	inform	remind	tell

Example: He **told** the reporters that he was planning to resign.

Verb + *wh-* **noun clause**

demonstrate	discover	explain	report	suggest

Example: She **described** what had happened.

EXERCISE 2

Revise the following sentences so that each verb is followed by a conventional complement.

1. The speaker discussed that applications had specific requirements.

2. He convinced that mass transit was affordable.

3. The two groups agreed how the problem could be solved.

4. Brown described that improvements had been made to the old house.

5. They wondered that such a catastrophe could happen.

24 Subordination and Coordination

Understanding subordination and coordination can help you indicate connections between ideas as well as add variety to your sentences (Chapter **30**). This chapter will help you

- use subordination and coordination effectively (**24a** and **24b**) and
- avoid faulty or excessive subordination and coordination (**24c**).

24a Using subordination effectively

Subordinate means "being of lower rank." A subordinate grammatical structure cannot stand alone; it is dependent on the main (independent) clause. The most common subordinate structure is the dependent clause (**1f(2)**), which usually begins with a subordinating conjunction or a relative pronoun.

(1) Subordinating conjunctions

A **subordinating conjunction** (**1g(3)**) specifies the relationship between a dependent clause and an independent clause. For example, it might signal a causal relationship.

The painters finished early **because they work well together.**

Here are a few of the most frequently used subordinating conjunctions:

Cause	*because*
Concession	*although, even though*
Condition	*if, unless*
Effect	*so that*
Sequence	*before, after*
Time	*when*

By using subordinating conjunctions, you can combine short sentences and indicate how they are related.

> _∧~~The~~ ^{After the} crew leader picked us up early on Friday.~~ We~~_∧ ^{, we} ate breakfast together at a local diner.

If the subjects of the two clauses are the same, the dependent clause can often be shortened to a phrase.

> After_∧ ^{eating}~~we ate~~ our breakfast, we headed back to the construction site.

(2) Relative pronouns

A **relative pronoun** (*who, whom, which, that,* or *whose*) introduces a dependent clause that, in most cases, modifies the pronoun's antecedent (**5a(3)**). By using this type of dependent clause, called an **adjectival clause**, or a **relative clause**, you can embed details into a sentence.

> The temple has a <u>portico</u> **that faces west**.

An adjectival clause can be shortened as long as the meaning of the sentence remains clear.

> The Parthenon is the Greek temple ~~that was~~ dedicated to the goddess Athena.

24b Using coordination effectively

Coordinate means "being of equal rank." Coordinate grammatical elements have the same form. For example, they may be two words that are both adjectives, two phrases that are both prepositional, or two clauses that are both independent.

> a **stunning** and **satisfying** conclusion [adjectives]
>
> **in the attic** or **in the basement** [prepositional phrases]
>
> **The company was losing money,** yet **the employees suspected nothing.** [independent clauses]

To indicate the relationship between coordinate words, phrases, or clauses, choose an appropriate coordinating or correlative conjunction (**1g**).

Addition	*and, both . . . and, not only . . . but also*
Alternative	*or, nor, either . . . or, neither . . . nor*
Cause	*for*
Contrast	*but, yet*
Result	*so*

By using coordination, you can avoid unnecessary repetition.

> The hike to the top of Angels Landing has countless
>
> switchbacks. ~~It also has~~ _{and} long drop-offs.

A semicolon can also be used to link coordinate independent clauses:

Hikers follow the path; climbers scale the cliff wall.

EXERCISE 1

Using subordination and coordination, revise the sentences in the following paragraph to emphasize ideas you consider important.

¹During the summer of 1998, many booklovers, young and old alike, spent hours reading about an adolescent wizard. ²The wizard had friends who were also wizards. ³Little time passed before Harry Potter, Ron Weasley, and Hermione Granger became familiar names. ⁴*Harry Potter and the Sorcerer's Stone* was the first of seven fantasy novels. ⁵All of these novels were penned by British author J. K. Rowling. ⁶Each of the novels has been a bestseller. ⁷Many readers consider the introductory novel their favorite. ⁸In *Sorcerer's Stone,* readers meet Harry. ⁹He is an orphan with extraordinary powers. ¹⁰Harry meets friends at a boarding school. ¹¹With them, he defeats the evil wizard. (¹²The wizard is so evil his name cannot be mentioned.) ¹³Some fans praise Rowling's treatment of the theme of friendship. ¹⁴They also appreciate her exploration of the theme of self-sacrifice. ¹⁵Other fans mention her plotting, humor, and straightforward style.

24c	**Avoiding faulty or excessive subordination and coordination**

(1) Choosing precise conjunctions

Effective subordination requires choosing subordinating conjunctions carefully. In the following sentence, the use of *as* is distracting because it can mean either "because" or "while."

    Because
    ~~As~~ time was running out, I randomly filled in the remaining circles on the exam sheet.

Your choice of coordinating conjunction should also convey your meaning precisely. For example, to indicate a cause-and-effect relationship, *so* is more precise than *and*.

    The timer rang, ~~and~~ *so* I turned in my exam.

(2) Avoiding excessive subordination and coordination

As you revise your writing, make sure that you have not over-used subordination or coordination. In the following ineffec-tive sentence, two dependent clauses compete for the reader's focus. The revision is clearer because it eliminates one of the dependent clauses.

    Although researchers used to believe that ancient Egyptians

    were the first to domesticate cats, they now think that

    cats may have provided company for humans 5,000 years

    earlier, *. They base their revised estimate on the discovery of* ~~because the intact skeleton of a cat has been~~
    an intact cat skeleton
    ~~discovered~~ in a Neolithic village on Cyprus.

Overuse of coordination results in a rambling sentence in need of revision.

    Because cats
    ~~Cats~~ hunt mice, and ~~they also hunt~~ other small rodents, ~~so~~
    they are popular pets.

The following strategies should help you avoid overusing coordinating conjunctions.

(a) Using a more specific subordinating conjunction or a conjunctive adverb

I worked all summer to earn tuition money, ~~and I didn't~~ _{so that I wouldn't} have to work during the school year.

OR

I worked all summer to earn tuition money, ~~and~~ _{; thus} I didn't have to work during the school year.

(b) Using a relative clause to embed information

Seafood, _{, which is nutritious and low in fat,} ~~is nutritious, and it is low in fat, and it~~ has become available in greater variety.

(c) Allowing two or more verbs to share the same subject

Marie quickly grabbed a shovel, ~~and then she~~ ran to the edge of the field, and ~~then she~~ put out the fire before it could spread to the trees.

(d) Placing some information in an appositive phrase

Karl Glazebrook, _{, a researcher in astronomy at Johns Hopkins University,} ~~is a researcher in astronomy at Johns Hopkins University, and he~~ has questioned the conventional theory of galaxy formations.

(e) Placing some information in a phrase

_{In the thick snow,} ~~The snow was thick, and~~ we could not see where we were going.

_{After pulling} ~~The plane pulled~~ away from the gate on time, _{the plane} ~~and then it~~ sat on the runway for two hours.

EXERCISE 2

Revise the following sentences to eliminate faulty or excessive coordination and subordination. Be prepared to explain why your sentences are more effective than the originals.

1. The Duct Tape Guys usually describe humorous uses for duct tape, providing serious information about the history of duct tape on their website.

2. Duct tape was invented for the US military during World War II to keep the moisture out of ammunition cases because it was strong and waterproof.

3. Duct tape was originally called "duck tape" as it was waterproof and ducks are like that too and because it was made of cotton duck, which is a durable, tightly woven material.

4. Duck tape was also used to repair jeeps and to repair aircraft, its primary use being to protect ammunition cases.

5. When the war was over, house builders used duck tape to connect duct work together, and the builders started to refer to duck tape as "duct tape" and eventually the color of the tape changed from the green that was used during the war to silver, which matched the ducts.

25 | Misplaced Modifiers

Modifiers are words, phrases, or clauses that modify; that is, they qualify or limit the meaning of other words. Modifiers enrich your writing with details and enhance its coherence— when they are correctly placed. This chapter will help you

- place modifiers effectively (**25a**) and
- revise dangling modifiers (**25b**) and strings of noun modifiers (**25c**).

25a | Placement of modifiers

Effective placement of modifiers will improve the clarity and coherence of your sentences. A **misplaced modifier** obscures the meaning of a sentence.

(1) Keeping related words together

Place the modifiers *almost, even, hardly, just,* and *only* before the words or word groups they modify. Altering placement can alter meaning.

The committee can **only** nominate two members for the position. [The committee cannot *appoint* the two members to the position.]

The committee can nominate **only** two members for the position. [The committee cannot nominate more than two members.]

Only the committee can nominate two members for the position. [No person or group other than the committee can nominate members.]

(2) Placing phrases and clauses near the words they modify

Readers expect phrases and clauses to modify the nearest grammatical element. The revision of the following sentence clarifies that the prosecutor, not the witness, was skillful:

> With great skill, the
> ∧The prosecutor cross-examined the witness ~~with great skill~~.

The following revision makes it clear that the phrase *crouched and ugly* describes the phantom, not the boy:

> The   crouched and ugly
> ∧~~Crouched and ugly, the~~ young boy gasped at the∧phantom
> moving across the stage.

The next sentence is fine as long as Jesse wrote the proposal, not the review. If he wrote the review, the sentence should be recast.

> I have not read the review of the proposal Jesse wrote.

> Jesse's
> I have not read∧~~the~~ review of the proposal ~~Jesse wrote~~.

(3) Revising squinting modifiers

A **squinting modifier** can be interpreted as modifying either what precedes it or what follows it. To avoid such lack of clarity, you can reposition the modifier or revise the entire sentence.

> Even though Erikson lists some advantages **overall** his vision of a successful business is faulty.

Revisions

> Even though Erikson lists some **overall** advantages, his vision of a successful business is faulty. [modifer repositioned; punctuation added]

> Erikson lists some advantages; **however, overall,** his vision of a successful business is faulty. [sentence revised]

EXERCISE 1

Improve the clarity of the following sentences by moving the modifiers. Not all sentences require editing.

1. Alfred Joseph Hitchcock was born the son of a poultry dealer in London.

2. Hitchcock was only identified with thrillers after making his third movie, *The Lodger*.

3. Hitchcock moved to the United States in 1939 and eventually became a naturalized citizen.

4. Hitchcock's most famous movies revolved around psychological improbabilities that are still discussed by movie critics today.

5. Although his movies are known for suspense sometimes moviegoers also remember Hitchcock's droll sense of humor.

6. Hitchcock just did not direct movie thrillers; he also produced two television series.

7. Originally a British citizen, Queen Elizabeth knighted Alfred Hitchcock in 1980.

25b Dangling modifiers

Dangling modifiers are phrases (**1e**) or **elliptical clauses** (clauses without a subject; see **1f(2)**) that lack an appropriate word to modify. To avoid including dangling modifiers in your writing, first look carefully at any sentence that begins with a phrase or an elliptical clause. If the phrase or clause suggests an action, be sure that what follows the modifier is the actor (the subject of the sentence). If there is no actor performing the action indicated in the phrase, the modifier is dangling. To revise this type of dangling modifier, name an actor—either in the modifier or in the main clause.

> Lying on the beach, time became irrelevant. [Time cannot lie on a beach.]

Revisions

> While **we** were lying on the beach, time became irrelevant. [actor in the modifier]

> Lying on the beach, **we** found that time became irrelevant. [actor in the main clause]

> While eating lunch, waves lapped at our toes. [Waves cannot eat lunch.]

Revisions

> While **we** were eating lunch, waves lapped at our toes. [actor in the modifier]

> While eating lunch, **we** noticed the water lapping at our toes. [actor in the main clause]

The following sentences illustrate revisions of other common types of dangling modifiers:

> To avoid getting sunburn, *you should apply* sunscreen ~~should be applied~~ before going outside. [Sunscreen cannot avoid getting sunburn.]

> *Because they were in* ^In a rush to get to the beach, an accident occurred. [An accident cannot be in a rush.]

Although you will most frequently find a dangling modifier at the beginning of a sentence, you may sometimes find one at the end of a sentence.

> Good equipment is important *for anyone* ^~~when~~ snorkeling. [Equipment cannot snorkel.]

Sentence modifiers and absolute phrases are *not* dangling modifiers.

The fog finally lifting, vacationers headed for the beach.

Marcus played well in the final game, **on the whole.**

EXERCISE 2

Revise the following sentences to eliminate misplaced and dangling modifiers.

1. Climbing a mountain, fitness becomes all-important.

2. In determining an appropriate challenge, considering safety precautions is necessary.

3. Taking care to stay roped together, accidents are less likely to occur.

4. Even when expecting sunny weather, rain gear should be packed.

5. Although adding extra weight, climbers should not leave home without a first-aid kit.

6. By taking pains at the beginning of a trip, agony can be averted at the end of a trip.

25c   Nouns as modifiers

Adjectives and adverbs are the most common modifiers, but nouns (**1a(2)**) can also be modifiers (***movie* critic, *reference* manual**). A string of noun modifiers can be cumbersome. The following example shows how a sentence with too many noun modifiers can be revised.

The Friday afternoon Student Affairs Committee meeting ^scheduled for Friday afternoon has been cancelled.

The infinitive marker *to* does not need to be repeated as long as the sentence remains clear.

She wanted her audience **to remember** the protest song and **understand** its origin.

To recognize parallelism in sentences that do not include repeated words, look for a coordinating conjunction (*and, but, or, yet, so, nor,* or *for*). The parts of a sentence that such a conjunction joins are parallel if they have similar grammatical forms (all nouns, all participial phrases, and so on).

Words	The young actor was **shy** <u>yet</u> **determined.** [two adjectives joined by *yet*]
Phrases	Her goals include **publicizing student and faculty research, increasing the funding for that research,** <u>and</u> **providing adequate research facilities.** [three gerund phrases joined by *and*]
Clauses	Our instructor explained **what the project had entailed** <u>and</u> **how the researchers had used the results.** [two noun clauses joined by *and*]

As you edit a draft, look for sentences that include two or three words, phrases, or clauses joined by a conjunction and make sure the grammatical forms being linked are parallel.

People all around me are **buying, remodeling,** or ~~they want to sell their~~ _{selling} houses.

| **26b** | **Linking parallel forms with correlative conjunctions** |

Correlative conjunctions (or **correlatives**) are pairs of words that link other words, phrases, or clauses (**1g(2)**).

both ... and

either ... or

neither ... nor

not only ... but also

whether ... or

Notice how the words or phrases following each of the paired conjunctions are parallel.

He will major in **either** <u>biology</u> **or** <u>chemistry</u>.

Whether <u>at home</u> **or** <u>at school</u>, he is always busy.

Be especially careful when using *not only ... but also*.

His team practices not only

~~Not only practicing~~ at 6 a.m. during the week, but ~~his team~~ also ~~scrimmages~~ on Sunday afternoons.

OR

does his team practice

Not only ~~practicing~~ at 6 a.m. during the week, but it ~~the team~~ also scrimmages on Sunday afternoons.

In the first revised example, each conjunction is followed by a prepositional phrase (**1e(4)**). In the second revised example, each conjunction accompanies a clause (**1f**).

26c Using parallelism to provide clarity and emphasis

Repeating a pattern emphasizes the relationship of ideas. The following two parallel sentences come from the conclusion of "Letter from Birmingham Jail":

> **If I have said anything** in this letter <u>that overstates the truth and indicates an unreasonable impatience</u>, **I beg you to forgive me. If I have said anything** <u>that understates the truth and indicates my having a patience</u> that allows me to settle for anything less than brotherhood, **I beg God to forgive me.**
>
> —MARTIN LUTHER KING, JR., "Letter from Birmingham Jail"
> © 1963, 1991. Reprinted by permission.

To create this parallelism, King repeats words and uses similar grammatical forms (sentences beginning with *if* and clauses beginning with *that*).

By expressing key ideas in parallel structures, you emphasize them. However, be careful not to overuse parallel patterns, or they will lose their impact. Parallelism is especially effective in the introduction or the conclusion of a paragraph or an essay. The preceding excerpt came from the conclusion of King's letter. The following passage from the introduction to a chapter of a book on advertising contains three examples of parallel forms:

> While **men are encouraged to fall in love with their cars, women are more often invited to have a romance,** indeed an erotic experience, with **something closer to home, something that truly does pump the valves of our hearts**—the food we eat. And the consequences become even more severe as we enter into the territory of **compulsivity** and **addiction.**
>
> —JEAN KILBOURNE, *Deadly Persuasion*

THINKING RHETORICALLY

PARALLELISM

Parallel elements make your writing easy to read. But consider breaking from the parallel pattern on occasion to emphasize a point. For example, to describe a friend, you could start with two adjectives and then switch to a noun phrase.

My friend Alison is **kind, modest,** and **the smartest mathematician in the state.**

EXERCISE 1

Make the structures in each sentence parallel. In some sentences, you may have to use different wording.

1. Helen was praised by the vice president, and her assistant admired her.

2. When she hired new employees for her department, she looked for applicants who were intelligent, able to stay focused, and able to speak clearly.

3. At meetings, she was always prepared, participating actively yet politely, and generated innovative responses to department concerns.

4. In her annual report, she wrote that her most important achievements were attracting new clients and revenues were higher.

5. When asked about her leadership style, she said that she preferred collaborating with others rather than to work alone in her office.

27 | Consistency

A consistent writing style will make it easier for readers to understand your message, your role in creating it, and their role as members of its audience. This chapter will help you maintain consistency

- in verb tense (**27a**),
- in point of view (**27b**), and
- in tone (**27c**).

27a | Verb tense

By using verb tenses consistently, you help your readers understand when the actions or events you are describing took place. Verb tenses convey information about time frames and grammatical aspect. *Time frame* refers to whether the tense is present, past, or future (refer to the columns of the chart on page 101). *Aspect* refers to whether it is simple, progressive, perfect, or perfect progressive (refer to the rows in the chart on page 101). Consistency in the time frame of a verb, though not necessarily in its aspect, ensures that any sequence of reported events is clearly and accurately portrayed. In the following paragraph, notice that the time frame remains the past, but the aspect varies among simple, perfect, and progressive:

past perfect

In the summer of 1983, I **had** just **finished** my third year

simple past

of architecture school and **had** to find a six-month internship.

past perfect (compound predicate)

I **had grown** up and **gone** through my entire education in the

past perfect

Midwest, but I **had been** to New York City once on a class

simple past   simple past

field trip and I **thought** it **seemed** like a pretty good place to

live. So, armed with little more than an inflated ego and my

simple past

school portfolio, I **was** off to Manhattan, oblivious to the bad

past progressive

economy and the fact that the city **was overflowing** with

young architects.   —PAUL K. HUMISTON, "Small World"

If you do need to shift to another time frame within a paragraph, you can use a time marker such as *now*, *then*, *during the 1920s*, or *after they leave*.

In the following paragraph, the time frame shifts back and forth between present and past—between today, when Edward O. Wilson is studying ants in the woods around Walden Pond, and the nineteenth century, when Thoreau lived there. The time markers are bracketed.

simple present   simple past

These woods **are** not wild; indeed, they **were** not wild

[in Thoreau's day]. [Today], the beach and trails of Walden

simple present

Pond State Reservation **draw** about 500,000 visitors a year.

simple present

Few of them **hunt** ants, however. Underfoot and under the leaf

simple present   simple past

litter there **is** a world as wild as it **was** [before human beings

simple past

came to this part of North America].

> —JAMES GORMAN, "Finding a Wild, Fearsome World beneath
> Every Fallen Leaf"

On occasion, a writer may change tenses without using any time marker, (1) to explain or support a general statement with information about the past, (2) to compare and contrast two different time periods, or (3) to comment on a topic. Why do you think the author of the following paragraph varies verb tenses?

Thomas Jefferson, author of the Declaration of Independence, **is** considered one of our country's most brilliant citizens. His achievements **were** many, as **were** his interests. Some historians **describe** his work as a naturalist, scientist, and inventor; others **focus** on his accomplishments as an educator and politician. Yet Jefferson **is** best known as a spokesman for democracy.

The author switches to the past tense in the second sentence to provide evidence from the past that supports the topic sentence.

Before you turn in a final draft, check the verb tenses you have used to ensure that they are logical and consistent. Revise any that are not.

The white wedding dress came ~~comes~~ into fashion when Queen Victoria wore a white gown at her wedding to Prince Albert of Saxe. Soon after, brides who could afford them bought stylish white dresses for their weddings. Brides of modest means, however, continued ~~continue~~ to choose dresses they could wear more than once.

EXERCISE 1

Revise the following paragraph so that there are no unnecessary shifts in verb tense.

I **had** already **been walking** for a half hour in the semidarkness of Amsterdam's early-morning streets when I **came** to a red light. I **am** in a hurry to get to the train station and no cars **were** out yet, so I **cross** over the cobblestones, passing a man waiting for the light to change. I never **look** back when he **scolds** me for breaking the law. I **had** a train to catch. I **was** going to Widnau, in Switzerland, to see Aunt Marie. I **have** not **seen** her since I **was** in second grade.

27b Point of view

Whenever you write, your point of view (perspective) will be evident in the pronouns you choose. *I* or *we* indicates a first-person point of view, which is appropriate for writing that includes personal views or experiences. If you decide to address the reader as *you*, you are adopting a second-person point of view. However, because a second-person point of view is rare in academic writing, avoid using *you* unless you need to address the reader. If you select the pronouns *he, she, it, one,* and *they*, you are writing with a third-person point of view. The third-person point of view is the most common point of view in academic writing.

Although you may find it necessary to use different points of view in a paper, be careful not to confuse readers by shifting perspective unnecessarily. The following paragraph has been revised to ensure consistency of point of view.

> To an observer, a sleeping person appears unresponsive, passive, and essentially isolated from the rest of the world and its barrage of stimuli. While it is true that ~~you are~~ someone asleep is unaware of most surrounding noises ~~when you are asleep, our~~ , that person's brain is far from inactive. In fact, the brain can be as active during sleep as it is ~~when you are awake.~~ in a waking state. When ~~our brains are~~ it is asleep, the rate and type of electrical activity change.

27c Tone

The tone of a piece of writing conveys a writer's attitude toward a topic (33a(3)). The words and phrases a writer chooses affect the tone he or she creates. Notice the difference in tone in the following excerpts describing the same scientific experiment. The first paragraph was written for the general public; the second was written for other researchers.

Imagine that I asked you to play a very simple gambling game. In front of you, are four decks of cards—two red and two blue. Each card in those four decks either wins you a sum of money or costs you some money, and your job is to turn over cards from any of the decks, one at a time, in such a way that maximizes your winnings. What you don't know at the beginning, however, is that the red decks are a minefield. The rewards are high, but when you lose on red, you lose *a lot*. You can really only win by taking cards from the blue decks, which offer a nice, steady diet of $50 and $100 payoffs. The question is: how long will it take you to figure this out? —**MALCOLM GLADWELL**, *Blink*

In a gambling task that simulates real-life decision-making in the way it factors uncertainty, rewards, and penalties, the players are given four decks of cards, a loan of $2000 facsimile U.S. bills, and asked to play so that they can lose the least amount of money and win the most (1). Turning each card carries an immediate reward ($100 in decks A and B and $50 in decks C and D). Unpredictably, however, the turning of some cards also carries a penalty (which is large in decks A and B and small in decks C and D). Playing mostly from the disadvantageous decks (A and B) leads to an overall loss. Playing from the advantageous decks (C and D) leads to an overall gain. The players have no way of predicting when a penalty will arise in a given deck, no way to calculate with precision the net gain or loss from each deck, and no knowledge of how many cards they must turn to end the game (the game is stopped after 100 card selections).

—**ANTOINE BECHARA, HANNA DAMASIO, DANIEL TRANEL, AND ANTONIO R. DAMASIO**, "**Deciding Advantageously before Knowing the Advantageous Strategy**"

In the excerpt from *Blink*, Malcolm Gladwell addresses readers directly: "Imagine that I asked you to play a very simple gambling game." In the excerpt aimed at an audience of researchers, Antoine Bechara and his coauthors describe their experiment without directly addressing the reader. Gladwell also uses less formal language than Bechara and his colleagues do. Finally, the scientists include a reference citation in their paragraph (the number *1* in parentheses), but Gladwell does not.

The tone of each excerpt is appropriate for its audience. However, shifts in tone can be distracting. The following paragraph was revised to ensure consistency of tone:

Scientists at the University of Oslo (Norway) ~~think they~~ [have evidence] ~~know why~~ [that] the common belief about the birth order of ~~kids has some truth to it.~~ [children carries some truth.] Using as data IQ tests taken from military records, the scientists found that older children ~~have~~ [score] significantly ~~more on the ball than kids in second or third place.~~ [higher than their siblings.] According to the researchers, the average variation in scores is large enough to account for differences in college admission.

EXERCISE 2

Revise the following paragraph so that there are no unnecessary shifts in tone.

[1]Many car owners used to grumble about deceptive fuel-economy ratings. [2]They often found, after they had already purchased a car, that their mileage was lower than that on the car's window sticker. [3]The issue remained pretty much ignored until our gas prices started to go up like crazy. [4]Because of increased pressure from consumer organizations, the Environmental Protection Agency reviewed and then changed the way it was calculating fuel-economy ratings. [5]The agency now takes into account factors such as quick acceleration, changing road grades, and the use of air conditioning, so the new ratings should reflect your real-world driving conditions. [6]Nonetheless, the ratings can never be right on target given that we all have different driving habits.

28 Pronoun Reference

The meaning of each pronoun in a sentence should be immediately obvious. In the following sentence, the pronouns *them* and *itself* clearly refer to their antecedents, *shells* and *carrier shell*, respectively.

The **carrier shell** gathers small empty **shells** and attaches **them** to **itself.**

As you draft and edit, be sure that your readers can easily determine the antecedents for the pronouns you use. This chapter will help you maintain clarity by avoiding pronoun references that are

- ambiguous or unclear (**28a**),
- remote or awkward (**28b**),
- broad or implied (**28c**), or
- nonspecific (**28d**).

28a Ambiguous or unclear pronoun references

When a pronoun can refer to either of two antecedents, replace the pronoun with a noun or rewrite the sentence. The following revised sentences clarify that Mr. Eggers, not Mr. Lee, is in charge of the project.

Mr. Lee told Mr. Eggers that ^Mr. Eggers he would be in charge of the project.

OR

Mr. Lee put Mr. Eggers in charge of the project.

28b Remote or awkward references

To help readers understand your meaning, place pronouns as close to their antecedents as possible. The following sentence needs to be revised so that the relative pronoun *that* is close to its antecedent, *poem.* Otherwise, the reader would wonder how a new book could be written in 1945.

The **poem** ^that was originally written in 1945 has been published in a new book ~~that was originally written in 1945~~.

A relative pronoun does not have to follow its antecedent directly when there is no risk of misunderstanding.

We began to notice **changes** in our lives **that** we had never expected.

28c Broad or implied references

Pronouns such as *it, this, that,* and *which* may refer to a specific word or phrase or to the sense of a whole clause or sentence. To avoid an ambiguous reference to the general idea of a preceding clause or sentence, clarify your pronoun reference. In the following sentence, *this* may refer to the class-attendance policy or to the students' feelings about it.

When class attendance is compulsory, some students feel that
                                                    perception
education is being forced on them. This ^is unwarranted.

In addition, remember to express an idea explicitly rather than
using a vague *it* or *they.*

My father is a music teacher. _{Teaching music} ~~It~~ is a profession that requires
much patience.

_{Former students}
~~They~~ say my father shows a great deal of patience with
everyone.

Be especially careful to provide clear antecedents when you
are referring to the work of others. The following sentence requires revision because *she* can refer to someone other than Jen
Norton:

In _{her} ~~Jen Norton's~~ new book, _{Jen Norton} ~~she~~ argues for election reform.

28d   The use of *it* without an antecedent

The expletive *it* does not have a specific antecedent (see **1b**).
Instead, it is used to postpone, and thus give emphasis to, the
subject of a sentence. A sentence that begins with the expletive
it can sometimes be wordy or awkward. Revise such a sentence
by replacing *it* with the postponed subject.

_{Trying to repair the car}   _{useless}
~~It was~~ ~~no use trying to repair the car.~~

EXERCISE 1

Edit the following sentences to make all references clear.

1. A champion cyclist, a cancer survivor, and a humanitarian, it is no wonder that Lance Armstrong is one of the most highly celebrated athletes in the world.

2. Armstrong's mother encouraged his athleticism, which led to his becoming a professional triathlete by age sixteen.

3. When he was twenty-five, he sought medical attention, and they told him he had testicular cancer.

4. Armstrong underwent dramatic surgery and aggressive chemotherapy; this eventually helped him recover.

5. For Lance Armstrong, it hasn't been only about racing bikes; he has become a humanitarian as well, creating the Lance Armstrong Foundation to help cancer patients and to fund cancer research around the world.

29 Emphasis

In any piece of writing, some of your ideas will be more important than others. You can direct the reader's attention to these ideas by emphasizing them. This chapter will help you

- place words where they receive emphasis (**29a**),
- use cumulative and periodic sentences (**29b**),
- arrange ideas in climactic order (**29c**),
- repeat important words (**29d**),
- invert word order in sentences (**29e**), and
- use an occasional short sentence (**29f**).

29a Placing words for emphasis

Words at the beginning or the end of a sentence receive emphasis. Notice how the revision of the following sentence adds emphasis to the beginning to balance the emphasis at the end:

~~In today's society, most good~~ Good jobs today require a college education.

You can also emphasize important words or ideas by placing them after a colon (**17d**) or a dash (**17e**).

At a later time [rocks and clay] may again become what they once were: dust.　　**—LESLIE MARMON SILKO, "Interior and Exterior Landscapes"**

By 1857, miners had extracted 760 tons of gold from these hills—and left behind more than ten times as much mercury, as well as devastated forests, slopes and streams.

—REBECCA SOLNIT, *Storming the Gates of Paradise: Landscapes for Politics*

EXERCISE 1

Below are three versions of a sentence. Decide which words receive more stress than the others in each version. Explain why.

1. The stunt double is essential to any action movie; that person may have to ride a horse backwards, jump from a tall building, or leap between speeding cars.

2. Essential to any action movie is the person riding a horse backwards, jumping from a tall building, or leaping between speeding cars: the stunt double.

3. The stunt double—essential to any action movie—may have to ride a horse backwards, jump from a tall building, or leap between speeding cars.

29b Using cumulative and periodic sentences

In a **cumulative sentence**, the main idea (the independent clause) comes first; less important ideas or supplementary details follow.

> **The day was hot for June,** a pale sun burning in a cloudless sky, wilting the last of the irises, the rhododendron blossoms drooping.
> —ADAM HASLETT, "Devotion"

In a **periodic sentence**, however, the main idea comes last, just before the period.

> In a day when movies seem more and more predictable, when novels tend to be plotless, baggy monsters or minimalist exercises in interior emotion, **it's no surprise that sports has come to occupy an increasingly prominent place in the communal imagination.** —MICHIKO KAKUTANI, "Making Art of Sport"

Both of these types of sentences can be effective. Because cumulative sentences are more common, however, the infrequently encountered periodic sentence tends to provide emphasis.

29c Ordering ideas from least to most important

By arranging your ideas in **climactic order**—from least important to most important—you build up suspense. If you place your most important idea at the end of the sentence, it will not only receive emphasis but also provide a springboard to the next sentence. In the following example, the writer emphasizes a doctor's desire to help the disadvantaged and then implies that this desire has been realized through work with young Haitian doctors:

> While he was in medical school, the soon-to-be doctor discovered his calling: to diagnose infectious diseases, to find ways of curing people with these diseases, and **to bring the lifesaving knowledge of modern medicine to the disadvantaged.** Most recently, he has been working with a small group of young doctors in Haiti.

29d Repeating important words

Although effective writers avoid unnecessary repetition, they also understand that deliberate repetition emphasizes key words or ideas.

> We **forget** all too soon the things we thought we could never **forget.** We **forget** the loves and betrayals alike, **forget** what we whispered and what we screamed, **forget** who we are.
>
> —JOAN DIDION, "On Keeping a Notebook"

In this case, the emphatic repetition of *forget* reinforces the author's point—that we do not remember many things that once seemed impossible to forget.

29e Inverting word order

Most sentences begin with a subject and end with a predicate. When you move words out of their normal order, you draw attention to them.

> <u>**At the back of the crowded room**</u> sat **a newspaper reporter.**
> [COMPARE: **A newspaper reporter** sat <u>**at the back of the crowded room.**</u>]

Notice the inverted word order in the second sentence of the following passage:

> ¹The Library Committee met with the City Council on several occasions to persuade them to fund the building of a library annex. ²So successful were their efforts that a new wing will be added by next year. ³This wing will contain archival materials that were previously stored in the basement.

The modifier *so successful* appears at the beginning of the sentence, rather than in its normal position, after the verb: *Their efforts were* so successful *that* . . . . The inverted word order emphasizes the committee's accomplishment.

MULTILINGUAL WRITERS

INVERTING WORD ORDER

English sentences are inverted in various ways. Sometimes the main verb in the form of a participle is placed at the beginning of the sentence. The subject and the auxiliary verb(s) are then inverted.

>   *part*   *aux*   *s*
> **Carved** into the bench **were someone's initials.**
> [COMPARE: Someone's initials were carved into the bench.]

For more information on English word order, see Chapter 45.

29f   Using an occasional short sentence

In a paragraph of mostly long sentences, try using a short sentence for emphasis. To optimize the effect, lead up to the short sentence with an especially long sentence.

> After organizing the kitchen, buying the groceries, slicing the vegetables, mowing the lawn, weeding the garden, hanging the decorations, and setting up the grill, I was ready to have a good time when my guests arrived. **Then the phone rang.**

EXERCISE 2

Add emphasis to each of the following sentences by using the technique indicated at the beginning. You may have to add some words and/or delete others.

1. (climactic order) In the 1960 Olympics, Wilma Rudolph tied the world record in the 100-meter race, she tied the record in the 400-meter relay, she won the hearts of fans from around the world, and she broke the record in the 200-meter race.

2. (periodic sentence) Some sports reporters described Rudolph as a gazelle because of her beautiful stride.

3. (inversion) Rudolph's Olympic achievement is impressive, but her victory over a crippling disease is even more spectacular.

4. (final short sentence) Rudolph was born prematurely, weighing only four and one-half pounds. As a child, she suffered from double pneumonia, scarlet fever, and then polio.

5. (cumulative sentence) She received help from her family. Her brothers and sister massaged her legs. Her mother drove her to a hospital for therapy.

6. (climactic order) After she built up strength and gained self-confidence, Rudolph set a scoring record in basketball, she set the standard for future track and field stars, and she set an Olympic record in track.

30 Variety

To make your writing lively and distinctive, include a variety of sentence types and lengths. Notice how the sentences in the following paragraph vary in length, form (simple, compound, and compound-complex), and function (statements, questions, and commands). The variety of sentences makes this paragraph about pleasure pleasurable to read.

> Start with the taste. Imagine a moment when the sensation of honey or sugar on the tongue was an astonishment, a kind of intoxication. The closest I've ever come to recovering such a sense of sweetness was secondhand, though it left a powerful impression on me even so. I'm thinking of my son's first experience with sugar: the icing on the cake at his first birthday. I have only the testimony of Isaac's face to go by (that, and his fierceness to repeat the experience), but it was plain that his first encounter with sugar had intoxicated him—was in fact an ecstasy, in the literal sense of the word. That is, he was beside himself with the pleasure of it, no longer here with me in space and time in quite the same way he had been just a moment before. Between bites Isaac gazed up at me in amazement (he was on my lap, and I was delivering the ambrosial forkfuls to his gaping mouth) as if to exclaim, "Your world contains *this?* From this day forward I shall dedicate my life to it." (Which he basically has done.) And I remember thinking, this is no minor desire, and then wondered: Could it be that sweetness is the prototype of *all* desire? —**MICHAEL POLLAN**, *The Botany of Desire*

This chapter will help you

- revise sentence length and form (**30a**);
- vary sentence openings (**30b**); and
- use an occasional question, command, or exclamation (**30c**).

30a Revising sentence length and form

To avoid the choppiness of a series of short sentences, combine some of them into longer sentences. You can combine sentences by using a coordinating conjunction (such as *and, but,* or *or*), a subordinating conjunction (such as *because, although,* or *when*), or a relative pronoun (such as *who, that,* or *which*).

Short	Americans typically eat popcorn at movie theaters. They also eat it at sporting events.
Combined	Americans typically eat popcorn at movie theaters **and** sporting events. [coordinating conjunction (**1g(1)**)]
Short	Researchers have found thousand-year-old popcorn kernels. These kernels still pop.
Combined	Researchers have found thousand-year-old popcorn kernels **that** still pop. [relative pronoun (**5a(3)**)]
Short	Popcorn was in demand during the Great Depression. Impoverished families could afford it.
Combined	**Because** impoverished families could afford it, popcorn was in demand during the Great Depression. [subordinating conjunction (**1g(3)**)]

You may sometimes be able to use both a subordinating and a coordinating conjunction.

Short	Sugar was sent abroad during World War II. Little sugar was left for making candy. Americans started eating more popcorn.
Combined	**Because** sugar was sent abroad during World War II, little was left for making candy, **so** Americans started eating more popcorn. [subordinating and coordinating conjunctions (**1g**)]

It is also possible to combine sentences by condensing one of them into a phrase (**1e**).

Short	Some colonial families ate popcorn for breakfast. They ate it with sugar and cream.
Combined	Some colonial families ate popcorn **with sugar and cream** for breakfast. [prepositional phrase (**1e(4)**)]

THINKING RHETORICALLY

SHORT SENTENCES

Occasionally, a series of brief sentences produces a special effect. The short sentences in the following passage capture the quick actions taking place as an accident is about to occur:

"There's a truck in your lane!" my friend yelled. I swerved toward the shoulder. "Watch out!" she screamed. I hit the brakes. The wheel locked. The back of the car swerved to the right.

30b Varying sentence openings

Most writers begin more than half of their sentences with a subject. Although this pattern is common, relying on it too heavily can make writing seem predictable. Experiment with the following alternatives for starting your sentences.

(1) Beginning with an adverb

Immediately, the dentist stopped drilling and asked me how I was doing.

(2) Beginning with a phrase

In the auditorium, voters waited in silence before casting their ballots. [prepositional phrase (1e(4))]

A tight contest, the gubernatorial election was closely watched by election officials. [appositive phrase (1e(5))]

Appealing to their constituents, candidates stated their positions. [participial phrase (1e(3))]

(3) Beginning with a transitional word or phrase

In each of the following examples, the transitional word or phrase shows the relationship between the ideas in the pair of sentences. (See also 33d.)

Many restaurants close within a few years of opening. **But** others, which offer good food at reasonable prices, become well established.

Independently owned restaurants struggle to get started for a number of reasons. **First of all,** they have to compete against successful restaurant chains.

(4) Beginning with a word that usually comes after the verb

I was an abysmal football player. **Soccer,** though, I could play well. [direct object]

Vital to any success I had were my mother's early lessons. [predicate adjective]

EXERCISE 1

Convert each set of short sentences into a single longer sentence.

1. On May 29, 1953, Edmund Hillary reached the summit of Mt. Everest. Hillary was a mountaineer from New Zealand. Tenzing Norgay was his Sherpa guide. Mt. Everest is the highest mountain in the world.

2. Hillary had been a member of a Swiss expedition. The Swiss expedition tried to reach the top of Mt. Everest in 1952. Bad weather stopped them eight hundred feet from the summit.

3. In March of 1953, Hillary joined an expedition from Great Britain. This expedition was led by John Hunt.

4. The expedition approached the peak. Conditions were worsening. Hunt directed Hillary and Norgay to continue to the summit.

5. Hillary thawed out his frozen boots on the morning of May 29. The two climbers then made the final ascent.

30c   Using questions, exclamations, and commands

You can vary sentences in a paragraph by introducing an occasional question, exclamation, or command (1i).

(1) Raising a question or two for variety

If people could realize that immigrant children are better off, and less scarred, by holding on to their first languages as they learn a second one, then perhaps Americans could accept a more drastic change. What if every English-speaking toddler were to start learning a foreign language at an early age, maybe in kindergarten? What if these children were to learn Spanish, for instance, the language already spoken by millions of American citizens, but also by so many neighbors to the South?

—ARIEL DORFMAN, "If Only We All Spoke Two Languages"

You can either answer the question you pose or let readers answer it for themselves, in which case it is called a **rhetorical question** (1i).

(2) Adding an exclamatory sentence for variety

But at other moments, the classroom is so lifeless or painful or confused—and I so powerless to do anything about it—that my claim to be a teacher seems a transparent sham. Then the enemy is everywhere: in those students from some alien planet, in the subject I thought I knew, and in the personal pathology that keeps me earning my living this way. What a fool I was to imagine that I had mastered this occult art—harder to divine than tea leaves and impossible for mortals to do even passably well!

—PARKER PALMER, *The Courage to Teach*

Although you can make sentences emphatic without using exclamation points (Chapter 29), the introduction of an exclamatory sentence can break up a regular pattern of declarative sentences.

(3) Including a command for variety

Now I stare and stare at people shamelessly. Stare. It's the way to educate your eye.   —WALKER EVANS, *Unclassified*

In this case, a one-word command, "Stare," provides variety.

EXERCISE 2

Explain how questions and commands add variety to the following paragraph. Describe other ways in which this writer varies his sentences.

[1]The gods, they say, give breath, and they take it away. [2]But the same could be said—couldn't it?—of the humble comma. [3]Add it to the present clause, and, of a sudden, the mind is, quite literally, given pause to think; take it out if you wish or forget it and the mind is deprived of a resting place. [4]Yet still the comma gets no respect. [5]It seems just a slip of a thing, a pedant's tick, a blip on the edge of our consciousness, a kind of printer's smudge almost. [6]Small, we claim, is beautiful (especially in the age of the microchip). [7]Yet what is so often used, and so rarely recalled, as the comma—unless it be breath itself?

—**PICO IYER, "In Praise of the Humble Comma"**

31 Reading, Writing, and the Rhetorical Situation

Rhetoric, the purposeful use of language, permeates your daily activities. You read rhetorically every time you review a syllabus, a newspaper, or a book. You write rhetorically when you draft and revise a purposeful message (aimed at a specific audience) to resolve a problem or address an opportunity for change, whether the message is a letter to the editor, an academic essay, or an e-mail. Because you have been reading and writing rhetorically most of your life, you are probably already knowledgeable about rhetoric.

In this chapter, you will see how reading and writing rhetorically are processes, each a series of sometimes overlapping steps. The chapter will help you

- understand the elements of any rhetorical situation (**31a**),
- recognize a rhetorical opportunity (**31b**),
- determine a purpose for a message (**31c**),
- consider the intended audience (**31d**), and
- think about the rhetorical effects of context (**31e**).

This chapter will also help you see exactly how reading and writing rhetorically can help you succeed with a variety of class assignments.

31a Understanding the rhetorical situation

Rhetoric is the *purposeful* use of language, whether for persuading, explaining, describing, informing, or some other purpose. The best way to establish a purpose for using language is to examine the **rhetorical situation** (Figure 31.1), which is composed of the writer (or speaker), the audience, the opportunity, the message, and the context. Any assignment that asks you to read or write rhetorically will be easier and more enjoyable if you keep the rhetorical situation in mind.

Writers (or speakers) enter a rhetorical situation when they identify an **opportunity**

Figure 31.1. The rhetorical situation.

© 2013 Cengage Learning

to propose change—in behavior, attitude, or perception—through the effective use of language. Once writers have identified a rhetorical opportunity, they prepare a **message** (using words and sometimes images) for a specific **audience**. Successful writers always link their purpose to their audience. The audience receives a writer's message within a specific **context** that includes what others have already said about the topic and how that topic was presented. Your primary role as a writer is to take into account all the elements of the rhetorical situation.

In your role as a rhetorical or critical reader, you also follow a series of steps, previewing an entire text—maybe jumping from the title to the table of contents and the author biography, then to the visuals, the final chapter, and the index—to see how much time and expertise reading the text will require. Often you preview a text **chronologically** (in order of occurrence), reading for content and responding with comments and questions, and then reread the text **recursively** (alternating between moving forward and looping back), maybe taking time to talk with your peers about their understanding of the content. Previewing a text means staying alert for the author's major points, for transitional words that reveal sequence (33d), and for developmental structures (or other clues) that indicate summary, causation, repetition, exemplification, or intensification. You may want to respond to those important points, as though you were carrying on a conversation with the author. To do so, you can use a pen, sticky notes, or online highlighting to underline, highlight, annotate, or question passages that interest or confuse you.

CHECKLIST for Reading Rhetorically

- What is your purpose for reading the text—pleasure, research, fulfillment of a course requirement, problem solving, inspiration?

- What is the author's purpose for writing? What do you know about this author's credibility, use of reliable (or unreliable) sources, experience, and biases?

- What knowledge or experience does the reading demand? Does your knowledge or experience meet that demand?

- What are the key parts of the text? How do those parts relate to your purpose for reading? What specific information from this text will help you achieve your purpose?

- What is your strategy for previewing, reading, and rereading? Are you reading online or on paper? How will you respond to the text?

- With which passages do you agree or disagree, and why?

- As you read, what do you understand clearly? What do you want to know more about?

- What questions do you have for your instructor or peers? What questions—and answers—might they have for you?

Reading and writing rhetorically allow you to consider each of the elements of the rhetorical situation separately as well as in combination: you evaluate the thesis statement (32c), the key points of the message, and the support provided for each point, as well as identify what needs to be said and what is purposefully left unsaid. When you *read* rhetorically, you read more effectively and thus are able to speak or write knowledgeably about what you have read. When you *write* rhetorically, you generate new ideas and communicate them clearly and concisely to your audience (Chapters **32** and **33**)—and you improve your understanding of what you have read.

31b   Responding to a rhetorical opportunity

A rhetorical situation offers you an opportunity to make a difference, often by solving a specific problem or addressing an issue for a specific audience. A college application, for instance, invites you to use words to address the problem of being accepted by a college. Once you engage the rhetorical opportunity—the reason that impels you to speak or write—you will be better able to gauge all the elements of your message (from word choice to organizational pattern) in terms of your intended audience and your purpose.

THINKING RHETORICALLY
OPPORTUNITY TO ADDRESS AN ISSUE

Historical events often serve as rhetorical opportunities. In 2010, for example, thirty-three Chilean miners were trapped nearly a mile underground for sixty-nine days, a tragic and near-fatal event that could have been prevented had the mining industry initially heeded the miners' warnings about unsafe conditions. The whole world watched—but also embraced the opportunity to advise, petition, and report—while Chilean officials planned the miners' successful rescue.

As a rhetorical reader, you need to determine the author's purpose for writing: to answer a question, solve a problem, address an issue, or entertain? The title of the text, the summary, or the abstract may provide that information. The cover of science

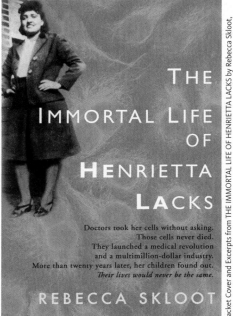

Jacket Cover and Excerpts from THE IMMORTAL LIFE OF HENRIETTA LACKS by Rebecca Skloot, copyright © 2010, 2011 by Rebecca Skloot. Used by permission of Crown Publishers, a division of Random House, Inc.

Figure 31.2. Cover of *The Immortal Life of Henrietta Lacks*.

writer Rebecca Skloot's book *The Immortal Life of Henrietta Lacks* (Figure 31.2) reveals that the book is the medical biography of an unwitting cell donor, whose harvested cells became the most commonly used line in biomedical research. Skloot addresses the complicated issue of patient consent in medical research.

31c   Determining a specific purpose

Once you realize that your words can spark change, modify a situation, address a need, or resolve a problem, you can determine your specific purpose. Writers choose their words carefully in order to clarify their purpose, always aligning that purpose with the potential to influence the thinking of their intended audience. The writer may want to evoke emotions, challenge beliefs, amuse or entertain, report information, explain or evaluate the significance of information, analyze a situation, clarify a point, invite the audience to consider alternative points of view, or argue for or against an attitude or a course of action.

Writers must identify their overall purpose in terms of their audience, keeping in mind that they can better achieve that purpose using various methods of development (such as narration, description, and cause-and-consequence analysis; see 32g). As a writer, your goal should be to respond to an opportunity to make a change and to provide a clear plan for effecting that change.

Readers need to identify the writer's purpose as well, assessing it to determine what the writer expects of the audience: to be entertained, informed, or persuaded. For example, the purpose of Rebecca Skloot's book is not clear from its title, but the blurb on the front cover provides helpful information: "Doctors took her cells without asking. Those cells never died. They launched a medical revolution and a multimillion-dollar industry. More than twenty years later, her children found out. *Their lives would never be the same.*" Skloot's purpose is to inform.

You can also look at a book's index to help orient yourself to the content and the writer's purpose. For example, the index of Skloot's book includes these entries, among others (Figure 31.3): "*A Conspiracy of Cells*," "Cell Culture Collection Committee," "cervical cancer," and "HeLa [or Henrietta Lacks] cells." Thus, you might be able to tell from the index whether the book contains the information you need to conduct research or satisfy your curiosity about African American medical history, cancer research, cell culture, or another related topic you are interested in.

Readers also need to establish their own purpose for reading: to summarize what they are learning, apply this information to solve a problem, analyze the constituent parts of the

Figure 31.3. Excerpt from the index of *The Immortal Life of Henrietta Lacks*.

text, make a decision, support a position, or combine the information in an original way.

Your challenge as a reader is to grasp the meaning the author wants to convey to you within the particular rhetorical situation. As a writer, your challenge is to make the purpose of your writing clear to your intended readers. Successful academic readers and writers always take time to talk with their instructor (and check the assignment sheet) to review the rhetorical opportunity, purpose, audience, context, and message of each reading and writing assignment. They also talk about their reading and writing with their peers, to make sure that they are on the right track. So, ask questions, listen to the answers, and try to answer the questions of your peers as you all work together to establish what is most significant about the writing and reading you are assigned.

> **CHECKLIST** for Assessing Purpose
>
> - How is the rhetorical purpose of the text linked to its intended audience? How might that audience help the writer fulfill the purpose of addressing an issue or resolving a problem?
>
> - What purpose does the writer want the message to fulfill: to evoke emotion, to entertain or inspire, to convey information, or to argue for or against a course of action or an attitude? Does the writer have more than one purpose?
>
> - How well do the topic and the audience connect to the rhetorical purpose? What examples or choice of words help fulfill that purpose?

Depending on the writer's overall purpose, writing can be classified as expressive, expository, or argumentative. Any of these types of writing can help a writer fulfill an overall purpose.

(1) Expressive writing

Expressive writing emphasizes the writer's feelings and reactions to people, objects, events, or ideas. Personal letters and journals are often expressive, as are many essays and short stories. As you read paragraph 1, which comes from a memoir, notice how philosopher Kathleen Dean Moore conveys her thoughts about what it takes to be happy. (For ease of reference, sample paragraphs in this chapter are numbered.)

1   So many people are telling me what should make me happy. Buy a cute car. Be thin. Get promoted or honored or given a raise. Travel: Baja! Belize! Finish the laundry. The voices may or may not be my own; they are so insistent that I can't distinguish them from the ringing in my ears. Maybe they are the voices of my mother and father, long dead and well intended, wanting only that I would be happy. Or my husband Frank, fully alive but ditto in all other respects. My colleagues. Maybe they're the voices of

advertisers, popular songwriters, even the president. Most of the time, I don't even think about making choices, plowing through my life as if I were pulled by a mule.

—KATHLEEN DEAN MOORE, "The Happy Basket"

(2) Expository writing

Much of the academic material you read—textbooks, news accounts, reports, books (such as the one by Rebecca Skloot featured in this chapter), and journal articles—is expository, as are most of the essays you will be asked to write in college. Expository writing focuses more on objects, events, or ideas than on the writer's feelings about them. Any time you report, explain, analyze, or assess, you are practicing exposition. Paragraph 2, an excerpt from Nina Jablonski's book, *Skin*, explains why the covering on our bodies varies.

2   [A] distinctive attribute of human skin is that it comes naturally in a wide range of colors, from the darkest brown, nearly black, to the palest ivory, nearly white. This exquisite sepia rainbow shades from darkest near the equator to lightest near the poles. This range forms a natural cline, or gradient, that is related primarily to the intensity of the ultraviolet radiation (UVR) that falls on the different latitudes of the earth's surface. Skin color is one of the ways in which evolution has fine-tuned our bodies to the environment, uniting humanity through a palette of adaptation. Unfortunately, skin color has also divided humanity because of its damaging association with concepts of race. The spurious connections made between skin color and social position have riven peoples and countries for centuries.

—NINA G. JABLONSKI, *Skin*

(3) Argumentative writing

Argumentative writing is intended to influence the reader's attitudes and actions. Most writing is to some extent an argument. Even something as apparently straightforward as a résumé can be seen as an argument for a job interview. However, writing is usually called argumentative if it clearly supports a specific position (Chapter 34). As you read paragraph 3, note how Rebecca Skloot demonstrates that Henrietta Lacks's story became an integral and ongoing part of Skloot's life.

3   I couldn't have imagined it then, but that phone call [an attempt to locate Lacks's husband] would mark the beginning of a decadelong adventure through scientific laboratories, hospitals, and mental institutions, with a cast of characters that would include Nobel laureates, grocery store clerks, convicted felons, and a professional con artist. While trying to make sense of the history of cell culture and the complicated ethical debate surrounding the use of human tissues in research, I'd be accused of conspiracy and slammed into a wall both physically and metaphorically, and I'd eventually find myself on the receiving end of something that looked a lot like an exorcism. I did eventually meet Deborah [Lacks's daughter], who would turn out to be one of the strongest and most resilient women I'd ever known. We'd form a deep personal bond, and slowly, without realizing it, I'd become a character in her story, and she in mine.

—REBECCA SKLOOT, *The Immortal Life of Henrietta Lacks*

EXERCISE 1

Write two paragraphs that begin to develop an expressive, expository, or argumentative essay on one of the following subjects.

1. your physical health
2. paying for college
3. your career goals
4. academic pressures
5. a good teacher
6. your living situation

31d   Considering audience

A clear understanding of the audience—its values, concerns, knowledge, and capabilities—helps writers convey their purpose through their message and helps readers orient themselves to that message. Writers fashion a message by considering quality and quantity of details, choice of words, and inclusion of effective examples and supporting details. Of course, the audience is anyone who reads a text, but the rhetorical, or intended, audience consists of those people whom the author considers capable of being influenced by the words or who are capable of bringing about change. Therefore, you need to think clearly about who exactly will be reading (or might end up reading) what you write and ask yourself whether your word choices and examples are appropriate for that audience.

(1) A specialized audience

A **specialized audience** has a demonstrated interest in the subject. If your sister is a Title I reading teacher, she might be interested in helping her colleagues build the school-district infrastructure necessary for implementing the *No Child Left Behind* (NCLB) Act of 2001. She would be part of the audience for paragraph 4.

4   Of course, the vast majority of districts . . . will be scrambling to find staff who can assume responsibility for NCLB. . . . Deciding how to assign new responsibilities for NCLB and how to restructure other duties should be predicated upon fully developed plans that identify the tasks that need to be accomplished. It may be a small comfort, but recognizing the trade-offs of different assignment decisions can help administrators and their staffs cope with them. Wherever NCLB coordination responsibility is housed, it is important to inform all district staff of where that is. Parents may contact the district at different entry points. They need to be directed efficiently to the NCLB authority.
                              —OFFICE OF INNOVATION AND IMPROVEMENT,
                                        US DEPARTMENT OF EDUCATION,
            "Innovations in Education: Creating Strong District School Choice Programs"

Although you can probably read every word in the preceding excerpt, the specialized terms and ideas might not be clear to you. A specialized audience has both interest in and knowledge of the topic under consideration.

Many of the essays you will be assigned to write in college—in the sciences, history, economics, English, and psychology, for example—will be aimed at your instructor, who comprises a specialized audience. Your job as an academic writer, then, is to weigh what your instructor knows against the specific information she or he needs and then determine how best to develop the needed information. Most of the materials you will be required to read in college are also aimed at a specialized audience. As part of such an audience, you will want to preview each text to assess what it demands of you in terms of time, effort, and knowledge. You will also want to review the directories within the text (table of contents, index, and bibliography) and the visual aids to orient yourself to the text and its context. You may also have to refer to a dictionary, an encyclopedia, or some other resource as you read.

(2) A diverse audience

A **diverse audience** consists of readers with differing levels of expertise and varying interest in the subject at hand. Paragraph 5 is taken from physician Atul Gawande's essay on blushing, which explains this universal human behavior.

5   Why we have such a reflex is perplexing. One theory is that the blush exists to show embarrassment, just as the smile exists to show happiness. This would explain why the reaction appears only in the visible regions of the body (the face, the neck, and the upper chest). But then why do dark-skinned people blush? Surveys find that nearly everyone blushes, regardless of skin color, despite the fact that in many people it is nearly invisible. And you don't need to turn red in order for people to recognize that you're embarrassed. Studies show that people detect embarrassment before you blush. Apparently, blushing takes between fifteen and twenty seconds to reach its peak, yet most people need less than five seconds to recognize that someone is

embarrassed—they pick it up from the almost immediate shift in gaze, usually down and to the left, or from the sheepish, self-conscious grin that follows a half second to a second later. So there's reason to doubt that the purpose of blushing is entirely expressive.   —ATUL GAWANDE, "Crimson Tide"

As a writer, you can easily imagine a diverse audience if you think of thoughtful, receptive, educated adults, with whom you may share some common ground. As a reader, you may often find yourself as a member of a diverse audience, one likely to include people with different beliefs, knowledge, and experience (34e(2)). Rarely will you write for or read something by someone who is exactly like you (19c), so remember that connections between writers and readers are made through the choice of language (Chapters 19–21), details, and examples (32f), which either invite readers into or exclude them from a written work.

(3) Multiple audiences

Writers often need to consider multiple audiences, tailoring word choice and tone to the primary audience, knowing that a secondary audience might have access to the text. For peer reviews (33e), for example, your primary audience is the peer whose writing you are reviewing. Knowing that your instructor (the secondary audience) will also be reading your commentary, you may respond to your peer's writing with more thoughtfulness and tact than you would otherwise. When you know that your rhetorical situation includes multiple audiences, you can better adjust your words and edit your information. And when you consciously read as a member of either a primary or a secondary audience, you can better evaluate your responses.

The following checklist may help you assess an audience, whether you are doing so as the writer or the reader.

CHECKLIST for Assessing Audience

- Who is the intended audience for this writing? Who else might read it? Has the writer identified the primary audience while also accommodating a secondary audience? What passages indicate that the writer has addressed the primary audience and also recognized the expectations of a secondary audience?

- What do you know about the backgrounds, values, and characteristics of the members of the intended audience? What do the audience members have in common? How are they different?

- What background, values, and characteristics do you (as either the writer or a reader) share with the members of the intended audience? How do you differ from them?

- How open are the members of this audience to views that are different from their own?

- What do you not know about this audience? In other words, what assumptions about its members might be risky to make?

- What kind of language, examples, and details are most appropriate (or inappropriate) for the members of this audience?

- What does this audience already know about the topic under consideration?

- What level of expertise will this audience expect from the writer?

31e Writing and reading a message within a context

Context includes the time and place in which a message is read or written, the writer and the intended audience, and the medium of delivery (print, online, spoken, or visual); in other words, context comprises the set of circumstances under which the writer and reader communicate. Social, political, religious, and other cultural factors as well as attitudes and beliefs influence context, as do the constraints (obstacles) or resources (positive influences) of the rhetorical situation. Whatever you read, write, or speak is always influenced (positively or negatively) by the context.

The medium of delivery is also part of the context. Writing material for a web page, for example, requires you to consider features of organization, design, and style that are related to onscreen presentation of material. Reading online also requires an adjustment to visual and audio elements that can enhance (or distract from) your experience. An online method of delivery, then, requires you to make different kinds of rhetorical decisions than you make for a text in a wholly static print medium (Chapter 35).

When you read the work of other writers, you will sometimes find the context for the work explicitly stated in a preface or an introduction. Often, however, the context must be inferred. Whether or not the context is announced, it is important that writers and readers identify and consider it.

CHECKLIST for Assessing Context

- What are the factors influencing the context in which you are writing: the time and place, the intended audience, and the medium of delivery (print, online, spoken, or visual)?

- What other events (personal, local, or global) are influencing the context for writing?

- What are the expectations concerning the length of this written message? If a length has not been specified, what seems appropriate in terms of purpose and audience?

- What document design (Chapter 8) is appropriate, given the context?

- Under what circumstances will this piece of writing be read? How can you help the intended audience quickly see the purpose of the text within these circumstances?

32 Planning and Drafting Essays

As an experienced writer, you already understand that writing is a process. Whether you are writing in or out of school, quickly or slowly, you revise and edit in light of your rhetorical opportunity, purpose, audience, and context (31c–e).

This chapter will help you understand your writing process by showing you how to

- recognize suitable topics (32a),
- focus your ideas (32b),
- write a clear thesis statement (32c),
- organize your ideas (32d),
- express your ideas in multiple drafts (32e), and
- use various strategies to develop effective paragraphs (32f) and essays (32g).

The writing process is **recursive**, which means that as you plan and draft an essay, you may need to return to a specific activity several times. For example, drafting may help you see that you need to go back and collect more ideas, modify your thesis statement, or maybe even start over. Experienced writers expect the writing process to lead to new ideas at the same time as it reveals passages in need of improvement. Despite the infinite variations of the writing process (generating, organizing, developing, and clarifying ideas, as well as polishing prose), writing usually involves four recursive stages, which are described in the following box.

STAGES OF THE WRITING PROCESS

- **Prewriting** is the initial stage of the writing process. As you begin thinking about a specific writing task, you consider the rhetorical opportunity, the audience and purpose, the context, and the medium of delivery. Then you start exploring your topic by talking with others working on the same assignment, keeping a journal, freewriting, asking questions, or conducting preliminary research. By now, you may already know the best ways to energize your thinking and jump-start your writing.

- **Drafting** involves writing down your ideas quickly, writing as much as you can without worrying about being perfect or staying on topic. The more ideas you get down on paper, the more options you will have as you begin to clarify your thesis, compose your next draft, and revise. Progress is your goal at this stage, not perfection.

- **Revising** offers you the opportunity to focus your purpose in terms of your audience, establish a clear thesis statement that conveys your main idea (**32c**), and organize your ideas toward those ends (**32d**). The revision stage is the time to stabilize the overall arrangement of your piece, develop the individual paragraphs (**32f** and **33c**), and reconsider your introduction and conclusion (**33b**). Remember that revising produces yet another draft meriting further revision and editing.

- **Editing** focuses on surface features: punctuation, spelling, word choice, grammar, sentence structure, and all the rest of the details of Standardized English (**33f**). As you prepare your work for final submission, consider reading it aloud to discover which sentence structures and word choices could be improved. You may even catch a few spelling errors in the process.

32a   Selecting a subject

If you are not assigned a subject and are free to choose your own, you can start by identifying a problem that your words can address or resolve. You can think about what you already know—or would like to learn—about the problem as well as what is likely to interest your intended audience (31d). The first step toward engaging an audience is being interested or experienced in the subject yourself. When subjects are important to you, they usually interest readers, especially when you write with a clear purpose and use well-chosen details and examples (31c and 32f).

More often, though, you will be asked to write about subjects that are outside your personal experience but are part of your academic coursework. Sometimes, you may be permitted to choose a subject that interests you, as long as it relates directly to a course. To find a subject that meets both those criteria, you can start by looking through your textbook, particularly in the sections listing suggestions for further reading. Go through your lecture notes, your reading journal, or any marginal annotations you have made in your textbook. Ask yourself whether any details of the subject have surprised, annoyed, or intrigued you—if you have discovered an opportunity for entering the scholarly conversation. Writing about a subject is one of the best ways to combine your own need to know more with the opportunity to deliver new information to an audience.

(1) Keeping a journal to explore subjects

Many experienced writers use journals to explore various subjects. In a **personal journal**, you can reflect on your experiences and inner life or focus on external events (such as political campaigns, sporting events, and new books and films), writing for your own benefit. Some writers prefer to keep a **reading journal**, where they record quotations, observations, and other material that they might use in future writing projects. Whatever type of journal you keep, write quickly, without worrying about spelling or grammar.

(2) Freewriting as a risk-free way to explore a subject

When **freewriting**, writers record whatever comes to mind about a subject, writing without stopping for a limited period of time—often no more than ten minutes. When they repeat themselves or get off track, writers keep going in order to generate ideas, make connections, create a tentative organization, and bring information and memories to the surface.

When Mary LeNoir's English instructor directed her to write for five minutes about how she chose a college, Mary produced the following directed freewriting as the first step toward her final essay (which appears in Chapter 33).

> I'm an athlete. I've always been an athlete. But being an athlete won't be my job when I graduate from college. So when I was thinking about which school to attend, I tried hard not to let high school athletics and the college recruiting process cloud my decision making. I also tried not to listen to all the people who seem to think that "money" is how we student-athletes make our decision. That's not true. In fact, many athletes place the athletic factor above the academic one, as though to say, "I'm going to X school to play Y sport and I'll take classes while I'm there." But play time is not the only factor that takes over the decision-making process. Some athletes commit to their school for reasons they haven't really thought through: they make their decision based on their emotions, on how much they like the school (whatever that means) or like the coach or like the other players. They can end up in a school located halfway across the country from home, a school with rigorous academics and a huge party scene. Sometimes, they don't find

out until too late that they cannot balance the academics with the athletics, not to mention the parties and big classes. And they're too far from home to get the support they need. If they're not partiers (or if they party too much) or if they need a small class and don't have that, then, they're out of luck. I'm also aware of athletes choosing a school based on potential for a professional career, even though the majority of them will move on from their sport, unless they venture into coaching or play it with friends. I know I'm throwing out a lot of things here and will need to narrow it down to one focus and come up with a thesis and an outline.

Mary's freewriting generated a number of possibilities for developing an essay about why she chose the college she did: she cites academics, emotional responses, and athletics as strong reasons. She was responding to the opportunity to explain the selection process, especially to people who believe that student-athletes think only of how much money they will receive if they attend a particular college. Notice, however, that her freewriting leads her to describe other athletes, not herself, and that she realizes she needs to think about what comes next in her writing process.

(3) Questioning to push the boundaries of a subject

You can also explore a subject by asking yourself some questions. The simplest questioning strategy for exploring a subject comes from journalism. **Journalists' questions**—*Who? What? When? Where? Why?* and *How?*—are easy to use and can help you generate ideas about any subject. Using journalists' questions to explore how a student-athlete chooses a college could lead you to the following: *Who* qualifies as a student-athlete? *What* criteria do and should student-athletes use in choosing a college? *When* should student-athletes expect to give up their sports? *Where* can student-athletes best succeed? *Why* is financial aid not the only or even the most important selection criterion? *How* might a student-athlete make the best decision, given his or her characteristics and circumstances?

32b   Focusing a subject into a specific topic

By exploring a subject, you can discover productive strategies for development as well as a specific focus for your topic. As you freewrite, you will decide that some ideas seem worth pursuing while others seem inappropriate for the rhetorical opportunity to which you are responding, your intended audience, or the context. Thus, some ideas will fall away as new ones arise and your topic comes into sharper focus.

After generating ideas through strategies such as freewriting and questioning, you can use various rhetorical methods for developing the ideas (32g). In responding to a rhetorical opportunity to explain how student-athletes choose (or should choose) a college, Mary LeNoir needed to focus this fairly broad subject into a more narrow (and manageable) topic. Therefore, she considered how she might use each of the rhetorical methods of development to sharpen her focus:

- *Narration.* What is a typical story about a student-athlete deciding on a college?
- *Description.* How do colleges distinguish themselves in terms of size, course offerings, location, and cost? How do student-athletes differ? What distinctive characteristics of colleges and students produce the best matches?
- *Process analysis.* What steps do student-athletes take as they choose a college? What are the most and least useful of those steps?

- *Cause-and-consequence analysis.* What considerations lead to the best choice of a college? What are the considerations that too many student-athletes overlook? What are the consequences of making the right choice? Of making the wrong one?
- *Comparison and contrast.* How does the process of choosing a college differ for an athlete and a nonathlete? How does the process differ for a student whose goal is to be only a college athlete and a student whose goal is to be a professional athlete?
- *Classification and division.* How might student-athletes' college-related needs and expectations be classified? How might colleges be categorized based on what they offer student-athletes?
- *Definition.* How can the "best" college be defined? What are the best reasons for choosing a college? Are these reasons defined by immediate or long-term benefits?

A combination of strategies soon led Mary to a tentative focus:

After interviewing five student-athletes about their personal goals and circumstances and considering my own situation, I discovered that despite our differences, we all used three basic criteria in choosing our college: the overall atmosphere of the school, the potential for our athletic development, and the material conditions associated with attending the college (costs and geographic location).

Whatever rhetorical method you use to bring a topic into focus, your final topic should be determined not only by the rhetorical opportunity but also by your intended audience, your purpose, and the context in which you are writing.

The following checklist may help you assess your topic.

CHECKLIST for Assessing a Topic

- What unresolved problem or issue related to this topic captures your interest? How can you use words to address this rhetorical opportunity for change?
- What audience might be interested in this topic?
- What is your purpose in writing about this topic for this audience?
- Can you address the topic in the time and space (page length) available to you? Or do you need to narrow the topic?
- Do you have the information you need to address this topic? If not, how will you acquire additional information?
- Are you willing to take the time to learn more about the topic in order to engage the rhetorical opportunity?

EXERCISE 1

Use the journalists' questions (page 323) to generate more ideas about a subject that interests you. Then identify a rhetorical opportunity that emerges from your answers to those questions. How does that opportunity connect with your subject to create a specific topic that is appropriate for an essay?

32c   Conveying a clearly stated thesis

Once you have identified a rhetorical opportunity, determined an intended audience, decided on a purpose (to entertain, explain, teach, analyze, persuade, or compare), and focused on an interesting topic, you have drawn close to settling on your controlling idea, or **thesis**, which may take several drafts to finalize.

Most academic writing has a **thesis statement**, an explicit declaration (usually a single, clearly focused, specifically worded sentence) of the main idea. A thesis can be thought of as an assertion, or **claim** (34d), which indicates what you believe to be true, interesting, or valuable about your topic.

An explicitly formulated thesis statement identifies the topic, the purpose, and, in some cases, the plan of development. Notice how the following thesis statements fulfill their purpose. The first is from a descriptive essay.

> If Lynne Truss were Catholic, I'd nominate her for sainthood.
>
> —FRANK McCOURT, Foreword, *Eats, Shoots & Leaves*

With this simple statement, McCourt establishes that the topic is Lynne Truss and indicates that he will describe why she should be a saint. He conveys enthusiasm and admiration for Truss's work.

The following thesis statement for a cause-and-consequence analysis sets the stage for the series of incidents that unfolded after surgeon and writer Richard Selzer was granted refuge in an Italian monastery when he had no hotel reservations:

> Wanderers know it—beggars, runaways, exiles, fugitives, the homeless, all of the dispossessed—that if you knock at the door of a monastery seeking shelter you will be taken in.
>
> —RICHARD SELZER, "Diary of an Infidel: Notes from a Monastery"

The main idea in an argumentative essay usually conveys a strong point of view, as in the following, which unmistakably argues for a specific course of action:

> Amnesty International opposes the death penalty in all cases without exception.
>
> —AMNESTY INTERNATIONAL, "The Death Penalty: Questions and Answers"

The following are possible thesis statements that Mary LeNoir might have written and that imply different rhetorical methods of development (**32g**). This sentence suggests a focus on comparison and contrast:

> Student-athletes who want to play only college sports and those who want to go pro should employ different criteria for selecting a college.

The following sentence focuses on cause and consequence:

> By establishing exactly what I wanted from my college experience, I was able to choose the best college for me.

It is just as important to allow your thesis statement to remain tentative in the early stages of writing as it is to allow your essay to remain flexible through the early drafts. Rather than sticking with an initial thesis, which you might have to struggle to support, you want to let your final thesis statement evolve as you think, explore, draft, and revise. The following tips might help you develop a thesis statement.

TIPS FOR DEVELOPING A THESIS STATEMENT

- Decide which feature of the topic opens up a rhetorical opportunity.
- Write down your opinion about that feature.
- Mark the passages in your freewriting, journal, or rough draft that support your opinion.
- Draft a thesis statement that connects the rhetorical opportunity, the rhetorical purpose, and the intended audience.
- After completing a draft, ask yourself whether your thesis should be adjusted to reflect the direction your essay has taken.
- If you are unhappy with the results, start again with the first tip, and be even more specific.

A clear, precise thesis statement helps unify your message; it directs your readers through the writing that follows. Therefore, as you write and revise, check your thesis statement frequently. It should influence your decisions about which details to keep and which to eliminate as well as guide your search for appropriate additional information to support your assertions.

A thesis statement is usually a declarative sentence with a single main clause—that is, either a simple or a complex sentence (**1h**). It most often appears in the first paragraph of an essay, although you can put yours wherever it best furthers your overall purpose (perhaps somewhere later in the introduction or even in the conclusion). The advantage of putting the thesis statement in the first paragraph is that readers know from the beginning what your essay is about, to whom you are writing, why you are writing, and how the essay is likely to take shape. This technique has proved to be especially effective in academic writing. If the thesis statement begins the opening paragraph, the rest of the sentences in the paragraph support or clarify it, as is the case in paragraph 1. (For ease of reference, each of the sample paragraphs in this chapter is numbered.)

1   *The cafeteria was a dreadful place in the basement.* Hundreds of kids at a time ate there, kids who'd spent all morning having to be quiet and sit still. Because of the room's low ceilings and hard surfaces, the sound bounced all over the place, creating a din, a roar so deafening you had to scream to be heard. It was a madhouse, the Hades of the school, a place where the Furies all ran wild.   —**SAM SWOPE**, "The Animal in Miguel"

If the thesis statement is the last sentence of the opening paragraph, the preceding sentences build toward it, as in paragraph 2.

2   The story of zero is an ancient one. Its roots stretch back to the dawn of mathematics, in the time thousands of years before the first civilization, long before humans could read and write. But as natural as zero seems to us today, for ancient peoples zero

was a foreign—and frightening—idea. An Eastern concept, born in the Fertile Crescent a few centuries before the birth of Christ, zero not only evoked images of a primal void, it also had dangerous mathematical properties. *Within zero there is the power to shatter the framework of logic.*
—**CHARLES SEIFE,** *Zero: The Biography of a Dangerous Idea*

Keep in mind that most academic writing features an easy-to-locate thesis statement. The following checklist may help you assess a thesis.

CHECKLIST for Assessing a Thesis

- Does your thesis respond to a rhetorical opportunity to create a change or to address a problem?

- Does your thesis accurately reflect your point of view about your topic?

- How does your thesis relate to the interests of your intended audience, your purpose, and the context in which you are writing?

- Where is your thesis located? Would your readers benefit from having it stated earlier or later?

- Does your thesis reflect your overall purpose? Does it clarify your focus and indicate your coverage of the topic?

- What are the two strongest assertions you can make to support your thesis?

- What specific examples, details, or experiences support your assertions?

32d Arranging or outlining ideas

Most writers benefit from a provisional organizational plan that helps them order their ideas and manage their writing. Other writers compose informal lists of ideas and then examine them for overlap, pertinence, and potential. While some ideas will be discarded, others might lead to a thesis statement, a provocative introduction, a reasonable conclusion, or an overall organizational plan. Some writers rely on more formal outlines, in which main points form the major headings and supporting ideas form the subheadings. Whatever method you choose for arranging your ideas, remember that you can always alter your plan to accommodate any changes your thinking undergoes as you proceed.

An outline of Mary LeNoir's essay might look something like the following:

TENTATIVE THESIS STATEMENT: No matter what their sport, student-athletes tend to choose a college using three criteria: (1) how much play time they will have; (2) material considerations, mainly geographic location and financial aid; and (3) emotional connection with the school.

I. Many student-athletes begin by dreaming about going pro after college, even though few will.
   A. They anticipate how much playing time they will have on the college field.
   B. They consider schools with the strongest teams.
II. Then, they consider the material reasons for attending a school.
   A. They consider the geographic location of the school.
   B. They try to negotiate the best financial aid package available to a student-athlete of their caliber.
III. Many student-athletes ultimately base their decision on an emotional connection with the school.
   A. They always dreamed about playing their sport at a particular college.
   B. They fell in love with the campus or the city—or really liked the coach or the other players.
IV. How I worked through these criteria toward my decision
V. The consequences (positive and negative) of my decision

I wanted more: a road to travel, a radio that actually worked, a destination and goal, a more finely tuned knowledge of navigation involving blinkers, lights, different driving conditions, and—most of all—the ability to travel and negotiate with others also on the road.

—BRENDA JO BRUEGGEMANN, "American Sign Language and the Academy"

Notice how the series of details in paragraph 3 supports the main idea, or topic sentence (33c), which has been italicized to highlight it. Also notice how one sentence leads into the next, creating a clear picture of the experience being described.

(2) Developing a paragraph with examples

Like details, examples contribute to paragraph development by making specific what otherwise might seem general and hard to grasp. **Details** describe a person, place, or thing; **examples** illustrate an idea with information that can come from different times and places. Both details and examples support the main idea of a paragraph.

The author of paragraph 4 uses several closely related examples (as well as details) to support the main idea with which she begins.

4   *It began with coveting our neighbor's chickens.* Lily would volunteer to collect the eggs, and then she offered to move in with them. Not the neighbors, the chickens. She said if she could have some of her own, she would be the happiest girl on earth. What parent could resist this bait? Our lifestyle could accommodate a laying flock; my husband and I had kept poultry before, so we knew it was a project we could manage, and a responsibility Lily could handle largely by herself. I understood how much that meant to her when I heard her tell her grandmother, "They're going to be just *my chickens,* grandma. Not even one of them will be my sister's." To be five years old and have some other life form entirely under your control—not counting goldfish or parents—is a majestic state of affairs.

—BARBARA KINGSOLVER, "Lily's Chickens"

32g   Employing rhetorical methods of development

When drafting an essay, you can develop a variety of paragraphs using **rhetorical methods**. These are approaches to
writing that help you address various rhetorical opportunities
by establishing boundaries (definition); investigating similarities or differences (comparison or contrast); making sense of
a person, place, or event (description and narration); organizing concepts (classification and division); thinking critically
about a process (process analysis or cause-and-consequence
analysis); or convincing someone (argumentation—see
Chapter 34). The strategies used for generating ideas, focusing a topic (32b), developing paragraphs and essays, and arranging ideas are already second nature to you. Every day,
you use one or more of them to define a concept, narrate a
significant incident, supply examples for an assertion, classify
or divide information, compare two or more things, analyze
a process, or identify a cause or a consequence. As a writer,
you have the option of employing one, or several, of the rhetorical methods to fulfill your overall purpose, which might
be to explain, entertain, argue, or evaluate.

(1) Narration
A **narrative** discusses a sequence of events, normally in
chronological order (the order in which they occur), to
develop a particular point or set a mood. This rhetorical
method, which often employs a setting, characters, dialogue,
and description, usually makes use of transition words or
phrases such as *first, then, later, that evening, the following
week,* and so forth to guide readers from one incident to the
next. Whatever its length, a narrative must remain focused

on the main idea. The narrative in paragraph 5 traces the history of the *Beaver,* a replica of the original Boston Tea Party ship:

5   In 1972, three Boston businessmen got the idea of sailing a ship across the Atlantic for the tea party's bicentennial. They bought a Baltic schooner, built in Denmark in 1908, and had her rerigged as an English brig, powered by an anachronistic engine that was, unfortunately, put in backwards and caught fire on the way over. Still, she made it to Boston in time for the hoopla. After that, the bicentennial *Beaver* was anchored at the Congress Street Bridge, next to what became the Boston Children's Museum. For years, it was a popular attraction. In 2001, though, the site was struck by lightning and closed for repairs. A renovation was planned. But that was stalled by the Big Dig, the excavation of three and a half miles of tunnel designed to rescue the city from the blight of Interstate 93, an elevated expressway that, since the 1950s, had made it almost impossible to see the ocean, and this in a city whose earliest maps were inked with names like Flounder Lane, Sea Street, and Dock Square. . . . In 2007, welders working on the Congress Street Bridge accidentally started another fire, although by then, the *Beaver* had already been towed, by tugboat, twenty-eight miles to Gloucester, where she'd been ever since, bereft, abandoned, and all but forgotten.

—JILL LEPORE, "Prologue: Party Like It's 1773"

(2) Description

By describing a person, place, object, or sensation, you can make your writing come alive. Even the most visual of descriptions can include the details of what you hear, smell, taste, or touch. Descriptions appeal to the senses.

Description should align with your rhetorical opportunity as well as with your purpose and audience. In paragraph 6,

The description of the candy and the appetizing image appeal to the reader's sense of taste.

Ishmael Beah employs vivid descriptive details to convey what he saw and heard as he walked through a small town in Sierra Leone that had been devastated by rebels.

6   I am pushing a rusty wheelbarrow in a town where the air smells of blood and burnt flesh. The breeze brings the faint cries of those whose last breaths are leaving their mangled bodies. I walk past them. Their arms and legs are missing; their intestines spill out through the bullet holes in their stomachs; brain matter comes out of their noses and ears. The flies are so excited and intoxicated that they fall on the pools of blood and die. The eyes of the nearly dead are redder than the blood that comes out of them, and it seems that their bones will tear through the skin of their taut faces at any minute. I turn my face to the ground to look at my feet. My tattered *crapes* [sneakers] are soaked with blood, which seems to be running down my army shorts. I feel no physical pain, so I am not sure whether I've been wounded. I can feel the warmth of my AK-47's barrel on my back; I don't

remember when I last fired it. It feels as if needles have been hammered into my brain, and it is hard to be sure whether it is day or night. The wheelbarrow in front of me contains a dead body wrapped in white bedsheets. I do not know why I am taking this particular body to the cemetery.

—ISHMAEL BEAH, *A Long Way Gone: Memoirs of a Boy Soldier*

(3) Process analysis

In explaining how something is done or made, process paragraphs often use both description and narration. You might describe the items used in a process and then narrate the steps of the process chronologically. By adding an explanation of a process to a draft, you could illustrate a concept that might otherwise be hard for your audience to grasp. In paragraph 7, Sam Swope explains the process by which an elementary school assistant principal tried (unsuccessfully) to intimidate students into identifying a fellow student who stole report cards.

7   Later that day, a frowning assistant principal appeared in the doorway, and the room went hush. Everyone knew why he was there. I'd known Mr. Ziegler only as a friendly, mild-mannered fellow with a comb-over, so I was shocked to see him play the

Time & Life Pictures/Getty Images

Sam Swope worked as an elementary school teacher in Queens, New York, for three years, interacting with children like these.

heavy. His performance began calmly, reasonably, solemnly. He told the class that the administration was deeply disappointed, that this theft betrayed the trust of family, teachers, school, and country. Then he told the children it was their duty to report anything they'd seen or heard. When no one responded, he added a touch of anger to his voice, told the kids no stone would go unturned, the truth would out; he vowed he'd find the culprit—it was only a question of time! When this brought no one forward, he pumped up the volume. His face turned red, the veins on his neck bulged, and he wagged a finger in the air and shouted, "I'm not through with this investigation, not by a long shot! And if any of you know anything, you better come tell me, privately, in private, because they're going to be in a lot of trouble, *a lot of trouble!*"

—SAM SWOPE, "The Case of the Missing Report Cards"

(4) Cause-and-consequence analysis

Writers who analyze cause or consequence often differentiate the **primary cause** (the most important one) from **contributory causes** (which add to but do not directly cause an event or situation) and the **primary consequence** (the most important result) from **secondary consequences** (which are less important than the primary consequence). In addition, they usually link a sequence of events along a timeline. Always keep in mind, though, that just because one event occurs before—or after—another event does not necessarily make it a cause—or a consequence—of that event. In paragraph 8, journalist Christopher Hitchens analyzes the consequences of his chemotherapy.

8   It's quite something, this chemo-poison. It has caused me to lose about 14 pounds, though without making me feel any lighter. It has cleared up a vicious rash on my shins that no doctor could ever name, let alone cure. . . . Let it please be this mean and ruthless with the alien and its spreading dead-zone colonies. But as against that, the death-dealing stuff and life-preserving stuff have also made me strangely neuter. I was fairly reconciled to the loss of my hair, which began to come out in the shower in the first two

weeks of treatment, and which I saved in a plastic bag so that it could help fill a floating dam in the Gulf of Mexico. But I wasn't quite prepared for the way that my razorblade would suddenly go slipping pointlessly down my face, meeting no stubble. Or for the way that my newly smooth upper lip would begin to look as if it had undergone electrolysis, causing me to look a bit too much like somebody's maiden auntie. (The chest hair that was once the toast of two continents hasn't yet wilted, but so much of it was shaved off for various hospital incisions that it's a rather patchy affair.) I feel upsettingly de-natured. If Penélope Cruz were one of my nurses, I wouldn't even notice. In the war against Thanatos, if we must term it a war, the immediate loss of Eros is a huge initial sacrifice.

—**CHRISTOPHER HITCHENS**, "Topic of Cancer"
© 2010. Reprinted by permission.

Writers also catalogue consequences, as Mark Orwoll does in paragraph 9, listing the results of a recent ruling in favor of air-passenger rights.

9   The turning point [in air-passenger rights] came last April when a new Department of Transportation (DOT) rule went into effect prohibiting lengthy tarmac delays on domestic flights at large and midsize hub airports. It requires that airlines provide food, water, and working toilets within two hours of delaying a plane on the ground and, after three hours, that passengers be allowed to safely leave the plane.

—**MARK ORWOLL**, "Revolution in the Skies"

(5) Comparison and contrast

A **comparison** points out similarities, and a **contrast** points out differences. When drafting, consider whether a comparison might help your readers see a relationship they might otherwise miss or whether a contrast might help them establish useful distinctions in order to better understand an issue or make a decision. In paragraph 10, Robert D. Putnam and David E. Campbell use descriptive details to compare two Catholic presidential candidates.

10   In 1960, presidential candidate John F. Kennedy had to
reassure Protestants that they
could safely vote for a Catholic.
(At the time 30 percent of
Americans freely told pollsters
that they would not vote for
a Catholic as president.) At
the same time, Kennedy won
overwhelming support from his
fellow Catholics, even though
he explicitly disagreed with his
church on a number of public
issues. In 2004, America had
another Catholic presidential

**Even though John Kerry shares
many characteristics with John
F. Kennedy, he was not able to
win the presidency.**

candidate—also a Democratic senator from Massachusetts,
also a highly decorated veteran, and also with the initials JFK.
Like Kennedy, John (Forbes) Kerry also publicly disagreed with
his church on at least one prominent issue—in this case, abortion. But unlike Kennedy, Kerry split the Catholic vote with his
Republican opponent, and lost handily among Catholics who
frequently attend church. Kennedy would likely have found
it inexplicable that Kerry not only lost to a Protestant, but to
George W. Bush, an evangelical Protestant at that. . . . In 1960,
religion's role in politics was mostly a matter of something akin
to tribal loyalty—Catholics and Protestants each supported their
own. In order to win, Kennedy had to shatter the stained glass
ceiling that had kept Catholics out of national elective office in a
Protestant-majority nation. By the 2000s, how religious a person
is had become more important as a political dividing line than
which denomination he or she belonged to. Church-attending
evangelicals and Catholics (and other religious groups too) have
found common political cause. Voters who are not religious have
also found common cause with one another, but on the opposite
end of the political spectrum.

—**ROBERT D. PUTNAM** AND **DAVID E. CAMPBELL,**
"Religious Polarization and Pluralism in America"

(6) Classification and division

Classification is a way to understand or explain something by establishing how it fits within a category or group of shared characteristics. For example, a book reviewer might classify a new novel as a mystery—leading readers to expect a plot based on suspense. **Division**, in contrast, separates something into component parts and examines the relationships among them. A novel can be discussed in terms of its components, such as plot, setting, and theme (Chapter 41).

Classification and division represent two different perspectives: ideas can be put into groups (classification) or split into subclasses (division). As strategies for organizing (or developing) an idea, classification and division often work together. In paragraph 11, for example, both classification and division are used to differentiate the two versions of the cowboy icon. Like many paragraphs, this one mixes rhetorical methods; the writer uses description, comparison and contrast, and classification to make her point.

KINGDOM	Animalia
	|
PHYLUM	Arthropoda
	|
CLASS	Insecta
	|
ORDER	Hymenoptera
	|
FAMILY	Apidae
	|
GENUS	*Apis*
	|
SPECIES	*mellifera*

© Scott T. Smith/Corbis

The scientific identification of the honeybee (*Apis mellifera*) requires a classification in the genus *Apis* and a division within that genus, the species *mellifera*.

11   First, and perhaps most fundamentally, the cowboy icon has two basic incarnations: the cowboy hero and the cowboy villain. Cowboy heroes often appear in roles such as sheriff, leader of a cattle drive, or what I'll call a "wandering hero," such as the Lone Ranger, who appears much like a frontier Superman wherever and whenever help is needed. Writers and producers most commonly place cowboy heroes in conflict either with "Indians" or with the cowboy villain. In contrast to the other classic bad guys of the Western genre, cowboy villains pose a special challenge because they are essentially the alter ego of the cowboy hero; the cowboy villain shares the hero's skill with a gun, his horse-riding maneuvers, and his knowledge of the land. What distinguishes the two, of course, is character: the cowboy hero is essentially good, while the cowboy villain is essentially evil.

—JODY M. ROY, "The Case of the Cowboy"

(7) Definition

By defining a concept or a term, you efficiently clarify your meaning and so develop an idea. Your readers will know what you are and are not talking about. Definitions are usually constructed in a two-step process: the first step locates a term by placing it in a class; the second step differentiates this particular term from other terms in the same class. For instance, "A concerto [the term] is a symphonic piece [the class] consisting of three movements performed by one or more solo instruments accompanied at times by an orchestra [the difference]." A symphony belongs to the same basic class as a concerto; it too is a symphonic piece. However, a symphony can be differentiated from a concerto in two specific ways: a symphony consists of four movements, and its performance involves an entire orchestra.

Paragraph 12 defines volcanoes by putting them into a class ("landforms") and by distinguishing them ("built of molten material") from other members of that class. The definition is then clarified by examples.

12   Volcanoes are landforms built of molten material that has spewed out onto the earth's surface. Such molten rock is called lava. Volcanoes may be no larger than small hills, or thousands of feet high. All have a characteristic cone shape. Some well-known mountains are actually volcanoes. Examples are Mt. Fuji (Japan), Mt. Lassen (California), Mt. Hood (Oregon), Mt. Etna and Mt. Vesuvius (Italy), and Paricutín (Mexico). The Hawaiian Islands are all immense volcanoes whose summits rise above the ocean, and these volcanoes are still quite active.

—JOEL AREM, *Rocks and Minerals*

Using definition and the other rhetorical methods just described will make your writing more understandable to your audience. Make sure that you use the method(s) best suited to your rhetorical situation, to supporting your thesis and making your purpose clear to your intended audience. As you draft and revise, check to see whether each rhetorical method you employ keeps your essay anchored to its thesis statement and helps you address your rhetorical opportunity. You may need to expand, condense, or delete paragraphs accordingly (33c and 33f).

33 Revising and Editing Essays

Revising, which literally means "seeing again," lies at the heart of all successful writing. When you are revising your writing, you resee it in the role of reader rather than writer. Revising involves considering a number of global issues: how successfully you have responded to the rhetorical opportunity, how clearly you have stated your thesis, how successfully you have communicated your purpose to your audience, how effectively you have arranged your information, and how thoroughly you have developed your assertions. **Editing**, on the other hand, focuses on local issues, which are smaller in scale. When you are editing, you polish your writing: you choose words more precisely (Chapter **20**), shape prose more distinctly (Chapter **21**), and structure sentences more effectively (Chapters **23–30**). While you are editing, you are also **proofreading**, focusing even more sharply to eliminate surface errors in grammar, punctuation, and mechanics. Revising and editing often overlap (just as drafting and revising do), and peer review can be helpful throughout these stages of the writing process. Usually revising occurs before editing, but not always. Edited passages may be redrafted, rearranged, and even cut as writers revise further.

As you revise and edit your essays, this chapter will help you

- consider your work as a whole (**33a(1)** and **33a(2)**),
- evaluate your tone (**33a(3)**),
- compose an effective introduction and conclusion (**33b**),
- strengthen the unity and coherence of paragraphs (**33c**),
- improve transitions (**33d**),

- benefit from a reviewer's comments (**33e**),
- edit to improve style (**33f**),
- proofread to eliminate surface errors (**33g**), and
- submit a final draft (**33h**).

33a   The essentials of revision

In truth, you are revising throughout the planning and drafting stages of the writing process, whether at the word, phrase, sentence, or paragraph level. A few writers prefer to start revising immediately after drafting, while their minds are still fully engaged by their topic. But most writers like to let a draft "cool off," so that when they return to it, they can assess it more objectively, with fresh eyes. Even an overnight cooling-off period will give you more objectivity as a reader and will reveal more options to you as a writer.

TECH SAVVY

Most word-processing programs enable you to track your revisions easily using a feature like Microsoft Word's Track Changes. Tracking changes is especially useful if your instructor requires you to submit all your drafts or if one or more peers are reviewing your drafts. If your word-processing program does not have this function, simply save and date each version of your work. You can open and compare these different versions as you write, and then submit all of them, if required.

(1) Revising purposefully

As you reread a draft, you need to keep in mind your audience, your purpose, and your thesis. Revision should enhance the development of your thesis while strengthening the connection between your rhetorical purpose and your intended audience (31d). In order to meet the needs, the expectations, and even the resistance of those in your audience, try to anticipate their responses (understanding, acceptance, or opposition) to your thesis statement, to each of your assertions, to the supporting examples and details you employ, and to the language you choose. In other words, revising successfully requires that you reread your work as both a writer and a reader. As a writer, ask yourself whether your words accurately reflect your intention and meaning. As a reader, ask yourself whether what seems clear and logical to you will also be clear to others.

(2) Adding essential information

Writers are always aware of what they have put on the page—but they seldom spend enough time considering what they may have left out. In order to ensure that you have provided all the information necessary for a reader to understand your points, consider the following questions: How might your audience be interested in addressing or resolving the rhetorical opportunity for change? What does your audience already know about this topic or issue? What information might your audience be expecting or be surprised by? What information might strengthen your thesis?

Keep in mind that your best ideas will not always surface in your first draft; you will sometimes come up with an important idea only after you have finished that draft, let it cool off, and then looked at it again.

(3) Creating the right tone

Tone reflects a writer's attitude toward a subject, so you will want to make sure that your tone is appropriate to your purpose, audience, and context (31a). Whether you are writing for school or work, your tone should reflect your confidence, preparation, fair-mindedness, and, perhaps most of all, your willingness to engage with your audience. If any of the passages in your draft sound defensive, self-centered, or apologetic to you or to a peer reviewer, revise them.

Consider the tone in paragraph 1, in which Dorothy Allison describes some of the positive and negative things she remembers about growing up in South Carolina. (For ease of reference, each of the sample paragraphs in this chapter is numbered.)

1   Where I was born—Greenville, South Carolina—smelled like nowhere else I've ever been. Cut wet grass, split green apples, baby shit and beer bottles, cheap makeup and motor oil. Everything was ripe, everything was rotting. Hound dogs butted my calves. People shouted in the distance; crickets boomed in my ears. That country was beautiful, I swear to you, the most beautiful place I've ever been. Beautiful and terrible. It is the country of my dreams and the country of my nightmares: a pure pink and blue sky, red dirt, white clay, and all that endless green—willows and dogwood and firs going on for miles.

—DOROTHY ALLISON, *Two or Three Things I Know for Sure*

When Mary LeNoir revised the first draft reprinted later in this chapter (pages 371–77), she adjusted her tone so that it was not so dry. Of course, Mary wanted to sound knowledgeable, but she also wanted to connect immediately with her audience. To meet her goals, she revised her introduction, striking a more natural and inviting tone that better aligned with her rhetorical opportunity (**31a**).

33b   Guiding readers with the introduction and the conclusion

Your introduction and conclusion orient your readers to the purpose of your essay as a whole. In fact, readers intentionally read these two sections for guidance and clarification.

(1) An effective introduction

Experienced writers know that the opening paragraph is their best chance to arouse the reader's interest; establish the rhetorical opportunity, the topic, and the writer as worthy of consideration; and set the overall tone. An effective introduction makes the intended audience want to read on. In paragraph 2, herpetologist Rick Roth introduces himself to a diverse audience, readers of *Sierra* magazine.

2   A lot of people know me as "Snake Man" now and don't know my real name. I've always been a critter person. My mother was never afraid of anything, and I used to actually get to keep snakes in the house. I'm 58, so this was a long time ago, when *nobody* got to keep snakes in the house. I've got 75 or so now at home—and a really cool landlord.   —RICK ROTH, "Snake Charmer"

Roth's friendly introduction immediately grabs readers' attention with his down-home language and unusual partiality for snakes. He then moves quickly to his childhood fascination with butterflies and dragonflies (thereby establishing common ground with those of his readers who are agitated by snakes) and goes on to explain his current occupation as the executive director of the Cape Ann Vernal Pond Team.

Introductions have no set length; they can be as brief as a couple of sentences or as long as two or more paragraphs, sometimes even longer. Although introductions always appear first, they are often drafted and revised after other parts of a work. Just like the thesis statements they often include, introductions evolve during the drafting and revising stages, as the material is shaped, focused, and developed toward fulfilling the writer's overall purpose.

You can arouse the interest of your audience by writing introductions in a number of ways.

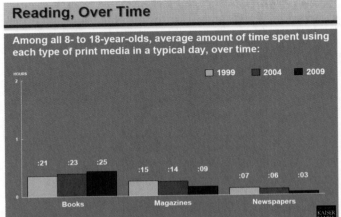

Opening with a thought-provoking statistic can be an effective introduction.

(a) Opening with an unusual fact or statistic

3   Americans aren't just reading fewer books, but are reading less and less of everything, in any medium. That's the doleful conclusion of "To Read or Not to Read," a report released last week by the National Endowment for the Arts.

—**JENNIFER HOWARD, "Americans Are Closing the Book on Reading, Study Finds"**

(b) Opening with an intriguing statement

4   I belong to a Clan of One-Breasted Women. My mother, my grandmothers, and six aunts have all had mastectomies. Seven are dead. The two who survive have just completed rounds of chemotherapy and radiation.

—**TERRY TEMPEST WILLIAMS, "The Clan of One-Breasted Women"**

(c) Opening with an anecdote or example

5   When I used to ask my mother which we were, rich or poor, she refused to tell me. I was then nine years old and of course what I was dying to hear was that we were poor. I was reading a book called *Five Little Peppers* and my heart was set on baking a cake for my mother in a stove with a hole in it. Some version of rich, crusty old Mr. King—up till that time not living on our street—was sure to come down the hill in his wheelchair and rescue me if anything went wrong. But before I could start a cake at all I had to find out if we were rich or poor, and poor *enough*; and my mother wouldn't tell me, she said she was too busy. I couldn't wait too long; I had to go on reading and soon Polly Pepper got into more trouble, some that was a little harder on her and easier on me.   —EUDORA WELTY, "A Sweet Devouring"

(d) Opening with a question

6   Fellow-Citizens—pardon me, and allow me to ask, why am I called upon to speak here today? What have I, or those I represent, to do with your national independence? Are the great principles of political freedom and of natural justice, embodied in that Declaration of Independence, extended to us? and am I, therefore, called upon to bring our humble offering to the national altar, and to confess the benefits, and express devout gratitude for the blessings, resulting from your independence to us?
      —FREDERICK DOUGLASS, "What to the Slave Is the Fourth of July?"

(e) Opening with an appropriate quotation

7   "My wife and I like the kind of trouble you've been stirring, Miss Williams," he said, with a smile and a challenge. He had an avuncular, wizardy twinkle, very Albus Dumbledore. It made me feel feisty and smart, like Hermione Granger. They *liked* my kind of trouble. But let this be a lesson: When a woman of my great dignity and years loses her sanity and starts imagining she's one

of Harry Potter's magical little friends, you can be sure that the
cosmic gyroscope is wobbling off its center . . . .

—PATRICIA J. WILLIAMS, *Open House: Of Family, Friends, Food,*
Piano Lessons, and the Search for a Room of My Own

(f) Opening with general information or background about the topic

8   Scientists have long touted the benefits of the Mediterranean
diet for heart health. But now researchers are finding more and
more evidence that the diet can keep you healthy in other ways,
too, including lowering the risk of certain cancers and easing the
pain and stiffness of arthritis.

—MELISSA GOTTHARDT, "The Miracle Diet"

(g) Opening with a thesis statement

9   When America first met her in 1992, Hillary Rodham
Clinton looked like what she was: a working mother. She had
recently chucked her Coke-bottle glasses but still sported head-
bands and weird amounts of ineptly applied makeup. Why
should it have been otherwise? Clinton was a busy woman when
her husband ran for president. Mind-bogglingly, she would be
the first first lady in American history to have maintained a full-
time career outside her husband's political life prior to his presidency. In short, Clinton was the first candidate for the job of first
lady to have a life that reflected [America after the feminist movement of the 1970s] and the many working women who made
their careers and raised their families here.

—REBECCA TRAISTER, *Big Girls Don't Cry*

However you open your essay, use your introduction to
specify your topic, engage your readers' attention, initiate an
appropriate tone, and establish your credibility (**34f(1)**).

(2) An effective conclusion

Just as a good introduction tantalizes readers, a good conclusion satisfies them. It helps readers recognize the significant points of your essay while wrapping up the essay in a meaningful, way. As you draft and revise, keep a list of ideas for your conclusion. Some suggestions for writing effective conclusions follow, beginning with the reliable method of simply restating the thesis and main points. This kind of conclusion can be effective for a long essay that includes several important points that the writer wants the reader to recall.

(a) Rephrasing the thesis and summarizing the main points

10   The Endangered Species Act should not take into account economic considerations. Economics doesn't know how to value a species or a forest. Its logic drives people to exploit resources to the point of extinction. The Endangered Species Act tells us that extinction is morally unacceptable. It was enacted by a Congress and president in a wise mood, to express a higher value than a bottom line.

—DONELLA MEADOWS, "Not Seeing the Forest for the Dollar Bills"

(b) Calling attention to larger issues

11   If tough breaks have not soured me, neither have my glory-moments caused me to build any altars to myself where I can burn incense before God's best job of work. My sense of humor will always stand in the way of my seeing myself, my family, my race or my nation as the whole intent of the universe. When I see what we really are like, I know that God is too great an artist for we folks on my side of the creek to be all of His best works. Some of His finest touches are among us, without doubt, but some more of His masterpieces are among those folks who live over the creek.

—ZORA NEALE HURSTON, *Dust Tracks on a Road: An Autobiography*

(c) Calling for a change in action or attitude

12   Although [Anna Julia] Cooper published *A Voice* in 1892, its political implications remain relevant to twenty-first-century scholars and activists. As our society grows increasingly multicultural, and the borders between colors and countries grow ever more porous, the strategies for organizing communities of resistance must necessarily follow suit. Academics and activists engaged in efforts to transform inequitable social relations benefit from thinking not only about what separates but also what unites humanity.

—**KATHY L. GLASS,** "Tending to the Roots"

(d) Concluding with a vivid image

13   At just past 10 A.M., farm workers and scrap-yard laborers in Somerset County looked up to see a large commercial airliner dipping and lunging as it swooped low over the hill country of southern Pennsylvania, near the town of Shanksville. A man driving a coal truck on Route 30 said he saw the jet tilt violently from side to side, then suddenly plummet "straight down." It hit nose first on the grassy face of a reclaimed strip mine at approximately 10:05 Eastern Daylight Time and exploded into a fireball, shattering windowpanes a half-mile away. The seventy-two-year-old man who was closest to the point of impact saw what looked to him like the yellow mushroom cloud of an atomic blast. Twenty-eight-year-old Eric Peterson was one of the first on the scene. He arrived to discover a flaming crater fifty feet deep. Shredded clothing hung from the trees, and smoldering airplane parts littered the ground. It did not look much like the site of a great American victory, but it was.

—**RANDALL SULLIVAN,** "Flight 93"

(e) Connecting with the introduction

In the following introduction and conclusion, Debra Utacia Krol focuses on an artist's achievements.

The introduction

14   Peterson Yazzie (Navajo) may be only 26, but this young contemporary painter from Greasewood Springs, Arizona has already

garnered impressive accolades and is considered one of the rising stars of the Native art realm.

The conclusion

15   Among other honors amassed over his meteoric career, Yazzie took home the best of class ribbon in painting from the Heard Museum Guild Indian Fair & Market in 2006, and looks forward to returning again this year as he continues to delve even further into expressing his worldview through art.

                              —DEBRA UTACIA KROL, "Peterson Yazzie"

Whatever technique you choose for your conclusion, provide readers with a sense of closure. Bear in mind that they may be wondering, "So what? Why have you told me all this?" Your conclusion gives you an opportunity to address that concern.

EXERCISE 1

Thumb through a magazine you enjoy, skimming the introductions of all the articles. Select two introductions that catch your attention. Consider the reasons *why* they interest you. What specific techniques for an introduction did the authors use? Next, look through the same or another magazine for two effective conclusions. Analyze their effectiveness as well. Be prepared to share your findings with the rest of the class.

33c   Revising for unified and coherent paragraphs

When revising the body of an essay, writers are likely to find opportunities for further development within each paragraph (**32f** and **32g**) and to discover ways to make each paragraph more **unified** by relating every sentence within the paragraph to a single main idea (**33c(2)**), which might appear in a topic sentence.

and Uncle Jack, both worked downtown, she at JC Penney and he at Jim Dugan's Menswear, so Babs and I would include visiting both of them at work in our day's activities. *Funny, now that I think back on those good summer afternoons, I think about how I haven't seen Babs since her mother's funeral, ten years ago.* On the days I didn't walk downtown, I usually swam in our neighborhood swimming pool, Fair Park pool, where all of us kids played freely and safely, often without any parents around but usually with our younger siblings trailing after us (including my own). Or we Fair Park kids might take a city bus out to the roller rink or, if something big was going on, walk out to the fairgrounds, sneaking under the fence to see what was happening.

Easy to delete, the italicized sentence about not having seen Babs for ten years violates the unity of a paragraph devoted to childhood activities in a small town.

As you revise your paragraphs for unity, the following tips may help you.

TIPS FOR IMPROVING PARAGRAPH UNITY

- **Identify.** Identify the topic sentence for each paragraph. Where is each located? Why is each one located where it is?

- **Relate.** Read each sentence in a paragraph and determine how (and if) it relates directly to or develops the topic sentence.

- **Eliminate.** Any sentence that does not relate to the topic sentence violates the unity of the paragraph—cut it or save it to use elsewhere.

- **Clarify.** If a sentence "almost" relates to the topic sentence, either revise it or delete it. As you revise, you might clarify details or add information or a transitional word or phrase to make the relationship clear.

- **Rewrite.** If more than one idea is being conveyed in a single paragraph, either rewrite the topic sentence so that it includes both ideas and establishes a relationship between them or split the single paragraph into two paragraphs, dividing up the information accordingly.

(3) Arranging ideas into coherent paragraphs

Some paragraphs are unified (33c(2)) but not coherent. In a unified paragraph, every sentence relates to the main idea of the paragraph. In a coherent paragraph, the relationship among the ideas is clear and meaningful, and the progression from one sentence to the next is easy for readers to follow. Paragraph 19 has unity but lacks coherence.

Lacks coherence

19   The land was beautiful, gently rolling hills, an old orchard with fruit-bearing potential, a small clear stream—over eleven acres. But the house itself was another story. It had sat empty for years. Perhaps not empty, though that's what the realtor told us. There were macaroni and cheese boxes, how-to-play the mandolin books and videos, extra countertops, a kitchen sink, single socks looking for their mates, a ten-year-old pan of refried beans, and all sorts of random stuff strewn all through the house. Had the owner stayed there until he gave up on remodeling it? Had homeless people squatted there? Or had it been a hangout for teenagers—until the hole in the roof got too big for comfort? Who had been living there, and what kind of damage had they brought to the house? *We looked at the house with an eye toward buying it.* The price was right: very low, just what we could afford. And the location and acreage were perfect, too. But the house itself was a wreck. It needed a new roof, but it also needed a kitchen, flooring, drywall, updated plumbing and electricity— and a great big dumpster. We didn't know if we had the energy, let alone the know-how, to fix it up. Plus it wasn't like it was just the two of us we had to think about. We had three children to consider. If we bought it, where would we start working to make it inhabitable?

Although every sentence in this paragraph has to do with the writer's reaction to a house offered for sale, the sentences themselves are not arranged coherently. The paragraph can easily be revised to allow the italicized topic sentence to control the meaningful flow of ideas—from the land and the condition of the house to the potential advantages and disadvantages of the purchase.

Revised for coherence

20   We looked at the house with an eye toward buying it. The land was beautiful, gently rolling hills, an old orchard with fruit-bearing potential, a small clear stream—over eleven acres. But the house itself was another story. It had sat empty for years. Perhaps not empty, though that's what the realtor told us. There were macaroni and cheese boxes, how-to-play the mandolin books and videos, extra countertops, a kitchen sink, single socks looking for their mates, a ten-year-old pan of refried beans, and all sorts of random stuff strewn all through the house. Had the owner stayed there until he gave up on remodeling it? Had homeless people squatted there? Or had it been a hangout for teenagers—until the hole in the roof got too big for comfort? By now, the house needed a new roof as well as a kitchen, flooring, drywall, updated plumbing and electricity—and a great big dumpster. We didn't know if we had the energy, let alone the know-how, to fix it up. Plus it wasn't like it was just the two of us we had to think about. We had three children to consider. Still, the price was right: very low, just what we could afford. And the location and acreage were perfect, too.

Paragraph 20 is coherent as well as unified.

To achieve coherence as well as unity in your paragraphs, study the following patterns of organization (chronological, spatial, emphatic, and logical), and consider which ones you might use in your own writing.

(a) Using chronological order

When you use **chronological order**, you arrange ideas according to the order in which things happened. This organizational pattern is particularly useful for narration.

21   The years following the Civil War laid the foundation for the division in the narrative. The war initially raised the hope that citizenship might replace race as the sole basis for American identity, but the realities of Reconstruction and its aftermath forced African Americans to redefine their collective history. . . . This memory of hope and sacrifice was given credence and promoted in the latter part of the nineteenth century through black-owned newspapers and black ministers, who reminded their audiences of the experiences and hopes of the past in connection with the struggles of the present. The journey, so filled with false promises and unrealized dreams, suggested greater rewards. The outline of this collective memory carried over into the twentieth century via slave narratives, autobiographies and adventure literature catering to the rising literacy rate for African Americans.

—K. BINDAS, "Re-remembering a Segregated Past: Race in American Memory"

(b) Using spatial order

When you arrange ideas according to **spatial order**, you orient the reader's focus from right to left, near to far, top to bottom, and so on. This organizational pattern is particularly effective in descriptions. Often the organization is so obvious that the writer can forgo a topic sentence, as in paragraph 22.

22   I went to see a prospective student, Steve, up on the North Branch Road. His mother, Tammi, told me to look for the blue trailer with cars in the yard. There were *lots* of junk cars—rusted, hoods up, and wheels off, a Toyota truck filled with bags of trash. The yard was littered with transmission parts, hubcaps, empty soda bottles, Tonka trucks, deflated soccer balls, retired chain saws and piles of seasoned firewood hidden in the overgrowth of jewelweed. A pen held an assortment of bedraggled, rain-soaked chickens and a belligerent, menacing turkey. A small garden of red and yellow snapdragons marked the way to the door.

—TAL BIRDSEY, *A Room for Learning: The Making of a School in Vermont*

(c) Using emphatic order

When you use **emphatic order**, you arrange information in order of importance, usually from least to most important. Emphatic order is especially useful in expository and persuasive writing, both of which involve helping readers understand logical relationships. The information in paragraph 23 leads up to the writer's conclusion—that rats raised in enriched environments are smarter rats.

23   Raising animals in a rich environment can result in increased brain tissue and improved performance on memory tests. Much of this work has been done with rats. The "rich" rat environment involved raising rats in social groups in large cages with exercise wheels, toys, and climbing terrain. Control "poor" rats were raised individually in standard laboratory cages without the stimulating objects the rich rats had. Both the rich and poor rats were kept clean and given sufficient food and water. Results of these studies were striking: Rich rats had substantially thicker cerebral cortex, the highest region of the brain and the substrate of cognition, with many more synaptic connections, than the poor rats. They also learned to run mazes better.

—**RICHARD F. THOMPSON** AND **STEPHEN A. MADIGAN**,
Memory: The Key to Consciousness

(d) Using logical order

Sometimes the movement within a paragraph follows a **logical order**, from specific to general (see paragraphs 17 and 24) or from general to specific (see paragraphs 20 and 25).

24   Whether one reads for work or for pleasure, comprehension is the goal. Comprehension is an active process; readers must interact and be engaged with a text. To accomplish this, proficient readers use strategies or conscious plans of action. Less proficient readers often lack awareness of comprehension strategies, however, and cannot develop them on their own. For adult literacy learners in particular, integrating and synthesizing information from any but the simplest texts can pose difficulties.

—**MARY E. CURTIS** AND **JOHN R. KRUIDENIER**, "Teaching Adults to Read"

25   It was not the only disappointment my mother felt in me. In the years that followed, I failed her so many times, each time asserting my own will, my right to fall short of expectations. I didn't get straight As. I didn't become class president. I didn't get into Stanford. I dropped out of college.

—AMY TAN, "Two Kinds"

33d   **Transitions within and between paragraphs**

Even if its sentences are arranged in a seemingly clear sequence, a single paragraph may lack internal coherence, and a series of paragraphs may lack overall coherence if transitions are abrupt or nonexistent. When revising, you can improve coherence by using pronouns, repetition, or conjunctions and transitional words or phrases (3c(5)).

(1) Using pronouns to establish links between sentences
In paragraph 26, the writer enumerates the similarities of identical twins raised separately. She mentions their names only once, but uses the pronouns *both, their,* and *they* to keep the references to the twins always clear.

26   Jim Springer and Jim Lewis were adopted as infants into working-class Ohio families. **Both** liked math and did not like spelling in school. **Both** had law enforcement training and worked part-time as deputy sheriffs. **Both** vacationed in Florida, **both** drove Chevrolets. Much has been made of the fact that **their** lives are marked by a trail of similar names. **Both** married and divorced women named Linda and had second marriages with women named Betty. **They** named **their** sons James Allan and James Alan, respectively. **Both** like mechanical drawing and carpentry. **They** have almost identical drinking and smoking patterns. **Both** chew **their** fingernails down to the nubs.

—CONSTANCE HOLDEN, "Identical Twins Reared Apart"

When revising, you must consider the effectiveness of the individual paragraphs in terms of the entire essay. Some writers like to start revising at the paragraph level, while others grapple first with larger issues concerning the thesis statement and the components of the rhetorical situation (opportunity, purpose, audience, and context.) Since there is no right way to revise, be guided by the principles and strategies discussed in this chapter—and trust your own good sense.

The following checklist can guide you in revising your paragraphs.

CHECKLIST for Revising Paragraphs

- Does the paragraph have a clear (or clearly implied) topic sentence (33c(1))?

- Do all the ideas in the paragraph relate to the topic sentence (33c(2))? Does each sentence link to previous and later ones? Are the sentences arranged in chronological, spatial, emphatic, or logical order, or are they arranged in some other pattern (33c(3))?

- Are sentences connected to each other with effective transitions (33d)?

- What rhetorical method or methods have been used to develop the paragraph (32g)?

- What evidence do you have that the paragraph is adequately developed (32f)? What idea or detail might be missing (33a(2))?

- How does the paragraph itself link to the preceding and following ones (33d)?

33e   Peer review

Because writing is a medium of communication, good writers check to see whether they have successfully conveyed their ideas to their readers. Instructors are one set of readers you need to think about. But often they are the last people to see your finished writing. Before you submit your work to an instructor, take advantage of other opportunities for getting responses to it. Consult with readers—at the writing center, in your classes, or in online writing groups—asking them for honest responses to your concerns about your writing.

(1) Establishing specific evaluation standards

Although you will always write within a rhetorical situation (31a), you will often do so in terms of an assigned task with specific evaluation standards. If your instructor has told you that your essay will be evaluated primarily in terms of whether you have a clear thesis statement (32c) and adequate support for it (32f and 32g), then those features should be your primary focus. Your secondary concerns may be the overall effectiveness of the introduction (33b(1)), sentence length and variety (Chapter 30), and mechanical correctness (Chapters 8–11, 18).

Evaluation standards cannot guarantee useful feedback, but they can help you focus as you write and can assist your reviewers as they respond to your draft.

A reviewer's comments should be based on the evaluation standards, pointing out what the writer has done well and suggesting how to improve particular passages. If a reviewer sees a problem that the writer did not mention, the reviewer should ask the writer if she or he wants to discuss it and should abide by the writer's decision. Ultimately, the success of the essay is the responsibility of the writer, who weighs the reviewer's advice, rejecting comments that might take the essay in a different direction and applying any suggestions that help fulfill the rhetorical purpose (31c).

If you are developing your own criteria for evaluation, the following checklist can help you get started. Based on the elements of the rhetorical situation, this checklist can be easily adjusted so that it meets your specific needs for a particular assignment.

CHECKLIST for Evaluating a Draft
of an Essay

- Does the essay fulfill all the requirements of the assignment?
- What rhetorical opportunity does the essay address (31a)?
- What is the specific audience for the essay (31d)? Is that audience appropriate for the assignment?
- What is the tone of the essay (33a(3))? How does the tone align with the overall purpose, the intended audience, and the context for the essay (31c–e)?
- Is the larger subject focused into a topic (32b)? What is the thesis statement (32c)?
- What assertions support the thesis statement? What specific evidence (examples or details) support these assertions?
- What pattern of organization is used to arrange the paragraphs (33c(3))? What makes this pattern effective for the essay? What other pattern(s) might prove to be more effective?
- How is each paragraph developed (32f and 32g)?
- What specifically makes the introduction effective (33b(1))? How does it address the rhetorical opportunity and engage the reader?
- How is the conclusion appropriate for the essay's purpose (33b(2))? How exactly does it draw the essay together?

(2) Informing reviewers about your purpose and your concerns

When submitting a draft for review, you can increase your chances of getting the kind of help you want by introducing your work and indicating what your concerns are. You can provide such an orientation orally or in writing. In either case, adopting the following model can help ensure that reviewers will give you useful responses.

SUBMITTING A DRAFT FOR REVIEW

Topic and Purpose

State your topic and the rhetorical opportunity for your writing (31b). Identify your thesis statement (32c), purpose (31c), and intended audience (31d). Such information gives reviewers useful direction.

Strengths

Mark the passages of your draft you are confident about. Doing so directs attention away from areas you do not want to discuss and saves time for all concerned.

Concerns

Put question marks by the passages you find troublesome and ask for specific advice wherever possible. For example, if you are worried about your conclusion, say so. Or if you suspect that one of your paragraphs may not fit the overall purpose, direct attention to that paragraph. You are most likely to get the kind of help you want and need when you ask for it specifically.

Mary submitted the draft on pages 371–77 for peer review in a first-year writing course. She worked with two classmates, giving them a set of criteria she had prepared. Because the reviewers were learning how to conduct peer evaluations, their comments are representative of responses you might receive in a similar situation. As members of writing groups gain experience and learn to employ the strategies outlined in this section, their advice usually becomes more helpful.

As you read the following assignment and then Mary's draft, remember that a first draft will not be a model of perfect writing—and also that this is the first time peer reviewers Ernie Lujan and Andrew Chama responded to it. Mary sent Ernie and Andrew her essay electronically, and they both used Track Changes to add a note summarizing their comments.

> **The assignment:** Draft a three- to four-page, double-spaced essay in which you analyze the causes or consequences of a choice you have had to make in the last year or two. Whatever choice you analyze, make sure that it concerns a topic you can develop with confidence and without violating your sense of privacy. Also, consider the expectations of your audience and whether the topic you have chosen will allow you to communicate something meaningful to readers. As you draft, establish an audience for your essay, a group that might benefit from or be interested in any recommendation or new knowledge that grows out of your analysis.

First Draft

The Search: Student-Athletes' Choice of College

Mary LeNoir

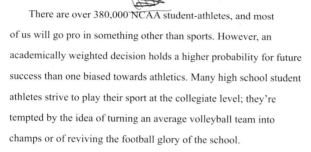

There are over 380,000 NCAA student-athletes, and most
of us will go pro in something other than sports. However, an
academically weighted decision holds a higher probability for future
success than one biased towards athletics. Many high school student
athletes strive to play their sport at the collegiate level; they're
tempted by the idea of turning an average volleyball team into
champs or of reviving the football glory of the school.

It all begins in the hearts of little leaguers and peewee soccer
players. Though we begin our athletic careers hitting off a tee or
paddling in the pool on a kickboard, many of us grow to hone
our athletic abilities and practice our skills to be the best and play
against the best. We play sports year round; for our schools, club
teams, all-star teams, AAU teams. We attend camps, separate
workouts, and practices. Our love of picking grass during peewee
soccer games evolves into the love of the competition and triumph
over our opponents. Aspirations of great collegiate careers and
even dreams of playing professional athletics consume our minds'
thoughts, driving the discipline behind our work ethic. College
prospects come knocking on our doors, prepared to enable our

Before revising, Mary considered the comments she received from Ernie and Andrew. Since she had asked them to respond to her introduction, conclusion, and organization, she had to weigh all of their comments—relevant and irrelevant—and use the ones that seemed to be most useful as she prepared her next draft.

EXERCISE 2

Reread Mary's first draft and the peer reviewers' responses. Identify the comments you think are the most useful, and explain why. Which comments seem to be less useful? Explain why. What additional comments would you make if Mary had asked you to review her draft?

After Mary had time to reconsider her first draft and to think about the responses she received, she made a number of large-scale changes, especially with regard to organization. She also strengthened her thesis statement and cleaned up the surface errors. After these and other revisions, more peer review, and some careful editing and proofreading, Mary was ready to submit her essay to her instructor. Her final draft is on pages 382–93.

33f   Editing for clarity

If you are satisfied with the revised structure of your essay and the content of your paragraphs, you can begin editing individual sentences for clarity, effectiveness, and variety (Chapters 23–30). The following checklist for editing contains cross-references to chapters or sections where you can find more specific information.

CHECKLIST for Editing

1 Sentences

- What is the unifying idea of each sentence (**23**)?
- How have you varied the lengths of your sentences? How many words are in your longest and shortest sentences?
- How have you varied the structure of your sentences? How many are simple? How many use subordination or coordination? If you overuse any one sentence structure, revise for variation (**30**).
- Does each verb agree with its subject (**6a**)? Does every pronoun agree with its antecedent (**6b**)?
- Which sentences have or should have parallel structure (**26**)?
- Do any sentences contain misplaced or dangling modifiers (**25a** and **25b**)?
- Do any of your sentences shift in verb tense or tone (**7b**)? Is the shift intentional?

2 Diction

- Have you repeated any words (**29**)? Is your repetition intentional?
- Are your word choices exact, or are some words vague or too general (**20**)?
- Have you used any language that is too informal (**19b**)?
- Is the vocabulary you have chosen appropriate for your audience, purpose, and context (**19** and **31c–e**)?
- Have you defined any technical or unfamiliar words for your audience (**19b(4)**)?

33g　Proofreading for an error-free essay

Once you have revised and edited your essay, it is your responsibility to format it properly and proofread it. Proofreading means making a special search to ensure that the final product you submit is free from error, or nearly so. An error-free essay allows your reader to read for meaning, without encountering incorrect spelling or punctuation that can interfere with meaning. As you proofread, you may discover problems that call for further revision or editing, but proofreading is usually the last step in the writing process.

Because the eye tends to see what it expects to see, many writers miss errors—especially minor ones, such as a missing comma or apostrophe—even when they think they have proofread carefully. To proofread well, then, you need to read your work more than once and read it aloud. Some people find it useful to read through a paper several times, checking for a different set of items on each pass. Other writers rely on peer editors to provide help with proofreading.

The proofreading checklist that follows refers to chapters and sections in this handbook where you will find detailed information to help you. Also, keep your dictionary (**19d**) at hand to look up any words whose meaning or spelling you are unsure about.

CHECKLIST for Proofreading

1 Spelling (18)

- Have you double-checked the words you frequently misspell and any the spell checker may have missed (for example, misspellings that still form words, such as *form* for *from*)?
- If you used a spell checker, did it overlook homophones (such as *there/their, who's/whose,* and *it's/its*) (18c)?
- Have you double-checked the spelling of all foreign words and all proper names?

2 Punctuation (12–17) and Capitalization (9)

- Does each sentence have appropriate closing punctuation, and have you used only one space after each end punctuation mark (17)?
- Is all punctuation within sentences—commas (12), semicolons (14), apostrophes (15), dashes (17e), and hyphens (18f)—used appropriately and placed correctly?
- Are direct quotations carefully and correctly punctuated (16a)? Where have you placed end punctuation with a quotation (16d)? Are quotations capitalized properly (16a and 9c(1))?
- Are all proper names, people's titles, and titles of published works correctly capitalized (9a and 9b)?
- Are titles of works identified with quotation marks (16b) or italics (10a)?

33h The final process-analysis essay

After her intensive revision, Mary edited and proofread her essay. The version that she ultimately submitted to her instructor follows.

LeNoir 1

Mary LeNoir

Professor Glenn

English 15

1 November 2010

How Student-Athletes Really Choose a College

The urge to win is a familiar feeling; mentioning it in the first sentence is a good way to hook readers.

It all begins in the hearts of little leaguers, peewee soccer players, and little swimmers with their kickboards—that yearning to play and win. Student-athletes may start small, but they grow and hone their athletic abilities in order to be the best and play against the best. They play sports year round—for their schools, club teams, all-star teams, Amateur Athletic Union (AAU) teams, and for themselves. They attend camps, workouts, and practices, lots of practices. Aspirations of great collegiate careers and even dreams of playing professional athletics begin to consume their dreams, driving their self-discipline and work ethic. Eventually, college prospects come knocking on their doors, the recruiters wearing blue and white or orange and maroon and talking to these young athletes of dreams that might come true. In those recruiters, student-athletes see their future. The smart ones know that their future as a competitive athlete might last through the college years, if they're lucky. They realize that only a very select few ever reach professional status.

LeNoir 2

So with the odds of becoming a professional against them, why do student-athletes agree to such a heavy commitment as playing college sports, knowing that the experience will take them only so far? Most people think student-athletes base their choice of college on financial gain and promised playing time. To find out how student-athletes actually made their choice of college, I decided to interview five college student-athletes who played various sports at five different kinds of schools. My goal was to uncover the personal reasons or criteria student-athletes used as they made their choices.

To answer my question, I selected both male and female student-athletes who play either a highly publicized or lesser-known sport. In addition, I chose athletes who attended big state universities, system schools, and smaller liberal arts colleges. Paige Wright represents lacrosse at Johns Hopkins University; Nick Huang, football at Capitol University; Bobby Dorsey, baseball at the University of Richmond; Marye Taranto, lacrosse at the University of Virginia; and Theresa Morales, lacrosse at Penn State University. I asked about their college search process (including why they wanted to play college athletics) and soon discovered that, despite their diverse goals and circumstances, the student-athletes all worked with three criteria when choosing their college: (1) the potential playing time at the school or how this school might help

To answer this question, the author interviewed student-athletes and thereby built her thesis statement.

The author explains how she went about answering the question posed in the previous paragraph.

a real possibility. Bobby and Nick both wanted a chance to play professionally, and their two sports happen to be two of the most popular professional sports.

This paragraph opens with a question that the writer goes on to answer.

    That said, what were student-athletes thinking when they signed on to play a college sport with no professional opportunity? Theresa admitted her athletic career would be over after four years of college:

> Even though there is no professional lacrosse league,
> I could always coach or help start a program at a school.
> But I wanted to do something else—that's one of the
> reasons Penn State appealed to me. The school offered
> so many majors and amazing academic support. I felt
> I needed that freedom to explore and figure out what I
> wanted to do with my life. (Morales)

Theresa went on to explain that her professional goals in no way diminished her competitive nature or desire to excel on the lacrosse field:

> Sports have been so much a part of my life that I could
> not imagine not playing in college. And because I'm
> competitive, there was no way I was going to commit to
> some school that was not serious about winning, that did
> not have a real chance, or where I would not have a chance

LeNoir 6

to play. Besides, college athletics is a good way to prepare

for any profession: I am competitive. I love to win, and

in order to win I train, work hard, practice discipline, and

budget my time—all qualities vital to success in life. Job

recruiters are well aware of an athlete's commitments, and

it looks great on a résumé to have played a varsity sport in

college while having kept up a good GPA. (Morales)

Marye and Paige, who also play lacrosse, also mentioned their

competitiveness and the importance of getting to play. They, too,

felt that being able to add varsity athletics to their résumés was

important. However, Marye and Paige went on to present criteria

other than "playing time" that had a strong influence on their final

decision about a college.

The second cluster of criteria that student-athletes use when

making their decision includes the material conditions associated

with a college, things that are the same for students and student-

athletes alike. Costs and the geographical location of the school—

these are logistics that do not disappear just because you are an

athlete. Being recruited does not imply the athlete has only to say

yes to the school and all these factors will fall into place. In Marye's

case, her athletic scholarship was the key to attending her dream

school, the University of Virginia (UVA). Marye loved UVA and its

Use of supporting details and explanation strengthens this paragraph.

LeNoir 7

team and had been accepted academically, but she had received no academic scholarship. She would have had to pay the full tuition, which she and her parents simply could not afford. Fortunately, Marye was a highly recruited lacrosse player, one UVA became very interested in after she had been admitted on academic grounds. The athletic scholarship offered to Marye made UVA affordable for her, giving her the opportunity to receive an education from one of the best academic institutions in the country and to live her dream of playing college lacrosse, neither of which would have been possible without her athletic scholarship. (See Fig. 1.)

Fig. 1. Playing lacrosse for a top college team is a dream of many high school players. Photograph © Paul A. Souders/Corbis.

LeNoir 8

Paige's situation presented an insurmountable obstacle: geographic location. Because her parents could not afford traveling expenses (neither for themselves nor for her), Paige simply had to attend a school close to home. Thus, Paige could not consider, let alone commit to, a school across the country. Johns Hopkins was particularly appealing because it was about a forty-minute car ride from her home. Bobby spoke about circumstances that combined Marye's and Paige's. Bobby did not want to go to school in state. He wanted to move away from home and experience another region of the country. However, he could not afford out-of-state tuition in addition to the traveling expenses. Bobby's parents told him that he would either have to attend a school in the state of Delaware or earn some kind of scholarship in order to go out of state. His baseball scholarship not only afforded him an opportunity to play ball and be drafted, but also fulfilled his desire to attend college away from home.

This is a strong paragraph, with good transitions, supporting details, and organization

All five athletes emphasized the importance of the "right" atmosphere in their college search, which actually meant an atmosphere that they connected with on an emotional level. When I asked each athlete to explain what she or he meant by "atmosphere," why it ranked higher than the other factors, and what exactly was the "right atmosphere" for each of them, they responded fully.

This transitional paragraph leads into the following ones.

For all of them, the atmosphere included athletic, academic, and social factors, the physical look and geographic location of the school, and the way all these features came together to enhance the student-athlete's emotional connection with the school.

Paige emphasized the size and social reputation of the school, the weather, and the look and layout of the campus. She claimed that visiting the school was crucial to her decision because the atmosphere is "felt," not described or quantified in statistics:

> A couple of days was what I needed to really test the feel
> [of the school]. What I had heard about the athletics and
> academics was what attracted me to the school and prompted
> me to visit in the first place. But, how I felt walking on
> campus, observing the students . . . this would be my home
> for the next four years! I wanted to fall in love. (Wright)

Nick and Theresa offered similar comments about the size and social aspects of their schools. Theresa knew she wanted a college town with a large student body. She wanted social options and a general camaraderie of school spirit among the faculty and students. In contrast, Nick desired a more intimate setting with a less intense social scene.

Having an idea of what "atmosphere" meant, I then asked them to touch on the academic and athletic aspects, explaining how those contributed to the emotional connection. Bobby wanted a smaller school with a solid academic reputation and the chance to play baseball. For him, it was important to play baseball at a school with a competitive reputation in the sport. Academically, Bobby had no specific career goal in mind. However, he did not totally disregard the academics; a university with a good academic reputation but a manageable coursework load was what he wanted. Paige, however, did have a specific career as a nurse in mind. This factor attracted her to Johns Hopkins, which is known for its medical program. Unlike Bobby, Paige emphasized the academic aspect of her college experience and desired a top-notch nursing program; thus, she knew she would be embarking on a rigorous academic schedule and was fully prepared for that commitment.

Marye offered me a slightly different definition of "atmosphere." Still including what the others mentioned, Marye simply named the atmosphere she already had in mind: her dream school, the University of Virginia. She had grown up loving UVA; she had a long history of an emotional connection. Playing lacrosse as a young girl, Marye imagined wearing the UVA uniform, which she thought "was the coolest-looking uniform ever," and winning national championships in the blue and orange. She had visited the school numerous times in high school to see games and visit her sister. For Marye, "atmosphere" meant attending her dream school

LeNoir 11

and playing lacrosse there. The reputation of UVA lacrosse ranks among the highest in the country; UVA teams are always in the top ten of polls and compete for ACC (Atlantic Coast Conference) championships. As for academics, Marye said,

> Yes, my love of the school was primarily based on this childhood dream of playing lacrosse there. UVA happens to have a stellar academic reputation too, which is a bonus. Not to say I would have thrown away academics and gone to any old school if it had great lacrosse. It's just, to be honest, I saw my future through this kind of tunnel-vision dream . . . I didn't think about what I wanted to do after college. Now, I feel a bit ashamed about my ignorance of UVA's academics. [We are] among the best in the country! I don't know how many kids dream of a UVA education, which I can't describe how much I now appreciate. But back in high school, I wanted to go to UVA to play lacrosse. (Taranto)

The essay ends with a reasonable and thoughtful conclusion

All these student-athletes described complex reasons for choosing a college, just as mine were. My love for lacrosse propelled my decision to take my sport into college, where I would pursue my professional goals, which would not include lacrosse. All my decisions were based on my wanting to achieve my goals of playing my sport in college and pursuing my career after college.

LeNoir 12

And just like all of these other student-athletes, I thought about other factors than the sport I would play. In some ways, whether we hope to play sports after college or not, we all made an emotional connection with the college of our choice.

LeNoir 13

Works Cited

Dorsey, Robert. Personal interview. 20 Oct. 2010.

Huang, Nick. Personal interview. 14 Oct. 2010.

Morales, Theresa. Personal interview. 22 Oct. 2010.

Taranto, Marye. Personal interview. 18 Oct. 2010.

Wright, Paige. Personal interview. 26 Oct. 2010.

EXERCISE 3

Compare the two versions of "How Student-Athletes Really Choose a College" that appear in this chapter and write a two-paragraph summary describing how Mary revised and edited her work. If she had shown her final draft to you, asking for your advice before submitting it for a grade, what would you have advised? Write a one-paragraph response to her draft.

34   Writing Arguments

You write arguments on a regular basis. When you send a memo reminding your colleague that a client needs to sign a contract, when you e-mail your parents to ask them for a loan, when you petition your academic advisor for a late drop, or when you demand a refund from a mail-order company, you are writing an argument in response to a rhetorical opportunity. You are expressing a point of view and using logical reasoning as you invite a specific audience to adopt that point of view or engage in a particular course of action.

Argument and *persuasion* are often used interchangeably, but they differ in two basic ways. **Persuasion** has traditionally referred to winning or conquering with the use of emotional reasoning, whereas **argument** has been reserved for the use of logical reasoning to convince listeners or readers that a particular course of action is the best one. But because writing often involves some measure of "winning" (even if that means just gaining the ear of a particular audience) and uses both emotion and reason to affect an audience, this book uses *argument* to cover the meanings of both terms.

When writing arguments, you follow the same process as for all your writing: you respond to a rhetorical opportunity and then begin planning, drafting, and revising (Chapters 32 and 33). Argumentative writing is distinctive, however, in its emphasis on audience and purpose. In order to achieve a rhetorical purpose, the writer of an argument must demonstrate a respectful acknowledgment of the beliefs, values, and expertise of the intended audience. Extending beyond mere victory over an opponent, the writer's purpose can be to invite exchange, understanding, cooperation, joint decision making, agreement, or negotiation of differences. Thus, an argument's purpose has three basic and sometimes overlapping components: to analyze a complicated issue or question an established belief, to express or defend a point of view, and to invite an audience to change a position or adopt a course of action.

This chapter will help you

- determine the purpose of an argument (34a),
- consider different viewpoints (34b),
- distinguish fact from opinion (34c),
- take a position or make a claim (34d),
- provide evidence to support a claim (34e),
- use the rhetorical appeals to ground an argument (34f),
- arrange an effective argument (34g),
- reason effectively and ethically (34h),
- avoid rhetorical fallacies (34i), and
- analyze an argument (34j).

As you proceed, you will understand the importance of determining your purpose, identifying your audience, marshaling your arguments, arguing ethically, and treating your audience with respect.

34a Determining the purpose of an argument

What opportunity for change calls to you? What topic is under discussion? What is at stake? What is likely to happen as a result of making this argument? How important are those consequences? Who is in a position to act or react in response to your argument?

When writing an argument, take care to establish the relationships among your topic, purpose, and audience. The relationship between audience and purpose is particularly significant because the audience often shapes the purpose.

- If there is little likelihood that you can convince members of your audience to change a strongly held opinion, you might achieve a great deal by inviting them to understand your position and offering to understand theirs.
- If the members of your audience are not firmly committed to a position, you might be able to convince them to agree with the opinion you are expressing or defending—or at least to consider it.
- If the members of your audience agree with you in principle, you might invite them to undertake a specific action—such as voting for a proposed school tax or supporting a particular candidate.

No matter how you imagine those in your audience responding to your argument, you must establish **common ground** with them, stating a goal toward which you both want to work or identifying a belief, assumption, or value that you both share. In other words, common ground is a necessary starting point, regardless of your ultimate purpose.

34b Considering differing viewpoints

Because people hold different points of view, much of the writing you do requires you to take an arguable position on a topic. The first step toward finding a topic for argumentation is to consider issues that inspire different opinions and offer opportunity for change.

Behind any effective argument is a question that can generate more than one reasonable answer that goes beyond "yes" or "no." If you ask "Are America's schools in trouble?" almost everyone will say "yes." But if you ask "In what ways can

America's schools be improved?" or "How can colleges better prepare teachers?" you will hear different answers. Answers differ because people approach questions with various backgrounds, experiences, and assumptions. As a consequence, they are often tempted to use reasoning that supports what they already believe. As a writer, you, too, will be tempted to employ such reasoning, but as you shape your argument, you need to demonstrate that you are well informed about your topic as well as aware and considerate of other views about it.

You write an argument in order to solve a problem, answer a question, or determine a course of action—with or for an audience. When you choose a topic for argumentation, you take a stance that allows you to question, while providing you an opportunity (or reason) for writing. You first focus on a topic, on the part of some general subject that you will address (32b), and then pose a question about it. As you formulate your question, consider (1) your values and beliefs with respect to the topic, (2) how your assumptions might differ from those of your intended audience, (3) what your ultimate purpose is for writing to this audience, and (4) how you might establish common ground with its members, while respecting any differences between your opinion and theirs. The question you raise will evolve into an arguable statement, your **thesis** (32c).

The most important criterion for choosing an arguable statement for an essay is knowledge of the topic; such knowledge makes you an informed writer, responsive to the expectations of your audience and faithful to your purpose. So stay alert for issues about which you have strong opinions or for ideas that circulate in classes, on television, on the Internet, in your reading, and in conversations.

To determine whether a topic might be suitable, make a statement about the topic ("I believe strongly that . . . " or "My view is that . . . ") and then check to see if that statement can be argued. If you can answer all the questions in the following box to your satisfaction, you should feel confident about your topic.

TIPS FOR ASSESSING AN ARGUABLE STATEMENT ABOUT A TOPIC

- What reasons can you state that support your belief (or point of view) about the topic? List those reasons. What else do you need to know?

- Who or what groups might disagree with your statement? Why? List those groups.

- What are other viewpoints on the topic and reasons supporting those viewpoints? List them. What else do you need to know?

- What is your purpose in writing about this topic?

- How does your purpose connect with your intended audience? Describe that audience.

- What do you want your audience to do in response to your argument? In other words, what do you expect from your audience? Write out your expectation.

As you move further into the writing process, researching and exploring your topic in the library and online (Chapter **36**), you will clarify your purpose and refine your thesis statement.

34c Distinguishing between fact and opinion

As you develop your thesis statement into an argument, you use both facts and opinions to establish your credibility (**34f(1)**). **Facts** are reliable pieces of information that can be verified through independent sources or procedures.

Opinions, on the other hand, are assertions or inferences that may or may not be based on facts. Opinions that are widely accepted, however, may seem to be factual when they are not.

Facts are significant only when they are used responsibly in support of an argument; otherwise, a thoughtful and well-informed opinion might have more impact. To determine whether a statement you have read is fact or opinion, ask yourself questions like these: Can it be proved? Can it be challenged? How often is the same result achieved? If a statement can consistently be proved true, then it is a fact. If it can be disputed, then it is an opinion, no matter how significant or reasonable it may seem.

Because the line between fact and opinion is not always clear, writers and readers of arguments must be prepared to assess the reliability of the information before them. They need to evaluate the beliefs supporting the argument's stance, the kinds of sources used, and the objections that could be made to the argument.

EXERCISE 1

Determine which of the following statements are facts and which are opinions. In each case, what kind of verification would you require in order to accept the statement as reliable?

1. China's Liu Xiaobo won the 2010 Nobel Peace Prize.

2. A college degree guarantees a higher income over the course of a lifetime.

3. Women who are overweight or who have a family history of diabetes have a higher risk of gestational diabetes.

4. Every American student can and should learn to write well in college.

5. Aerobic exercise is good for your health.

34d Taking a position or making a claim

When making an argument, a writer takes a position on a particular topic. Whether the argument analyzes, questions, expresses, defends, invites, or convinces, the writer's position needs to be clear. That position, which is called the **claim**, or **proposition**, clearly states what the writer wants the audience to do with the information being provided. The claim is the thesis of the argument and usually appears in the introduction and sometimes again in the conclusion.

(1) Extent of a claim

Claims vary in extent; they can be absolute or moderate, large or limited. Absolute claims assert that something is always true or false, completely good or bad; moderate claims make less sweeping assertions.

Absolute claim	Any great college athlete can go pro.
Moderate claim	Most pro athletes went to college.
Absolute claim	Harry Truman was the best president the United States has ever had.
Moderate claim	Truman's domestic policies helped advance civil rights.

Moderate claims are not necessarily superior to absolute claims. After all, writers frequently need to take a strong position for or against something. But the stronger the claim, the stronger the evidence needed to support it. Be sure to consider the quality and the significance of the evidence you use—not just its quantity.

(2) Types of claims

(a) Substantiation claims

Without making a value judgment, a **substantiation claim** asserts that something exists or is evident. This kind of point can be supported by evidence.

> The job market for those with a liberal arts degree is limited.

> The post office is raising rates and losing money again.

(b) Evaluation claims

According to an **evaluation claim**, something has a specific quality: it is good or bad, effective or ineffective, successful or unsuccessful.

> The high graduation rate for athletes at Penn State is a direct result of the school's supportive academic environment.

> The public transportation system in Washington DC is reliable and safe.

Sometimes, writers use an evaluation claim as a way to invite their audience to consider an issue.

> It is important for us to consider the graduation rate of all Big Ten athletes, regardless of the sport they play.

(c) Policy claims

When making **policy claims**, writers call for specific action.

> We must establish the funding necessary to hire the best qualified high school teachers.

> We need to build a light-rail system linking downtown with the western suburbs.

Much writing involves substantiation, evaluation, and policy claims. When writing about the job market for engineers with recent degrees, you might tap your ability to substantiate a claim; when writing about literature (Chapter 41), you might need to evaluate a character. Policy claims are commonly found in arguments about social or political issues such as health care, social security, affirmative action, or defense spending. These claims often grow out of substantiation or evaluation claims: first,

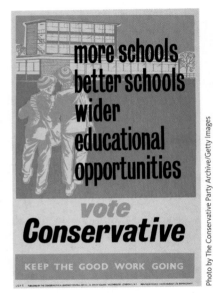

Photo by The Conservative Party Archive/Getty Images

Policy claims, such as the one made by this 1961 British political poster, call for specific action. In this case, the requested action is to vote for Conservative candidates.

you demonstrate that a problem exists; then, you establish the best solution for that problem.

TIPS FOR MAKING A CLAIM ARGUABLE

- Write down your opinion.
- Describe the situation and experiences that produced your opinion.
- Decide who constitutes the audience for your opinion and what you want that audience to do about your opinion.

- Write down the verifiable and reliable facts that support your opinion.

- Transform your initial opinion into a thoughtful claim that reflects those facts and considers at least two sides of the issue.

- Ask yourself, "So what?" If the answer to this question shows that your claim leads nowhere, start over, beginning with the first tip.

34e   Providing evidence for an effective argument

Effective arguments are well developed and supported. You should explore your topic in enough depth that you have the evidence to support your position intelligently and ethically, whether that evidence is based on personal experience or on research (Chapters **32** and **36**). You want to consider the reasons others might have to disagree with you and be prepared to respond to those reasons.

(1) Establishing the claim

If you want readers to take your ideas seriously, you must establish the reasons that have led to your claim and the opinions, values, and assumptions that underlie your thinking. So, as you explore your topic, make a list of the reasons that have led to your belief (**32d** and **32f**). When Anna Seitz was working on her argumentative essay (at the end of this chapter; see pages 423–29), she listed the following reasons for her belief that universities should not allow individuals or corporations to buy naming rights to campus buildings:

1. By purchasing naming rights, donors gain influence over educational policy decisions, even though they are not qualified to make such decisions.
2. Significant donations can adversely affect overall university finances by replacing existing funding sources.
3. Donors who purchase naming rights are associated with the university, in spite of the fact that they or their corporations may subscribe to a different set of values.

Although it is possible to base an argument on one good reason (such as "The selling of naming rights distracts from the educational purposes of universities"), doing so can be risky. If your audience does not find this reason convincing, you have no other support for your position. When you show that you have more than one reason for believing as you do, you increase the likelihood that your audience will find merit in your argument. Sometimes, however, one reason is stronger—and more appropriate for your audience—than several others you could advance. To develop an argument for which you have only one good reason, explore the opinions, values, and assumptions that led you to take such a stand. By revealing the thinking behind the single reason on which you are building your case, you can create a well-developed argument.

Whether you have one reason or several, be sure to provide sufficient evidence from credible sources to support your claim:

- facts,
- statistics,
- examples, and
- testimony (based on personal experience or professional expertise).

This evidence must be accurate, representative, and sufficient. Accurate information should be verifiable by others (34c). Recognize, however, that even if the information a writer provides is accurate, it may not be representative or sufficient if it was drawn from an exceptional case, a biased sample, or a one-time occurrence. If, for example, you are writing an argument about the advantages of using Standardized English but you draw all of your supporting evidence from a proponent of the English-Only movement, your evidence represents only the views of that movement. Such evidence is neither representative of all the support for the use of Standardized English nor sufficient to support a thoughtful argument. In order to better represent your viewpoint, you should gather supporting evidence from sociolinguists, speakers of other dialects and languages, education specialists, professors, and other experts. In other words, consult more than a single source (Chapter 36).

When gathering evidence, be sure to think critically about the information you find. If you are using the results of polls or other statistics or statements by authorities, determine how recent and representative the information is and how it was gathered. Consider, too, whether the authority you plan to quote is qualified to address the topic under consideration and is likely to be respected by your readers.

Whatever form of evidence you use—facts, statistics, examples, or testimony—you need to make clear to your audience exactly *why* and *how* the evidence supports your claim. After all, even accurate information has to be interpreted by the writer and the reader. As soon as the relationship between your claim and your evidence is clear to you, make that connection explicit to your readers, helping them understand your thinking.

(2) Responding to diverse views

Issues are controversial because good arguments can be made on all sides. Therefore, effective arguments consider and respond to other points of view. Fairness, respect, and acknowledgment of other points of view are crucial for connecting with your audience. When you introduce diverse views and then respectfully demonstrate why you disagree with each of them, you are using **refutation**, the most common strategy for addressing opposing points of view. As you consider opposing points of view, you are likely to discover some you cannot refute, perhaps because they are based in a belief system markedly different from your own. You are also likely to discover that some of those other views have real merit. If you understand the reasons behind opposing viewpoints but remain unconvinced, you will need to demonstrate why.

When you find yourself agreeing with a point that supports another side of an issue, you can benefit from offering a **concession**. By openly admitting that you agree with opponents on one or more specific points, you demonstrate that you are fair-minded and credible (34f(1)). Your concessions also increase the likelihood that your opponents will find merit in parts of your argument.

Whether you agree or disagree with other positions, you must recognize and assess them. It is hard to persuade people to agree with you if you insist that they are entirely wrong. If you admit that they are partially right, they are more likely to admit that you could be partially right as well. In this sense, then, argument involves working with an audience as much as getting them to work with you.

EXERCISE 2

The following paragraph is the conclusion of an argument written by Tucson writer Debra Hughes shortly after the January 2011 shooting of Representative Gabrielle Giffords and eighteen other people at a local political rally. Hughes's piece connects inflammatory rhetoric with such acts of violence. Write a short analysis of this paragraph in which you note (a) an opposing viewpoint to which she is responding, (b) a refutation she offers to this viewpoint, (c) a concession she makes, and (d) any questions this excerpt raises for you.

[1]Habits are hard to break. [2]Only days after the shooting, Arizona passed laws prohibiting picketing within three hundred feet of any home, cemetery, funeral home, or house of worship before, during, or after a ceremony or burial. [3]As the shooting victims were being laid to rest, a group called Angel Action donned white wings and stood with hundreds of others dressed in white, to shield mourners from potential protests by fanatics, such as the Westboro Baptist Church congregants, who have been disrupting services for US soldiers slain in Iraq and Afghanistan. [4]The Westboro group is emblematic of the flashpoints of opinion and controversy across the land. [5]In the weeks and months ahead, only determined national leadership and conscientious, principled activity and restraint on the part of major media, and on the part of each citizen, can restore humanity and civility to American discourse.

—**DEBRA HUGHES**, **"The Tucson Shootings: Words and Deeds"**

34f Using the rhetorical appeals to ground an argument

Human beings do not form their beliefs or act on the basis of facts or logic alone; if we did, we would all agree and would act accordingly. Therefore, your best chance at shaping an effective argument is by incorporating a combination of persuasive strategies, which include the **rhetorical appeals** of ethos, logos, and pathos. **Ethos** (an ethical appeal) establishes the speaker's or writer's credibility and trustworthiness. An ethical appeal demonstrates goodwill toward the audience, good sense or knowledge of the subject at hand, and good character. Establishing common ground with the audience is another feature of ethos. However, ethos can rarely carry an argument by itself; therefore, you also need to use **logos** (a logical appeal). Logos demonstrates an effective use of reason and judicious use of evidence, whether that evidence consists of facts, statistics, comparisons, anecdotes, expert opinions, or observations. You employ logos when you are supporting claims, drawing reasonable conclusions, and avoiding rhetorical fallacies (34i). But logic may not be sufficient to persuade an audience, unless the audience feels emotionally stirred by the topic under discussion. Therefore, **pathos** (an emotional appeal) involves using language that will connect with the beliefs and feelings of the audience. If you misuse pathos in an attempt to manipulate your audience (as sentimental movies and manipulative speakers often do), your attempt can easily backfire. Still, pathos can be used successfully when it establishes empathy, authentic understanding, and a human connection with the audience. The most effective arguments are those that combine these three persuasive appeals—ethos, logos, and pathos—responsibly and knowledgeably.

In the next three subsections, additional passages from Debra Hughes's "Tucson Shootings: Words and Deeds" illustrate how a writer can use all three of the classical rhetorical appeals.

(1) Ethical appeals

The ethical appeal, ethos, establishes a writer's credibility, moral character, and goodwill. In her introductory paragraphs, Debra Hughes captures the atmosphere of downtown Tucson on the evening of the shootings and demonstrates her knowledge of the day's events, her credibility as a local Tucsonan, and her goodwill toward her readers, who have "their own heavy feelings." She also establishes her thesis.

> The night of the mass shootings in Tucson, a downtown art gallery hosted an already scheduled [exhibition of images] from François Robert's photography series *Stop the Violence* . . . of human bones arranged in the shapes of a handgun, grenade, knife, Kalashnikov, fighter jet, and other symbols of violence, all starkly set on black backgrounds. Those images confronted viewers with their own heavy feelings. That morning six people had been killed and thirteen wounded in the shooting rampage at Gabrielle Giffords's political rally at a local Safeway. Jared Lee Loughner had tried to assassinate the Arizona congresswoman, using a Glock 19 semiautomatic pistol and firing thirty-one rounds into the crowd in about fifteen seconds.
>
> The shooting took place at a small shopping center in my neighborhood. . . . Our bank is there, along with the stores where we mail our packages, buy pastries, toothpaste, and paper towels, and where we regularly run errands. That morning people had gathered to hear what their state representative had to say. She called the event "Congress at Your Corner."
>
> In the afternoon, . . . Pima County Sheriff Clarence Dupnik, a seventy-five-year-old with the sagging cheeks and drooping eyes of a bulldog, spoke his mind. "People tend to pooh-pooh this business about all the vitriol that we hear inflaming the American

public . . . ." He was alluding to talk-show hosts and politicians who use inflammatory rhetoric, and he added that the effect of their words should not be discounted. "That may be free speech, but it's not without consequences." Almost immediately a heated public debate began over whether or not political rhetoric had spurred Jared Lee Loughner to kill.

(2) Logical appeals

Logos, the logical appeal, is considered to be especially trustworthy, as it is rooted in a writer's reliance on reason and supporting evidence (facts, statistics, observations, interviews with authorities, survey results, and so on) to build an argument. Logical appeals, however, need to be examined closely to determine whether facts are accurate, testimony has been considered within its context, and sources are reliable. To help her audience appreciate the connection she's trying to make, Debra Hughes builds on the sheriff's testimony, building her logos with facts, expert opinions, observations, and vivid examples, while recognizing all sides of the controversy. She ends the following excerpt by quoting an obvious (and ironic) logical fallacy:

> Tucson forensic psychologist Dr. Gary Perrin, a professional familiar with violent crime, was asked if a mentally disturbed person might distort strong messages into a belief that violent acts are noble. Perrin replied, "In . . . the past few years, rhetoric has increased. Words are powerful, and certainly words can make a [mentally unstable] person act in a certain way." But he emphasized that violent acts are "situational, and many things contribute. Words can be one of the factors."
>
> . . . Within a week of the shooting a Google search produced 55 million results about the event, including myriad bloggers arguing over free speech. Stephen Colbert, on his Comedy Central TV show, aired a segment entitled "The Word: Life, Liberty, and the Pursuit of Angriness," in which he deadpanned, "If incendiary rhetoric isn't connected to the Arizona tragedy, it logically follows that it must be good."

(3) Emotional appeals

Emotional appeals, pathos, stir feelings and help a writer connect with the audience. Although emotional appeals can be manipulative (like unethical appeals to faulty logic or to false authority), they can be used ethically and logically to move the audience to a new way of thinking or acting.

In the following passage, which moves toward her conclusion (see page 407), Debra Hughes continues to use logos while alluding to moderation and personal responsibility, thus evoking feelings that people on all sides of the issue can share.

> The night before Gabrielle Giffords was shot, she sent an email offering congratulations to Kentucky Secretary of State Trey Grayson, . . . [newly] named director of Harvard University's Institute of Politics. . . . "After you get settled, I would love to talk about what we can do to promote centrism and moderation. I am one of twelve Democrats in a GOP district (the only woman) and we need to figure out how to tone our rhetoric and partisanship down." . . .
>
> The Tucson shooting and its aftermath are being followed around the world, especially in places where violence is a problem. Venezuelan magazine editor Sergio Dahbar wrote, "The Giffords shooting is being followed very closely in Latin America because we also have this illness. We have the illness of intolerance." A well-known maxim from Victor Hugo commenting on unrest in France in the 1830s runs, "The guilty one is not he who commits the sin, but the one who causes the darkness."
>
> —**DEBRA HUGHES, "The Tucson Shootings: Words and Deeds"**

Although ethos is often developed in the introduction to an argument, logos in the body, and pathos in the conclusion, these classical rhetorical appeals can overlap and appear throughout an argument.

Sometimes the premise is not stated, for the simple reason that the writer assumes that an audience shares the belief.

Imogen has graduated from medical school, so she must complete a residency.

In this sentence, the unstated premise is that all doctors must complete a residency. A syllogism with an unstated premise—or even an unstated conclusion—is called an **enthymeme**. Frequently found in written arguments, enthymemes can be very effective because they presume shared beliefs or knowledge. For example, the argument "The college needs to build a new dormitory because the present overcrowded dorms are unsafe" contains the unstated premise that the college has a responsibility to reduce unsafe conditions.

34i   Avoiding rhetorical fallacies

Logical reasoning fortifies the overall effectiveness of an argument as well as builds the ethos of the speaker or writer. Constructing an argument effectively means avoiding errors in logic, known as **rhetorical fallacies**, which weaken an argument as well as the writer's ethos. These fallacies signal to your audience that your thinking is not entirely trustworthy and that your argument is not well reasoned or researched.

Therefore, you need to recognize and avoid several kinds of rhetorical fallacies. As you read the arguments of others (37a) and revise the arguments you draft (Chapter 33), keep the following common fallacies in mind.

(1) *Non sequitur*
A *non sequitur,* the basis for most of the other rhetorical fallacies, attempts to make a connection where none actually exists

(the phrase is Latin for "it does not follow"). Just because the first part of a statement is true does not mean that the second part is true, will become true, or will necessarily happen.

Faulty　Heather is married and will start a family soon.

This assertion is based on the faulty premise that *all* women have children soon after marrying (**34h(2)**).

(2) Ad hominem

The *ad hominem* fallacy refers to a personal attack that draws attention away from the issue under consideration (the Latin phrase translates to "toward the man himself").

Faulty　With his penchant for expensive haircuts, that candidate cannot relate to the common people.

The fact that a candidate pays a lot for a haircut may say something about his vanity but says nothing about his political appeal. When private or personal information about political candidates is used in criticizing them, the focus is on the person, not the political issues in play.

(3) Appeal to tradition

The appeal to tradition argues that because things have always been done a certain way, they should continue that way. Assuming that a mother will stay home and take care of children or that a father will be the sole breadwinner is an appeal to tradition.

Faulty　Because they are a memorable part of the pledge process, fraternity hazings should not be banned.

Times change; what was considered good practice in the past is not necessarily considered acceptable now.

Maiming and pillaging have a long history, but that does not mean they should continue.

(4) Bandwagon

The bandwagon fallacy implies that because everyone is doing, saying, or thinking something, you should too. It makes an irrelevant and disguised appeal to the human desire to be part of a group.

> **Faulty**   Everyone uses cell phones while driving, so why won't you answer my calls?

Even if the majority of people talk on the phone while driving, doing so has proven to be dangerous. The majority is not automatically right.

(5) Begging the question

The begging-the-question fallacy presents the conclusion as though it were a major premise. What is assumed to be fact actually needs to be proved.

> **Faulty**   If we get rid of the current school board, we will see an end to all the school district's problems.

Any connection between the current school board and the school district's problems has not been established.

(6) Equivocation

The rhetorical fallacy of equivocation falsely relies on the use of one word or concept in two different ways.

Faulty Today's students are illiterate; they do not know the characters in Shakespeare's plays.

Traditionally, *literacy* has meant knowing how to read and write, how to function in a print-based culture. Knowing about Shakespeare's characters is not the equivalent of literacy; someone lacking this special kind of knowledge might be characterized as uneducated or uninformed but not as illiterate.

(7) False analogy

A false analogy assumes that because two things are alike in some ways, they are alike in others as well.

Faulty The United States lost credibility with other nations during the war in Vietnam, so we should not get involved in the Middle East, or we will lose credibility again.

The differences between the war in Southeast Asia in the 1960s and 1970s and the current conflict in the Middle East may well be greater than their similarities.

(8) False authority (or appeal to authority)

The fallacy of false authority assumes that an expert in one field is credible in another. Every time you see a movie star selling cosmetics, a sports figure selling shirts, or a talk-show host selling financial advice, you are the target of an appeal to authority.

Faulty We must stop sending military troops into Iran, as Zack de la Rocha has argued.

De la Rocha's membership in the politically engaged band Rage Against the Machine does not qualify him as an expert in foreign policy.

(9) False cause

Sometimes called *post hoc, ergo propter hoc* (meaning "after this, so because of this"), the fallacy of false cause is the assumption that because one event follows another, the first is the cause of the second.

Faulty   If police officers did not have to carry carry guns and batons as part of their job, there would be no incidents of violence by off-duty officers.

The assumption is that because police officers carry weapons, they use those weapons indiscriminately. Making such a connection is like announcing that you are not going to wash your car any more because every time you do, it rains (as though a clean car causes rain).

(10) False dilemma

Sometimes called the *either/or fallacy*, a false dilemma is a statement that only two alternatives exist, when in fact there are more than two.

Faulty   We must either build more nuclear power plants or be completely dependent on foreign oil.

Other possibilities for generating energy without using foreign oil exist.

(11) Guilt by association

An unfair attempt to besmirch a person's credibility by linking that person with untrustworthy people or suspicious actions is the fallacy of guilt by association.

Faulty   You should not vote for her for class treasurer because her mother was arrested for shoplifting last year.

The mother's behavior should not be held against the daughter.

Joseph Sohm/Corbis

This sign exemplifies a false dilemma, as though only two alternatives exist when, in reality, there are more than two.

(12) Hasty generalization

A hasty generalization is a conclusion based on too little evidence or on exceptional or biased evidence.

> **Faulty**   Ellen is a poor student because she failed her first history test.

Ellen's performance may improve in the weeks ahead. Furthermore, she may be doing well in her other subjects.

(13) Oversimplification

A statement or argument that implies a single cause or solution for a complex problem, leaving out relevant considerations and complications, relies on the oversimplification fallacy.

> **Faulty**   We can eliminate unwanted pregnancies by teaching birth control and abstinence.

Teaching people about birth control and abstinence does not guarantee the elimination of unwanted pregnancies.

(14) Red herring

Sometimes called *ignoring the question*, the red herring fallacy dodges the real issue by drawing attention to a seemingly related but irrelevant one.

Faulty   Why worry about violence in schools when we ought to be worrying about international terrorism?

International terrorism has no direct link to school violence.

(15) Slippery slope

The slippery slope fallacy assumes that one thing will inevitably lead to another—that if one thing is allowed, it will be the first step in a downward spiral.

Faulty   Handgun control will lead to a police state.

Handgun control has not led to a police state in England.

Be alert for rhetorical fallacies in your writing. When you find such a fallacy, be sure to moderate your claim, clarify your thinking, or, if necessary, eliminate the fallacious statement. Even if your argument as a whole is convincing, rhetorical fallacies can damage your credibility (37a).

34j   An argument essay

The following argumentative essay was Anna Seitz's response to an assignment asking her to identify a specific problem in her living quarters, on her campus, in her town, or in the world at large and then recommend a solution for that problem. As you read Anna's essay (which she formatted according to MLA guidelines; see Chapter 39), consider how she argued her case and whether she argued effectively. Note her use of the rhetorical appeals (ethos, logos, and pathos) and classical arrangement and her inductive reasoning. Also, identify the kinds of evidence she uses (facts, examples, testimony, and authority).

Anna Seitz

Professor Byerly

Library Science 313

30 November 2007

The writer's last name and the page number appear as the running head on each page of the paper.

Naming Opportunities: Opportunities for Whom?

All over the nation, football stadiums, business schools, law schools, dining halls, and even coaching positions have become naming opportunities (also known as "naming rights" and "legacy opportunities"). Since 1979, when Syracuse University signed a deal with the Carrier Corporation for lifetime naming rights to its sports stadium—the Carrier Dome—naming has become a common practice with an alleged twofold payoff: universities raise money and donors get their names writ large. Universities use the money obtained from naming opportunities to hire more faculty, raise salaries, and support faculty research. Reser Stadium (Oregon State), The Donald Bren School of Law (University of California–Irvine), or the Malloy Paterno Head Football Coach Endowment (Penn State University)—all these naming opportunities seem like a good solution for raising money, especially at a time when state legislatures have cut back on university funding and when wealthy alumni are being besieged for donations from every college they have ever attended. Naming opportunities seem like a good

In her introduction (the first paragraph), the writer uses concrete examples to catch her readers' attention and set up her argument.

Seitz 2

solution for donors, too, because their donations will be broadly

recognized. While naming opportunities may seem like a perfect

solution for improving colleges and universities and simplifying

funding, in reality they are not. In this paper, I argue against naming

opportunities on college and university campuses because they

create more problems than they solve.

> The last sentence in the first paragraph is the thesis statement.

The naming of sports stadiums is a familiar occurrence.

But naming opportunities in other spheres of academic life

are unfamiliar to most people, even though such naming is an

established practice. A quick search of the Web pages of university

libraries reveals that many of them, especially those in the midst

of major development campaigns, have created a price list just for

naming opportunities. Entire buildings are available, of course.

For example, a $5 million donation earns the right to name the

music library at Northwestern University (Northwestern). But parts

of buildings are also available these days. North Carolina State

University will name an atlas stand according to the donor's wishes

for only $7,500 or put a specific name on a lectern for $3,500 (North

Carolina).

> The writer provides background information in the second, third, and fourth paragraphs.

> The writer's use of specific information helps establish her ethos as knowledgeable and trustworthy.

Naming opportunities can clearly bring in a good deal of

money. It has become commonplace for schools to offer naming

opportunities on planned construction in exchange for 51 percent

Seitz 3

of the cost of the building! That's a big head start to a building project, and naming opportunities may be what allow some schools to provide their students with better facilities than their counterparts that do not offer naming opportunities. In fact, donors are often recruited for the opportunity to pay for named faculty chairs, reading rooms, or major library or art collections—all of which enhance student life.

> The writer's use of the phrase *enhance student life* at the end of the third paragraph demonstrates an effort to establish common ground with her readers.

Clearly the more opportunities and resources any university can offer current and potential students and even alumni, the more that university enhances its own growth and that of its faculty. Library donors and recipients say that if it is possible for a library to pay for a new computer lab just by adding a sign with someone's name over the door, the advantages often seem to outweigh the disadvantages. Proponents of naming opportunities point out that small donors are often hailed as library supporters, even when big donors are maligned as corporate flag-wavers.

> The writer lays out specific opposing arguments without losing her focus on her own argument or alienating her readers. She is maintaining her ethos while using logos.

Few would argue that these donations necessarily detract from the educational mission of the institution. However, selling off parts of a university library, for example, does not always please people, especially those whose responsibility includes managing that donation. The curator of rare books and manuscripts at a prominent state university told me that one of the most frustrating parts of her

> The writer's refutation of opposing arguments is fair-minded.

Seitz 4

job is dealing with "strings-attached" gifts, which is what too many library donations turn out to be. Some major donors like to make surprise visits, during which they monitor the prominence of their "legacy opportunity." Others like to create rules which limit the use of their funds to the purchase of certain collections or subjects; still others just need constant personal maintenance, including lunches, coffees, and regular invitations to events. But this meddling after the fact is just a minor inconvenience compared to some donors' actions.

This paragraph and the next introduce the proposition.

Donors who fund an ongoing educational program and who give money on a regular basis often expect to have regular input. Because major donors want major prestige, they try to align themselves with successful programs. Doing that can result in damage to university budgets. First of all, high-profile programs can become increasingly well funded, while less prominent, less glamorous ones are continually ignored. Second, when corporate or private funds are regularly available, existing funding sources can erode. Simply put, if a budgeted program becomes funded by donation, the next time the program needs funding, the department or unit will likely be told that finding a donor is its only option. Essentially, once donor-funded, always donor-funded.

The writer uses cause-and-consequence analysis in these paragraphs to support her thesis.

Seitz 5

Additionally, many academics believe that selling off naming rights can create an image problem for a university. While buildings, schools, endowed chairs, even football stadiums were once named for past professors, university presidents, or others with strong ties to the university, those same facilities are now named for virtually anyone who can afford to donate, especially corporations. Regular input from a corporation creates the appearance of a conflict of interest in a university, which is exactly the reason such arrangements are so often vehemently opposed by the university community. Boise State University in Idaho received such negative press for negotiating a deal with labor-unfriendly Taco Bell that it was finally pressured to terminate the $4 million contract (Langrill 1).

Given these drawbacks, many universities are establishing guidelines for the selection of appropriate donors for named gifts. To that end, fundraising professional and managing director of Changing Our World, Inc. Robert Hoak suggests that naming opportunities should be mutually beneficial for the donor (whether a corporation or an individual) and the university and that these opportunities should be viewed as the start of a long-term relationship between the two. Additionally, he cautions that even if the donor seems the right fit for the organization, it is in the best interest of both parties to add an escape clause to the contract in

This paragraph provides the proof, or confirmation. Here the writer describes a reasonable middle ground, a way to encourage endowments, enhance student life, and protect the integrity of the university.

Seitz 6

order to protect either side from potential embarrassment or scandal.
He provides the example of Seton Hall University, which regrettably
had both an academic building and the library rotunda named for
Tyco CEO Dennis Kozlowski. When Kozlowski was convicted of
grand larceny, the university pulled the names (Hoak).

Although the writer uses the final paragraph for a refutation of opposing arguments, she also emphasizes her common ground with her readers.

Although many people prefer that naming be an honor given to
recognize an accomplished faculty member or administrator, most
realize that recruiting major donors is good business. Whether it is
"good education" is another question. Naming university property
for major donors is not a recent phenomenon. New College in
Cambridge, Massachusetts was just that—until local clergyman
John Harvard died and left half his estate and his entire library
to what would soon become Harvard College. Modern naming
opportunities, however, do not necessarily recognize and remember
individuals who had significant influence on university life; rather,
they create obligations for the university to operate in such a way

The conclusion repeats the writer's main point.

as to please living donors or their descendants. Pleasing wealthy
donors should not replace educating students as a university's
primary goal.

Seitz 7

Works Cited

Hoak, Robert. "Making the Most of Naming Opportunities." *on Philanthropy*. Changing Our World, 28 Mar. 2003. Web. 5 Nov. 2007.

Langrill, Chereen. "BSU Faculty Says 'No Quiero' to Taco Bell." *Idaho Statesman* [Boise] 27 Oct. 2004: 1+. Print.

North Carolina State University Libraries. "NCSU Libraries East Wing Renovation: Naming Opportunities." *NCSU Libraries*. North Carolina State U, n.d. Web. 5 Nov. 2007. <http://www.lib.ncsu.edu/renovation/namingOp/>.

Northwestern University Library. "Making a Gift: Naming Opportunities." *Naming Opportunities: Library Development Office*. Northwestern U, 2007. Web. 20 Nov. 2007.

Following MLA guidelines, the writer includes the URL for an online source that would be difficult for readers to locate without that information.

35 Online Writing

In addition to word-processing capabilities, computers also offer you the opportunity to communicate with a wider, often global, audience. Online writing is often **interactive** (that is, a writer is linked to other writers, and a document is linked to other documents), dramatically expanding a work's audience and context. Because composing in this medium differs somewhat from writing essays or research papers delivered in hard copy, online writing calls for different skills—many of which you already have. This chapter will help you

- assess the rhetorical situation for online writing (**35a**),
- participate in online discussions (**35b**),
- understand conventions for online communication (**35c**),
- compose effective documents in an online environment (**35d**), and
- manage the visual elements of a website (**35e**).

35a Assessing the online rhetorical situation

Whenever you compose an e-mail message, create a web page, engage in an online discussion, or post a note on Facebook or an update on a blog, you are using rhetoric, or purposeful language, to influence the outcome of an interaction (**31a**). Some online communication is so quick, casual, and visual that you may forget that you are responding to a rhetorical situation.

The key difference between online communication and other kinds of communication is the former's instant access to many different audiences. You may have already learned the hard way how easily your so-called private e-mail can be forwarded to, accidentally copied to, or printed out for an unintended audience. Thus, online communication offers enormous challenges in terms of audience, which are evident whether you are composing within an online learning platform, contributing to a campus club's listserv, or updating your status on Facebook or Twitter. For example, if your instructor asks you to contribute regularly to a class-related blog, the instructor and your classmates comprise your primary audience. But as soon as you post an entry on the blog, your work is available to a variety of secondary audiences (**31d(3)**) via the Internet. Therefore, as you compose online, you will want to consider the responses of all possible readers and pay careful attention to appropriateness, accuracy, and tone.

Purpose is another important feature of any online rhetorical opportunity, whether you are responding to an e-mail or creating a website. You will always want to identify your purpose, connecting it with your audience, just as you do in print. Whether you wish to express your point of view, create a mood, or amuse or motivate your audience, you need to make your purpose clear. In an e-mail message, you can often state a purpose in the subject line. In a course-related blog, you can announce the subject, so that your audience has a sense of your purpose and of how your comments connect to it. You might state the purpose of a web page as a mission statement. For instance, the website for English 202C, Technical Writing (Figure 35.1), a course designed for students in the sciences, states that one of the site's primary purposes is to "serve as a repository for the syllabus, assignment sheets, and resources for the course." However readers encounter your online composition, they want to be able to identify your purpose without effort. Therefore, you need to take extra care to clarify and make apparent your purpose.

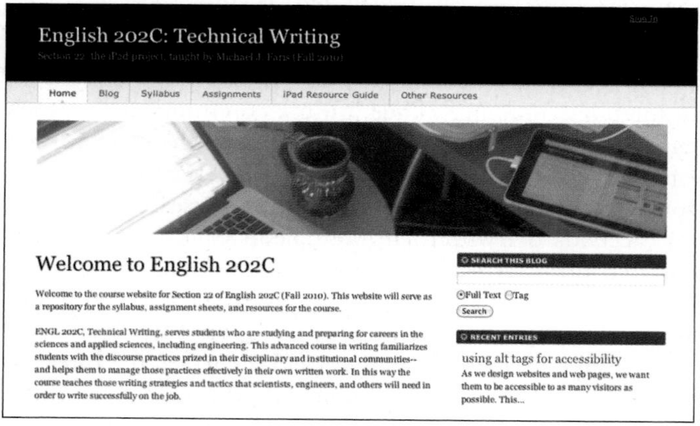

Figure 35.1. This home page of a website for a technical writing course states the purpose of the site.

In addition to attention to purpose and audience, online composing requires a sensitivity to rhetorical context that sometimes exceeds that demanded by conventional academic writing projects (**31e**). In an online context, the boundary between writer and audience can often become blurred, as writer and audience are both participants in the rhetorical situation, responding to one another nearly simultaneously. In addition, the accessibility of online discussion communities (**35b**) encourages many people to add to or comment on what has already been written and posted. This flow of new material contributes to an always evolving rhetorical context, requiring you to be familiar with the preceding discussion and to understand the conventions of the forum in order to communicate effectively.

Timeliness is another important feature of an online rhetorical situation. Internet users expect online compositions to be up-to-date, given the ease of altering an electronic document.

Just think of the constant updates you receive via Twitter and Facebook. Web pages, too, must be constantly updated in order to be useful. As you know, it is frustrating to access a web page only to discover that it is months or even years out of date. Whether you are reading or composing online material, you want the information to be current, detailed, and correct.

35b Participating in online discussion communities

Participating in an online discussion community is a good way to learn more about a topic that interests you or to network with friends, classmates, and online acquaintances while developing your writing skills. However, just as you evaluate information in print sources (Chapter 37), you need to evaluate the information and advice you receive from this kind of online source.

The two main types of online discussion groups are asynchronous forums and synchronous forums. **Asynchronous forums**, such as blogs, listservs, and various discussion forums, allow easy access, regardless of time and place, because participants do not need to be online at the same time. The delay between the posting and viewing of messages can lead to thoughtful discussions because it emphasizes the importance of *responding* to the existing rhetorical situation. On the other hand, **synchronous forums**, provided by electronic meeting software and instant-messaging (IM) programs (such as iChat, Skype, and Gchat), allow users to view text (and any multimedia elements) in real time—that is, as it is being posted—and to respond immediately. Such discussions resemble face-to-face interactions. Both asynchronous and synchronous forums are used by a variety of groups—social groups, students, scholarly or special-interest groups, and business groups—as a convenient way to communicate across geographic distances.

Whenever you are involved in an online forum, you need to pay attention to your rhetorical situation, taking care to post and respond respectfully (see Figure 35.2).

Whenever—and however—you participate in an online forum, be sure to present yourself as a trustworthy, credible writer and person. To do so, start by reading what has already been said about the topic before adding your comments to an

---

Heather Brook Adams | January 18, 2011 8:12 PM | Reply

Your points are well taken. I am most interested in your feeling like there is a level of political engagement you "should" embrace. Where does that sense of responsibility and obligation come from? Do you really feel that you "should" follow politics and current events? Or do you think that these are the sorts of things that we are supposed to feel? After all, you articulate some level of hopelessness in light of an "underlying power struggle."

Do public speakers hold themselves to a higher standard of this obligation? If so, where and when does that obligation break down?

These are just questions for thinking. Of course, anyone is welcome to respond!

JOSEPHINE SEUNGAH LEE | January 19, 2011 9:41 AM | Reply

Well, being a college student who is thinking about making some kind of difference in this world or in someone's life, I know that there really is not much I can do to help anyone as I would like to if I don't know what their problems are. I am a sociology major, and we learn that people are influenced not only by one another directly but also by larger groups or powers indirectly. Even if I want to help people on a more personal level, often times, the source of their problems stem from what is happening around them, and this includes political issues. Part of me wants to just keep my head in the sand, but I feel as though I am missing out on something big if I continue to do that.

Perhaps I do see politics as an endless power struggle, but as I said, I do not really know much about it. Debate is just so complex to me. It is an exchange that does not work if someone does not listen. Without listening, it is just an exchanging of words. I am more concerned about action to back up words. Without a following action, I don't know what the purpose of debate was.

I think everyone is equal in their obligation to speak. Some people actually speak in public, others speak out visually or in other ways. Some people can not speak, but they are there crying out silently. It all depends on who listens and responds with actions.

Figure 35.2. A thread from a course blog, *Rhetoric and Civic Life*, featuring messages between an instructor (Adams) and her student (Lee).

existing thread or starting a new one. Keep in mind both the specific information in and the overall tone of the messages posted by others, and monitor your own messages for tone and clarity. For instance, in the exchange in Figure 35.2, the instructor draws inferences from the student's previous post and poses several pointed questions to spark further reflection by the student.

Because tone is difficult to convey in online postings, take care when responding to others, making sure to stay on topic or to announce when you are changing the topic. Avoid jokes, criticism, or any other kinds of comments that might be misinterpreted as rude or mean-spirited. If you have a question about a previous post, raise it respectfully. If you detect a factual error, diplomatically present what you believe to be the correct information. And if someone criticizes you online (an attack referred to as **flaming**), try to remember how difficult tone is to convey and give that person the benefit of the doubt.

Except in the most informal of e-mail correspondence, stay alert to all the conventions of correct English. If you use all lowercase letters, misspell words, or make usage errors, readers of your online writing may come away with a negative impression of you, especially if the writing is business, academic, or professional correspondence. Finally, given that friends, teachers, and professional colleagues can easily access your online writing, take care to establish a professional relationship with the multiple audiences in your online groups and always monitor your privacy settings.

35c   Netiquette and online writing

Netiquette (from the phrase *Internet etiquette*) is a set of social practices developed by Internet users in order to regulate online interactions so that they are always conducted respectfully.

TIPS FOR USING NETIQUETTE IN ONLINE INTERACTIONS

Audience

- Keep in mind the potential audience(s) for your message: those for whom it is intended and others who may read it. If privacy is important, do not use online communication.

- Make the subject line of your message as descriptive as possible so that your reader(s) will immediately recognize the topic.

- Keep your message focused and limit it to one screen, if possible. If you want to attach a text or graphic file, keep its size under 1 MB. Readers' time or bandwidth may be limited, or their mobile device may be subject to an expensive usage plan.

- Before uploading a large file, consider reducing the resolution (and file size) of an image, which rarely affects quality.

- Avoid using fancy fonts and multiple colors unless you are certain that they will appear on your audience's screen.

- Give people adequate time to respond, remembering that they may be away from their computers or may be contemplating what to say.

- Consider the content of your message, making sure that it pertains to the interests and needs of your audience.

- Respect copyright. Never post something written by someone else or pass it off as your own.

Style and Presentation

- Maintain a respectful tone, whether your message is formal or informal.

- Be sure of your facts, especially when you are offering a clarification or a correction.

- Present ideas clearly and logically, using bullets or numbers if doing so will help.

- Pay attention to spelling and grammar. If your message is a formal one, you will certainly want to proofread it (perhaps even in hard copy) and make corrections before sending it out.

- Use emoticons (such as ☺) and abbreviations (such as IMHO for "in my humble opinion" or LOL for "laughing out loud") only when you are sure your audience will understand them and find them appropriate.

- Use all capital letters only when you want to be perceived as SHOUTING.

- Use boldface only if you wish the reader to be able to quickly locate a key item in your message, such as the due date for a report or the name of someone to contact.

- Abusive, critical, or profane language is never appropriate.

Context

- Observe what others say and how they say it before you engage in an online discussion; note what kind of information participants find appropriate to exchange.

- If someone is abusive, ignore that person or change the subject. Do not respond to flaming.

- Tone is difficult to convey online, and thus gentle sarcasm and irony may inadvertently come across as personal attacks.

- Do not use your school's or employer's network for personal business.

Credibility

- Use either your real name or an appropriate online pseudonym to identify yourself to readers. Avoid suggestive or inflammatory pseudonyms.

- Be respectful of others even when you disagree, and be welcoming to new members of an online community.

35d   Composing in an online environment

The Internet offers you the chance to communicate with many different audiences for a variety of purposes. More than an electronic library for information and research, it is also a kind of global marketplace, allowing people all over the world to exchange ideas as well as goods.

As you know, websites are sets of electronic pages, anchored to a home page. Instead of the linear arrangement of print texts (in which arguments, passages, and paragraphs unfold sequentially, from start to finish), websites rely on **hypertext** (electronic text that includes **hyperlinks**, or **links**, to other online text, graphics, and animations) to emphasize arrangement and showcase content. That is, websites are created and delivered with text, graphics, and animations integrated into their content. You are probably accustomed to navigating websites by clicking on hyperlinks. The Memphis Zoo's home page (Figure 35.3) illustrates both the integration of text and graphics on a web page and the use of hyperlinks for navigation within a website.

Courtesy of the Memphis Zoo

Figure 35.3. The Memphis Zoo's website illustrates successful integration of text and graphics and features easy-to-use hyperlinks for navigation.

Another important online tool is the navigation bar that stretches across the top of a web page. The home page of the White House's website (Figure 35.4) features a well-organized navigation bar that includes tabs for current issues on its left side and tabs for more stable information about the administration, the White House and its staff, and the US government on the right side. Each of these tabs redirects users to other web pages with more information. The **arrangement** (the pattern of organization of the ideas, text, and visual elements in a composition) of the site is clear because information is grouped meaningfully. The site is thus easy to use: at the top of each page, below a purpose statement, is a list of main topics, each of which links to more detailed information. Arrangement also involves the balance of visual elements and text. The White House's home page is unified by the use of several shades of blue for accent boxes, links, and headings, and the entire website is given coherence by

Figure 35.4. The navigation bar on the home page of the White House's website provides coherence.

the navigation bar, which appears on every page. Finally, the White House seal, prominent in the middle of the navigation bar, and the American flag, at the left corner of the navigation bar, are visual cues that remind users of the official nature of this site. Visual links—such as the current images that are presented as a slide show on the home page and that link to in-depth blog posts and videos—combine arrangement and **delivery** (the presentation and interaction of visual elements with content).

Your web documents will likely be less elaborate than the White House site. Nevertheless, given the flexible nature of electronic composition, you can be creative when planning, drafting, and revising web documents.

(1) Planning a website

As you develop any web document, including a website, you need to keep all the elements of the rhetorical situation in mind: audience, message, context, and purpose. Considering your audience and purpose, you must decide which ideas or information to emphasize and how to arrange your document to achieve that emphasis. You also need to consider the overall impression you want the document to make. Do you want it to be motivational, informative, entertaining, or analytical? Do you want it to look snazzy, soothing, fun, or serious?

While you are generating the content (with your overall purpose in mind), you need to consider the supplementary links that will help you achieve your purpose. But you do not have to do everything at once; fine-tuning the visual design can wait until the content is in place.

When you are planning a website, you may find it helpful to create a storyboard or other visual representation of the site's organization. You can sketch a plan on paper or in a word-processing file if your site is fairly simple. If you have some time to devote to the planning process, you may want to learn how

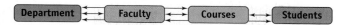

Figure 35.5. Linear pattern for organizing a website. (© 2013 Cengage Learning)

to use a program such as Web Studio or Dreamweaver to help you map out your site.

The possibilities for organizing a website are endless. As starting points, you can consider three basic arrangement patterns—linear, hierarchical, and radial. A linear site (Figure 35.5) is easy to set up, as it is presented with a narrative structure, running from beginning to end. Hierarchical arrangements and radial arrangements are more complex to develop and may be better suited to group projects. The hierarchical arrangement (Figure 35.6) branches out at each level, and the radial arrangement (Figure 35.7) features individual pages that

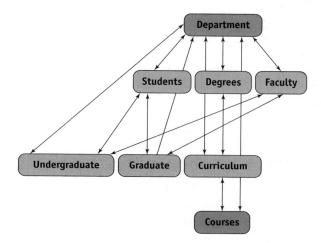

Figure 35.6. Hierarchical pattern for organizing a website. (© 2013 Cengage Learning)

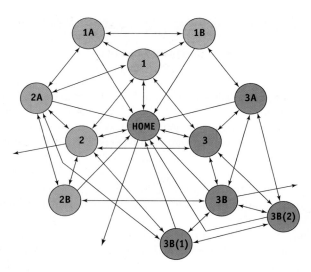

Figure 35.7. Radial pattern for organizing a website. (© 2013 Cengage Learning)

can be linked and viewed in a variety of sequences. Whatever arrangement you choose, keep in mind how your arrangement will affect a user's experience in navigating the website. However you decide to organize your site, be sure to represent each main element in your plan. A good plan will be invaluable to you as you draft text, incorporate visual and multimedia elements, and refine your arrangement.

(2) Creating effective online documents

When you plan and compose a web page or website, you will rely on hypertext. You will also rely on design or visual elements (such as background and color) and links (to the home page as well as to all other pages of the website) to establish consistency and orienting guideposts for your readers. Those design and visual elements and links create important associations among the concepts and ideas in your web document and serve as valuable tools for its development. Your inclusion of links allows the website users (your audience) to read the information in whatever sequence is most productive for each of them. Therefore, you will want to consider how users' different approaches may affect the intended purpose of your document and try to arrange your document accordingly.

Some basic principles can help you use hyperlinks effectively in your web documents.

(a) Enhancing coherence with hyperlinks

The choice and placement of hyperlinks should be a vital part of your organizational plan. A site map, located on the home page, is essential for a large site and helpful for a compact one, as it provides a snapshot of the site's content and arrangement as well as direct access to its various pages. Hyperlinks to the individual pages of a website not only indicate logical divisions of the document but also provide transitions based on key words or ideas. As navigational signposts, hyperlinks serve as powerful rhetorical tools that provide coherence and reflect an effective arrangement.

(b) Taking advantage of the flexibility of hyperlinks

You can use individual words, phrases, or even sentences as textual hyperlinks. Hyperlinks can also be icons or other graphical elements, such as pictures or logos that appropriately reflect the information contained in each link (see Figure 35.8). Remember that you must get permission to use text, graphics, or multimedia elements taken from other sources. Even though such material is often free, its source must be acknowledged (38g).

Internal hyperlinks are those that take the user between pages or sections of the website in which they appear. When

of graphical elements, facilitating the downloading of materials by using low-resolution images (which have smaller file sizes), and avoiding animated graphics. To accommodate users with physical disabilities or different means of accessing web documents (for example, visually impaired users who employ talking computer programs that read web pages), incorporate basic accessibility features such as **alt tags** (descriptive lines of text for each visual image that can be read by screen-reading software). Such accommodations will make your online writing accessible to the greatest number of users.

The following checklist will help you plan a website and develop ideas for each page.

CHECKLIST for Planning and Developing a Website

- What information, ideas, or perspective should a user take away from your site?
- How does the arrangement of your site reflect your overall purpose? How does it assist your intended users in understanding your purpose?
- Ideally, how would a user navigate your website? What are the other options for navigating within your site?
- Should you devote each page to a single main idea or combine several ideas on one page?
- How will you help users return to the home page and find key information quickly?
- What key connections between ideas or pieces of information might be emphasized through the use of hyperlinks?
- Will a user who follows external links be able to get back to your site?

- To ensure that your website has more impact than a paper document, have you used web-specific resources—such as hyperlinks, sound and video clips, and animations—in creating it? How do those multimedia elements help you achieve your purpose?

- Do you need graphics—charts, photos, cartoons, clip art, logos, and so on—to enhance the site so that it will accomplish your purpose? Where should key visual elements be placed to be most effective?

- How often will you update your site?

- How will you solicit feedback for revisions to your site?

- Will your site be accessible to users with slow Internet access and those with physical limitations?

EXERCISE 1

Plan and compile information for a web page that supports a paper you are writing for one of your classes. If you have access to software that converts documents to web pages, start by converting your document. Make adjustments to it based on the criteria in the preceding checklist. Finally, critique your web page.

35e Visual elements and rhetorical purpose

Visual design sends messages to users: an effective design not only invites them to explore a website but also conveys the designer's rhetorical purpose (Chapter 8). All the design elements of an online document, like the tone and style of a printed one, are rhetorical tools that help you achieve your purpose and reach your intended audience.

(1) Basic design principles for easy navigation

A number of basic principles apply to the visual design of web documents.

- **Balance** involves the way in which the design elements used in a document are related to one another spatially. Web pages with a symmetrical arrangement of elements convey a formal, static impression, whereas asymmetrical arrangements are informal and dynamic.
- **Proportion** has to do with the relative sizes of design elements. Large elements attract more attention than small ones and will be perceived as more important.
- **Movement** concerns the way in which our eyes scan a page for information. Most of us look at the upper-left corner of a page first and the lower-right corner last. Therefore, the most important information on a web page should appear in those locations. Vertical or horizontal arrangement of elements on a page implies stability; diagonal and zigzagging arrangements suggest movement.
- **Contrast** between elements can be achieved by varying their focus or size. For instance, a web page about the Siberian Husky might show a photo of one of these dogs in sharp focus against a blurred background; the image of the dog might also be large relative to other elements on the page to enhance contrast. In text, you can emphasize an idea by presenting it in a contrasting font—for example, a playful display font such as Marker Felt Thin or an elegant script font such as *Edwardian Script*. An easy-to-read (sans serif) font such as Arial or Helvetica, however, should be used for most of the text on a web page.
- **Unity** refers to the way all the elements (and pages) of a site combine to give the impression that they are parts of a complete whole. For instance, choose a few colors and fonts to reflect the tone you want to convey, and use them consistently throughout your site. Creating a new design for each page of a website makes the site seem chaotic and thus is ineffective.

(2) Using color and background in online composition

Like the other elements of a web document, color and background are rhetorical tools that can be used to achieve various visual effects (Figure 35.9). Current web standards allow the display of a wide array of colors for backgrounds, text, and frames. You can find thousands of background graphics on the Internet or create them with software.

Designers recommend using no more than three main colors for a document, although you may use varying intensities, or shades, of a color to connect related materials. Besides helping to organize your site, color can have other specific effects.

Courtesy of the International Federation of Organic Agriculture Movements (IFOAM)

Figure 35.9. The use of a consistent color palette on this web page enhances the purpose of the website. The Organic World Foundation promotes organic farming; the visual composition suggests simplicity and purity.

Bright colors, such as red and yellow, are more noticeable and can be used on a web page to emphasize a point or idea. In addition, some colors have associations you may wish to consider. For instance, reds can indicate danger or an emergency, whereas brown shades such as beige and tan suggest a formal atmosphere. Textual hyperlinks usually appear in a color different from that of the surrounding text on a web page so that they are more visible to users. Also, such links generally appear in one color before they are clicked and change to a different color when a user clicks on them. Select colors for textual hyperlinks that fit in with the overall color scheme of your document and help readers navigate between pages on your site.

Background, too, contributes to a successful website. Although a dark background can create a dramatic appearance, it often makes text difficult to read and hyperlinks difficult to see. A dark background can also cause a printout of a web page to be blank or unreadable. If you do use a dark background, be sure that the color of the text is bright enough to be readable on screen and that you provide a version that will print clearly. Similarly, a background with a pattern can be dramatic but can obscure the content of a web page or other online document. If you want to use a pattern for your background, check the readability of the text. You may need to change the color of the text or adjust the background to make the page easier to read.

Use different background colors or patterns for different pages of your online document only if you have a good rhetorical reason for doing so. When you do this, adhere closely to the other design principles in **35e(1)** so that your site appears coherent to your audience.

CHECKLIST for Designing an Online Document

- Have you chosen background and text colors that allow users to print readable copies of your pages if they wish?

- Have you used no more than three colors, perhaps varying the intensity of one or more of them?

- Does a background pattern on your page make the text difficult or easy to read?

- Have you chosen a single, easy-to-read font such as Arial or Helvetica for most of your text? Are the type styles (bold, italic, and so on) used consistently throughout the document?

- Have you used visual elements sparingly? Are any image files larger than 4 or 5 MB, making it likely that they will take a long time to transfer? If so, can you reduce their size using a lower resolution or by cropping?

- Have you indicated important points graphically by using bullets or numbers or visually by dividing the text into short blocks?

- Is any page or section crowded? Can users scan the information on a single screen quickly?

- Does each page include adequate white space for easy reading?

- Have you made sure that all links work?

- Have you identified yourself as the author and noted when the site was created or last revised?

- Have you run a spell checker and proofread the site yourself?

R

RESEARCH and DOCUMENTATION

36 | Finding Sources Online, in Print, and in the Field

When research is part of a writing assignment, students some-times ask, "Why do we have to do research? Why can't we just use our own opinions?" A personal opinion is a good starting point, but research can help you broaden and even challenge your initial perspective. Doing research is a way of joining a larger conversation on a topic—a conversation that often holds surprises.

This chapter will help you

- use the rhetorical situation to frame your research (**36a**),
- find online sources (**36b**),
- find books (**36c**),
- find articles (**36d**), and
- conduct field research (**36e**).

36a | Research and the rhetorical situation

To make the most of the time you spend doing research, think carefully about your rhetorical situation early in the research process.

(1) Identifying a research question
The starting point for any writing project is the rhetorical op-portunity—the issue or problem that has prompted you to write. For research assignments, it is helpful to turn the issue or problem into a question that can guide your research. This question will help you choose relevant articles, books, and other materials. Research questions often arise when you try to relate what you are studying to your own experience. For

instance, you may start wondering about voting regulations while reading about past elections for a history class and, at the same time, noticing news stories about the role technology plays in elections or the unfair practices reported in some states. Each observation, however, may give rise to a different question. Focusing on the influence of technology may prompt you to ask, "What are the possible consequences of having only electronic ballots?" However, if you focus on unfair voting practices, you may ask, "How do voting procedures differ from state to state?" Because you can ask a variety of research questions about any topic, choose the one that interests you the most and also allows you to fulfill the assignment.

To generate research questions, you may find it helpful to ask yourself about causes, consequences, processes, definitions, or values.

Questions about causes

What are the causes of low achievement in our schools?

What causes power outages in large areas of the country?

Questions about consequences

What are the consequences of taking antidepressants for a long time?

How might stronger gun-control laws affect the frequency of public shootings?

Questions about processes

How can music lovers prevent corporations from controlling the development of new music?

How are presidential campaigns funded?

Questions about definitions

How do you know if you are addicted to something?

What is the opportunity gap in the American educational system?

Questions about values

Should the result of a parental DNA test be the deciding factor in a custody case?

Would the construction of wind farms be detrimental to the environment?

If you have trouble coming up with a research question, you may need a jump start. The following tips can help you.

TIPS FOR FINDING A RESEARCH QUESTION

- What rhetorical opportunity (problem or issue) presented in one of your classes would you like to address?
- What have you observed recently (on television, in the newspaper, on campus, or online) that piqued your curiosity?
- What widely discussed local or national problem would you like to help solve?
- Is there anything that you find unusual or intriguing and would like to explore? Consider lifestyles, fashion trends, political views, and current news stories.

EXERCISE 1

Each of the following subjects would need to be narrowed down for a research paper. Compose two questions about each subject that could be answered in a ten-page paper (refer to the list that begins on page 455 for examples of questions).

1. college education
2. water supply
3. environment
4. extreme sports
5. body image
6. social networking

(2) Keeping your purpose and audience in mind

A research paper often has one of the following rhetorical purposes:

- *To inform an audience.* The researcher reports current thinking on a specific topic, including opposing views, without analyzing them or siding with a particular position.

    Example To inform an audience about current nutritional guidelines for children

- *To analyze and synthesize information and then offer possible solutions.* The researcher analyzes a topic and synthesizes the available information about it, looking for points of agreement and disagreement as well as gaps in coverage. After presenting the analysis and synthesis, the researcher sometimes offers possible ways to address any problems found.

    Example To analyze and synthesize various proposals for alternative energy sources

- *To convince or issue an invitation to an audience.* The researcher states a position and supports it with data, statistics, examples, testimony, and/or relevant experience. The researcher's purpose is to persuade or invite readers to take the same position.

    Example To persuade people to support a political candidate

Some writing assignments may require the researcher to achieve all these purposes.

(3) Preparing a working thesis

During the research process, you may find it beneficial to state a **working thesis**—essentially, an answer to your question, which you will test against the research you do. Because you form a working thesis during the early stages of writing, you will need to revise it during later stages. Note how the following research question, working thesis, and final thesis statement differ:

Research question: What is happiness?

Working thesis: Being happy is more than feeling cheerful.

Final thesis statement: Although most people think of cheerfulness when they hear the word *happiness,* they should not exclude contentment and confidence.

Clearly, the final thesis statement is more specific than the working thesis.

(4) Using primary and secondary sources

As you proceed with research, be aware of whether your sources are primary or secondary. **Primary sources** for researching topics in the humanities are generally documents such as archived letters, records, and papers, as well as literary, autobiographical, and philosophical texts. In the social sciences, primary sources may be field observations, case histories, or survey data. In the natural sciences, primary sources may be field observations, measurements, discoveries, or experimental results. **Secondary sources** are commentaries on primary sources. For example, a review of a new novel is a secondary source, as is a discussion of adolescence based on survey data. Experienced researchers usually consult both primary and secondary sources.

36b Finding online sources

You are probably well acquainted with search engines such as Google and Bing. Unlike these search engines, **subject directories** are collections of Internet sources arranged topically. They include categories such as "Arts," "Health," and "Education." Some useful subject directories for academic and professional research are Academic Info, Internet Public Library, and WWW Virtual Library.

Although searching the Internet is a popular research technique, it is not the only technique you should use. You will not find library books or database materials through an Internet search because library and database services are available only to paid subscribers (students fall into this category).

CAUTION

AllTheWeb, AltaVista, Ask, Bing, Google, Yahoo!, and other search engines list both reliable and unreliable sources. Choose only sources that have been written and reviewed by experts (see Chapter 37). Entries in *Wikipedia* include links to useful information, but avoid using *Wikipedia* itself, or any other wiki, as a research source. Because nonexperts can write or alter entries, the information on a wiki is not considered reliable.

(1) Keeping track of online sources

As you click from link to link, you can keep track of your location by looking at the **URL (uniform resource locator)** at the top of the screen. These addresses generally include the server name, domain name, directory (and perhaps subdirectory) name, file name, and file type.

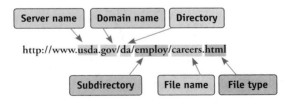

If you find that a URL of a site you are looking for has changed, you may still be able to find the site by dropping the last part of the address and trying again. If this strategy does not work, you can also run a search or look at the links on related websites.

A convenient way to keep track of any useful website you find is to create a **bookmark**—a record of a URL that allows you to go to the website with one click. The bookmarking function of a browser can usually be opened by selecting Bookmarks or Favorites on the brower's main menu bar.

Because sites change and even disappear, scholarly organizations such as the Modern Language Association (Chapter 39) require that bibliographic entries for websites include both the **access date** (the date on which the site was visited) and the **publication date** (the date when the site was published or last modified). When you print out material from the Internet, the access date usually appears at the top or bottom of the printout. The publication date generally appears on the site itself. If a site does not have a publication date, note that it is undated; doing so will establish that you did not accidentally omit a piece of information.

(2) Finding US government documents

If you need information on particular federal laws, court cases, or population statistics, US government documents may be your best sources. You can find these documents by using online databases such as Congressional Universe, MARCIVE, LexisNexis Academic, and STAT-USA. In addition, the following websites may be helpful:

FedStats	www.fedstats.gov
FirstGov	www.firstgov.gov
US Courts	www.uscourts.gov

(3) Finding images

If your rhetorical situation calls for the use of images, the Internet offers you billions from which to choose. However, if an image you choose is copyrighted, you need to contact the author, artist, or designer for permission to use it. Figure 36.1 is an example of an image with a caption and a credit line, which signifies that the image is used with permission. You do not need to obtain permission to use images that are in the public domain (meaning that they are not copyrighted) or those that are cleared for reuse.

Figure 36.1. Genetically modified foods look like naturally produced foods. (Photo © Tom Grill/Getty Images)

Many search engines allow you to search for images. Collections of specific types of images are also available at the following Internet sites:

Advertisements

Ad*Access   scriptorium.lib.duke.edu/adaccess

Adflip   www.adflip.com

Art

Artchive   www.artchive.com

The Web Gallery of Art   www.wga.hu

Clip art

Microsoft Office   microsoft.office.com

Clipart.com   www.clipart.com

Photography

National Geographic   photography.nationalgeographic.com

Smithsonian Images   smithsonianimages.com

36c   Finding books

Three types of books are commonly used in the research process. **Scholarly books** are written by experts to advance knowledge of a certain subject and usually include original research. Before being published, these books are reviewed by scholars in the same field as the author(s). **Trade books** are also written by experts or scholars and often by journalists or freelance writers as well. Authors of trade books write to inform a general audience of research that has been done by others. **Reference books** such as encyclopedias and dictionaries provide factual information in short articles or entries written and reviewed by experts in the field. The audience for these books includes both veteran scholars and those new to a field of study.

(1) Locating sources through an online catalog

The easiest way to find books related to your research question is to consult your library's online catalog, doing either a keyword search or a subject search. To perform a **keyword search**, choose a word or phrase that you think might be found in the title of a book or in notes in the catalog's records. Some online catalogs allow users to be quite specific. Figure 36.2 shows the keyword search page Marianna Suslin used to begin researching her paper on genetically modified foods (**39c**). Notice that

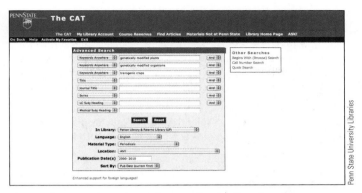

Figure 36.2. Keyword search page from a university library's website.

a user of this online catalog can list a number of related items and specify a library (Penn State has more than one), a language, the type of material desired (periodicals, books, or archival materials, for example), publication date(s), and the way the results should be organized.

By inserting into the keyword search box a word or part of a word followed by asterisks, you can find all sources that include that root, even when suffixes have been added. For example, if you entered *environment***, your search would yield not only sources with *environment* in the title, subject headings, and content notes but also sources with *environments, environmental,* or *environmentalist* in those locations. This search technique is called **truncation**.

The keyword search page at most libraries also allows the user to select a **logical** (or **Boolean**) **operator**—*and, or,* or *not.* These words narrow or broaden an electronic search.

LOGICAL OPERATORS

The words *and, or,* and *not* are the most common logical operators. However, online catalogs and periodical databases have various instructions for using them. If you have trouble following the guidelines presented here, check the instructions for the particular search box you are using.

and   narrows a search (Entering "genetically modified **and** food" returns only those records that contain both keywords.)

or   broadens a search (Entering "genetically modified **or** food" finds all records that contain information about either keyword.)

not   excludes specific items (Entering "genetically modified **and** food **not** humans" excludes any records that mention genetic modification of human beings.)

Once you find the online catalog record for a book you would like to use, write down its **call number**. This number appears on the book itself and indicates where the book is shelved. The online record will reveal the status of the book, letting you know whether it is currently checked out or has been moved to a special collection. To find the book, consult the key to your library's shelving system, usually posted throughout the library. Library staff can also help you find books.

(2) Locating specialized reference books

A specialized encyclopedia or dictionary can often provide background information about people, events, and concepts related to the topic you are researching. To find such sources using a library's search page, enter the type of reference book and one or two keywords identifying your topic. For example, entering "encyclopedia of alcoholism" resulted in the following list of titles:

> *Encyclopedia of Drugs, Alcohol, and Addictive Behavior*
>
> *Encyclopedia of Drugs and Alcohol*
>
> *The Encyclopedia of Alcoholism*

(3) Consulting books not listed in a library's online catalog
If you cannot find a particular book in your school's library, you have several options. Frequently, library websites have links to the catalogs of other libraries. By using such links, you can determine whether another library has the book you want and order it directly from that library or through the interlibrary loan service. In addition, your library may offer access to the database WorldCat, which locates books as well as images, sound recordings, and other materials.

EXERCISE 2

Choose a research question, perhaps one you composed in Exercise 1. Find the titles of a scholarly book, a trade book, and a reference book related to your choice.

36d Finding articles

Because articles offer information that is often more recent than that found in books, they can be crucial to your research. Articles can be found in various **periodicals** (publications that appear at regular intervals). **Scholarly journals** contain reports of original research written by experts for an academic audience. **Professional** (or **trade**) **magazines** feature articles written by staff writers or industry specialists who address on-the-job concerns. **Popular magazines** and **newspapers**, generally written by staff writers, carry a combination of news stories that attempt to be objective and essays that reflect the opinions of editors or guest contributors. The following are examples of the various types of periodicals:

Scholarly journals: *The Journal of Developmental Psychology, The Journal of Business Communication*

Trade magazines: *Farm Journal, Automotive Weekly*

Magazines (news): *Time, Newsweek*

Magazines (public affairs): *The New Yorker, National Review*

Magazines (special interest): *National Geographic, Discover*

Newspapers: *The New York Times, The Washington Post*

A library's online catalog lists the titles of periodicals; however, it does not provide the titles of individual articles within these periodicals. The best strategy for finding print articles is to use an **electronic database**, which is a collection of articles compiled by a company that indexes them according to author, title, date, subject, keywords, and other features. The electronic databases available in libraries are sometimes called **database subscription services**, **licensed databases**, or **aggregated databases**.

A database search will generally yield an **abstract**, a short summary of an article. By scanning the abstract, you can determine whether to locate the complete text of the article, which can often be downloaded and printed. You can access your library's databases by using its computers or, if you have a password, by linking from a remote computer. College libraries subscribe to a wide variety of database services, but the following carry articles on the widest range of topics: EBSCO Academic Search Complete, LexisNexis Academic, JSTOR, and CQ Researcher. You may also be able to search databases that focus on a single field, for example, ERIC for education and PsycINFO for psychology.

Your school's library is likely to provide access to a number of databases. To find those related to your research, go to the library's database search page and enter a word or a part of a word that might be in the titles of relevant databases. You may also find relevant databases by entering a description, category, type, or database vendor. During her research on genetically modified foods, Marianna Suslin inserted "food" in the title box and selected "Agriculture + Biology" as a category (see the boxes at the left on the screen in Figure 36.3).

Most database search pages also allow you to view an alphabetical listing of the various databases available. If you were using such a list of databases to research the status of genetically modified foods in the United States, as Marianna did for her paper (**39c**), you could select Agropedia (agriculture encyclopedias), Consumer Health, or Engineered Materials.

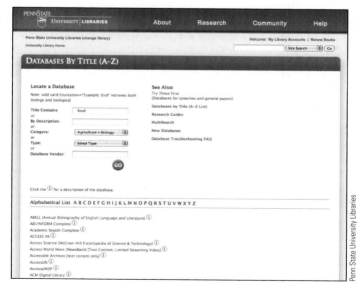

Figure 36.3. A database search page from a university's website.

TIPS FOR CONDUCTING A DATABASE SEARCH

- Identify keywords that clearly represent the topic.
- Determine the databases you want to search.
- Perform your search, using logical operators (36c(1)).
- Refine your search strategy if the first search returns too many or too few citations or (worse) irrelevant ones.
- Download and print the relevant articles or save them to a folder you have created for your research project.
- Be sure that the name of the database is on the printout or electronic copy. If it is not, jot down the name so that you will have it when you prepare your bibliography.

EXERCISE 3

Perform a database search and an Internet search, using the same keywords for each. Print the first screen of the hits (results) you get for each type of search. Compare the two printouts and describe how the results of the two searches differ.

36e   Field research

Although much of your research will be done in a library or online, you may also find it helpful to conduct **field research**—to gather information in a natural setting. Interviews, questionnaires, and observations are the most common methods for such research.

(1) Interviewing an expert

After you have consulted articles, books, or other sources on your topic, you may find that you still have questions that might best be answered by someone who has firsthand experience in the area you are researching—a teacher, government official, business owner, or other person who may be able to provide the information you are seeking. If your assignment calls for an expert, be sure the person you contact has educational or professional credentials in the relevant area.

To arrange an interview, introduce yourself, briefly describe your project, and then explain your reasons for requesting the interview. Most people are busy, so try to accommodate the person you hope to interview by asking him or her to suggest an interview date. If you intend to record your interview, ask for permission ahead of time.

Start preparing your list of questions before the day of the interview. Effective interviews usually contain a blend of open (or broad) questions and focused (or narrow) questions.

Rather than posing a question that elicits just "yes" or "no," reformulate the question so that it begins with *why, when, what, who, where,* or *how.* By doing so, you give your interviewee a chance to elaborate. If you know that the person you are interviewing has published articles or a book on your topic, ask questions that will advance your knowledge, rather than questions that the author has already answered in print.

After the interview, review and expand your notes. If you recorded the interview, transcribe the relevant parts of the recording. The next step is to write extensively about the interview. Ask yourself what you found most important, most surprising, and most puzzling. You will find this writing especially worthwhile when you are able to use portions of it in your final paper.

(2) Using a questionnaire

Whereas an interview elicits information from one person whose name you know, a questionnaire provides information from a number of anonymous people. To be effective, a questionnaire should be short and focused. If the list of questions is too long, people may not be willing to take the time to answer them all. If the questions are not focused on your research topic, you will find it difficult to integrate the results into your paper.

The first four types of questions in the following box are the easiest for respondents to answer. Open questions, which require much more time to answer, should be asked only when the other types of questions cannot elicit the information you want.

EXAMPLES OF TYPES OF SURVEY QUESTIONS

Questions that require a simple yes-or-no answer:
Do you commute to work in a car? (Circle one.)
Yes   No

Multiple-choice questions:
How many people do you commute with? (Circle one.)
0   1   2   3   4

Questions with answers on a checklist:
How long does it take you to commute to work? (Check one.)
___ 0–30 minutes   ___ 30–60 minutes
___ 60–90 minutes   ___ 90–120 minutes

Questions with a ranking scale:
If the car you drive or ride in is not working, which of the following types of transportation do you rely on? (Rank the choices from 1 for most frequently used to 4 for least frequently used.)
___ bus   ___ shuttle van   ___ subway   ___ taxi

Open questions:
What feature of commuting do you find most irritating?

Begin your questionnaire with an introduction stating what the purpose of the questionnaire is, how many questions it contains or approximately how long it should take to complete, and how the results will be used. In the introduction, you should also assure participants that their answers will remain confidential. To protect survey participants' privacy, colleges and universities have **institutional review boards (IRBs)** set up to review questionnaires. Before you distribute your questionnaire, check with the IRB on your campus to make certain that you have followed its guidelines.

If you decide to mail your questionnaire, provide a self-addressed envelope and directions for returning it. It is a good idea to send out twice as many questionnaires as you would like returned because the proportion of responses is generally low. Questionnaires can sometimes be distributed in college dormitories or in classes, but this procedure must be approved by school officials.

Once the questionnaires have been returned, tally the results for all but the open questions on an unused copy. To find patterns in the responses to the open questions, first read through them all; you might find that you can create categories for the responses. For example, the open question "What feature of commuting do you find most irritating?" might elicit answers that fall into such categories as "length of time," "amount of traffic," and "bad weather conditions." By first creating categories, you will find it easier to tally the answers to open questions.

CHECKLIST for Creating a Questionnaire

- Does each question relate directly to the purpose of the survey?
- Are the questions easy to understand?
- Are they designed to elicit short, specific responses?
- Are they designed to collect concrete data that can be analyzed easily?
- Have respondents been given enough space to write their answers to open questions?
- Do you have access to the group you want to survey?
- Have you asked a few classmates to "test-drive" your questionnaire?

37 Evaluating Print and Online Sources

As you find sources that seem to address your research question, you have to evaluate them to determine how, or even whether, you can use them in your paper. This chapter will help you

- assess an author's credibility (**37a**),
- evaluate a publisher's credibility (**37b**),
- evaluate online sources (**37c**), and
- determine the relevance and timeliness of a source (**37d**).

37a Credibility of authors

Credible (or trustworthy) authors present facts accurately, support their opinions with evidence, connect their ideas reasonably, and demonstrate respect for any opposing views. Evaluating the credibility of authors involves determining what their credentials are, what beliefs and values they hold, and how other readers respond to their work.

(1) Evaluating an author's credentials

When evaluating sources, consider whether the authors have credentials that are relevant to the topics they address. Be sure to take into account the credentials of all the authors responsible for the material in the sources you use. Credentials include academic or professional training, publications, and experience. To find information about the credentials of an author whose work you want to use, look

- on the jacket of a book,
- on a separate page near the front or back of the book,

- in the preface of the book,
- in a note at the bottom of the first or last page of an article in print, or
- on a separate page of a periodical or a web page devoted to providing background on contributors.

CHECKLIST for Assessing an Author's Credentials

- Does the author's education or profession relate to the subject of the work?
- With what institutions, organizations, or companies does the author affiliate?
- What awards has the author won?
- What other works has the author produced?
- Do other experts speak of the author as an authority?

---

(2) Examining an author's values and beliefs

An author's values and beliefs underpin his or her research and publications. To determine what these values and beliefs are, consider the author's purpose and intended audience. For example, a lawyer may write an article about malpractice suits to convince patients to sue health providers, a doctor may write a presentation for a medical convention to highlight the frivolous nature of malpractice claims, and a linguist might prepare a conference paper proposing that miscommunication is at the core of malpractice suits.

As you read and use sources, keep in mind that they reflect the views of the authors and often of the audience for which they were written. By identifying the underlying values and beliefs, you can responsibly report the information you retrieve from various sources. When you find source material that suggests economic, political, religious, or social biases, you should feel free to point out such flaws.

CHECKLIST for Determining an Author's Beliefs and Values

- What is the author's educational and professional background?
- What are the author's and publisher's affiliations?
- What is the editorial slant of the organization publishing the author's work?
- Can you detect any signs of bias on the part of the author?
- Is the information purported to be factual? objective? personal?
- Who advertises in the source?
- To what types of websites do any links lead?
- How can you use the source—as fact, opinion, support, authoritative testimony, or material to be refuted?

37b   Credibility of publishers

When doing research, you need to consider not only the credibility of authors but also the credibility of the media through which their work is made available to you. Some publishers hold authors accountable to higher standards than others do. When evaluating books, you can usually assume that publishers associated with universities demand a high standard of scholarship—because the books they publish are reviewed by experts before publication. Books published by commercial (or trade) presses typically do not undergo the same scrutiny. To determine how a trade book has been received by others writing in the same area, you may have to rely on book reviews.

Similarly, journals that carry scholarly articles are considered more credible than magazines that publish articles for a general audience. Authors of journal articles must include both in-text citations and bibliographies so that expert reviewers and other researchers can consult the sources used (Chapters **39** and **40**). Articles that appear in magazines and newspapers may also be reliable, but keep in mind that they are usually written by someone on the periodical's staff—not by an expert in the field. Because magazines and newspapers often discuss research initially published elsewhere, try to find the original source to ensure the accuracy of their reports. Because in-text citations and bibliographies are rarely provided in these periodicals, your best bet for finding the original source is to search databases (**36d**) or the Internet (**36b**).

37c   Evaluation of online sources

If you are evaluating a source from a periodical that you obtained online, you can follow the guidelines for print-based sources (**37a** and **37b**). But if you are evaluating a website, you also need to consider the nature of the site and its sponsor. Although many sites are created by individuals working on their own, many others are sponsored by colleges or universities, professional or nonprofit organizations, and commercial enterprises. The type of sponsor is typically indicated in the site's address, or URL, by a suffix that represents the domain (**36b**). As you evaluate the content of websites, remember that every site has been created to achieve a specific purpose and to address a target audience.

You can find out more about the sponsor of a website by clicking on a navigational link such as About Us or Our Vision. Figure 37.1 shows a page from the website of the American Red Cross, reached by clicking on About Us on the site's home page. In the text on the page, the Red Cross establishes its credibility by explaining its history, its mission, and the scope of its services. It also discloses information about the donations it receives.

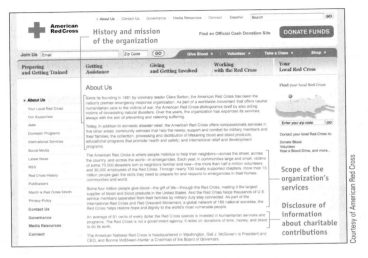

Figure 37.1. The American Red Cross establishes its credibility on its "About Us" page.

EXERCISE 1

Find three websites that have different kinds of sponsors but contain material about a specific subject, such as global warming, energy conservation, or disaster relief efforts. Explain the differences and similarities among the three sites you choose.

37d   Relevance and timeliness of sources

A source is useful only when it is relevant to your research question. Given the ever-growing amount of information available on most topics, you should be prepared to put aside a source that will not help you answer your research question or achieve your rhetorical purpose.

As you conduct research, draft, and revise, you may reject some sources altogether and use only parts of others. Seldom will an entire book, article, or website be useful for a specific research paper. A book's table of contents can lead you to relevant chapters or sections, and its index can lead you to relevant pages. Websites have links that you can click on to locate relevant information. Once you find potentially useful material, read it with your research question and rhetorical purpose in mind.

Useful sources are also up to date. An advanced search, which is an option offered by search engines and database search pages, allows you to select the dates of articles you would like to review. If you are writing about a specific era in the past, you should also consult **contemporary sources**— sources written during that period.

To determine when a source was published, look for the date of publication. In books, it appears with other copyright information on the page following the title page. (See the example on page 529.) Dates of periodicals appear on their covers and frequently at the top or bottom of pages throughout each issue (see page 523). The publication date on a website (the date when the site was published or last modified) frequently appears at the bottom of each screen on the site.

CHECKLIST for Establishing Relevance and Timeliness

- Does the table of contents, index, or directory of the work include key words related to your research question?
- Does the abstract of a journal article contain information on your topic?
- If an abstract is not available, are any of the article's topic sentences relevant to your research question?
- Do the section heads of the source include words connected to your topic?
- On a website, are there links that lead to relevant information?
- Is the work recent enough to provide useful information?
- If you need a source from another time period, is the work from the right period?

EXERCISE 2

Using the questions in the preceding checklist, evaluate the relevance and timeliness of the sources you found for Exercise 1.

38 | Integrating Sources and Avoiding Plagiarism

To use sources effectively, you need to think and write about them critically. To use sources responsibly, you must acknowledge the ideas and words of other writers as you incorporate them into your paper. This chapter will help you

- consider the rhetorical situation for a research paper (**38a**);
- take notes and organize them effectively (**38b**);
- compile a working bibliography or an annotated bibliography (**38c**);
- quote, paraphrase, and summarize sources (**38d**);
- analyze and respond to sources (**38e**);
- synthesize sources (**38f**); and
- avoid plagiarism (**38g**).

38a | The rhetorical situation and the research paper

Although you might think that a research paper is simply a paper that reports research on a topic, it is much more than that. A research paper not only describes what others have discovered but also points out the connections between those discoveries and explains the writer's response to them.

The following introduction to a research article from the *Journal of Film and Video* reveals how the author, Marsha Orgeron, has addressed her rhetorical situation. In the first paragraph, she mentions a problem: a gap in the historical research on the effects of movie magazines on their readers. To help fill this gap is Orgeron's purpose. In the last line of the excerpt, she states that her intention is to determine a specific effect that movie magazines published between 1910 and 1950 had on their readers. Orgeron shows her understanding of the journal's academic audience and context by discussing the work of others, by indicating endnotes with superscript numbers, and by appropriately identifying the work of another researcher she has quoted.

There exists a significant critical literature about motion picture marketing and advertisement, especially concerning the related subject of American movie fan magazines. Much of this scholarship revolves around the gendering of discourse aimed at the fan magazine reader, especially over the course of the 1910s and 20s, and the degree to which these magazines increasingly spoke to women who were confronted with a range of entertainment options and related forms of consumerism.[2] However, there have been few attempts by scholars to account for the ways that the readers of movie magazines both were encouraged to behave and, indeed, responded to this institutionalizing of fan culture. Jane Gaines makes a point akin to this in her 1985 essay "War, Women, and Lipstick": "Our most sophisticated tools of structural analysis can't tell us who read fan magazines, in what spirit or mood, or in what social context. Were they read on magazine stands next to bus stops, in waiting rooms, or under the dryers at beauty parlors? Or maybe they were never read at all, but purchased only for images, to cut up, tack on walls, or paste into scrapbooks" (46).

Where Gaines abandons this quest, casting it aside as an ancillary and perhaps even futile pursuit, I want to investigate one relatively unexplored avenue for understanding how fans both read and responded to movie magazines and the culture they created. Although this article begins somewhat conventionally with a discussion of how fan magazines from Hollywood's heyday (the 1910s through the 40s) were encoded, its ultimate aim is to assess how the magazines shaped their readers' understanding of their own relation to star culture.

—MARSHA ORGERON, "'You Are Invited to Participate':
Interactive Fandom in the Age of the Movie Magazine"

By conducting research and acknowledging sources, you demonstrate that you have

- educated yourself about your topic,
- drawn accurately on the work of others (including diverse points of view),
- understood what you have discovered,
- integrated published research into a paper that is clearly your own, and
- provided all the information readers need to consult the sources you have used.

The rest of this chapter and Chapters 39 and 40 will help you fulfill these responsibilities.

38b Taking and organizing notes

Taking accurate notes and organizing those notes are both critical when you are preparing to write a research paper in which you attribute specific words and ideas to others while adding your own ideas. Some researchers are most comfortable taking notes in notebooks. Others write notes on index cards (three-by-five cards) or type them into computer files—two methods that allow notes to be rearranged easily. Still others like to write notes directly on pages they have photocopied or printed out from an online source. Many researchers rely on web-based research tools, such as Zotero, which allow them not only to collect and organize materials but also to take notes and keep track of bibliographic information. Choose the method that best meets the requirements of your research project and fits your working style. However, your notes will be most useful for drafting a paper if you include the elements identified in the following box.

TIPS ON TAKING NOTES AND ORGANIZING THEM

- **Subject heading.** Use a short descriptive phrase to summarize the content of the note. This phrase will help you retrieve the information later.

- **Type of note.** Indicate whether the note is a quotation (**38d(2)**), a paraphrase (**38d(3)**), a summary (**38d(4)**), or your own thoughts. Place quotations between quotation marks (**16a**). Indicate any changes you made to quoted material with square brackets (**17g**) or ellipsis points (**17h**). If you are using a computer to take notes, you can use a different font color to highlight your own thoughts.

- **Bibliographic information.** Jot down the author's name and/ or the title of the source. If the source has page numbers, indicate which pages your notes refer to. You can provide complete bibliographic information in a working bibliography (**38c**).

- **Computer folders.** If you are using a computer, create a master folder (or directory) for the paper. Within that folder, create separate folders for your notes, drafts, and bibliography. In your notes folder, create files or documents that correspond to each source you use.

Another way to take notes is to use photocopies of articles and excerpts from books or printouts of sources from the Internet. On a printout or a photocopy, you can highlight quotable material and jot down your own ideas in the margins. The example in Figure 38.1 comes from the work Marianna Suslin did for her research paper (**39c**). Make sure to record the bibliographic information if it is not shown on the photocopy or printout. If you have downloaded an article from a database (**36d**) as a PDF file, consider using the commenting feature of Adobe Acrobat or Adobe Reader to make notes.

Genetic tinkering is the process of adding a gene or genes (the transgene) to plant or animal DNA (the recipient genome) to confer a desirable trait, for example, inserting the genes of an arctic flounder into a tomato to give antifreeze properties, or inserting human genes into fish to increase growth rates.

Author defines "genetic engineering"; his use of the word "tinkering" reveals how he feels about the technology.

examples of genetic modification

But, as we are about to discover, this is a technology that no one wants, that no one asked for, and that no one but the biotech companies will benefit from. This is why the biotech lobby has such a vast, ruthless, and well-funded propaganda machine. If they can reinvent our food and slap a patent on it all, they have just created an unimaginably vast new market for themselves.

Author believes no one but big corporations will benefit from this technology.

And to try to convince a suspicious public, they have given us dozens of laudable reasons why the world will benefit from this tinkering. The companies who so enthusiastically produce millions of tons of pesticides every year are now telling us that GMOs will help reduce pesticide use. The companies who have so expertly polluted the world with millions of tons of toxic chemicals are now telling us that GM will help the environment. The companies who have so nonchalantly used child labor in developing countries, and exported dangerous pesticides that are banned in the developed countries to the developing countries, are now telling us that they really do care about people and that we must have GM to feed the world.

Author seeks to discredit biotech companies.

Rees, Andy. *Genetically Modified Food: A Short Guide for the Confused.* Ann Arbor: Pluto, 2006. 8.

Figure 38.1. Photocopied source with notes.

38c Working bibliography and annotated bibliography

A **working bibliography**, or preliminary bibliography, contains information about the materials you think you might use for your research paper. Creating a working bibliography can help you evaluate the quality of your research. If you find that your most recent source is five years old, for example, or that you have relied exclusively on information from magazines or websites, you may need to find some other sources.

Some researchers find it convenient to compile a working bibliography using a web-based research tool or a word processor, which can sort and alphabetize automatically, making it easier to move material directly to the final draft. Others prefer to put each bibliographic entry on a separate index card.

It is a good idea to use the bibliographical format your instructor prefers when compiling your working bibliography. This book covers the most common formats: those preferred by the Modern Language Association (MLA; see Chapter **39**) and the American Psychological Association (APA; see Chapter **40**). The examples given in the rest of this chapter follow the MLA's bibliographical and citation style.

If you are asked to prepare an **annotated bibliography** (also called an **annotated list of works cited**), you should list all your sources alphabetically according to the last name of the author. Then, at the end of each entry, summarize the content of the source in one or two sentences.

Zimmer, Carl. *Soul Made Flesh: The Discovery of the Brain—and How It Changed the World.* New York: Free, 2004. Print. This book is a historical account of how knowledge of the brain developed and influenced ideas about the soul. It covers a span of time and place, beginning four thousand years ago in ancient Egypt and ending in Oxford, England, in the seventeenth century.

38d	Quoting, paraphrasing, and summarizing sources

You can integrate sources into your writing in a number of ways: quoting exact words, paraphrasing sentences, and summarizing longer sections of text or even entire texts. Whenever you borrow others' ideas in these ways, be careful to integrate the material—properly cited—into your own sentences and paragraphs.

(1) Introducing sources

When you borrow textual material, introduce it to readers by establishing the source, usually by providing an auther's name. You may also need to include additional information about the author, especially if he or she is unfamiliar to your audience. For example, in a paper on the origins of literacy, the following statement becomes more credible if it includes the added information about Oliver Sacks's background:

> professor of neurology and psychiatry at Columbia University,
> According to Oliver Sacks, ˄ "[t]he origin of writing and reading cannot be understood as a direct evolutionary adaptation" (27).

Phrases such as *According to Oliver Sacks* and *from the author's perspective* are called **signal phrases** (or *attributive tags*) because they indicate the source from which information was taken. The following box suggests phrases to use when you quote, paraphrase, or summarize information, as well as verbs to use in variations of some of the phrases.

SIGNAL PHRASES FOR QUOTING, PARAPHRASING, AND SUMMARIZING

- According to ___author's name___ , ...
- In ___author's name___ 's view, ...
- In ___title of article or book___ , ___author's name___ states that . . . .
- The author points out that . . . . She [or he] also stresses that . . . .

You can vary the last two of the preceding signal phrases by using one of the following verbs instead of *state, point out, or stress*. For a list of the types of complements that follow such verbs, see **23d(5)**.

admit	conclude	find	propose
advise	deny	imply	reject
argue	disagree	indicate	reply
believe	discuss	insist	report
claim	emphasize	note	suggest
concede	explain	observe	think

Signal phrases often begin a sentence, but they can also appear in the middle or at the end:

According to Jim Cullen, "The American Dream would have no drama or mystique if it were a self-evident falsehood or a scientifically demonstrable principle" (7).

"The American Dream," **claims Jim Cullen,** "would have no drama or mystique if it were a self-evident falsehood or a scientifically demonstrable principle" (7).

"The American Dream would have no drama or mystique if it were a self-evident falsehood or a scientifically demonstrable principle," **asserts Jim Cullen in his book *The American Dream: A Short History of an Idea That Shaped a Nation*** (7).

If you decide to integrate visual elements such as photos or graphs as source material, you must label them as figures and assign them arabic numerals. You can then refer to them within the text of your paper, as in this example: "The markings on the sixspine butterfly fish (*Parachaetodon ocellatus*) resemble those on the wings of some butterflies (see fig. 38.2)." Under the visual element, include both the figure number and a title or caption.

Figure 38.2. Sixspine butterfly fish.

Offscreen/Shutterstock.com

(2) Using direct quotations

Direct quotations draw attention to key passages. Include a direct quotation in a paper only if

- you want to retain the beauty or clarity of someone's words,
- you need to reveal how the reasoning in a specific passage is flawed or insightful, or
- you plan to discuss the implications of the quoted material.

Keep quotations as short as possible and make them an integral part of your text.

Any quotation of another person's words should be placed in quotation marks or, if longer than four lines, set off as an indented block (**39a(2)**). If you need to clarify a quotation by changing it in any way, place square brackets around the added or changed words.

"In this role, he [Robin Williams] successfully conveys a diverse range of emotion."

If you want to omit part of a quotation, replace the deleted words with ellipsis points (**17h**).

"Overseas markets **. . .** are critical to the financial success of Hollywood films."

When modifying a quotation, be sure not to alter its essential meaning.

Each quotation you use should be accompanied by a signal phrase to help readers understand why the quotation is important. A sentence that consists of only a quotation is called a **dropped quotation**. Notice how the signal phrase improves the following passage:

Joel Achenbach recognizes that compromises
~~Compromises~~ must be made to promote safer sources of energy: "To accommodate green energy, the grid needs not only more storage but more high-voltage power lines" (~~Achenbach~~ 137).

Readers will also want to know how a quotation is related to the point you are making. When the connection is not readily apparent, provide an explanation in a sentence or two following the quotation.

Joel Achenbach recognizes that compromises must be made to promote safer sources of energy: "To accommodate green energy, the grid needs not only more storage but more high-voltage power lines" (137). If we are going to use green energy to avoid dependence on types of energy that cause air pollution, we may have to tolerate visual pollution in the form of power lines strung between huge towers.

CHECKLIST for Using Direct Quotations

- Have you copied all the words and punctuation accurately?
- Have you attributed the quotation to a specific source?
- Have you used square brackets around anything you added to or changed in a direct quotation (**17g**)?
- Have you used ellipsis points to indicate anything you omitted (**17h**)?
- Have you included a signal phrase with the quotation?
- Have you included a sentence or two after a quotation to indicate its relevance?
- Have you used quotations sparingly? Rather than using too many quotations, consider paraphrasing or summarizing the information instead.

(3) Paraphrasing another person's ideas

A **paraphrase** is a restatement of someone else's ideas in approximately the same number of words. Paraphrasing allows you to demonstrate that you have understood what you have read; it also enables you to help your audience understand it. Paraphrase when you want to

- clarify difficult material by using simpler language,
- use someone else's idea but not his or her exact words,
- create a consistent tone (**33a(3)**) for your paper as a whole, or
- interact with a point that a source has made.

Your paraphrase should be almost entirely in your own words and should accurately convey the content of the original passage.

(a) Using your own words and sentence structure

As you compare the source below with the paraphrases that follow, note the similarities and differences in both sentence structure and word choice.

Source

Zimmer, Carl. *Soul Made Flesh: The Discovery of the Brain—and How It Changed the World*. New York: Free, 2004. Print.

> The maps that neuroscientists make today are like the early charts of the New World with grotesque coastlines and blank interiors. And what little we do know about how the brain works raises disturbing questions about the nature of our selves. (page 7)

Inadequate paraphrase

The maps used by neuroscientists today resemble the rough maps of the New World. Because we know so little about how the brain works, we must ask questions about the nature of our selves (Zimmer 7).

If you simply change a few words in a passage, you have not adequately restated it. You may be committing plagiarism (**38g**) if the wording of your version follows the original too closely, even if you provide a page reference for the source.

Adequate paraphrase

Carl Zimmer compares today's maps of the brain to the crude maps made of the New World. He believes that the lack of knowledge about the workings of the brain makes us ask serious questions about human nature (7).

In the second paraphrase, both vocabulary and sentence structure differ from those in the original. This paraphrase also includes a signal phrase ("Carl Zimmer compares").

(b) Maintaining accuracy

Any paraphrase must accurately maintain the sense of the original. If you unintentionally misrepresent the original because you did not understand it, you are being *inaccurate*. If you deliberately change the gist of what a source says, you are being *unethical*. Compare the original statement below with the paraphrases.

Source

Hanlon, Michael. "Climate Apocalypse When?" *New Scientist* (17 Nov. 2007): 20.

> Disastrous images of climate change are everywhere. An alarming graphic recently appeared in the UK media showing the British Isles reduced to a scattered archipelago by a 60-metre rise in sea level. Evocative scenes of melting glaciers, all-at-sea polar bears and forest fires are routinely attributed to global warming. And of course Al Gore has just won a Nobel prize for his doomsday flick *An Inconvenient Truth,* starring hurricane Katrina.
>
> . . . There is a big problem here, though it isn't with the science. The evidence that human activities are dramatically modifying the planet's climate is now overwhelming—even to a former paid-up sceptic like me. The consensus is established, the fear real and justified. The problem is that the effects of climate change mostly haven't happened yet, and for journalists and their editors that presents a dilemma. Talking about what the weather may be like in the 2100s, never mind the 3100s, doesn't sell.

Inaccurate or unethical paraphrase

Evocative scenes of melting glaciers, landless polar bears, and forest fires are attributed to global warming in Al Gore's *An Inconvenient Truth.* The trouble is that Gore cannot predict what will happen (Hanlon 20).

Accurate paraphrase

According to Michael Hanlon, the disastrous images of climate change that permeate the media are distorting our understanding of what is actually happening globally and what might happen in the future (20).

Although both paraphrases include a reference to an author and a page number, the first focuses misleadingly on Al Gore, whereas the second paraphrase notes the much broader problem, which can be blamed on the media's focus on selling a story.

(4) Summarizing an idea

Although a summary omits much of the detail used by the writer of the original source, it accurately reflects the essence of that work. In most cases, then, a **summary** reports a writer's main idea (32c) and the most important support given for it.

A summary is shorter than the material it reports. When you summarize, you present just the gist of the author's ideas, without including background information and details. Summaries can include short quotations of key words or phrases, but you must always enclose another writer's exact words in quotation marks when you blend them with your own.

Source

Marshall, Joseph M., III. "Tasunke Witko (His Crazy Horse)." *Native Peoples* (Jan./Feb. 2007): 76-79. Print.

> The world knows him as Crazy Horse, which is not a precise translation of his name from Lakota to English. *Tasunke Witko* means "his crazy horse," or "his horse is crazy." This slight mistranslation of his name seems to reflect the fact that Crazy Horse the man is obscured by Crazy Horse the legendary warrior. He was both, but the fascination with the legendary warrior

hides the reality of the man. And it was as the man, shaped by his family, community and culture—as well as the events in his life—that he became legend.

Summary

The Lakota warrior English speakers refer to as "Crazy Horse" was actually called "his crazy horse." That mistranslation may distort impressions of what Crazy Horse was like as a man.

EXERCISE 1

Find a well-developed paragraph in one of your recent reading assignments. Rewrite it in your own words, varying the sentence structure of the original. Make your paraphrase approximately the same length as the original. Next, write a one-sentence summary of the same paragraph.

38e Analyzing and responding to sources

Though quotations, paraphrases, and summaries are key to academic writing, thinking critically involves more than referring to someone else's work. Quotations, paraphrases, and summaries call for responses. Readers of your papers will want to know what you think about an article, a book, or another source. They will expect you to indicate its strengths and weaknesses and to mention the impact it has had on your own ideas.

Your response to a source will be based on your analysis of it. You can analyze a source in terms of its rhetorical situation (31a), its use of rhetorical appeals (34f), or its reasoning (34h and 34i). You can also evaluate a source by using some common criteria: currency, coverage, and reliability.

(1) Considering the currency of sources

Depending on the nature of your research, the currency of sources may be an important consideration. Using up-to-date sources is crucial when researching most topics. (Historical research may call for sources from a specific period in the past.) When you consider the currency of a source, start by looking for the date of its publication. Then, examine any data reported. Even a source published in the same year that you are doing research may include data that are several years old and thus possibly irrelevant. In the following example, the writer questions the usefulness of an out-of-date statistic mentioned in a source:

According to Jenkins, only 50% of all public schools have web pages (23). However, this statistic is taken from a report published in 1997; a more recent count would likely yield a much higher percentage.

(2) Noting the thoroughness of research

Coverage refers to the comprehensiveness of research. The more comprehensive a study is, the more convincing are its findings. Similarly, the more examples an author provides, the more compelling are his or her conclusions. Claims or opinions that are based on only one instance are often criticized for being merely anecdotal or otherwise unsubstantiated. The writer of the following response suggests that the author of the source in question may have based his conclusion on too little information:

Johnson concludes that middle-school students are expected to complete an inordinate amount of homework given their age, but he bases his conclusion on research conducted in only three schools (90). To be more convincing, Johnson needs to conduct research in more schools, preferably located in different parts of the country.

(3) Checking the reliability of findings

Research, especially when derived from experiments or surveys, must be reliable. Experimental results are considered **reliable** if they can be reproduced by researchers using a similar methodology. Results that cannot be replicated in this way are not reliable because they are supported by only one experiment.

Reliability is also a requirement for reported data. Researchers are expected to report their findings accurately and honestly, not intentionally excluding any information that weakens their conclusions. When studies of the same phenomenon give rise to disputes, researchers should discuss conflicting results or interpretations. The writer of the following response focuses on the problematic nature of her source's methodology:

Jamieson concludes from her experiment that a low-carbohydrate diet can be dangerous for athletes (73), but her methodology suffers from lack of detail. No one would be able to confirm her experimental findings without knowing exactly what and how much the athletes consumed.

Researchers often use certain phrases when responding to sources. The following list presents a few examples.

COMMON PHRASES FOR RESPONDING TO SOURCES

Agreeing with a source

- Recent research confirms that ___author's name___ is correct in asserting that....
- ___Author's name___ aptly notes that....
- I agree with ___author's name___ that....

(continued on page 496)

(continued from page 495)

Disagreeing with a source

- Several of the statements made by ___author's name___ are contradictory. She [or he] asserts that..., but she [or he] also states that....
- In stating that..., ___author's name___ fails to account for....
- I disagree with ___author's name___ on this point. I believe that....

Expressing both agreement and disagreement with a source

- Although I agree with ___author's name___ that ..., I disagree with his [or her] conclusion that....
- In a way, ___author's name___ is correct: .... However, from a different perspective, one can say that....
- Even though ___author's name___ may be right that..., I must point out that....

38f   Synthesizing sources

The word *synthesis* may remind you of the word *thesis*. The two words are, of course, related. A *thesis* is typically defined as a claim, a proposition, an informed opinion, or a point of view; a *synthesis* refers to a combination of claims, propositions, opinions, or points of view. When you synthesize sources, you combine them, looking for similarities, differences, strengths, weaknesses, and so on. Like summarizing and responding, synthesizing is not only a writing skill but also a critical-thinking skill.

In the following excerpt, a writer reports two similar views on the topic of ecotourism.

The claim that ecotourism can benefit local economies is supported by the observations of Ellen Bradley, tour leader in

Cancun, Mexico, and Rachel Collins, county commissioner in Shasta County, California. Bradley insists that ecotourism is responsible for creating jobs and improved standards of living in Mexico (10). Similarly, Collins believes that ecotourism has provided work for people in her county who had formerly been employed by the timber industry (83).

Notice that the writer uses the transitional word *similarly* (33d(3)) to indicate a comparison. In the next excerpt on the topic of voting fraud, a writer contrasts two different views, using the transitional word *although*.

Although Ted Kruger believes voting fraud is not systematic (45), that does not mean there is no fraud at all. Kendra Berg points out that voter rolls are not updated often enough (18), which leaves the door open for cheaters.

In both of the previous examples, the writers not only summarize and respond to sources but synthesize them as well. The box below suggests phrases you can use when synthesizing sources.

COMMON PHRASES FOR SYNTHESIZING SOURCES

- The claim that...is supported by the observations of _author 1's name_ and _author 2's name_ . _Author 1's name_ insists that.... Similarly, _author 1's name_ believes that....

- _Author 1's name_ asserts that.... _author 2's name_ supports this position by arguing that....

- Although _author 1's name_ believes that..., this interpretation is not accepted universally. For example, _author 2's name_ notes that....

- _Author 1's name_ asserts that...; however, he [or she] fails to explain why [or how].... _author 2's name_ points out that....

38g   Avoiding plagiarism

To use the work of other writers responsibly, give credit for all the information you gather through research. Always ensure that your audience can distinguish between the ideas of other writers and your own contributions. It is not necessary, however, to credit information that is **common knowledge**, which includes well-known facts such as the following: "The *Titanic* hit an iceberg and sank on its maiden voyage." This event has been the subject of many books and movies, so some information about it has become common knowledge.

If, however, you are writing a research paper about the *Titanic* and wish to include the ship's specifications, such as its overall length and gross tonnage, you will be providing *un*common knowledge, which must be documented. After you have read about a given subject in a number of sources, you will be able to distinguish between common knowledge and the distinctive ideas or interpretations of specific writers. If you have been scrupulous about identifying your own thoughts while taking notes, you should have little difficulty distinguishing between what you knew to begin with and what you learned through your research.

Taking even part of someone else's work and presenting it as your own leaves you open to criminal charges. In the film, video, music, and software businesses, this sort of theft is called **piracy**. In publishing and education, it is called **plagiarism**. Whatever it is called, it is illegal, and penalties range from failing a paper or a course to being expelled from school. Never compromise your integrity or risk your future by submitting someone else's work as your own.

CAUTION

Although it is fairly easy to copy material from a website or even to purchase a paper on the Internet, it is just as easy for a teacher or employer to locate that same material and determine that it has been plagiarized. Many teachers routinely use Internet search tools such as Google or special services such as Turnitin if they suspect that a student has submitted a paper that was plagiarized.

To review how to draw responsibly on the words and ideas of others, consider the following examples:

Source

McConnell, Patricia B. *The Other End of the Leash*. New York: Ballantine, 2002. 142. Print.

> Status in male chimpanzees is particularly interesting because it is based on the formation of coalitions, in which no single male can achieve and maintain power without a cadre of supporting males.

Paraphrase with documentation

Patricia B. McConnell, an authority on animal training, notes that by forming alliances with other male chimpanzees, a specific male can enjoy status and power (142).

This example includes not only the author's name but also a parenthetical citation, which marks the end of the paraphrase and provides the page number where the original material can be found.

Quotation with documentation

Patricia B. McConnell, an authority on animal training, argues that male chimpanzees achieve status "based on the formation of coalitions, in which no single male can achieve and maintain power without a cadre of supporting males" (142).

Quotation marks show where the copied words begin and end; the number in parentheses indicates the exact page in McConnell's book on which those words appear. Again, the author is identified at the beginning of the sentence. However, the quoted material can instead be completely documented in a parenthetical reference at the end of the sentence:

Male chimpanzees achieve status "based on the formation of coalitions, in which no single male can achieve and maintain power without a cadre of supporting males" (McConnell 142).

If, after referring to the following checklist, you cannot decide whether you need to cite a source, the safest policy is to cite it.

CHECKLIST of Sources That Should Be Cited

- Writings, both published and unpublished
- Opinions and judgments that are not your own
- Statistics and other facts that are not widely known
- Images and graphics, such as works of art, drawings, charts, graphs, tables, photographs, maps, and advertisements
- Personal communications, such as interviews, letters, and e-mail messages
- Public electronic communications, including television and radio broadcasts, motion pictures and videos, sound recordings, websites, and posts to online discussion groups or blogs

EXERCISE 2

After reading the source material, decide which of the quotations and paraphrases that follow it are written correctly and which would be considered problematic. Be prepared to explain your answers.

Source

Despommier, Dickson D. "A Farm on Every Floor." *New York Times.* New York Times, 23 Aug. 2009. Web. 19 July 2010.

If climate change and population growth progress at their current pace, in roughly 50 years farming as we know it will no longer exist. This means that the majority of people could soon be without enough food or water. But there is a solution that is surprisingly within reach: Move most farming into cities, and grow crops in tall, specially constructed buildings. It's called vertical farming.

© 2009. Reprinted by permission.

1. Vertical farming is a way to provide food in the future.

2. Dickson D. Despommier believes that vertical farming may be the solution to the growing demand for food and water.

3. According to Dickson D. Despommier, in fifty years "farming...will no longer exist."

4. Dickson D. Despommier claims that the farming we are accustomed to will no longer be in existence in fifty years.

5. Vertical farming is the use of specially designed city buildings to grow crops (Despommier).

6. "If climate change and population growth progress at their current pace, in roughly 50 years farming as we know it will no longer exist" (Despommier).

39 │ MLA Documentation

The Modern Language Association (MLA) provides guidelines for documenting research in literature, languages, linguistics, history, philosophy, and composition studies. The *MLA Handbook for Writers of Research Papers* is published specifically for college writers. This chapter includes

- guidelines for citing sources within the text of a paper (**39a**),
- guidelines for documenting sources in a works-cited list (**39b**), and
- a sample student paper (**39c**).

39a │ MLA-style in-text citations

(1) Citing material from other sources

The citations you use within the text of a research paper refer your readers to the list of works cited at the end of the paper, tell them where to find the borrowed material in the original source, and indicate the boundaries between your ideas and those you have borrowed. In the following example, the parenthetical citation guides the reader to page 88 of the book by Pollan documented in the works-cited list:

In-text citation

Since the 1980s virtually all the sodas and most of the fruit drinks sold in supermarkets have been sweetened with high-fructose corn syrup (HFCS)—after water, corn sweetener is their principal ingredient (Pollan 88).

Works-cited entry

Pollan, Michael. *The Omnivore's Dilemma: A Natural History of Four Meals.* New

    York: Penguin, 2006. Print.

The MLA suggests reserving numbered notes for supplementary comments—for example, when you wish to explain a point further but the subject matter is tangential to your topic. When numbered notes are used, superscript numbers are inserted in the appropriate places in the text, and the notes are gathered at the end of the paper on a separate page with the heading *Notes* (not italicized). The first line of each note is indented one-half inch. You can create a superscript number in Microsoft Word by typing the number, highlighting it, pulling down the Format menu, clicking on Font, and then clicking in the box next to Superscript.

In-text note number

Most food found in American supermarkets is ultimately derived from corn.[1]

Notes entry

    1. Nearly all farm animals—from cows and chickens to various kinds of

farmed fish—are fed a diet of corn.

An in-text citation usually provides two pieces of information about borrowed material: (1) information that directs the reader to the relevant source on the works-cited list and (2) information that directs the reader to a specific page or section within that source. An author's last name and a page number generally suffice. To create an in-text citation, either place both the author's last name and the page number in parentheses or

MLA

introduce the author's name in the sentence and supply just the page number in parentheses.

A "remarkably narrow biological foundation" supports the variety of America's

supermarkets (Pollan 18).

Pollan explains the way corn products "feed" the familiar meats, beverages, and

dairy products that we find on our supermarket shelves (18).

When referring to information from a range of pages, separate the first and last pages with a hyphen: (34-42). If the page numbers have the same hundreds or thousands digit, do not repeat it when listing the final page in the range: (234-42) or (1350-55) but (290-301) or (1395-1402). If you refer to an entire work or a work with only one page, no page numbers are necessary.

The following examples are representative of the types of in-text citations you might be expected to use. For more details on the placement and punctuation of citations, including those following long quotations, see pages 510–14.

Directory of MLA Parenthetical Citations

1. Work by one author

Set on the frontier and focused on characters who use language sparingly, Westerns often reveal a "pattern of linguistic regression" (Rosowski 170).

OR

Susan J. Rosowski argues that Westerns often reveal a "pattern of linguistic regression" (170).

2. More than one work by the same author(s)

When your works-cited list includes more than one work by the same author(s), provide a shortened title in your in-text citation that identifies the relevant work. Use a comma to separate the name (or names) from the shortened title when both are in parentheses. For example, if you listed two books by Antonio Damasio, *Looking for Spinoza* and *The Feeling of What Happens*, on your works-cited page, then you would cite one of them within your text as follows:

According to one neurological hypothesis, "feelings are the expression of human flourishing or human distress" (Damasio, *Looking* 6).

OR

Antonio Damasio believes that "feelings are the expression of human flourishing or human distress" (*Looking* 6).

3. Work by two or three authors

Some environmentalists seek to protect wilderness areas from further development so that they can both preserve the past and learn from it (Katcher and Wilkins 174).

Use commas to separate the names of three authors: (Bellamy, O'Brien, and Nichols 59).

4. Work by more than three authors

Use either the first author's last name followed by the abbreviation *et al.* (from the Latin *et alii,* meaning "and others") or all the last names. (Do not italicize the abbreviated Latin phrase, which ends with a period.)

In one important study, women graduates complained more frequently about "excessive control than about lack of structure" (Belenky et al. 205).

OR

In one important study, women graduates complained more frequently about "excessive control than about lack of structure" (Belenky, Clinchy, Goldberger, and Tarule 205).

5. Works by different authors with the same last name

When your works-cited list includes works by different authors with the same last name, provide a first initial, along with the last name, in parenthetical citations, or use the author's first and last name in the text. For example, if your works-cited list included entries for works by both Richard Enos and Theresa Enos, you would cite the work of Theresa Enos as follows.

Pre-Aristotelian rhetoric still has an impact today (T. Enos 331-43).

OR

Theresa Enos mentions the considerable contemporary reliance on pre-Aristotelian rhetoric (331-43).

If two authors have the same last name and first initial, spell out each author's first name in a parenthetical citation.

6. Work by a corporate author

A work has a corporate author when individual members of the group that created it are not identified. If the corporate author's name is long, you may use common abbreviations for parts of it—for example, *Assn.* for "Association" and *Natl.* for "National." Do not italicize the abbreviations.

Strawbale constructions are now popular across the nation (Natl. Ecobuilders Group 2).

7. Two or more works in the same citation

When two sources provide similar information or when you combine information from two sources in the same sentence, cite both sources, listing them in alphabetical order by the first author of each one and separating them with a semicolon.

Agricultural scientists believe that crop productivity will be adversely affected by solar dimming (Beck and Watts 90; Harris-Green 153-54).

8. Multivolume work

When you cite material from more than one volume of a multi-volume work, include the volume number (followed by a colon and a space) before the page number.

Katherine Raine claims that "true poetry begins where human personality ends" (2: 247).

You do not need to include the volume number in a paren-thetical citation if your list of works cited includes only one volume of a multivolume work.

9. Anonymous work

The Tehuelche people left their handprints on the walls of a cave, now called Cave of the Hands ("Hands of Time" 124).

Use the title of an anonymous work in place of an author's name. If the title is long, provide a shortened version, begin-ning with the word by which it is alphabetized in the list of works cited. For example, the shortened title for "Chasing Down the Phrasal Verb in the Discourse of Adolescents" is "Chasing Down."

10. Indirect source
If you need to include material that one of your sources quoted from another work because you cannot obtain the original source, use the following format (*qtd.* is the abbreviation for "quoted"):

The critic Susan Hardy Aikens has argued on behalf of what she calls "canonical multiplicity" (qtd. in Mayers 677).

A reader turning to the list of works cited should find a bibliographic entry for Mayers, the source consulted, but not for Aikens.

11. Poetry, drama, and sacred texts
When you refer to poetry, drama, or sacred texts, you should give the numbers of lines, acts, and scenes or of chapters and verses, rather than page numbers. This practice enables readers to consult an edition other than the one you have used. Act, scene, and line numbers (all arabic numerals) are separated by periods with no space before or after them. The MLA suggests that biblical chapters and verses be treated similarly, although some writers prefer to use colons instead of periods in such citations. In all cases, the progression is from larger to smaller units.

The following citation refers to lines of a poem:

Emily Dickinson alludes to her dislike of public appearance in "I'm Nobody! Who Are You?" (5-8).

The following citation shows that the famous "To be, or not to be" soliloquy from *Hamlet* appears in act 3, scene 1, lines 56–89:

In *Hamlet*, Shakespeare presents the most famous soliloquy in the history of the English theater: "To be, or not to be . . ." (3.1.56-89).

Citations of biblical material identify the book of the Bible, the chapter, and the pertinent verses. In the following example, the writer refers to the creation story in Genesis, which begins in chapter 1 with verse 1 and ends in chapter 2 with verse 22:

The Old Testament creation story, told with remarkable economy, culminates in the arrival of Eve (New American Standard Bible, Gen. 1.1-2.22).

Mention in your first citation which version of the Bible you are using; list only book, chapter, and verse in subsequent citations. Note that the names of biblical books are neither italicized nor enclosed in quotation marks.

The *MLA Handbook* provides standard abbreviations for the parts of the Bible, as well as for the works of Shakespeare and Chaucer and certain other literary works.

12. Constitution

When referring to the US Constitution, use the full title in the list of works cited. For in-text citations, use the following common abbreviations:

United States Constitution	US Const.
article	art.
section	sec.

The testimony of two witnesses is needed to convict someone of treason (US Const., art. 3, sec. 3).

13. Online sources

If paragraphs or sections in an online source are numbered, cite the number(s) of the paragraph(s) or section(s) after the abbreviation *par.* (or *pars.* for more than one paragraph) or *sec.* (or *secs.* for more than one section).

Alston describes three types of rubrics for evaluating customer service (pars. 2-15).

Hilton and Merrill provide examples of effective hyperlinks (sec. 1).

PDFs (stable files that can be viewed on and downloaded from the Internet) usually have numbered pages, which you should cite.

If an online source includes no numbers that distinguish one part from another, either indicate an approximate location of the cited passage within the sentence that introduces the material or treat the source as unpaginated in the parenthetical citation, as in the following examples:

Raymond Lucero's *Shopping Online* offers useful advice for consumers who are concerned about transmitting credit-card information over the Internet.

OR

Shopping Online offers useful advice for consumers who are concerned about transmitting credit-card information over the Internet (Lucero).

If an electronic source is only one page long, you may omit the page number in your citation. However, including a page number demonstrates to your readers that you did not unintentionally omit it and gives the citation the proper form.

(2) Guidelines for in-text citations and quotations

(a) Placement of in-text citations

When you acknowledge your use of a source by placing the author's name and a relevant page number in parentheses,

MLA

insert this parenthetical citation directly after the information you used, generally at the end of a sentence but *before* the final punctuation mark (a period, question mark, or exclamation point).

Oceans store almost half the carbon dioxide released by humans into the atmosphere (Wall 28).

However, you may need to place a parenthetical citation earlier in a sentence to indicate that only the first part of the sentence contains borrowed material. Place the citation after the clause containing the material but before a punctuation mark (a comma, semicolon, or colon).

Oceans store almost half the carbon dioxide released by humans into the atmosphere (Wall 28), a fact that provides hope for scientists studying global warming but alarms scientists studying organisms living in the oceans.

If you cite the same source more than once in a paragraph, with no intervening citations of another source, you can place one parenthetical citation at the end of the last sentence in which the source is used: (Wall 28, 32).

(b) Lengthy quotations

When a quotation is more than four lines long, set it off from the surrounding text by indenting all lines one inch from the left margin. Such quotations (sometimes referred to as **block quotations**) are usually introduced by a colon, but other punctuation marks or none at all may be more appropriate. The first line should not be indented more than the others. The right margin should remain the same as it is for the surrounding text. Double-space the entire quotation and do not enclose it in quotation marks.

In *Nickel and Dimed*, Barbara Ehrenreich describes the dire living conditions of the working poor:

> The lunch that consists of Doritos or hot dog rolls, leading to faintness before the end of the shift. The "home" that is also a car or a van. The illness or injury that must be "worked through," with gritted teeth, because there's no sick pay or health insurance and the loss of one day's pay will mean no groceries for the next. These experiences are not part of a sustainable lifestyle, even a lifestyle of chronic deprivation and relentless low-level punishment. They are, by almost any standard of subsistence, emergency situations. And that is how we should see the poverty of millions of low-wage Americans—as a state of emergency. (214)

A problem of this magnitude cannot be fixed simply by raising the minimum wage.

Note that the period precedes the parenthetical citation at the end of an indented (block) quotation. Note, too, how the writer introduces and then comments on the quotation from Ehrenreich.

Rarely will you need to quote more than a paragraph, but if you do, indent the first line of each paragraph an extra quarter of an inch.

(c) Punctuation within citations and quotations

Punctuation marks clarify meaning in quotations and citations. The following list summarizes their common uses:

- A colon separates volume numbers from page numbers in a parenthetical citation.

  (Raine 2: 247)

- A comma separates the author's name from the title when it is necessary to list both in a parenthetical citation.

  (Kingsolver, *Animal Dreams*)

- A comma also indicates that page or line numbers are not sequential.

  (44, 47)

- Ellipsis points indicate an omission within a quotation.

  "They lived in an age of increasing complexity and great hope; we in an age of . . . growing despair" (Krutch 2).

  When an ellipsis indicates that the end of a sentence has been omitted, the final punctuation follows the in-text citation.

  "They lived in an age of increasing complexity and great hope . . ." (Krutch 2).

- A hyphen indicates a continuous sequence of pages or lines.

  (44-47)

- A period separates acts, scenes, and lines of dramatic works.

  (3.1.56)

- A period also distinguishes chapters from verses in biblical citations.

  (Gen. 1.1)

- A question mark placed inside the final quotation marks indicates that the quotation itself is a question. Notice that the period after the parenthetical citation marks the end of the sentence.

  Peter Elbow asks, "What could be more wonderful than the pleasure of creating or appreciating forms that are different, amazing, outlandish, useless—the opposite of ordinary, everyday, pragmatic?" (542).

  When placed outside the final quotation marks, a question mark indicates that the quotation has been incorporated into a question posed by the writer of the paper.

  What does Kabat-Zinn mean when he advises people to practice mindfulness "as if their lives depended on it" (305)?

- Square brackets enclose words that have been added to the quotation as clarification and are not part of the original material.

  "The publication of this novel [*Beloved*] establishes Morrison as one of the most important writers of our time" (Boyle 17).

MLA

39b   MLA list of works cited

All of the works you cite should be listed at the end of your paper, beginning on a separate page that has the heading *Works Cited*. Use the following tips as you prepare your list.

TIPS FOR PREPARING A LIST OF WORKS CITED

- Center the heading *Works Cited* (not italicized) one inch from the top of the page.
- Arrange the list of works alphabetically by the authors' last names.
- If a source has more than one author, alphabetize the entry according to the last name of the first author.
- If you use more than one work by the same author, alphabetize the works by the first major word in each title. For the first entry, provide the author's complete name (last name given first), but substitute three hyphens (---) for the author's name in subsequent entries. If the author is also the first of two or more authors of another work in the list, do not use three hyphens for the author's name but instead write it out in full. A multiple-author entry follows all entries for the first author's works.
- For a work without an author or editor, alphabetize the entry according to the first important word in the title.
- Type the first line of each entry flush with the left margin and indent subsequent lines one-half inch (a hanging indent).
- Double-space equally throughout—between lines of an entry and between entries as well as between the heading *Works Cited* and the first entry.

MLA

Directory of MLA-Style Entries for a Works-Cited List

When writing down source information for your bibliography, be sure to copy the information directly from the source (e.g., the table of contents of a journal or the title page of a book). (See Figure 39.1 on page 523 for an example of a journal's table of contents and Figure 39.2 on page 528 for an example of a book's title page.)

GENERAL DOCUMENTATION GUIDELINES FOR PRINT-BASED SOURCES

Author or Editor

One author. Place the last name before the first, separating them with a comma. Add any middle name or initial after the first name. Use another comma before any abbreviation or number that follows the name. Titles, affiliations, and degrees should be omitted. Indicate the end of this unit of the entry with a period.

Halberstam, David.

Johnston, Mary K.

King, Martin Luther, Jr.

MLA

Two or three authors. List names in the same order used in the original source. The first person's name is inverted (that is, the last name appears first); the others are not. Separate all names with commas, placing the word *and* before the final name.	West, Nigel, and Oleg Tsarev. Green, Bill, Maria Lopez, and Jenny T. Graf.
Four or more authors. List the names of all the authors or provide just the first person's name (inverted) and follow it with the abbreviation *et al.* (for *et alii,* meaning "and others").	Quirk, Randolph, Sidney Greenbaum, Geoffrey Leech, and Jan Svartvik. OR Quirk, Randolph, et al.
Corporate or group author. Omit any initial article (*a, an,* or *the*) from the name.	Institute of Medicine. Department of Justice.
Editor. If an editor or editors are listed instead of an author or authors, include the abbreviation *ed.* for "editor" or *eds.* for "editors."	Espinoza, Toni, ed. Gibb, Susan, and Karen Enochs, eds.

Title

Italicized titles. Italicize the titles of books, magazines, journals, newspapers, films, plays, and screenplays. Capitalize all major words (nouns, pronouns, verbs, adjectives, adverbs, and subordinating conjunctions). Do not use a period after the title of a periodical.	*Hamlet.* *Weird English.* *The Aviator.* *Newsweek*

(continued on page 520)

MLA

Title *(continued from page 519)*

Titles in quotation marks. Use quotation marks to enclose the titles of short works such as journal or magazine articles, short stories, poems, and songs (**16b**).	"Three Days to See." "Selling the Super Bowl." "Generations."
Subtitles. Always include a subtitle if the work has one. Use a colon to separate a main title and a subtitle. However, if the main title ends in a question mark or exclamation point, no colon is used.	*Lost in Translation: Life in a New Language.* "Silence: Learning to Listen."
Titles within titles. When an italicized title includes the title of another work normally italicized, do not italicize the embedded title.	*Essays on* Death of a Salesman. BUT *Death of a Salesman.*
If the embedded title normally requires quotation marks, it should be italicized as well as enclosed in quotation marks.	*Understanding "The Philosophy of Composition" and the Aesthetic of Edgar Allan Poe.* BUT "The Philosophy of Composition."
When a title in quotation marks includes the title of another work normally italicized, retain the italics.	"A Salesman's Reading of *Death of a Salesman.*"
If the embedded title is normally enclosed in quotation marks, use single quotation marks.	"The European Roots of 'The Philosophy of Composition.'"

Publication Data

City of publication. If more than one city is listed on the title page, mention only the first. Place a colon after the name of the city.	Boston: New York:
Publisher's name. Provide a shortened form of the publisher's name, and place a comma after it. To shorten the name of the publisher, use the principal name. For books published by university presses, abbreviate *University* and *Press* without periods or italics.	Knopf (for Alfred A. Knopf) Random (for Random House) Harvard UP (for Harvard University Press)
If two publishers are listed, provide the city of publication and the name of the publisher for each. Use a semicolon to separate the two.	Manchester: Manchester UP; New York: St. Martin's
Publisher's imprint. You will sometimes need to identify both a publisher and an imprint. The imprint is usually listed above the publisher's name on the title page. In a works-cited entry, the imprint is listed first with a hyphen to separate the two names.	Quill-Harper Vintage-Random

(continued on page 522)

MLA

Publication Data *(continued from page 521)*

Copyright date. Although the copyright date may be found on the title page, it is usually found on the next page—the copyright page (see Figure 39.3 on page 529). Place a period after the date.

Medium of publication. Entries for all print publications— books, newspapers, magazines, journals, maps, articles, reviews, editorials, letters to the editor, pamphlets, published dissertations, and so on—must include the medium of publication: *Print.* Do not italicize the medium of publication; follow it with a period.

PRINT PUBLICATIONS

Print Articles

A **journal** is a publication written for a specific discipline or profession. **Magazines** and **newspapers** are written for the general public. You can find most of the information required for a works-cited entry for a journal article in the table of contents for the issue (Figure 39.1) or at the bottom of the first page of the article.

ENGLISH JOURNAL

Name of journal

VOL. 98 NO. 1 SEPTEMBER 2008

Volume number — Issue number — Date of publication

7 **Call for Manuscripts**

11 **From the Editor**
Ken Lindblom

13 **From the Secondary Section**
Meet me in San Antonio
Dianne Watt

16 **In Memoriam: Margaret J. Early**
Ben F. Nelms

102 **Speaking My Mind**
Dear Mr. Shakespeare: Please Stay Out of My Middle School English Class!
Mike Roberts

"ONCE MORE UNTO THE BREACH, DEAR FRIENDS": BEGINNING ANOTHER YEAR

18 ***EJ* in Focus**
Energy and Enthusiasm: Don't Start the School Year without Them
Nancy Mack
An esteemed teacher-scholar offers good-humored advice for staying fresh in the classroom.

26 **Teacher to Teacher**
What Are Some Innovative Assignment Ideas for a Good Start to a New School Year?
Sandi Marinella; Lisa Winkler; Jonathan Tsui

29 **Incorporating Student Choice: Reflective Practice and the Courage to Change**
Katie Dredger
Can a veteran teacher of AP English survive the risk when she opens the class to any student who wants to try it?

36 **My New Teaching Partner? Using the Grammar Checker in Writing Instruction**
Reva Potter and Dorothy Fuller
A teacher and professor collaborate to show that students' grammatical awareness improves when their computer's grammar checker is incorporated into their learning.

Title of article

42 **Everything I Need to Know About Teaching I Learned from Beowulf**
Name of author — **Ruth R. Caillouet**
A veteran teacher uses an enduring classic to give humorous and useful advice for contemporary teaching. Are you ready for the battle?

continued

THE JOURNAL OF THE SECONDARY SECTION OF THE NATIONAL COUNCIL OF TEACHERS OF ENGLISH. PUBLISHED SINCE 1912.

Printed on recycled paper.

Figure 39.1. Table of contents of a journal.

Title of article and name of periodical

Put the article title in quotation marks with a period inside the closing quotation marks. Italicize the name of the periodical, but do not add any punctuation following the name. Capitalize all major words (nouns, pronouns, verbs, adjectives, adverbs, and subordinating conjunctions). Omit the word *A*, *An*, or *The* from the beginning of the name of a periodical.

"Activities to Create Yearlong Momentum." *English Journal*

Volume and issue numbers

In an entry for an article from a journal, provide the volume number. If the issue number is available, put a period after the volume number and add the issue number.

Contemporary Review 194   *Studies in the Literary Imagination* 26.3

Date

For journals, place the year of publication in parentheses after the volume or issue number. For magazines and newspapers, provide the date of issue after the name of the periodical. Note the day first (if provided), followed by the month (abbreviated except for May, June, and July) and year.

Journal	*American Literary History* 20.1-2 (2008)
Magazine	*Economist* 13 Aug. 2005
Newspaper	*Chicago Tribune* 24 July 2002

Page numbers

Use a colon to separate the date from the page number(s). Note all the pages on which the article appears, separating the first and last page with a hyphen: 21-39. If the page numbers have the same hundreds or thousands digit, do not repeat it when listing the final page in the range: 131-42 or 1680-99. Magazine and newspaper articles are often interrupted by advertisements or other articles. If the first part of an article appears on pages 45 through 47 and the rest on pages 92 through 94, give only the first page number followed by a plus sign: 45+.

Medium of publication

Identify the medium of publication, *Print* (not italicized), at the end of the entry.

1. Article in a journal

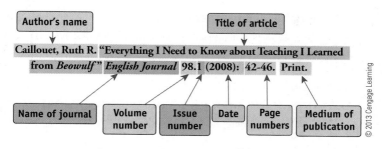

2. Article in a monthly magazine

Keizer, Garret. "How the Devil Falls in Love." *Harper's* Aug. 2002: 43–51. Print.

3. Article in a weekly magazine or newspaper

Chown, Marcus. "Into the Void." *New Scientist* 24 Nov. 2007: 34-37. Print.

4. Article in a daily newspaper

Moberg, David. "The Accidental Environmentalist." *Chicago Tribune* 24 Sept. 2002, final ed., sec. 2: 1+. Print.

When the name of the city is not part of a locally published newspaper's name, it should be given in brackets after the name of the paper: *Star Telegram* [Fort Worth]. If an edition is specified on the newspaper's masthead, include this information (using *ed.* for "edition") after the date. If a specific edition is not identified, put a colon after the date and then provide the page number(s). Specify a section by using the number and/or letter that appears in the newspaper (7, A, A7, or 7A, for example).

5. Unsigned article or wire service article

"View from the Top." *National Geographic* July 2001: 140. Print.

6. Editorial in a newspaper or magazine

Beefs, Anne. "Ending Bias in the Human Rights System." Editorial. *New York Times* 22 May 2002, natl. ed.: A27. Print.

7. Book or film review in a magazine

Denby, David. "Horse Power." Rev. of *Seabiscuit*, dir. Gary Ross. *New Yorker* 4 Aug. 2003: 84–85. Print.

Include the name of the reviewer, the title of the review (if any), the phrase *Rev. of* (for "Review of"), the title of the work being reviewed, and the name of the editor (preceded by the abbreviation *ed.*, not italicized), or the author (preceded by the word *by*) of the book or that of the director (preceded by the abbreviation *dir.*) of the film.

8. Book or film review in a journal

Graham, Catherine. Rev. of *Questionable Activities: The Best*, ed. Judith Rudakoff. *Canadian Theatre Review* 113 (2003): 74–76. Print.

Print Books

9. Book by one author

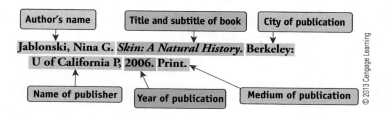

The title page and copyright page of a book (see Figures 39.2 and 39.3) provide the information needed to create a bibliographic entry. Be sure to include the medium of publication at the end of the entry.

10. Book by two authors

West, Nigel, and Oleg Tsarev. *The Crown Jewels: The British Secrets at the Heart of the KGB Archives*. New Haven: Yale UP, 1999. Print.

11. Book by three authors

Spinosa, Charles, Ferdinand Flores, and Hubert L. Dreyfus. *Disclosing New Worlds: Entrepreneurship, Democratic Action, and the Cultivation of Solidarity*. Cambridge: MIT P, 1997. Print.

12. Book by more than three authors

Bullock, Jane A., George D. Haddow, Damon Cappola, Erdem Ergin, Lissa Westerman, and Sarp Yeletaysi. *Introduction to Homeland Security*. Boston: Elsevier, 2005. Print.

OR

Bullock, Jane A., et al. *Introduction to Homeland Security*. Boston: Elsevier, 2005. Print.

MLA

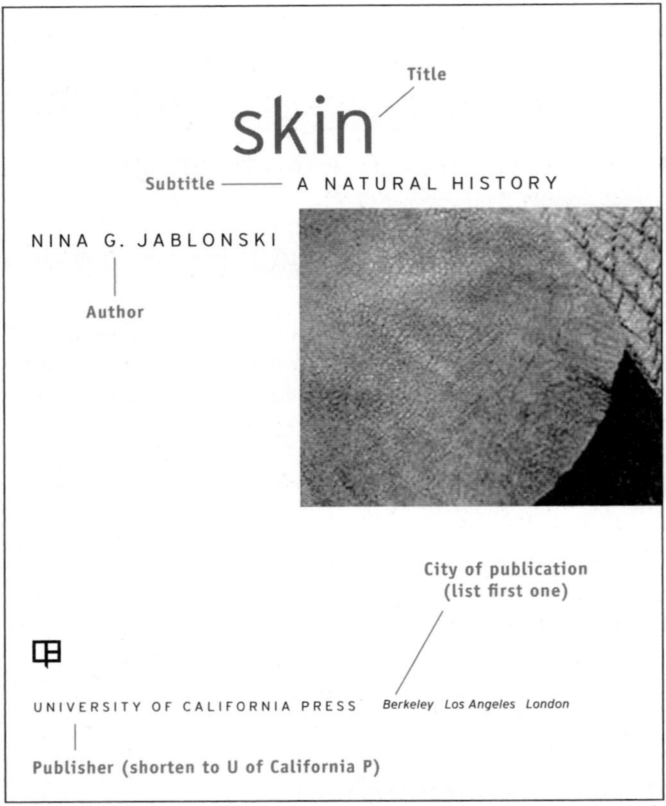

Figure 39.2. A title page includes most, if not all, of the information needed for a bibliographic entry. In this case, the title page omits the publication date.

University of California Press, one of the most distinguished university presses in the United States, enriches lives around the world by advancing scholarship in the humanities, social sciences, and natural sciences. Its activities are supported by the UC Press Foundation and by philanthropic contributions from individuals and institutions. For more information, visit www.ucpress.edu.

University of California Press
Berkeley and Los Angeles, California

University of California Press, Ltd.
London, England

——————————————————— Copyright year

©2006 by Nina G. Jablonski

From SKIN: A Natural History, by Nina G. Jablonski; © 2006 by Nina G. Jablonski. Published by the University of California Press. Used by permission.

Figure 39.3. If the title page does not give the book's date of publication, turn to the copyright page, which is usually the page following the title page.

13. Book by a corporate author

Institute of Medicine. *Blood Banking and Regulation: Procedures, Problems, and Alternatives*. Washington: Natl. Acad., 1996. Print.

14. Book by an anonymous author

Primary Colors: A Novel of Politics. New York: Warner, 1996. Print.

Begin the entry with the title. Do not use *Anonymous* or *Anon.*

15. Book with an author and an editor

Stoker, Bram. *Dracula*. 1897. Ed. Glennis Byron. Peterborough: Broadview, 1998. Print.

Include both the name of the author and the name of the editor (preceded by *Ed.*). The original publication date, followed by a period, can be included after the title.

MLA

16. Book with an editor instead of an author

Kachuba, John B., ed. *How to Write Funny*. Cincinnati: Writer's Digest, 2000. Print.

17. Edition after the first

Murray, Donald. *The Craft of Revision*. 4th ed. Boston: Heinle, 2001. Print.

18. Introduction, preface, foreword, or afterword to a book

Olmos, Edward James. Foreword. *Vietnam Veteranos: Chicanos Recall the War*. By

Lea Ybarra. Austin: U of Texas P, 2004. ix-x. Print.

The name that begins the entry is that of the author of the section of the book, not the author of the entire book. The section author's name is followed by the title of the section (Introduction, Preface, Foreword, or Afterword).

19. Anthology (a collection of works by different authors)

Buranen, Lisa, and Alice M. Roy, eds. *Perspectives on Plagiarism and Intellectual*

Property in a Postmodern World. Albany: State U of New York P, 1999. Print.

Include the name(s) of the editor(s), followed by the abbreviation *ed.* (or *eds.*). (For documenting individual works within an anthology, see items 20–22.)

20. A work originally published in an anthology

Rowe, David. "No Gain, No Game? Media and Sport." *Mass Media and Society*.

Ed. James Curran and Michael Gurevitch. 3rd ed. New York: Oxford UP,

2000. 346-61. Print.

Use this form for an article, essay, story, poem, or play that was published for the first time in the anthology. Place the title of the anthology after the title of the individual work. Provide the name(s) of the editor(s) after the abbreviation *Ed.* for "edited by," and note the edition if it is not the first. List the publication data for the anthology and the range of pages on which the work appears. (See pages 504 and 524–25 for information on inclusive page numbers.)

If you cite more than one work from an anthology, provide only the name(s) of the author(s), the title of the work, the name(s) of the editor(s), and the inclusive page numbers in an entry for each work. Also provide an entry for the entire anthology, in which you include the relevant publication data (see the sample entry for an anthology in item 19).

Clark, Irene L. "Writing Centers and Plagiarism." Buranen and Roy 155-67.

Howard, Rebecca Moore. "The New Abolitionism Comes to Plagiarism."

Buranen and Roy 87-95.

21. A work from a journal reprinted in a textbook or an anthology

Selfe, Cynthia L. "Technology and Literacy: A Story about the Perils of Not Paying

Attention." *College Composition and Communication* 50.3 (1999): 411-37.

Rpt. in *Views from the Center: The CCCC Chairs' Addresses 1977-2005.* Ed.

Duane Roen. Boston: Bedford; Urbana: NCTE, 2006. 323-51. Print.

Use the abbreviation *Rpt.* (not italicized) for "Reprinted." Two cities and publishers are listed in the sample entry because the collection was copublished.

22. A work from an edited collection reprinted in a textbook or an anthology

Brownmiller, Susan. "Let's Put Pornography Back in the Closet." *Take Back the*

Night: Women on Pornography. Ed. Laura Lederer. New York: Morrow, 1980.

252-55. Rpt. in *Conversations: Readings for Writing.* By Jack Selzer. 4th ed.

New York: Allyn, 2000. 578-81. Print.

See item 20 for information on citing more than one work from the same anthology.

MLA

23. Translated book

Garrigues, Eduardo. *West of Babylon*. Trans. Nasario Garcia. Albuquerque: U of

New Mexico P, 2002. Print.

Place the abbreviation *Trans.* (not italicized) for "Translated by" before the translator's name.

24. Republished book

Alcott, Louisa May. *Work: A Story of Experience*. 1873. Harmondsworth: Penguin,

1995. Print.

After the title of the book, provide the original publication date, followed by a period.

25. Multivolume work

Young, Ralph F., ed. *Dissent in America*. 2 vols. New York: Longman-Pearson,

2005. Print.

Cite the total number of volumes in a work when you have used material from more than one volume. If all the volumes were not published in the same year, provide inclusive dates: 1997–99 or 1998–2004. If publication of the work is still in progress, include the words *to date* (not italicized) after the number of volumes. If you have used material from only one volume of a multivolume work, include that volume's number (preceded by the abbreviation *Vol.*) in place of the total number of volumes.

Young, Ralph F., ed. *Dissent in America*. Vol. 1. New York: Longman-Pearson,

2005. Print.

Note that the publisher's name in this entry is hyphenated: the first name is the imprint, and the second is the publisher.

26. Article in a multivolume work

To indicate a specific article in a multivolume work, provide the author's name and the title of the article in quotation marks. Provide the page numbers for the article after the date of publication.

Baxby, Derrick. "Jenner, Edward." *Oxford Dictionary of National Biography*. Ed.

H. C. G. Matthew and Brian Harrison. Vol. 30. Oxford: Oxford UP, 2004.

4-8. Print.

If required by your instructor, include the number of volumes and the inclusive publication dates after the medium of publication: 382-89. Print. 23 vols. 1962-97.

27. Book in a series

Sumner, Colin, ed. *Blackwell Companion to Criminology*. Malden: Blackwell,

2004. Print. Blackwell Companions to Sociology 8.

When citing a book that is part of a series, add the name of the series after the medium of publication. If one is listed, include the number designating the work's place in the series. The series name is not italicized. Abbreviate words in the series name according to the MLA guidelines; for example, the word *Series* is abbreviated *Ser*.

Other Print Texts

28. Encyclopedia entry

Robertson, James I., Jr. "Jackson, Thomas Jonathan." *Encyclopedia of the American*

Civil War: A Political, Social, and Military History. Ed. David S. Heidler and

Jeanne T. Heidler. Santa Barbara: ABC-CLIO, 2000. 1058-66. Print.

When the author of an encyclopedia article is indicated by initials only, check the table of contents for a list of contributors. If an article is anonymous, begin the entry with the article title.

MLA

Page numbers and full publication information are not necessary in an entry for an article from a well-known reference work that is organized alphabetically. After the author's name, the title of the article, and the name of the encyclopedia, provide the edition and/or year of publication, for example, *5th ed. 2004.* or *2002 ed.* (not italicized). Conclude with the medium of publication.

Petersen, William J. "Riverboats and Rivermen." *The Encyclopedia Americana.*

1999 ed. Print.

29. Dictionary entry

When citing a specific dictionary definition for a word, use the abbreviation *Def.* (for "Definition"), and indicate which definition you used if the entry has two or more.

"Reactive." Def. 2a. *Merriam-Webster's Collegiate Dictionary.* 10th ed. 2001. Print.

30. Sacred text

Begin your works-cited entry for a sacred text with the title of the work, rather than information about editors or translators, and, if appropriate, end the entry with the name of the version of that work.

The Bible. Anaheim: Foundation, 1997. Print. New American Standard Bible.

The Qur'an. Trans. Muhammad A. S. Abdel Haleem. Oxford: Oxford UP, 2004.

Print.

31. Government publication

United States. Office of Management and Budget. *A Citizen's Guide to the Federal Budget.* Washington: GPO, 1999. Print.

When citing a government publication, list the name of the government (e.g., United States or Minnesota) and the agency that issued the work. Italicize the title of a book or pamphlet. Indicate the city of publication. Federal publications are usually printed by the Government Printing Office (GPO) in Washington, DC, but be alert for exceptions.

When the name of an author, editor, or compiler appears on a government publication, you can begin the entry with that name, followed by the abbreviation *ed.* or *comp.* if the person is not the author. Alternatively, insert that name after the publication's title and introduce it with the word *By* or the abbreviation *Ed.* or *Comp.* to indicate the person's contribution.

32. Law case

Chavez v. Martinez. 538 US 760. Supreme Court of the US. 2003. *United States Reports*. Washington: GPO, 2004. Print.

Include the last name of the first plaintiff, the abbreviation *v.* for "versus," the last name of the first defendant, data on the law report (volume, abbreviated name, and page or reference number), the name of the deciding court, the year of the decision, and appropriate publication information for the medium consulted. Although names of law cases are italicized in the text of a paper, they are *not* italicized in works-cited entries.

33. Public law

No Child Left Behind Act of 2001. Pub. L. 107-10. 115 Stat. 1425-2094. 8 Jan. 2002. Print.

Include the name of the act, its Public Law number (preceded by the abbreviation *Pub. L.*), its Statutes at Large volume number and page numbers (separated by the abbreviation *Stat.*), the date it was enacted, and the medium of publication.

Although no works-cited entry is needed for familiar sources such as the US Constitution, an in-text citation should still be included (see page 509).

MLA

34. Pamphlet or bulletin

Stucco in Residential Construction. St. Paul: Lath & Plaster Bureau, 2000. Print.

If the pamphlet has an author, begin with the author's name, as you would in an entry for a book.

35. Published dissertation

Fukuda, Kay Louise. *Differing Perceptions and Constructions of the Meaning*

of Assessment in Education. Diss. Ohio State U, 2001. Ann Arbor: UMI,

2002. Print.

After the title of the dissertation, include the abbreviation *Diss.,* the name of the university granting the degree, the date of completion, and the publication information. In the example, *UMI* stands for "University Microfilms International," which publishes many dissertations. If a dissertation was published by its author, use *privately published* (not italicized) instead of a publisher's name.

36. Published letter

In general, treat a published letter like a work in an anthology, adding the date of the letter and the number (if the editor assigned one).

Jackson, Helen Hunt. "To Thomas Bailey Aldrich." 4 May 1883. *The Indian*

Reform Letters of Helen Hunt Jackson, 1879-1885. Ed. Valerie Sherer Mathes.

Norman: U of Oklahoma P, 1998. 258-59. Print.

Print Cartoons, Maps, and Other Visuals

37. Cartoon or comic strip

Cheney, Tom. Cartoon. *New Yorker* 9 June 2003: 93. Print.

Trudeau, Garry. "Doonesbury." Comic strip. *Daily Record* [Ellensburg] 21 Apr.

2005: A4. Print.

After the creator's name, place the title of the work (if given) in quotation marks and include the descriptor *Cartoon* or *Comic strip*.

38. Map or chart

Cincinnati and Vicinity. Map. Chicago: Rand, 2008. Print.

Include the title and the appropriate descriptor, *Map* or *Chart*.

39. Advertisement

Nu by Yves Saint Laurent. Advertisement. *Allure* June 2003: 40. Print.

The name of the product and/or that of the company being advertised is followed by the designation *Advertisement*.

ONLINE PUBLICATIONS

Many of the MLA guidelines for documenting online sources are similar to those for print sources. For sources you find online, provide electronic publication information and access information. (However, if you conduct all your research online, you may want to consult *The Columbia Guide to Online Style*, which offers formatting guidelines, sample in-text citations, and sample bibliographic entries.)

Electronic publication information

Indicate the author's name, the title of the work, the title of the website (and version or edition used), the site's sponsoring organization (usually found at the bottom of the site's home page; see Figure 39.4), the date of publication, and the medium of publication (*Web*). All of this information precedes the access information.

Figure 39.4. A page on the website of the Centers for Disease Control and Prevention indicates the name of the sponsoring organization and the date of publication (most recent update).

Access information

When you document an online source, you must include the date of access: the day, month, and year on which you consulted the source. Either keep track of the date of access or print out the source so that you have a record (see Figure 39.5).

You are not required to include the URL if your readers can easily locate the online source by searching for the author's

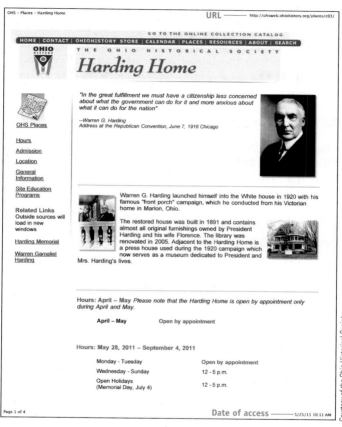

Figure 39.5. When you print a page from a website, the URL and the date of access usually appear at the top or bottom of the page.

name and the title of the work. For cases in which your readers cannot easily locate a source, you should provide the complete URL (between angle brackets), including the protocol (*http, ftp, telnet,* or *news*). When a URL does not fit on a single line, break it only after a slash or a double slash. Make sure that the URL is accurate. Take care to distinguish between uppercase and lowercase letters and to include hyphens and underscores. The URL follows the date of access, appearing after a period and a space. The closing angle bracket should also be followed by a period.

Online Articles

The following formats apply to articles available only online. For articles available through online databases, see pages 542–44. If you need to include a URL, follow the instructions that begin on page 539.

40. Scholarly journal article

Harnack, Andrea, and Gene Kleppinger. "Beyond the *MLA Handbook*:

Documenting Sources on the Internet." *Kairos* 1.2 (1996): n. pag. Web. 14

Aug. 1997.

If no page numbers are provided for an online journal, write *n. pag.* (for "no pagination"). If page numbers are provided, place them after the publication date and a colon. The entry ends with the date of access.

41. Popular magazine article

Plotz, David. "The Cure for Sinophobia." *Slate.com.* Newsweek Interactive, 4

June 1999. Web. 15 June 1999.

42. Newspaper article

"Tornadoes Touch Down in S. Illinois." *New York Times.* New York Times,

16 Apr. 1998. Web. 20 May 1998.

When no author is identified, begin with the title of the article. If the article is an editorial, include *Editorial* (not italicized)

after the title: "America's Promises." Editorial. (In the sample entry, the first mention of *New York Times* is the title of the website, and the second, which is not italicized, is the name of the site's sponsor.)

Online Books

43. Book available only online

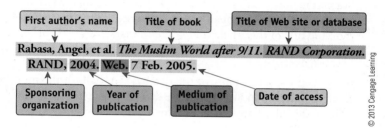

Rabasa, Angel, et al. *The Muslim World after 9/11. RAND Corporation.* RAND, 2004. Web. 7 Feb. 2005.

Because there are more than three authors, the abbreviation *et al.* has been used in the example entry, but listing all names is also acceptable: Rabasa, Angel, Cheryl Benard, Peter Chalk, C. Christine Fair, Theodore W. Karasik, Rollie Lal, Ian O. Lesser, and David E. Thaler.

44. Book available online and in print

Rohrbough, Malcolm J. *Days of Gold: The California Gold Rush and the American Nation.* Berkeley: U of California P, 1997. *History E-book Project.* Web. 17 Feb. 2005.

Begin the citation with print citation information: author's name, title of the work, city of publication, publisher, and date. Follow this information with the title of the database or website (italicized) where the book was accessed, the medium of publication (*Web*), and the date of access.

45. Part of an online book

Strunk, William, Jr. "Elementary Rules of Usage." *The Elements of Style*. Ithaca:

Humphrey, 1918. n. pag. *Bartleby.com*. Web. 6 June 2003.

Online Databases

Many print materials are available online through databases such as JSTOR, Project MUSE, ERIC, PsycINFO, Academic Search Premier, LexisNexis, ProQuest, InfoTrac, and Silver Platter. To cite material from an online database, begin with the author, the title of the article (in quotation marks), the title of the publication (in italics), the volume and issue numbers, the year of publication, and the page numbers (or the abbreviation *n. pag.*). Then add the name of the database (in italics), the medium of publication (*Web*), and the date of access. You can find most of the information you need for a works-cited entry for an article on the abstract page from the database (see Figure 39.6).

46. ERIC

Taylor, Steven J. "Caught in the Continuum: A Critical Analysis of the Principle

of the Least Restrictive Environment." *Research and Practice for Persons with*

Severe Disabilities 29.4 (2004): 218-30. *ERIC*. Web. 3 Mar. 2009.

47. Academic Search Premier

Folks, Jeffrey J. "Crowd and Self: William Faulkner's Sources of Agency in *The*

Sound and the Fury." *Southern Literary Journal* 34.2 (2002): 301. *Academic*

Search Premier. Web. 6 June 2003.

For sources that list only the page number on which a work begins, include that number and a plus sign.

48. LexisNexis

Suggs, Welch. "A Hard Year in College Sports." *Chronicle of Higher Education* 19

Dec. 2003: 37. *LexisNexis*. Web. 17 July 2004.

Name of database

Title of article

Author

Name of journal

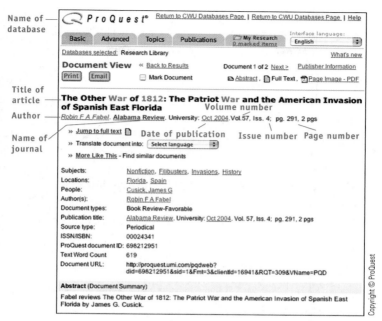

Volume number

Date of publication   Issue number   Page number

Figure 39.6. Abstract page from an online subscription database.

49. ProQuest

Fabel, Robin F. A. "The Other War of 1812: The Patriot War and the American
Invasion of Spanish East Florida." *Alabama Review* 57.4 (2004): 291-92.
ProQuest. Web. 8 Mar. 2005.

50. InfoTrac

Priest, Ann-Marie. "Between Being and Nothingness: The 'Astonishing Precipice'
of Virginia Woolf's *Night and Day*." *Journal of Modern Literature* 26.2 (2002-03):
66-80. *InfoTrac*. Web. 12 Jan. 2004.

51. JSTOR

Blum, Susan D. "Five Approaches to Explaining 'Truth' and 'Deception' in
Human Communication." *Journal of Anthropological Research* 61.3 (2005):
289-315. *JSTOR*. Web. 3 Mar. 2009.

52. Project MUSE

Muñoz, Alejandro Anaya. "Transnational and Domestic Processes in the
Definition of Human Rights Policies in Mexico." *Human Rights Quarterly*
31.1 (2009): 35-58. *Project MUSE*. Web. 3 Mar. 2009.

53. Encyclopedia entry from a subscription database

Turk, Austin T. "Terrorism." *Encyclopedia of Crime and Justice*. Ed. Joshua
Dressler. 2nd ed. Vol. 4. New York: Macmillan Reference USA, 2002. *Gale
Virtual Reference Library*. Web. 7 Feb. 2005.

54. Abstract from a subscription database

Landers, Susan J. "FDA Panel Findings Intensify Struggles with Prescribing of
Antidepressants." *American Medical News* 47.37 (2004): 1-2. *ProQuest Direct*.
Web. 7 Feb. 2005.

Online Communications and Web Sites

55. Web site

McGann, Jerome, ed. *The Complete Writings and Pictures of Dante Gabriel
Rossetti*. Inst. for Advanced Technology in the Humanities, U of Virginia, n.d.
Web. 16 Mar. 2009.

Include the name of the author, editor, or compiler, followed
by the title of the site (italicized), the version or edition (if
given), the publisher or sponsor (if not available, use *N.p.*), the

date of publication (if not available, use *n.d.*), the medium of publication (*Web*), and the date of access.

56. Web site with incomplete information

Breastcancer.org. N.p., 2 Feb. 2008. Web. 5 Feb. 2008.

If a site does not provide all the information usually included in a works-cited entry, list as much as is available.

57. Section of a Web site

Altman, Andrew. "Civil Rights." *Stanford Encyclopedia of Philosophy*. Ed. Edward

  N. Zalta. Center for the Study of Lang. and Information, Stanford U, 3 Feb.

  2003. Web. 12 June 2003.

Mozart, Wolfgang Amadeus. "Concerto No. 3 for Horn, K. 447." *Essentials of*

  Music. Sony Music Entertainment, 2001. Web. 3 Mar. 2009.

58. Personal home page

Gladwell, Malcolm. Home page. N.p., 8 Mar. 2005. Web. 2 Mar. 2009.

After the name of the site's creator, provide the title or include the words *Home page* (not italicized).

59. E-mail

Peters, Barbara. "Scholarships for Women." Message to Rita Martinez. 10 Mar.

  2003. E-mail.

The entry begins with the name of the person who created the e-mail. Put the subject line of the e-mail message in quotation marks. The recipient of the message is identified after the words *Message to*. If the message was sent to you, use *the author* rather than your name. The date of the message and the medium complete the citation.

60. Discussion group or forum

Schipper, William. "Re: Quirk and Wrenn Grammar." *Ansaxnet*. N.p., 5 Jan.

1995. Web. 12 Sept. 1996.

Provide the name of the forum (in this case, *Ansaxnet*) between the title of the work and the sponsor (use *N.p.* if no sponsor is identified). If the posting is untitled, note the genre (for example, *Online posting*) in place of the title.

61. Newsgroup

May, Michaela. "Questions about RYAs." *Generation X.* N.p., 19 June 1996.

Web. 29 June 1996.

The name of the newsgroup (for example, *Generation X*) takes the place of the title of the website.

62. Blog

Cuthbertson, Peter. "Are Left and Right Still Alright?" *Conservative Commentary*.

N.p., 7 Feb. 2005. Web. 18 Feb. 2005.

Other Online Documents
63. Online encyclopedia entry

"Iran." *Encyclopaedia Britannica Online*. Encyclopaedia Britannica, 2002. Web. 6

Mar. 2004.

64. Online congressional document

United States. Cong. Senate. Special Committee on Aging. *Global*

Aging: Opportunity or Threat for the U.S. Economy? 108th Cong.,

1st sess. S. Hrg. 108-30. Washington: GPO, 2003. *GPO Access*. Web.

7 Jan. 2005.

Provide the number and session of Congress and the type and number of publication. (*S* stands for "Senate"; *H* or *HR* stands for "House of Representatives.")

Bills	S 41, HR 82
Reports	S. Rept. 14, H. Rept. 18
Hearings	S. Hrg. 23, H. Hrg. 25
Resolutions	S. Res. 32, H. Res. 52
Documents	S. Doc. 213, H. Doc. 123

65. Online document from a government office

United States. Dept. of State. Bureau of Democracy, Human Rights, and Labor.

Guatemala Country Report on Human Rights Practices for 1998. Feb. 1999.

Web. 1 May 1999.

Begin with the name of the country, state, or city whose government is responsible for the document and the name of the department or agency that issued it. If a subdivision of the larger organization is responsible, also name the subdivision. If an author is identified, provide his or her name, preceded by the word *By*, between the title and the date of issue of the document.

66. Online law case

Tennessee v. Lane. 541 US 509. Supreme Court of the US. 2004. *Supreme Court*

Collection. Legal Information Inst., Cornell U Law School, n.d. Web. 28 Jan.

2005.

MLA

67. Online public law

Individuals with Disabilities Education Act. Pub. L. 105-17. 104 Stat. 587-698.

> *Thomas.* Lib. of Cong., 4 June 1997. Web. 29 Jan. 2005.

Thomas is an online government resource that makes federal legislative information available to the public.

68. Online sacred text

Sama Veda. Trans. Ralph T. H. Griffith. 1895. *Sacred-Texts.com.* Ed. John B.

> Hare. N.p., 2008. Web. 6 Mar. 2008.

Online Recordings and Images

69. Online music

Moran, Jason. "Jump Up." *Same Mother.* Blue Note, 2005. *Blue Note.* Blue Note

> Records. Web. 7 Mar. 2005.

In this entry, the first mention of "Blue Note" identifies the manufacturer of the CD, *Blue Note* is the title of the website where the song was accessed, and "Blue Note Records" identifies the sponsor of that site.

70. Online speech

Malcolm X. "The Ballot or the Bullet." Detroit. 12 Apr. 1964. *American Rhetoric:*

> *Top One Hundred Speeches.* Ed. Michael E. Eidenmuller. N.p., 2005. Web. 14

> Jan. 2005.

In this entry, "12 Apr. 1964" identifies the date the speech was originally delivered, "2005" specifies the year of the speech's electronic publication, and "14 Jan. 2005" gives the date of access.

71. Online video

Riefenstahl, Leni, dir. *Triumph of the Will*. Reichsparteitag-Film, 1935. *Movieflix.*

  com. MovieFlix, 2005. Web. 17 Feb. 2005.

In this entry, "1935" specifies the year in which the movie was originally released, "2005" identifies the year in which it was made available online, and "17 Feb. 2005" gives the date of access. An entry like this one can begin with either the name of the director or the title of the work, depending on the emphasis of the discussion of the work.

72. Online television or radio program

"Religion and the American Election." Narr. Tony Hassan. *The Religion Report*.

  ABC Radio National, 3 Nov. 2004. Web. 18 Feb. 2005.

73. Online interview

McLaughlin, John. Interview by Wolf Blitzer. *CNN.com*. Cable News Network,

  14 July 2004. Web. 21 Dec. 2004.

74. Online work of art

Vermeer, Johannes. *Young Woman with a Water Pitcher*. c. 1660. Metropolitan

  Museum of Art, New York. *The Metropolitan Museum of Art*. Web. 2 Oct. 2002.

75. Online photograph

Marmon, Lee. *Engine Rock*. 1985. *Lee Marmon Gallery*. Web. 9 Feb. 2009.

76. Online map or chart

"Virginia 1624." Map. *Map Collections 1544-1996*. Lib. of Cong. Web. 26 Apr. 1999.

United States. Dept. of Health and Human Services. Centers for Disease Control

  and Prevention. "Daily Cigarette Smoking among High School Seniors."

  Chart. 27 Jan. 2005. *National Center for Health Statistics*. Web. 25 Feb. 2005.

77. Online advertisement

Adflip LLC. "Got Milk?" Advertisement. *Adflip.com.* May 2001. Web. 16 Feb. 2005.

78. Online cartoon or comic strip

Cagle, Daryl. "Social Security Pays 3 to 2." Cartoon. *Slate.com.* Newsweek
    Interactive, 4 Feb. 2005. Web. 5 Feb. 2005.

OTHER COMMON SOURCES
Live and Recorded Performances
79. Play performance

Proof. By David Auburn. Dir. Daniel Sullivan. Walter Kerr Theater, New York.
    8 Oct. 2002. Performance.

Cite the date of the performance you attended.

80. Lecture or presentation

Guinier, Lani. Barbara Jordan Lecture Ser. Schwab Auditorium, Pennsylvania
    State U, University Park. 4 Oct. 2004. Address.

Scharnhorst, Gary. English 296.003. Dane Smith Hall, U of New Mexico,
    Albuquerque. 30 Apr. 2008. Class lecture.

Identify the site and the date of the lecture or presentation.
Use the title if available; otherwise, provide a descriptive label.

81. Interview

Furstenheim, Ursula. Personal interview. 16 Jan. 2003.

Sugo, Misuzu. Telephone interview. 20 Feb. 2003.

For an interview you conducted, give only the name of the
person you interviewed, the type of interview, and the date of

the interview. If the interview was conducted by someone else, add the name of the interviewer, a title or a descriptive label, and the name of the source.

Harryhausen, Ray. Interview by Terry Gross. *Fresh Air*. Natl. Public Radio.

    WHYY, Philadelphia. 6 Jan. 2003. Radio.

82. Film

My Big Fat Greek Wedding. Dir. Joel Zwick. IFC, 2002. Film.

The name of the company that produced or distributed the film (IFC, in this case) appears before the year of release. It is not necessary to cite the city in which the production or distribution company is based.

When you want to highlight the contribution of a specific person, list the contributor's name first. Other supplementary information may be included after the title.

Gomez, Ian, perf. *My Big Fat Greek Wedding*. Screenplay by Nia Vardalos. Dir.

    Joel Zwick. IFC, 2002. Film.

83. Radio or television program

When referring to a specific episode, place quotation marks around its title. Italicize the title of the program.

"'Barbarian' Forces." *Ancient Warriors*. Narr. Colgate Salsbury. Dir. Phil Grabsky.

    Learning Channel. 1 Jan. 1996. Television.

To highlight a specific contributor or contributors, begin the entry with the name or names and note the nature of the contribution.

Abumrad, Jad, and Robert Krulwich, narrs. "Choice." *Radiolab*. New York Public

    Radio. WNYC, New York, 14 Nov. 2008. Radio.

MLA

Works of Visual Art

84. Painting

Gauguin, Paul. *Ancestors of Tehamana*. 1893. Oil on canvas. Art Inst. of Chicago,

Chicago.

Identify the artist's name, the title of the work (italicized), the date of composition (if known; otherwise, write *N.d.*), the medium of composition, the organization or individual holding the work, and the city in which the work is located. For a photograph or reproduction of a work of art, provide the preceding information followed by complete publication information for the source, including medium of publication.

85. Photograph

Marmon, Lee. *White Man's Moccasins*. 1954. Photograph. Native American

Cultural Center, Albuquerque.

An entry for a photograph parallels one for a painting; see item 84.

Digital Sources

86. CD-ROM

"About *Richard III*." *Cinemania 96*. Redmond: Microsoft, 1996. CD-ROM.

Indicate which part of the CD-ROM you are using, and then provide the title of the CD-ROM. Begin the entry with the name of the author if one has been provided.

Jordan, June. "Moving towards Home." *Database of Twentieth-Century African*

American Poetry on CD-ROM. Alexandria: Chadwyck-Healey, 1999.

CD-ROM.

87. Work from a periodically published database on CD-ROM

Parachini, John V. *Combating Terrorism: The 9/11 Commission Recommendations and the National Strategies.* CD-ROM. *RAND Electronically Distributed Documents.* RAND. 2004. Disc 8.

If the work was issued in print at the same time, the print publication information appears before the CD-ROM information.

88. DVD

A River Runs through It. Screenplay by Richard Friedenberg. Dir. Robert Redford. 1992. Columbia, 1999. DVD.

Cite relevant information about the title and director as you would for a film. Provide both the original release date of the film and the release date for the DVD. If the company that originally produced the film did not release the DVD, list the company that released the DVD instead.

89. Sound recording on CD

Franklin, Aretha. *Amazing Grace: The Complete Recordings.* Atlantic, 1999. CD.

For a sound recording on another medium, identify the type (*Audiocassette* or *LP*).

Raitt, Bonnie. *Nick of Time.* Capitol, 1989. Audiocassette.

When citing a recording of a specific song, begin with the name of the performer or composer (depending on your emphasis) and place the song title in quotation marks. Identify the composer(s) or performer after the song title.

MLA

If the performance is a reissue from an earlier recording, provide the original date of recording (preceded by *Rec.* for "Recorded").

Horne, Lena. "The Man I Love." By George Gershwin and Ira Gershwin. Rec. 15

Dec. 1941. *Stormy Weather*. BMG, 1990. CD.

39c   MLA-style research paper

(1) Title page

The MLA recommends omitting a title page and instead providing the title of the paper and your name and other pertinent information on the first page of the paper (see page 557). If your instructor requires a title page but does not supply specific instructions for one, include the title of the paper, your name, the instructor's name, the course title with its section number, and the date—all centered on the page. A sample title page is shown in Figure 39.7 (on page 556). (If your instructor requires a title page, omit the heading on the first page of your paper.) Some instructors require a final outline with a paper, which serves as a table of contents. If you are asked to include an outline, prepare a title page as well.

(2) Sample paper

Interested in the controversy surrounding genetically modified foods, Marianna Suslin explores both sides of the debate as she comes to her conclusion. As you study her paper, notice how she develops her thesis statement, considers more than one point of view, and observes the conventions for an MLA-style paper.

TIPS FOR PREPARING AN MLA-STYLE PAPER

- Number all pages (including the first one) with an arabic numeral in the upper-right corner, one-half inch from the top. Put your last name before the page number.

- On the left side of the first page, one inch from the top, type a heading that includes your name, the name of your professor, the course number, and the date of submission.

- Double-space between the heading and the title of your paper, which should be centered on the page. If your title consists of two or more lines, double-space them and center each line.

- Double-space between your title and the first line of text.

- Indent the first paragraph, and every subsequent paragraph, one-half inch.

- Double-space throughout.

MLA

Genetically Modified Foods and Developing Countries

Marianna Suslin

Professor Squier
Sociology 299, Section 1
27 November 2007

Figure 39.7. Sample title page for an MLA-style paper.

One inch

One-half inch

Marianna Suslin

Professor Squier

Sociology 299, Section 1

27 November 2007

A header consisting of writer's name, instructor's name, course title, and date is aligned at the left side.

The writer's last name and the page number appear as the running head on each page.

Genetically Modified Foods and Developing Countries

Center the title.

Genetic engineering first appeared in the 1960s. Since then,

Double-space throughout.

thousands of genetically modified plants, also referred to as

"genetically modified organisms" (GMOs) and "transgenic crops,"

have been introduced to global markets. Those who argue for

Use one-inch margins on all sides of the page.

continued support of genetic modification claim that the crops

have higher yield, grow in harsher conditions, and benefit the

ecology. Some experts even argue that genetic engineering has the

potential to benefit poor farmers in developing countries, given

that genetically modified plants increase the production of food,

thereby alleviating world hunger. Despite these claims, the practice

of genetic engineering—of inserting genetic material into the DNA

of a plant—continues to be controversial, with no clear answers

as to whether genetically engineered foods can be the answer for

developing countries, as proponents insist.

The last sentence in the first paragraph is the thesis statement.

One of the most important potential benefits of the technology

The second paragraph provides background information.

to both proponents and opponents of genetic engineering is its

potential to improve the economies of developing countries.

MLA

One-inch bottom margin

Suslin 2

Direct quotations are used as evidence.

According to Sakiko Fukuda-Parr, "Investing in agricultural technology increasingly turns up these days on the lists of the top ten practical actions the rich world could take to contribute to reducing global poverty" (3). Agriculture is the source of income for the world's poorest—70 percent of those living on less than a dollar a day support themselves through agriculture. These farmers could benefit greatly from higher yielding crops that could grow in nutrient-poor soil. Genetic modification "has shown how high-yielding varieties developed at international centers can be adapted to local conditions, dramatically increasing yields and farm incomes" (Fukuda-Parr 3).

Indent each paragraph one-half inch.

The writer describes some advantages of growing genetically modified crops.

Theoretically, genetic engineering can bring about an increase in farm productivity that would give people in developing countries the chance to enter the global market on better terms. Developing countries are often resource poor and thus have little more than labor to contribute to the world economy. Farming tends to be subsistence level as farmers can grow only enough on the land—which tends to be nutrient poor—to feed themselves. But the higher yield of genetically modified crops along with the resistance to pests and ability to thrive in nutrient-poor soil can enable the farmers to produce more crops, improve the economy, and give their countries something more to contribute globally by exporting extra crops not

MLA

Suslin 3

needed for subsistence (Fukuda-Parr 1). Genetic modification can
also help poor farmers by delaying the ripening process. If fruits
and vegetables don't ripen as quickly, the farmer is able to store
the crops longer and thus have more time in which to sell the crops
without fear of spoilage. Small-scale farmers often "suffer heavy
losses because of uncontrolled ripening and spoiling of fruits and
vegetables" (Royal Society et al. 238).

Today, eighteen percent of people living in developing
countries do not have enough food to meet their needs (Royal
Society et al. 235). "Malnutrition plays a significant role in half
of the nearly 12 million deaths each year of children under five
in developing countries" (UNICEF, qtd. in Royal Society et al.
235). Genetically modified foods that produce large yields even
in nutrient-poor soils could potentially help to feed the world's
increasing population. Moreover, scientists are working on ways
to make the genetically modified foods more nutritious than
unmodified crops, which would feed larger numbers of people with
less food while, at the same time, combating malnutrition. The
modification of the composition of food crops has already been
achieved in some species to increase the amount of protein, starch,
fats, or vitamins. For example, a genetically modified rice
has already been created, one that "exhibits an increased production

A work by an
organization is
cited.

of beta-carotene," which is a precursor to vitamin A (Royal Society et al. 240). Because vitamin A deficiencies are common in developing countries and contribute to half a million children becoming partially or totally blind each year, advances in genetic engineering offer hope for millions of people who live with nutrient deficiencies (Royal Society et al. 239).

Proponents of genetic engineering have also argued that genetically modified crops have the potential to decrease the amount of damage modern farming technologies inflict on ecology, thereby improving the economy of developing countries without the ecological damage many developed countries have suffered. For example, genetically modified plants with resistance to certain insects would decrease the amount of pesticides that farmers have to use. Genes for insect resistance have already been introduced into cotton, making possible a huge decrease in insecticide use (Royal Society et al. 238). A decrease in the amount of pesticides used is good from an ecological perspective.[1] Not only can pesticides be washed into streams and be harmful to wildlife, but they have also been known to appear in groundwater, thus potentially causing harm to humans.

Scientists have argued that genetic engineering is only the latest step in the human involvement in plant modification that has been going on for thousands of years.[2] Since the dawn of the agricultural

A superscript number indicates an endnote.

MLA

Suslin 5

revolution, people have been breeding plants for desirable traits
and thus altering the genetic makeup of plant populations. The key
advantage of genetic engineering over traditional plant breeding
is that genetic engineering produces plants with the desirable trait
much faster (Fukuda-Parr 5).

Many benefits may come from genetic engineering for farmers
in developing countries and even in the United States, but many
people remain skeptical about this new technology. Research
shows that many Americans are uneasy about consuming foods
that have been genetically enhanced. That same research points out
potential risks of consuming GMOs, which some believe outweigh
the benefits (Brossard, Shanahan, and Nesbitt 10). Considering the
risks of genetically modified foods, people in developing countries
are likely to feel the same way: that the risks outweigh the benefits.
No matter how many potential benefits genetically modified crops
may bring, if they are not safe for consumption, they will hurt the
economies of developing countries.

In "Genetically Modified Food Threatens Human Health,"
Jeffrey Smith argues that inserting foreign genetic material into
food is extremely dangerous because it may create unknown toxins
or allergens. Smith argues that soy allergies increased significantly
after genetically modified soybean plants were introduced in the

The writer describes the disadvantages of eating genetically modified foods.

MLA

Suslin 6

United Kingdom (103). Smith also points to the fact that gene
insertion could damage a plant's DNA in unpredictable ways. For
example, when scientists were working with the soybean plant,
the process of inserting the foreign gene damaged a section of
the plant's own DNA, "scrambling its genetic code" (105). The

A direct quotation of a phrase from a cited work is integrated into the text.

sequence of the gene that was inserted had inexplicably rearranged
itself over time. The protein the gene creates as a result of this
rearrangement is likely to be different, and since this new protein
has not been evaluated for safety, it could be harmful or toxic (105).

   In *Genetically Modified Food: A Short Guide for the Confused*,
Andy Rees argues a similar point: genetically modified foods carry
unpredictable health risks. As an example, he cites the 1989 incident
in which bacteria genetically modified to produce large amounts
of the food supplement L-tryptophan "yielded impressively toxic
contaminants that killed 37 people, partially paralyzed 1,500 and
temporarily disabled 5,000 in the US" (75). Rees also argues that
genetically modified foods can have possible carcinogenic effects.
He states that "given the huge complexity of genetic coding, even
in very simple organisms such as bacteria, no one can possibly
predict the overall, long-term effects of GM [genetically modified]
foods on the health of those who eat them" (78). Rees cites a 1999
study on male rats fed genetically modified potatoes to illustrate the

MLA

Suslin 7

possible carcinogenic effect. The study found that the genetically
modified potatoes had "a powerful effect on the lining of the gut
(stomach, small bowel, and colon)" leading to a proliferation of
cells. According to histopathologist Stanley Ewen, this proliferation
of cells caused by genetically modified foods is then likely to "act
on any polyp present in the colon ... and drastically accelerate the
development of cancer in susceptible persons" (qtd. in Rees 78).

Three ellipsis points mark an omission in quoted material.

In addition to the health risks involved in consuming
genetically modified foods, some experts also argue that such foods
will not benefit farmers in developing countries but will aid big
corporations here in the United States. Brian Halweil, author of
"The Emperor's New Crops," brings up the fact that global sales for
genetically modified crops grew from seventy-five million dollars
in 1995 to one and a half billion dollars in 1998, which is a twenty-
fold increase. Genetically modified crops are obviously lucrative
for large companies. In addition, of the fifty-six transgenic products
approved for commercial planting in 1998, thirty-three belong to
just four corporations (Halweil 256).

The writer cites statistical evidence.

The spread of genetic engineering can change power relations
between nations (Cook 3). The big American corporations that sell
genetically modified seeds can hold power over the governments of
developing countries, hindering their further economic development.

MLA

Suslin 10

reasons (49). Rees also argues that genetically modified crops have
not increased farmers' incomes, regardless of what proponents
of genetic engineering may claim. He points to a 2003 study by
Professor Caroline Saunders at Lincoln University, New Zealand,
which found that "GM food releases have not benefited producers
anywhere in the world" and that "the soil association's 2002 'Seeds
of Doubt' report, created with feedback from farmers and data from
six years of commercial farming in North America, shows that GM
soy and maize crops deliver less income to farmers (on average)
than non-GM crops" (50-51). The potential benefit of genetically
modified crops thus remains uncertain.

The writer's
conclusion is
based on her
own insights
as well as
on research
reported on
the previous
pages.

While proponents of genetic engineering insist that genetically
modified crops can increase yield and help feed the hungry,
opponents point to health risks and challenge the research that
appears to prove that genetically modified foods are beneficial.
However, even if these foods do prove to be as beneficial as
proponents claim, there is nothing to ensure that this technology
will benefit poor farmers in developing countries. Since large
corporations hold patents on all genetically modified seeds, poor
farmers may not have access to these seeds. Therefore, it is far
from certain whether this new technology will benefit developing
nations in the dramatic way its proponents assert.

One inch

Suslin 11

Notes

1. There is some concern, however, about the long-term effects of crops genetically engineered for pest resistance. Since these plants are engineered to continually produce a form of the pesticide used to combat the best problem, insects are constantly exposed to the chemical used to kill them. Such exposure increases the likelihood that the insects will develop a tolerance for this chemical, making the pesticide ineffective.

2. The main difference between genetic engineering and the breeding of plants for desired traits that people have practiced for thousands of years is that genetic engineering actually alters the DNA of a particular plant. Traditional breeding cannot alter the DNA of an individual plant but instead seeks to increases the number of plants that have a trait that occurs naturally. While the end product of both genetic engineering and selective breeding is similar in that both produce plants with desirable traits, the actual processes are radically different.

One inch

Suslin 12

Center the
heading.

Works Cited

Alphabetize
the entries
according to
the authors'
last names.

Brossard, Dominique, James Shanahan, and T. Clint Nesbitt,

eds. *The Public, the Media, and Agricultural Biotechnology.*

Cambridge: CABI, 2007. Print.

Cook, Guy. *Genetically Modified Language: The Discourse of*

Arguments for GM Crops and Food. New York: Routledge,

Indent the
second and
subsequent
lines of
each entry
one-half
inch.

2005. Print.

Easton, Thomas A., ed. *Taking Sides: Clashing Views on*

Controversial Environmental Issues. 11th ed. Dubuque:

McGraw, 2005. Print.

Federal Register 54.104 (1992): 22991. Print.

Fukuda-Parr, Sakiko, ed. *The Gene Revolution: GM Crops and*

Unequal Development. London, England: Earthscan, 2007. Print.

Halweil, Brian. "The Emperor's New Crops." Easton 249-59.

Huffman, W. E. "Production, Identity Preservation, and Labeling in

a Marketplace with Genetically Modified and Non-Genetically

Modified Foods." *Plant Physiology* 134 (2004): 3-10. Web.

5 Nov. 2007.

Newell, Peter. "Corporate Power and 'Bounded Autonomy' in the

Global Politics of Biotechnology." *The International Politics*

of Genetically Modified Food: Diplomacy, Trade, and Law. Ed.

Robert Falkner. Hampshire: Palgrave, 2007. 67-84. Print.

MLA

Suslin 13

Rees, Andy. *Genetically Modified Food: A Short Guide for the Confused.* Ann Arbor: Pluto, 2006. Print.

Royal Society et al. "Transgenic Plants and World Agriculture." Easton 234-45.

Smith, Jeffrey M. "Genetically Modified Food Threatens Human Health." *Humanity's Future.* Ed. Louise I. Gerdes. Detroit: Gale, 2006. 103-08. Print.

Weirich, Paul, ed. *Labeling Genetically Modified Food: The Philosophical and Legal Debate.* New York: Oxford UP, 2007. Print.

40 APA Documentation

The American Psychological Association (APA) publishes a style guide entitled *Publication Manual of the American Psychological Association.* Its documentation system (called an *author-date system*) is used for work in psychology and many other disciplines, including education, economics, sociology, and business management. Updates to the style guide are provided at www.apastyle.org. This chapter presents

- guidelines for citing sources within the text of a paper (**40a**),
- guidelines for documenting sources in a reference list (**40b**), and
- a sample student paper (**40c**).

40a APA-style in-text citations

(1) Citing material from other sources

APA-style in-text citations usually include just the last name(s) of the author(s) of the work and the year of publication. However, be sure to specify the page number(s) for any quotations you use in your paper. The abbreviation *p.* (for "page") or *pp.* (for "pages") precedes the number(s). If you do not know the author's name, use a shortened version of the source's title instead. If your readers want to find more information about a source, they will look for the author's name or, in its absence, the title of the work in the bibliography at the end of your paper.

You will likely consult a variety of sources for a research paper. The following examples are representative of the types of in-text citations you can expect to use.

APA

Directory of APA-Style Parenthetical Citations

1. Work by one author

Yang (2006) admits that speech, when examined closely, is a "remarkably messy means of communication" (p. 13).

OR

When examined closely, speech is "a remarkably messy means of communication" (Yang, 2006, p. 13).

Use commas within a parenthetical citation to separate the author's name from the date and the date from the page number(s). Include a page number or numbers only when you are quoting directly from the source.

2. Work by two authors

Waldron and Dieser conclude from their interview data that the media are responsible for shaping student perceptions of health and fitness (2010).

OR

The media are greatly responsible for shaping student perceptions of health and fitness (Waldron & Dieser, 2010).

APA

When the authors' names are in parentheses, use an ampersand (&) to separate them.

3. Work by more than two authors

Students have reported the benefits of talking with a teacher about their academic progress (Komarraju, Musulkin, & Bhattacharya, 2010).

For works with three, four, or five authors, cite all the authors the first time the work is referred to, but in subsequent references give only the last name of the first author followed by *et al.* (which means "and others" and is not italicized).

According to Komarraju et al. (2010), when students find a teacher accessible, they "are more likely to report being confident of their academic skills and being motivated, both intrinsically and extrinsically" (p. 339).

For works with six or more authors, provide only the last name of the first author followed by *et al.* in the first and subsequent citations.

4. Anonymous work

Use a shortened version of the title to identify an anonymous work.

Chronic insomnia often requires medical intervention ("Sleep," 2009).

This citation refers to an article listed in the bibliography as "Sleep disorders: Standard methods of treatment."

If the word *Anonymous* is used in the source itself to designate the author, it appears in place of an author's name.

The documents could damage the governor's reputation (Anonymous, 2009).

5. Two or more works by different authors in the same parenthetical citation

Smokers frequently underestimate the long-term effects of smoking (O'Conner, 2005; Polson & Truss, 2007).

Arrange the citations in alphabetical order, using a semicolon to separate them.

6. Two or more works by the same author in the same parenthetical citation

The amygdala is active when a person experiences fear or anger (Carey, 2001, 2002).

Jameson (2007a, 2007b) has proposed an anxiety index for use by counselors.

Order the publication dates of works by the same author from earliest to most recent; however, if two or more works have the same publication date, distinguish the dates with lowercase letters (*a, b, c,* and so on) assigned according to the order in which the entries for the works are listed in your bibliography (see page 578).

7. Personal communication

State educational outcomes are often interpreted differently by teachers in the same school (J. K. Jurgensen, personal communication, May 4, 2009).

Personal communications include letters, memos, e-mail messages, interviews, and telephone conversations. These sources are cited in the text only; they do not appear in the reference list.

APA

8. Indirect source

Korovessis (2002, p. 63) points out Toqueville's description of the "strange
melancholy" exhibited by citizens living amid abundance.

Toqueville (as cited in Korovessis, 2002, p. 63) observed the "strange melancholy"
exhibited by citizens living amid abundance.

In the reference list, include a bibliographic entry for the
source you read, not for the original source. (Use an indirect
source only when you are unable to obtain the original.)

9. Electronic source

Cite an electronic source such as an online newspaper or a
website according to the guidelines already mentioned. If there
is no date, use the abbreviation *n.d.* If no page numbers are
provided in a source, give the number of the paragraph containing the words you are quoting, preceded by the abbreviation *para.*

Researchers believe that athletes should warm up before exercising, but
according to Kolata (2010), "what's missing is evidence showing actual
effects on performance" (para. 18).

If the source is divided into sections, use the section heading and the number of the paragraph following that heading:
(Methods, para. 2).

(2) Guidelines for in-text citations and quotations

(a) Placement of in-text citations

According to APA guidelines, there are two ways to cite a
source: one focuses on the researcher and the other on the researcher's findings. If you focus on the researcher, use that per-

son's name in the sentence and place the publication date of the source in parentheses directly after the name.

Diaz (2011) reported that all-night cram sessions do not improve performance.

When making subsequent references to a researcher within the same paragraph, you do not need to repeat the date.

If you decide to focus on the researcher's findings, place the researcher's name and the date (separated by a comma) in parentheses at the end of the sentence.

All-night cram sessions do not improve performance (Diaz, 2011).

Parenthetical citations that include a researcher's name must also include a publication date.

(b) Punctuation for citations and quotations

Quotations should be incorporated into the text when they include fewer than forty words. Use double quotation marks to enclose the quotation. Then cite the source and page number in parentheses. Use *p.* for "page" and *pp.* for "pages." Place a period after the last parenthesis.

According to recent research on the effects of birth order, "laterborns are 1.5 times more likely than first-borns to engage in such activities, including football, soccer, rugby, bobsledding, and skydiving" (Sulloway & Zweigenhaft, 2010, p. 412).

OR

Sulloway and Zweigenhaft (2010) report that "laterborns are 1.5 times more likely than first-borns to engage in such activities, including football, soccer, rugby, bobsledding, and skydiving" (p. 412).

If a quotation has forty or more words, format it as a block quotation. Because a block quotation is set off from the rest of the text, quotation marks are not used. Notice that the parenthetical citation is placed at the end of the paragraph *after* the end punctuation (in this case, a period).

Sulloway and Zweigenhaft (2010) report the effect of birth order on individuals' decisions to take more risks in sports:

> Data on 700 brothers whose major league careers ended by 2008, and who collectively played in more than 300,000 baseball games, reveal significantly heterogeneous results for birth order and its relationship to specific abilities in baseball, including skill, power, self-restraint, and risk taking. As predicted, younger brothers were more likely to engage in the risky business of stealing bases; they attempted more steals per game, and they were more likely to succeed in doing so. (p. 412)

(3) Headings

In research reports, headings set off sections and subsections. The APA specifies five levels of headings:

- Level 1 headings are centered and boldfaced, with each major word capitalized:

Method

- Level 2 headings are flush with the left margin and boldfaced, with each major word capitalized:

Materials and Procedure

- Level 3 headings are boldfaced, begin on a paragraph indent, have only the first word capitalized, and end with a period:

   Sampling procedures.

■ Level 4 headings are boldfaced and italicized, begin on a paragraph indent, have only the first word capitalized, and end with a period:

> ***Use of a random generator.***

■ Level 5 headings are italicized, begin on a paragraph indent, have only the first word capitalized, and end with a period:

> *Problems with generated data points.*

Most papers that students write have two or three levels of headings. For a paper with two levels, use levels 1 and 2; for a paper with three levels, use levels 1, 2, and 3.

40b APA-style reference list

All of the works you cite should be listed at the end of your paper, beginning on a separate page with the heading *References* (not italicized). The following tips will help you prepare your list.

TIPS FOR PREPARING A REFERENCE LIST

■ Center the heading *References* one inch from the top of the page.

■ Include in your reference list only the sources you explicitly cite in your paper. Do not, however, include entries for personal communications or for original works cited in indirect sources.

■ Arrange the list of works alphabetically by the author's last name. If a source has more than one author, alphabetize by the last name of the first author.

(continued on page 578)

APA

(continued from page 577)

- If you use more than one work by the same author(s), arrange the entries according to the date of publication, placing the entry with the earliest date first. If two or more works by the same author(s) have the same publication date, the entries are arranged so that the titles of the works are in alphabetical order, according to the first important word in each title; lowercase letters are then added to the date (for example, 2008a and 2008b) to distinguish the works.

- When an author's name appears both in a single-author entry and as the first name in a multiple-author entry, place the single-author entry first.

- For a work without an author, alphabetize the entry according to the first important word in the title.

- Type the first line of each entry flush with the left margin and indent subsequent lines one-half inch or five spaces (a hanging indent).

- Double-space throughout—between lines in each entry and between entries.

Whether you are submitting an APA-style paper in a college course or preparing a manuscript for publication, you can be guided by the format of the following sample entries.

Directory of APA-Style Entries for a Reference List

APA

The following guidelines are for books, articles, and most electronic sources. For additional guidelines for documenting electronic sources, see pages 593–99.

When preparing entries for your reference list, be sure to copy the bibliographic information directly from each source (for example, the title page of a book). (See Figure 40.1, on page 587.)

APA

GENERAL DOCUMENTATION GUIDELINES
FOR PRINT-BASED SOURCES

Author or Editor

One author. Use the author's first initial and middle initial (if given) and his or her last name. Invert the initials and the last name; place a comma after the last name. Include a space between the first and middle initials. Any abbreviation or number that is part of a name, such as *Jr.* or *II*, is placed after a comma following the initials. Indicate the end of this information unit with a single period.	Walters, D. M. Thayer-Smith, M. S. Villa, R. P., Jr.
Two to seven authors. Invert the last names and initials of all authors. Use commas to separate last names from initials. Use an ampersand (&) (in addition to the comma) before the last name of the last author.	Vifian, I. R., & Kikuchi, K. Kempf, A. R., Cusack, R., & Evans, T. G.
Eight or more authors. List the first six names, add three ellipsis points, and include the last author's name.	Bauer, S. E., Berry, L., Hacket, N. P., Bach, R., Price, T., Brown, J. B., . . . Green, J.
Corporate or group author. Provide the author's full name.	Hutton Arts Foundation. Center for Neuroscience.

(continued on page 582)

Author or Editor *(continued from page 581)*

Editor. If a work has an editor or editors instead of an author or authors, include the abbreviation *Ed.* for "editor" or *Eds.* for "editors" in parentheses after the name(s).	Harris, B. E. (Ed.). Stroud, D. F., & Holst, L. F. (Eds.).

Publication Date

Books and journals. Provide the year of publication in parentheses, placing a period after the closing parenthesis. For books, this date can be found on the copyright page, which is the page following the title page (see Figure 40.2, on page 588). The publication date of a journal article can be found at the bottom of the first page of the article (see Figure 40.3, on page 591). For a work that has been accepted for publication but has not yet been published, place *in press* in parentheses. For a work without a date of publication, use *n.d.* in parentheses.	(2008). (in press). (n.d.).
Magazines and newspapers. For monthly publications, provide both the year and the month, separated by a comma. For daily publications, provide the year, month, and day. Use a comma between the year and the month.	(2007, January). (2008, June 22).

Conferences and meetings. If a paper presented at a conference, symposium, or professional meeting is published, the publication date is given as the year only, in parentheses. For unpublished papers, provide the year and the month in which the gathering occurred, separated by a comma.	(2008) (2009, September).

Title

Books. Capitalize only the first word and any proper nouns in a book title. Italicize the entire title and place a period at the end of this information unit.	*An introduction to Vygotsky.* *Avoiding work-related stress.*
Journals, magazines, or newspapers. In the name of a journal, magazine, or newspaper, capitalize all major words, as well as any other words consisting of four or more letters. Italicize the entire name and place a comma after it.	*Journal of Child Psychology,* *Psychology Today,* *Los Angeles Times,*
Articles and chapters. Do not italicize the titles of short works such as journal articles or book chapters. The title of an article or chapter appears before the book title or the name of the journal, magazine, or newspaper. Capitalize only the first word of the title and any proper nouns.	Treating posttraumatic stress disorder.

(continued on page 584)

APA

Title *(continued from page 583)*

Subtitles. Always include any subtitle provided for a source. Use a colon to separate a main title and a subtitle. Capitalize only the first word of the subtitle and any proper nouns.	*Reading images: The grammar of visual design.* Living in Baghdad: Realities and restrictions.

Volume, Issue, Chapter, and Page Numbers

Journal volume and issue numbers. A journal paginated *continuously* designates only the first page of the first issue in a volume as page 1. The first page of a subsequent issue in the same volume is given the page number that follows the last page number of the previous issue. In contrast, each issue of a journal paginated *separately* begins with page 1. When you use an article from a journal paginated continuously, provide only the volume number (italicized). When you use an article from a journal paginated separately, provide the issue number (placed in parentheses) directly after the volume number. Do not insert a space between the volume and issue numbers. Italicize only the volume number. Place a comma after this unit of information.	*Journal of Applied Social Psychology, 32,* *Behavior Therapy, 33*(2),

Book chapters. Provide the numbers of the first and last pages of the relevant chapter preceded by the abbreviation *pp.* (for "pages"). Place this information in parentheses. Use an en dash (a short dash; see **11g(2)**) between the page numbers.	*New communitarian thinking* (pp. 126–140).
Articles. List the page numbers after the comma that follows the volume or issue number.	*TESOL Quarterly, 34*(2), 213–238.

Publication Data

City and state. Identify the city in which the publisher of the work is located, including the two-letter US Postal Service abbreviation for the state. If two or more cities are given on the title page, use the first one listed. If the publisher is a university press whose name mentions a state, do not include the state abbreviation. When a work has been published in a city outside the United States, include the name of the country.	Boston, MA: Lancaster, PA: University Park: Pennsylvania State University Press. Oxford, England:

(continued on page 586)

Publication Data *(continued from page 585)*

Publisher's name. Provide only enough of the publisher's name so that it can be identified clearly. Omit words such as *Publishers* and abbreviations such as *Inc.* However, include *Books* and *Press* when they are part of the publisher's name. The publisher's name follows the city and state or country, after a colon. A period ends this unit of information.

New Haven, CT: Yale University Press.

New York, NY: Harcourt.

Cambridge, England: Cambridge University Press.

BOOKS

1. Book by one author

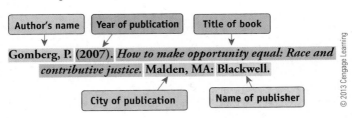

Gomberg, P. (2007). *How to make opportunity equal: Race and contributive justice.* Malden, MA: Blackwell.

© 2013 Cengage Learning

2. Book by two or more authors

Edwards, M., & Titman, P. (2010). *Promoting psychological well-being in children with acute and chronic illness.* London, England: Jessica Kingsley.

If there are eight or more authors, list the first six names followed by three ellipsis points and the last author's name.

Title——**How to Make Opportunity Equal**

Subtitle———— *Race and Contributive Justice*

Author——————— Paul Gomberg

Publisher
(shorten to————— **Blackwell**
Blackwell) Publishing

Figure 40.1. The title page of a book provides most of the information necessary for creating a bibliographic entry for a research paper.

Year of — © 2007 by Paul Gomberg
publication

BLACKWELL PUBLISHING
Cities of ⌈350 Main Street, Malden, MA 02148-5020, USA
publication ⎮9600 Garsington Road, Oxford OX4 2DQ, UK
(use Malden, ⌊550 Swanston Street, Carlton, Victoria 3053, Australia
MA)

The right of Paul Gomberg to be identified as the Author of this
Work has been asserted in accordance with the UK Copyright,
Designs, and Patents Act 1988.

All rights reserved. No part of this publication may be reproduced,
stored in a retrieval system, or transmitted, in any form or by any
means, electronic, mechanical, photocopying, recording or
otherwise, except as permitted by the UK Copyright, Designs, and
Patents Act 1988, without the prior permission of the publisher.

First published 2007 by Blackwell Publishing Ltd.

Figure 40.2. The year in which a book was published and the city where it was published can be found on the copyright page, which follows the title page.

3. Book with editor(s)

Wolfe, D. A., & Mash, E. J. (Eds.). (2005). *Behavioral and emotional disorders in adolescents: Nature, assessment, and treatment.* New York, NY: Guilford Press.

4. Book with a corporate or group author

U.S. War Department. (2003). *Official military atlas of the Civil War.* New York, NY: Barnes & Noble.

5. Edition after the first

Lycan, W., & Prinz, J. (Eds.). (2008). *Mind and cognition* (3rd ed.). Malden, MA: Blackwell.

Identify the edition in parentheses immediately after the title. Use abbreviations: *2nd, 3rd,* and so on for the edition number and *ed.* for "edition."

6. Translation

Rank, O. (2002). *Psychology and the soul: A study of the origin, conceptual evolution, and nature of the soul* (G. C. Richter & E. J. Lieberman, Trans.). Baltimore, MD: Johns Hopkins University Press. (Original work published 1930)

A period follows the name of the publisher but not the parenthetical note about the original publication date.

7. Republished book

Petersen, J. (2009). *Our street.* (B. Rensen, Trans.) London, England: Faber. (Original work published 1938)

8. Multivolume work

Fitzduff, M., & Stout, C. (Eds.). (2006). *The psychology of resolving global conflicts: From war to peace* (Vols. 1–3). Westport, CT: Praeger.

9. Government report

Executive Office of the President. (2003). *Economic report of the President, 2003* (GPO Publication No. 040-000-0760-1). Washington, DC: U.S. Government Printing Office.

10. Selection from an edited book

Empson, R. (2007). Enlivened memories: Recalling absence and loss in Mongolia. In J. Carsten (Ed.), *Ghosts of memory: Essays on remembrance and relatedness* (pp. 58–82). Malden, MA: Blackwell.

Italicize the book title but not the title of the selection.

11. Selection from a reference book

Wickens, D. (2001). Classical conditioning. In *The Corsini encyclopedia of psychology and behavioral science* (Vol. 1, pp. 293–298). New York, NY: John Wiley.

ARTICLES IN PRINT

12. Article with one author in a journal with continuous pagination

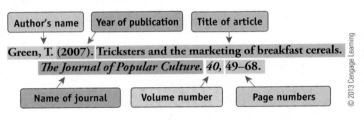

Figure 40.3 shows where the information for this type of entry is found on the first page of an article.

13. Article with two authors in a journal with each issue paginated separately

Rudisill, J. R., & Edwards, J. M. (2002). Coping with job transitions. *Consulting Psychology Journal, 54*(1), 55–62.

14. Article with three to seven authors

Frost, R. O., Steketee, G., & Williams, L. (2002). Compulsive buying, compulsive hoarding, and obsessive-compulsive disorder. *Behavior Therapy, 33*(2), 201–213.

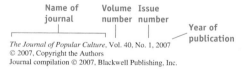

Title —— Tricksters and the Marketing of
Breakfast Cereals

Author —— THOMAS GREEN

BREAKFAST CEREALS ARE SOLD BY TRICKSTERS. FROM LUCKY THE
Leprechaun to the Cookie Crook to the mischievous live-action
squirrels who vend General Mills Honey Nut Clusters, an
astounding number of Saturday morning television commercials feature
30-second dramatizations of trickster tales that are designed to promote
breakfast cereals. True, breakfast cereals are not the only products sold by
tricksters, and not all cereals are sold by tricksters—especially in the last
decade. But the association is common enough to persist as an unexamined
assumption that seems obvious to most Americans once it is pointed out.
Naturally, breakfast cereals are often sold by animated tricksterish mascot
characters, and naturally such commercials feature motifs and narrative
patterns that are common in trickster tales. But the perception of an inherent
internal logic in this scheme overlooks a couple of key questions. Why, for
example, are tricksters considered a particularly appropriate or effective
means of marketing breakfast cereals? And why breakfast cereals in
particular (and a few other breakfast products), almost to the exclusion of
tricksters in other types of marketing campaigns? The answers to these
questions, it turns out, may lie back in the semi-mystical, pseudoreligious
origins of prepared breakfast foods and the mating of the mythology of
those foods with the imperatives of the competitive, prepared-foods
marketplace.

Name of   Volume Issue
journal   number number
                                    Year of
                                    publication
The Journal of Popular Culture, Vol. 40, No. 1, 2007
© 2007, Copyright the Authors
Journal compilation © 2007, Blackwell Publishing, Inc.

**Figure 40.3. The first page of a journal article provides the information
needed to complete a bibliographic entry for that source.**

APA

15. Article with eight or more authors

Lockenhoff, C. E., De Fruyt, F., Terracciano, A., McCrae, R. R., De Bolle, M.,

Costa, P. T., Jr., …Yik, M. (2009). Perceptions of aging across 26 cultures

and their culture-level associates. *Psychology and Aging, 24,* 941–954.

16. Article in a monthly, biweekly, or weekly magazine

Winson, J. (2002, June). The meaning of dreams. *Scientific American, 12,* 54–61.

For magazines published weekly or biweekly, add the day of the issue: (2003, May 8).

17. Article in a newspaper

Simon, S. (2007, October 14). Winning hearts, minds, homes. *Los Angeles Times,* p. A1.

Include the letter indicating the section with the page number.

18. Letter to the editor

Budington, N. (2010, July 20). Social class and college admissions [Letter to the editor]. *The New York Times,* p. A26.

After the title, indicate within brackets that the work is a letter to the editor.

19. Book review

If the review lacks its own title, use a descriptive phrase (like that shown below) in brackets.

Orford, J. (2007, November). [Review of the book *Drug addiction and families, by M. Barnard*]. *Addiction, 102,* 1841–1842.

If the review has a title, include that title before the bracketed information.

Herman, O. (2011, April 10). A little help from your friends: How evolution explains altruism. [Review of the book *Supercooperators,* by M. A. Nowak with R. Highfield]. *The New York Times Book Review,* p. 18.

20. Conference paper

Dodgen, L. (2010, May). *Perceptions of coercion.* Paper presented at the Symposium on University Research and Creative Expression, Central Washington University, Ellensburg, WA.

SOURCES PRODUCED FOR ACCESS BY COMPUTER

The APA guidelines for electronic sources are similar to those for print sources. Many scholarly journals assign a digital object identifier (DOI) to each article so that the article can be accessed easily. The DOI is listed on the first page of the article, which usually contains the abstract. Figure 40.4 shows the location of a DOI and other pertinent bibliographic information on the first page of an online journal. Whenever possible, end a reference list entry for a journal article with the DOI (without a period following it). In an entry for an article without a DOI, use the URL for the periodical's home page. If the URL has to continue on a new line, break it before a punctuation mark or other special character. Do not include a period at the end of the URL.

21. Online journal article with a digital object identifier (DOI)

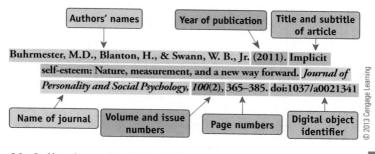

Buhrmester, M.D., Blanton, H., & Swann, W. B., Jr. (2011). Implicit self-esteem: Nature, measurement, and a new way forward. *Journal of Personality and Social Psychology, 100*(2), 365–385. doi:1037/a0021341

Authors' names — Year of publication — Title and subtitle of article — Name of journal — Volume and issue numbers — Page numbers — Digital object identifier

© 2013 Cengage Learning

22. Online journal article without a DOI

Tuladhar-Douglas, W. (2007). Leaf blowers and antibiotics: A Buddhist stance for science and technology. *Journal of Buddhist Ethics, 14,* 200–238. Retrieved from http://blogs.dickinson.edu/buddhistethics/

Instead of a DOI, the URL for the journal's home page, preceded by the phrase *Retrieved from*, appears at the end of the entry.

APA

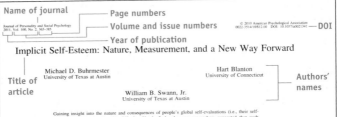

Name of journal

Journal of Personality and Social Psychology
2011, Vol. 100, No. 2, 365–385
© 2010 American Psychological Association
0022-3514/10/$12.00   DOI: 10.1037/a0021341

— Page numbers
— Volume and issue numbers
— Year of publication
— DOI

Title of article

Implicit Self-Esteem: Nature, Measurement, and a New Way Forward

Michael D. Buhrmester
University of Texas at Austin

Hart Blanton
University of Connecticut

William B. Swann, Jr.
University of Texas at Austin

Authors' names

Gaining insight into the nature and consequences of people's global self-evaluations (i.e., their self-esteem) has been fraught with difficulty. Nearly 2 decades ago, researchers suggested that such difficulties might be addressed by the development of a new class of measures designed to uncover implicit self-esteem. In this article, we evaluate the construct validity of the 2 most common measures of implicit self-esteem, the Implicit Association Test (IAT) and Name-Letter Test (NLT). Our review indicates that the research literature has not provided strong or consistent support for the validity of either measure. We conclude that both tests are impoverished measures of self-esteem that are better understood as measures of either generalized implicit affect (IAT) or implicit egotism (NLT). However, we suggest that there surely are aspects of self-esteem that people are unwilling or unable to report and suggest a general approach that may allow researchers to tap these unspoken aspects of self-esteem.

Keywords: self-esteem, implicit measures, Implicit Association Test, name-letter effect, implicit egotism

Proper self-esteem [is] a state of mind that ought to be. Those, moreover, who estimate their own worth correctly, do so on the basis of their past deeds, and so, what they have done, they dare to try again. Those who estimate their worth too highly, or who pretend to be what they are not, or who believe flatterers, become disheartened when dangers actually confront them.

—Thomas Hobbes, *De Homine*

As Hobbes's remarks testify, the construct of self-esteem has a long history in Western culture. But his remarks also acknowledge the difficulty of accurately assessing and reporting one's value. And so, although self-esteem has soared in popularity among both laypersons (e.g., Branden, 1994) and academics (e.g., Bosson & Swann, 2009), some have registered gnawing doubts regarding the capacity of people to report faithfully their true self-evaluations. These concerns have recently dovetailed with growing interest in implicit, nonconscious processes to produce an exciting new construct: Implicit self-esteem. Here, we focus on the nature and measurement of this new construct, what is known about its relation to theoretical correlates of self-esteem, and how researchers might approach this new construct in the future. To place our analysis in context, we begin with a brief discussion of challenges associated with measuring self-esteem using traditional, explicit methods.

This article was published Online First November 1, 2010.
Michael D. Buhrmester and William B. Swann, Jr., Department of Psychology, University of Texas at Austin; Hart Blanton, Department of Psychology, University of Connecticut.
Correspondence concerning this article should be addressed to William B. Swann, Jr., Department of Psychology, University of Texas at Austin, 1 University Station, Austin, TX 78712. E-mail: swann@mail.utexas.edu

Measuring Self-Esteem

Throughout most of the 6 decades since the introduction of the first measure of self-esteem (Raimy, 1948), theorists agreed that explicit self-esteem refers to feelings of self-worth or the global evaluation of the self (Bosson & Swann, 2009; Rosenberg, 1965; Shavelson, Hubner, & Stanton, 1976). But, if there is consensus regarding the nature of self-esteem, there is less agreement regarding the most appropriate means of assessing it. To date, the most widespread approach has been to explicitly ask people a series of questions regarding their global self-evaluation. Critics have faulted such measures on at least three counts. First, the verbal questionnaires designed to assess explicit self-esteem may fail to capture self-views of which people are unaware (e.g., Greenwald & Farnham, 2000). Second, even if people are aware of a given self-view, they might fail to express it due to self-presentational pressures that tempt them to inflate their self-evaluations (e.g., Paulhus, 1991, 2002). Third, perhaps due to the foregoing limitations, researchers have failed to uncover strong and consistent support for the predictive validity of measures of explicit self-esteem (see Baumeister, Campbell, Krueger, & Vohs, 2003).

Although some have countered the foregoing critiques (e.g., Marsh & Craven, 2006; Swann, Chang-Schneider, & McClarty, 2007), it is certainly possible that measures of explicit self-esteem may fail to capture important aspects of self-knowledge. With this possibility in mind, researchers developed several measures of implicit self-esteem (see Bosson, Swann & Pennebaker, 2000). By far the most popular of these measures have been a pair dubbed the Implicit Association Test (IAT) and the Name-Letter Test (NLT). Both measures are based on the assumption that implicit self-esteem is a valenced association that a person has toward himself or herself. Although researchers differ in how they characterize this association, some consensus has emerged regarding its nature

Figure 40.4. First page of an online journal article.

23. Online magazine article based on a print source

Acocella, J. (2008, May 26). A few too many. *The New Yorker, 84*(15), 32–37.

    Retrieved from http://www.newyorker.com

24. Online magazine article not found in print

Saletan, W. (2008, August 27). Unfinished race: Race, genes, and the future of

    medicine. *Slate.* Retrieved from http://www.slate.com

25. Article in an online newspaper

Redden, J. (2011, April 18). Portland, Lake Oswego councils endorse

    streetcar proposal. *Portland Tribune.* Retrieved from http://www

    .portlandtribune.com

In Figure 40.5, you can see where to find the author's name, the date, the title of the article, and the name of the newspaper. Be sure to use the URL for the newspaper's home page (for example, http://www.portlandtribune.com), not the URL for the page on which you found the article.

26. Article from a database

Include the article's DOI if it has one; if it does not have a DOI, list the URL of the journal's home page (see items 21 and 22). Note that the name of the database is not included.

Hill, E. J., Erickson, J. J., Holmes, E. K., & Ferris, M. (2010). Workplace

    flexibility, work hours, and work-life conflict: Finding an extra day

    or two. *Journal of Family Psychology, 24*(3), 349–358. doi:10.1037

    /a0019282

Shellenbarger, S. (2010, July 21). Kids quit the team for more family time. *The*

    Wall Street Journal. Retrieved from http://online.wsj.com/home-page

Figure 40.5. Article from an online newspaper.

27. Online book

Pine, R. C. (2004). *Science and the human prospect*. Retrieved from http://home
.honolulu.hawaii.edu/~pine/book1-2.html

If access to the online book is not free, use *Available from* instead of *Retrieved from*.

28. Online book chapter

Brady, V. (2006). A flaw in the nation-building process: Negotiating the sacred in
our multicultural society. In E. B. Coleman & K. White (Eds.), *Negotiating
the sacred: Blasphemy and sacrilege in a multicultural society* (pp. 43–49).
Retrieved from http://epress.anu.edu.au/nts_citation.html

If page numbers are not provided in the online book, simply omit the parenthetical identification of such numbers.

29. Message posted to a newsgroup, forum, or discussion group

Vellenzer, G. (2004, January 24). Synonyms of entreaty [Msg 2]. Message posted
    to http://groups.google.com/groups?selm=MPG.1a7cacccd54e9c27989b95
    %40news.CIS.DFN.DE&output=gplain

If the message has been archived at an address different from the one it was posted to, a comma should be placed after the first URL, followed by *archived at* (not italicized) and the URL for the archived version.

30. Blog posting

Chatham, C. (2008, August 29). Action without intention: Parietal damage
    alters attention awareness. Message posted to http://scienceblogs.com
    /developingintelligence/2008/08/action_without_intention_parie.php

31. Lecture notes posted online

Wolfe, J. (2004). *Lecture 18: Freud and fairy tales.* Retrieved from Massachusetts
    Institute of Technology OpenCourseWare website: http://ocw.mit.edu
    /OcwWeb/Brain-and-Cognitive-Sciences/9-00Fall-2004/LectureNotes
    /index.htm

32. Authored document from a website

Ennis, R. H. (2002, July 20). *An outline of goals for a critical thinking curriculum
    and its assessment.* Retrieved from http://faculty.ed.uiuc.edu/rhennis
    /outlinegoalsctcurassess3.html

APA

33. Online document with no identified author

American School Counselor Association. (2006). *Position statement: Equity for all students*. Retrieved from http://asca2.timberlakepublishing.com/content .asp?contentid=503

Use the name of the organization sponsoring the website as the author of the document.

34. Personal communication

Entries for personal communications such as e-mail messages, letters, interviews, and telephone conversations are not included in the reference list. Such sources should be cited in the text as follows: (S. L. Johnson, personal communication, September 3, 2003).

35. Online encyclopedia

Dowe, P. (2007). Causal processes. In E. N. Zalta (Ed.), *The Stanford encyclopedia of philosophy*. Retrieved from http://plato.stanford.edu/archives/sum2007 /entries/cognitive-science/

36. Online dictionary

Paranormal. (2000). In *The American Heritage dictionary of the English language* (4th ed.). Retrieved from http://www.bartleby.com/61/35 /P0063500.html

37. Online consumer brochure

American Psychological Association. (2008). *Elder abuse and neglect: In search of solutions* [Brochure]. Retrieved from http://www.apa.org/pi/aging /eldabuse.html

38. Online government document

Pashler, H., Bain, P., Bottge, B., Graesser, A., Koedinger, K., McDaniel, M., &
Metcalfe, J. (2007, September). *Organizing instruction and study to improve
student learning: IES practice guide* (NCER 2007-2004). Washington,
DC: National Center for Education Research, Institute of Education
Sciences, U.S. Department of Education. Retrieved from http://ies.ed.gov
/pubsearch/pubsinfo.asp?pubid=NCER20072004

If no authors are identified, an entry for an online government
document is formatted as follows:

U.S. Department of Health and Human Services, National Institutes of
Health, National Institute on Alcohol Abuse and Alcoholism. (2010).
Beyond hangovers: Understanding alcohol's impact on your health (NIH
Publication No. 10-7604). Retrieved from http://www.niaaa.nih.gov
/Publications/PamphletsBrochuresPosters/English/Pages/default.aspx

39. Online audio or video file

Begin the entry with the name of the contributor whose role is
most relevant to the topic of your paper. Identify the contributor's role (for example, *Producer, Director, Writer, Host,* or *Presenter*) in parentheses. The medium (for example, *Audio file,
Video file, Audio podcast,* or *Video webcast*) is placed in square
brackets after the title of the work.

Davies, D. (Host). (2010, July 13). A psychiatrist's prescription for his profession
[Audio file]. In T. Gross & D. Miller (Executive producers), *Fresh air.*
Retrieved from http://www.npr.org/templates/rundowns/rundown.php
?prgrid=13&prgDate=7-13-2010

Gopnik, A. (Presenter). (2009, July 28). Moments of absolute absorption [Video
file]. In D. McGee & P. W. Kunhardt (Executive producers), *This emotional
life: The meaning of happiness* (Chapter 6). Retrieved from http://www.pbs
.org/thisemotionallife/perspective/meaning-happiness

OTHER SOURCES

40. Motion picture

Gaviria, M. (Producer/Director), & Smith, M. (Producer/Writer). (2001).

Medicating kids [Motion picture]. United States: Public Broadcasting Service.

Begin with the relevant contributor(s), identifying the nature of the contribution in parentheses following each name. Then provide the release date and the title, followed by a descriptive label in square brackets. The entry ends with the name of the country where the film was produced and the name of the studio or organization.

41. Television program

Holt, S. (Producer). (2002, October 1). *The mysterious lives of caves* [Television

program]. Alexandria, VA: Public Broadcasting Service.

Give the title of the program in italics. If citing an entire series (for example, *Nova* or *The West Wing*), use the name of the producer of the series as a whole and the descriptive label *Television series* in the square brackets. If the program is a single episode of a series, its title is not italicized, and the descriptive label in the brackets is *Television series episode*.

42. Advertisement

Rosetta Stone [Advertisement]. (2010, July). *National Geographic, 218*(1), 27.

| 40c | APA-style research paper |

The APA recognizes that a paper may have to be modified so that it adheres to an instructor's requirements. The following boxes offer tips for preparing a title page, an abstract page, and the body of a typical student paper. For tips on preparing a reference list, see **40b**.

TIPS FOR PREPARING THE TITLE PAGE OF AN APA-STYLE PAPER

- The title page includes both the full title of the paper and a shortened version of it. The shortened version appears in the header. On the left side of the header, place the words *Running head*: (not italicized, followed by a colon) and the shortened version of your title. The shortened version should consist of no more than fifty characters (including punctuation and spaces), and all letters should be capitals. On the right side of the header, insert the page number. The title page is page 1 of your paper.

- Place the full title in the upper half of the page, with your name and your institutional affiliation below it. You may include the course name and number instead of the affiliation if your instructor requests it. Double-space and center these lines.

TIPS FOR PREPARING THE ABSTRACT AND THE BODY OF AN APA-STYLE PAPER

- The header for the abstract page and each page in the body of the paper consists of the shortened title on the left and the page number on the right. The abstract is on page 2; the body of the paper begins on page 3.

- Center the word *Abstract* (not italicized or boldfaced) one inch from the top of the paper.

- Be sure that the abstract is no more than 250 words. For advice on summarizing, see **38d(4)**.

- Double-space throughout the body of the abstract. Do not indent the first line of the abstract.

- Center the title of the paper one inch from the top of the page on page 3.

- Use one-inch margins on both the left and right sides of all pages.

- Double-space throughout the body of the paper, indenting each paragraph one-half inch, or five to seven spaces.

- If there are headings within the body of the paper, format them according to their levels (**40a(3)**).

APA

Use 1-inch
margins on
both sides
of the page.

1/2 inch

The shortened title
in the page header
should consist of
no more than 50
characters.

Place the
page number
in the upper
right corner.

The Social Status of an Art:

Historical and Current Trends in Tattooing

Rachel L. Pinter and Sarah M. Cronin

Central Washington University

If required
by the
instructor,
the course
name and
number
replace the
institutional
affiliation.

APA

↕ 1/2 inch
SOCIAL STATUS OF AN ART | 1 inch   2

Abstract

Center the heading.

Current research demonstrates that the social practice of tattooing

has changed greatly over the years. Not only have the images

chosen for tattoos and the demographic of people getting tattoos

changed, but the ideology behind tattooing itself has evolved. This

paper first briefly describes the cross-cultural history of the practice.

It then examines current social trends in the United States and

related ideological issues.

The maximum length for an abstract is 250 words.

APA

SOCIAL STATUS OF AN ART   1 inch   3

Center the title.

The Social Status of an Art:

Historical and Current Trends in Tattooing

Tattoos, defined as marks made by inserting pigment into

the skin, have existed throughout history in countless cultures.

Use 1-inch margins on both sides of the page.

Currently, tattoos are considered popular art forms. They can

be seen on men and women from all walks of life in the United

States, ranging from a trainer at the local gym to a character on a

television show or even a sociology professor. Due to an increase

in the popularity of tattooing, studies of tattooing behavior have

proliferated as researchers attempt to identify trends. This paper

The writers' thesis statement forecasts the content of the essay.

seeks to explore both the history of tattooing and its current practice

in the United States.

Tattooing has a long history in most of the world, though

its origin is currently unknown. Krcmarik (2003) provides a

The writers provide historical and cultural information about tattooing.

helpful geographical overview. In Asia, tattooing has existed for

thousands of years in Chinese, Japanese, Middle Eastern, and

Indian cultures. Evidence of its existence can be seen on artifacts

such as 7,000-year-old engravings. In Europe, tattooing flourished

during the 19th century, most notably in England. Many of the

sailors traveling with Captain James Cook returned with tales of

exotic tattooing practices and sometimes with tattoos themselves.

The Samoans in the South Pacific are famous for their centuries-

APA

SOCIAL STATUS OF AN ART   4

old tattooing practice, known as *tatau*—the word from which *tattoo* is said to have originated. The Maori of New Zealand are also well known for their hand-carved facial tattoos, known as *Moko* (see Figure 1).

Figure 1. A Maori man with a facial tattoo. *Note.* Photo © Tim Graham/Getty Images.

In Africa, tattoos can be found on Egyptian and Nubian mummies, which date back to approximately 2000 BCE. Tattooing is noted in the written accounts of Spanish explorers' encounters with tattooed Mayans in Central America. Finally, in North America, tattooing became popular in the early part of the 20th century and has experienced advances and retreats in social acceptance since then. Starting in the 1960s, its popularity rose dramatically.

Clearly, the history of tattooing spans generations and cultures. The practice has gained and lost popularity, often as a result of rather extreme changes in the ideologies supporting or discouraging it. This roller-coaster pattern of acceptance is well demonstrated in the United States. Since the 19th century, the wearing of tattoos has allowed for subculture identification among such persons as sailors, bikers,

The writers discuss changing perspectives on the appropriateness of tattoos.

APA

The writers include a photograph to support a point.

Figure 2. Tattoos are becoming more common among middle-class professionals. *Note.* Photo © Eric Anthony Johnson/Photolibrary.com

circus "freak" performers, and prison inmates (DeMello, 1995). As a collective group behavior indicating deviant subculture membership, tattooing flourished during the 20th century but remained plagued by negative stereotypes and associations. In the last 10 years, however, the practice has represented a more individualistic yet mainstream means of body adornment. As Figure 2 illustrates, it is not unusual to see a white-collar worker sporting a tattoo.

Tattooing is now common among both teenagers and older adults, men and women, urbanites and suburbanites, the college-educated and the uneducated, and the rich as well as the poor (Kosut, 2006). Table 1 indicates the wide range of Americans wearing tattoos in 2003 and 2008.

Table 1

Percentages of American Adults with One or More Tattoos

Category	Year 2003	2008
All adults	16	14
Region		
East	14	12
Midwest	14	10
South	15	13
West	20	20
Age range		
18–24	13	9
25–29	36	32
30–39	28	25
40–49	14	12
50–64	10	8
65+	7	9
Sex		
Male	16	15
Female	15	13

Note. Adapted from "Three in Ten Americans with a Tattoo Say Having One Makes Them Feel Sexier," by R. A. Corso, 2008, *Harris Interactive.* Copyright 2008 by Harris Interactive.

The trend toward acceptance of tattoos may be a result of how American society views the people who wear them. Earlier, tattoos were depicted in mainstream print and visual media as worn by people with low socioeconomic or marginal status; now, they are

SOCIAL STATUS OF AN ART 7

considered to be a means of self-expression among celebrities as well

as educated middle- and upper-class individuals (Kosut, 2006). This

shift in the symbolic status of tattoos—to a form of self-expression

among the social elite rather than a deviant expression among the

lower classes—has allowed tattoos to be obtained in greater numbers,

owing in great part to the importance placed on self-expression in the

United States. Even in the workplace, where employees had often

been forbidden to display tattoos, employers now "take advantage

of the open-mindedness and innovation that younger [tattooed]

employees bring into the workplace" (Org, 2003, p. D1).

As the popularity and acceptability of tattoos have increased,

tattooing has become part of the greater consumer culture and has

thus undergone the process of commercialization that frequently

occurs in the United States. Tattoos are now acquired as status

symbols, and their prevalence helps to sell tattoo maintenance

products, clothing, and skateboards (Kosut, 2006). This introduction

into the consumer culture allows tattoos to gain even more

popularity; they are now intertwined with mainstream culture.

Researchers have been tracking the popularity of tattoos,

though no one seems able to agree on exact numbers (Libbon,

2000). In 2000, MSNBC aired an investigative piece called

Tattoos—Skin Deep, which cited the tattooing rate at 20% of the

SOCIAL STATUS OF AN ART   8

U.S. population (Rosenbaum, 2000). In 2003, citing a lower number,
Harris Interactive reported that 16% of all adults in the United
States have at least one tattoo (Sever, 2003). The actual number of
individuals with tattoos is unknown, but most researchers believe
the trend has been consistently gaining ground since the 1960s.
Statistics on the frequency of tattooing among specific age groups
generally show increases (Armstrong, Owen, Roberts, & Koch,
2002; Mayers, Judelson, Moriarty, & Rundell, 2002) although one
study (Corso, 2008) showed a slight decrease. However, because of
the limitations of the various research designs, more research on a
national level is needed to obtain truly representative figures.

Significantly, the increase in acceptance of tattoos has
resulted in trends concerning the images and locations of tattoos,
which appear to be divided along lines of gender. Many of the
tattoo images commonly found on men include, but are not
limited to, death themes, various wildlife, military insignia,
tribal armbands, and family crests or last names. During the
1980s, cartoon images such as Bugs Bunny and the Tasmanian
Devil were also popular for males. Males choose various
locations for tattoos, but the most popular sites are the upper
back, back of the calves, and the upper arm, according to tattoo
artist Ben Reames (personal communication, July 12, 2007).

The writers
list statistics
to support a
claim.

Two citations
of articles,
both written
by four
authors, are
separated by
a semicolon.

APA

Citation of an
interview with
a tattoo artist

Conversely, females often obtain tattoos that symbolize traditional femininity, such as flowers, stars, hearts, and butterflies. A noticeable trend for females in the 1980s was the rose tattoo, which was often located on the breast or ankle. Stars and butterflies now rival the rose in popularity. The ankle continues to be a popular location for females today. Other popular spots for tattoos include the hip, the foot, and the lower back. In fact, the lower back experienced a huge surge in popularity during the 1990s (B. Reames, personal communication, July 12, 2007).

The last paragraph is the conclusion.

The art of tattooing has existed in many culturally determined forms throughout human history, and its current manifestations are as varied as the cultures themselves. However, based on the current literature, the social behavior of tattooing is still quite common in the United States. In fact, Kosut (2006) argues, "New generations of American children are growing up in a cultural landscape that is more tattoo-friendly and tattoo-flooded than any other time in history" (p. 1037). Because today's children see tattoos and tattoo-related products everywhere, usually in neutral or positive situations, they will likely be more accepting of tattoos than earlier generations were. Certainly, the tattooing trend shows no signs of decreasing significantly.

APA

References

Armstrong, M. L., Owen, D. C., Roberts, A. E., & Koch, J. R. (2002).

   College students and tattoos: Influence of image, identity, family,

   and friends. *Journal of Psychosocial Nursing, 40*(10), 20–29.

Corso, R. A. (2008, February 12). *Three in ten Americans with

   a tattoo say having one makes them feel sexier*. Retrieved

   from http://www.harrisinteractive.com/harris_poll/index

   .asp?PID=868

DeMello, M. (1995). Not just for bikers anymore: Popular

   representations of American tattooing. *Journal of Popular

   Culture, 29*(3), 37–53. Retrieved from http://www.wiley.com

   /bw/journal.asp?ref=0022-3840

Kosut, M. (2006). An ironic fad: The commodification and

   consumption of tattoos. *Journal of Popular Culture*, 39(6),

   1035–1049. Retrieved from http://www.wiley.com/bw/journal

   .asp?ref=0022-3840

Krcmarik, K. L. (2003). *History of tattooing*. Retrieved from

   Michigan State University website: http://www.msu

   .edu/~krcmari1/individual /history.html

Libbon, R. P. (2000). Dear data dog: Why do so many kids sport

   tattoos? *American Demographics, 22*(9), 26. Retrieved from

   http://amiga.adage.com/de

Center the
heading.

Alphabetize
the entries
according to
the author's
(or the first
author's) last
name.

Indent
second and
subsequent
lines of each
entry one-half
inch or five
spaces.

APA

SOCIAL STATUS OF AN ART   11

Mayers, L. B., Judelson, D. A., Moriarty, B. W., & Rundell, K. W.
    (2002). Prevalence of body art (body piercing and tattooing)
    in university undergraduates and incidence of medical
    complications. Mayo Clinic Proceedings, 77, 29–34.

Org, M. (2003, August 28). The tattooed executive. *The Wall Street
    Journal*. Retrieved from http://online.wsj.com/public/us

Rosenbaum, S. (Executive Producer). (2000, August 20). *MSNBC
    investigates: Tattoos—skin deep* [Television program]. New
    York and Englewood Cliffs, NJ: MSNBC.

Sever, J. (2003, October 8). *A third of Americans with tattoos say
    they make them feel more sexy*. Retrieved from http://www
    .harrisinteractive.com/Insights/HarrisVault.aspx

Identification
of the type
of medium is
placed in square
brackets.

No period follows
a URL at the end
of an entry.

APA

41 | Writing about Literature

You have been interpreting and writing about literature ever since you wrote your first book report. When you write about literature in college, you will still discuss plot, characters, and setting. You will also establish a rhetorical opportunity for writing, explore and focus your subject, formulate a purposeful thesis statement that is supported by reference to the literary work itself, address an audience, and arrange your thoughts in the most effective way. In short, when you write about literature, you respond to the rhetorical situation. This chapter will help you

- recognize the various genres of literature (**41a**),
- realize the value of careful reading (**41b**),
- use the specialized vocabulary for discussing literature (**41c**),
- employ various critical approaches for interpreting literature (**41d**), and
- apply the conventions for writing about literature (**41e** and **41f**).

41a | Literature and its genres

Works of literature can be divided into categories, or **genres**. A genre is identified by its particular features and conventions. Some genres are timeless and universal (drama and poetry, for instance); others are context-specific and develop within particular cultures (the graphic novel, for instance, is a fairly recent cultural phenomenon).

Some of the most widely studied literary genres are fiction, drama, and poetry, though many forms of nonfiction (including personal essays and memoirs, literacy narratives, and manifestos) are being studied in college courses on literature. All imaginative literature can be characterized as fictional, but the term **fiction** is applied specifically to novels and short stories. **Drama** differs from all other imaginative literature in one specific way: it is meant to be performed. In a novel, you often find extensive descriptions of characters and setting as well as passages revealing what characters are thinking. In drama, however, characters reveal what they are thinking through either spoken dialogue with other characters or a **dramatic soliloquy** (a speech delivered to the audience by an actor alone on the stage). Poetry shares the components of both fiction and drama. But poetry distinguishes itself from the other literary genres with its extensive use of connotative language, imagery, allusions, figures of speech, symbols, sound, meter, and rhythm.

41b Rhetorical reading and literary interpretation

The most successful writing about literature starts with rhetorical (or active) reading. As you read, note your reactions. Were you amused, moved, or confused? Which characters interested you? Were you able to follow the plot? Did the work remind you of any experience of your own or other works you have read? Did it introduce you to a different historical or geographical setting, or did you encounter a familiar setting and cast of characters? These first impressions can provide the seeds from which strong essays will grow, especially when they are later modified as you consider the work further.

(1) Your personal response to reading literature

When reflecting on your response to some element in a work of literature, consider how your reading might be shaped by the factors that define who you are. If you feel yourself responding positively or negatively to a character, a plot twist, or the setting in a novel or play, you might ask yourself whether this response has anything to do with your psychological makeup, political beliefs, gender or sexual orientation, cultural or ethnic group, social class, religion, or geographic location.

Thinking about what your individual reading contributes to a work of literature helps you focus on a rhetorical opportunity for writing and may suggest a theoretical approach to use as a basis for your interpretation (41d). Keep in mind, though, that just as your life experiences and values can enhance your understanding of a literary work, they can also limit that understanding—yet another way your identity can shape your response.

(2) Developing your topic using evidence in the text

If you are choosing your own topic, your first step is to reflect on your personal response as you formulate a tentative thesis statement. Next, consider what specific evidence from the text will best explain and support your interpretation and thesis statement.

Most readers (including your instructor) will be interested in what *you* think, so you need to discover a way to demonstrate your originality by focusing on a topic you can develop adequately and then applying one or more rhetorical methods (32g). You might explain why you consider a character heroic, classify a play as a comedy of manners, or describe a setting that anchors a literary work's meaning. Perhaps you can compare and contrast two poems on a similar subject or explore cause-and-consequence relationships in a novel. What circumstances, for instance, lead to an otherwise intelligent character's monumentally bad decision? Or you might trace the repeated appearance of an object (hands, dark skies, or a cat, for example) throughout a story and explain how that repetition serves to remind the reader of some particular idea or theme.

(3) Researching what other readers have said about a literary work

You will undoubtedly anchor your essay in your own interpretation, but you enrich that interpretation with the sometimes conflicting responses of others, from literary experts to classmates. Every time you read works of literary criticism, visit online discussion groups (**35b**), participate in class discussions, or take part in a book club or reading group, you are engaging in literary-based dialogue with others. Although it is tempting to lean heavily on the interpretations of scholarly experts, use them only to enrich your own interpretation and support your own points. And be sure to cite any sources you do use.

To locate scholarly material on a specific writer, work, or literary theory, you can start by consulting your library's resources. Your library's catalog (**36b** and **36c**) and certain reference books are the best starting points. For instance, *The MLA International Bibliography,* an index of books and articles about literature, is an essential resource for literary studies and is available in print and online. Works such as *Contemporary Authors, The Oxford Companion to English Literature,* and *The New Princeton Handbook of Poetic Terms* can be useful when you are beginning your research or when you have encountered terms you need to clarify.

(4) Types of literary interpretation

Writing about a literary work requires you to focus on the work itself and to demonstrate that you have read it carefully—a process known as **close reading**. (Compare close reading with reading rhetorically, discussed in Chapter **31**.) Close reading allows you to offer an **interpretation**, an explanation of what you see in a work. When your interpretation explains the contribution of one feature of a literary work (such as the setting or main character) to the work's overall meaning, it is called an **analysis**. A common form of analysis is **character analysis**, in which a writer interprets the significance of one or more features of a single character. An analysis can also focus on a single scene, symbol, or theme.

Explication, usually used only with poetry, is an interpretation that attempts to explain every element in a literary work. When explicating William Wordsworth's "A Slumber Did My Spirit Seal," a writer might note that the *s* sound reinforces the hushed feeling of sleep and death in the poem. But it would also be necessary to consider the meanings of *slumber, spirit,* and *seal.*

An **evaluation** of a literary work gauges how successfully the author communicates meaning to readers. The most common types of evaluation are book, theater, and film reviews. Like any other interpretation, an evaluation is a type of argument in which a writer cites both positive and negative textual evidence to persuade readers to accept a clearly formulated thesis. (See Chapters 32 and 34.)

41c Vocabulary for discussing literature

Like all specialized fields, literature has its own vocabulary, which describes the various features of literary texts and the concepts of literary analysis. As you learn this vocabulary, you will learn more than just a list of terms: you will learn how to understand, interpret, and write about literature.

(1) Characters
The **characters** are the humans or humanlike personalities (aliens, robots, animals, and other creatures) who carry the plot forward; they usually include a main character, called a **protagonist**, who is in external conflict with another character

Understanding how a particular character moves the plot forward will help you interpret a work as a whole.

or an institution or in internal conflict with himself or herself. This conflict usually reveals the **theme**, or the central idea of the work (41c(7)).

Because you need to understand the characters in any work you read, pay close attention to their appearance, their language, and their actions. You also need to pay attention to what the narrator or other characters say about them and how the other characters treat and react to them.

(2) Imagery

The imagery in a piece of literature is conveyed by **descriptive language**, or words that describe a sensory experience. Notice the images in the following excerpt from a prose poem by Pinkie Gordon Lane that focuses on the death—and life—of a mother.

> My mother died walking along a dusty road on a Sunday morning in New Jersey. The road came up to meet her sinking body in one quick embrace. She spread out like an umbrella and dropped into oblivion before she hit the ground. In that one swift moment all light went out at the age of forty-nine. Her legacy: the blackened knees of the scrub-woman who ransomed her soul so that I might live, who bled like a tomato whenever she fought to survive, who laughed fully when amused—her laughter rising in one huge crescendo—and whose wings soared in dark despair....
>
> —**PINKIE GORDON LANE, "Prose Poem: Portrait"**
> © 1991. Reprinted by permission.

The dusty road, the sinking body, the quick embrace—these images convey the loneliness and swiftness of death. The blackened knees, tomato-like bleeding, and rising laughter are, in contrast, images of a life's work, struggle, and joy.

(3) Narrator

The **narrator** of a literary work tells the story. The voice doing the telling can seem to be that of the author himself or herself (but such a voice is that of the author's **persona**, which is a fictional construction and not actually the author). The voice might instead be that of a specific character (or one of several characters who are taking turns telling the story). Or the voice can be that of an all-knowing presence (referred to as an **omniscient narrator**) that transcends characters and author alike. Whatever the voice, the narrator's tone reveals his or her attitude toward events and characters and even, in some circumstances, toward readers.

(4) Plot

The **plot** is what happens in the story, the sequence of events (the narrative)—and more. The plot establishes how events are patterned or related in terms of conflict and resolution. Narrative answers "What comes next?" and plot answers "Why?" Consider this example:

> **Narrative**
> A woman is confined to a room with yellow wallpaper.

> **Plot**
> The physician husband of a highly imaginative woman moves her into a room with yellow wallpaper, where she is restricted to silence and idleness.

A plot usually begins with a conflict, an unstable situation that sets events in motion. In what is called the **exposition**, the author introduces the characters, setting, and background—the elements that not only constitute the unstable situation but also relate to the events that follow. The subsequent series

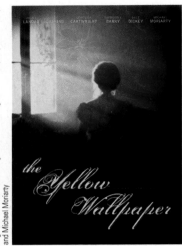

In "The Yellow Wallpaper," a doctor confines his wife to an upstairs bedroom in an attempt to restore her mental health by means of a rest cure.

of events leads to the **climax**, the most intense event in the narrative. The climax is also referred to as the **turning point** because what follows is **falling action** (or **dénouement**) that leads to a resolution of the conflict and a more stable situation, though not necessarily a happy ending.

(5) Setting

Setting involves place—not just the physical setting, but also the social setting (the morals, manners, and customs of the characters). Setting also involves time— not only historical time, but also the length of time covered by the narrative. Setting includes **atmosphere**, or the emotional response to the situation, often shared by the reader with the characters. Being alert to the features of the setting will help you better understand a story, whether it is written as fiction, drama, or poetry.

(6) Symbols

Frequently used by writers of literature, a **symbol** is usually a physical object that stands for something else, usually something abstract. For example, at the beginning of *A Streetcar Named Desire,* a play by Tennessee Williams, one of the main characters buys a paper lantern to cover a naked lightbulb. During the scenes that follow, she frequently talks about light, emphasizing her preference for soft lighting. At the end of the play, another character tears off the lantern, and a third character tries to return the ruined lantern to the main character as she is being taken away to a mental hospital. Anyone seeing this play performed or reading it carefully would note that the paper lantern is a symbol. It is an object that is part of the setting and the plot, but it also stands for something more—a character's avoidance of harsh truths.

When you write about a particular symbol, first note where it appears in the literary work. To determine what the symbol might mean, consider why it appears in those places and to what effect. Once you have an idea about the meaning, trace the incidents in the literary work that reinforce that interpretation.

(7) Theme

The main idea of a literary work is its **theme**. Depending on how they interpret a work, different readers may identify different themes. To test whether an idea is central to the work in question, check to see if the idea is supported by the setting, plot, characters, and symbols. If you can relate these components to the idea you are considering, then it can be considered the work's theme. The most prominent literary themes arise out of external or internal conflict: character versus character, character versus herself or himself, character versus nature, or character versus society.

When you believe you have identified the theme of a literary work, state it as a sentence—and be precise. A theme conveys a specific idea; it should not be confused with a topic.

Topic	a physician's care of his ill wife
Vague theme	the subordination of nineteenth-century married women
Specific theme	"The Yellow Wallpaper" deals with a conflict between an imaginative woman and a society that insists that she abandon her artistic endeavors.

CHECKLIST for Interpreting a Literary Work

- From whose point of view is the story told?
- Who is the protagonist? How is his or her character developed?
- With whom or what is the protagonist in conflict?
- How are the other characters depicted and distinguished through dialogue?
- What symbols, imagery, or figures of speech (20a(3)) does the author use? To what effect?
- What is the theme of the work? How does the author use setting, plot, characters, and symbols to establish that theme?

41d Approaches to interpreting literature

An interpretation of a literary work can be shaped by your personal response to what you have read, by the views of other readers whom you wish to support or challenge, or by a specific type of literary theory.

Literary theory, the scholarly discussion of how the nature and function of literature can be determined, ranges from approaches that focus almost exclusively on the text itself (its language and structure) to approaches that show how the text relates to author, reader, language, society, culture, economics, or history. Familiarity with literary theory enriches your reading of literature as well as your understanding of the books and essays about literature that you will discover when you do research (see Chapter 36).

(1) Reader-response theory

According to **reader-response theory**, readers construct meaning as they read and interact with the elements within a text. Thus, meaning is not fixed *on* the page but rather depends on what each reader brings *to* the page. Furthermore, the same reader can have different responses to the same literary work when rereading it later: a father of teenagers, for example, might find Gwendolyn Brooks's "we real cool" more disturbing now than when he first read it in high school. Although a reader-response approach to literature encourages diverse interpretations, you cannot simply say, "Well, that's what this work means to me," or "That's my interpretation." You must demonstrate to your audience how the elements of the work support your interpretation.

(2) Feminist and gender-based literary theories

The significance of sex, gender, or sexual orientation within a particular social context is the interpretive focus of **feminist** and **gender-based literary theories**. These theories enable a reader to analyze the ways in which a work (through its characters, theme, or plot) promotes or challenges the prevailing intellectual or cultural assumptions of its day regarding issues related to gender and sexuality, such as patriarchy and compulsory heterosexuality. For instance, Edith Wharton's *The Age of Innocence* compares two upper-class nineteenth-century women with respect to the specific social pressures that shaped and constricted their lives and loves. A feminist critic might emphasize the oppression of these women and the repression of their sexuality. Using a gender-based approach, another critic might read Henry James's *The Bostonians* and focus on the positive features of the domestic relationship between the financially independent Olive Chancellor and Verena Tarrant, the feminist activist she supports.

(3) Race-based literary theory

A useful form of race-based literary criticism, **critical race theory**, focuses on the significance of race relations within a specific historical and social setting in order to explain the

zora neale hurston

THEIR EYES WERE WATCHING GOD
A NOVEL

Harper Perennial Modern Classics

In *Their Eyes Were Watching God*, a mature and complex black woman recounts her life in three "acts," based on her relationships with three very different men.

experience and literary production of any people whose history is characterized by political, social, or psychological oppression. Previously neglected works such as Zora Neale Hurston's *Their Eyes Were Watching God,* Rudolfo Anáya's *Bless Me, Ultima,* and Frederick Douglass's *Narrative,* which demonstrate how racism affects the characters' lives, have taken on considerable cultural value in the last thirty years. **African American literary criticism**, for example, has been particularly successful in invigorating the study of great African American writers. Closely associated with critical race theory is **postcolonial theory**, which takes into account the relationship of the colonized with the colonizer. Joseph Conrad's *Heart of Darkness,* Jean Rhys's *Wide Sargasso Sea,* Daniel Defoe's *Robinson Crusoe,* and E. M. Forster's *A Passage to India* can all be read productively through the lens of postcolonial theory.

(4) Class-based literary theory

Class-based literary theory draws on the work of Karl Marx, Terry Eagleton, and others who have addressed the implications of social hierarchies and the accompanying economic tensions, which divide people in profoundly significant ways.

Thus, a class-based approach can be used to explain why Emma Bovary is unhappy, despite her "good" (that is, financially advantageous) marriage, in Gustave Flaubert's *Madame Bovary*, why Bigger Thomas gets thrown into such a confused mental state in Richard Wright's *Native Son*, or why a family loses its land in John Steinbeck's *The Grapes of Wrath*.

(5) Text-based literary theory

Text-based literary theory demands concentration on the piece of literature itself; that is, it calls for a close reading. With this approach, only the use of concrete, specific examples from the text itself validates an interpretation. Nothing more than what is contained within the text itself—not information about the author's life, culture, or society—is needed to understand and appreciate the text's unchanging meaning. Readers may change, but the meaning of the text does not.

(6) Context-based literary theory

Context-based literary theory considers the historical period during which a work was written and the cultural and economic patterns that prevailed during that period. For example, recognizing that Willa Cather published *My Ántonia* during World War I can help account for the darker side of that novel about European immigrants' harsh life in the American West. Critics who use a context-based and class-based approach known as **cultural studies** consider how a literary work interacts with economic conditions, socioeconomic classes, and other cultural artifacts (such as songs or fashion) of the period in which it was written.

(7) Psychoanalytic theories

Psychoanalytic theories seek to explain human experience and behavior in terms of sexual impulses and unconscious motivations (drives, desires, fears, needs, and conflicts). When applied to literature, these theories (based on the work of Hélène Cixous, Sigmund Freud, Jacques Lacan, and others) help readers discern the motivations of characters, envision the psychological state of the author as implied by the text, and evaluate the psychological reasons for their own interpretations. Readers may apply a psychoanalytic approach to explain why Hamlet is deeply disturbed by his mother's remarriage, why Holden Caulfield rebels at school (in J. D. Salinger's *The Catcher in the Rye*), or why Rochester is blinded (in Charlotte Brontë's *Jane Eyre*).

Theorists who apply the work of psychiatrist Carl Jung to explore **archetypes** (meaningful images that arise from the human unconscious and that appear as recurring figures or patterns in literature) are also using a psychoanalytic approach to interpret literature. Archetypal figures include the hero, the earth mother, the warrior, the outcast, and the cruel stepmother. Archetypal patterns include the quest, the initiation, the test, and the return.

EXERCISE 1

Attend a film, a play, or a poetry reading at your school or in your community. Write a two- to three-page essay evaluating the work, using one of the theoretical approaches discussed in this section.

41e   Conventions for writing about literature

When writing about literature, you need to adhere to several conventions.

(1) Using first person

When writing an analysis of a piece of literature, you may use the first-person singular pronoun, *I*.

Although some critics believe Rudolfo Anáya's novel to be about witchcraft, I think it is about the power of belief.

By using *I*, you indicate that you are presenting your opinion about a work. When you propose or argue for a particular belief or interpretation or offer an opinion, you must support it with specific evidence from the text itself.

(2) Using present tense

Use the present tense when discussing a literary work, since the author of the work is communicating to the reader at the present time (**7b(1)**).

In "A Good Man Is Hard to Find," the grandmother reaches out to touch her killer just before he pulls the trigger.

Similarly, use the present tense when reporting how other writers have interpreted the work you are discussing.

As Toni Morrison demonstrates in her analysis of the American literary tradition, black Americans continue to play a vital role.

(3) Documenting sources

When writing about a work assigned by your instructor, you may not need to give the source and publication information. However, if you are using an edition or translation that may be different from the one your reader(s) will use, you should indicate this. You can document the version of the work you are discussing by using the MLA format for listing works cited (**39b**), although your bibliography in this case will consist of only a single entry.

An alternative way of providing documentation for a single source is by acknowledging the first quotation from or reference to the work using a superscript number and then providing an explanatory note on a separate page at the end of your paper.

In-text citation

... as Toni Morrison states (127).[1]

OR

... tendency to misread texts by African American writers (Morrison 127).[1]

Note

1. Toni Morrison, *Playing in the Dark: Whiteness and the Literary Imagination* (New York: Vintage, 1992). All subsequent references to this work will be identified with page numbers in parentheses within the text.

If you use this note form, you do not need to repeat the bibliographical information in a separate entry or include the author's name in subsequent parenthetical references to page numbers. Check with your instructor about the format he or she prefers.

When you use a bibliography to provide publication data, you must indicate specific references whenever you quote a line or passage. According to MLA style, such bibliographic information should be placed in the text in parentheses directly after the quotation. A period, a semicolon, or a comma should follow the parentheses (**39a(1)** and **16d(1)**).

Quotations from short stories and novels are identified by the author's name and page number:

"A man planning to spend money on me was an experience rare enough to feel odd" (Gordon 19).

Quotations from poems are referred to by line number:

"O Rose, thou are sick!" (Blake 1).

Quotations from Shakespeare's plays are identified using abbreviations of the titles; the following line is from act I, scene I, line 28 of Shakespeare's play *Much Ado about Nothing*:

"How much better it is to weep at joy than to joy at weeping" (*Ado* 1.1.28).

(4) Quoting poetry

When quoting from poems and verse plays, type quotations involving three or fewer lines in the text and insert a slash (see 17i) with a space on each side to separate the lines.

"Does the road wind uphill all the way? / Yes, to the very end" (Rossetti 1-2).

Christina Rossetti opens her poem "Uphill" with this two-line question and answer.

Quotations of more than three lines should be indented one inch from the left-hand margin and double-spaced. Do not use slashes at the ends of lines, and make sure to follow the original text for line breaks, special indenting, or spacing. For this type of block quotation, place your citation after the final punctuation mark.

(5) Referring to authors' names

Use the full name of the author of a work in your first reference and only the last name in all subsequent references. For instance, write "Charles Dickens" or "Willa Cather" the first time and use "Dickens" or "Cather" after that. Never refer to a female author differently than you do a male author. For example, use "Robert Browning and Elizabeth Barrett Browning" or "Browning and Barrett Browning" (not "Browning and Mrs. Browning" or "Browning and Elizabeth").

41f   Literary interpretation of a short story

In the following literary interpretation, English major Kristin Ford focuses on the political and personal implications of a woman's mental illness as portrayed in Charlotte Perkins Gilman's short story "The Yellow Wallpaper."

Ford 1

Kristin Ford

Professor Glenn

English 232

19 November 2010

The Role of Storytelling in Fighting

Nineteenth-Century Chauvinism

The writer provides a critical overview of the story, demonstrating her understanding of it.

Widely considered to be one of the most influential pieces of early feminist literature, "The Yellow Wallpaper," published in 1892 by Charlotte Perkins Gilman, illustrates nineteenth-century men's patronizing treatment of and abusive power over women, exploring the smudged line between sanity and insanity, men's alleged ability to distinguish between the two, and women's inability to pull themselves out of depression or any form of mental illness without seeming to further demonstrate their insanity. The protagonist of Gilman's story descends into madness, a mental state unnecessarily exacerbated, if not caused, by her husband's prescribed "rest cure,"

The writer defines *double bind*, which is the operative term for her thesis.

which entailed total inactivity and isolation. Such was her double bind: the stronger the constraints of the cure, the worse her mental illness. She had no way to resolve her problem.

During Gilman's time, women were understood largely in relation to the "Cult of True Womanhood," which prescribed women's "proper" place in society, especially within the middle

Ford 2

and upper classes. Piety, purity, submissiveness, and domesticity
were not merely encouraged but demanded in order for a woman
to avoid breaking this strict social code (Lavender). Such virtues
meant that a "true woman" of that time was a wife, housewife, and
mother—always yielding to the demands of her husband and her
family. Any woman who went against these norms risked being cast
out or labeled insane (Mellor 156). Men dominated medicine, and
mental illness remained largely unexplored and thus misunderstood.
Many doctors still feared it and thus ignorantly tried to pass off
serious psychological disorders as cases of "nervousness" or
"hysteria" or "fragile constitutions" (Tierny 1456). One of the most
influential doctors at that time, Silas Weir Mitchell, made popular
his "rest cure," which was thought to be especially effective for such
disorders.

> The writer includes historical background for the story. She uses past tense to refer to these actions and beliefs.

These societal views are reflected in "The Yellow Wallpaper."
The physician husband of the main character imposes the "rest
cure" on her. She is forced to obey her husband and has no choice
in her treatment. Furthermore, her husband does not listen when
she tries to tell him more about her condition, her fears, and her
aspirations. This feature of the story—men not listening to their
wives—accurately reflects the social climate of the late nineteenth
century, when husbands could impose their rules on their wives,

> The writer uses the literary present tense to describe the action in the story itself.

Ford 3

with little (if any) thought given to what the women knew, felt, or wanted.

Such a male-centered ideology fostered the development of the "rest cure," initiated by Weir Mitchell in the late 1880s. He describes his "Rest Treatment for Nervous Disorders" (Tierny 1456) as well as the temperament of women in his book *Fat and Blood: and How to Make Them*:

The writer uses information from a physician's writings to support her interpretation and bolster her historical connection.

> The American woman is, to speak plainly, too often physically unfit for her duties as woman, and is perhaps of all civilized females the least qualified to undertake those weightier tasks which tax so heavily the nervous system of man. She is not fairly up to what nature asks from her as wife and mother. How will she sustain herself under the pressure of those yet more exacting duties which nowadays she is eager to share with the man? (13)

Because of this general belief about American women's fragility (or weakness), Weir Mitchell often diagnosed patients as having neurasthenia, a catch-all term for any nervous disorder that affected mainly women. Many cases, like the one depicted in "The Yellow Wallpaper," were what would now be considered postpartum depression, a legitimate psychological disorder requiring medication and therapy.

Ford 4

Conversely, Weir Mitchell's theory was that neurasthenia was all in a woman's head. His rest treatment, prescribed only for women, involved complete rest, little mental stimulation, and overfeeding. A woman was not allowed to leave her bed for months at a time, and she was certainly never allowed to read or write (Weir Mitchell 39). This tendency to diagnose women as "hysterical," coupled with the era's chauvinism, made it easy for doctors like Weir Mitchell to simply, almost flippantly, dismiss the protesting pleas of mentally ill women.

Gilman herself was prescribed this treatment. In "Why I Wrote 'The Yellow Wallpaper,' " Gilman describes how she tried the "rest cure" for three months and "came so near the border of mental ruin that I could see over" (820). In the end, in order to save herself from insanity, Gilman had to ignore what society told her. She could not lead a domestic, sedentary life without falling into insanity. However, according to Weir Mitchell, such a life was considered sane for a woman, a prime indicator of her mental stability. The resulting conflict between Gilman's personal experience and Weir Mitchell's impersonal theory begs the question "What is true sanity?" For Gilman, the only way to cure herself of her madness was the very thing she was told she could not do: write and engage in mental stimulation. This is the double

The writer presents relevant biographical information about the author of the story.

bind that women of the day faced. What Gilman was prescribed to do caused her to fall further into mental illness, but doing what she needed to do to get over the illness was considered a symptom of insanity. This is the same double bind trapping Gilman's protagonist throughout the story.

The rest cure is a tool to suppress all mental activity in women (Tierny 1457). At the beginning of the story, the struggle is among competing factors: what the protagonist is told, what she knows is right, and what she feels she should do. She wants to listen to her husband, but she senses that her illness will not be cured by his proposed remedy. All the while her husband assures her that she only needs the "rest cure" and she will be the wife and mother she should be. Throughout "The Yellow Wallpaper," the wife repeatedly says that although she may be getting physically better, mentally she is not. Her husband repeatedly replies, "Never for one instant let that idea enter your mind! There is nothing so dangerous, so fascinating to a temperament like yours" (Gilman, "Yellow Wallpaper" 814). In addition, he often admonishes her to get well. Gilman juxtaposes what men believed at this time with the actual implications of this cure for the female mind. Although her husband remarks that she seems to be getting better and better, the woman slowly descends further into her madness, showing just how oblivious men, even renowned physicians, were to the struggles of women.

Gilman's goal in this story is to expose this "rest cure" for what it truly is and make clear the struggle women have in a society in which they are expected to be entirely domestic and submissive to men. Gilman makes a particular yet subtle argument when she demonstrates the "domesticated" woman's double bind: If she uses her imagination in an "unsuitable" way, she is exhibiting mental illness. The cure for that illness is constraint, a prohibition on imagination and activity, which only worsens her mental condition. Gilman experienced another double bind as a female author functioning within a realm of male control and expectations. Any woman who published, particularly if her stories dealt with mentally ill women, was revealing her own mental instability. Of course, if an author was not able to write and publish, she would feel even worse.

Gilman portrays the feminist challenge to society's standards through character development and the interactions between the physician husband and his wife. When developing the character of the husband, Gilman illustrates his dominance over his wife through much of their dialogue. The physician speaks to his wife much like an adult speaking to a child. Gilman juxtaposes the husband's view of the woman's improving health against what the reader actually sees happening: the woman creeps around the room becoming completely involved in the pattern of the wallpaper, clearly a sign that she is becoming increasingly ill. This disconnect between what the husband wants to believe and the reality of his wife's condition exemplifies the disconnect in their marital life. It demonstrates the lack of understanding men had toward women and the lack of concern with which they reacted to women's problems.

In "The Yellow Wallpaper," Gilman produced an insightful work using the symbolism of a room turned jail cell to express her views on the way women were treated in her society. Gilman masterfully crafted a story that describes a woman's descent into madness, using that descent as an allegory for the oppression of women of the late nineteenth century. Beyond its importance as a powerful piece of feminist literature, "The Yellow Wallpaper" made a profound impact on its society. After the publication of "Why I Wrote 'The Yellow Wallpaper,' " Weir Mitchell quietly changed his "rest cure." For a respected physician in the late nineteenth century to change his practice based on the literary work of a woman is powerful testimony to the impact of "The Yellow Wallpaper."

Ford 8

Works Cited

Gilman, Charlotte P. "Why I Wrote 'The Yellow Wallpaper.'" *The Norton Anthology of American Literature*. 7th ed. Vol. C. New York: Norton, 2007. Print.

---. "The Yellow Wallpaper." *The Norton Anthology of American Literature*. 7th ed. Vol. C. New York: Norton, 2007. Print.

Lavender, Catherine. "The Cult of Domesticity & True Womanhood." *Women in New York City, 1890–1940*. The College of Staten Island of CUNY, Fall 1998. Web. 21 Nov. 2008. <http://www.csi.cuny.edu/dept/history/lavender/386/truewoman.html>.

Mellor, Ann K. *Romanticism and Feminism*. Bloomington: Indiana UP, 1988. Print.

Tierny, Helen. *Women's Studies Encyclopedia*. Westport: Greenwood, 1997. Print.

Weir Mitchell, Silas. *Fat and Blood: and How to Make Them*. Philadelphia: Lippincott, 1882. Print.

EXERCISE 2

Based on your reading of Kristin Ford's essay on "The Yellow Wallpaper," what personal or political values do you think she brought to her interpretation of that text? Which of the theoretical approaches to literature did she use as the basis for her interpretation (**41d**)? Write a one- or two-page paper analyzing her interpretation of the story.

42 | Writing in Business

Writing in business, like writing in any other environment, requires close attention to rhetorical opportunity, audience, purpose, message, and context. It differs, however, in the nature of authorship: as a business writer, you need to present yourself and your employer as credible and reliable. To do so, you need to follow the conventions and formats expected by the business community.

On the job or in business courses, you will receive a variety of writing assignments: letters, memos and e-mails, PowerPoint presentations, oral reports, and business reports. This chapter will help you

- recognize the stylistic conventions of standard business writing (**42a**),
- draft a business letter (**42b**),
- produce business memos and e-mails (**42c**),
- compose a résumé (**42d**) and a letter of application (**42e**),
- prepare an oral report including a PowerPoint presentation (**42f**), and
- research and write a formal business report (**42g**).

42a | Conventions of language and organization

Whether you are using a word processor or writing on paper, preparing a memo or a business plan, you will face both anticipated and unexpected deadlines for business documents. The following strategies for effective business communication will help you produce comprehensive, concise, and well-organized documents on time.

STRATEGIES FOR EFFECTIVE BUSINESS COMMUNICATION

Be direct.

- Know who your audience members are and consider their needs.
- State the purpose of your document in your opening sentence or paragraph.
- Write straightforward sentences, beginning with a subject and including an active verb (**7c(2)**).
- Use technical language sparingly, unless the document is intended for a specialized audience (**19b(4)**).

Be concise.

- Compose direct, uncomplicated sentences.
- Include only necessary details.
- Use numbers, bullets, or descriptive headings that guide readers to information.
- Use graphs, tables, and other visual elements that convey information succinctly.

Use conventional formatting.

- Follow the standard formats that have been established within a business or industry or use the formats outlined in this chapter (**42b–f**).
- Avoid informal language unless you know that a casual tone is acceptable.
- Edit and proofread your documents carefully. Typos, grammatical mistakes, sentence fragments, and missing words detract significantly from your ethos.

Business letters serve a variety of purposes—to inquire, to inform, to complain, or to respond, for example. (For letters of application, see **42e**.) Regardless of its purpose, a business letter is usually single-spaced and fits on one sheet of paper. It also follows a standard block format: each element is aligned flush with the left margin, with double spacing between paragraphs.

ELEMENTS OF A STANDARD BUSINESS LETTER

- **Return address.** Your employer may require you to use stationery with a letterhead. If not, type your mailing address one inch from the top of the paper, flush left on a one-inch margin, and single-spaced.

- **Date.** Type the date beneath your return address. If you are using letterhead stationery, type the date one or two lines below the letterhead's last line.

- **Recipient's name and address.** Provide the full name and address of the recipient. Single-space these lines, and allow an extra line space above them. If you do not know the person's name, try to find it by checking the company's website or phoning the company. If you cannot find the recipient's name, use an appropriate title such as *Personnel Director* or *Customer Service Manager* (not italicized).

- **Greeting.** Type your greeting two lines below the last line of the recipient's address. The conventional greeting is *Dear* (not italicized) followed by the recipient's name and a colon. If you and the recipient use first names to address each other, use the person's first name. Otherwise, use *Mr., Ms., Mrs.,* or *Miss* and the last name. (Choose *Ms.* when you do not know a woman's preference.) Avoid the sexist *Dear Sir, Gentlemen,* or

Dear Madam and the stilted *To Whom It May Concern* or *Dear Sir or Madam.*

- **Body of the letter.** Begin the first paragraph two lines below the greeting. Single-space lines within a paragraph; double-space between paragraphs. If your letter must continue on a second page, include the recipient's last name, the date, and the page number in three single-spaced lines at the top left on the second page.

- **Closing.** Close your letter two lines after the end of the body with an expression such as *Sincerely* or *Cordially* (not italicized) followed by a comma.

- **Signature.** Type your full name four lines below the closing. Then, in the space above your typed name, sign your full name, using blue or black ink. If you have addressed the recipient by his or her first name, sign just your first name.

- **Additional information.** If you are enclosing extra material such as a résumé, type the word *Enclosure* or the abbreviation *Encl.* (not italicized) two lines below your name. You may also note the number of enclosures or the identity of the document(s): for example, *Enclosures (3)* or *Encl.: 2002 Annual Year-End Report.* If you would like the recipient to know the names of people receiving copies of the letter, use the abbreviation *cc* (for "carbon copy") and a colon followed by the other recipients' names. Place this element on the line directly below the enclosure line or, if there is no enclosure, two lines below your name.

The sample **letter of inquiry** (a letter intended to elicit information) in Figure 42.1 illustrates the parts of a typical business letter.

Return address and date	550 First Avenue Ellensburg, WA 98926 February 4, 2009
Name and address of recipient	Mr. Mark Russell Bilingual Publications 5400 Sage Avenue Yakima, WA 98907
Greeting	Dear Mr. Russell:
Body of letter	I am a junior in the Bilingual Education Program at Central Washington University. For my coursework, I am investigating positions in publishing that include the use of two languages. Your name and address were given to me by my instructor, Marta Cole, who worked for you from 2003 through 2007. I have learned something about your publications on your website. I am most interested in dual documents—those in both English and Spanish. Could you please send me samples of such documents so that I can have a better idea of the types of publications you produce? I am also interested in finding out what qualifications I would need to work for a business like yours. I am fluent in both Spanish and English and have taken a course in translation. If possible, I would like to ask you a few questions about your training and experience. Would you have time for an interview some day next week?
Closing	Sincerely,
Signature	*Chris Humphrey* Chris Humphrey

Figure 42.1. A sample letter of inquiry. (© 2013 Cengage Learning)

42c   Business memos and e-mails

A **memo** (short for *memorandum*) is a brief document, usually focusing on one topic, sent within a business to announce a meeting, explain an event or situation, set a schedule, or request information or action (see Figure 42.2). E-mail messages are used for internal communication, but they are also used for external communication, for initiating and maintaining relationships with clients, prospective employees, or people at other companies. The basic guidelines for writing memos also apply to e-mail messages.

Because it is circulated internally, a memo or e-mail is usually less formal than a letter. Nonetheless, it should still be direct and concise: a memo should be no longer than a page, and an e-mail

---

To: Intellectual Properties Committee   *Heading*
From: Leo Renfrow, Committee Chair
Date: March 15, 2010
Subject: Review of Policy Statement

At the end of our last meeting, we decided to have our policy   *Body of*
statement reviewed by someone outside our university. Clark   *memo*
Beech, chair of the Intellectual Properties Committee at Lincoln
College, agreed to help us. Overall, as his review shows, the
format of our policy statement is sound. Dr. Beech believes that
some of the content should be further developed, however. It
appears that we have used some ambiguous terms and included
some conditions that would not hold up in court.

My assistant will deliver a copy of Dr. Beech's review to each of you.
Please look it over before our next meeting, on March 29. If you have
any questions or comments before then, please call me at ext. 1540.
I look forward to seeing all of you at the meeting.

Figure 42.2. A sample business memo. (© 2013 Cengage Learning)

no longer than a screen. The following guidelines for formatting these kinds of documents are fairly standard, but a particular company or organization may establish its own format.

ELEMENTS OF A STANDARD BUSINESS MEMO OR E-MAIL

- **Heading.** On four consecutive lines, type *To* (not italicized) followed by a colon and the name(s) of the recipient(s), *From* followed by a colon and your name and title (if appropriate), *Date* followed by a colon and the date, and *Subject* followed by a colon and a few words identifying the memo's subject. (The abbreviation *Re*, for "regarding," is sometimes used instead of *Subject*.) This information should be single-spaced. If you are sending copies to individuals whose names are not included in the *To* line, place those names on a new line beginning with *cc* ("carbon copy") and a colon. Most e-mail software supplies these header lines on any new message.

- **Body.** Use the block format (42b), single-spacing lines within each paragraph and double-spacing between paragraphs. Double-space between the heading and the body of the memo. Open your memo with the most important information and establish how it affects your audience. Use your conclusion to establish goodwill.

The effectiveness of memos and e-mails depends on several essential features: tone, length, and directness. A conversational tone is acceptable for an internal message to a coworker, but a more formal tone is required for a memo or an e-mail to a supervisor or a larger group of associates. One way to enhance the professional tone of your e-mails is to use an e-mail signature: a set of information that identifies you and your institution and is appended to the end of all your outgoing messages. Tone also includes the content of a message, so take care not to mention, let alone forward, any information that you or other correspondents might prefer to keep private. And keep in mind that anything you send in an e-mail can easily be forwarded by others. (For information on writing to multiple audiences, see 31d.)

Excessive length also detracts from the effectiveness of your memos and e-mails. Keep your messages to one page for a memo, one screen (or twenty lines) for an e-mail. Because people tend to read only one rhetorical unit, compose a message that fits on a single page or screen, yet has enough white space to allow for easy reading.

Regular e-mail users receive a large volume of messages every day, scanning messages, deleting many without reading them, and responding quickly to the rest. To ensure that an e-mail receives the attention it merits, announce your topic in the subject line and then arrange and present your message in concise, readable chunks (perhaps bulleted or numbered lists) that incorporate white space and guide recipients to important information. Short paragraphs also allow for white space, which helps readers to maintain their attention and absorb the key points.

TIPS FOR SENDING ATTACHMENTS WITH E-MAIL MESSAGES

- Before you send any attachment, consider the size of the file—many inboxes have limited space and cannot accept large files or multiple files (totalling over 1000 KB) or files that contain streaming video, photographs, or sound clips. If there is a chance that a large file might crash a recipient's e-mail program, call or e-mail the recipient to ask permission before sending it.

- When you do not know the type of operating system or software installed on a recipient's computer, send text-only documents in rich text format (indicated by the file suffix **.rtf**), which preserves most formatting and is recognized by many word-processing programs.

- Attachments are notorious for transmitting computer viruses; therefore, never open an attachment sent by someone you do not know or any attachment if your computer does not have active antivirus software. You can get virus-related updates and alerts on the website for your computer's operating system or from suppliers such as McAfee and Symantec.

42d Résumés

A **résumé** is essentially an argument (Chapter 34) designed to emphasize a person's job qualifications by highlighting his or her experience, education, and abilities. If you create and save your résumé in a word-processing file, you can easily tailor it for each job application.

You want your résumé to be easy to read, with clear headings, adequate white space, and traditional formatting. It should establish a strong link between you and the organization to which you are applying and should include your contact information, your work or educational experience (whichever is more suitable), your honors, your extracurricular interests, and any other relevant information.

Your résumé can be organized in either of two ways. A **chronological résumé** lists positions and activities in *reverse* chronological order; that is, your most recent experience comes first. This format works well if you have a steady job history and want to emphasize your most recent experience because it is closely related to the position for which you are applying. An alternative way to organize a résumé is to list experience in terms of job skills rather than jobs held. This format, called a **functional résumé**, is especially useful when you have the required skills, but your work history in the particular field is modest or you are just starting your career.

Regardless of the format you choose, remember that your résumé is, in effect, going to someone's office for a job interview. Make sure that it is dressed for success. Effective résumés are brief, so try to design your résumé to fit on a single page. Use good-quality paper (preferably white or off-white) and a laser printer. Choose a standard format and a traditional typeface, applying them consistently throughout. Use boldface or italic type only for headings. Resist the impulse to make the design unnecessarily complicated: when in doubt, opt for simplicity.

Joe Delaney's résumé (Figure 42.3) incorporates features of both the chronological and the functional formats. He starts with his education, then describes his computer and technical skills because those relate directly to the position he is applying for; his library job, which he lists later, is less relevant.

TIPS FOR RÉSUMÉ WRITING

- Include your name, address, telephone number, e-mail address, and a fax number, if available.

- Identify your career or job objective if you have a compelling one. You can provide details about your future plans during an interview.

- Whenever possible, establish links between jobs you have had and the job you are seeking, describing tasks and responsibilities that relate directly to the position you seek.

- List your college or university degree and any pertinent areas in which you have had special training, highlighting educational details that best demonstrate your qualifications for the job. Consider including your GPA (your overall GPA or your GPA in your major, whichever is higher), particular coursework (list specific classes or note areas of specialization, such as twenty hours of coursework in accounting), and relevant class projects.

- Do not include personal data such as age, marital status, race, religion, or ethnicity.

- Even if an advertisement or posting asks you to state a salary requirement, avoid any mention of salary in your résumé. That topic is usually deferred until an interview or a job offer.

- To show that you are well organized and thoughtful, use a clean, clear format.

- Meticulously proofread your résumé before sending it, and have others read it carefully as well. Errors in a résumé can destroy your chances of getting an interview.

| Return address and date | Joseph F. Delaney III
138 Main Street, Apt 10D
Cityville, PA 16800
June 4, 2007 |
| --- | --- |
| Name and address of recipient | Mr. Jim Konigs, Human Resource Director
E. G. Hickey Technical Enterprise
333 Cumberville State Road, Suite 110
West Cumberville, PA 19428-2949 |
Greeting	Dear Mr. Konigs:
Body of letter	I am applying for the position of project manager advertised on Monster.com. I graduated on May 15 with a B.S. degree in information sciences and technology from Pennsylvania State University. I believe that my in-depth research and education in information technology make me an ideal candidate for this position.
	I have completed the required coursework and an internship in information technology, consulting, and security, working under such distinguished professors as James Wendle and David Markison. I am currently a teaching instructor with Dr. Markison, responsible for student evaluation and advising. I have served as a project team leader in database management; my team created Rabble Mosaic Creator, a website that allows users to create mosaics out of images.
	In addition, I have applicable experience as a member of the student government's Academic Committee, which analyzes students' problems in light of policy before presenting the issues to the dean and professor in charge. I have also worked in the Penn State libraries.
	I would appreciate the opportunity to talk with you about the position and my interest in risk and project management. I am available for an interview and can be reached at the phone number or e-mail address at the top of my résumé.
Closing	Sincerely,
Signature	*Joseph F. Delaney*
	Joseph F. Delaney III
Enclosure line	Encl.: résumé

Figure 42.4. Model letter of application.

TIPS FOR WRITING LETTERS OF APPLICATION

- Address your letter to a specific person. If you are responding to an ad that mentions a department without giving a name, call the company and find out who will be doing the screening. If you cannot obtain a specific name, use an appropriate title such as *Human Resources Director* (not italicized).

- Be brief. You can assume that the recipient will be screening many applications, so keep your letter to one easy-to-read page.

- Mention that you are enclosing a résumé or refer to it, but do not summarize it. Your goal is to attract the attention of a busy person (who will not want to read the same information in both your letter and your résumé).

- Indicate why you are interested in working for the company or organization to which you are applying. Demonstrating that you already know something about the company and the position, that you can contribute to it, indicates your seriousness and motivation. If you want more information about the company, locate an annual report and other information by searching the Internet (36d).

- In your closing, be sure to specify how and where you can be reached and emphasize your availability for an interview.

42f Oral presentations with PowerPoint

Oral reports accompanied by PowerPoint presentations are commonplace in business. Such reports can be either internal (for supervisors and colleagues) or external (for clients or investors). They may take the form of project status reports, demonstrations of new equipment or software, research reports, or recommendations.

Keep in mind the following guidelines as you compose an oral report and create PowerPoint slides to accompany it.

ELEMENTS OF A STANDARD ORAL PRESENTATION

- **Introduction.** Taking no more than one-tenth of your overall presentation time (for example, one minute of a ten-minute presentation), your introduction should indicate who you are, your qualifications, your topic, and the relevance of that topic to your audience. The introduction provides an outline of your main points so that listeners can easily follow your presentation.

- **Body.** Make sure the organization of your presentation is clear through your use of transitions. You can number each point (*first, second, third,* and so on) and use cause-and-consequence transitions (*therefore, since, due to*) and chronological transitions (*before, following, next, then*). Provide internal summaries to remind your listeners where you have been and where you are going and offer comments to help your audience sense the weight of various points (for example, "Not many people realize that . . ." or "The most important thing I have to share is . . .").

- **Conclusion.** Anyone can simply restate the main ideas in the conclusion to an oral presentation; you will want to consider ways to make your conclusion memorable. To do so, you may want to end with a proposal for action, a final statistic, recommendations, or a description of the benefits of a certain course of action. In general, conclusions should be even shorter than introductions.

**TIPS FOR INCORPORATING POWERPOINT
INTO AN ORAL PRESENTATION**

- Design your slides for your audience, not for yourself. If you need speaking notes for your talk, write them on note cards or type them into the notes section provided below each slide in the PowerPoint program.

- Use text and visuals on the PowerPoint slides that complement the oral part of your presentation and do not repeat what you have said. If everything you are saying is on the slides, your audience will skim each slide and then become bored listening to you catch up.

- Be aware of the limitations of PowerPoint. For example, PowerPoint slides do not accommodate large amounts of text. In general, use no more than five lines of text per slide. Because PowerPoint tends to encourage oversimplification of information, be sure to tell your audience whenever you had to simplify the information presented on a slide (for example, in order to fit time constraints) and let them know where they can find more details.

- In general, keep text and visuals separate. Alternating predominantly visual slides with slides of text will keep your audience's attention. Let visuals (charts, pictures, or graphs) stand alone with just a heading or a title. Use text slides to define terms, to present block quotes that might be difficult to follow orally, and to list the main points you will be making.

- Time your speaking with your presentation of the slides so that the two components are synchronized. Make sure to give your audience enough time to absorb complex visuals.

Student Emily Cohen and fellow group members created PowerPoint slides to accompany an oral presentation in a business class in which they were writing a business plan. The group created a draft set of slides, then realized that the slides did not follow basic guidelines for effective PowerPoint presentations. Figure 42.5 shows two slides from

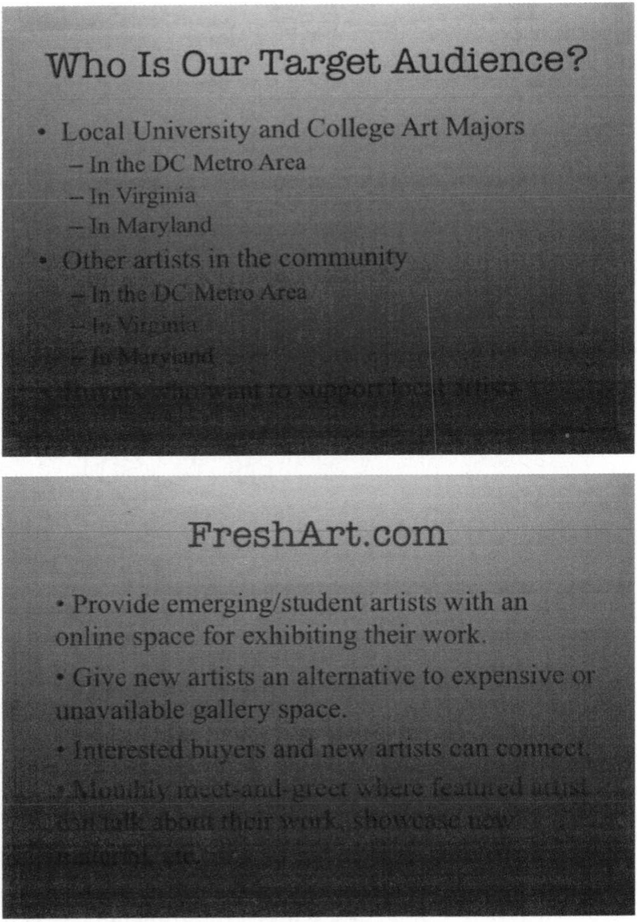

Courtesy of Heather Adams

Figure 42.5. These PowerPoint slides incorporate too much text, and the blue-on-blue design makes the text hard to read. Unnecessary repetition of the subpoints in the lower slide adds clutter and distracts listeners from what the presenter is saying.

Courtesy of Heather Adams

Figure 42.6. A more precise title and concise bulleted points will help audience members skim, not read, the upper slide, letting them focus more on what the presenter is saying. Reducing the amount of text and using contrasting colors make slides easier to read.

the first draft of the presentation. Figure 42.6 shows the same slides, revised so that they better serve as visuals accompanying an oral presentation. Emily's group revised the slides in several ways:

- Using a solid background color and a contrasting typeface to make reading easier
- Limiting the slides to one sans-serif typeface
- Writing more concise bulleted points to cover main ideas
- Removing subpoints that were discussed in the presentation
- Making all bulleted items parallel in structure (by starting each point with an active verb)
- Incorporating simple graphics created with PowerPoint software

42g   Business reports

Business reports take many forms, including periodic reports, sales reports, progress reports, incident reports, and longer reports that assess relocation plans, new lines of equipment or products, marketing schemes, and so on. The following box describes elements of such reports.

ELEMENTS OF A STANDARD BUSINESS REPORT

- **Front matter.** Depending on the audience, purpose, and length of a given report, the front matter materials may include a letter of transmittal (explaining the relevance of the report), a title page, a table of contents, a list of illustrations, and/or an abstract.

- **Introduction.** This section should identify the problem addressed by the report (the rhetorical opportunity), present background information about it, and include a purpose statement and a description of the scope of the report (a list of the limits that framed the investigation). In a long report, each of these elements may be several paragraphs long, and some may have their own subheadings. An introduction should not take up more than ten to fifteen percent of the length of a report.

- **Body or discussion.** This, the longest section of the report, presents the research findings. It often incorporates charts and graphs to help make the data easy to understand. This section should be subdivided into clear subsections by subheadings or, for a shorter report, paragraph breaks.

- **Conclusion(s).** This section summarizes any conclusions and generalizations deduced from the data presented in the body of the report.

- **Recommendation(s).** Although not always necessary, a section that outlines what should be done with or about the findings is included in many business reports.

- **Back matter.** Like the front matter, the back matter of a report depends on the audience, purpose, and length of the report. Back matter may include a glossary, a list of the references cited, and/or one or more appendixes.

In the following sample business report, a student makes recommendations for changes in business communication curricula based on research into day-to-day business practices.

Changing Forms of Business Communication:

Implications for Business Communication Curricula

Joseph F. Delaney III

Penn State University

June 11, 2007

2

Table of Contents

3

Abstract

American businesses gain a competitive advantage in today's global economy by properly and effectively using various means of business communication. The use of different communication strategies in the workplace is essential to the success of businesses ranging from large corporations to on-campus student technology centers. Thus, any successful business education program must also prepare students to use and choose between the diverse media available for business communication today. This report discusses the strengths and weaknesses of different communication modes such as telephone, e-mail, and face-to-face communication, relying largely on an in-depth study of the Computer Store at Penn State University. Based on research and observations of staff, managers, and executives of the organization, the research team developed some strategies for successful business communication, in particular a critique of the overuse of e-mail in business settings. Finally, this report proposes changes to the business communication curriculum at Penn State University, in light of the team's research into real-world business communication practices.

Introduction

Background

Business communication is constantly evolving as technology provides new methods of communication. In the past two decades,

4

e-mail, instant messaging, video conferencing, and cell phone technology have allowed colleagues to cooperate with unparalleled efficiency. The number of e-mails sent daily in 2006 was estimated at around 62 billion. The International Data Corporation has also reported 600 billion minutes of usage per year by mobile phone users (Berkeley School of Information Management and Systems, 2003, p. 10). However, these new methods have not replaced more traditional forms of communication. The average office worker uses about 12,000 sheets of paper per year. With so many options available, it is sometimes difficult to choose the best one for a given situation. Furthermore, the range of options for communication necessitates changes to business communication curricula.

Problem

In the world of modern business, it can be difficult to choose between memos, telephone calls, voice mails, e-mails, meetings, and other modes of communication. All too often employees choose a form of communication that may not be properly suited to their purpose. More specifically, e-mail is easy to use, but at the same time it leads to impersonal interactions between the sender and the receiver. Many employees use e-mail when contacting individuals they are not closely acquainted with, such as new clients. The problem with this strategy is that, in business, building relationships

5

and networking are valuable activities and can increase productivity and quality throughout an organization.

Jeremy Burton, Vice-President of VERITAS Software Corporation in Silicon Valley, like many other executives, has taken this issue to heart. Burton banned e-mails on Fridays, imposing a small fine for each e-mail sent. Though his 240 employees initially resisted the change, they began to think more critically about their communication with others, and productivity increased (Walker, 2004). The real problem, however, is that this kind of critical thinking should be encouraged before employees reach the workforce. Business communication skills like those that innovative employers like Burton are teaching their employees should be addressed in college coursework, not just on the job. The research team believes that the Penn State College of Business is not currently addressing this need.

Purpose

The object of this report is to encourage the College of Business at Penn State to consider changes to the business communication curriculum. Specifically, this report examines a few major forms of communication, evaluates their strengths and weaknesses, and gives suggestions for when and how to use them effectively. This analysis shows that such topics need to be more

6

thoroughly addressed in business classes at Penn State. The analysis
is followed by a set of recommendations for the college.

Scope

This report includes a general analysis of various types of
communication and their uses. A case study of communication
practices at Penn State's Computer Store examines, in particular,
e-mail overuse and some factors that may contribute to it. The report
shows the value of employees who can properly identify appropriate
communication strategies and media. The recommendations suggest
the relevance and consequences of the findings for the business
curriculum at Penn State.

Discussion

Successful Business Communication

Like those at many other contemporary businesses, the
employees of the Computer Store at Penn State University use many
forms of communication. As Robin Becker, the Director of Sales
and Marketing, said, "There is no one ideal form of communication"
(personal communication, June 5, 2007). Besides e-mail, the Computer
Store uses three effective forms of communication for communicating
with clients. During in-house technical consulting, employees assist
students and parents in person as they purchase and learn to use a
laptop. Second, consultants use the telephone to advise individuals

about purchasing laptops and accessories. The last medium, which has taken off recently, is the store's website. The Computer Store website has become a very useful tool for assisting people in their decision making and preparing them for computer ownership, and it complements the first two methods for communicating with clients. The site helps the meeting with a consultant run more smoothly, whether it takes place face to face or over the phone.

As for communication between workers at the Computer Store, no one form is ideal. Table 1 shows the different modes of communication and ranks their appropriateness for different purposes. Generally, e-mail is used for follow-ups or quick notes. For other tasks, such as negotiating with a corporation to make its software available to all Penn State students, a face-to-face conference or a conference call is preferred. For passing quick notes and bits of information within the Computer Store, Becker explains, e-mail or instant messaging works best (personal communication, June 5, 2007). James Murphy, team leader for the consultants, advised, "Use a mix [of communication], try not to limit yourself to one style, and try to cater to other people. If I need to get a lot of information across, I'll write a memo and send it via e-mail, but if I just need to get across one thing to one person, I will walk over there and tell that person face to face" (personal communication, June 5, 2007).

8

Table 1

Business Communication Methods

How Well Method Is Suited To:	Hard Copy	Phone Call	Voice Mail	E-mail	Meeting	Web-site
Assessing commitment	3	2	3	3	1	3
Building consensus	3	2	3	3	1	3
Mediating a conflict	3	2	3	3	1	3
Resolving a misunderstanding	3	1	3	3	2	3
Addressing negative behavior	3	2	3	2	1	3
Expressing support/ appreciation	1	2	2	1	2	3
Encouraging creativity	2	3	3	1	3	3
Making an ironic statement	3	2	2	3	1	3
Conveying a document	1	3	3	3	3	2
Reinforcing one's authority	1	2	3	3	1	1
Providing a permanent record	1	3	3	1	3	3
Maintaining confidentiality	2	1	2	3	1	3
Conveying simple information	3	1	1	1	2	3
Asking an informational question	3	1	1	1	3	3
Making a simple request	3	1	1	1	3	3
Giving complex instructions	3	3	2	2	1	2
Addressing many people	2	2	2	2	3	1

Key: 1 = Excellent   2 = Adequate   3 = Inappropriate
Note. From *Email Composition and Communication (EMC2)* by
T. Galati. Practical Communications, Inc. (www.praccom.com), 2001.

9

Research into business communication confirms the necessity of the Computer Store's reliance on modes of communication other than e-mail. When working with a large number of people, businesses need to be able to get a message across accurately and promptly. If a message is misinterpreted by just one individual, a large-scale problem might result. When an e-mail message requires immediate attention, some businesses, like the Computer Store, address the problem of delayed response by using instant messaging.

Though instant messaging may alleviate in-house communication lapses, other media, like websites, are more suited to reaching foreign audiences. Foreign investment by U.S. companies was approximately $9 trillion in 1966 but had grown to $300 trillion in 2002 (Blalock, 2005). This rising foreign investment by U.S. companies has increased the need for better global communication media. Though the Penn State Computer Store is, for the most part, not communicating on a global scale, its practices demonstrate the same need for a variety of communication media, without an uncritical overreliance on e-mail.

E-mail Overuse

Many factors go into choosing the proper medium for a given situation, such as privacy concerns, size of the group, type of

10

information, and desired level of immediacy. People tend to overuse e-mail because it is an easy, inexpensive way to send information to several people. When asked why sending an e-mail to a large group is not as effective as sending one to a small group, CIO Carol Hildenbrand, answered:

> As a group increases in size, you have a whole slew of management challenges. Communicating badly exponentially increases the possibility of making fatal mistakes. A large-scale project has a lot of moving parts, which makes it that much easier to break down. Communication is the oil that keeps everything working properly. It's much easier to address an atmosphere of distrust among a group of five team members than it is with a team of 500 members. (Schwalbe, 2006, p. 399)

Information distribution, therefore, involves more than creating and sending status reports, and different media are suited to different contexts, purposes, and audiences. Table 1 shows some findings about the suitability of forms of communication to particular business goals, but each situation should be assessed individually.

Implementation of Communication Strategies

One of the most important functions of business communication is to transfer information from one level in a hierarchy to another: from employee to manager, for example, or vice versa. After examining the functioning of the Penn State Computer Store, the research team found that it had a very effective communication structure. Employees shared their insights as to how different kinds of information are conveyed through different media. Ideas, proposals, and other important information are generally communicated in person or over the phone. E-mail is relied on for notifications and follow-ups, but not for complex tasks where interactivity and collaboration are desired.

Personal Relationships

It is vital for a business to build successful relationships that might prove beneficial in the future. Yet in today's workplace, it is becoming harder to develop a network of relationships because of the overuse of e-mail. The Penn State Computer Store avoids this problem by having employees within the office interact face to face. The store also maintains good relationships with customers at other campuses by sending out consultants to meet with them. Face-to-face communication creates perceived added value for the store's services and products. When a consumer buys a product or a warranty from the Computer Store, he or she reaps the benefits of the store's strong network of coworkers who have good communication skills.

12

Conclusion

Over the past few decades, modes of communication have changed rapidly. When we have something to say, we have the option of sending an instant message or an e-mail, making a phone call, sending a text message, posting information on a web site or a blog, or even creating a podcast. The rise of these new means of communication has posed some problems for businesses that are not operating with a high level of efficiency. Without employees who can both use the technologies and, more importantly, choose the most fitting technology for a given situation, businesses will not be able to communicate efficiently either internally or externally, with clients and other businesses.

Research at the Penn State Computer Store showed that each form of communication has its own benefits and drawbacks. If an important message needs to be conveyed, a face-to-face meeting is recommended. If an employee wants to check with a supervisor before leaving the office for an hour, an instant message is sufficient. However, there is an overuse of e-mail, which has become a mode used between people who are unfamiliar with each other and are likely, if using this medium, to remain that way. This impersonality may hurt businesses because networking and building relationships are crucial to business success.

13

Recommendation

Overview of the Problem

Business curricula are not devoting enough time to teaching potential new employees strategies and procedures for communicating properly through different media. This issue needs to be addressed immediately by the College of Business at Penn State.

Possible Solutions

The way to correct this problem is to increase the amount of business communication courses offered in the College of Business. The inclusion of more communication-oriented material will increase graduates' abilities to begin and maintain careers in the business world. The curriculum should cover topics such as the use and misuse of e-mail and the advantages and disadvantages thereof. Courses should cover networking within businesses and explore how this networking can create successful relationships among employees. The curriculum should also present students with multiple opportunities to work in groups using various media as well as the opportunity to develop their communication skills before entering the business community.

Benefits

Implementation of this new curriculum will put Penn State business students in an enviable position for future employment

14

with successful companies. Businesses need employees well versed in communication techniques crucial to a global market. Penn State students will be well placed with these revisions to the business communication curriculum. Ultimately, as Penn State students achieve more success in business, the prestige and reputation of the College of Business will continue to grow.

15

References

Berkeley School of Information Management and Systems. (2003). *Executive summary: How much information?* Retrieved from http://www2.sims.berkeley.edu/research/projects/how-much-info-2003/execsum.htm

Blalock, M. (2005, December 23). *Why good communication is good business.* Retrieved from http://www.bus.wisc.edu/update/winter05/business_communication.asp

Schwalbe, K. (2006). *Information technology project management* (4th ed.). Waterloo, Ontario, Canada: Thomson Course Technology.

Walker, M. (2004, August 26). The day the e-mail dies. *The Wall Street Journal Online*. Retrieved from http://www.lucid-minds.com/public/p66.htm

43 Determiners, Nouns, and Adjectives

If you grew up speaking a language other than English, consider yourself fortunate: knowing two languages provides many cultural and linguistic resources. You may, however, encounter difficulties as you write in English, because it is impossible to translate word for word from one language to another. Studying the topics in this chapter and in Chapters 44 and 45 may help you overcome typical challenges faced by multilingual writers. This chapter will help you

- use determiners with proper nouns (43a),
- use determiners with count nouns (43b),
- use determiners with noncount nouns (43c),
- use determiners with adjectives (43d), and
- shift from nonspecific to specific references (43e).

43a Determiners and proper nouns

A **proper noun** is the specific name of someone or something (for example, *Benito, Mexico,* or *Museum of Popular Art*). Proper nouns are capitalized (9a). A **determiner** is a noun marker—a word that comes *before* a noun. The most common determiners are **articles** (*a, an,* and *the*), **demonstratives** (*this, that, these,* and *those*), **possessives** (*my, your, his, her, its, our,* and *their*), and **quantifiers** (*many, few, much,* and *less*). Choosing the right determiner may be difficult for you if your primary language does not have determiners or if its determiners function differently than those in English.

Proper nouns are preceded either by *the* or by no article at all. To decide whether to use *the* with a proper noun, first ask yourself this question: **Is the proper noun singular or plural?** If the proper noun is plural, use *the* (*the Browns, the United States*). If the proper noun is singular, you can usually omit the article (*Jeff, Mr. Brown, Atlanta, China*). However, the following singular proper nouns *are* preceded by *the*.

SINGULAR PROPER NOUNS PRECEDED BY *THE*

- **Historical documents, periods, and events:** *the* Magna Carta, *the* Renaissance, *the* Velvet Revolution

- **Buildings, hotels, museums, and some bridges:** *the* Burj Khalifa, *the* Sheraton, *the* Prado, *the* Brooklyn Bridge

- **Oceans, seas, rivers, and deserts:** *the* Pacific Ocean, *the* Aegean Sea, *the* Amazon River, *the* Sahara Desert

- **Names that include *of:*** *the* Gulf of Mexico, *the* Statue of Liberty, *the* University of Tennessee, *the* Fourth of July

---

43b Determiners and count nouns

A **common noun** is a general label (such as *president, country, museum*). Common nouns are not capitalized. A common noun may or may not be preceded by a determiner (43a). To decide whether to use a determiner with a common noun, begin by asking yourself this question: **Is the noun a count noun or a noncount noun?** A **count noun** names something that can be counted and has a singular and a plural form (one *car,* two *cars*). If the common noun is a count noun, ask yourself another question: **Is the count noun referring to someone or something specific?** If so, the following guidelines indicate appropriate determiners to use with the noun.

DETERMINERS WITH COUNT NOUNS: SPECIFIC REFERENCES

- A singular count noun is preceded by *the, this, that,* or a possessive.

    The/This/That/My book is heavy.

- A plural count noun is preceded by *the, these, those,* or a possessive.

    The/These/Those/My books are heavy.

- Count nouns that refer to unique individuals or entities (that have only one possible referent) are preceded by *the.*

    The president of the organization will arrive soon.

    The sun sets in the west.

If a count noun does not refer to someone or something specific, but instead to a type of individual or entity, use the following guidelines.

DETERMINERS WITH COUNT NOUNS: NONSPECIFIC REFERENCES

- The article *a* is used before a singular count noun or an adjective beginning with a consonant *sound* (not necessarily the letter representing the sound).

    We went to *a* **c**afe.

    He lives in *a* **s**mall apartment.

- The article *an* is used before a singular count noun or an adjective beginning with a vowel *sound* (not necessarily the letter representing the sound).

    They had *an* **a**rgument.

    The problem had *an* **e**asy solution.

- Plural count nouns are preceded by *some, many,* or *few* when the quantity is a consideration.

    Some/Many/Few students have volunteered.

- Plural count nouns take no determiner at all when quantity is not a consideration.

    Students have volunteered.

    Potatoes are grown in Idaho.

43c   **Determiners and noncount nouns**

A **noncount noun** names something that cannot be counted; it has neither a singular nor a plural form. Some noncount nouns never take determiners.

TYPES OF NONCOUNT NOUNS THAT TAKE NO DETERMINERS

- **Games and sports:** baseball, basketball, chess, football, poker, tennis

    Soccer is my favorite sport.

- **Subjects of study:** biology, chemistry, economics, English, history, mathematics, psychology, sociology

    English is my favorite subject.

Other types of noncount nouns may or may not take determiners.

TYPES OF NONCOUNT NOUNS THAT MAY OR MAY NOT TAKE DETERMINERS

- **Abstractions:** democracy, education, health, knowledge, love

  An **education** is of utmost importance.

  Education is crucial to economic security.

- **Groups of things:** clothing, equipment, garbage, homework, money, traffic

  The **homework** for French class is time-consuming.

  I spend a lot of time doing **homework.**

- **Substances:** air, blood, coffee, ice, rice, tea, water, wood

  This **tea** is watery.

  She prefers **tea** for breakfast.

To decide which determiner to use with a noun referring to an abstraction, a group, or a substance, begin with the question you asked about count nouns: **Is the noncount noun referring to something specific?** If it is, use the determiners in the following list.

DETERMINERS WITH NONCOUNT NOUNS THAT REFER TO SPECIFIC ABSTRACTIONS, GROUPS OF THINGS, OR SUBSTANCES

Use *the, this, that,* or a possessive before a noncount noun making a specific reference.

The/This/That/Our information is important.

If the noncount noun is *not* referring to something specific, use the following guidelines.

DETERMINERS WITH NONCOUNT NOUNS THAT DO NOT REFER TO SPECIFIC ABSTRACTIONS, GROUPS OF THINGS, OR SUBSTANCES

A noncount noun is preceded by the determiner *some, much,* or *little* when quantity is a consideration.

    We drank *some/much/little* water.

A noncount noun takes no determiner at all when quantity is not a consideration.

    We drank only water.

Because noncount nouns do not have singular and plural forms, sentences like the following should be edited:

We learned to use a lot of equipments. [An *s* is not added to a noncount noun.]

I finished two homeworks_∧today. ^assignments^ [Numbers are not used with noncount nouns.]

The job requires a~~ special machinery. [*A* and *an* are not used with noncount nouns.]

The vocabulary_∧~~are~~ ^is^ difficult. [Use a singular verb with a noncount noun.]

Some words can be used as either a count noun or a noncount noun.

They believed **life** was sacred. [noncount noun]

He led *an* interesting **life.** [count noun]

43d   Determiners and adjectives

Some adjectives add specificity to nouns. Use *the* before the following types of adjectives.

ADJECTIVES AND DETERMINERS: SPECIFIC REFERENCES

- Adjectives indicating sequence, such as *first, next, last,* and so forth

  The **first/next/last** person in line will win a prize.

- Adjectives indicating a single person or item, such as *right, only,* and so forth

  She had *the* **right/only** answer.

When describing how one of two individuals or entities differs from or surpasses the other, use the comparative form of an adjective (**4b**). The comparative form has the suffix *-er* or the word *more* or *less*:

Cars are cheap*er* here. Cars are *more* expensive there.

Use the article *the* before the comparative form in this phrase: *the* [comparative form] *of the two* [plural noun].

The older *of the two* sons is now a doctor.

When describing how one of three or more individuals or entities surpasses all the others, use the superlative form of an adjective (**4b**). There are two superlative forms: (1) the adjective has the suffix *-est* and is preceded by the article *the*, or

(2) the adjective does not have that suffix and is preceded by *the most* or *the least*.

Cars are *the* cheap*est* here. Cars are *the most* expensive there.

Use the following guidelines to help you choose which form to use.

**GUIDELINES FOR FORMING COMPARATIVES
AND SUPERLATIVES**

- One-syllable words generally take the ending *-er* or *-est*: *fast, faster, fastest*.

- Two-syllable words ending in a consonant and *-y* also generally take the ending *-er* or *-est*, with the *y* changed to an *i*: *noisy, noisier, noisiest*.

- Two-syllable adjectives ending in *-ct, -nt,* or *-st* are preceded by *more/less* or *most/least*: *less exact, least exact; more recent*. Two-syllable adjectives with a suffix such as *-ous, -ish, -ful, -ing,* or *-ed* are also preceded by *more/less* or *most/least*: *more/most famous; less/least careful*.

- Two-syllable adjectives ending in *-er, -ow,* or *-some* either take the ending *-er* or *-est* or are preceded by *more/less* or *most/least*: *narrower, more narrow, less narrow, narrowest, most narrow, least narrow*.

- Words of three or more syllables are preceded by *more/less* or *most/least*: *less/least fortunate; more/most intelligent*.

- Some modifiers have irregular comparative and superlative forms:

  little, less, least

  good/well, better, best

| 43e | **Shifting from nonspecific to specific references** |

In writing, you usually introduce an individual or an entity with a nonspecific reference. After you have mentioned the individual or entity, you can use specific references.

First mention

A tsunami <u>warning</u> was issued last night.

Subsequent mention

This <u>warning</u> affected all low-lying areas.

A subsequent mention does not have to repeat the word used in the first mention. However, the word chosen must be closely related to the one introduced first.

The weather service <u>warned</u> people about possible flooding.
The <u>warning</u> included possible evacuation routes.

EXERCISE 1

Edit the following common sayings so that determiners are used correctly.

1. The absence makes a heart grow fonder.
2. The actions speak louder than the words.
3. The bad news travels fast.
4. Best things come in the small packages.
5. The blood is thicker than the water.
6. Don't cry over the spilled milk.

44 | Verbs and Verb Forms

Learning how to use verbs effectively involves more than looking up their meanings in a dictionary. You must also understand how the *form* of a verb affects its meaning. Building on the discussion in Chapter 7, this chapter gives more information about

- verb tenses (44a),
- auxiliary verbs (44b),
- prepositional and phrasal verbs (44c), and
- participles used as adjectives (44d).

44a | Verb tenses

English verbs are either regular verbs (7a(1)) or irregular verbs (7a(2)). This distinction is based on the forms of a verb. The forms of irregular verbs do not follow the set pattern that the forms of regular verbs do. If you have trouble choosing the right verb forms, study the charts on pages 94 and 95–97. As you become more familiar with English verb forms, you will understand how they provide information about time. Keep in mind that, although the words *present, past,* and *future* may lead you to think that these tenses refer to actions or events occurring now, in the past, and in the future, respectively, this strict separation is not always the case.

(1) Simple tenses

Simple tenses have many uses, which are not limited to indicating specific times. The conjugation of the **simple present tense** of a regular verb includes two forms of the verb: the base form and the *-s* form. Notice that the third-person singular form is the only form with the *-s* ending.

Simple Present Tense		
	Singular	**Plural**
First person	I **work**	We **work**
Second person	You **work**	You **work**
Third person	He, she, it **works**	They **work**

Use the simple present tense for the following purposes.

USES OF THE SIMPLE PRESENT TENSE

- To indicate a current state: We **are** ready.
- To report a general fact: The sun **rises** in the east.
- To describe a habitual action: Dana **uses** common sense.
- To add a sense of immediacy to a description of a historical event: **In** 1939, Hitler's armies **attack** Poland.
- To discuss literary and artistic works: Joseph Conrad **writes** about what he sees in the human heart.
- To refer to future events: The festival **begins** next month.

The simple past tense of a regular verb has only one form: the base form with the *-ed* ending. The past tense forms of irregular verbs vary (see **7a(2)**).

Simple Past Tense
I, you, he, she, it, we, they **worked**

The simple past tense is used to refer to completed actions or past events.

USES OF THE SIMPLE PAST TENSE

- To indicate a completed action: He **traveled** to the Philippines.
- To report a past event: The accident **occurred** several weeks ago.

The simple future tense of a regular verb also has only one form: the base form accompanied by the auxiliary *will.*

Simple Future Tense
I, you, he, she, it, we, they **will work**

The simple future tense refers to future actions or states.

USES OF THE SIMPLE FUTURE TENSE

- To promise to perform an action: I **will call** you tonight.
- To predict a future action: They **will finish** the project soon.
- To predict a future state of being: Everyone **will be** weary.

It is also possible to use a form of *be going to* when referring to the future (44b(2)).

I **am going to** study in Russia next year.

(2) Progressive tenses

Progressive tenses indicate that actions or events are repetitive, ongoing, or temporary. The present progressive tense consists of a present-tense form of the auxiliary verb *be* and the present participle (*-ing* form) of the main verb, whether that verb is regular or irregular. Notice that the present participle remains the same regardless of person and number, but the auxiliary *be* appears in three forms: *am* for first-person singular, *is* for third-person singular, and *are* for other person-number combinations.

Present Progressive Tense		
	Singular	**Plural**
First person	I **am working**	We **are working**
Second person	You **are working**	You **are working**
Third person	He, she, it **is working**	They **are working**

The present progressive tense signals an activity in progress or a temporary situation.

USES OF THE PRESENT PROGRESSIVE TENSE

- To show that an activity is in progress: The doctor **is seeing** another patient right now.
- To indicate that a situation is temporary: We **are living** in a yurt right now.
- To refer to an action that will occur at a specific time in the future: They **are leaving** for Alaska next week.

Like the present progressive, the past progressive tense is a combination of the auxiliary verb *be* and the present participle (*-ing* form) of the main verb. However, the auxiliary verb is in the past tense, rather than in the present tense.

Past Progressive Tense		
	Singular	**Plural**
First person	I **was working**	We **were working**
Second person	You **were working**	You **were working**
Third person	He, she, it **was working**	They **were working**

The past progressive tense signals that an action or event occurred in the past and was repeated or ongoing.

USES OF THE PAST PROGRESSIVE TENSE

- To indicate that a past action was repetitive: The new member **was** constantly **interrupting** the discussion.
- To signal that a past action was occurring when something else happened: We **were eating** dinner when we heard the news.

A verb in the future progressive tense has only one form. Two auxiliaries, *will* and *be,* are used with the *-ing* form of the main verb.

Future Progressive Tense
I, you, he, she, it, we, they **will be working**

The future progressive tense indicates that actions will occur over some period of time in the future.

USE OF THE FUTURE PROGRESSIVE TENSE

- To indicate that an action will occur over a span of time in the future: She **will be giving** her report at the end of the meeting.

CAUTION

Some verbs do not express actions but rather mental states, emotions, conditions, or relationships. These verbs are not used in progressive forms; they include *believe, belong, contain, cost, know, like, own, prefer,* and *want.*

The book ∧ <s>is containing</s> many Central American folktales.
      contains

He ∧ <s>is knowing</s> many old myths.
   knows

(3) Perfect tenses

Perfect tenses indicate actions that were performed or events that occurred before a particular time. The present perfect tense is formed by combining the auxiliary *have* with the past participle of the main verb. The participle remains the same regardless of person and number; however, the auxiliary has two forms: *has* for third-person singular and *have* for the other person-number combinations.

	Present Perfect Tense	
	Singular	**Plural**
First person	I **have worked**	We **have worked**
Second person	You **have worked**	You **have worked**
Third person	He, she, it **has worked**	They **have worked**

The present perfect tense is used for the following purposes.

USES OF THE PRESENT PERFECT TENSE

- To signal that a situation originating in the past is continuing into the present: They **have lived** in New Zealand for twenty years.
- To refer to a past action that has current relevance: I **have read** that book already, but I could certainly read it again.

The past perfect tense is also formed by combining the auxiliary *have* with the past participle. However, the auxiliary is in the past tense. There is only one form of the past perfect.

Past Perfect Tense
I, you, he, she, it, we, they **had worked**

The past perfect tense specifies that an action was completed at a time in the past prior to another time or before another past action.

USES OF THE PAST PERFECT TENSE

- To indicate that a past action occurred prior to a given time in the past: By the time he turned forty, he **had earned** enough money for retirement.

- To indicate that a past action occurred prior to another past action: She **had** already **mailed** the letter when she realized her mistake.

- To emphasize the point of preceding discourse: I spent the morning in my office. I shelved all my books, arranged the furniture, hung a few photographs, and learned to use the computer. My new job **had begun**.

The future perfect tense consists of two auxiliaries, *will* and *have*, along with the past participle of the main verb. There is only one form of the future perfect tense.

Future Perfect Tense
I, you, he, she, it, we, they **will have worked**

The future perfect tense refers to an action that is to be completed prior to a future time.

USE OF THE FUTURE PERFECT TENSE

■ To refer to future completion of an action: By this time next year, I **will have finished** medical school.

(4) Perfect progressive tenses

Perfect progressive tenses combine the forms and meanings of the progressive and the perfect tenses. The present perfect progressive form consists of two auxiliaries, *have* and *be*, plus the present participle (*-ing* form) of the main verb. The form of the auxiliary *have* varies with person and number. The auxiliary *be* appears as the past participle, *been*.

Present Perfect Progressive Tense		
	Singular	Plural
First person	I **have been working**	We **have been working**
Second person	You **have been working**	You **have been working**
Third person	He, she, it **has been working**	They **have been working**

The present perfect progressive signals that an action, state, or event originating in the past is ongoing or incomplete.

USES OF THE PRESENT PERFECT PROGRESSIVE TENSE

■ To signal that a state of being is ongoing: I **have been feeling** tired for a week.

■ To indicate that an action is incomplete: We **have been organizing** the conference since April.

The past perfect progressive tense follows the pattern *had* + *been* + present participle (*-ing* form) of the main verb. (The auxiliary *have* is in the past tense.)

Past Perfect Progressive Tense
I, you, he, she, it, we, they **had been working**

The past perfect progressive tense refers to a situation or an action occurring over a period of time in the past and prior to another past action.

USE OF THE PAST PERFECT PROGRESSIVE TENSE

- To indicate that an ongoing action occurred prior to a past action: She **had been living** so frugally all year that she saved enough money for a new car.

The future perfect progressive form has the pattern *will* + *have* + *been* + present participle (*-ing* form) of the main verb.

Future Perfect Progressive Tense
I, you, he, she, it, we, they **will have been working**

The future perfect progressive tense refers to an action that is occurring in the present and will continue to occur for a specific amount of time.

USE OF THE FUTURE PERFECT PROGRESSIVE TENSE

- To indicate that an action will continue until a specified time: In one month, I **will have been working** on this project for five years.

EXERCISE 1

Explain how the meaning of each sentence changes when the verb tense changes.

1. In "Fiji's Rainbow Reef," Les Kaufman (describes/described) the coral reefs of Fiji and (discusses/discussed) the factors affecting their health.

2. Rising water temperatures (damaged/have damaged/did damage) the reefs.

3. The lack of algae (has left/had left) the coral "bleached."

4. Strangely, though, new life (is flourishing/was flourishing/has been flourishing) in some of these areas.

5. Scientists (study/will study) this area to understand its resilience.

(5) Using verb tenses to convey the duration or time sequence of actions and events

When you use more than one tense in a single sentence, you give readers information about how actions or events are related in time and duration.

> When the speaker **had finished,** everyone **applauded.**
> [The past perfect tense *had finished* indicates a time before the action expressed by *applauded.*]

Infinitives and participles can be used to express time relations within a sentence. The present infinitive (*to* + base form) of a verb expresses action occurring later than the action expressed by the main verb.

> They **want to design** a new museum. [The action of designing will take place in the future.]

The perfect infinitive (*to* + *have* + past participle) signals that an action, state, or event is potential or hypothetical or that it did not occur.

> She **hopes to have earned** her degree by the end of next year. [Earning the degree has the potential to occur.]
>
> The governor **would like to have postponed** the vote. [The postponement did not occur.]

The present participle (*-ing* form) indicates simultaneous or previous action.

> **Laughing** loudly, the old friends **left** the restaurant arm in arm. [The friends were laughing as they were leaving.]
>
> **Hearing** that she was ill, I **rushed** right over. [The action of hearing occurred first.]

The perfect participle (*having* + past participle) expresses action completed before the action conveyed by the main verb.

> **Having learned** Spanish at an early age, she **spoke** to the Mexican diplomats in their native language.

The past participle can be used to express either simultaneous action or previous action.

> **Led** by a former Peace Corps worker, the volunteers **provided** medical assistance. [The two actions occurred simultaneously.]
>
> **Encouraged** by job prospects, he **moved** to Atlanta. [The encouragement preceded the move.]

EXERCISE 2

Revise the following sentences so that all verbs express logical time sequences.

1. We expected the storm to have bypassed our town, but it made a direct hit.

2. We would like to have prior notice; however, even the police officers were taken by surprise.

3. Not having known much about flooding, the emergency crew was at a disadvantage.

4. Having thrown sandbags all day, the volunteers had been exhausted by 5 p.m.

5. They went home, succeeding in preventing a major disaster.

44b   Auxiliary Verbs

Auxiliary verbs add nuances of meaning to main verbs (7a(4)). Some provide information about time (44a), while others are used to provide emphasis, to form questions, or to indicate ability, certainty, obligation, and so on.

(1) The auxiliary verb *do*

Unlike *be* and *have,* the auxiliary verb *do* does not occur with other verbs to indicate tense. Instead, it is used in questions, negations, and emphatic sentences.

> **Do** you have any questions? [question]
>
> I **do** not have any questions. [negation]
>
> I **do** have a few comments. [emphatic sentence]

The auxiliary *do* is used only in the simple present (*do, does*) and the simple past (*did*).

(2) Modal verbs

The modal auxiliary verbs in English are *can, could, may, might, must, shall, should, will,* and *would* (7a(4)). English also has **phrasal modals**, which are modal auxiliaries consisting of more than one word. They have meanings similar to those of one-word modals.

> **be able to** (ability): We **were able to** find the original document.

> **have to** (obligation): You **have to** report your test results.

Other common phrasal modals are *be going to, be supposed to, had better, need to,* and *used to.*

Both modal auxiliaries and phrasal modals indicate a variety of meanings, including obligation, permission, and probability. Modal auxiliaries have only two forms: the base form and the perfective form (base form + *have* + past participle). Most phrasal modals have more than two forms (*am able to, is able to, were able to, has been able to*). Only *had better* and *used to* have a single form. The following box shows the most common uses of modal verbs in academic writing.

USING MODAL VERBS

Modal Auxiliaries

Verb	Meaning	Example(s)
can	Ability	New legislation **can** change tax rates.
could	Possibility	The announcement **could** cause unrest.
may	Possibility	Funding **may** be the problem.
must	Obligation	Judges **must** be neutral.
should	Obligation	Dissent **should** be acknowledged.
will	Certainty	A statistical analysis **will** be done.
would	Prediction	All **would** benefit from better roads.

(continued on page 696)

(continued from page 695)

Modal Auxiliaries with *Have* + Past Participle

Verb	Meaning	Example
might have	Conjecture	The accident **might have** caused the delay.
must have	Conjecture	The police **must have** known about the protest.
should have	Criticism	Monitors **should have** reported the incident.

Phrasal Modals in the Present Tense

Verb	Meaning	Example(s)
be able to	Ability	They **are able to** respond quickly.
have to	Obligation	The president **has to** attend the meeting.
need to	Necessity	A good summary **needs to** be clear.

Phrasal Modals in the Past Tense

Verb	Meaning	Example
was/were able to	Ability	They **were able to** finish on time.
had to	Obligation	The journalist **had to** divulge his sources.

EXERCISE 3

Fill in the blank with a modal auxiliary or a phrasal modal and describe the meaning it conveys.

1. Nations at war _____ follow the Geneva Conventions.

2. Everyone _____ be treated humanely.

3. In various wars, humanitarian groups _____ protect innocent victims.

4. Without humanitarian groups, many more people _____ lost their lives.

5. Critics of past wars state that more aid _____ been provided.

44c Prepositional verbs and phrasal verbs

A prepositional verb consists of a verb followed by a preposition; a phrasal verb consists of a verb followed by a particle.

(1) Prepositional verbs
Some verbs are typically followed by prepositions. Following are ten prepositional verbs that commonly occur in academic writing. Some are more often used in the active voice; others are more often used in the passive voice (7c).

TEN PREPOSITIONAL VERBS COMMON IN ACADEMIC WRITING

Active Voice	Passive Voice
depend on	be applied to
lead to	be derived from
look at	be divided into
refer to	be known as
result in	be used in

(2) Phrasal verbs
A **phrasal verb** consists of a verb and a particle such as *up, out,* or *on.* A phrasal verb is often idiomatic, conveying a meaning that differs from the common meanings of the individual words. For example, the definitions that first come to mind for the words *blow* and *up* are not likely to help you understand the phrasal verb *blow up* when it means "to enlarge."

She **blew up** the photograph so that she could see the faces better.

Phrasal verbs may have more than one meaning. *To blow up* means not only "to enlarge" but also "to inflate" or "to explode."

A few phrasal verbs retain the common meanings of the verb and the particle.

The protesters **hung up** a banner.

The verb and particle in most phrasal verbs may be separated by a short noun phrase (**1e(1)**).

She **called** the meeting **off.**

OR

She **called off** the meeting.

If you use a pronoun with a phrasal verb, always place it between the verb and the particle.

The student **turned** <u>it</u> **in** yesterday.

Some phrasal verbs are not separable, however.

The group **went over** the proposal.

I **came across** an interesting fact.

You should be able to find definitions of phrasal verbs in a conventional dictionary; however, a specialized dictionary (**19d**) will also provide information about these verbs' separability.

EXERCISE 4

Insert an appropriate preposition or particle after the verb in each sentence.

1. Overpopulation has brought _____ great changes on earth.

2. Deforestation often leads _____ extinctions.

3. High levels of carbon dioxide result _____ increased global temperatures.

4. Proposals for curbing emissions should be looked _____ closely.

5. Scientists have taken _____ the challenge of slowing the destruction.

44d Participles used as adjectives

Both present participles and past participles can be used as adjectives; however, they are not interchangeable. When you want to indicate an emotion, use a present participle with a noun referring to someone or something that is the cause of the emotion. In the phrase *the exciting tennis match,* the tennis match is the cause of the excitement. Use the past participle with a noun referring to someone who experiences an emotion. In the phrase *the excited crowd,* the crowd is experiencing the excitement.

EXERCISE 5

Choose the correct form of each participle.

1. My uncle is interesting/interested in most but not all sports.
2. He was exciting/excited by the World Cup matches in South Africa.
3. However, he did not like the annoying/annoyed sound of the vuvuzelas.
4. Soccer is his favorite sport; baseball he finds boring/bored.
5. He jokes that being a sports fan is a tiring/tired job.

45 Word Order

The general order in an English sentence is subject-verb-object; however, few sentences consist of just three words. This chapter discusses

- the appropriate sequence for adjectives (**45a**),
- the placement of adverbs of frequency (**45b**),
- the order of adverbs and direct objects (**45c**), and
- the order of words within certain clauses (**45d**).

45a Ordering adjectives

In English, two or three adjectives modifying the same noun are used in a particular order based on their meanings. The following list shows the usual order for adjectives of different types and gives examples of each type:

Evaluator	*fascinating, painful, content*
Size	*large, long, small, short*
Shape	*square, round, triangular*
Age	*young, old, aged, newborn, antique*
Color	*black, white, green, brown*
Origin	*Arabian, Cuban, Peruvian, Slavic*
Material	*silk, paper, pine, rubber*

We visited a **fascinating Italian** village. [evaluator, origin]

An **old black** dog stared at us. [age, color]

45b   Placing adverbs of frequency

Adverbs of frequency (such as *always, never, sometimes,* and *often*) appear before one-word verbs.

He **rarely** goes to horror movies.

However, these adverbs appear after a form of *be* when it is the main verb.

Novels written by Stephen King are **always** popular.

When a sentence contains more than one verb in a verb phrase, the adverb of frequency is placed after the first auxiliary verb.

My friends have **never** read *The Shining.*

45c   Placing adverbs and direct objects

An adverb may be placed after a verb when the verb has no direct object (**1c**).

They worked **efficiently**.

Revise any sentence that includes an adverb before a direct object.

I read ~~quickly~~ the letter ^quickly^.

OR

I ^quickly^ read ~~quickly~~ the letter.

45d   Ordering words within clauses

The word order of embedded questions and adjectival clauses differs from the standard subject-verb-object order of clauses.

(1) Embedded questions

The word order of questions and embedded questions is not the same. Notice the difference in each of the following pairs of sentences:

Is the source reliable? [question]

I do not know whether **the source is** reliable. [embedded question]

How **was the source evaluated?** [question]

He explained how **the source was evaluated**. [embedded question]

Does the author make a good argument? [question]

We should decide whether **the author makes** a good argument. [embedded question]

In the question in each pair, the subject and the verb (or the auxiliary verb if there is one) are inverted; in the embedded question, they are not. The auxiliary verb *do* is not used in embedded questions.

If a question begins with an interrogative pronoun such as *who* or *what* as the subject, the order of the question and the embedded question are the same.

s
Who worked on the project? [question]

s
They did not mention **who worked on the project.**
[embedded question]

(2) Adjectival clauses

If a relative pronoun is the subject of an adjectival clause (**1f(2)**), the word order of the clause is standard.

s   v
Twitter is based in San Francisco, California.

s   v
Twitter, **which is based** in San Francisco, California, has users from around the world.

If the relative pronoun is the object of the adjectival clause, no direct object follows the verb.

do
Protestors sent **tweets.**

Tweets **that** protestors sent ~~them~~ to journalists were highly effective.

EXERCISE 1

Revise the word order of each sentence.

1. The Human Genome Project, which Francis Collins initiated it, turned into an international effort.

2. Prior to 1953, scientists did not know for certain was the structure of DNA a double helix.

3. The discovery of the double helix, which James D. Watson and Francis Crick described it in 1953, eventually led to the study of human genetics.

4. Without an understanding of gene sequences, scientists would not know how do people inherit traits.

5. No one can predict with certainty how will people in the future use this knowledge.

Glossary of Usage

The term *usage* refers to the ways words are used in specific contexts. As you know from speaking and writing every day, the words you choose depend on your audience and your purpose. By consulting the entries in this glossary, you will increase your ability to use words effectively. Many of the entries describe the contexts in which words are used; others distinguish between words that sound or look similar.

The labels below will help you choose appropriate words for your rhetorical situation. Be aware that the idea of standard usage may carry with it the assumption that words not considered standard are inferior. Words labeled *nonstandard* are commonly condemned, even though they may be words some people have grown up hearing and using. A better way to describe usage is to identify what is conventional, or accepted practice, for a specific rhetorical situation.

Conventional   Words or phrases listed in dictionaries without special usage labels; generally considered appropriate in academic and professional writing.

Conversational   Words or phrases that dictionaries label *informal, slang,* or *colloquial;* although often used in informal speech and writing, not generally appropriate for formal writing assignments.

Unconventional   Words or phrases not generally considered appropriate in academic or professional writing and often labeled *nonstandard* in dictionaries; best avoided in formal contexts.

Agreement on usage occurs slowly, often after a period of debate. In this glossary, entries are marked with an asterisk (*) when new usages have been reported by dictionary editors but may not yet be accepted by everyone.

a lot of *A lot of* is conversational for *many, much,* or *a great deal of:* They do not have ~~a lot of~~ **much** time. *A lot* is sometimes misspelled as *alot.*

a while, awhile *A while* means "a period of time." It is often used with the prepositions *after, for,* and *in:* We rested for **a while.** *Awhile* means "a short time." It is not preceded by a preposition: We rested **awhile.**

accept, except The verb *accept* means "to receive": I **accept** your apology. The verb *except* means "to exclude": The policy was to have everyone wait in line, but parents with small children were **excepted.** The preposition *except* means "other than": All **except** Joe will attend the conference.

advice, advise *Advice* is a noun: They asked their attorney for **advice.** *Advise* is a verb: The attorney **advised** us to save all relevant documents.

affect, effect *Affect* is a verb that means "to influence": The lobbyist's pleas did not **affect** the politician's decision. The noun *effect* means "a result": The **effect** of his decision on the staff's morale was positive and long lasting. When used as a verb, *effect* means "to produce" or "to cause": The activists believed that they could **effect** real political change.

all ready, already *All ready* means "completely prepared": The rooms are **all ready** for the conference. *Already* means "by or before the time specified": She has **already** taken her final exams.

****all right** *All right* means "acceptable": The students asked whether it was **all right** to use dictionaries during the exam. *Alright* is not yet a generally accepted spelling of *all right,* although it is becoming more common in journalistic writing.

all together, altogether *All together* means "as a group": The cast reviewed the script **all together.** *Altogether* means "wholly, thoroughly": That game is **altogether** too difficult.

allude, elude *Allude* means "to refer to indirectly": The professor **alluded** to a medieval text. *Elude* means "to evade" or "to escape from": For the moment, his name **eludes** me.

allusion, illusion An *allusion* is a casual or indirect reference: The **allusion** was to Shakespeare's *Twelfth Night.* An *illusion* is a false idea or an unreal image: His idea of college is an **illusion.**

alot See **a lot of.**

already See **all ready, already.**

alright See **all right.**

altogether See **all together, altogether.**

*****among, between** To follow traditional usage, use *among* with three or more entities (a group): The snorkelers swam **among** the fish. Use *between* when referring to only two entities: The rivalry **between** the two teams is intense. Current dictionaries also note the possibility of using *between* to refer to more than two entities, especially when these entities are considered distinct: We have strengthened the lines of communication **between** the various departments.

amount of, number of Use *amount of* before nouns that cannot be counted: The **amount of** rain that fell last year was insufficient. Use *number of* with nouns that can be counted: The **number of** students attending college has increased.

and/or This combination denotes three options: one, the other, or both. These options can also be presented separately with *or:* The student's application should be signed by a parent **and/or** a teacher. The student's application should be signed by a parent, a teacher, **or** both.

*****angry at, angry with** Both *at* and *with* are commonly used after *angry,* although according to traditional guidelines, *with* should be used when a person is the cause of the anger: She was **angry with** me because I was late. Many voters were **angry at** the newspaper's coverage of the debate.

another, other, the other *Another* is followed by a singular noun: **another** book. *Other* is followed by a plural noun: **other** books. *The other* is followed by either a singular or a plural noun: **the other book, the other books.**

anymore, any more *Anymore* meaning "any longer" or "now" most frequently occurs in negative sentences: Sarah doesn't work here **anymore.** Its use in positive sentences is considered conversational; *now* is generally used instead: All he ever does ~~anymore~~ **now** is watch television. As two words, *any more* appears with *not* to mean "no more": We do not have **any more** time.

anyone, any one *Anyone* means "any person at all": We did not know **anyone.** *Any one* refers to one of a group: **Any one** of the options is better than the current situation.

as, because The use of *as* to signal a cause may be vague; if it is, use *because* instead: ~~As~~ **Because** we were running out of gas, we turned around.

****as, like** According to traditional usage, *as* begins either a phrase or a clause; *like* begins only a phrase: My brother drives too fast, just ~~like~~ **as** my father did. Current dictionaries note the informal use of *like* to begin clauses, especially after verbs such as *look, feel,* and *sound.*

assure, ensure, insure *Assure* means "to state with confidence, alleviating any doubt": The flight attendant **assured** us that our flight would arrive on time. *Ensure* and *insure* are usually interchangeable to mean "make certain," but only *insure* means "to protect against loss": The editor **ensured** [OR **insured**] that the reporter's facts were accurate. Physicians must **insure** themselves against malpractice suits.

awhile See **a while, awhile.**

bad Unconventional as an adverb; use *badly* instead. The team played **badly.** However, the adjective *bad* is used after sensory verbs such as *feel, look,* and *smell:* I feel **bad** that I forgot to return your book yesterday.

because See **as, because.**

being as, being that Unconventional; use *because* instead. ~~Being as~~ **Because** the road was closed, traffic was diverted to another route.

beside, besides According to traditional usage, these two words have different meanings. *Beside* means "next to": The president sat **beside** the prime minister. *Besides* means "in addition to" or "other than": She has written many articles **besides** those on political reform. Current dictionaries report that professional writers regularly use *beside* to convey this meaning, as long as there is no risk of ambiguity.

between See **among, between.**

can, may *Can* refers to ability, and *may* refers to permission: You **can** [are able to] drive seventy miles an hour, but you **may** not [are not permitted to] exceed the speed limit. Current dictionaries report that in contemporary usage *can* and *may* are used interchangeably to denote possibility or permission, although *may* is used more frequently in formal contexts.

can't hardly, can't scarcely Unconventional. Use *can hardly* or *can scarcely:* The students **can't hardly** wait for summer vacation.

capital, capitol As a noun, *capital* means either "a governing city" or "funds": The **capital** of Minnesota is St. Paul. An anonymous donor provided the **capital** for the project. As a modifier, *capital* means "chief" or "principal": This year's election is of **capital** importance. It may also refer to the death penalty: **Capital** punishment is legal in some states. A *capitol* is a statehouse; the *Capitol* is the US congressional building in Washington, DC.

censor, censure, sensor As a verb, *censor* means "to remove or suppress because of immoral or otherwise objectionable ideas": Do you think a ratings board should **censor** films? As a noun, *censor* refers to a person who is authorized to remove material considered objectionable: The **censor** recommended that the book be banned. The verb *censure* means "to blame or criticize"; the noun *censure* is an expression of disapproval or blame. The Senate **censured** Joseph McCarthy. He received a **censure** from the Senate. A *sensor* is a device that responds to a stimulus: The **sensor** detects changes in light.

cite, site, sight *Cite* means "to mention": Be sure to **cite** your sources. *Site* is a location: The president visited the **site** for the new library. As a verb, *site* also means "to situate": The builder **sited** the factory near the freeway. The verb *sight* means "to see": The crew **sighted** land. As a noun, *sight* refers to the ability to see or to a view: Her **sight** worsened as she aged. What an incredible **sight!**

climactic, climatic *Climactic* refers to a climax, or high point: The actors rehearsed the **climactic** scene. *Climatic* refers to the *climate:* Many environmentalists are worried about the recent **climatic** changes.

coarse, course *Coarse* refers to roughness: The jacket was made of **coarse** linen. *Course* refers to a route: Our **course** to the island was indirect. *Course* may also refer to a plan of study: I want to take a **course** in nutrition.

compare to, compare with *Compare to* means "to regard as similar," and *compare with* means "to examine for similarities and/or differences": She **compared** her mind **to** a dusty attic. The student **compared** the first draft **with** the second.

complement, complementary; compliment, complimentary *Complement* means "to complete" or "to balance": Their personalities **complement** each other. They have **complementary** personalities. *Compliment* means "to express praise": The professor **complimented** the students on their first drafts. Her remarks were **complimentary.** *Complimentary* may also mean "provided free of charge": We received **complimentary** tickets.

*****compose, comprise** *Compose* means "to make up": That collection **is composed** of medieval manuscripts. *Comprise* means "to consist of": The anthology **comprises** many famous essays. Dictionary editors have noted the increasing use of *comprise* in the passive voice to mean "to be composed of."

conscience, conscious, consciousness *Conscience* means "the sense of right and wrong": He examined his **conscience** before deciding whether to join the protest. *Conscious* means "awake": After an hour, the patient was fully **conscious.** After an hour, the patient regained **consciousness.** *Conscious* may also mean "aware": We were **conscious** of the possible consequences.

continual, continually; continuous, continuously *Continual* means "constantly recurring": **Continual** interruptions kept us from completing the project. Telephone calls **continually** interrupted us. *Continuous* means "uninterrupted": The job applicant had a record of ten years' **continuous** employment. The job applicant worked **continuously** from 2000 to 2009.

could of *Of* is often mistaken for the sound of the unstressed *have:* They **could of have** [OR might **have,** should **have,** would **have**] gone home.

couldn't care less *Couldn't care less* expresses complete lack of concern: She **couldn't care less** about her reputation. *Could care less* is considered unconventional in academic writing.

council, counsel A *council* is an advisory or decision-making group: The student **council** supported the new regulations. A *counsel* is a legal adviser: The defense **counsel** conferred with the judge. As a verb, *counsel* means "to give advice": She **counsels** people with eating disorders.

criteria, criterion *Criteria* is a plural noun meaning "a set of standards for judgment": The teachers explained the **criteria** for the assignment. The singular form is *criterion:* Their judgment was based on only one **criterion.**

***data** *Data* is the plural form of *datum,* which means "piece of information" or "fact": When the **data are** complete, we will know the true cost. However, current dictionaries also note that *data* is frequently used as a mass entity (like the word *furniture*), appearing with a singular verb.

desert, dessert As a noun, *desert* means "a barren land": Gila monsters live in the **deserts** of the Southwest. As a verb, *desert* means "to leave": I thought my friends had **deserted** me. *Dessert* refers to something sweet eaten at the end of a meal: They ordered apple pie for **dessert.**

differ from, differ with *Differ from* means "to be different": A bull snake **differs from** a rattlesnake in a number of ways. *Differ with* means "to disagree": Senator Brown has **differed with** Senator Owen on several issues.

different from, different than *Different from* is generally used with nouns, pronouns, noun phrases, and noun clauses: This school was **different from** most others. The school was **different from** what we had expected. *Different than* is used with adverbial clauses; *than* is the conjunction: We are no **different than** they are.

disinterested, uninterested *Disinterested* means "impartial": A **disinterested** observer will give a fair opinion. *Uninterested* means "lacking interest": She was **uninterested** in the outcome of the game.

****due to** Traditionally, *due to* was not synonymous with *because of:* ~~Due to~~ Because of holiday traffic, we arrived an hour late. However, dictionary editors now consider this usage of *due to* acceptable.

effect See **affect, effect.**

e.g. Abbreviation of *exempli gratia,* meaning "for example." Use only within parentheses: Digestive problems may be treated with herbs (**e.g.,** peppermint and fennel). Otherwise, replace *e.g.* with the English equivalent, *for example:* Social media differ from traditional media, ~~e.g.,~~ for example, television and newspapers. Do not confuse *e.g.* with *i.e.,* meaning "that is."

elicit, illicit *Elicit* means "to draw forth": He is **eliciting** contributions for a new playground. *Illicit* means "unlawful": The newspaper reported their **illicit** mishandling of public funds.

elude See **allude, elude.**

emigrate from, immigrate to *Emigrate* means "to leave one's own country": My ancestors **emigrated from** Ireland. *Immigrate* means "to arrive in a different country to settle": The Ulster Scots **immigrated to** the southern United States.

eminent, imminent *Eminent* means "distinguished": An **eminent** scholar in physics will be giving a public lecture tomorrow. *Imminent* means "about to happen": The merger of the two companies is **imminent.**

ensure See **assure, ensure, insure.**

enthuse Many readers object to the use of *enthuse*. Use *enthusiastic* or *enthusiastically* instead: Students ~~enthused~~ spoke **enthusiastically** about the new climbing wall. They were ~~enthused~~ **enthusiastic** about the new climbing wall.

especially, specially *Especially* emphasizes a characteristic or quality: Some people are **especially** sensitive to the sun. *Especially* also means "particularly": Wildflowers are abundant in this area, **especially** during May. *Specially* means "for a particular purpose": The classroom was **specially** designed for music students.

etc. Abbreviation of *et cetera,* meaning "and others of the same kind." Use only within parentheses: Be sure to bring appropriate camping gear (tent, sleeping bag, mess kit, **etc.**). Because *and* is part of the meaning of *etc.,* avoid using the combination *and etc.*

everyday, every day *Everyday* means "routine" or "ordinary": These are **everyday** problems. *Every day* means "each day": I read the newspaper **every day.**

everyone, every one *Everyone* means "all": **Everyone** should attend. *Every one* refers to each person or item in a group: **Every one** of you should attend.

except See **accept, except.**

explicit, implicit *Explicit* means "expressed clearly and directly": Given his **explicit** directions, we knew how to proceed. *Implicit* means "implied or expressed indirectly": I mistakenly understood his silence to be his **implicit** approval of the project.

farther, further Generally, *farther* refers to geographic distance: We will have to drive **farther** tomorrow. *Further* means "more": If you need **further** assistance, please let me know.

*__feel__ Traditionally, *feel* was not synonymous with "think" or "believe": I ~~feel~~ **think** that more should be done to protect local habitat. Dictionary editors now consider this use of *feel* to be a standard alternative.

fewer, less *Fewer* occurs before nouns that can be counted: **fewer** technicians, **fewer** pencils. *Less* occurs before nouns that cannot be counted: **less** milk, **less** support. *Less than* may be used with measurements of time or distance: **less than** three months, **less than** twenty miles.

*__first, firstly; second, secondly__ Many college instructors prefer the use of *first* and *second*. However, dictionary editors state that *firstly* and *secondly* are also well-established forms.

former, latter Used together, *former* refers to the first of two; *latter* to the second of two: John and Ian are both English. The **former** is from Manchester; the **latter** is from Birmingham.

further See **farther, further.**

get Considered conversational in many common expressions: The weather ~~got better~~ **improved** overnight. I did not know what he ~~was getting at~~ **meant.**

good, well *Good* is an adjective, not an adverb: He pitched ~~good~~ **well** last night. *Good* in the sense of "in good health" may be used interchangeably with *well:* I feel **good** [OR **well**] this morning.

half *A half a* or *a half an* is unconventional; use *half a, half an,* or *a half:* You should be able to complete the questionnaire in **a half ~~an~~** hour.

hanged, hung *Hanged* means "put to death by hanging": The prisoner was **hanged** at dawn. For all other meanings, use *hung:* He **hung** the picture above his desk.

hardly See **can't hardly, can't scarcely.**

has got, have got Conversational; omit *got*: I **have ~~got~~** a meeting tomorrow.

he/she, his/her As a solution to the problem of sexist language, these combinations are not universally accepted. Consider using *he or she* and *his or her.* See **19c(1)**.

herself, himself, myself, yourself Unconventional as subjects in a sentence. Joe and ~~myself~~ I will lead the discussion. See **5a(2)**.

hopefully According to traditional usage, *hopefully* means "with hope," not "it is hoped": **Hopefully,** the negotiators discussed the proposed treaty. However, dictionary editors have accepted the use of *hopefully* as a sentence modifier: **Hopefully,** the treaty will be ratified. If your instructor prefers you to follow traditional usage, use *I hope* in such a sentence: **I hope** the treaty will be ratified.

hung See **hanged, hung.**

i.e. Abbreviation of *id est,* meaning "that is." Use only within parentheses: All participants in the study ran the same distance (**i.e.,** six kilometers). Otherwise, replace *i.e.* with the English equivalent, *that is:* Assistance was offered to those who might have difficulty boarding, ~~i.e.,~~ **that is,** the elderly, the disabled, and parents with small children. Do not confuse *i.e.* with *e.g.,* meaning "for example."

illicit See **elicit, illicit.**

illusion See **allusion, illusion.**

immigrate See **emigrate from, immigrate to.**

imminent See **eminent, imminent.**

****impact** Though *impact* is commonly used as a verb in business writing, many college teachers still use it as a noun only: The new tax ~~impacts~~ **affects** everyone.

implicit See **explicit, implicit.**

imply, infer *Imply* means "to suggest without actually stating": Though he never mentioned the statistics, he **implied** that they were questionable. *Infer* means "to draw a conclusion based on evidence":

Given the tone of his voice, I **inferred** that he found the work substandard.

in regards to Unconventional; see **regard, regarding, regards.**

inside of, outside of Drop *of* when unnecessary: Security guards stood **outside ~~of~~** the front door.

insure See **assure, ensure, insure.**

irregardless Unconventional; use *regardless* instead.

its, it's *Its* is a possessive form: The committee forwarded **its** recommendation. *It's* is a contraction of *it is* or *it has:* **It's** a beautiful day. **It's** been sunny for days.

-ize Some readers object to using this ending to create new verbs, such as *calendarize.* Some of these new verbs, however, have already entered into common usage: *computerize.*

kind of a, sort of a The word *a* is unnecessary: This **kind of a** book sells well. *Kind of* and *sort of* are not conventionally used to mean "somewhat": The report was **~~kind of~~ somewhat** difficult to read.

later, latter *Later* means "after a specific time" or "a time after now": The concert ended **later** than we had expected. *Latter* refers to the second of two items: Of the two versions described, I prefer the **latter.**

lay, lie *Lay* (*laid, laying*) means "to put" or "to place": He **laid** the book aside. *Lie* (*lay, lain, lying*) means "to rest" or "to recline": I had just **lain** down when the alarm went off. *Lay* takes an object (to **lay** something); *lie* does not. These verbs may be confused because the present tense of *lay* and the past tense of *lie* are spelled the same way.

lead, led As a noun, *lead* means "a kind of metal": The paint had **lead** in it. As a verb, *lead* means "to conduct": A guide will **lead** a tour of the ruins. *Led* is the past tense of the verb *lead:* He **led** the country from 1949 to 1960.

less, less than See **fewer, less.**

lie See **lay, lie.**

like See **as, like.**

literally Conversational when used to emphasize the meaning of another word: I was ~~literally~~ **nearly** frozen after I finished shoveling the sidewalk. *Literally* is conventionally used to indicate that an expression is not being used figuratively: My friend **literally** climbs the walls after work; his fellow rock climbers join him at the local gym.

lose, loose *Lose* is a verb: She does not **lose** her patience often. *Loose* is chiefly used as an adjective: A few of the tiles are **loose.**

lots, lots of Conversational for *many* or *much:* He has ~~lots of~~ **many** friends. We have ~~lots~~ **much** to do before the end of the quarter.

many, much *Many* is used with nouns that can be counted: **many** stores, too **many** assignments. *Much* is used with nouns that cannot be counted: **much** courage, not **much** time.

may See **can, may.**

may of, might of See **could of.**

maybe, may be *Maybe* is an adverb: **Maybe** the negotiators will succeed this time. *May* and *be* are verbs: The rumor **may be** true.

***media, medium** According to traditional definitions, *media* is a plural word: The **media** have sometimes created the news in addition to reporting it. The singular form is *medium:* The newspaper is one **medium** that people seem to trust. Dictionary editors note the frequent use of *media* as a collective noun taking a singular verb, but this usage is still considered conversational.

might could Conversational for "might be able to": The director **might** ~~could~~ **be able to** review your application next week.

most Unconventional to mean "almost": We watch the news ~~most~~ **almost** every day.

much See **many, much.**

myself See **herself, himself, myself, yourself.**

neither . . . or *Nor,* not *or,* follows *neither:* The book is **neither** as funny ~~or~~ **nor** as original as critics have reported.

nothing like, nowhere near Unconventional; use *not nearly* instead: Her new novel is ~~nowhere near~~ not nearly as mysterious as her previous one.

number of When the expression *a number of* is used, the reference is plural: **A number of** positions **are** open. When *the number of* is used, the reference is singular: **The number of** possibilities **is** limited. See also **amount of, number of.**

off of Conversational; omit *of:* He walked **off ~~of~~** the field.

on account of Conversational; use *because of:* The singer canceled her engagement ~~on account of~~ because of a sore throat.

on the other hand Use *however* instead or make sure that the sentence or independent clause beginning with this transitional phrase is preceded by one starting with *on the one hand.*

other See **another, other, the other.**

passed, past *Passed* is the past tense of the verb *pass:* Deb **passed** the other runners right before the finish line. *Past* means "beyond a time or location": We walked **past** the high school.

per In ordinary contexts, use *a* or *an:* You should drink at least six glasses of water **per** a day.

percent, percentage *Percent* (also spelled *per cent*) is used with a specific number: **Sixty percent** of the students attended the ceremony. *Percentage* refers to an unspecified portion: The **percentage** of high school graduates attending college has increased in recent years.

perspective, prospective *Perspective* means "point of view": We discussed the issue from various **perspectives.** *Prospective* means "likely to become": **Prospective** journalists interviewed the editor in chief.

phenomena, phenomenon *Phenomena* is the plural form of *phenomenon:* Natural **phenomena** were given scientific explanations.

plus *Plus* joins nouns or noun phrases to make a sentence seem like an equation: Her endless curiosity **plus** her boundless energy makes her the perfect camp counselor. Note that a singular form of the verb is required (e.g., *makes*). *Plus* is not used to join clauses: I telephoned ~~plus~~ **and** I sent flowers.

precede, proceed To *precede* is to "go before": A moment of silence **preceded** the applause. To *proceed* is to "go forward": After stopping for a short rest, we **proceeded** to our destination.

prejudice, prejudiced *Prejudice* is a noun: They were unaware of their **prejudice.** *Prejudiced* is an adjective: She accused me of being **prejudiced.**

pretty *Pretty* means "attractive," not "rather" or "fairly": We were ~~pretty~~ **fairly** tired after cooking all day.

principal, principle As a noun, *principal* means "chief official": The **principal** greeted the students every day. It also means "capital": The loan's **principal** was still quite high. As an adjective, *principal* means "main": Tourism is the country's **principal** source of income. The noun *principle* refers to a rule, standard, or belief: She explained the three **principles** supporting the theory.

proceed See **precede, proceed.**

prospective See **perspective, prospective.**

quotation, quote In academic writing, *quotation,* rather than *quote,* refers to a sentence or passage repeated or copied from another source: She began her speech with a ~~quote~~ **quotation** from *Othello. Quote* expresses an action: My friend sometimes **quotes** lines from television commercials.

raise, rise *Raise* (*raised, raising*) means "to lift or cause to move upward, to bring up or increase": Retailers **raised** prices. *Rise* (*rose, risen, rising*) means "to get up" or "to ascend": The cost of living **rose** sharply. *Raise* takes an object (to **raise** something); *rise* does not.

real, really *Really* rather than *real* is used to mean "very": He is from a ~~real~~ **really** small town. To ensure this word's effectiveness, use it sparingly.

*****reason why** Traditionally, this combination was considered redundant: No one explained **the reason ~~why~~** the negotiations failed.

[OR No one explained ~~the reason~~ **why** the negotiations failed.] However, dictionary editors report its use by highly regarded writers.

regard, regarding, regards These forms are used in the following expressions: *in regard to, with regard to, as regards,* and *regarding* [NOT *in regards to, with regards to,* or *as regarding*].

***relation, relationship** According to traditional definitions, *relation* is used to link abstractions: We studied the **relation** between language and social change. *Relationship* is used to link people: The **relationship** between the two friends grew strong. However, dictionary editors now label as standard the use of *relationship* to connect abstractions.

respectfully, respectively *Respectfully* means "showing respect": The children learned to treat one another **respectfully.** *Respectively* means "in the order designated": We discussed the issue with the chair, the dean, and the provost, **respectively.**

rise See **raise, rise.**

sensor See **censor, censure, sensor.**

sensual, sensuous *Sensual* refers to gratification of the physical senses, often those associated with sexual pleasure: Frequently found in this music are **sensual** dance rhythms. *Sensuous* refers to gratification of the senses in response to art, music, nature, and so on: **Sensuous** landscape paintings lined the walls of the gallery.

set, sit *Sit* means "to be seated": Jonathan **sat** in the front row. *Set* means "to place something": The research assistant **set** the chemicals on the counter. *Set* takes an object (to **set** something); *sit* does not.

shall, will Traditionally, *shall* was used with *I* or *we* to express future tense, and *will* was used with the other personal pronouns, but *shall* has almost disappeared in contemporary American English. *Shall* is still used in legal writing to indicate an obligation.

should of See **could of.**

sight See **cite, site, sight.**

site See **cite, site, sight.**

so *So* intensifies another word when it is used with *that:* He was **so** nervous **that** he had trouble sleeping. Instead of using *so* alone, find a precise modifier: She was **so intensely** focused on her career. See **22c.**

sometime, sometimes, some time *Sometime* means "at an unspecified time": They will meet **sometime** next month. *Sometimes* means "at times": **Sometimes** laws are unfair. *Some time* means "a span of time": They agreed to allow **some time** to pass before voting on the measure.

sort of a See **kind of a, sort of a.**

specially See **especially, specially.**

supposed to, used to Be sure to include the frequently unsounded *d* at the end of the verb form: We are **supposed to** leave at 9:30 a.m. We **used to** leave earlier.

than, then *Than* is used in comparisons: The tape recorder is smaller **than** the radio. *Then* refers to a time sequence: Go straight ahead for three blocks; **then** turn left.

*__**that, which**__ *Which* occurs in nonessential (nonrestrictive) clauses: Myanmar, **which** borders Thailand, was formerly called Burma. Both *that* and *which* occur in essential (restrictive) clauses, although traditionally only *that* was considered acceptable: I am looking for an atlas **that** [OR **which**] includes demographic information. (For more information on essential and nonessential clauses, see **12d(1)** and **13d.**)

*__**that, which, who**__ In essential (restrictive) clauses, *who* and *that* refer to people. We want to hire someone **who** [OR **that**] has had experience programming. Traditionally, only *who* was used to refer to people. *That,* as well as *which,* refers to things: He proposed a design **that** [OR **which**] will take advantage of solar energy.

their, there, they're *Their* is the possessive form of *they:* They will give **their** presentation tomorrow. *There* refers to location: I lived **there** for six years. *There* is also used as an expletive (see **21a(3)**): **There** is no explanation for the phenomenon. *They're* is a contraction of *they are:* **They're** leaving in the morning.

theirself, theirselves Unconventional; use *themselves.* The students finished the project by ~~theirself~~ themselves.

then See **than, then.**

to, too, two *To* is an infinitive marker: She wanted **to** become an actress. *To* is also used as a preposition, usually indicating direction: They walked **to** the memorial. *Too* means either "also" or "excessively": I voted for her **too.** They are **too** busy this year. *Two* is a number: She studied abroad for **two** years.

toward, towards Although both are acceptable, *toward* is preferred in American English.

try and Conversational for *try to:* The staff will **try ~~and~~ to** finish the project by Friday.

uninterested See **disinterested, uninterested.**

*****unique** Traditionally, *unique* meant "one of a kind" and thus was not preceded by a qualifier such as *more, most, quite,* or *very:* Her prose style is ~~quite~~ **unique.** However, dictionary editors note that *unique* is also widely used to mean "extraordinary."

use, utilize In most contexts, *use* is preferred to *utilize:* We ~~utilized~~ **used** a special dye in the experiment. However, *utilize* may suggest an effort to employ something for a purpose: We discussed how to **utilize** the resources we had been given.

used to See **supposed to, used to.**

very To ensure this word's effectiveness, use it sparingly. Whenever possible, choose a stronger word: She was **very ~~satisfied~~ delighted** with her new digital camera.

ways Conversational when referring to distance; use *way* instead: It's a long ~~ways~~ **way** from home.

well See **good, well.**

where Conversational for *that:* I noticed ~~where~~ that she had been elected.

where ... at, where ... to Conversational; omit *at* and *to:* **Where** is the library ~~at~~? **Where** are you moving ~~to~~?

which See **that, which** and **that, which, who.**

*****who, whom** *Who* is used as the subject or subject complement in a clause: We have decided to hire Marian Wright, ~~whom~~ who I believe is currently finishing her degree in business administration. [*Who* is the subject in *who is currently finishing her degree in business administration.*] See also **that, which, who.** *Whom* is used as an object: Jeff Kruger, ~~who~~ whom we hired in 2007, is now our top sales representative. [*Whom* is the object in *whom we hired.*] Dictionary editors note that in conversation *who* is commonly used as an object as long as it does not follow a preposition. See **5b(5).**

whose, who's *Whose* is a possessive form: **Whose** book is this? The book was written by a young Mexican American woman **whose** family still lives in Chiapas. *Who's* is the contraction of *who is:* **Who's** going to run in the election? See **5b(3).**

will See **shall, will.**

with regards to Unconventional; see **regard, regarding, regards.**

would of See **could of.**

your, you're *Your* is a possessive form: Let's meet in **your** office. *You're* is a contraction of *you are:* **You're** gaining strength.

yourself See **herself, himself, myself, yourself.**

Glossary of Terms

This glossary provides brief definitions of frequently used terms. Consult the index for references to terms not listed here.

absolute phrase A sentencelike structure containing a subject and its modifiers. Unlike a sentence, an absolute phrase has no verb marked for person, number, or tense: *The ceremony finally over,* the graduates tossed their mortarboards in the air. See **1e(6)**.

active voice See **voice**.

adjectival clause A dependent clause, also called a **relative clause,** that modifies a noun or a pronoun. See **1f(2)**.

adjectival phrase A phrase that modifies a noun or a pronoun.

adjective A word that modifies a noun or a pronoun. Adjectives typically end in suffixes such as *-al, -able, -ant, -ative, -ic, -ish, -less, -ous,* and *-y.* See **1a(4)** and **4a**. **Coordinate adjectives** are two or more adjectives modifying the same noun and separated by a comma: a *brisk, cold* walk. See **12c(2)**.

adverb A word that modifies a verb, a verbal, an adjective, or another adverb. Adverbs commonly end in *-ly.* Some adverbs modify entire sentences: *Perhaps* the meeting could be postponed. See **1a(5)** and **4a**.

adverbial clause A dependent clause that modifies a verb, an adjective, or an adverb. See **1f(2)**.

adverbial conjunction See **conjunctive adverb.**

antecedent A word or group of words referred to by a pronoun. See **1a(3)** and **6b**.

appositive A pronoun, noun, or noun phrase that identifies, describes, or explains an adjacent pronoun, noun, or noun phrase. See **1e(5)** and **5b(4)**.

article A word used to signal a noun. *The* is a definite article; *a* and *an* are indefinite articles. See **1a(4)** and **43a**.

auxiliary verb A verb that combines with a main verb. Also called an **auxiliary** or a **helping verb**. *Be, do,* and *have* are auxiliary verbs when they are used with main verbs. **Modal auxiliaries** include *could, should,* and *may* and are used for such purposes as expressing doubt or obligation and making a request. See **1a(1), 7a(4),** and **44b**.

Boolean operators See **logical operators**.

case The form of a noun or a pronoun that indicates its relationship to other words in a sentence. Nouns and pronouns can be subjects or subject complements (**subjective case**), objects (**objective case**), or markers of possession and other relations (**possessive case**). See **5b**.

claim A statement that a writer wants readers to accept; also called a **proposition**. See **34d**.

clause A sequence of related words forming an independent unit (**independent clause,** or **main clause**) or an embedded unit (**dependent clause** used as an adverb, adjective, or noun). A clause has both a subject and a predicate. See **1f**.

cliché An expression that has lost its power to interest readers because of overuse. See **20b**.

collective noun A noun that refers to a group: *team, faculty, committee*. See **1a(2)**.

collocation Common word combination such as *add to, adept at,* or *admiration for*. See **20c**.

colloquial A label for any word or phrase that is characteristic of informal speech. *Kid* is colloquial; *child* is used in formal contexts. See **19b**.

common noun A noun referring to any or all members of a class or group (*woman, city, holiday*) rather than to specific members (*Susan, Reno, New Year's Day*). COMPARE: **proper noun**. See **1a(2)** and **43b**.

complement   A word or words used to complete the meaning of a verb. A **subject complement** is a word or phrase that follows a linking verb and categorizes or describes the subject. An **object complement** is a word or phrase that categorizes or describes a direct object when it follows such verbs as *make, paint, elect,* and *consider.* See **1c**.

complex sentence   A sentence containing one independent clause and at least one dependent clause. See **1h(3)**.

compound-complex sentence   A sentence containing at least two independent clauses and one or more dependent clauses. See **1h(4)**.

compound predicate   A predicate that has two parts joined by a connecting word such as *and, or,* or *but;* each part contains a verb: Clara Barton *nursed the injured during the Civil War* and *later founded the American Red Cross.* See **1b**.

compound sentence   A sentence containing at least two independent clauses and no dependent clauses. See **1h(2)**.

compound subject   Two subjects joined by a connecting word such as *and, or,* or *but: Students* and *faculty* are discussing the issue of grade inflation. See **1b**.

compound word   Two or more words functioning as a single word: *ice cream, double-check.* See **18f(1)**.

conditional clause   An adverbial clause, usually beginning with *if,* that expresses a condition: *If it rains,* the outdoor concert will be postponed.

conjunction   A word used to connect other words, phrases, clauses, or sentences. **Coordinating conjunctions** (*and, but, or, nor, for, so,* and *yet*) connect and relate words and word groups of equal grammatical rank. See **1a(7)** and **1g(1)**. A **subordinating conjunction** such as *although, if,* or *when* begins a dependent clause and connects it to an independent clause. See **1a(7)** and **1g(3)**.

conjunctive adverb   A word such as *however* or *thus* that joins one independent clause to another; also known as an **adverbial conjunction.** See **1g(4)**.

convention, conventional Refers to language or behavior that follows the customs of a community such as the academic, medical, or business community.

coordinate adjective See **adjective.**

coordinating conjunction See **conjunction.**

coordination The use of grammatically equivalent constructions to link or balance ideas. See **24b.**

correlative conjunctions, correlatives Two-part connecting words such as *either ... or* and *not only ... but also.* See **1a(7)** and **1g(2).**

count nouns Nouns naming things that can be counted (*word, student, remark*). See **1a(2)** and **43b.** COMPARE: **noncount nouns.**

dangling modifier A word or phrase that does not clearly modify another word or word group. See **25b.** COMPARE: **misplaced modifier.**

deductive reasoning A form of logical reasoning in which a conclusion is formed after relating a specific fact (minor premise) to a generalization (major premise). See **34h(2).** COMPARE: **inductive reasoning.**

demonstratives Four words (*this, that, these,* and *those*) that distinguish one individual, thing, event, or idea from another. Demonstratives may occur with or without nouns: *This* [demonstrative determiner] *law* will go into effect in two years. *This* [demonstrative pronoun] will go into effect in two years. See **5a(5)** and **43a.**

dependent clause See **clause.**

determiner A word that signals the approach of a noun. A determiner may be an article, a demonstrative, a possessive, or a quantifier: *a reason, this reason, his reason, three reasons.* See **43a–c.**

direct address See **vocative.**

direct object See **object.**

direct quotation See **quotation.**

elliptical clause A clause missing one or more words that are assumed to be understood. See **1f(2)**.

essential element A word or word group that modifies another word or word group, providing information that is essential for identification. Essential elements are not set off by commas, parentheses, or dashes: The woman *who witnessed the accident* was called to testify. Also called a **restrictive element.** COMPARE: **nonessential element.** See **1f(2)** and **12d**.

ethos One of the three classical appeals; the use of language to demonstrate the writer's trustworthy character, good intentions, and substantial knowledge of a subject. Also called an **ethical appeal.** See **34f(1)**. See also **logos** and **pathos.**

expletive A word signaling a structural change in a sentence, usually used so that new or important information is given at the end of the sentence: *There* were over four thousand runners in the marathon. See **1b(1)** and **21a(3)**.

faulty predication A sentence error in which the predicate does not logically belong with the given subject. See **23d**.

figurative language The use of words in an imaginative rather than in a literal sense. See **20a(3)**.

first person See **person.**

gender The grammatical label that distinguishes nouns or pronouns as masculine, feminine, or neuter. In English, grammatical gender usually corresponds to biological gender. Gender also describes how people see themselves, or are seen by others, as either male or female. See **6b(2)** and **19c(1)**.

genre A literary category, such as drama or poetry, identified by its own conventions. See **41a**.

gerund A verbal that ends in *-ing* and functions as a noun: *Snowboarding* is a popular winter sport. See **1e(3)**.

gerund phrase A verbal phrase that employs the *-ing* form of a verb and functions as a noun: Some students prefer *studying in the library.* See **1e(3)**.

helping verb See **auxiliary verb.**

homophones Words that have the same sound and sometimes the same spelling but differ in meaning: *their, there,* and *they're* or *capital* meaning "funds" and *capital* meaning "the top of a pillar." See **18c.**

idiom An expression whose meaning often cannot be derived from its elements. *Burning the midnight oil* means "staying up late studying." See **20c.**

imperative mood See **mood.**

indefinite article See **article.**

indefinite pronoun A pronoun such as *everyone* or *anything* that does not refer to a specific person, place, thing, or idea. See **5a(6).**

independent clause See **clause.**

indicative mood See **mood.**

indirect object See **object.**

indirect question A sentence that includes an embedded question, punctuated with a period instead of a question mark: My friends asked me *why I left the party early.* See **17a(1).**

indirect quotation See **quotation.**

inductive reasoning The reasoning process that begins with facts or observations and moves to general principles that account for those facts or observations. See **34h(1).** COMPARE: **deductive reasoning.**

infinitive A verbal that consists of the base form of the verb, usually preceded by the infinitive marker *to.* An infinitive is used chiefly as a noun, less frequently as an adjective or adverb: My father likes *to golf.* See **1e(3)** and **44a(5).**

infinitive phrase A verbal phrase that contains the infinitive form of a verb: They volunteered *to work at the local hospital.* See **1e(3).**

inflection A change in the form of a word that indicates a grammatical feature such as number, person, tense, or degree. For example, *-ed* added to a verb indicates the past tense, and *-er* indicates the comparative degree of an adjective or adverb.

intensive pronoun See **reflexive pronoun.**

interjection A word expressing a simple exclamation: *Hey! Oops!* When used at the beginnings of sentences, mild interjections are set off by commas. See **1a(8).**

intransitive verb A verb that does not take an object: Everyone *laughed.* See **1d.** COMPARE: **transitive verb.**

inversion A change in the usual subject-verb order of a sentence: *Are you* ready? See **1d.**

linking verb A verb that relates a subject to a subject complement. Examples of linking verbs are *be, become, seem, appear, feel, look, taste, smell,* and *sound.* See **1a(1)** and **1c(3).**

logical operators Words used to broaden or narrow electronic database searches. These include *or, and, not,* and *near.* Also called **Boolean operators.** See **36c(1).**

logos One of the three classical appeals; the use of language to show clear reasoning. Also called a **logical appeal.** See **34f(2).** See also **ethos** and **pathos.**

major premise See **premise.**

main clause Also called **independent clause.** See **clause.**

minor premise See **premise.**

misplaced modifier A descriptive or qualifying word or phrase placed in a position that confuses the reader: I read about a wildfire that was out of control *in yesterday's paper.* [The modifier belongs after *read.*] See **25a.**

mixed construction A confusing sentence that is the result of an unintentional shift from one grammatical pattern to another: When police appeared who were supposed to calm the crowds showed up, most people had already gone home. [The sentence should be recast with either *appeared* or *showed up,* not with both.] See **23c.**

mixed metaphor A construction that includes parts of two or more unrelated metaphors: Her *fiery* personality *dampened* our hopes of a compromise. See **23b**.

modal auxiliary See **auxiliary verb.**

modifier A word or word group that describes, limits, or qualifies another. See **4a**.

mood A set of verb forms or inflections used to indicate how a speaker or writer regards an assertion: as a fact or opinion (**indicative mood**); as a command or instruction (**imperative mood**); or as a wish, hypothesis, request, or condition contrary to fact (**subjunctive mood**). See **7d**.

nominalization Formation of a noun by adding a suffix to a verb or an adjective: *require, requirement; sad, sadness.*

nominative case Also called **subjective case.** See **case.**

noncount nouns Nouns naming things that cannot be counted (*architecture, water*). See **1a(2)** and **43c**. COMPARE: **count nouns.**

nonessential element A word or word group that modifies another word or word group but does not provide information essential for identification. Nonessential elements are set off by commas, parentheses, or dashes: Carol Murphy, *president of the university,* plans to meet with alumni representatives. Also called a **nonrestrictive element.** See **1f(2)** and **12d**. COMPARE: **essential element.**

nonrestrictive element See **nonessential element.**

nonstandard, nonstandardized Refers to speech forms that are not considered conventional in many academic and professional settings. See the Glossary of Usage.

noun A word that names a person, place, thing, idea, animal, quality, event, and so on: *Alanis, America, desk, justice, dog, strength, departure.* See **1a(2)**. See also **collective noun, common noun, count noun, noncount noun,** and **proper noun.**

noun clause A dependent clause used as a noun. See **1f(2)**.

noun phrase A noun and its modifiers. See **1e(1)**.

number The property of a word that indicates whether it refers to one (**singular**) or to more than one (**plural**). Number is reflected in the word's form: *river/rivers, this/those, he sees/they see.* See Chapter **6**.

object A noun, pronoun, noun phrase, or noun clause that follows a preposition or a transitive verb or verbal. A **direct object** names the person or thing that receives the action of the verb: I sent the *package.* An **indirect object** usually indicates to whom the action was directed or for whom the action was performed: I sent *you* the package. See **1c(1)** and **1c(2)**. The **object of a preposition** follows a preposition: I sent the package to *you.* See **1e(4)**.

object complement See **complement.**

object of a preposition See **object.**

objective case See **case.**

parenthetical element Any word, phrase, or clause that adds detail to a sentence or any sentence that adds detail to a paragraph but is not essential for understanding the core meaning. Commas, dashes, or parentheses separate these elements from the rest of the sentence or paragraph. See **12d**, **17e**, and **17f**.

participial phrase A verbal phrase that includes a participle: The stagehand *carrying the trunk* fell over the threshold. See **1e(3)**. See also **participle** and **phrase.**

participle A verb form that may function as part of a verb phrase (had *determined,* was *thinking*) or as a modifier (a *determined* effort; the couple, *thinking* about their past). A **present participle** is formed by adding *-ing* to the base form of a verb. A **past participle** is usually formed by adding *-ed* to the base form of a verb (*walked, passed*); however, many verbs have irregular past-participle forms (*written, bought, gone*). See **7a(1)** and **7a(5)**.

particle A word such as *across, away, down, for, in, off, out, up, on,* or *with* that combines with a main verb to form a phrasal verb: *write down, look up.* See **7a(3)** and **44c(2)**.

parts of speech The classes into which words may be grouped according to their forms and grammatical relationships. The traditional parts of speech are verbs, nouns, pronouns, adjectives, adverbs, prepositions, conjunctions, and interjections. See **1a**.

passive voice See **voice**.

past participle See **participle**.

pathos One of the three classical appeals; the use of language to stir the feelings of an audience. Also called an **emotional appeal** or a **pathetic appeal**. See **34f(3)**. See also **ethos** and **logos**.

person The property of nouns, pronouns, and their corresponding verbs that distinguishes the speaker or writer (**first person**), the individuals addressed (**second person**), and the individuals or things referred to (**third person**). See **5c** and **7b**.

personal pronoun A pronoun that refers to a specific person, place, thing, and so on. Pronoun forms correspond to three cases: subjective, objective, and possessive. See **5a(1)**.

phrasal modal A modal auxiliary verb consisting of more than one word: *be able to*. See **44b(2)**.

phrasal verb A grammatical unit consisting of a verb and a particle such as *after, in, up, off*, or *out: fill in, sort out*. See **7a(3)** and **44c(2)**.

phrase A sequence of grammatically related words that functions as a unit in a sentence but lacks a subject, a predicate, or both: *in front of the stage*. See **1e**.

point of view The vantage point from which a topic is viewed; also, the stance a writer takes: objective or impartial (third person), directive (second person), or personal (first person). See **5c**.

possessive case See **case**.

predicate The part of a sentence that expresses what a subject is, does, or experiences. It consists of the main verb, its auxiliaries, and any complements and modifiers. The **simple predicate** consists of only the main verb and any accompanying auxiliaries. See **1b** and **1c**. COMPARE: **subject**.

premise An assumption or a proposition on which an argument or explanation is based. In logic, premises are either **major**

(general) or **minor** (specific); when combined correctly, they lead to a conclusion. See **34h(2)**. See also **syllogism.**

preposition A word such as *at, in, by,* or *of* that relates a pronoun, noun, noun phrase, or noun clause to other words in the sentence. See **1a(6)**.

prepositional phrase A preposition with its object and any modifiers: *at* the nearby airport, *by* the sea. See **1e(4)**.

prepositional verb A frequently occurring combination of a verb and a preposition: *rely on, think about.* See **7a(3)** and **44c(1)**.

present participle See **participle.**

primary source A source that provides firsthand information. See **36a(4)**. COMPARE: **secondary source.**

pronoun A word that takes the position of a noun, noun phrase, or noun clause and functions as that word or word group does: *it, that, he, them.* See **1a(3)** and **5a**.

proper adjective An adjective that is derived from the name of a person or place: *Marxist* theories. See **9a(9)**.

proper noun The name of a specific person, place, organization, and so on: *Dr. Pimomo, Fargo, National Education Association.* Proper nouns are capitalized. See **1a(2)** and **43a**. COMPARE: **common noun.**

proposition See **claim.**

qualifier A word that intensifies or moderates the meaning of an adverb or adjective: *quite* pleased, *somewhat* reluctant.

quotation A **direct quotation** is the exact repetition of someone's spoken or written words. An **indirect quotation** is a report of someone's written or spoken words not stated in the exact words of the writer or speaker. See **16a** and **38d**.

reflexive pronoun A pronoun that ends in *-self* or *-selves* (*myself* or *themselves*) and refers to a preceding noun or pronoun in the sentence: *He* added a picture of *himself* to his web page. When used to provide emphasis, such a pronoun is called an **intensive pronoun:** The president *herself* awarded the scholarships. See **5a(2)**.

refutation A strategy for addressing opposing points of view by discussing those views and explaining why they are unsatisfactory. See **34e(2)** and **34g.**

relative clause See **adjectival clause.**

relative pronoun A word (*who, whom, that, which,* or *whose*) used to introduce an **adjectival clause**, also called a **relative clause.** An antecedent for the relative pronoun can be found in the main clause. See **1f(2)** and **5a(3).**

restrictive element See **essential element.**

rhetorical appeal The means of persuasion in argumentative writing, relying on reason, authority, or emotion. See **34f.**

rhetorical opportunity A chance for a writer to resolve an issue or problem with the purposeful use of language. See **31b.**

secondary source A source that analyzes or interprets firsthand information. See **36a(4).** COMPARE: **primary source.**

sentence modifier A modifier related to a whole sentence, not to a specific word or word group within it: *All things considered,* the committee acted appropriately when it approved the amendment to the bylaws.

signal phrase A short phrase that identifies the source of a quotation: *according to Jones, Jones claims.* See **3d** and **38d(1).**

simple predicate See **predicate.**

simple subject See **subject.**

split infinitive The separation of the two parts of an infinitive form by at least one word: *to completely cover.* See **1e(3).**

squinting modifier A modifier that is unclear because it can refer to words either preceding it or following it: Proofreading *quickly* results in missed spelling errors. See **25a(3).**

Standardized English The usage expected in most academic and business settings. See the Glossary of Usage.

subject The general idea addressed in a piece of writing. See **32a.** COMPARE: **topic.** Also, the pronoun, noun, or noun phrase that carries out the action or assumes the state described in the predicate of a sentence. Usually preceding the predicate, the subject includes the main noun or pronoun and all modifiers. A **simple subject** consists of only the main noun or pronoun. See **1b** and **1d**. COMPARE: **predicate.**

subject complement   See **complement.**

subjective case   See **case.**

subjunctive mood   See **mood.**

subordinating conjunction   See **conjunction.**

subordination   The connection of a grammatical structure to another, usually a dependent clause to an independent clause: *Even though customers were satisfied with the product,* the company wanted to improve it. See **24a**.

syllogism   Method for deductive reasoning consisting of two premises and a conclusion. See **34h(2)**. See also **premise.**

tense   The form of a verb that indicates when and for how long an action or state occurs. See **7b** and **44a**.

theme   The main idea of a literary work. See **41c(7)**.

thesis   The central point or main idea of an essay. See **32c**.

tone   The writer's attitude toward the subject and the audience, usually conveyed through word choice and sentence structure. See **33a(3)**.

topic   The specific, narrowed main idea of an essay. See **32b**. COMPARE: **subject.**

topic sentence   A statement of the main idea of a paragraph. See **33c(1)**.

transitions   Words, phrases, sentences, or paragraphs that relate ideas by linking sentences, paragraphs, or larger segments of writing. See **3c(5)** and **33d**.

transitive verb   A verb that takes an object. The researchers *reported* their findings. See **1d**. COMPARE: **intransitive verb.**

verb   A word denoting action, occurrence, or existence (state of being). See **1a(1)** and Chapter 7.

verb phrase A main verb and any auxiliaries. See **1e(2)** and **7a(4)**.

verbal A verb form functioning as a noun, an adjective, or an adverb. See **1e(3)**. See also **gerund, infinitive,** and **participle.**

vocative Set off by commas, the name of or the descriptive term for a person or persons being addressed. See **12b(3)**.

voice A property of a verb that indicates the relationship between the verb and its subject. The **active voice** is used to show that the subject performs the action expressed by the verb; the **passive voice** is used to show that the subject receives the action. See **7c**.

Credits

These pages constitute an extension of the copyright page. We have made every effort to trace the ownership of all copyrighted material and to secure permission from copyright holders. In the event of any question arising as to the use of any material, we will be pleased to make the necessary corrections in future printings.

p. 35: Bill McKibben, "Small World," *Harper's*, December 2003, 46–54.

p. 43: Bill Holm, *Eccentric Islands: Travels Real and Imaginary* (Minneapolis, MN: Milkweed Editions, 2001), 7.

p. 51: Ken Carey, *Flat Rock Journal: A Day in the Ozark Mountains* (San Francisco: Harper, 1995), 1.

p. 60: Elaine Scarry, "On Beauty and Being Just" (The Tanner Lectures on Human Values, Yale University, New Haven, CT, October 25–26, 1998), www.tannerlectures.utah.edu/lectures /documents/scarry00.pdf.

p. 60: Rebecca Solnit, *Wanderlust: A History of Walking* (New York: Penguin, 2000), 289.

p. 61: Derek Bok, *The Politics of Happiness: What Government Can Learn from the New Research on Well-Being* (Princeton, NJ: Princeton University Press), 212.

p. 131: Excerpt by Nadine Gordimer in *For the Love of Books: 115 Celebrated Writers on the Books They Most Love* by Ronald B. Shwartz (New York: Grosset/Putnam, 1999), 102.

p. 131: Excerpt by Ved Mehta in *For the Love of Books: 115 Celebrated Writers on the Books They Most Love* by Ronald B. Shwartz (New York: Grosset/Putnam, 1999), 181.

p. 138: Gary Snyder, *Back on the Fire: Essays* (Berkeley, CA: Counterpoint Press, 2008), 107.

p. 173: Richard F. Thompson and Stephen A. Madigan, *Memory: The Key to Consciousness* (Washington, DC: Joseph Henry Press, 2005), 83.

p. 187: Martha Graham, "An Athlete of God" in *This I Believe: The Personal Philosophies of Remarkable Men and Women* (New York: Holt, 2007), 84.

p. 189: Tracy Kidder, *Mountains Beyond Mountains: The Quest of Dr. Paul Farmer, a Man Who Would Cure the World* (New York: Random House, 2009), 218.

p. 193: Diane Ackerman, *A Natural History of Love* (New York: Vintage/Random House, 1995), 86.

p. 204: Oliver Sacks, "The Last Hippie" in *An Anthropologist on Mars* (New York: Vintage/Random House, 1995), 59.

p. 205: David Weinberger, *Small Pieces Loosely Joined: A Unified Theory of the Web* (New York: Basic Books, 2003), 10, 170.

pp. 206, 207, 208: Patricia Gadsby, "Endangered Chocolate," *Discover*, August 1, 2002.

p. 209: Wallace Stevens, "The Snow Man," in *The Collected Poems of Wallace Stevens* (New York: Random House/Vintage, 1990), 10.

p. 233: By permission. From *Merriam-Webster's Collegiate Dictionary*, 11th Edition © 2011 by Merriam-Webster, Incorporated (www.MerriamWebster.com).

p. 237: Charles Wohlforth, "Conservation and Eugenics," *Orion*, July/August 2010, 28.

p. 237: Loida Maritza Pérez, *Geographies of Home* (New York: Penguin, 2000).

p. 237: Hisaye Yamamoto, "Las Vegas Charley," in *Seventeen Syllables* (San Diego, CA: Harcourt, 1994).

p. 237: Denise Chávez, *A Taco Testimony: Meditations on Family, Food and Culture* (Tuscon, AZ: Rio Nuevo, 2006).

p. 238: Jane Kramer, "The Reporter's Kitchen," *New Yorker,* August 19, 2002.

p. 239: Gayle Pemberton, "The Zen of Bigger Thomas," in *The Hottest Water in Chicago: Notes of a Native Daughter* (Middletown, CT: Weslyan University Press, 1998), 168.

p. 276: Letter from Birmingham Jail by Dr. Martin Luther King, Jr. Reprinted by arrangement with The Heirs to the Estate of Martin Luther King Jr., c/o Writers House as agent for the proprietor New York, NY. Copyright 1963 Dr. Martin Luther King Jr; copyright renewed 1991 Coretta Scott King.

p. 276: Jean Kilbourne, *Deadly Persuasion* (New York: Free Press, 1999).

p. 279: Paul K. Humiston, "Small World," in *I Thought My Father Was God* by Paul Auster (New York: Holt, 2001), 183.

p. 280: James Gorman, "Finding a Wild, Fearsome World beneath Every Fallen Leaf," in *The Best American Science and Nature Writing 2003*, edited by Richard Dawkins (Boston: Houghton, 2003), 67.

p. 283: Malcolm Gladwell, *Blink: The Power of Thinking without Thinking* (Boston: Little, Brown: 2005), 8–9.

p. 283: Antoine Bechara, Hanna Damasio, Daniel Tranel, and Antonio R. Damasio, "Deciding Advantageously before Knowing the Advantageous Strategy," *Science* 275 (February 28, 1997), 1293.

p. 290: Leslie Marmon Silko, "Interior and Exterior Landscapes," in *Yellow Woman and a Beauty of the Spirit* (New York: Simon & Schuster, 1997), 27.

p. 290: Rebecca Solnit, *Storming the Gates of Paradise: Landscapes for Politics* (Berkeley, CA: University of California Press, 2008), 115.

p. 291: Adam Haslett, "Devotion," in *You Are Not a Stranger Here: Stories* (New York: Doubleday/Anchor, 2003), 65.

p. 291: Michiko Kakutani, "Making Art of Sport," *New York Times Magazine*, December 15, 1997.

p. 292: Joan Didion, "On Keeping a Notebook," in *Slouching towards Bethlehem* (New York: Farrar, Straus & Giroux, 2008), 139.

p. 295: Michael Pollan, *The Botany of Desire: A Plant's-Eye View of the World* (New York: Random House, 2002), 18.

p. 300: Ariel Dorfman, "If Only We All Spoke Two Languages," *New York Times*, June 24, 1998.

p. 300: Parker Palmer, *The Courage to Teach: Exploring the Inner Landscape of a Teacher's Life* (San Francisco: Jossey Bass, 2007).

p. 300: Walker Evans, *Unclassified* (Zurich, Switzerland: *Scalo*, 2000).

p. 301: Pico Iyer, "In Praise of the Humble Comma, *Time*, June 13, 1988.

pp. 311–12: Kathleen Dean Moore, "The Happy Basket" in *Wild Comfort: The Solace of Nature* (Boston: Trumpeter, 2010), 21–22.

p. 312: Nina G. Jablonski, *Skin: A Natural History* (Berkeley: University of California Press, 2006), 3.

p. 313: Rebecca Skloot, *The Immortal Life of Henrietta Lacks* (New York: Crown/Random House, 2010), 6–7.

p. 314: US Department of Education, Office of Innovation and Improvement, "Innovations in Education: Creating Strong District School Choice Programs," Washington DC (May 2004), http://www2.ed.gov/admins/comm/choice/choiceprograms/report.pdf.

pp. 315–16: Atul Gawande, "Crimson Tide" in *Complications: A Surgeon's Notes on an Imperfect Science* (New York: Picador, 2002), 150.

pp. 322–23, 325, 328, 331: Freewriting, thesis statements, and outline used by permission of Mary LeNoir.

p. 327: Frank McCourt, "Foreword," in *Eats, Shoots & Leaves* by Lynne Truss (New York: Gotham, 2003), xi.

p. 327: Richard Selzer, "Diary of an Infidel: Notes from a Monastery" in *Taking the World in for Repairs* (New York: Morrow, 1986), 13.

p. 327: Amnesty International, "Abolish the Death Penalty," http://www.amnesty.org/en/death-penalty.

p. 329: Sam Swope, "The Animal in Miguel" in *I Am a Pencil: A Teacher, His Kids, and Their Wonderful World of Stories* (New York: Holt/Owl, 2005), 114.

pp. 329–30: Charles Seife, *Zero: The Biography of a Dangerous Idea* (New York: Penguin, 2000), 5.

pp. 333–34: Brenda Jo Brueggemann, "American Sign Language and the Academy" in *Deaf Subjects: Between Identities and Places* (New York: New York University Press, 2009), 29.

p. 334: Barbara Kingsolver, "Lily's Chickens," in *Small Wonder: Essays* (New York: HarperCollins, 2002), 110.

p. 336: Jill Lepore, "Prologue: Party Like It's 1773" in *The Whites of Their Eyes: The Tea Party's Revolution and the Battle over American History* (Princeton, NJ: Princeton University Press, 2010), 2.

pp. 337–38: Ishmael Beah, *A Long Way Gone: Memoirs of a Boy Soldier* (New York: Crichton, 2007), 18.

pp. 338–39: Sam Swope, "The Case of the Missing Report Cards," in *I Am a Pencil: A Teacher, His Kids, and Their Wonderful World of Stories* (New York: Holt/Owl, 2005), 140.

pp. 339–40: From "Topic of Cancer" by Christopher Hitchens from *Vanity Fair*, September 2010, p. 204. Used by permission of the author.

p. 340: Mark Orwoll, "Revolution in the Skies," *Travel & Leisure*, November 2010, 169.

p. 341: Reprinted with the permission of Simon & Schuster, Inc. from *American Grace* by Robert D. Putnam and David E. Campbell, p. 2. Copyright © 2010 Robert D. Putnam and David E. Campbell.

p. 343: Jody M. Roy, "The Case of the Cowboy," in *Love to Hate: America's Obsession with Hatred and Violence* (New York: Columbia University Press, 2002).

p. 344: Joel Arem, *Rocks and Minerals* (Totnes, England: Geoscience Press, 1994).

p. 348: Dorothy Allison, *Two or Three Things I Know for Sure* (New York: Penguin/Plume, 1996), 6.

p. 349: Rick Roth, "Snake Charmer," interview by Sara Martel, *Sierra,* January–February 2011, 11.

p. 350: Jennifer Howard, "Americans Are Closing the Book on Reading, Study Finds," *Chronicle of Higher Education* 54.14 (November 19, 2007), A12.

p. 350: Terry Tempest Williams, "The Clan of One-Breasted Women" in *Refuge* (New York: Random House/Vintage, 1992), 281.

p. 351: Eudora Welty, "A Sweet Devouring" first published in *Mademoiselle*, 1957. Collected in *The Eye of the Story* (New York: Penguin/Random House, 1977).

p. 351: Frederick Douglass, "What to the Slave Is the Fourth of July?" Reprinted in his 1855 autobiography, *My Bondage and My Freedom.*

pp. 351–52: Patricia J. Williams, *Open House: Of Family, Friends, Food, Piano Lessons, and the Search for a Room of My Own* (New York: Farrar, 2004), 3.

p. 352: Melissa Gotthardt, "The Miracle Diet," *AARP,* January/February 2007, 26.

p. 352: Rebecca Traister, *Big Girls Don't Cry* (New York: Free Press, 2010), 15.

p. 353: From "Not Seeing the Forest for the Dollar Bills" by Donella Meadows, *Valley News*, June 30, 1990. Reprinted by permission of Sustainability Institute.

p. 353: Zora Neale Hurston, *Dust Tracks on a Road: An Autobiography* (Philadelphia: Lippincott, 1942), 288.

p. 354: Kathy L. Glass, "'Tending to the Roots': Anna Julia Cooper's Sociopolitical Thought and Activism," *Meridians* 6.1 (2005), 23–55.

p. 354: Excerpt from "September 11, 2001 Somewhere over Pennsylvania" by Randall Sullivan from *Rolling Stone*, 4/11/02. © Rolling Stone LLC 2002. All Rights Reserved. Reprinted by permission.

pp. 354–55: Debra Utacia Krol, "Peterson Yazzie," *Native Peoples: Arts and Lifeways*, January/February 2007, 45.

p. 356: Malcolm Gladwell, *Blink: The Power of Thinking without Thinking* (Boston: Little, Brown, 2005), 13–14.

p. 357: Patricia Marx, "Dressin' Texan," *New Yorker*, March 19, 2007, 68.

p. 361: K. Bindas. "Re-remembering a Segregated Past: Race in American Memory." *History and Memory* 22.1 (2010), 113–35. Used by permission of Indiana University Press.

p. 361: Tal Birdsey, *A Room for Learning: The Making of a School in Vermont* (New York: St. Martin's, 2009), 34.

p. 362: Richard F. Thompson and Stephen A. Madigan, *Memory: The Key to Consciousness* (Washington, DC: Joseph Henry Press, 2005), 83.

p. 362: Mary E. Curtis and John R. Kruidenier, "Teaching Adults to Read: A Summer of Scientifically Based Research Principles" (Jessup, MD: National Institute for Literacy, 2005), 9, accessed October 27, 2010, http://www.eric.ed.gov/PDFS/ED493064 .pdf.

p. 363: Amy Tan, "Two Kinds" in *The Joy Luck Club* (New York: Penguin, 2006), 142.

p. 363: Constance Holden, "Identical Twins Reared Apart," *Science*, 207 (March 1980), 1323–25.

p. 364: "Spellings Addresses PTA Convention," *The Achiever* 4.10 (September 2005), n.p., accessed October 27, 2010, http:// www2.ed.gov/news/newsletters.achiever.2005/090105.html.

pp. 371–75: First draft by Mary LeNoir. Used by permission of the author.

pp. 382–93: "How Student-Athletes Really Choose a College" by Mary Lenoir. Used by permission of the author.

pp. 407, 409–10, 411: "The Tucson Shootings: Words and Deeds" by Debra Hughes originally appeared in *Narrative* magazine. Reprinted by permission of *Narrative* magazine and the author.

pp. 404, 423–29: List of reasons and "Naming Opportunities: Opportunities for Whom" by Anna Seitz. Used by permission of the author.

p. 480: Marsha Orgeron, "'You Are Invited to Participate': Interactive Fandom in the Age of the Movie Magazine," *Journal of Film and Video* 61.3 (Fall 2009), 3–23.

p. 483: Andy Rees, *Genetically Modified Food: A Short Guide for the Confused* (Ann Arbor, MI: Pluto, 2006), 8.

p. 485: Oliver Sacks, "A Man of Letters," in *The Mind's Eye* (New York: Knopf, 2010), 75.

p. 486: Jim Cullen, *The American Dream: A Short History of an Idea That Shaped a Nation* (New York: Oxford University Press, 2004), 7.

p. 488: Joel Achenbach, "Electrical Grid," *National Geographic*, July 2010, 137.

p. 490: Carl Zimmer, *Soul Made Flesh: The Discovery of the Brain—and How It Changed the World* (New York: Free Press, 2004), 7.

p. 491: Michael Hanlon, "Climate Apocalypse When?" *New Scientist*, November 17, 2007, 20.

pp. 492–93: Joseph M. Marshall, III, "Tasunke Witko (His Crazy Horse)," *Native Peoples,* January/February 2007, 76–79.

p. 499: Patricia McConnell, *The Other End of the Leash* (New York: Ballantine, 2002), 142.

p. 501: Excerpt from "A Farm on Every Floor" by Dickson Despommier from *New York Times*, 23 Aug. 2009. Copyright © 2009 The New York Times. Reprinted by permission.

p. 512: Barbara Ehrenreich, *Nickel and Dimed: On (Not) Getting By in America* (New York: Holt/Metropolitan, 2001), 214.

p. 523: From *English Journal: The Journal of the Secondary Section of the National Council of Teachers of English* 98.1 (September 2008). Used by permission of National Council of Teachers of English.

pp. 528, 529: From *Skin: A Natural History*, by Nina G. Jablonski, © 2006 by Nina G. Jablonski. Published by the University of California Press. Used by permission.

pp. 556, 557–69: "Genetically Modified Foods and Developing Countries" by Marianna Suslin. Used by permission of the author.

p. 576: F. J. Sulloway and R. L. Zweigenhaft, "Birth Order and Risk Taking in Athletics: A Meta-Analysis and Study of Major League Baseball," *Personality and Social Psychology Review* 14 (April 30, 2010), 412.

p. 587: Title page from *How to Make Opportunity Equal* by Paul Gomberg. Used by permission of Blackwell Publishing Ltd.

p. 588: Copyright page from *How to Make Opportunity Equal* by Paul Gomberg. Reprinted by permission of Blackwell Publishing Ltd.

p. 591: Excerpt from "Tricksters and the Marketing of Breakfast Cereals" by Thomas Green from *Journal of Popular Culture,* Vol. 40, No. 1, 2007. Reprinted by permission of Blackwell Publishing Ltd.

p. 594: M. D. Buhrmester, H. Blanton, and W. B. Swann, Jr., "Implicit Self-Esteem: Nature, Measurement, and a New Way Forward," *Journal of Personality and Social Psychology*, 100.2 (2011), 365, doi:10.1037/a0021341.

pp. 602–12: "The Social Status of an Art: Historical and Current Trends in Tattooing" by Rachel L. Pinter and Sarah M. Cronin. Used by permission of the authors.

p. 618: "Prose Poems: Portrait" from *Girl at the Window* © 1991 by Pinkie Gordon Lane. Reprinted by permission of Louisiana State University Press.

Index

Numbers and letters in color refer to chapters and sections in the handbook; other numbers refer to pages.

(cont.)

(cont.)

(cont.)

(cont.)